ANNUAL REPORTS
ON THE PROGRESS OF CHEMISTRY

SECTION A

Annual Reports

on the Progress of Chemistry

Volume 83, 1986

SECTION A

Inorganic Chemistry

The Royal Society of Chemistry

Burlington House, London, W1V 0BN

ISBN 0-85186-180-6
ISSN 0260-1818

Annual Reports on the Progress of Chemistry Section 'A' is published annually by The Royal Society of Chemistry, Burlington House, London W1V 0BN, England. 1987 Annual Subscription price U.K. £57.00, Rest of World £64.00, U.S.A. $113.00, including air speed delivery. Post-publication prices: U.K. £60.00, Rest of World £67.00, U.S.A. $118.00. (Combined subscription package to Annual Reports on the Progress of Chemistry Sections A, B, and C, details available from The Distribution Centre.) Change of address and orders with payment in advance to The Royal Society of Chemistry, The Distribution Centre, Blackhorse Road, Letchworth, Herts. SG6 1HN, England. Air Freight and mailing in the U.S.A. by Publications Expediting Inc., 200 Meacham Avenue, Elmont, New York 11003. Second class postage paid at Jamaica, New York 11431. All other despatches outside the U.K. by Bulk Airmail, and Accelerated Surface Post outside Europe. PRINTED IN THE U.K.

Set in Times on Digiset and printed offset by
J. W. Arrowsmith Ltd., Bristol, England

Made in Great Britain

Contributors

Senior Reporter

J. D. Donaldson, B.Sc., Ph.D., D.Sc., C.Chem., F.R.S.C., *The City University, London*

Reporters

F. J. Berry, B.Sc., Ph.D., C.Chem., F.R.S.C., *University of Birmingham*
B. W. Fitzsimmons, B.A., Ph.D., D.Sc., *Birkbeck College, University of London*
R. Greatrex, B.Sc., M.Sc., Ph.D., *University of Leeds*
S. M. Grimes, B.Sc., Ph.D., C.Chem., M.R.S.C., *The City University, London*
P. G. Harrison, Ph.D., D.Sc., *University of Nottingham*
J. D. Miller, M.A., Ph.D., *University of Aston in Birmingham*
J. E. Newbery, B.Sc., M.A.., Ph.D., *Goldsmiths' College, University of London*
S. D. Robinson, B.Sc., Ph.D., C.Chem., F.R.S.C., *King's College (KQC), University of London*
J. Silver, B.Sc., Ph.D., C.Chem., *University of Essex*
R. Snaith, B.Sc., Ph.D., *University of Cambridge*
M. J. K. Thomas, B.Sc., Ph.D., C.Chem., M.R.S.C., *Goldsmiths' College, University of London*
C. E. Urch, *University of Southampton*
D. S. Urch, B.Sc., Ph.D., *Queen Mary College, University of London*

Contents

Section B, Organic Chemistry, contains the following items

Section C, Physical Chemistry, contains the following items

1 Introduction

By J. D. DONALDSON

Department of Chemistry, The City University, Northampton Square, London EC1V 0HB

Volume 83 of *Annual Reports* covers the highlights of the work published during 1986 dealing with various aspects of inorganic chemistry. The format of the volume is the same as that of Volume 82.

It is once again my duty and pleasure, as Senior Reporter, to thank the authors of the various chapters in this volume. I am particularly grateful to Dr. Robinson who provided an article for the Report for the first time. This introductory chapter is followed by six chapters that deal with the chemistry of the *s*- and *p*-block elements written by Dr. Ron Snaith, Dr. Bob Greatrex, Dr. Susan Grimes, Dr. Philip Harrison, Dr. Frank Berry, and Dr. Mike Thomas. Chapters 8 to 12 deal with various aspects of *d*- and *f*-block element chemistry and have been written by Dr. John Newbery, Dr. Brian Fitzsimmons, Dr. Stephen Robinson, Dr. Jack Silver, and Dr. David Miller. Chapter 13 on radiochemistry has been provided by Dr David Urch in conjunction with Miss Catherine Urch.

Since this is my last year as Senior Reporter I would like to thank all the authors who have provided chapters for the six volumes of *Annual Reports A* with which I have been associated. Reading their chapters each year has made me aware of some developing areas of inorganic chemistry that I might not otherwise have noticed. The strength and value of *Annual Reports* lies in its ability to draw the reader's attention to the more important publications on the inorganic chemistry of any of the elements. I have always been impressed by the efficient way in which all of the authors have tackled the difficult task of describing the published work for one year in so few pages of the Reports. In so doing they have, in my opinion, both stressed important developments and produced interesting and informative articles.

I leave the task of Senior Reporter to my successor, Professor Jon McCleverty, with the hope that he too will find the preparation of *Annual Reports A* to be a stimulating and informative exercise.

2 Li, Na, K, Rb, Cs, Fr; Be, Mg, Ca, Sr, Ba, Ra

By R. SNAITH

University Chemical Laboratory, Lensfield Road, Cambridge CB2 1EW

1 Introduction

Although this report employs the same headings as the 1985 one, the *extent* of coverage within each section has altered somewhat. Of course, useful work is going on in all areas but, as the following brief resumé tries to emphasize, this is particularly so for solid-state inorganic chemistry and for lithium-Group 5 and -Group 6 coordination compounds.

Within inorganic chemistry, there is much interest in alkali- and alkaline earth-metal clusters as bulk metal prototypes, in the mechanism of 'simple' reactions and the electronic structures of products, in insertion reactions and fast-ion conductors, and in the use of the metals or their compounds as catalysts *per se* or as promotion agents. Pioneering work has come within alkali-metal solid-state n.m.r. spectroscopy and in crystal structure determinations of, for example, metal silicides and phosphides, intermetallics and Zintl-type species. It is worth noting that the above are *fundamental* studies, yet ones directed decidedly towards applications, and that, although relying on physical methods, their subjects are truly 'inorganic'. Furthermore, unlike most other areas covered here, they are not lithium-dominated, but call on most of the Group 1 and 2 elements quite equitably.

There has also been some fascinating work done for M^I-X coordination compounds where M^I = Li especially, and X = Hal; O and S; N, P, and As; and Si. X-Ray diffraction has revealed a wealth of structural types – polymers, clusters, rings, and monomers – and such results, alongside those from MO calculations and multinuclear n.m.r. spectroscopy, have been used to rationalize such diversity. Lithium$\cdots$H bridges/interactions, some maybe 'agostic' in origin, have also been implicated in several structures. Once more, such studies deserve the label 'fundamental' since many of the compounds investigated are of course widely used as reagents particularly, as a recent compilation shows,[1] in organic syntheses.

The organometallic compounds of these elements continue to be probed widely. Highlights in their synthetic applications pinpoint an increasing tendency to employ 'mixed' reagents, while organopotassiums have been shown to be very powerful metallating agents. Mechanisms of the formation of organometallics and of their subsequent operation have also been looked at intensely, especially by MO calculations of transition states. Macrocyclic-ligand chemistry has concerned mainly the syntheses and metal ion-complexing abilities of ingenious variations on crowns and crypts, including double side-arm or 'bibracchial' lariat ethers (BiBLEs), redox- or

---

[1] M. Fieser, 'Fieser's Reagents for Organic Synthesis', Vol. 12, Wiley-Interscience, New York, 1986.

photochemically-switched ionophores, bis(crowns) and aza- and transition metal-functionalized ones, spherands of various degrees of pre-organization, and a totally new class, torands. Important developments in bioinorganic chemistry have included increasing applications of metal n.m.r. spectroscopy and of computer simulations to bio-systems, and some significant improvements of methods for essential metal analysis.

One review will be of use to chemists working in all the areas mentioned in this introduction: it gives guidelines for the handling of air-sensitive species of Groups 1 and 2, including the metals themselves, their hydrides and alkyls, and strong bases such as metal amides and alkoxides.[2]

2 Inorganic Chemistry

The recovery or preparation, isotopes, and analyses of the metals themselves have attracted attention. For lithium in particular, a world-wide increase in demand is expected through its projected use as a nuclear fusion tritium fuel source and reactor blanket material. Such considerations prompted a successful study of lithium recovery from geothermal waters using sodium aluminate as co-precipitating agent,[3] and determinations of $^6Li/^7Li$ isotopic ratios by n.m.r. spectroscopy[4] and by an improved a.a.s. method incorporating an ultimate absorbance-ratio technique.[5] For caesium, the metal has been obtained in highly pure form by thermolysis of the azide, CsN_3.[6] Human $^{134}Cs/^{137}Cs$ radioisotope levels in Scotland after the Chernobyl disaster were shown, thankfully, to be only *ca.* 1% up on the natural background dose.[7] Tandem accelerator mass spectrometry can, unlike conventional detection methods, measure cosmogenic ^{41}Ca in a meteorite so making this radioisotope, $t_{1/2}$ 1.03×10^5 years, a good candidate for dating terrestial ages of $<300\,000$ years.[8] In analysis, Be has been determined in natural waters using electron-capture detection gas chromatography,[9] while, important for hydrogeochemical prospecting, a multi-element determination of nine elements (including Li, Na, K, Ca, Mg) in water using inductively coupled plasma atomic fluorescence spectrometry has been described.[10]

There have been numerous experimental and theoretical contributions to the cluster chemistry of these metals, such interest reflecting the intermediacy of clusters as possible models of more extended metallic systems. One review in particular surveys critically the contributions made by theoretical methods to the understanding of the shapes and properties of clusters such as M_4-M_9 (M = Li, Na) and Be_4.[11] Ionic clusters M_4^{3+} (M = Na, K) have been formed within zeolite cages by exposure of M^+-exchanged zeolite-Y to M vapour, MO calculations with Raman spectroscopy

[2] G. B. Gill and D. A. Whiting, *Aldrichimica Acta*, 1986, **19**, 31.
[3] T. Yoshinaga, K. Kawano, and H. Imoto, *Bull. Chem. Soc. Jpn.*, 1986, **59**, 1207.
[4] K. J. Franklin, J. D. Halliday, L. M. Plante, and E. A. Symons, *J. Magn. Reson.*, 1986, **67**, 162.
[5] K. Kushita, *Anal. Chim. Acta*, 1986, **183**, 225.
[6] F. Blatter and E. Schumacher, *J. Less-Common Met.*, 1986, **115**, 307.
[7] W. S. Watson, *Nature (London)*, 1986, **323**, 763.
[8] P. W. Kubik, D. Elmore, N. J. Conard, K. Nishiizumi, and J. R. Arnold, *Nature (London)*, 1986, **319**, 568.
[9] C. I. Measures and J. M. Edmond, *Anal. Chem.*, 1986, **58**, 2065.
[10] R. F. Sanzolone and A. L. Meier, *Analyst (London)*, 1986, **111**, 645.
[11] J. Koutecký and P. Fantucci, *Chem. Rev.*, 1986, **86**, 539.

indicating tetrahedral geometries.[12] Two sets of calculations have probed Be_{13}, the smallest cluster that possesses a central atom coordinated by nearest-neighbour atoms corresponding to the bulk environment. In the first, both all-electron and effective core potential treatments concluded that Be_{13} prefers a D_{3d} geometry (face-centred cubic-like), albeit much distorted;[13] in the second, a chosen D_{3h} geometry (hexagonal close-packed-like) was subjected to *ab initio* methods which showed that the valence electrons have high mobility, producing small binding energies thus low ionization potentials.[14] Outwith theory, co-deposition of vapours of Ca and Mg with argon and argon/CH_3Hal mixtures at 9K has given growth rates for Ca_n ($n = 1, 2$, and higher) and Mg_n ($n = 1$—4, and higher), and has intimated that clusters are *more* reactive than atoms (and particularly so the larger the cluster).[15] Some way between such clusters and the 'bulk', Mg, Ca, and Ba slurries, prepared by metal vapour cocondensation with toluene, have been reacted with some bis-cyclopentadienyl transition-metal complexes,[16] while ultrasound has been shown to accelerate greatly the room-temperature Li reduction of metal halides to metal powders (including Mg).[17]

Turning to compounds, several 'simple' binary and ternary ones have been synthesized for the first time or by improved methods, while others have found important synthetic uses. Thus, cleavage of white P with MPH_2 (M = Li, Na, K) produces M_3P_{19}, the first salts with nonadecaphosphide ions,[18] and mixed crystals Rb_3 $(P_{7-x}As_x)$ can be obtained from the elements at 900K: in solution, a series of compounds with $x = 1$—5, all interconvertible by valence tautomerism, are observed.[19] A faster and cheaper route to $Ca(AlH_4)_2$.2THF involves first adding initiator $NaAlH_4$ to a $NaCl$–CaH_2 mixture, heating to 60 °C, and then treating with $AlCl_3$ in THF.[20] An important paper records the syntheses, then applications, of 'superactive' MH metallation reagents (M = Li, Na, K) by hydrogenation of organometallics in the presence of a Lewis base (equation 1).[21] Still in the area of activated reagents,

$$Bu^nM + H_2 \xrightarrow{\text{TMEDA, hexane}} Bu^nH + MH \text{ suspension} \qquad (1)$$

two groups have reported on the use of CaF_2 as a support for KF,[22] the combination being a much enhanced fluorinating agent; both conclude, from solid-state n.m.r. spectroscopic and powder diffraction data, that there is no *chemical* interaction between the two fluorides. An attempt to obtain an arene-soluble fluoride complex, $LiF.HMPA$, failed, reaction of $Bu^nLi.HMPA$ with BF_3 giving instead the crystalline and highly soluble $LiBF_4.4HMPA$ which, in solution, is thought to exist as tight

[12] P. Sen, C. N. R. Rao, and J. M. Thomas, *J. Mol. Struct.*, 1986, **146**, 171.
[13] C. W. Bauschlicher and L. G. M. Pettersson, *J. Chem. Phys.*, 1986, **84**, 2226.
[14] W. C. Ermler, C. W. Kern, R. M. Pitzer, and N. W. Winter, *J. Chem. Phys.*, 1986, **84**, 3937.
[15] K. J. Klabunde and A. Whetten, *J. Am. Chem. Soc.*, 1986, **108**, 6529.
[16] W. E. Lindsell and R. A. Parr, *Polyhedron*, 1986, **5**, 1259.
[17] P. Boudjouk, D. P. Thompson, W. H. Ohrbom, and B.-H. Han, *Organometallics*, 1986, **5**, 1257.
[18] M. Baudler, D. Duester, and J. Germeshausen, *Z. Anorg. Allg. Chem.*, 1986, **534**, 19.
[19] W. Hönle and H.-G. von Schnering, *Angew. Chem., Int. Ed. Engl.*, 1986, **25**, 352.
[20] P. Shen, Y. Zhang, S. Chen, X. Feng, and T. Li, *Huaxue Tongbao*, 1986, 31.
[21] P. A. A. Klusener, L. Brandsma, H. D. Verkruijsse, P. von R. Schleyer, T. Friedl, and R. Pi, *Angew. Chem., Int. Ed. Engl.*, 1986, **25**, 465.
[22] J. H. Clark, A. J. Hyde, and D. K. Smith, *J. Chem. Soc., Chem. Commun.*, 1986, 791; J. Ichihara, T. Matsuo, T. Hanafusa, and T. Ando, *ibid.*, 1986, 793.

Li^+ and BF_4^- units held by $Li\cdots F$ interactions.[23] Also concerning halides, the conversion of alkyl chlorides into synthetically more useful bromides by their reaction with MBr (M = Li, Na, K) or $CaBr_2$ under phase-transfer conditions appears dependent on the amount of water present and on the identity of the metal; specifically, the conversion is favoured by using LiBr containing 1% water, a result attributed to the stronger hydration of this salt *cf.* the LiCl produced.[24] Finally, anhydrous $MgBr_2$ in benzene/ether has been found to cleave selectively HO-protecting benzyl ether groups, perhaps *via* the intermediate complexes shown in equation 2.[25]

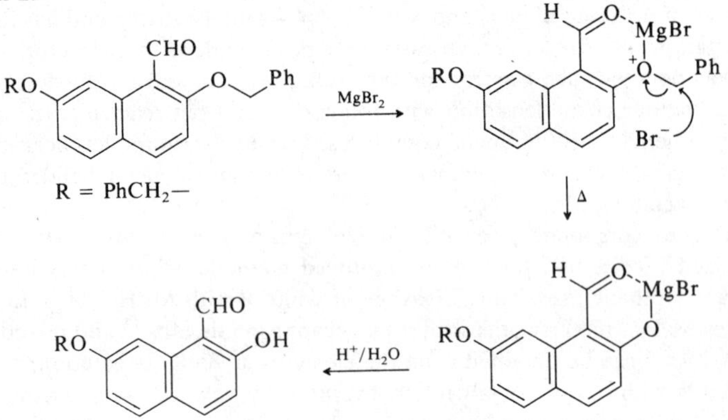

$$R = PhCH_2-$$

(2)

The thermal decompositions of two industrially important sodium salts, $NaHCO_3$,[26] whose decomposition is a key stage of the Solvay process, and $Na_2S.5H_2O$,[27] used in chemical heat pumping and paper technology, have been investigated. For the former, below 390 K first-order kinetics apply with E_{act} in nitrogen *ca.* 64 kJ mol^{-1}, though much higher (*ca.* 130 kJ mol^{-1}) under high partial pressures of CO_2. For the latter, thermogravimetry and powder diffractometry have identified an intermediate dihydrate phase produced *via* a topotactic structural transformation. Molten salt studies have included the detection, by ^7Li n.m.r. and potentiometric measurements, of $LiCl_2^-$ and $Li_2Cl_4^{2-}$ ions in 'solutions' of Li salts in the room-temperature melt $AlCl_3$-n-butylpyridinium chloride,[28] and Raman spectroscopic evidence for $Mg^{2+}\cdots Cl^-$ contacts in molten hydrates of $MgCl_2$,[29] this being contrary to earlier conclusions that these were ionic liquids consisting of separated $Mg(H_2O)_6^{2+}$ and Cl^-.

There have been some very significant investigations of the essential structures of inorganic materials and of the energetics, kinetics, and mechanisms of reactions involving them, such studies encompassing product analysis, rate measurements,

[23] D. Barr, K. B. Hutton, J. H. Morris, R. E. Mulvey, D. Reed, and R. Snaith, *J. Chem. Soc., Chem. Commun.*, 1986, 127.

[24] Y. Sasson, M. Weiss, A. Loupy, G. Bram, and C. Pardo, *J. Chem. Soc., Chem. Commun.*, 1986, 1250.

[25] J. E. Baldwin and G. G. Haraldsson, *Acta Chem. Scand., Ser. B*, 1986, **40**, 400.

[26] M. C. Ball, C. M. Snelling, A. N. Strachan, and R. M. Strachan, *J. Chem. Soc., Faraday Trans. 1*, 1986, **82**, 3709.

[27] J. Y. Andersson and M. Azoulay, *J. Chem. Soc., Dalton Trans.*, 1986, 469.

[28] R. R. Rhinebarger, J. W. Rovang, and A. I. Popov, *Inorg. Chem.*, 1986, **25**, 4430.

[29] W. Voigt and P. Kloeboe, *Acta Chem. Scand., Ser. A*, 1986, **40**, 354.

spectroscopy of many kinds, and MO calculations. The kinetics of the room-temperature reaction of atomic sodium with ozone, a process vital to understanding thermo/mesosphere chemistry, have been measured (equation 3).[30]

$$Na + O_3 \rightarrow NaO + O_2; \qquad \Delta H = -167 \pm 42 \text{ kJ mol}^{-1} \tag{3}$$

$$k = (3.1 \pm 0.4) \times 10^{-10} \text{ cm}^3 \text{ molecule}^{-1} \text{ s}^{-1}$$

Of likely interest regarding modification of transition-metal catalysts by alkali metals, krypton-matrix co-deposition of Li with CO leads to, among other products, $Li(CO)_n$ with $n = 1$—3, and 4 or 6; SCF calculations predict linear structures when $n = 1, 2$, with lowered $\nu(CO)$ force constants.[31] Matrix i.r. studies have also shown that Li co-condensed with C_2H_4 and N_2 produces $Li^+N_2^-$, a species not observed with N_2 alone and so implying that preliminary complexation of C_2H_4 obviates the energy barrier for the fixation reaction.[32] The 'lithium bond', suggested over the years by analogy with the hydrogen bond, has again been discussed. A brief review on the topic has appeared,[33] and force constants for $H_3N\cdots LiF$ and $H_3N\cdots HF$ calculated, a prime conclusion being that $X-F$ bond-lengthening ($X = Li$ or H) and dipole moment changes on complexation are much the same in both species.[34] Also concerning fluoro-species, SCF–MO calculations have provided good estimates of decomposition energies, MF_n to M^{n+} and nF^- ($M = Li$, Na, Be, Mg), thereby allowing compilations of fluoride ion affinities.[35] For hydride reduction mechanisms, reaction of LiH with formaldehyde is calculated to proceed *via* a linear complex (1) then a bridged, 'early' transition state, (2), about 10% along the reaction pathway;

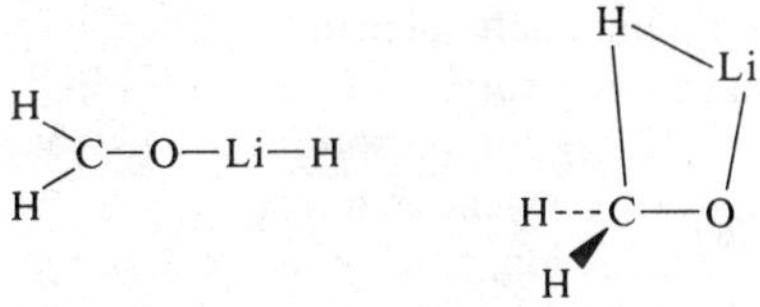

(1) (2)

in both, Li retains its ionicity, bearing a charge of *ca.* $+0.85$.[36] The $LiAlH_4$ reductions of certain amino acids in esters have been shown by ^7Li and ^{27}Al n.m.r. spectroscopy to involve alkoxyaluminates (equation 4) and their disproportionation equilibria (*e.g.*, equation 5).[37]

$$LiAlH_4 + nROH \rightarrow LiAlH_{4-n}(OR)_n + nH_2 \tag{4}$$

$$2LiAlH_2(OR)_2 \rightleftharpoons LiAlH_3OR + LiAlH(OR)_3 \tag{5}$$

[30] J. A. Silver and C. E. Kolb, *J. Phys. Chem.*, 1986, **90**, 3263.
[31] O. Ayed, A. Loutellier, L. Manceron, and J. P. Perchard, *J. Am. Chem. Soc.*, 1986, **108**, 8138; B. Silvi, O. Ayed, and W. B. Person, *ibid.*, 1986, **108**, 8148.
[32] L. Manceron, M. Hawkins, and L. Andrews, *J. Phys. Chem.*, 1986, **90**, 4987.
[33] A. B Sannigrahi, *J. Chem. Educ.*, 1986, **63**, 843. .
[34] Z. Latajka, K. Morokuma, H. Ratajczak, and W. J. Orville-Thomas, *J. Mol. Struct. (Theochem.)*, 1986. **135**, 429.
[35] M. O'Keeffe, *J. Am. Chem. Soc.*, 1986, **108**, 4341.
[36] S. M. Bachrach and A. Streitwieser, *J. Am. Chem. Soc.*, 1986, **108**, 3946.
[37] A.-C. Malmvik, U. Obenius, and U. Henriksson, *J. Chem. Soc., Perkin Trans. 2*, 1986, 1899.

Small heteronuclear cluster molecules and cations of the metals with hydrogen and the halides have been looked at intensively by theoretical methods and mass spectrometry, and the use of the latter technique in following the generation and growth of such species has been reviewed.[38] For hydrides, the structures of all possible Li_nH_m molecules and mono-cations with $n + m \leq 4$ were optimized by *ab initio* calculations, stable species predicted for the first time including planar LiH_3 (3) and $Li_2H_2^+$ (4), and kite-shaped Li_3H^+ (5).[39] Concerning alkali halides,

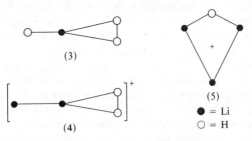

(3)

(4)

(5)

● = Li
○ = H

an interesting study, albeit one concerned more with the analysis of such salts, has shown the feasibility of using an unmodified commercial mass spectrometer with various ionization modes to produce then detect such ions and clusters.[40] Calculations on $M^I(M^IHal)_4^+$ species reveal that, rather than having the cubic structures proposed to explain the occurrence of 'magic numbers' for alkali halide clusters, these larger and more polarizable clusters form in a slightly puckered plane; moreover, the Born-Mayer model appears to give a level of accuracy comparable to that from far more computationally-demanding ones.[41] Spectroscopic parameters and dipole moments have been calculated for the Group 2 monohalides, MHal (M = Be, Mg, Ca; Hal = F, Cl), the Ca species being predominantly ionic though the Be ones showing signs of covalent character.[42]

As mentioned in the Introduction, much of current inorganic chemistry is centred firmly 'in the solid-state', especially as it concerns insertion and conduction studies, alkalides, MAS n.m.r., catalytic properties, and, of course, crystal structures. Huge numbers of publications are appearing in these five areas, and some highlights and representative examples from 1986 are now reported.

Insertion reactions in electrode materials have been reviewed.[43] Of the wide variety of methods for effecting insertion, direct chemical ones have included the use of LiI to form LiV_2O_5 and Li_2ReO_3,[44] and the partial de-intercalation of $NaCoO_2$ by Br_2 and I_2, giving Na_xCoO_2 (x = 0.5 and 0.6 respectively).[45] Electrochemical methods have been used to insert up to eight Li atoms into $Mo_{15}Se_{19}$, giving a material of potential importance in secondary Li batteries,[46] and such a battery has

[38] T. P. Martin, *Angew. Chem., Int. Ed. Engl.*, 1986, **25**, 197.
[39] B. H. Cardelino, W. H. Eberhardt, and R. F. Borkman, *J. Chem. Phys.*, 1986, **84**, 3230.
[40] F. A. Bencsath and F. H. Field, *Anal. Chem.*, 1986, **58**, 679.
[41] B. I. Dunlap, *J. Chem. Phys.*, 1986, **84**, 5611.
[42] S. R. Langhoff, C. W. Bauschlicher, H. Partridge, and R. Ahlrichs, *J. Chem. Phys.*, 1986, **84**, 5025.
[43] D. W. Murphy, *Solid State Ionics*, 1986, **18—19**, 847.
[44] D. W. Murphy and S. M. Zahurak, *Inorg. Synth.*, 1986, **24**, 200.
[45] S. Kikkawa, S. Miyazaki, and M. Koizumi, *J. Solid State Chem.*, 1986, **62**, 35.
[46] J. M. Tarascon and D. W. Murphy, *Phys. Rev. B: Condens. Matter*, 1986, **33**, 2625.

been obtained by the uptake of 1.7 Li into NbS_3.[47] An intriguing observation, maybe of analytical value, is that alkali-metal ions can be separated by the electrochemical reduction (intercalation), then oxidation (release), cycles of a Pt electrode modified with Prussian Blue.[48] As examples of spectroscopic investigations of intercalated materials, ^7Li n.m.r. relaxation rates have been measured in Li_5AlO_4, so giving information on Li diffusion,[49] and i.r. spectra analysis of $Li—S_n$ vibrations in Li_xFeS_2 systems ($0 \leq x \leq 2$) has shown that when $x \geq 1$, Li^+ ions occupy both octahedral and tetrahedral sites, though the former positions are favoured when $0 \leq x < 1$.[50] Insertions into spinels have attracted particular attention since the products seem to hold especial promise as Li battery cathode materials. In the case of $LiFe_5O_8$ insertion at 400 °C causes Fe extrusion to produce the rock-salt phase $Li_2Fe_3O_5$.[51] For MMn_2O_4 (M = Zn, Ni, Cd), on Li insertion Mn^{3+} is reduced to Mn^{2+} and it seems that the ease of reducibility of the transition metal ions may partly control the extent and kinetics of such reactions.[52] Several fast-ion conductors have been scrutinized. A structural study over the range $133 \leq T \leq 919K$ on the hollandite $K_{1.54}Mg_{0.77}Ti_{7.23}O_{16}$ helped reveal the conduction mechanism at various temperatures, *e.g.*, at room temperature, K^+ ions diffuse at a high rate across the bottleneck between the neighbouring cavities.[53] The search for an alternative to the unviable Li_3N as a cell material for energy storage devices has come up with polycrystalline LiI.4MeOH [formulated $Li(MeOH)_4^+$.I^- from the crystal structure], which gives purely ionic conductance with the highest conductivity so far observed for Li (presumably due to mobile complexed cations).[54] By contrast, coumarin(C)–alkali-halide–iodine complexes, *e.g.*, C_4RbI_4, conduct by the migration of M^+ *and* I^- ions through the organic lattice framework.[55]

A notable first is the crystal structure of the electride $Cs^+(18\text{-crown-}6)_2.e^-$ which shows that each electron is trapped and localized in a near-spherical, otherwise empty, hole of radius *ca.* 2.4 Å, surrounded by eight complexed Cs^+ ions.[56] Similar isolation has been noted for the Na^- ion in 12-crown-4 solution, measured ^{23}Na nuclear spin relaxation rates being essentially independent of the anion environment, so showing that this really is a 'gas-like' ionic moiety.[57] Such solution studies have been backed up by calculations of the difference between the shieldings of the gaseous alkali metal anions and those of the gaseous metal atoms; for $^{23}Na^-$, theory and experimental solution data concur, though for $^{87}Rb^-$ and $^{133}Cs^-$ they diverge, these heavier anions being perturbed by solvation effects.[58] Anions apart, cations, solvated electrons, and various cation–electron aggregates have been detected by

[47] T. Yamamoto, S. Kikkawa, and M. Koizumi, *J. Electrochem. Soc.*, 1986, **133**, 1558.

[48] T. Ikeshoji, *J. Electrochem. Soc.*, 1986, **133**, 2108.

[49] T. Matsuo, T. Shibuya, and M. Ohno, *J. Chem. Soc., Faraday Trans. 2*, 1986, **82**, 1.

[50] P. Gard, C. Sourisseau, G. Ouvrard, and R. Brec, *Solid State Ionics*, 1986, **20**, 231.

[51] L. A. De Picciotto and M. M. Thackeray, *Mater. Res. Bull.*, 1986, **21**, 583.

[52] C. J. Chen, M. Greenblatt, and J. V. Waszczak, *Mater. Res. Bull.*, 1986, **21**, 609.

[53] H.-P. Weber and H. Schulz, *J. Chem. Phys.*, 1986, **85**, 475.

[54] W. Weppner, W. Welzel, R. Kniep, and A. Rabenau, *Angew. Chem., Int. Ed. Engl.*, 1986, **25**, 1087.

[55] C. Wu and M. M. Labes, *J. Phys. Chem.*, 1986, **90**, 4199.

[56] S. B. Dawes, D. L. Ward, R. H. Huang, and J. L. Dye, *J. Am. Chem. Soc.*, 1986, **108**, 3534.

[57] A. S. Ellabboudy, D. M. Holton, N. C. Pyper, P. P. Edwards, B. Wood, and W. McFarlane, *Nature (London)*, 1986, **321**, 684; D. M. Holton, A. Ellabboudy, N. C. Pyper, and P. P. Edwards, *J. Chem. Phys.*, 1986, **84**, 1089.

[58] N. C. Pyper and P. P. Edwards, *J. Am. Chem. Soc.*, 1986, **108**, 78.

optical, e.s.r., and n.m.r. spectroscopy in solutions of K, Rb, Cs in 12-crown-4 and 15-crown-5.[59] More widely, the thermodynamic stabilities of gas-phase anions throughout the Periodic Table (including Groups 1 and 2) have been calculated, along with ΔG_f^0 values of these anions in water and in liquid ammonia; one conclusion is that an impossibly high concentration of Na metal would be needed to produce Na^- under standard-state conditions in liquid NH_3.[60] Interestingly, though, optical spectroscopic evidence has been presented for Na^- in a solid NH_3 matrix.[61]

The emergence of solid-state n.m.r. spectroscopy continues, and, linking to the last section, it has been applied to sodium anion salts[62] and, for the first time, to crystalline rubidides, which show chemical shifts of -186 ± 2 p.p.m. for Rb^- relative to Rb^+_{aq}.[63] The power of the technique is further demonstrated by ^{23}Na studies of simple sodium salts and ones complexed with small ligands or ionophores; the range of chemical shifts is large, about 60 p.p.m., but the real value is how ion–ion and ion–ligand interactions are revealed, now being free from interference by the exchange processes encountered in solution.[64] Solid Group 1 fluorides, on their own, as hydrates, or supported, have also been examined by ^{19}F n.m.r., and here again a far wider range of shifts is found compared to that in solution; this maybe reflects the observation of ion-pairing effects and coordinative unsaturation of F^-.[65] Clearly, 'simplification' of the system by removing solvents is, albeit perhaps paradoxically, going to provide much more information on the environment and bonding of the nuclide under study. The structural value of static and MAS n.m.r. spectroscopy has been further demonstrated by ^{29}Si work on belinite, $4Ca_2SiO_4.MgCl_2$, which has identified the presence of the new 4-coordinate magnesium silicate anion (6).[66]

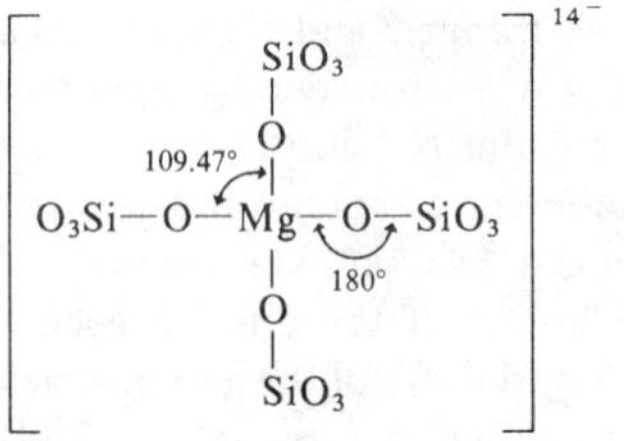

(6)

The catalytic uses of Group 1 and 2 metals and compounds continue to expand, and tremendous efforts are being made to understand how such catalyses work. Potassium-containing ruthenium on active carbon is the first heterogeneous system capable of giving NH_3 from N_2 and H_2 under ambient conditions; the activity is

[59] R. N. Edmonds, D. M. Holton, and P. P. Edwards, *J. Chem. Soc., Dalton Trans.*, 1986, 323; A. Ellaboudy, D. M. Holton, R. N. Edmonds, and P. P. Edwards, *J. Chem. Soc., Chem. Commun.*, 1986, 1444.

[60] S. G. Bratsch and J. J. Lagowski, *Polyhedron*, 1986, **5**, 1763.

[61] D. Smith, B. E. Williamson, and P. N. Schatz, *Chem. Phys. Lett.*, 1986, **131**, 457.

[62] A. Ellaboudy and J. L. Dye, *J. Magn. Reson.*, 1986, **66**, 491.

[63] M. L. Tinkham, A. Ellaboudy, J. L. Dye, and P. B. Smith, *J. Phys. Chem.*, 1986, **90**, 14.

[64] R. Tabeta, M. Aida, and H. Saitô, *Bull. Chem. Soc. Jpn.*, 1986, **59**, 1957.

[65] J. H. Clark, E. M. Goodman, D. K. Smith, S. J. Brown, and J. M. Miller, *J. Chem. Soc., Chem. Commun.*, 1986, 657.

[66] A.-R. Grimmer and F. von Lampe, *J. Chem. Soc., Chem. Commun.*, 1986, 219.

10^5 times lower than that of certain nitrogenases, but at least a start has been made.[67] Syn-gas conversions have been extensively studied. For example, K_2O-promotion of Ru-K-Al_2O_3 catalysts has a dramatic effect on the hydroformylation of propene, f.t.-i.r. spectroscopic studies implying that surface KH is formed (equation 6), subsequently reacting to stabilize intermediate formyl species (equation 7).[68]

$$K_2O(s) + 2H_2(g) \rightleftharpoons 2KH(s) + H_2O(g) \qquad (6)$$

$$KH(s) + CO(g) \rightleftharpoons K-C{\overset{\overset{H}{\diagup}}{\underset{\diagdown}{\diagdown O}}} \quad (s) \qquad (7)$$

In similar vein, doping of Cu + ZnO catalyst with Cs doubles the methanol yield from H_2-CO mixes yet keeps higher oxygenate formation very low ($\leqslant 1.1$ mol %), effects attributed in part to intermediate formation of surface formate and acetate species.[69] Partial oxidation of methane to C_2-compounds C_2H_4 and C_2H_6 is enhanced specifically towards the former by $LiCl$-Sm_2O_3 catalyst and by addition of various alkali metal salts to NiO catalyst;[70] these salts may poison active sites which cause the burning of CH_4 or C_2-products, and create new sites effective to dehydrogenation and coupling of CH_4. A wide-ranging paper in the same area has concerned the marked influence of KNO_3 addition to Ni + SiO_2 catalysts on the kinetic parameters of several organic transformations, the general conclusion being that K-doping blocks certain Ni atoms and electronically modifies others, leading to 'electron-enrichment' (acknowledged to be an unequivocal concept) of the metallic phase.[71] Spanning Groups 1 and 2, $BaCO_3$ and $SrCO_3$ have been found to be among the most effective catalysts for CH_4-O_2 to C_2-hydrocarbon processes, doping with alkali cations (Li, Na, K, Rb) improving activity and selectivity still further.[72] Powdered MgO alone has photocatalytic activity in isomerizing butenes, the rate of isomerism being paralleled by the intensity of photoluminescence, indicating that coordinatively unsaturated surface ions play a major role in such activity.[73] Although perhaps away from catalysis as such, mention must be made of a remarkable 'salting' effect whereby a whole range of non-ionic guests [*e.g.*, benzene, phenol, even $Fe(CO)_5$] can be co-condensed into an alkali halide matrix at 77K; the incorporation is permanent even at 25 °C, and in some cases for up to two years or after heating to 450 °C, and can lead to host dimerization after u.v. irradiation.[74]

The crystal structures of many interesting binary and ternary compounds have been reported. Among them, $Li_{12}Si_7$ has a complex 3D-structure with Si_4 and Si_5 clusters which are enveloped by Li atoms; however, to bring it in line with amended Zintl electron-counting rules, the structure is best treated in terms of two 1D-partial

[67] K. Aika, *Angew. Chem., Int. Ed. Engl.*, 1986, **25**, 558.
[68] I. L. C. Freriks, P. C. de Jong-Versloot, A. G. T. G. Kortbeek, and J. P. van den Berg, *J. Chem. Soc., Chem. Commun.*, 1986, 253.
[69] J. Nunan, K. Klier, C.-W. Young, P. B. Himelfarb, and R. G. Herman, *J. Chem. Soc., Chem. Commun.*, 1986, 193.
[70] K. Otsuka, Q. Liu, and A. Morikawa, *J. Chem. Soc., Chem. Commun.*, 1986, 586; *Inorg. Chim. Acta*, 1986, **118**, L23.
[71] H. Praliaud, J. A. Dalmon, C. Mirodatos, and G. A. Martin, *J. Catal.*, 1986, **97**, 344.
[72] K. Aika, T. Moriyama, N. Takasaki, and E. Iwamatsu, *J. Chem. Soc., Chem. Commun.*, 1986, 1210.
[73] M. Anpo, Y. Yamada, and Y. Kubokawa, *J. Chem. Soc., Chem. Commun.*, 1986, 714.
[74] E. Kirkor, J. Gebicki, D. R. Phillips, and J. Michl, *J. Am. Chem. Soc.*, 1986, **108**, 7106.

ones, the 26e⁻ fragment $Li_5(LiSi_5)$ and the 28e⁻ fragment $Li_{12}Si_4$.[75] In Li_3P_7, the $P_7{}^{3-}$ anions are surrounded by no fewer than 12 Li cations, these having coordination numbers of 3 and 4 to P.[76] For Group 2, the structures of several hydrates have been determined, including those of $Ba(OH)_2.3H_2O$, whose Ba atoms are coordinated to $6H_2O$ and $2OH^-$ oxygens forming a bicapped trigonal prism (Figure 1a), of γ-$Ba(OH)_2.H_2O$, having a tricapped trigonal prismatic Ba environment of $3H_2O$ and $6OH^-$ oxygens (Figure 1b), and of isotypic β-$Ba(OH)_2.H_2O$ and $Sr(OH)_2.H_2O$,

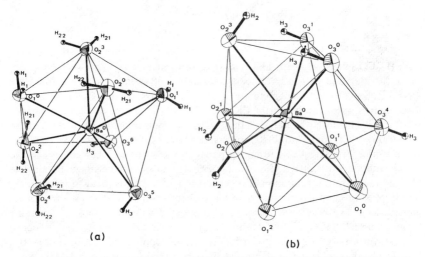

(a)   (b)

Figure 1 *Coordination sphere of barium* (a) *in* $Ba(OH)_2.3H_2O$ *and* b) *in* γ-$Ba(OH)_2.H_2O$ (Reproduced by permission from *Z. Anorg. Allg. Chem.*, 1986, **538**, 131)

with 8-coordinate metal atoms.[77] A hitherto unknown structural type has been found for $SrI_2.2H_2O$ which has face-shared $SrI_5(H_2O)_4$ tricapped trigonal prisms forming columns and channels along [010].[78] Among the several fascinating mixed-metal silicide structures to have emerged, the novel Zintl compound Li_8MgSi_6 contains planar Si_5 rings which form $\frac{1}{\infty}[LiSi_5]$ sandwich stacks,[79] while red K_3LiSi_4 has a $\frac{1}{\infty}[LiSi_4]^{3-}$ chain structure with Li atoms functioning twice as μ_3-ligands, with one K also μ_3 to the Si_4 tetrahedron, the other K atoms bridging only edges or coordinating only corners (Figure 2a); a further variation occurs in $K_7Li(Si_4)_2$ which has a dumb-bell-shaped unit $[Li(Si_4)_2]^{7-}$, with six of the K atoms serving as further μ_3-ligands (Figure 2b).[80] The unintentional use of excess Rb during the synthesis of Rb_3As_7 in a niobium tube led, after extraction with crypt, to the remarkable compound $[Rb(2, 2, 2,-crypt)]_2.[Rb(NbAs_8)]$, whose third Rb has been complexed by the Nb^V-centred crown-shaped anion $(As^-)_8$ to give a 1—D chain

75 R. Nesper, H. G. von Schnering, and J. Curda, *Chem. Ber.*, 1986, **119**, 3576.
76 V. Manriquez, W. Hönle, and H. G. von Schnering, *Z. Anorg. Allg. Chem.*, 1986, **539**, 95.
77 W. Buchmeier and H. D. Lutz, *Z. Anorg. Allg. Chem.*, 1986, **538**, 131.
78 W. Buchmeier and H. D. Lutz, *Acta Crystallogr., Sect.C*, 1986, **42**, 651.
79 R. Nesper, J. Curda, and H. G. von Schnering *J. Solid State Chem.*, 1986, **62**, 199.
80 H. G. von Schnering, M. Schwarz, and R. Nesper, *Angew. Chem., Int. Ed. Engl.*, 1986, **25**, 566.

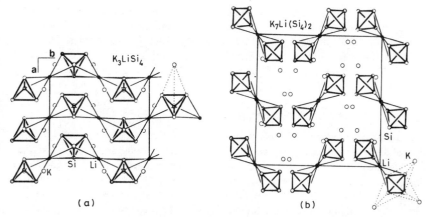

Figure 2 *Crystal structures of* (a) K_3LiSi_4 *with* $^1_\infty[Li(Si_4)]^{3-}$ *polymeric chains, and* (b) $K_7Li(Si_4)_2$ *with dumb-bell* $[Li(Si_4)_2]^{7-}$ *groups*
(Reproduced by permission from *Angew. Chem., Int. Ed. Engl.*, 1986, **25**, 566)

$^1_\infty[Rb(NbAs_8)]^{2-}$.[81] In the first structurally well-characterized triclinic bronze, $LiMo_3O_9$, layers containing distorted MoO_6 and LiO_6 octahedra are interconnected through edge- and corner-sharing to give a 3-D network.[82] Within Group 2, the known $CaBe_2Ge_2$ structure, (7), has been interpreted in terms of the donor–acceptor concept, each component $(BeGe^-)_n$ being a 2-D layer, (8); such a treatment allows

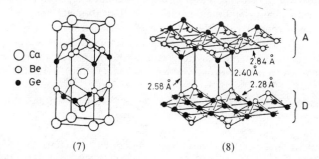

(7)   (8)

calculations of the detailed features of the electronic structure.[83] New structures elucidated included $Sr_2P_4O_{12}.6H_2O$ whose P_4O_{12} ring anions are linked by $SrO_5.(H_2O)_3$ polyhedra and hydrogen bridges,[84] and $Ba_5(Mo_4O_6)_8$, whose formula reflects an ordering of Ba^{2+}, each having a coordination environment of eight O atoms, in 5 of the 8 sites in the channels of the $NaMo_4O_6$ structural type.[85] Other yet more complicated structures worth a note are: $Sr_2[UO_2(CO_3)_3].8H_2O$ whose asymmetric unit contains one 8- and three 9-coordinate Sr atoms, two $[UO_2(CO_3)_3]$

[81] H. G. von Schnering, J. Wolf, D. Weber, R. Ramirez, and T. Meyer, *Angew. Chem., Int. Ed. Engl.*, 1986, **25**, 353.

[82] P. P. Tsai, J. A. Potenza, M. Greenblatt, and H. J. Schugar, *J. Solid State Chem.*, 1986, **64**, 47.

[83] C. Zheng and R. Hoffmann, *J. Am. Chem. Soc.*, 1986, **108**, 3078.

[84] A. Durif and M. T. Averbuch-Pouchot, *Acta Crystallogr., Sect. C*, 1986, **42**, 927.

[85] C. C. Torardi and R. E. McCarley, *J. Less-Common Met.*, 1986, **116**, 169.

units, thirteen Sr-bonded H_2O molecules, and three lattice-bonded ones;[86] $BaCs_4(PO_3)_6$ consisting of $(PO_3)_\infty$ chains with all the Ba atoms and one Cs atom located on the internal threefold axis and being 9-coordinate to O;[87] $Ba_2AsSe_4(OH).2H_2O$, with isolated $AsSe_4^{3-}$ tetrahedra and Ba^{2+} ions coordinated by six Se and three O atoms;[88] and $Ba_{1.14}K_{0.72}VS_4$, whose Ba and K atoms are coordinated to 9 and 11 S atoms respectively, forming capped trigonal prisms.[89]

3 Organometallic Chemistry

Key areas of concern within this discipline remain the preparations and synthetic uses of organometallics and the mechanisms of such reactions, and structural studies drawing heavily on MO and n.m.r. spectroscopic methods as well as on X-ray diffraction. For syntheses, relevant compilations are: a book including preparative details for close on twenty Group 1 and Group 2 organometallics,[90] a review surveying the applications of organometallic sonochemistry and discussing usefully the physics of the technique,[91] and a reported lecture dealing with aspects of di-Grignard reagent chemistry.[92] For mechanisms, an authoritative review focuses on intramolecularly assisted (by O, N) lithiations and discusses relationships between the resulting internally coordinated products and the more usual intermolecular adducts of organolithiums and Lewis bases;[93] along a similar line, 'complex-induced proximity effects', involving a transition state formed from the initial complex of RLi.(heteroatom-functionalized reactant), have been shown to be important in numerous RLi reactions.[94] The increasing structural value of high-resolution metal-n.m.r. spectroscopy (including ^{25}Mg) of organometallics has been reviewed.[95]

Several synthetically important mono- and di-alkali-metal species have been described. A new route to terminal and exocyclic olefins from carbonyl compounds involves addition of $ClCH_2Li$ (generated *in situ* from $ClCH_2I$ + MeLi) then lithiation of the resulting intermediate to give a dilithium species which β-eliminates (equation 8).[96] A much improved synthesis of $(MeO)(SPh)CHLi$ (for use in the

$$RR'C=O \xrightarrow{ClCH_2Li} \underset{OLi}{RR'C-CH_2Cl} \xrightarrow{Li} \underset{OLi}{RR'C-CH_2Li} \rightarrow RR'C=CH_2 \quad (8)$$

preparation of γ-ketoesters) has been described, reaction of PhSH and $MeOCH_2OMe$ in the presence of $BF_3.OEt_2$ giving $MeOCH_2SPh$ which is then readily lithiated by $Bu^sLi.TMEDA$.[97] Allyl- and vinyl-lithiums continue to be sought after

[86] K. Mereiter, *Acta Crystallogr., Sect. C*, 1986, **42**, 1678.
[87] M. T. Averbuch-Pouchot and A. Durif, *Acta Crystallogr., Sect. C*, 1986, **42**, 928.
[88] J. Kaub, *Z. Naturforsch., B*, 1986, **41**, 436.
[89] H. Vincent, M. Anne, A. Chang, and J. Marcus, *J. Solid State Chem.*, 1986, **61**, 332.
[90] 'Organometallic Syntheses', ed. R. B. King and J. J. Eisch, Vol. 3, Elsevier, Amsterdam, 1986; see especially pp. 352—406.
[91] K. S. Suslick, *Adv. Organomet. Chem.*, 1986, **25**, 73.
[92] F. Bickelhaupt, *Pure Appl. Chem.*, 1986, **58**, 537.
[93] G. W. Klumpp, *Recl. Trav. Chim. Pays-Bas*, 1986, **105**, 1.
[94] P. Beak and A. I. Meyers, *Acc. Chem. Res.*, 1986, **19**, 356.
[95] R. Benn and A. Rufińska, *Angew. Chem., Int. Ed. Engl.*, 1986, **25**, 861.
[96] J. Barluenga, J. L. Fernández-Simón, J.-M. Concellón, and M. Yus, *J. Chem. Soc., Chem. Commun.*, 1986, 1665.
[97] S. Hackett and T. Livinghouse, *J. Chem. Soc., Chem. Commun.*, 1986, 75.

and used. A general method for generating the former involves treatment of allyl phenyl sulphides with Li naphthalenide-radical anions,[98] while the useful synthon trifluorovinyllithium has been obtained by treatment of $F_2C=CFCl$ with Bu^sLi in ether at $-60\,°C$.[99] Vinyllithium itself reacts with MCl_4 (M = Ti, Zr, Hf) with reduction of the metal compounds and formation of butadiene, such reactions probably proceeding *via* an outer-sphere redox mechanism.[100] Addition of Li to several simple cyclic and acyclic alkynes has been found to proceed surprisingly smoothly, *e.g.* with cyclooctyne, excess Li powder in Et_2O at $-35\,°C$ over 2h produces the dilithio compound (9) with lesser amounts of the dimerization product (10), both perhaps having the doubly bridged structures shown.[101]

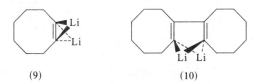

(9)   (10)

Last year's prediction that organo-Na,K chemistry was about to be opened up has been justified. Previous neglect of this area was due to low thermal stability and limited solubility, so the hexane 'solubilization' of unsolvated butylsodium by complexation with magnesium 2-ethoxyethoxide is noteworthy: both 2:1 and 1:1 Na:Mg complexes are available, and have moderate to good thermal stability and pronounced reducing ability.[102] Like Bu^nNa in 1985, Bu^nK, prepared by metal–metal exchange between Bu^nLi and potassium t-amylate, will dissolve in hexane after addition of TMEDA; the solution, probably containing a monomeric $Bu^nK.TMEDA$ complex, is a powerful metallating agent, though it must be stored below $-70\,°C$ to limit K-attack on the Lewis base.[103] A related study has shown that Bu^nK does not in fact need to be prepared separately, simple mixing of Bu^tOK, Bu^nLi, TMEDA (1:1:1) in hexane or pentane at $< -40\,°C$ giving the first reagent which can efficiently metallate ethene.[104] A like-system, Bu^tOK, Bu^nLi in THF-hexane, has been used to metallate and subsequently derivatize norbornene and norbornadiene,[105] so another fairly safe prediction would be that 'mixed' or composite reagents are here to stay. Interestingly, this final comment extends to Group 2 organometallics, as in the use of CuI with Grignard reagents in Et_2O–THF at $-60\,°C$ to open β-epoxy ring systems rapidly.[106] The future synthetic application of alternative organocalcium halides may also have received a boost through a report of a much improved method for their preparation from organic halides and calcium ultrafine particles produced by a heated tungsten surface.[107]

[98] T. Cohen and B.-S. Guo, *Tetrahedron*, 1986, **42**, 2803.
[99] J. P. Gillet, R. Sauvêtre, and J. F. Normant, *Synthesis*, 1986, 355.
[100] R. Beckhaus and K.-H. Thiele, *J. Organomet. Chem.*, 1986, **317**, 23.
[101] A. Maercker, T. Graule, and U. Girreser, *Angew. Chem., Int. Ed. Engl.*, 1986, **25**, 167.
[102] C. G. Screttas and M. Micha-Screttas, *J. Organomet. Chem.*, 1986, **316**, 1.
[103] R. Pi, W. Bauer, B. Brix, C. Schade, and P. von R. Schleyer, *J. Organomet. Chem.*, 1986, **306**, Cl.
[104] L. Brandsma, H. D. Verkruijsse, C. Schade, and P. von R. Schleyer, *J. Chem. Soc., Chem. Commun.*, 1986, 260.
[105] H. D. Verkruijsse and L. Brandsma, *Recl. Trav. Chim. Pays-Bas*, 1986, **105**, 66.
[106] R. Tanikaga, K. Hosoya, and A. Kaji, *J. Chem. Soc., Chem. Commun.*, 1986, 836.
[107] K. Mochida, S. Ogura, and T. Yamanishi, *Bull. Chem. Soc. Jpn.*, 1986, **59**, 2633.

The mechanisms of numerous reactions of organometallics, particularly those of Li, have been probed by diverse techniques. MO calculations have shown that complexation of ethene with Li^+ results in a strong kinetic acceleration of methyl radical addition; thus complex (11), stabilized by 24.3 kcal mol^{-1} relative to C_2H_4 + Li^+, adds $\dot{C}H_3$ *via* transition state (12), the final radical-cation product (13) being

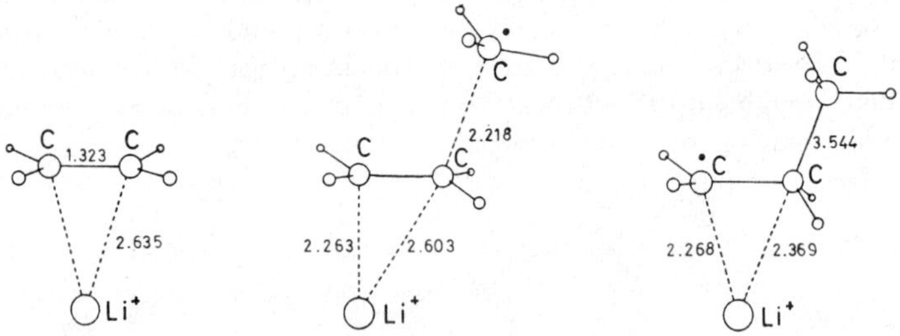

(11)   (12)   (13)

23.3 kcal mol^{-1} more stable than Li^+ + $CH_3CH_2\dot{C}H_2$.[108] The interconversion of *cis*-1,2-dilithioethylene (14) into the more stable (experimentally and by theory) *trans*-isomer (15) has also been investigated by *ab initio* methods and shown to proceed by in-plane inversion of a CH group through a planar *cis* form (16) and transition state (17).[109] No fewer than seven mechanisms were considered for the

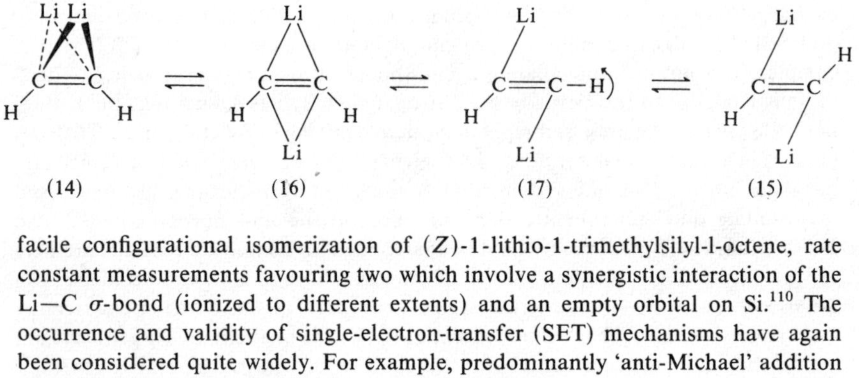

(14)   (16)   (17)   (15)

facile configurational isomerization of (*Z*)-1-lithio-1-trimethylsilyl-l-octene, rate constant measurements favouring two which involve a synergistic interaction of the Li—C σ-bond (ionized to different extents) and an empty orbital on Si.[110] The occurrence and validity of single-electron-transfer (SET) mechanisms have again been considered quite widely. For example, predominantly 'anti-Michael' addition of Bu^{t}Li to (*E*)-3-(trimethylsilyl)- and (*E*)-3-phenylpropenoic acids has been so explained.[111] For reactions of various 1-arylpropenyllithiums with t-alkyl bromides, however, analysis of alkylation products implies that two alternative pathways, SET and nucleophilic substitution, can operate. Their relative contributions depend on the electronic stability of the organolithium precursor and the polarity of the solvent

[108] T. Clark, *J. Chem. Soc., Chem. Commun.*, 1986, 1774.
[109] P. von R. Schleyer, E. Kaufmann, A. J. Kos, T. Clark, and J. A. Pople, *Angew. Chem., Int. Ed. Engl.*, 1986, **25**, 169.
[110] E. Negishi and T. Takahashi, *J. Am. Chem. Soc.*, 1986, **108**, 3402.
[111] K. J. H. Kruithof, A. Mateboer, M. Schakel, and G. W. Klumpp, *Recl. Trav. Chim. Pays-Bas*, 1986, **105**, 62.

system.[112] In contrast, no evidence has been found for the intermediacy of free-radicals in metal–halogen exchanges of 1-iodo-5-hexenes with Bu^{t}Li, and in particular and contrary to earlier results, no products attributable to cyclization of 5-hexen-1-yl radicals could be detected.[113] The Li—Sn exchange reaction between MeLi and Me$_4$Sn in HMPA has been followed by ^{119}Sn n.m.r. spectroscopy, the data pinpointing an intermediate hypervalent pentaorgano tin 'ate complex Li$^+$SnMe$_5^-$; interestingly, given the report above, these findings suggest that analogous species may be generally involved in Li-metalloid (including halide) exchanges.[114] Other possible mechanistic effects are also proving contentious, not least the role of metal cation$\cdots$electron-rich-group interactions in assisting and/or directing the course of reactions. For example, $E : Z$ ratios of Ph$_2$CH.CH=C(OLi)Mes formed by reactions of PhLi with PhCH=CHCOMes and Ph$_2$CHCH$_2$COMes vary according to the Li$^+$-solvating ability of the medium in a way which suggests that $\geq$C=O$\cdots$Li$^+$ coordination helps direct stereoselective formation of the E-enolate.[115] Further, metal-assisted ionization of fluoride on treatment of optically active fluorocyclo-hexylidenes with Bu^{t}Li has been invoked to explain configuration retention after LiF elimination.[116] Notwithstanding this, the lithiation of various tricarbonyl(fluoroanisole)chromium complexes, including (OC)$_3$Cr.p-MeOC$_6$H$_4$F, has been shown to occur exclusively *ortho* to F rather than to the MeO-group, this despite the fact that in uncomplexed arenes the latter has the greater 'directing ability'; the crystal structure of the above complex indicates that the O is not sterically unavailable to the incoming organolithium, the implication being that it is the inductive effect of F, rather than intermolecular coordinations, which determines kinetic acidity in these cases.[117] Finally, for mechanisms concerning Li species, 1,1'-dilithioferrocene.2TMEDA when added to ethylene oxide promotes its ring-opening anionic polymerization probably, since Fe occurs in the polymer, *via* ferrocenylalkoxy Li complexes.[118] Two interesting and related papers have addressed the mechanistics of Grignard reagent formation. In one, the alkyl radicals responsible for alkyl isomerizations during reagent formation have been shown, by analysis of isomerization kinetics, to be freely diffusing intermediates in solution, and not Mg-surface adsorbed.[119] In the other, a study of the chemisorption and subsequent decomposition of MeBr on a Mg(0001) single-crystal surface, neat or in Me$_2$O, has shown that products are a surface bromide and gas-phase hydrocarbons, *i.e.* no metal–carbon bond is formed, and no stable surface alkyls are detectable.[120]

Structural investigations of organometallics continue to rely on a wide range of techniques. Emphasizing this, the following report is organized according to the type of compound under study.

For dilithiated species, the stability associated with the ion triplet model (based on electrostatic interactions between a dianion and two bridging cations) has been

[112] J. Tanaka, M. Nojima, and S. Kusabayashi, *J. Chem. Soc., Chem. Commun.*, 1986, 242.
[113] W. F. Bailey, J. J. Patricia, T. T. Nurmi, and W. Wang, *Tetrahedron Lett.*, 1986, **27**, 1861.
[114] H. J. Reich and N. H. Phillips, *J. Am. Chem. Soc.*, 1986, **108**, 2102.
[115] E. P. Ignatova-Avramova and I. G. Pojarlieff, *J. Chem. Soc., Perkin Trans. 2*, 1986, 69.
[116] J. Rachon, V. Goedken, and H. M. Walborsky, *J. Am. Chem. Soc.*, 1986, **108**, 7435.
[117] J. P. Gilday and D. A. Widdowson, *J. Chem. Soc., Chem. Commun.*, 1986, 1235.
[118] K. Gonsalves and M. D. Rausch, *J. Polym. Sci., Polym. Chem.*, 1986, **24**, 1419.
[119] J. F. Garst, J. E. Deutch, and G. M. Whitesides, *J. Am. Chem. Soc.*, 1986, **108**, 2490.
[120] R. G. Nuzzo and L. H. Dubois, *J. Am. Chem. Soc.*, 1986, **108**, 2881.

further confirmed by MO calculations on 1,3-dilithiocyclobutane, the results implying that the second pK_a of cyclobutane is actually 5.3 units lower than the first.[121]
A distinctly different, non-equivalent bridging mode was observed for $Li_2C_2H_4$,
produced along with $Li(C_2H_4)_n$, $n = 1, 2, 3$, on condensation of the constituents in
an argon matrix; i.r. spectral analysis pinpoints a structure (18), while Li's partial
covalency is emphasized by the formation of merely van der Waals complexes
between C_2H_4 and heavier alkali metals.[122] The novel dilithiated dibenzyl compound
(19) has been shown by multinuclear n.m.r. spectroscopy to incorporate two very
different benzyllithium moieties,[123] and non-equivalent Li nuclei have also been
observed in THF solutions of 3,4-dilithio-2,5-dimethyl-2,4-hexadiene, two ^{6}Li n.m.r.
signals being apparent at -70 °C: ^{6}Li, ^{6}Li COSY, performed for the first time, proves
that these signals represent different environments within just one cluster, preliminary
X-ray results suggesting structure (20) with four inner and four outer Li atoms.[124]

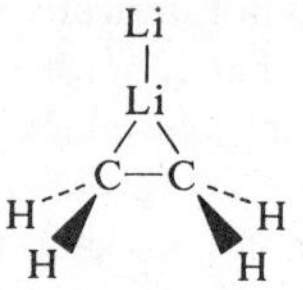

(18) 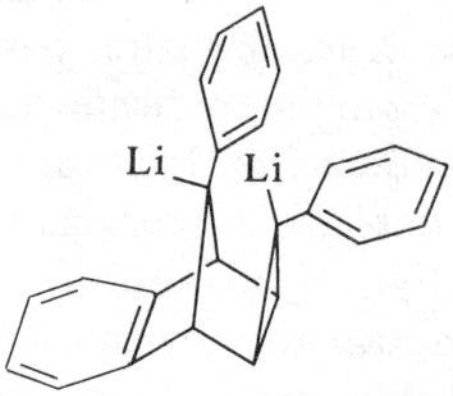   (19)

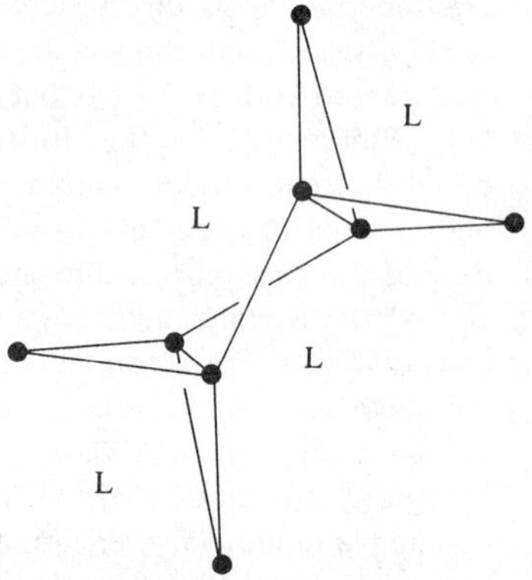

● = Li
L = $[Me_2C=C-C=CMe_2]^{2-}$
(20)

A further new n.m.r. method, ^{6}Li–^{1}H 2D HOESY, has detected the spatial proximity
of each Li to one *ortho* phenyl proton in the dimer of 2-Li-1-phenylpyrrole, so tying
in with an X-ray structure analysis showing a similar disposition and with earlier
experimental observations that second lithiation does indeed occur at this particular,
presumably activated, C—H bond.[125] Similar Li···HC stabilizing interactions may
be present also within the crystal structure of the hexamer $(Me_3SiCH_2Li)_6$.[126] Without
doubt, one of the most interesting structures of 1986 was that of the dilithium salt
of a diboratabenzene, $[Li(TMEDA)]_2 \cdot [1, 2\text{-}C_4H_4 (BNMe_2)_2]$, Figure 3, whose 6π-
electron C_4B_2 ring is near-planar, with the Li units slipped away from the B atoms
towards C-2 and C-2'.[127]

[121] S. M. Bachrach, *J. Am. Chem. Soc.*, 1986, **108**, 6406.
[122] L. Manceron and L. Andrews, *J. Phys. Chem.*, 1986, **90**, 4514.
[123] D. Hoell, J. Lex, and K. Müllen, *J. Am. Chem. Soc.*, 1986, **108**, 5983.
[124] H. Günther, D. Moskau, R. Dujardin, and A. Maercker, *Tetrahedron Lett.*, 1986, **27**, 2251.
[125] W. Bauer, G. Müller, R. Pi, and P. von R. Schleyer, *Angew. Chem., Int. Ed. Engl.*, 1986, **25**, 1103.
[126] B. Tecle, A. F. M. M. Rahman, and J. P. Oliver, *J. Organomet. Chem.*, 1986, **317**, 267.
[127] G. E. Herberich, B. Hessner, and M. Hostalek, *Angew. Chem., Int. Ed. Engl.*, 1986, **25**, 642.

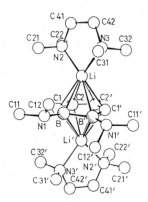

Figure 3 *Molecular structure of* [Li(TMEDA)]$_2$.[1,2-C$_4$H$_4$(BNMe$_2$)$_2$]
(Reproduced by permission from *Angew. Chem., Int. Ed. Engl.*, 1986, **25**, 642)

Turning to monolithiated compounds, the mass spectra of several alkyllithiums $(RLi)_n$ have been recorded in the range 10–70 eV, mass peak data showing that when R = Me, Pr^i, Bu^{s,t} gas-phase tetramers occur, while R = Pr^n, Bu^{n,i} are mixes of tetramers and hexamers.[128] Several structures of alkyllithiums bearing an intramolecular Lewis base function have appeared, including those of the tetrameric pseudo-cubanes $(LiCH_2CH_2CH_2NMe_2)_4$[129] and $(LiCH_2CH_2CH_2OMe)_4$[130] in which each internal donor atom coordinates as a fourth ligand to one of the Li atoms. The enthalpies of intramolecular etheration of the latter compound and of several related γ-methoxy $(MeO—R—Li)_n$ species were also determined[131] (see also ref. 93). Similar, but tripled, coordinations occur in the dimer $[(MeO.Me_2Si)_3CLi]_2$ (21), each of the component monomers having one of its own OMe-groups internally coordinated to its own Li atom, and its other two OMe groups bridging to the Li atom of the other monomer.[132] The ubiquitous bicyclo[3.2.1]octa-2,6-dienyl anion and its Li salt (22) have concerned three groups in efforts to resolve finally the stabilizing role, if any, of homoaromaticity. The crystalline complex of (22) with TMEDA is monomeric with Li almost centrally placed below the 7-membered ring so engaging both the allylic and olefinic functions, which seem to move towards each other.[133] MO calculations have shown, it seems quite conclusively, that homoaromaticity is a negligible factor in stabilizing the gas-phase anion compared to hyperconjugation and inductive effects, and that in solutions of the salt alkenyl–Li$^+$ interactions contribute also to stability.[134] Anticipated η^3-allyl Li moieties have been found for the first time in the crystal structure of 1,3-diphenylallylLi.OEt$_2$ (23),

[128] D. Plavšić, D. Srzić, and L. Klasinc, *J. Phys. Chem.*, 1986, **90**, 2075.

[129] K. S. Lee, P. G. Williard, and J. W. Suggs, *J. Organomet. Chem.*, 1986, **299**, 311.

[130] G. W. Klumpp, P. J. A. Geurink, N. J. R. van Eikema Hommes, F. J. J. de Kanter, M. Vos, and A. L. Spek, *Recl. Trav. Chim. Pays-Bas*, 1986, **105**, 398.

[131] P. J. A. Geurink and G. W. Klumpp, *J. Am. Chem. Soc.*, 1986, **108**, 538.

[132] N. H. Buttrus, C. Eaborn, S. H. Gupta, P. B. Hitchcock, J. D. Smith, and A. C. Sullivan, *J. Chem. Soc., Chem. Commun.*, 1986, 1043.

[133] N. Hertkorn, F. H. Köhler, G. Müller, and G. Reber, *Angew. Chem., Int. Ed. Engl.*, 1986, **25**, 468.

[134] P. von R. Schleyer, E. Kaufmann, A. J. Kos, H. Mayr, and J. Chandrasekhar, *J. Chem. Soc., Chem. Commun.*, 1986, 1583; R. Lindh, B. O. Roos, G. Jonsall, and P. Ahlberg, *J. Am. Chem. Soc.*, 1986, **108**, 6554.

(21)   (22)   (23)

polymeric chains forming with Li atoms lying above and below the near-planar allyl ligands.[135]

Three especially interesting crystal structures of organo-derivatives of heavier alkali metals have been reported. The benzylsodium complex, $(PhCH_2Na.TMEDA)_4$, is the first organo alkali-metal tetramer shown to prefer an 8-membered ring over a tetrahedral arrangement, the four Na atoms forming a square edge-bridged by benzyl α-C atoms and possibly, more weakly, by *ipso* carbons.[136] The structure of the allenylsodium complex, $Na(TMEDA)_2^+.[Bu^tC^-=C=C(Me)-C\equiv CBu^t]$, Figure 4, provides a superb

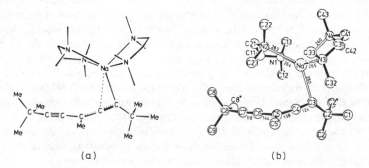

(a)   (b)

Figure 4 *Crystal structure of* $Na(TMEDA)_2^+.Bu^t{}^-C=C=C(Me)-C\equiv C-Bu^t$: (a) *schematic representation, and* (b) *SCHAKAL drawing with key bond lengths* (pm) (Reproduced by permission from Angew. Chem., Int. Ed. Engl., 1986, **25**, 902)

example of why supposedly 'free' carbanions just aren't this, at least in the solid state: the Na^+ ('counterion' seems a somewhat dismissive term), aided by the presence of two TMEDA ligands, localizes the negative charge so maximizing its electrostatic interaction with just the terminal C and thereby over-riding the normal preference for delocalization.[137] In the diglyme adduct of dipotassium bis(t-butyl[8]annulene)ytterbate, Yb is sandwiched between two parallel annulene rings, the O-complexed K atoms capping the other sides of these rings.[138]

[135] G. Boche, H. Etzrodt, M. Marsch, W. Massa, G. Baum, H. Dietrich, and W. Mahdi, *Angew. Chem., Int. Ed. Engl.*, 1986, **25**, 104.
[136] C. Schade, P. von R. Schleyer, H. Dietrich, and W. Mahdi, *J. Am. Chem. Soc.*, 1986, **108**, 2484.
[137] C. Schade, P. von R. Schleyer, M. Geissler, and E. Weiss, *Angew. Chem., Int. Ed. Engl.*, 1986, **25**, 902.
[138] S. A. Kinsley, A. Streitwieser, and A. Zalkin, *Acta Crystallogr., Sect. C*, 1986, **42**, 1092.

In Group 2 organometallic chemistry, the Mg alkyl $\{\overline{Mg[NC_5H_4C}(SiMe_3)_2 - 2]_2\}$ has, like Zn and Cd analogues, a monomeric structure; metal 4-coordination is provided by two anionic C and two dative pyridyl N centres.[139] Various O-, N-, and P- complexes of Cp_2Mg have been prepared and relative base strengths estimated by $^{13}C/^{25}Mg$ shift correlations; in the same study, $CpMgOEt$ was shown to have a $(MgO)_4.(exo-\eta^5-Cp)_4$ cubane structure.[140] The solution fluxionality of Be in beryllocene between one η^5-ring and one peripherally bonded ring has been investigated by observation of partially relaxed coupling to ^9Be in the ^{13}C n.m.r. spectrum, a rate of molecular inversion of *ca.* $10^{10} s^{-1}$ being implied.[141] In a similar area, gasphase reactions of Ca or Sr with CpH have allowed laser spectroscopic detection of CaCp and SrCp free-radicals with open-faced sandwich C_{5V} symmetry; this dual laser chemistry seems particularly promising, both prompting the chemistry by exciting metal atoms and allowing investigations of ensuing products.[142]

4 Coordination Chemistry

Several calculational and spectroscopic methods have been used to explore ion hydration, and solvation in general. The Born model was extended by incorporating a distance-dependent permittivity into the expression for the electrostatic energy between M^I, M^{II} ions and surrounding water molecules, so-calculated hydration energies agreeing well with experimental values.[143] A similarly accurate empirical relationship allowing calculation of enthalpies of formation of Group 1 and 2 hydrates from their anhydrous salts has also been described.[144] The apparent requirement of Na^+ to maintain octahedral coordination during Na^+Cl^- hydration was demonstrated by computer simulations, the shell comprising $5 H_2O$, $1 Cl^-$ in the contact ion-pair, with slight separation causing H_2O rotation to form bridges to the Cl^-, and with entry of a sixth H_2O on formation of the solvent-separated pair.[145] Matrix-isolated gaseous ion-pairs $M^+ClO_3^-$ and $H_2O.M^+ClO_3^-$ have been compared at 700K by f.t.-i.r. spectroscopy, results showing that, certainly for M = Li, bidentate ligand binding is preferred.[146] Far-i.r. spectra for LiX (X = Hal, BF_4, ClO_4) in DMSO indicate fivefold solvent primary coordination of Li^+ in a trigonal bipyramidal geometry, though this solvated cation is also associated with an external anion.[147] Kinetic data from ultrasonic relaxation spectra have been linked with structural information from i.r. spectra in a study of MSCN (M = Li, Na) in THF, a solvent-separated species, $LiNCS\cdots LiNCS$, and a contact dimer, $(NaNCS)_2$, being observed.[148] For Group 2 salts $M(ClO_4)_2$ (M = Mg, Ca, Sr), i.r., Raman, and n.m.r. measurements indicate little ion-association when M = Mg, but formation of both

[139] M. J. Henderson, R. I. Papasergio, C. L. Raston, A. H. White, and M. F. Lappert, *J. Chem. Soc., Chem. Commun.*, 1986, 672.
[140] H. Lehmkuhl, K. Mehler, R. Benn, A. Rufińska, and C. Krüger, *Chem. Ber.*, 1986, **119**, 1054.
[141] K. W. Nugent and J. K. Beattie, *J. Chem. Soc., Chem. Commun.*, 1986, 186.
[142] L. C. O'Brien and P. F. Bernath, *J. Am. Chem. Soc.*, 1986, **108**, 5017.
[143] M. Bucher and T. L. Porter, *J. Phys. Chem.*, 1986, **90**, 3406; M. Bucher, *ibid.*, 1986, **90**, 3411.
[144] P. Vieillard and H. D. B. Jenkins, *J. Chem. Res. (S)*, 1986, 444, 446.
[145] A. C. Belch, M. Berkowitz, and J. A. McCammon, *J. Am. Chem. Soc.*, 1986, **108**, 1755.
[146] G. Ritzhaupt and J. P. Devlin, *J. Phys. Chem.*, 1986, **90**, 6764.
[147] S. Chang, P. P. Schmidt, and M. W. Severson, *J. Phys. Chem.*, 1986, **90**, 1046.
[148] D. Saar and S. Petrucci, *J. Phys. Chem.*, 1986, **90**, 3326.

solvent-shared and contact anion–cation complexes for the other two salts.[149] An X-ray structure is also relevant here, that of $LiNO_3.(diacetamide)_2$ demonstrating conclusive cation–anion coordination with Li^+ 5-coordinate to the four neutral donor O atoms and to one O of the NO_3^-.[150]

Regarding non-cyclic polyethers, the electrochemical reduction of (24) to its anion radical 'switches on' Li^+ transport across a membrane, e.s.r. work showing formation of a strong $(24)^{\overline{\cdot}}$ Li^+ ion pair.[151] 3,7-Dioxaazelaamides like (25) are effective

(24)   (25)

ionophores with Li^+/Na^+ selectivities up to 80, and sensor membranes based on them allow reliable measurements of blood serum Li^+ levels.[152] A new class of open-ionophores, organometallics of type $CpCo[P(O)(OR)_2]_3^-$, have also been described,[153] and f.a.b. mass spectrometry has been used to monitor complex formation between M^+ ions and tris(3-6-dioxaheptyl)amine in solution; results showing that Li^+ is most strongly bound were substantiated by i.r. and n.m.r. spectra, so giving confidence in the method.[154]

Aside from macrocyclic ligand complexes (dealt with later in their own subsection), the remainder of this section summarizes some of the major 1986 publications on metal-anion systems with the primary atom of the anionic ligand being drawn from Groups 7, 6 (O and S), 5 (N, P, As), 4 (Si); the Introduction noted that much significant work is going on in this area.

Several lithium derivatives of the above types have been used as reagents in key organic syntheses. Chiral lithium amide bases such as (26) have given similarly chiral products (for use as synthons) *via* enantioselective deprotonations of ketones[155] and of 4-alkylcyclohexanones;[156] the latter paper also discusses the steric and

[149]   D. W. James and P. G. Cutler, *Aust. J. Chem.*, 1986, **39**, 137.
[150]   D. D. Bray and N. F. Bray, *Inorg. Chim. Acta*, 1986, **111**, L39.
[151]   L. Echeverria, M. Delgado, V. J. Gatto, G. W. Gokel, and L. Echegoyen, *J. Am. Chem. Soc.*, 1986, **108**, 6825.
[152]   E. Metzger, R. Aeschimann, M. Egli, G. Suter, R. Dohner, D. Ammann, M. Dobler, and W. Simon, *Helv. Chim. Acta*, 1986, **69**, 1821.
[153]   H. Shinar, G. Navon, and W. Klaui, *J. Am. Chem. Soc.*, 1986, **108**, 5005.
[154]   J. M. Miller, S. J. Brown, R. Theberge, and J. H. Clark, *J. Chem. Soc., Dalton Trans.*, 1986, 2525.
[155]   N. S. Simpkins, *J. Chem. Soc., Chem. Commun.*, 1986, 88.
[156]   R. Shirai, M. Tanaka, and K. Koga, *J. Am. Chem. Soc.*, 1986 **108**, 543.

structural (*e.g.* N-lone pair orientation, carbonyl$\cdots$Li$^+$ coordination) factors governing such selectivity. Similarly mechanistic, reactions of LiN^{i}Pr$_2$ with *N,N*-dialkylpyridinecarboxamides are thought to proceed either through SET to give radical anions or through unstable lithiated amide intermediates.[157] The role of 'additives' (*e.g.* LiCl, HMPA, TMEDA) in increasing the metallating ability of various lithium amides has been probed by measuring pK_a values of five carbon acids in the presence of such mixed systems, the conclusion being that thermodynamic acidity is little altered *i.e.* the additives are beneficial by increasing *kinetic* acidity, with TMEDA being the most effective.[158] The aminating reagent generated from MeLi and MeO.NH$_2$ (probably straightforward MeO.NHLi rather than nitrene :$\ddot{\text{N}}$H *via* MeOLi elimination) has been used to prepare successfully a polystyrene polymer with amine end-group functionality.[159] Phospha- and arsa-ethynes, R-C≡E (E = P, As), the latter being the first stable compound with a C≡As triple bond, have been obtained *via* treatment of various acid chlorides RCOCl with LiE(SiMe$_3$)$_2$.2THF reagents.[160] For M—O systems (M = Li, K), ring opening of epoxides to give β-metallated alkoxides (27) has been effected using arene radical anion salts,[161] while treatment of fluoronitroalkanes with Bu^{n}Li produced the first relatively stable (up to −70 °C) dilithiofluoro-derivatives such as (28) which, through treatment with carbonyl organics, make accessible a wide variety of new functionalized organofluorine compounds.[162]

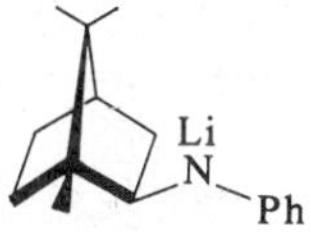

(26) 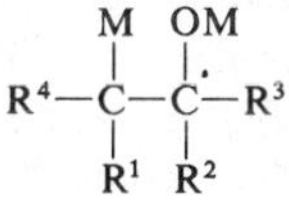   (27) 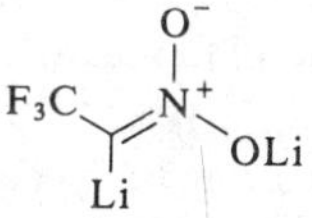   (28)

Structural studies on M—X bonded species have been impressively wide, employing many techniques and with X being drawn from Main Groups 7 → 4. This report follows the descending periodic order, starting with M = Li, then to the heavier alkali metals, and finally to Group 2 metals. For X = halogen, MeLi with the aminofluorosilane (Me$_2$CH)$_2$Si(F).NH.C$_6$H$_2$Bu$^t_3$ gave the most unusual LiF iminosilane adduct (29).[163] Equally novel are the *X*-ray-characterized products from reactions of PhSi(NLiSiMe$_3$)$_3$ with CpTiCl$_3$, *viz.* (30) with a titanabicycloheptane structure incorporating a Ti—Cl—Li(OEt$_2$)—N—Si bridge,[164] and of (Me$_3$Si)$_3$CLi.2THF (made *via* MeLi containing LiCl) with PMDETA [L = Me$_2$N(CH$_2$)$_2$.NMe.(CH$_2$)$_2$NMe$_2$], this being [L.Li(Cl)Li.L]$^+${Li[C(SiMe$_3$)$_3$]$_2$}$^-$ with a linear Li—Cl—Li linkage within the cation.[165] Diverse Li—O systems have been

[157] J. Epsztajn, J. Z. Brzeziński, and A. Jóźwiak, *J. Chem. Res. (S)*, 1986, 18.
[158] R. R. Fraser and T. S. Mansour, *Tetrahedron Lett.* 1986, **27**, 331.
[159] R. P. Quirk and P.-L. Cheng, *Macromolecules*, 1986, **19**, 1291.
[160] G. Märkl and H. Sejpka, *Angew. Chem., Int. Ed. Engl.*, 1986, **25**, 264.
[161] E. Bartmann, *Angew. Chem., Int. Ed. Engl.*, 1986, **25**, 653.
[162] D. Seebach, A. K. Beck, and P. Renaud, *Angew. Chem., Int. Ed. Engl.*, 1986, **25**, 98.
[163] R. Boese and U. Klingebiel, *J. Organomet. Chem.*, 1986, **315**, C17.
[164] D. J. Brauer, H. Bürger, and G. R. Liewald, *J. Organomet. Chem.*, 1986, **307**, 177.
[165] N. H. Buttrus, C. Eaborn, P. B. Hitchcock, J. D. Smith, J. G. Stamper, and A. C. Sullivan, *J. Chem. Soc., Chem. Commun.*, 1986, 969.

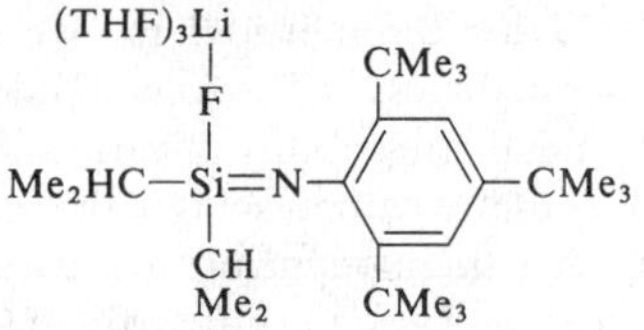

(29) 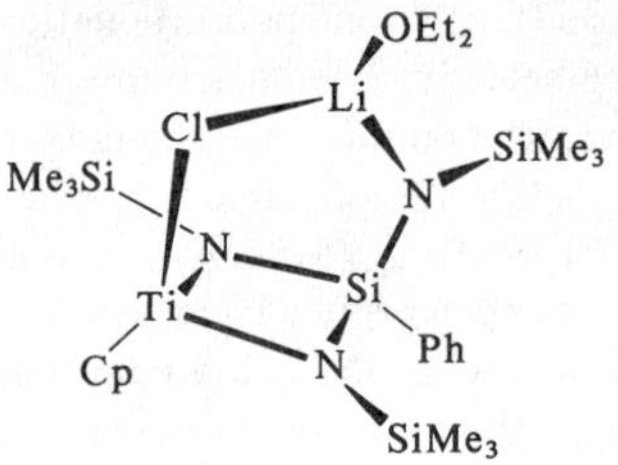   (30)

probed, among them the salt of the dianion of 3-methylbut-2-enoic acid, $H_2\bar{C}.CH=CH.COO^-$, for which ^{13}C n.m.r. spectroscopy and *ab initio* calculations suggest a delocalized structure with one Li^+ in the plane of the carboxy group but with the other above the allyl C_3 backbone plane.[166] The simplest possible enolate has been shown by multinuclear n.m.r. spectroscopy to exist in THF as two similar complexed tetramers, $(H_2C=CHOLi.THF)_4$, with a barrier of only *ca.* 6.6 kJ mol^{-1} for rotation of the vinyl group.[167] The alkoxide $[(Me_3Si)_3CSiMe_2OLi.THF]_2$ is also tetrameric in solution, although it exists as a dimer in the solid with, unusually, a *non*-planar Li_2O_2 ring—suggesting considerable steric strain in the molecule, this in turn presumably preventing further association to the usual cubane.[168] Dimers have also been found for the α-sulphinyllithium $\{[PhCMe.S(O)Ph]^-Li^+.TMEDA\}_2$ which has a Li_2O_2 4-membered ring, but no C—Li bond,[169] and for the α-sulphonylallyllithium, $[CH_2=CH.CH.S(Ph)(O)OLi.diglyme]_2$, now with a LiOSO.LiOSO 8-membered ring but yet again with the C atoms of the allyl lying outside Li coordination spheres;[170] both results are crucial to understanding the operation of these supposed 'carbanions' in asymmetric organic syntheses. Contrastingly, the α-sulphonimidoylmethyl carbanion, $\{[H_2C—S(O)(Ph)NMe]^-Li^+\}_4.2TMEDA$ deserves the label in part, 2Li being coordinated by one TMEDA and two O atoms each, but each of the other two by 3N atoms of the sulphoximide groups and also by one C of each of the CH_2 groups.[171] Although the above three (S—O)Li species show no coordinative involvement of their S atoms, that of the monothiobenzoate $(PhCOSLi.TMEDA)_2$ certainly does so, having an 8-membered ring composed of co-planar $(COS)_2$ units and of two TMEDA-complexed Li atoms above and below this plane; the deviation from planarity seems required to allow donor attachment, *ab initio* optimized $(HCOSLi)_2$ being planar but showing excessively large ring angles at Li centres (SLiO 163°, *cf. ca.* 124° in the crystal structure).[172] Helping to lead on to Li-Group 5 chemistry, three interesting γ-P functionalized alkoxides result from addition of $R_2PCH_2Li.TMEDA$ across $Bu^t_2C=O$, two being of type (31), R = Me or Ph, with fused tricyclic structures involving both $P \rightarrow Li$ and Li—O

[166] A. Bongini, M. Orena, and S. Sandri, *J. Chem. Soc., Chem. Commun.*, 1986, 50.
[167] J. Q. Wen and J. B. Grutzner, *J. Org. Chem.*, 1986, **51**, 4220.
[168] P. B. Hitchcock, N. H. Buttrus, and A. C. Sullivan, *J. Organomet. Chem.*, 1986, **303**, 321.
[169] M. Marsch, W. Massa, K. Harms, G. Baum, and G. Boche, *Angew. Chem., Int. Ed. Engl.*, 1986, **25**, 1011.
[170] H.-J. Gais, J. Vollhardt, and H. J. Lindner, *Angew. Chem., Int. Ed. Engl.*, 1986, **25**, 939.
[171] H.-J. Gais, I. Erdelmeier, H. J. Lindner, and J. Vollhardt, *Angew. Chem., Int. Ed. Engl.*, 1986, **25**, 938.
[172] D. R. Armstrong, A. J. Banister, W. Clegg, and W. R. Gill, *J. Chem. Soc., Chem. Commun.*, 1986, 1672.

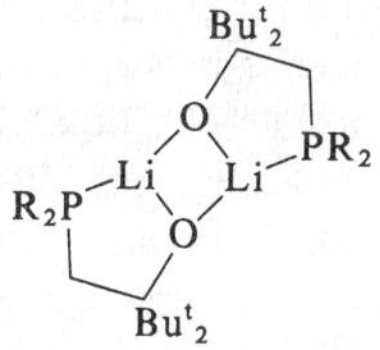

(31) 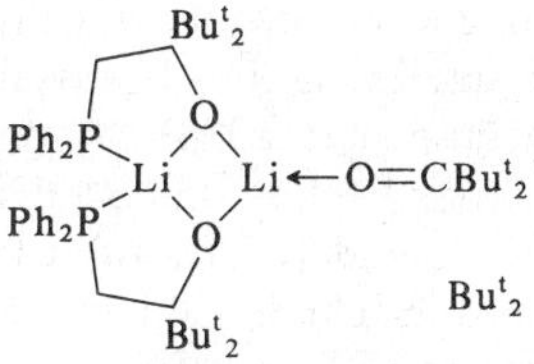   (32)

coordination, and the third being a ketone adduct (32) isolated as an intermediate on the way to (31), R = Ph.[173]

For solely Group 5 anionic ligands, the first X-ray structure of an α-nitrile 'carbanion' has appeared, $(PhCH\overset{..}{-}C\overset{..}{=}NLi.TMEDA)_2$ being a slightly puckered $(NLi)_2$ ring dimer with near-linear CCN units.[174] In contrast, clustered species are found for bis- and tris-(lithioamino)silanes; $[(Bu^tNLi)_2SiMe_2]_2$ has a Li_4 bis-phenoidal core whose two pairs of fused Li_3 faces are linked by the two NSiN bridges, and $[(Bu^tNLi)_3SiPh]_2$ has a Li_6 trigonal antiprismatic core with N atoms of the two ligand moieties in bridging positions over the six Li_3 faces. In both compounds the Li—N bonding is clearly electron-deficient and appreciably ionic; Li$\cdots$Li contacts are deemed electrostatically repulsive (not bonding, despite short Li$\cdots$Li distances) and there are indications that Li$\cdots$H$_3$C interactions aid cluster stability.[175] Attempts to rationalize and delineate cluster and ring formation in Li—N chemistry, and indeed more widely for Li—O (see *e.g.*, refs. 167, 168) and Li—C species, have come through ring-stacking and ring-laddering principles. Concerning the former, analysis of the dimensions of imino-Li hexamers $(RR'C{=}NLi)_6$ (R = Ph, R' = Bu^t or Me$_2$N) has shown that they are best regarded as pairs of stacked cyclic trimers; thus the flat N-ligands permit approach of two such rings, alternate electron-precise N—Li bonds of each transforming into 3-centre N$\cdots$Li$_2$ ones both within and between trimers, and thereby raising Li's coordination number from 2 to 3 (equation 9).[176] Contrastingly, amidolithium rings, $(RR'NLi)_n$, cannot stack since R groups project above and below the $(NLi)_n$ ring plane; instead, further association of such rings must occur laterally in a ladder-like manner, and such a ladder has

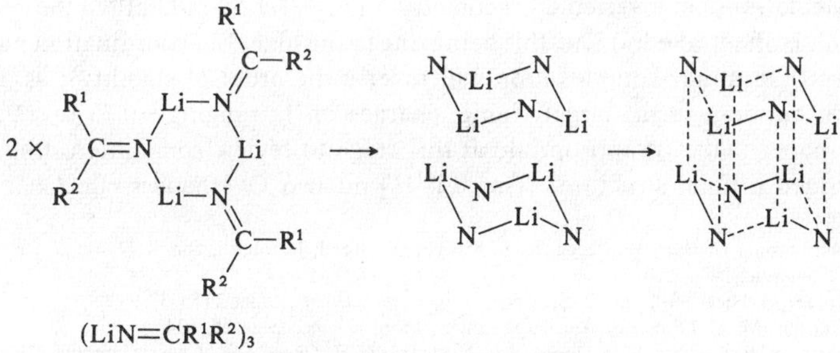

$(LiN{=}CR^1R^2)_3$

(9)

[173] L. M. Engelhardt, J. M. Harrowfield, M. F. Lappert, I. A. MacKinnon, B. H. Newton, C. L. Raston, B. W. Skelton, and A. H. White, *J. Chem. Soc., Chem. Commun.*, 1986, 846.

[174] G. Boche, M. Marsch, and K. Harms, *Angew. Chem., Int. Ed. Engl.*, 1986, **25**, 373.

[175] D. J. Brauer, H. Bürger, and G. R. Liewald, *J. Organomet. Chem.*, 1986, **308**, 119.

[176] D. Barr, W. Clegg, R. E. Mulvey, R. Snaith, and K. Wade, *J. Chem. Soc., Chem. Commun.*, 1986, 295.

been donor-intercepted in the structure of $[(\text{pyrrolididoLi})_3 \cdot \text{PMDETA}]_2$, Figure 5.[177] Several interesting Li—P structures have been solved, including that of monomeric $(\text{Ph}_2\text{P})_2\text{CHLi.TMEDA}$ which has no Li—C link, but rather a P—C—P—Li ring.[178] For lithium diorganophosphides proper, $(\text{Ph}_2\text{PLi.OEt}_2)_\infty$, $(\text{Ph}_2\text{PLi.2THF})_\infty$, and $[(\text{C}_6\text{H}_{11})_2\text{PLi.THF}]_\infty$ all have R_2P units alternating with Li.donor groups in infinite —Li—P—Li—P—chains.[179] Interesting 'ate complexes $\text{Bu}^t_2\text{PE.}(\text{PBu}^t_2)_2\text{Li.THF}$ with EP_2Li rings have resulted from reactions of Bu^t_2PLi in THF with ECl_2 (E = Sn or Pb),[180] and a similar bimetallic ring system, $\text{Lu}(\mu_2\text{-}P)_2\text{Li}$, was found for $\text{Cp}_2\text{Lu}(\text{PPh}_2)_2\text{Li.TMEDA}$.[181] The first structures of diorganoarsenides have also materialized, for $[\text{Bu}^t_2\text{As-As}^*(\text{Bu}^t)\text{Li.THF}]_2$ which has a central planar Li_2As^*_2 ring,[182] and for dimeric $(\text{Ph}_2\text{AsLi.2Et}_2\text{O})_2$ and monomeric $\text{Ph}_2\text{AsLi.}(1,4\text{-dioxane})_3$;[183] the latter paper also details the first two Li diorganostibinide structures, consisting of $[\text{Li}(12\text{-crown-4})_2]^+$ cations with separated $(\text{Ph}_2\text{Sb})^-$ and $(\text{Ph}_4\text{Sb}_3)^-$ anions.

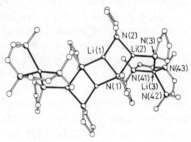

Figure 5 *Molecular structure of* $\{[\text{H}_2\overline{\text{C(CH}_2)_3\text{N}}\text{Li}]_3 \cdot \text{PMDETA}\}_2$
(Reproduced from *J. Chem. Soc., Chem. Commun.,* 1986, 869)

Reaching Group 4, Li—C compounds have of course been dealt with in Section 3 of this report but there is a developing interest in organosilyllithiums. Thus Ph_3SiLi and Ph_2MeSiLi mixtures in THF have been shown, on the basis of ^7Li n.m.r. data, to undergo bimolecular exchange between monomeric structures; the spectra also support essentially ionic Li—Si bonding.[184] The ion-pair $\text{Li}^+\text{SiH}_3^-$ prefers, according to MO calculations, an inverted C_{3v} geometry, $\text{Li}(\mu_2 - \text{H})_3\text{Si}$, rather than the conventional van't Hoff tetrahedral one, this being due to the higher Li coordination number so acquired (so donor complexation will reverse the order of stability); as above, the Li—Si bonding seems largely ionic, charges on Li ranging +0.73 to +0.93.[185] This last paper makes it appropriate at this stage to review some interesting work on H-bridged lithium structures. The $^6\text{Li}[^1\text{H}]$ nuclear Overhauser effect has been

[177] D. R. Armstrong, D. Barr, W. Clegg, R. E. Mulvey, D. Reed, R. Snaith, and K. Wade, *J. Chem. Soc., Chem. Commun.,* 1986, 869.

[178] D. J. Brauer, S. Hietkamp, and O. Stelzer, *J. Organomet. Chem.,* 1986, **299**, 137.

[179] R. A. Bartlett, M. M. Olmstead, and P. P. Power, *Inorg. Chem.,* 1986, **25**, 1243.

[180] A. M. Arif, A. H. Cowley, R. A. Jones, and J. M. Power, *J. Chem. Soc., Chem. Commun.,* 1986, 1446.

[181] H. Schumann, E. Palamidis, G. Schmid, and R. Boese, *Angew Chem., Int. Ed. Engl.,* 1986, **25**, 718.

[182] A. M. Arif, R. A. Jones, and K. B. Kidd, *J. Chem. Soc., Chem. Commun.,* 1986, 1440.

[183] R. A. Bartlett, H. V. R. Dias, H. Hope, B. D. Murray, M. M. Olmstead, and P. P. Power, *J. Am. Chem. Soc.,* 1986, **108**, 6921.

[184] E. Buncel, T. K. Venkatachalam, and U. Edlund, *Can. J. Chem.,* 1986, **64**, 1674.

[185] P. von R. Schleyer and T. Clark, *J. Chem. Soc., Chem. Commun.,* 1986, 1371.

used to show that the H-bridged systems of solid-state $(Me_{3-n} Ph_nSi)_3CB(\mu - H)_3Li.3THF$, $n = 0$, 1, are probably retained in solution, and measurement of the build-up rate of the n.O.e. even made it possible to estimate a Li$\cdots$H distance of 2.2 Å.[186] Studies such as these are obviously going to be extremely important in probing the reactivity and regioselectivity of metal borohydrides and related reducing agents. Two crystal structures have implicated Li$\cdots$H interactions. In $[(Me_3P)_3WH_5^-.Li^+]_4$, each W atom is connected to two neighbouring Li atoms *via* two normal $W(\mu_2$-H)Li bridges and then by one trifurcated (LiHWHLi) bridge,[187] and an agostic Li$\cdots$H contact (2.07 Å) with a Bu^t group – presumably prompted by the Li being just 2-coordinate – has been observed in the Li$(\mu_2$-O)$_2$Co ring alkoxide, $(Bu^t_3CO)_2Co[N(SiMe_3)_2]Li$;[188] in this same paper, two other structures are noteworthy, that of $(Bu^t_3CO)_2Co.ClLi(THF)_3$ for its Co—Cl—Li unit, and that of $[Li(THF)_{4.5}]^+.\{Co[N(SiMe_3)_2].(OCBu^t_3)_2\}^-$ for its novel pseudo-five-coordinate lithium cation.

Turning to include the heavier alkali metals, enolates of pinacolone, $[Bu^tC(=CH_2)O^-M^+]_n.mTHF$ with M = Na and K, as well as Li, have been structurally characterized; while the Li derivative is an unsolvated $(OLi)_6$ hexamer-viewed incidentally as two stacked and puckered $(OLi)_3$ rings (*cf.* ref. 176) – the Na one aggregates as a tetramer ($n = 4$) solvated not by THF ($m = 0$) but by unenolized ketone (*cf.* ref. 173), and the K complex is a THF-solvated hexamer ($m = 6$, $n = 6$).[189] A further contrast between Li and Na coordination chemistry is provided by the unusual neutral O-ligand, $N(CH_2CH_2POPh_2)_3.3H_2O[L.3H_2O]$ which with MBPh$_4$ (M = Li, Na) in butanol gives $[Li_2L_2]^{2+}.[LiL]_2^{2+}.4BPh_4$, with tetrahedral LiO$_4$ and trigonal LiO$_3$ environments in the different cations, and $(NaL)^+.(H_2O).(BuOH).BPh_4^-$, with the Na surrounded tetrahedrally by four O atoms, two from L and one each from water and butanol.[190] A cubane cluster accommodating *both* Li and Na has been isolated, $LiNa_3(HMPA)_3.[N=C(NMe_2)_2]_4$ being built up of novel μ_3-N—LiNa$_2$ and μ_3-N—Na$_3$ pyramids, though interestingly (see previous paragraph) the HMPA-bare Li may be involved in Li$\cdots$HC interactions to Me$_2$N-units within the imine ligand.[191] Studies on Na alone have included the structure of NaI.diglyme which forms as infinite zig-zag chains of *intact* NaI bonds linked together by bridging diglyme molecules.[192] The structure of the synthetically useful compound chloramine-T (33) as its trihydrate, shows Na$^+$ as having interactions with one sulphonyl oxygen and a chlorine from a neighbouring molecule, the usually assumed Na$^+\cdots$N contacts being absent.[193] Sodium reduction of $[Fe(acacen)]_2O$, acacen being $[MeC(O^-)CH=CMe=NCH_2]_2$, has given the Fe—Na—oxo aggregate $\{[Fe(acacen)]_2ONa\}_2$ (34) whose Na cations achieve hexacoordination within a cage provided by the O atoms of the two ligands.[194]

[186] A. G. Avent, C. Eaborn, M. N. A. El-Kheli, M. E. Molla, J. D. Smith, and A. C. Sullivan, *J. Am. Chem. Soc.*, 1986, **108**, 3854.
[187] A. R. Barron, M. B. Hursthouse, M. Motevalli, and G. Wilkinson, *J. Chem. Soc., Chem. Commun.*, 1986, 81.
[188] M. M. Olmstead, P. P. Power, and G. Sigel, *Inorg. Chem.*, 1986, **25**, 1027.
[189] P. G. Williard and G. B. Carpenter, *J. Am. Chem. Soc.*, 1986, **108**, 462.
[190] C. A. Ghilardi, P. Innocenti, S. Midollini, and A. Orlandini, *J. Chem. Soc., Dalton Trans.*, 1986, 2075.
[191] W. Clegg, R. E. Mulvey, R. Snaith, G. E. Toogood, and K. Wade, *J. Chem. Soc., Chem. Commun.*, 1986, 1740.
[192] R. E. Mulvey, W. Clegg, D. Barr, and R. Snaith, *Polyhedron*, 1986, **5**, 2109.
[193] M. M. Olmstead and P. P. Power, *Inorg. Chem.*, 1986, **25**, 4057.
[194] F. Arena, C. Floriani, A. Chiesi-Villa, and C. Guastini, *J. Chem. Soc., Chem. Commun.*, 1986, 1369.

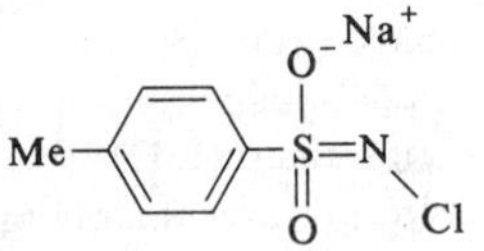

(33) 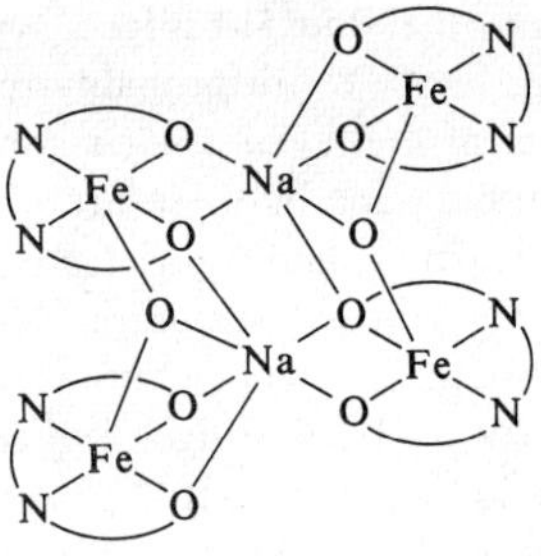   (34)

In Group 2 coordination chemistry, the mass and ^9Be n.m.r. spectra of several fluorinated β-diketonate–Be complexes have been reported.[195] The bimetallic dioctahedral cation of $[Mg_2Cl_3(THF)_6]^+.[MoOCl_4.THF]^-$ has been shown by X-ray diffraction to have the $(\mu-Cl)_3$-bridged structure shown in (35).[196] Four-coordination of Mg occurs in the first structure of a bromomagnesium ketone enolate; obtained from $Bu^t.MeCHBr.C{=}O$ and Mg in Et_2O, the dimeric complex has a slightly puckered $(MgO)_2$ ring and Z-alkene geometry (36), and it seems probable that such

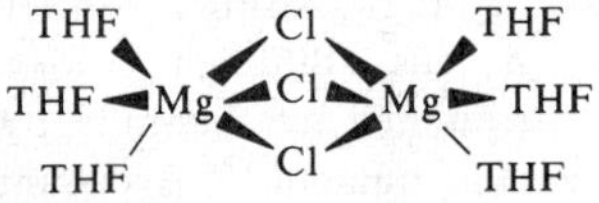

(35) 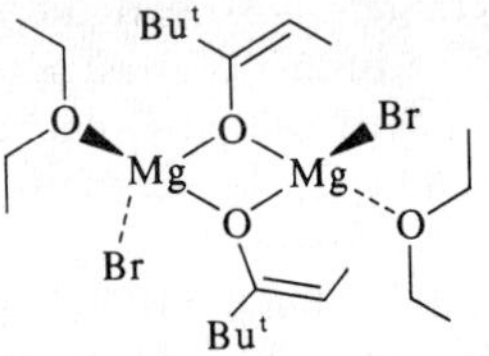   (36)

dimers may operate in aldol reactions.[197] Two especially interesting hetero-bimetallic complexes have resulted from partial or full displacement of ethene from $CpCo(C_2H_4)_2$ by Grignard reagents: the first, using PhMgBr in TMEDA, contains an intact complexed Grignard in the coordination sphere of Co (Figure 6a), while the second, derived via allylMgBr in THF, exhibits total C_2H_4-displacement along with transfer of the allyl group from Mg to Co (Figure 6b).[198] The crystal structure of the complex $Ca[S_2CC(CN)_2].5H_2O$ reveals each Ca with bicapped trigonal-prismatic coordination of six O and two N atoms, two of these O atoms being shared by a further Ca so giving a dimeric unit $[CaO_4(\mu-O)_2N_2]_2$ and with these dimers in turn being linked via their N atoms.[199] Two remarkable arene-soluble molecular alkoxostannates, $[Sn(OBu^t)_3.M.(OBu^t)_3Sn]$ with M = Sr and Ba, have been made; the structure of the former consisting of two trigonal bipyramids $Sn(\mu_2-OR)_3Sr$ sharing a common Sr vertex.[200] Alkaline earth metal monoalkoxide

[195] J. C. Kunz, M. Das, and D. T. Haworth, *Inorg. Chem.*, 1986, **25**, 3544.
[196] P. Sobota, T. Pluziński, and T. Lis. *Z. Anorg. Allg. Chem.*, 1986, **533**, 215.
[197] P. G. Williard and J. M. Salvino, *J. Chem. Soc., Chem. Commun.*, 1986, 153.
[198] K. Jonas, G. Koepe, and C. Krüger, *Angew. Chem., Int. Ed. Engl.*, 1986, **25**, 923.
[199] C. Wolf and H.-U. Hummel, *J. Chem. Soc., Dalton Trans.*, 1986, 43.
[200] M. Veith, D. Käfer, and V. Huch, *Angew. Chem., Int. Ed. Engl.*, 1986, **25**, 375.

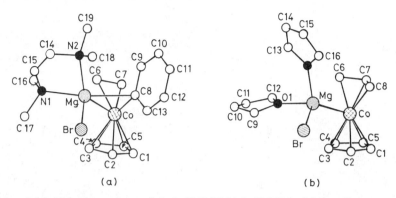

Figure 6 (a) *Molecular structure of* CpCo(C₂H₄)PhMgBr.TMEDA; (b) *crystal structure of* CpCo(η^3-C₃H₅)MgBr.(THF)₂
(Reproduced by permission from *Angew. Chem., Int. Ed. Engl.*, 1986, **25**, 924)

radicals (M = Ca, Sr, Ba) have resulted from reactions of the metal vapours with various alcohols, and their structures have been studied by laser spectroscopy.[201]

Macrocyclic Ligands.—Selected lectures from the 11th International Symposium on Macrocyclic Chemistry (Florence, September 1986) have been published,[202] as has an authoritative review on strategies for the molecular pre-organization of hosts prior to complexation of, among others, M^I ions.[203] Several papers have concerned the likely uses of crown-complexed species in analysis, catalysis, and ion-transport separation. For example, fluorescence quantum yields of anthracene monoaza-crowns (37), n = 0 or 1, are low in methanol due to quenching by electron transfer from the N lone pair, but are enhanced by factors of up to 47 on addition of Na^+ or K^+: such 'switching on' of an optical property promises realization of quantitative remote signalling of the presence of such ions in micro-environments.[204] Insertion of Na^+ or K^+ into the cavity of 18-crown-6 having a pendant thiazolium ion increases the rate of oxidative decarboxylation of pyruvic acid, probably by attracting the substrate anion into the vicinity of the thiazolium catalytic site, (38).[205] New active transport systems, with selectivities of K^+ over Na^+ of *ca.* 10:1, have been based on 18-crown-6 with amino-CH₂O(CH₂)ₙNRR′ side arms (*e.g.* n = 2 or 5, R = R′ = H).[206] The mechanism of Na^+ transport across an organic (nitrobenzene)–H₂O interface in the presence of dibenzo-18-crown-6 has been probed by kinetic analysis of polarographic data which indicates that complex formation and its dissociation occur at the phase interface.[207] Chromatographic behaviours of M^I and M^{II} ions on crown-modified silica gels and polymer matrices were investigated and retention sequences were found to mirror crown-complex stability sequences.[208] Resins containing dibenzo-18-

[201] C. R. Brazier, L. C. Ellingboe, S. Kinsey-Nielsen, and P. F. Bernath, *J. Am. Chem. Soc.*, 1986, **108**, 2126.
[202] *Pure Appl. Chem.*, 1986, **58**, pp. 1429—1534.
[203] D. J. Cram, *Angew. Chem., Int. Ed. Engl.*, 1986, **25**, 1039.
[204] A. P. de Silva and S. A. de Silva, *J. Chem. Soc., Chem. Commun.*, 1986, 1709.
[205] Y. Yano, M. Kimura, K. Shimaoka, and H. Iwasaki, *J. Chem. Soc., Chem. Commun.*, 1986, 160.
[206] Y. Nakatsuji, H. Kobayashi, and M. Okahara, *J. Org. Chem.*, 1986, **51**, 3789.
[207] T. Kakutani, Y. Nishiwaki, T. Osakai, and M. Senda, *Bull. Chem. Soc. Jpn.*, 1986, **59**, 781.
[208] T. Iwachido, H. Naito, F. Samukawa, K. Ishimaru, and K. Tôei, *Bull. Chem. Soc. Jpn.*, 1986, **59**, 1475.

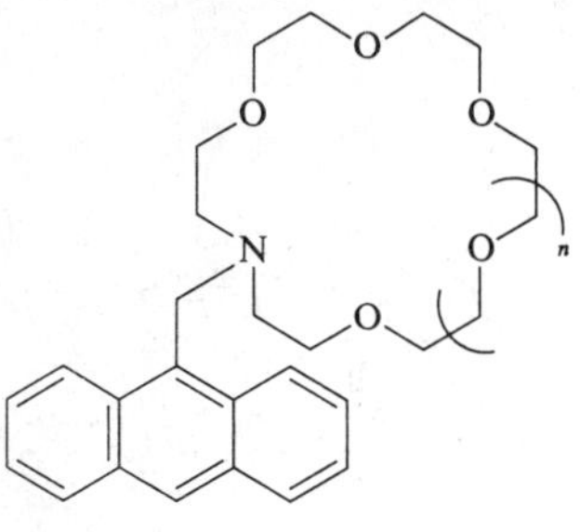

(37) 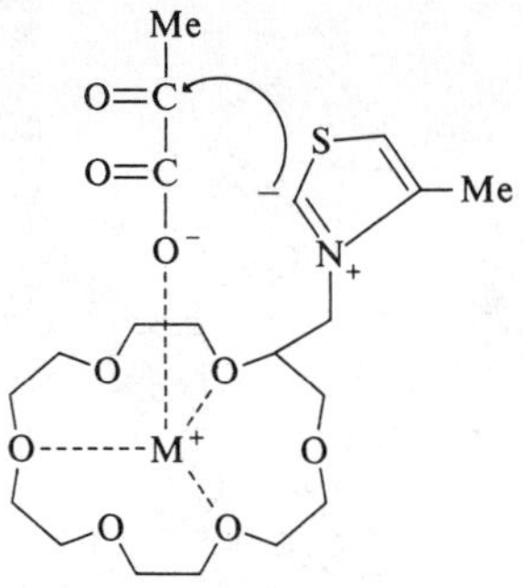   (38)

crown-6 as anchor groups have been used to extract M^I electrolytes from aqueous solutions, analysis for anions giving equilibrium constants for the equilibration process.[209]

The thermodynamics of the formation of crown complexes, their identities, and their bonding continue to be widely investigated. Calorimetric and potentiometric titrations have given ΔH, $T \Delta S$, and stability constants for complexation of M^+ ions by 18-crown-6 and by some azacrowns and some cryptands, the conclusion being that the 'macrocyclic' and the 'cryptate effect' are caused only by favourable entropic changes.[210] Analogous thermodynamic data for various Cs^+–crown species in DMF–MeCN were obtained from variable-temperature ^{133}Cs n.m.r. measurements, and here complexes were found to be enthalpy stabilized but entropy destabilized.[211] The thermal stabilities of solids KX.dibenzo-18-crown-6 (X = NCS, Br, I, NO_3) were followed by differential scanning calorimetry. The complex with X = NCS melted congruently but the others were found to undergo partial dissociation into their components.[212] Analysis of the Raman spectra of 18-crown-6, 15-crown-5, and their complexes with some M^+ and M^{2+} ions has given information on complex stoicheiometries and on changes in crown ring conformations: the D_{3d} state of the former crown seems largely retained, though 15-crown-5 complexes show four types of ring structure depending on the size and charge of the cation.[213] Complexes of four chiral 18-crown-6 ligands with M^INO_3 and $M^{II}(N^-_3)_2$ ion pairs have been characterized in $CHCl_3$ solution, the induced circular dichroism of the $NO_3 n \rightarrow \pi^*$ transition being especially sensitive to structural differences; thus the presence of two NO_3^- in a M^{2+} complex constrains the crown in a flat conformation whereas it can fold around the M^+ cations which have just one coordinated NO_3^-.[214]

As noted in the Introduction, work on the syntheses and complexing behaviours of modified crown ligands occupies many research groups. Thermodynamic and kinetic data on complexes of Na^+ with four crowns based on dibenzo-16-crown-5

[209] A. F. Danil de Namor and E. Sigstad, *Polyhedron*, 1986, **5**, 839.
[210] H.-J. Buschmann, *Inorg. Chim. Acta*, 1986, **125**, 31.
[211] G. Rounaghi and A. I. Popov, *Inorg. Chim. Acta*, 1986, **114**, 145.
[212] A. Bianchi, J. Giusti, P. Paoletti, and S. Mangani, *Inorg. Chim. Acta*, 1986, **117**, 157.
[213] H. Takeuchi, T. Arai, and I. Harada, *J. Mol. Struct.*, 1986, **146**, 197.
[214] R. B. Dyer, D. H. Metcalf, R. G. Ghirardelli, R. A. Palmer, and E. M. Holt, *J. Am. Chem. Soc.*, 1986, **108**, 3621.

but with carboxylic, sulphonic, or phosphonic mono-ionizable side-arms were consistent with significant ion–side-arm interaction;[215] for the 5-oxyacetic acid ligand, it was further shown that M^+ binding can indeed be tuned by simply adjusting the pH of the medium, Na^+ having a higher formation constant than K^+ when the ligand is deprotonated.[216] The acronym BiBLE has been applied to two-armed or bibracchial lariat ethers. Many such N,N'-disubstituted derivatives of diazacrowns have been synthesized by new procedures (*e.g.* as in equation 10) and their M^+

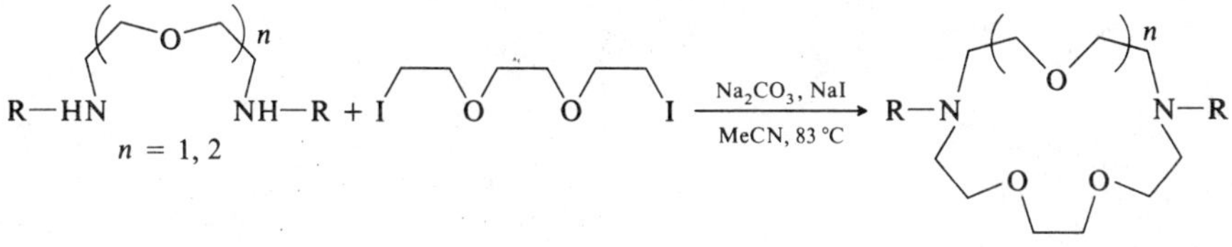

$$(10)$$

binding constants measured; the data indicate that binding relies on cooperative ring and side-arm interactions when the latter contains donor groups.[217] The crystal structures of five K^+ complexes with BiBLE or single lariat ethers have been determined and compared with those of crowns and cryptands: 'cavity-size' concepts fail to explain differences gleaned, but rather the M^+-donor distances coincide with the effective ionic radius defined by Shannon for various ions of differing coordination numbers.[218] The first example of M^+ binding enhancement by electrochemical switching in a lariat ether has been reported, this being accomplished by one- or two-electron reduction of a quinone side-arm on 15-crown-5.[219] In a related study the e.s.r. spectra of the radical anion of N-(2-nitrobenzyl)aza-15-crown-5 change most markedly when Na^+ is added; significantly this cation is closest in size to the cavity formed by the cooperating macro-ring and side-arm.[220]

For bis(crown ether) ligands, E- and Z-isomers of benzo-crown molecules linked by vinylene $-CH=CH-$ units (*i.e.* stilbene crowns) have been prepared and shown to interconvert on irradiation, the two isomers having different complexing properties.[221] Where the connecting group is a dithia-Schiff base, so containing recognition sites for both main group and transition-metal ion guests, ^{13}C n.m.r. studies have indicated that the prior presence of an ion such as Ag^+ or Cu^{2+} can prevent or allow, respectively, complexation of alkali-metal ions to give an intramolecular sandwich complex: for example, (39) shows how the preference of Cu^{2+} for square planar/tetragonal geometry allows such 1:1 complexation for K^+.[222] Similarly, several species having a redox-active 16-electron $Mo(NO)^{3+}$ core alongside a crown

[215] R. J. Adamic, B. A. Lloyd, E. M. Eyring, S. Petrucci, R. A. Bartsch, M. J. Pugia, B. E. Knudsen, Y. Liu, and D. H. Desai, *J. Phys. Chem.*, 1986, **90**, 6571.
[216] C. A. Chang, J. Twu, and R. A. Bartsch, *Inorg. Chem.*, 1986, **25**, 396.
[217] V. J. Gatto, K. A. Arnold, A. M. Viscariello, S. R. Miller, C. R. Morgan, and G. W. Gokel, *J. Org. Chem.*, 1986, **51**, 5373.
[218] R. D. Gandour, F. R. Fronczek, V. J. Gatto, C. Minganti, R. A. Schultz, B. D. White, K. A. Arnold, D. Mazzocchi, S. R. Miller, and G. W. Gokel, *J. Am. Chem. Soc.*, 1986, **108**, 4078.
[219] L. Echegoyen, D. A. Gustowski, V. J. Gatto, and G. W. Gokel, *J. Chem. Soc., Chem. Commun.*, 1986, 220.
[220] M. Delgado, L. Echegoyen, V. J. Gatto, D. A. Gustowski, and G. W. Gokel, *J. Am. Chem. Soc.*, 1986, **108**, 4135.
[221] G. Lindsten, O. Wennerström, and B. Thulin, *Acta Chem. Scand., Ser. B*, 1986, **40**, 545.
[222] P. D. Beer, *J. Chem. Soc., Chem. Commun.*, 1986, 1678.

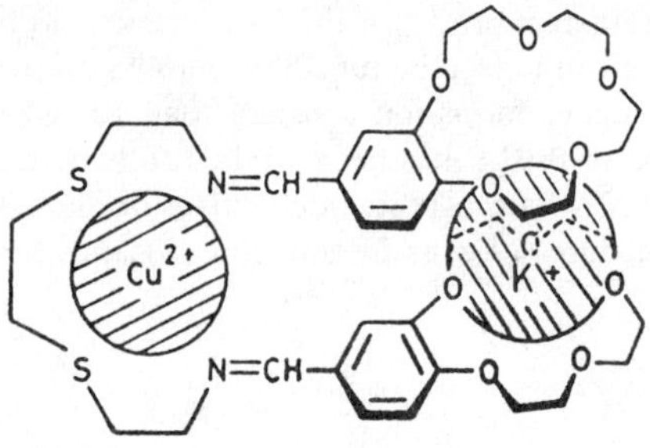

(39) 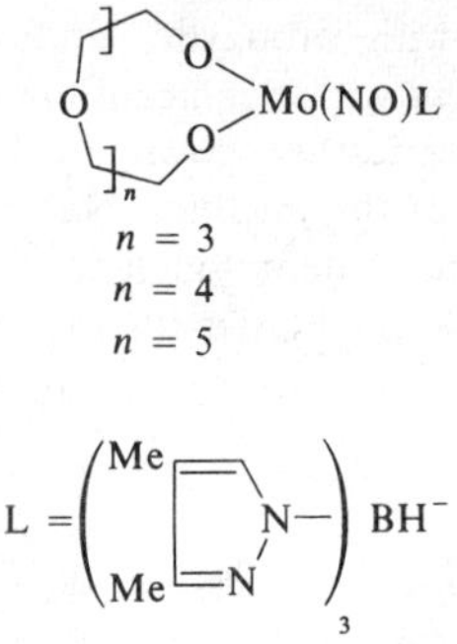   (40)

moiety *e.g.* (40) have the reduction potentials of their Mo-centres significantly affected by Na^+ complexation, so promising future reagents which may effect stereospecific electron transfers on bound organic substrates.[223] A liquid membrane containing carrier pentaoxa[13]ferrecenophane has been shown to undergo electrochemical redox-driven Na^+-transport.[224] Two papers illustrating the differing standpoints of modifying ligand photochemistry by M^+ complexation (*cf.* ref. 204) or, conversely, of photocontrol of ion complexing behaviour have appeared. The former approach has been illustrated by observation of greatly affected fluorescence emission of a decaoxa-anthracenophane system caused by large conformational changes on Na^+ complexation,[225] and the latter by alterations in M^+-complexing abilities of crown-linked spirobenzopyran after u.v./visible irradiation.[226]

All-aza crowns have also merited attention, among them the 18-crown-6 analogue (41) which has been isolated metal-free for the first time and structurally character-ized by *X*-ray diffraction; ^1H n.m.r. data implied quite drastic conformational changes within this flexible host on K^+ or Sr^{2+} complexation.[227] The structure of the first Li complex of a neutral ligand containing only sp^2-N donor atoms, [Li(L)-(MeOH)]$^+$.PF_6^- with L = (42), shows the Li^+ complexed only by the pyridine and imine donors, its pentagonal pyramidal environment being completed by MeOH.[228] The first synthesis of a rigidly toroidal macrocycle (christened a 'torand') has been reported (43); it has been shown to complex Ca^{2+}.[229]

Spherand-type ligands with functional groups like OMe or OCH_2OMe on the outer sphere have been devised and synthesized, so providing anchoring points for binding these ligands and their M^+ complexes to biological carriers such as pro-teins.[230] The pre-organized host (44) binds ArN_2^+.BF_4^- salts well, but the cation is instantaneously released on Na^+ addition so that, in the presence of anilines, a dye

[223] N. Al Obaidi, P. D. Beer, J. P. Bright, C. J. Jones, J. A. McCleverty, and S. S. Salam, *J. Chem. Soc., Chem. Commun.*, 1986, 239.
[224] T. Saji and I. Kinoshita, *J. Chem. Soc., Chem. Commun.*, 1986, 716.
[225] H. Bouas-Laurent, A. Castellan, M. Daney, J.-P. Desvergne, G. Guinand, P. Marsau, and M.-H. Riffaud, *J. Am. Chem. Soc.*, 1986, **108**, 315.
[226] H. Sasaki, A. Ueno, T. Anzai, and T. Osa, *Bull. Chem. Soc. Jpn.*, 1986, **59**, 1953.
[227] T. W. Bell and F. Guzzo, *J. Chem. Soc., Chem. Commun.*, 1986, 769.
[228] E. C. Constable, L.-Y. Chung, J. Lewis, and P. R. Raithby, *J. Chem. Soc., Chem. Commun.*, 1986, 1719.
[229] T. W. Bell and A. Firestone, *J. Am. Chem. Soc.*, 1986, **108**, 8109.
[230] P. J. Dijkstra, B. J. van Steen, and D. N. Reinhoudt, *J. Org. Chem.*, 1986, **51**, 5127.

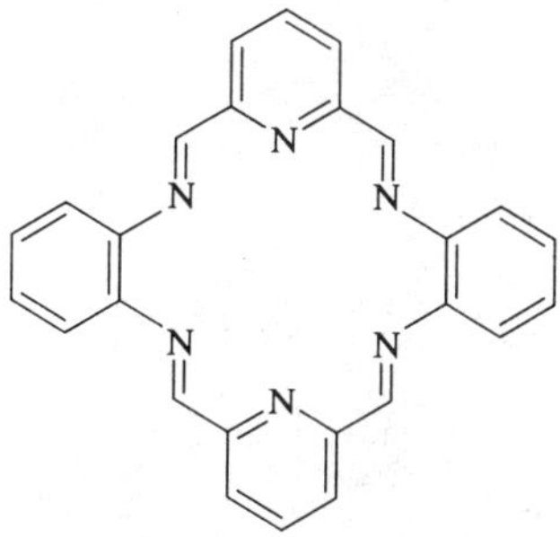

(41)

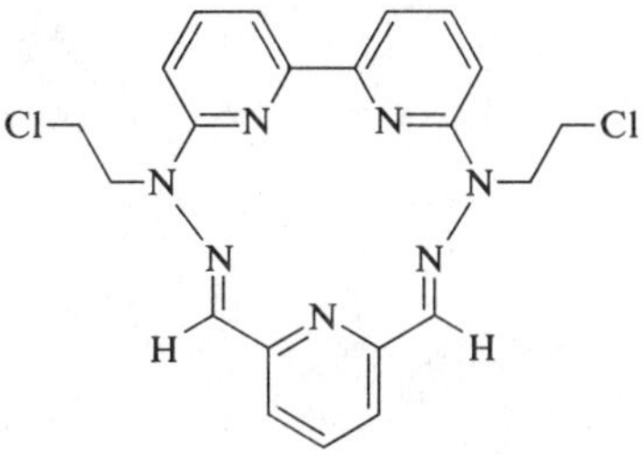

(42)

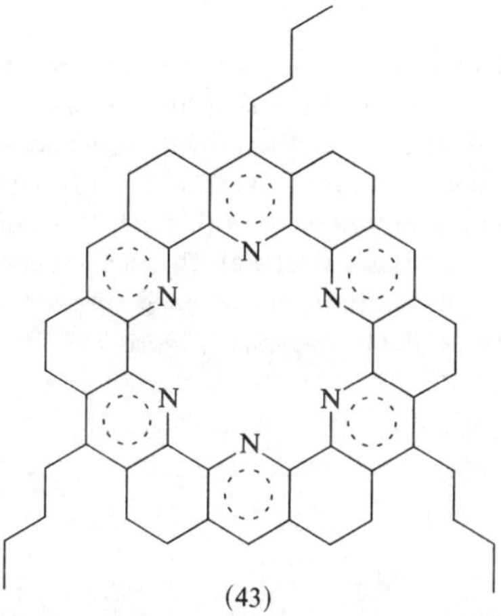

(43)

forms *i.e.* the whole system acts as an indicator for the presence of M^+ ions.[231] The syntheses and crystal structures of a whole new class of hosts *e.g.* (45), half spherand and half cryptand (so 'cryptahemispherands'), have been reported and shown to be highly selective and highly effective in complexing alkali-metal ions.[232] Other X-ray structures have been those of bis(benzo-15-crown-5)M^+.NO_3^-.H_2O, $M^+ = K^+$ or Rb^+, which are charge-separated sandwich complexes containing just (!) 10-coordinate M^+ ions with no NO_3^- or H_2O involvement.[233] In contrast, a complex of a (2S,6S)-dimethylhexaoxacyclooctadecane with $K^+NO_3^-$ exists as a contact ion-pair with bidentate NO_3^- to K^+ coordination in the crystal and, by f.t.-i.r. spectroscopy, in $CHCl_3$ solution.[234] Several 1:2 and 1:3 complexes of 18-crown-6 and its dicyclohexano- and dibenzo-derivatives with Na^+ and K^+ salts of PhO^- and NCS^- have been synthesized and identified.[235] From these, the 1:2 *cis-anti-cis* dicyclohexano-18-

[231] D. J. Cram and K. N. Doxsee, *J. Org. Chem.*, 1986, **51**, 5068.
[232] D. J. Cram, S. P. Ho, C. B. Knobler, E. Maverick, and K. N. Trueblood, *J. Am. Chem. Soc.*, 1986, **108**, 2989; D. J. Cram and S. P. Ho, *ibid.*, 1986, **108**, 2998.
[233] W. S. Sheldrick and N. S. Poonia, *J. Inclusion Phenom.*, 1986, **4**, 93; C. Momany, K. Clinger, M. L. Hackert, and N. S. Poonia, *ibid.*, 1986, **4**, 61.
[234] R. B. Dyer, R. G. Ghirardelli, R. A. Palmer, and E. M. Holt, *Inorg. Chem.*, 1986, **25**, 3184
[235] A. Rodrigue, J. W. Bovenkamp, B. V. Lacroix, R. A. B. Bannard, and G. W. Buchanan, *Can. J. Chem.*, 1986, **64**, 808.

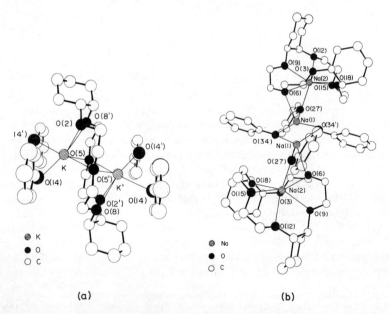

(44)   (45)

crown-6:2KOPh complex has each K^+ mid-way between the plane formed by the crown oxygens (which outline an elliptical cavity and are near-planar) and the PhO⁻ anions, and coordinated by four of the crown oxygens and by two of the anion oxygens (Figure 7a); in contrast, for the most unusual Na^+ complex of the *cis-syn-cis* ligand, a tetramer is being complexed, with half the sodium atoms individually coordinated by all the six highly irregular crown oxygens but by just one PhO⁻ oxygen, and with the other sodium atoms each interacting with just one crown oxygen but now with three PhO⁻ oxygens (Figure 7b).[236]

Figure 7 (a) *Crystal structure of* dicyclohexano-18-crown-6(isomerB).2KOPh; (b) *crystal structure of* dicyclohexano-18-crown-6(isomer A).2NaOPh
(Reproduced by permission from *Can. J. Chem.*, 1986, **64**, 816)

[236] M. E. Fraser, S. Fortier, A. Rodrigue, and J. W. Bovenkamp, *Can. J. Chem.*, 1986, **64**, 816.

Cryptand-complexation studies have included ^{133}Cs n.m.r. measurements of formation constants between Cs$^+$ and cryptand-222 in binary solvent systems.[237] A conductance stopped-flow examination of the kinetics of complexation of this same ligand with Ca^{2+}, Sr^{2+}, and Ba^{2+} indicated that after very rapid conformational changes in the ligand a fast step first cleaves H$_2$O from the metal ions, then a slower step includes M^{2+} in the cryptand cavities.[238] Sodium thiocyanate complexes of C21C$_5$ (with three O donors) and C211 (with four) have been investigated by both ^{23}Na n.m.r. and X-ray methods: the former gives a decomplexation rate constant for [Na.C211]$^+$ in pyridine and DMF 500—2000 times less than for the [Na.C21C$_5$]$^+$ complex, though the latter shows that both complexes are 'exclusive', with Na$^+$ located above the cryptand-O planes.[239] In contrast, the Li$^+$ cryptate of a benzo-C22 cryptand isolated as its AlCl$_4^-$ salt from a n-butylpyridinium chloride–AlCl$_3$ melt is inclusive, with six Li—O interactions but Li—N interactions weak or absent and with a well-separated AlCl$_4^-$ anion.[240] Finally, f.a.b.-m.s. has demonstrated the formation of ferrocene and ruthenocene cryptands with selectivity for K$^+$ complexation.[241]

5 Bioinorganic Chemistry

N.m.r. techniques have continued to play a major role in the investigation of these metal ions with naturally occurring molecules. As examples, a ^{7}Li n.m.r. study at various phospholipid membranes has found effective competition between Li$^+$ and Mg^{2+}, Ca^{2+} ions at binding sites,[242] while, of more technical interest, a novel r.f. coil for imaging the short T_2 relaxation component of intracellular ^{23}Na has been designed and constructed.[243] For the ^{87}Rb nucleus, n.m.r. measurements have given binding constants for the interaction of Rb$^+$ with the gramicidin transmembrane channel and information on conductance rates through the channel.[244] An interesting paramagnetic n.m.r. approach used Dy^{3+} as an isomorphous replacement for Ca^{2+} to probe the binding of the latter to bile salts: the data indicated that Ca^{2+} prefers the carboxylate sites of taurocholate to the sulphonate sites of glycholate.[245] Also for calcium, MNDO-based calculations on antagonists like nifedipine (46) and molecularly similar agonists like Bay K 8644 (47) implicated a crucial differential feature, in that there is a conformation of the latter which is not available to the former, *viz.* one with the nitro and C=O groups in the plane of the dihydropyridine ring.[246] The possible direct structural involvement of Ca in histamine release in inflammatory reactions has been examined through the crystal structure of [CaCl$_4$(H$_2$O)$_2$CaCl$_2$(H$_2$O)$_2$(histamine)$_2$] which contains the protonated 4-(2-

[237] G. Rounaghi and A. I. Popov, *Polyhedron*, 1986, **5**, 1935.
[238] H. Kitano, J. Hasegawa, S. Iwai, and T. Okubo, *J. Phys. Chem.*, 1986, **90**, 6281.
[239] S. F. Lincoln, I. M. Brereton, and T. M. Spotswood, *J. Am. Chem. Soc.*, 1986, **108**, 8134; S. F. Lincoln, E. Horn, M. R. Snow, T. W. Hambley, I. M. Brereton, and T. M. Spotswood, *J. Chem. Soc., Dalton Trans.*, 1986, 1075.
[240] D. L. Ward, R. R. Rhinebarger, and A. I. Popov, *Inorg. Chem.*, 1986, **25**, 2825.
[241] P. D. Beer, C. G. Crane, A. D. Keefe, and A. R. Whyman, *J. Organomet. Chem.*, 1986, **314**, C9.
[242] E. T. Fossel, M. M. Savasua, and K. A. Koehler, *J. Magn. Reson.*, 1986, **64**, 536.
[243] P. M. Joseph and R. M Summers, *J. Magn. Reson.*, 1986, **68**, 198.
[244] D. W. Urry, J. L. Trapane, C. M. Venkatachalam, and K. U. Prasad, *J. Am. Chem. Soc.*, 1986, **108**, 1448.
[245] E. Mukidjam, S. Barnes, and G. A. Elgavish, *J. Am. Chem. Soc.*, 1986, **108**, 7082.
[246] M. Mahmoudian and W. G. Richards, *J. Chem. Soc., Chem. Commun.*, 1986, 739.

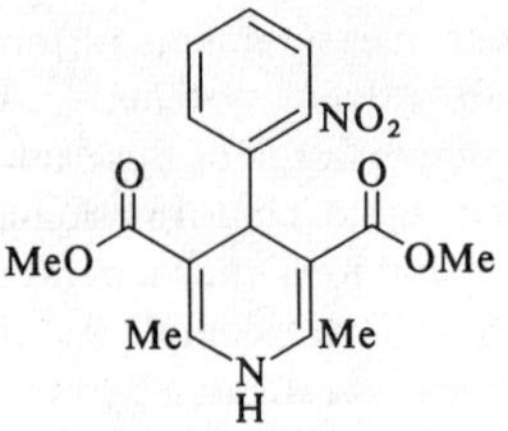

(46) 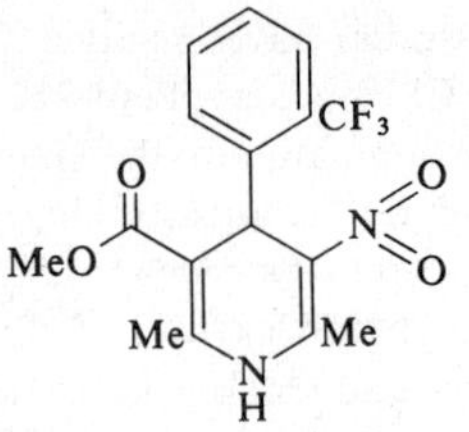   (47) 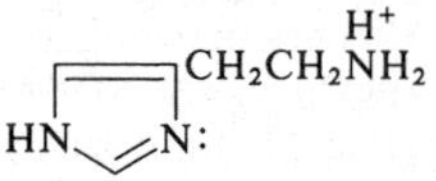   (48)

aminoethyl)imidazole tautomeric forms of the ligand (48), each bound in unidentate fashion to the metal *via* the N atom alpha to the side chain.[247]

Studies concerned with metal-ion binding to peptides and proteins have included MO calculations on model M^+ (M = Li, Na) complexes of *N*-acetylalanine methyl ester, the chelate geometry involving both peptide and ester carbonyl groups being the preferred conformation,[248] and the *X*-ray characterization of $M(NO_3)_2$ [M = Ca, Mg] complexes of *N*-methylacetamide: it seems that with smaller cations the preferred geometry of binding is in the amide plane, *trans* to the amide N and with a $M-C=O$ angle of $135 \pm 15°$.[249] Biological carboxylic acid ionophores such as monensin and lasalocid A have been shown to mediate biomimetic transport of guest cations (*e.g.*, amino-acid ester salts, Pb^{2+}, Co^{2+}) across liquid membranes, the driving force being the concentration gradient of the counter-transported cation (*e.g.* Li^+, Na^+);[250] inversion of transport selectivity of Na^+ *vs.* K^+ has also been found on introduction of the 4-chloropolyurethane group into the monensin chain.[251] The magnesium aspartate trihydrate (49), obtained from basic media, contains a tridentate acid dianion with octahedral coordination of the Mg atom completed by $2H_2O$ and an O atom of a carboxy group of a neighbouring complex, so giving strands of (49) running parallel to the *a* axis; on neutralization with HCl, protonated NH_3^+ units seem to hold Cl^- ions, as in (50), a finding of significance since $1:1:1$ Mg:Cl:Asp species are known to be especially active pharmacologically in Mg therapy.[252]

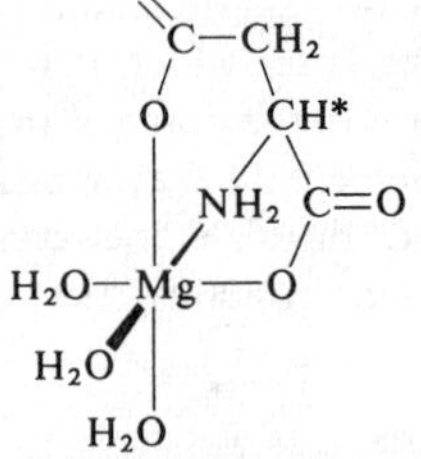

(49) 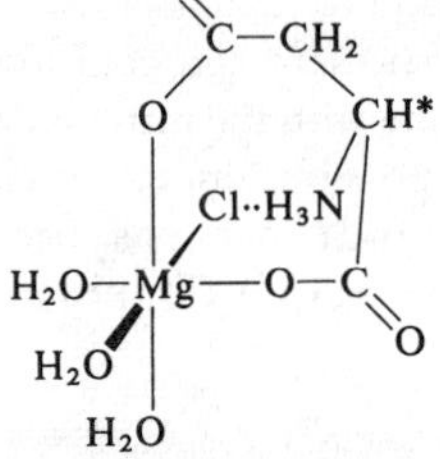   (50)

[247] L. B. Cole and E. M. Holt, *J. Chem. Soc., Perkin Trans. 1*, 1986, 151.
[248] S. U. Kokpol, S. V. Hannongbua, B.-M. Rode, and J. Limtrakul, *Inorg. Chim. Acta*, 1986, **125**, 107.
[249] K. Lewinski and L. Lebioda, *J. Am. Chem. Soc.*, 1986, **108**, 3693.
[250] H. Tsukube, K. Takagi, T. Higashiyama, T. Iwachido, and N. Hayama, *J. Chem. Soc., Chem. Commun.*, 1986, 448.
[251] J. Bolte, S. Caffarel-Mendes, G. Dauphin, C. Demuynck, and G. Jeminet, *Bull. Soc. Chim. Fr.*, 1986, 370.
[252] H. Schmidbaur, G. Müller, J. Riede, G. Manninger, and J. Helbig, *Angew. Chem., Int. Ed. Engl.*, 1986, **25**, 1013.

The identities and thermodynamic parameters of several ATP complexes have been elucidated from potentiometric measurements *e.g.* [M(ATP)H](M = any alkali metal, Ca^{2+}), $Li_2(ATP)$, $Ca_2(ATP)$, and $Ca(ATP)H_2$.[253] Spectroscopic (^1H n.m.r., f.t.-i.r.) studies on H_2ATP salts of, among others, Na^+, Mg^{2+}, and Ca^{2+} have shown that metal binding occurs through the phosphate oxygen atoms with no direct metal–base involvement.[254] In porphyrins and related chemistry, an interesting study has shown that under pulse radiolytic conditions Br_2^- oxidizes magnesium porphyrins to radical cations whose stabilities are controlled by the nature of the water-solubilizing groups: with *meso*-tetrakis(4-sulphonatophenyl)(TSP) groups for example the $t_{1/2}$ is 37s, and this Mg(TSP)porphyrin$^+$ oxidizes water to O_2 at the surface of colloidal $RuO_2.2H_2O$.[255] Methyl bacteriochlorophyllide *d* [Et, Et] has been shown by n.m.r. studies to form a dimer in $CHCl_3$ solution, probably with a 'piggy-back' structure involving coordination of the OH of the 1-hydroxyethyl substituent with the Mg atom of the adjacent molecule.[256] The *X*-ray structure of the tetraphenylporphyrin (TPP) complex $MgTPP.H_2O.(2\text{-picoline})_2$ shows that Mg is coordinated to H_2O and the four chelating N atoms though it is displaced out of the ring plane of these atoms, in part it seems due to strengthened $Mg\cdots OH_2$ interactions caused by $OH_2\cdots$picoline hydrogen-bonding.[257] Resonance Raman and i.r. bands of chlorophyll *a* (Chl) have identified solvent-dependent shifts between 5- and 6-coordinate Mg.[258] Raman spectra have also helped chart the Chl forms available in $DMSO-H_2O$ mixtures, *e.g.* Chl monomer, $(Chl)_n$ aggregate, $(Chl.2H_2O)_n$, and a new form, $(Chl.DMSO)_n$,[259] and an analysis of fluorescence spectra has allowed assignments of monomeric and aggregated Chl in dry and wet hexane.[260] Laser-desorption f.t.-m.s. has been applied to Chl*a* and Chl*b* and the results suggest that this technique permits the detection of some molecules found to be highly labile when analysed by time-of-flight m.s.[261] Native chlorophylls have been studied by a new m.s. technique involving separate laser-vaporizing and multiphoton-ionizing processes; a unique feature is the 'softness' of the former process which precludes hydrogen transfer or abstraction during heating.[262]

In polysaccharide chemistry, an EXAFS investigation on calcium poly(α-D-galacturonate) in solid and gel forms and on solid calcium poly(α-L-guluronate) has given Ca—O bond distances and their distributions: in the former solid, Ca^{2+} is ten coordinate with a $2:6:1:1$ partitioning of Ca—O lengths, while in the latter (of known structure) and in the gel there is a $3:3:2$ arrangement, so establishing that a polymorphic phase-transition occurs on drying the galacturonate.[263] The enthalpies, heat capacities, and volumes of aqueous solutions of Ca^{2+}-D-ribose and

[253] A. De Robertis, C. De Stefano, S. Sammartano, R. Cali, R. Purrello, and C. Rigano, *J. Chem. Res. (S)*, 1986, 164.
[254] H. A. Tajmir-Riahi, M. J. Bertrand, and T. Theophanides, *Can. J. Chem.*, 1986, **64**, 960.
[255] A. Harriman, P. Neta, and M.-C. Richoux, *J. Phys. Chem.*, 1986, **90**, 3444.
[256] K. M. Smith, F. W. Bobe, D. A. Goff, and R. J. Abraham, *J. Am. Chem. Soc.*, 1986, **108**, 1111.
[257] C. C. Ong, V. McKee, and G. A. Rodley, *Inorg. Chim. Acta*, 1986, **123**, L11.
[258] M. Fujiwara and M. Tasumi, *J. Phys. Chem.*, 1986, **90**, 250.
[259] Y. Koyama, Y. Umemoto, A. Akamatsu, K. Uchara, and M. Tanaka, *J. Mol. Struct.*, 1986, **146**, 273.
[260] N. E. Binnie, L. V. Haley, T. A. Mattioli, D. L. Thibodeau, W. Wang, and J. A. Koningstein, *J. Phys. Chem.*, 1986, **90**, 4938.
[261] R. S. Brown and C. L. Wilkins, *J. Am. Chem. Soc.*, 1986, **108**, 2447.
[262] J. Grotemeyer, U. Bosel, K. Walter, and E. W. Schlag, *J. Am. Chem. Soc.*, 1986, **108**, 4233.
[263] L. Alagna, T. Prosperi, A. A. G. Tomlinson, and R. Rizzo, *J. Phys. Chem.*, 1986, **90**, 6853.

Ca^{2+}-D-arabinose have been measured and used to determine Ca^{2+}—OH interaction parameters.[264]

Suitable sensors for monitoring Li levels in blood serum during lithium therapy remain an attractive goal. A Li^+-selective electrode based on a liquid membrane of a 14-crown-4 ether has been tested in a flow injection analysis system; for the first time, hospital serum samples were analysed, results matching favourably with those from a.a.s.[265] High sodium intake is being increasingly linked with hypertension, and the Na^+-selective electrode has been used to determine the ion in various salted foods.[266] Computer modellings of bio-systems have included one used to monitor the precipitation of stone-forming solids from urine e.g. calcium oxalate monohydrate, calcium hydroxyapatite.[267] A computer simulation has also been used to predict *in vivo* equilibrium concentrations of species formed during food digestion from metal ions (including Na^+, Mg^{2+}, Ca^{2+}) and amino acids.[268]

[264] J.-P. Morel, C. Lhermet, and N. Morel-Desrosiers, *Can. J. Chem.*, 1986, **64**, 996.

[265] R. Y. Xie, and G. D. Christian, *Anal. Chem.*, 1986, **58**, 1806.

[266] E. Florence, *Analyst (London)*, 1986, **111**, 571.

[267] P. W. Linder and J. C. Little, *Inorg. Chim. Acta*, 1986, **123**, 137.

[268] P. Robb, D. R. Williams, and D. J. McWeeny, *Inorg. Chim. Acta*, 1986, **125**, 207.

3 Boron

By R. GREATREX

Department of Inorganic and Structural Chemistry, University of Leeds, Leeds LS2 9JT

1 Introduction

This chapter follows the pattern adopted in recent years. After a brief section on borides and related materials, there are four larger sections dealing with different aspects of boron hydride chemistry; binary boranes and their derivatives are discussed first, followed by metallaboranes (including metal borohydride complexes), carbaboranes, and metallacarbaboranes. The synthesis, structure, and bonding of these materials remains a dominant interest in inorganic boron chemistry, and it seem likely that this trend will continue. Section 7 covers a miscellaneous selection of other species containing boron–carbon bonds, ranging from a novel graphite-like material of composition BC_3, through small-ring boron–carbon systems, to multi-decker metal sandwich complexes containing boron heterocyclic ligands; organoboranes in general are beyond the scope of this article. There is a steady growth in the volume of work on boron–nitrogen compounds, particularly by German chemists, and this subject is covered in section 8, together with the first reported compounds containing B=P multiple bonds. The final two sections are less substantial by comparison and deal mainly with boron–oxygen compounds and boron halides, respectively.

2 Borides and Related Materials

In a search for new ionic conductors, a monoclinic polymorph of Li_3BN_2 has been synthesized by slow-cooling of a mixture of Li_3N and BN from 1200 K. The structure has two kinds of alternating layers parallel to (100). One layer contains Li and B atoms, and the other only N atoms. The boron is linearly coordinated by two N atoms at distances of 133.93(5) and 133.61(5) pm.[1]

A detailed single-crystal X-ray diffraction analysis of SiB_6 has revealed a new polyhedral arrangement for a boron-rich phase. The novel feature is a silicon-containing icosihexahedron (15 atoms, 26 faces) similar to that found previously in BeB_3. Four such entities are present in the 290-atom unit cell, together with 18 icosahedra, eight single atoms, and about four interstitial atoms.[2] A critical review of the available data has shown that the homogeneity range of the important thermionic emitter LaB_6 is narrow, with a variation in the B/La ratio of less than

[1] H. Yamane, S. Kikkawa, H. Horiuchi, and M. Koizumi, *J. Solid State Chem.*, 1986, **65**, 6.
[2] M. Vlasse, G. A. Slack, M. Garbauskas, J. S. Kasper, and J. C. Viala, *J. Solid State Chem.*, 1986, **63**, 31.

2% up to 2000 K; contradictory results in the literature were attributed to deficiencies in the various classical techniques employed.[3] The solubility of vanadium in β-rhombohedral boron has been studied by single-crystal diffractometry for two samples of composition $VB_{\sim 65}$ and $VB_{\sim 165}$. The boron framework was found to be essentially unaltered, though there was a slight expansion of the unit cell in $VB_{\sim 165}$.[4a] On the basis of a combined EXAFS/XANES study, the vanadium atoms in this particular sample were found to be twelvefold co-ordinated by B atoms at a distance of 217(2) pm.[4b]

The isotypic monoclinic borides $Ca_2Os_3B_5$ and $Eu_2Os_3B_5$ have been prepared by reaction of the elemental components in sealed tantalum tubes, and found to contain puckered Os—B layers connected by B atoms with very short Os—B distances.[5a] The new compound Mg_2IrB_{2-x} (homogeneity range $0.2 \leq x \leq 0.35$) is related to the W_2CoB_2-type structure, and contains distorted Mg_4Ir_2 prisms. The boron atoms, which are partly disordered, are situated either inside the prisms forming pairs, or with very short Ir—B distances (199 pm) on an elongated prism edge formed by two iridium atoms.[5b] The electronic structure has been studied for the systems MB_2C_2 (M = Ca, La, etc.),[6a] and for transition-metal borides with the AlB_2 structure; It is suggested that the puckering of the graphite-like net of boron atoms in ReB_2 and RuB_2 is due to strong metal–metal repulsions perpendicular to the non-metal sheets.[6b]

In last year's report the bonding of boron within a variety of isolated $Zr_6Cl_{12}{}^{n-}$ clusters ($0 < n < 4$) was noted. This work has now been extended to include the synthesis and structural characterization of well-faceted crystals of $Zr_6I_{12}B$ and $MZr_6I_{14}B$ (M = Cs or K) in good yield from reactions of stoicheiometric amounts of Zr, ZrI_4, CsI, or KI (when appropriate), and B at 850 °C in sealed tantalum containers.[7a] The phases $Sc_7Cl_{12}B$ and $Sc_4Cl_{12}B$ have been similarly prepared and characterized; the former contains discrete $Sc_6Cl_{12}B$ clusters together with isolated Sc^{3+} ions, whereas the latter, which is predicted to be metallic, consists of infinite chains of the condensed clusters sharing Sc—Sc edges.[7b] Extended-Hückel calculations on all these materials indicate the importance of strong bonding between the boron interstitial and the cluster. The analogous species $Sc_7I_{12}B$ has also been synthesized.[7c]

Studies on the mechanism of transition-metal-assisted $NaBH_4$ reductions of nitriles, alkenes, and alkyl halides have established unequivocally that metal borides are involved as catalysts. The work suggests that, in future, it should be possible to design specific borides for particular reactions, and thereby improve the efficiency of key industrial chemical processes.[8] The mechanism of the thermochemical process

[3] T. Lundström, Z. Anorg. Allg. Chem., 1986, **540/541**, 163.
[4] (a) M. F. Garbauskas, J. S. Kasper, and G. A. Slack, J. Solid State Chem., 1986, **63**, 424; (b) J. Wong and G. A. Slack, 1986, ibid., **61**, 203.
[5] (a) K. Schweitzer and W. Jung, Z. Anorg. Allg. Chem., 1986, **533**, 30; (b) W. Jung and B. Schmidt, ibid., 1986, **543**, 89.
[6] (a) J. K. Burdett, E. Canadell, and T. Hughbanks, J. Am. Chem. Soc., 1986, **108**, 3971; (b) J. K. Burdett, E. Canadell, and G. J. Miller, ibid., p. 6561.
[7] (a) J. D. Smith and J. D. Corbett, J. Am. Chem. Soc., 1986, **108**, 1927; (b) S.-J. Hwu and J. D. Corbett, J. Solid State Chem., 1986, **64**, 331; (c) D. S. Dudis, J. D. Corbett, and S.-J. Hwu, Inorg. Chem., 1986, **25**, 3434.
[8] (a) J. O. Osby, S. W. Heinzman, and B. Ganem, J. Am. Chem. Soc., 1986, **108**, 67; (b) B. Ganem and J. O. Osby, Chem. Rev., 1986, **86**, 763.

for the boriding of steels has been studied *via* electrochemical studies on molten systems based on $Na_2B_4O_7$.[9]

3 Boranes and Derivatives

In a very readable and informed overview entitled 'Boranes and Heteroboranes', Fehlner and Housecroft have attempted to show the various links between borane cluster structure, bonding, energetics, and reactivity. The article provides an excellent summary of the field and recognizes the central role of these materials as 'pattern makers' for other cluster systems, emphasizing the importance of continued development of all aspects of borane chemistry.[10] In this respect there have been several important developments during the year in the general area of cluster bonding. Following on from the recent work of Fowler (see last year's report), but incorporating a previously unrecognized group-theoretical restriction based on the Pairing Principle in Stone's Tensor Surface Harmonic Theory, Johnston and Mingos have demonstrated that there is a general class of so-called *polar deltahedra* with n atoms and n skeletal electron pairs.[11a] The bonding consequences for *closo* clusters of having n or indeed $n + 2$, instead of the usual $n + 1$, skeletal electron pairs have been probed quantitatively by Wade and co-workers using the CNDO-based Molecular Orbital Bond Index (MOBI) method. The particular species considered were the real and hypothetical borane anions $[B_nH_n]^{c-}$ ($n = 5$—12; $c = 0$, 2, or 4). The results confirm earlier qualitative predictions and show for example that for the normal *closo*-$[B_nH_n]^{2-}$ species, addition or removal of an electron pair leads to the same type of polyhedral distortion, because, whereas the HOMO is bonding for a particular polyhedral edge, the LUMO is antibonding.[11b] On the basis of the well-known electron-counting rules developed by Wade, the known molecular structures of several *closo*-boranes have been used to derive a relationship between bond length and bond order for boron–boron two-centre bonds. The procedure is analogous to that of a previous approach by Wade *et al.*, and provides a useful means of deducing the distribution of electronic charge over the B—B and B—H—B bonds. For the *arachno*-clusters B_4H_{10} and B_5H_{11} the average charges per B—H—B bond were erroneously large, and doubts were raised about the commonly assumed role of the terminal *endo*-hydrogen atoms in contributing an electron to the framework.[11c] Other notable theoretical studies include an attempt to extend Wade's rules to include both single and conjunctopolyhedral boranes and heteroboranes,[12] and detailed *ab initio* calculations on the bonding and electronic structure in $[B_6H_6]^{2-}$ [13] and $[B_{12}H_{12}]^{2-}$.[13,14] A normal co-ordinate analysis has also been carried out on the anions $[B_{12}X_{12}]^{2-}$ (X = H, D, F, Cl, Br, I).[15]

[9] M. Makyta, M. Chrenková, P. Fellner, and K. Matiasovský, *Z. Anorg. Allg. Chem.*, 1986, **540/541**, 169.
[10] T. P. Fehlner and C. E. Housecroft, 'Boranes and Heteroboranes', Chapter 6, in 'Molecular Structure and Energetics, Vol. 1. Chemical Bonding Models', ed. J. F. Liebman and A. Greenberg, VCH Publishers Inc., FRG, 1986.
[11] (a) R. L. Johnston and D. M. P. Mingos, *Polyhedron*, 1986, **5**, 2059; (b) R. E. Mulvey, M. E. O'Neill, K. Wade, and R. Snaith, *ibid.*, p. 1437; (c) N. W. Thomas, *ibid.*, p. 1207.
[12] A.-C. Tang and Q.-S. Li, *Int. J. Quantum Chem.*, 1986, **29**, 579.
[13] P. W. Fowler, *J. Chem. Soc., Faraday Trans. 2*, 1986, **82**, 61.
[14] A. Goursot, E. Pénigault, H. Chermette, and J. G. Fripiat, *Can. J. Chem.*, 1986, **64**, 1752.
[15] S. J. Cyvin, B. N. Cyvin, and T. Mogstad, *Spectrochim. Acta, Part A*, 1986, **42**, 985.

Ab initio MO calculations have been carried out for the small molecules BH and B_2H_4, both of which may exist as reaction intermediates. The former was shown to have a triplet state with an energy 133.5 kJ mol^{-1} above the singlet ground-state;[16] the latter was studied at the HF/6-31G and MP2/6-31G** level and it was found that the classical H_2BBH_2 (D_{2d}) structure was more stable by only 6.3 kJ mol^{-1} than the non-planar doubly-bridged structure.[17] It has been suggested that p-doped amorphous silicon arising from the decomposition of silane–borane mixtures may involve silylboranes as intermediates. The mechanisms of such reactions may be of importance in the development of inexpensive and efficient thin-film solar cells, and have therefore been studied by *ab initio* methods; the specific reaction systems chosen for study were SiF_2 with BH_3, and SiHF with BFH_2 to produce $F_2HSi-BH_2$.[18] The transient species HBO and $MeOBH_2$ have been observed; the former was detected by means of diode laser spectroscopy in an a.c. discharge plasma of $B_2H_6-O_2$ or B_2H_6-NO mixtures,[19a] and the latter by microwave spectroscopy in the reaction of B_2H_6 with MeOH or HCHO.[19b] The radical intermediates H_4B· and Ph_4B· have been observed in the absorption spectra of pulse-irradiated N_2O-saturated aqueous mixtures of NaX (X = N_3, Br, or SCN) and either $NaBH_4$ or $NaBPh_4$. The sequence of events is thought to involve an initial interaction between N_2O and the hydrated electron to produce the OH radical, which in turn reacts with the anion X^- to generate the radicals N_3·, Br_2^-, or $(SCN)_2^-$ responsible for the final oxidation.[20] Pyrolysis of Et_4NBH_4 in refluxing decane–dodecane mixtures at temperatures of 175—190 °C has been found to yield $[B_{10}H_{10}]^{2-}$, $[B_9H_9]^{2-}$, $[B_{12}H_{12}]^{2-}$, and $[B_{11}H_{14}]^-$ as the only products after a period of 12 h. Et_3NBH_3 is formed initially but reacts rapidly with the starting material to give $[B_3H_8]^-$, which appears to be the only stable intermediate.[21]

The quantum yield Φ_{BH_3} for BH_3 production from ArF-laser-excited B_2H_6 at 193 nm has been determined to be 2.00 ± 0.25 by trapping with PF_3. From this it was concluded that the primary photochemical step is the same as that proposed in thermolysis: $B_2H_6 + h\nu \rightarrow 2BH_3$.[22] Microwave spectroscopy has been used to study a linear B_2H_6-H/DF dimer, held together by weak van der Waals hydrogen-bonding interactions.[23] A neutron diffraction study of $[N(PPh_3)_2]^+[B_2H_7]^-\cdot CH_2Cl_2$ at 80 K has revealed that the anion has a non-crystallographic C_s symmetry with a bent geometry (B—H—B 127(2)°] and staggered terminal B—H bonds. The bridging hydride is significantly displaced off the pseudo-threefold axis defined by each terminal BH_3 group, which is interpreted as evidence that the B—H—B bond is of the 'closed' type.[24]

[16] J. A. Pople and P. von R. Schleyer, *Chem. Phys. Lett.*, 1986, **129**, 279.

[17] R. R. Mohr and W. N. Lipscomb, *Inorg. Chem.*, 1986, **25**, 1053.

[18] C. W. Bock, M. Trachtman, and G. J. Mains, *J. Phys. Chem.*, 1986, **90**, 51.

[19] (a) Y. Kawashima, K. Kawaguchi, and E. Hirota, *Chem. Phys. Lett.*, 1986, **131**, 205; (b) Y. Kawashima, *J. Mol. Spectrosc.*, 1986, **116**, 23.

[20] H. Horii and S. Taniguchi, *J. Chem. Soc., Chem. Commun.*, 1986, 915.

[21] (a) M. Colombier, J. Atchekzai, and H. Mongeot, *Inorg. Chim. Acta*, 1986, **115**, 11; (b) H. Mongeot, B. Bonnetot, J. Atchekzai, M. Colombier, and C. Vigot-Vieillard, *Bull. Chim. Soc. Fr.*, 1986, 385.

[22] M. P. Irion and K. L. Kompa, *J. Photochem.*, 1986, **32**, 139.

[23] H. S. Gutowsky, T. Emilsson, J. D. Keen, T. D. Klots, and C. Chuang, *J. Chem. Phys.*, 1986, **85**, 683.

[24] S. I. Khan, M. Y. Chiang, R. Bau, T. F. Koetzle, S. G. Shore, and S. H. Lawrence, *J. Chem. Soc., Dalton Trans.*, 1986, 1753.

The first bridge-cyanide derivative containing *two* polyborane substituents, $[B_3H_7(NC)B_3H_7]^-$, has been synthesized by simultaneous oxidation and substitution of $[B_3H_8]^-$ by the reagent BrCN. Treatment with HCl yielded $[B_3H_6Cl(NC)B_3H_7]$, in which substitution was shown by ^{11}B n.m.r. data to have occurred at the nitrogen-bonded triborane fragment. Disubstitution to give $[B_3H_6Cl(NC)B_3H_6Cl]$ could only be achieved electrochemically. Other new species included the ions $[B_3H_7(NC)BH_2(NC)BH_3]^-$, $[B_3H_7(NC)BPh_3]^-$, $[B_3H_7(NC)BH_2(CN)]^-$, $[B_3H_7(NC)BH_2(CN)B_3H_7]^-$, and the nonaborate derivatives $[B_9H_{13}(NC)BPh_3]^-$, $[B_9H_{13}(NC)BH_2(NC)BH_3]^-$, $[B_9H_{13}(NC)BH_2(CN)]^-$, and $[B_9H_{13}(NC)BH_2(CN)B_9H_{13}]^-$.[25a] ^{11}B—^{11}B COSY n.m.r. studies have been reported for some of these compounds.[25b]

To resolve conflicting reports in the literature, the kinetics of thermal decomposition of B_4H_{10} have been investigated by a mass-spectrometric technique in the pressure range 0.86—38.69 mmHg and temperature range 40.3—77.9 °C. The initial rate of consumption of B_4H_{10} was shown to be first-order, with no evidence for the 3/2-order dependence claimed in an earlier report. Significant induction periods were observed in the build-up of species other than B_5H_{11} (*i.e.* B_2H_6, B_6H_{12}, and $B_{10}H_{14}$), suggesting that they were not formed in the initial reaction. The activation energy for the decomposition was found to be $E_a = 99.2 \pm 0.76$ kJ mol^{-1}, with a pre-exponential factor of $\sim 6 \times 10^{11}$ s^{-1}. The results were discussed in terms of a mechanism involving elimination of H_2 from B_4H_{10} to generate the reactive intermediate $\{B_4H_8\}$ in an initial rate-determining step.[26] The potential energy surface connecting the known 'butterfly' C_{2v} structure of B_4H_{10} and the unknown bis(diboranyl) C_2 structure has been explored theoretically at various levels of sophistication, including MP2/6-31G*. It was suggested that the bis(diboranyl) structure may be formed as a very minor product during the pyrolysis of B_2H_6 under conditions that favour B_4H_{10} formation, and that it may be detectable *via* its asymmetric bridge stretching frequencies in the i.r. spectrum.[27] The u.v. absorption spectra of B_4H_{10} and B_5H_9 in the gas-phase have been reported.[28]

The use of B_5H_9 as a source for higher boron hydride systems such as $B_9H_{13}L$ (L = Me$_2$S, Et$_2$S, Bu$_2^n$S, Bu$_2^i$S, Ph$_2$S, or Ph$_3$P), n-$B_{18}H_{22}$, $B_{10}H_{14}$, *nido*-5,6-$C_2B_8H_{12}$, and *nido*-5,6-Me$_2$-5,6-$C_2B_8H_{10}$ has been described. The synthesis of $B_{10}H_{14}$ is an improved procedure over the earlier reported method involving B_5H_9. All the reactions involve the anion $[B_9H_{14}]^-$, which can be generated *in situ* by the reaction of B_5H_9 with KH or NaH in THF. This particular reaction was studied in detail to obtain evidence for reaction pathways, and it was found that the $[B_5H_8]^-$ intermediate reacted surprisingly rapidly with B_5H_9 to give $[B_9H_{14}]^-$ as the predominant species in solution.[29a] The anion $[B_9H_{14}]^-$ and other important higher borane species such as $B_9H_{13}.SMe_2$, $B_8H_{10}.NEt_3$, 1,2-$R_2C_2B_8H_8$, and $[B_{12}H_{12}]^{2-}$ can also be pro-

[25] (a) D. G. Meina and J. H. Morris, *J. Chem. Soc., Dalton Trans.*, 1986, 2645; (b) D. G. Meina, J. H. Morris, and D. Reed, *Polyhedron*, 1986, **5**, 1639.

[26] R. Greatrex, N. N. Greenwood, and C. D. Potter, *J. Chem. Soc., Dalton Trans.*, 1986, 81.

[27] M. L. McKee, *Inorg. Chem.*, 1986, **25**, 3545.

[28] M. P. Irion, M. Seitz, and K. L. Kompa, *J. Mol. Spectrosc.*, 1986, **118**, 64.

[29] (a) S. H. Lawrence, J. R. Wermer, S. K. Boocock, M. A. Banks, P. C. Keller, and S. G. Shore, *Inorg. Chem.*, 1986, **25**, 367; (b) J. J. Briguglio, P. J. Carroll, E. W. Corcoran, and L. G. Sneddon, *ibid.*, p. 4618; (c) D. F. Gaines and D. E. Coons, *ibid.*, p. 364; (d) J. H. Osborne, R. C. P. Hill, and D. M. Ritter, *ibid.*, p. 372.

Scheme 1

(Reproduced by permission from *Inorg. Chem.*, 1986, **25**, 364)

duced in good yields from 1:2'-$[B_5H_8]_2$ (the crystal structure of which has now been determined) *via* coupled-cage to single-cage condensations.[29b] A new class of bridge-substituted pentaborane derivatives has been synthesized by reaction of the iminium salt $[Me_2NCH_2]I$ with salts of the $[B_5H_8]^-$, $[1\text{-}Et_2B_5H_7]^-$, and $[1\text{-}BrB_5H_7]^-$ anions. The new compounds μ-$(Me_2NCH_2)B_5H_8$, 1-Et_2-μ-$(Me_2NCH_2)B_5H_7$, and 1-Br-μ-$(Me_2NCH_2)B_5H_7$ are suggested to have *arachno* structures, in which a bridging hydrogen atom of *nido*-B_5H_9 has been replaced by a C—N two-atom bridge. The formation is though to occur in two steps as shown in Scheme 1.[29c] High-field n.m.r. measurements (500.1 MHz for ^1H and 160.4 MHz for ^{11}B) have revealed that preparations considered previously to be 2-MeB_5H_{10} and 3-MeB_5H_{10} are in fact identical mixtures of the two isomers; temperature-dependent equilibrium constants and associated thermodynamic quantities were determined for the isomerization process.[29d]

The migration of the proton in the $[B_6H_7]^-$ anion has been studied by ^1H and ^{11}B n.m.r. spectroscopy, and shown to take place chiefly through an edge of the octahedral core;[30a] i.r. spectra have also been reported for this anion.[30b] The observed similarity in the ^{11}B—^{11}B COSY n.m.r. spectra of $[B_9H_{12}]^-$ and 7-CB_8H_{12} has established that they have similar nine-vertex structures (1).[31] The anion $[B_9H_9]^{2-}$ and the isoelectronic carborane $C_2B_7H_9$ have the alternative nine-vertex structure (2) based on a tricapped trigonal prism, and possible mechanisms for their isomerization have been discussed. It was shown that a mechanism involving a single diamond–square–diamond rearrangement is forbidden by the principle of conservation of orbital symmetry, but a double DSD process is allowed. This involves

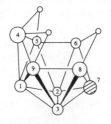

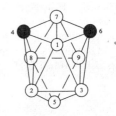

(1) ○ = BH, ◍ = BH or CH   (2) ○ = BH, ● = BH or C

[30] (a) K. A. Solntsev, Yu. A. Buslaev, and N. T. Kuznetsov, *Russ. J. Inorg. Chem.*, 1986, **31**, 633; (b) D. M. Vinitskii, K. A. Solntsev, N. T. Kuznetsov, and L. V. Goeva, *ibid.*, p. 1340.

[31] S. Heřmánek, J. Fusek, B. Štíbr, J. Plešek, and T. Jelínek, *Polyhedron*, 1986, **5**, 1873.

simultaneous rupture of the 1,7 and 2,3 bonds in (2), and the formation of two new bonds, 4,6 and 1,5.[32]

In recent years little progress has been made in studying the chemistry of nitrogen-containing boranes. In taking up this topic, Todd and co-workers have shown that the reaction at room temperature between $B_{10}H_{14}$ and $NaNO_2$ in THF is more interesting and complex than previously thought. The major products were found to be $Na[B_9H_{12}NH]$, $Na[B_9H_{14}]$, and $Na[B_{10}H_{15}]$, whereas $Na[B_{10}H_{12}NO_2]$ exists only as a transient intermediate. The Lewis-base adduct $B_9H_{11}NH.CN(C_6H_{11})$ was shown by a single-crystal X-ray study to have an *arachno* open-cage structure (Figure 1), similar to that reported previously for the isoelectronic compound 9-NEt_3-6-SB_9H_{11}. Slow-passage of $B_9H_{11}NH$ through an evacuated hot tube (400 °C) produced H_2 and the first *closo*-azaborane, B_9H_9NH, for which n.m.r. data suggested an axial placement of the NH unit in a bicapped archimedian antiprismatic structure.[33]

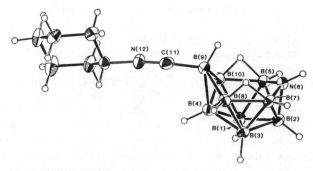

Figure 1 *ORTEP diagram of* 9-[$(C_6H_{11})NC$]-6-NHB_9H_{11}[33]
(Reproduced by permission from *Inorg. Chem.*, 1986, **25**, 2758)

The *arachno*-decaborane derivatives 2-X-6,9-$(SMe_2)_2$-$B_{10}H_{11}$ have been synthesized in almost quantitative yield from the reaction of *nido*-2-$XB_{10}H_{13}$ (X = Cl, Br, or I) with SMe_2. When refluxed with ROH (R = Me or Et) the bis(ligand) decaboranes gave moderate yields of *arachno*-1-X-4-(SMe_2)-B_9H_{12} together with the corresponding *arachno*-4-(SMe_2)-7-(OR)-B_9H_{12}. The use of some of these complexes in metallaborane synthesis is discussed in the next section.[34a] The *arachno*-decaboranyl structure suggested by n.m.r. spectroscopy for the anionic species $[B_{10}H_{12}(PPh_2)]^-$ has been confirmed by a single-crystal X-ray diffraction analysis of the $[PMePh_3]^+$ salt. The ten-boron cage resembles that of $[B_{10}H_{14}]^{2-}$, but it has a bridging PPh_2 group between the B(6) and B(9) positions instead of *endo*-terminal hydrides.[34b]

The formation of non-volatile white or yellow solids during the thermolysis of the boranes is well known and is generally regarded as an undesirable feature because it reduces the yields of 'useful' volatile boranes in interconversion reactions.

[32] B. M. Gimarc and J. J. Ott, *Inorg. Chem.*, 1986, **25**, 2708.
[33] A. Arafat, J. Baer, J. C. Huffman, and L. J. Todd, *Inorg. Chem.*, 1986, **25**, 3757.
[34] (a) R. Ahmad, J. E. Crook, N. N. Greenwood, and J. D. Kennedy, *J. Chem. Soc., Dalton Trans.*, 1986, 2433; (b) M. Thornton-Pett, M. A. Beckett, and J. D. Kennedy, *ibid.*, p. 303.

Very little is known about these materials because their intractable nature renders them unsuitable for study by the more common spectroscopic techniques. However, by use of field desorption mass spectrometry it has been possible to identify and tentatively assign high molecular weight products in the thermolysis of $B_{10}H_{14}$. These include not only known species such as $B_{20}H_{26}$ and $B_{30}H_{38}$, but also materials with mass cut-offs at m/z 364—366, 420, and 496, assignable to condensed boranes having two or more common boron atoms.[35]

Other topics worthy of mention include a new preparation of $[B_{11}H_{14}]^-$ from $NaBH_4$ and MeI at room temperature in diglyme,[36] the synthesis and structural characterization of $[NMe_4][Et_3NB_{12}H_{11}]$,[37] derivative chemistry of $[B_{12}H_{10}(CH_2OH)_2]^{2-}$,[38] and a report of more general interest dealing with rules for predicting the ^{11}B n.m.r. spectra of closo-boranes and -heteroboranes.[39]

4 Metallaboranes

This year has seen the publication of an authoritative review entitled 'The polyhedral metallaboranes Part II. Metallaborane clusters with eight vertices and more'. Together with Part I, this impressive achievement provides essentially comprehensive coverage of the structural, preparative, and behavioural chemistry of this burgeoning area up to late 1984.[40]

There is continuing interest in the field of metal tetrahydroborate chemistry, largely because of the versatile bonding modes of the BH_4^- ligands. For example in $FeH(dmpe)_2(BH_4)$ an X-ray diffraction study has confirmed the presence of a singly-bridged BH_4 ligand with a bent Fe—H—B bond.[41] Of much greater interest, however, are the structures of $Ti(BH_4)_3(PMe_3)_2$ (3)[42] and $[(tripod)HRu(\mu-\eta^2-BH_4)RuH(tripod)]BPh_4$ [tripod = $MeC(CH_2PPh_2)_3$] (4)[43] which are reproduced in Figures 2 and 3, respectively. The structure of (3) reveals that two of the BH_4 groups possess unusual geometries involving interaction of the titanium centre with one B—H bond in a 'side-on' manner. This geometry resembles one proposed transition state for the approach of a molecule of methane to a transition metal leading to C—H bond cleavage, and it is suggested that the study provides the first experimentally-determined structural parameters for a chemical system closely related to this 'side-on' transition state for the activation of methane.[42] Compound (4) is the first example of a bimetallic group 8 tetrahydroborate complex with a $\mu,\eta^2\text{-}BH_4^-$ ligand. Interestingly, the Ru—B distances [208(3) and 212(3) pm] are comparable with the sum of the covalent radii of Ru and B (213 pm), which may indicate the presence of a direct Ru—B interaction.[43] A single-crystal X-ray study has revealed that the uranium atom in $U(BH_4)_3.3THF$ is surrounded by three $\eta^3\text{-}BH_4^-$ and three

[35] V. V. Volkov, L. I. Brezhneva, and G. S. Voronina, Izskab. Isv. Sib. Otd. Akad. Nauk SSSR Ser. Khim., 1986, 61.

[36] A. Ouassas and B. Frange, Bull. Soc. Chim. Fr., 1986, 22.

[37] G. F. Mitchell and A. J. Welch, Acta Crystallogr., Sect. C, 1986, 42, 101.

[38] G. T. King and N. E. Miller, Inorg. Chem., 1986, 25, 4309.

[39] F. Teixidor, C. Viñas, and R. W. Rudolph, Inorg. Chem., 1986, 25, 3339.

[40] J. D. Kennedy, Prog. Inorg. Chem., 1986, 34, 211.

[41] R. Bau, H. S. H. Yuan, M. V. Baker, and L. D. Field, Inorg. Chim. Acta, 1986, 114, L27.

[42] J. A. Jensen and G. S. Girolami, J. Chem. Soc., Chem. Commun., 1986, 1160.

[43] L. F. Rhodes, L. M. Venanzi, C. Sorato, and A. Albinati, Inorg. Chem., 1986, 25, 3335.

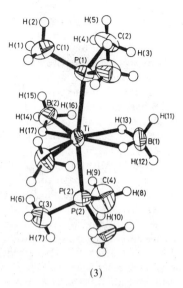

(3)

Figure 2 *Molecular structure of* Ti(BH$_4$)$_3$(PMe$_3$)$_2$ (3)[42]
(Reproduced from *J. Chem. Soc., Chem. Commun.*, 1986, 1160)

THF ligands in a distorted *facial*-octahedral arrangement,[44] and cell parameters have been reported for the rare-earth complexes Ln(BH$_4$)$_3$.3THF.[45]

The borohydride derivative (η^5-C$_5$Me$_5$)$_2$NbBH$_4$ has been obtained in 30—50% yield from NbCl$_5$, LiC$_5$Me$_5$, and NaBH$_4$ in 1,2-dimethoxyethane, and has proved

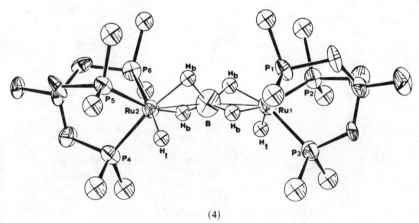

(4)

Figure 3 *ORTEP view of* [(tripod)HRu(μ-η^2-BH$_4$)RuH(tripod)]BPh$_4$ [tripod = MeC(CH$_2$PPh$_2$)$_3$] (4)[43]
(Reproduced by permission from *Inorg. Chem.*, 1986, **25**, 3335)

[44] D. Männig and H. Nöth, *Z. Anorg. Allg. Chem.*, 1986, **543**, 66.
[45] U. Mirsaidov, I. B. Shaimuradov, and M. Khikmatov, *Russ. J. Inorg. Chem.*, 1986, **31**, 753.

to be a useful starting material for the synthesis of other pentamethylniobocene derivatives. It features bidentate co-ordination of the BH_4^- ligand (i.r. evidence), and by far the largest yet reported barrier for bridge $\rightleftharpoons$ terminal hydride exchange (68.62 kJ mol^{-1}).[46] The complex *cis*-[Re(η^2-BH_4)(dppe)$_2$] also contains a bidentate BH_4^- which exhibits static n.m.r. behaviour, but temperature-dependence studies were not carried out to determine activation energies.[47] Other relevant reports include observations on copper(I) borohydride, which appears to retain a persistent chloride impurity;[48] brief reference to the compound Me_2GaBH_4, formed in the reaction of $(Me_2GaH)_2$ and B_2H_6 at $-23\,°C$;[49] quasirelativistic Xα-SW electronic structure calculations on the series $M(BH_4)_4$ (M = Zr, Hf, Th, U);[50a] and photoelectron spectra of the related compounds $M(BH_3Me)_4$ (M = Zr, Hf, Th, U).[50b]

Fehlner and Housecroft and co-workers have continued their fascinating work on ferraborane cluster chemistry.[51] For example, the full characterization of the two related ferraboranes $HFe_3(CO)_9BH_3R$ [R = H (5) and Me (6)] and the conjugate bases $[HFe_3(CO)_9BH_2R]^-$ [R = H (7) and Me (8)] has allowed features of their geometrical and electronic structures to be contrasted with those of the derivatives $(\mu\text{-}H)_3Fe_3(CO)_9(\mu_3\text{-}CR)$ [R = H (9) and Me (10)] with which they are isoelectronic. This relationship together with the observation that the ferraborane anions (7) and (8) are isoprotonic with the hydrocarbyls (9) and (10) has suggested the description of these metallaboranes as 'inorganometallic' compounds, and has led to insights into the factors important in determining stable hydrogen positions on these tetranu-

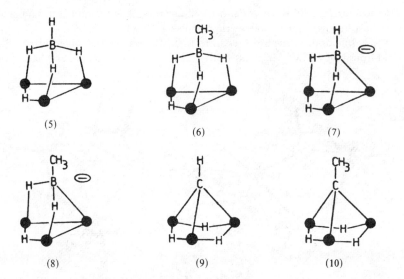

(5)   (6)   (7)

(8)   (9)   (10)

[46] R. A. Bell, S. A. Cohen, N. M. Doherty, R. S. Threlkel, and J. E. Bercaw, *Organometallics*, 1986, **5**, 972.
[47] A. J. L. Pombeiro and R. L. Richards, *J. Organomet. Chem.*, 1986, **306**, C33.
[48] R. J. Spokas and B. D. James, *Inorg. Chim. Acta*, 1986, **118**, 99.
[49] P. L. Baxter, A. J. Downs, M. J. Goode, D. W. H. Rankin, and H. E. Robertson, *J. Chem. Soc., Chem. Commun.*, 1986, 805.
[50] (*a*) D. Hohl and N. Rösch, *Inorg. Chem.*, 1986, **25**, 2711; (*b*) J. C. Green, R. Shinomoto, and N. Edelstein, *ibid.*, p. 2718.

clear systems.[51a] In recognizing that (7) represents a novel example of BH_3 supported on a multinuclear metal framework, experiments were designed to assess the effect of this unusual environment on the reactivity of the borane fragment. It was shown that in the presence of Lewis bases (*e.g.* PMe_2PH), (7) was subject to competition between cluster retention (*i.e.* substitution *via* H_2, rather than CO, elimination) and cluster fragmentation, with the former being favoured at low ligand concentrations. A kinetic study revealed a first-order dependence on both substrate and ligand for fragmentation, suggesting an associative mechanism.[51b,c] Compound (7) has also been shown to take part in a rational and high-yield cluster expansion reaction (equation 1) to generate the butterfly anionic cluster $[(\mu\text{-}H)Fe_4(CO)_{12}BH]^-$ (11).

$$[(\mu\text{-}H)Fe_3(CO)_9BH_3]^- + 2Fe_2(CO)_9 \rightarrow [(\mu\text{-}H)Fe_4(CO)_{12}BH]^- + H_2 + 3Fe(CO)_5 \quad (1)$$

Elimination of the *endo* hydrogens from (7) as dihydrogen is believed to be the driving force for this reaction, and it is suggested that hydrogen-rich clusters may constitute important cluster building materials in the future.[51d] Phosphine substitution in (11) was shown to be specific, being 'wing-tip' for monosubstitution (12) and 'wing-tip'–'wing-tip' for disubstitution (13). Monosubstitution was found to increase the barrier to proton scrambling, whereas disubstitution is seen (13) to convert an Fe—H—Fe interaction into an Fe—H—B interaction. These observations were rationalized on the basis of extended Hückel calculations, which suggest that the phosphine substituent attacks the most positive centre and the hydrogens move to positions of maximum negative charge.[51e] The dynamic structure of $Fe_2(CO)_6B_3H_7$ (14) has also been discussed, and Fenske–Hall quantum chemical calculations have been used to suggest reasons why its deprotonation to give $[Fe_2(CO)_6B_3H_6]^-$ (15) occurs *via* loss of an Fe—H—B rather than a B—H—B proton.[51f] Related studies deal with the synthesis and structural aspects of the iron–gold borido cluster $Fe_4(CO)_{12}[Au(PPh_3)]_2BH$ (16), in which the boron atom is virtually encapsulated by six metal atoms;[51g] and with u.v.-photoelectron spectroscopic and quantum-chemical analyses of $H_3Os_3(CO)_9BCO$, which suggest that boron is acting as a pseudo metal cluster atom.[51h]

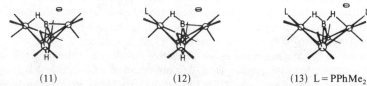

(11)   (12)   (13) L = PPhMe$_2$

The pale yellow-green solid $(C_5H_5)Mo(CO)_2\{P(BH_3)(Ph)[N(SiMe_3)_2]\}$ (17) has been synthesized in the reaction of the metallophosphenium complex $(C_5H_5)Mo(CO)_2P(Ph)[N(SiMe_3)_2]$ with either B_2H_6 or BH_3.THF, and shown by X-ray diffraction analysis to contain the unanticipated B—H—Mo interaction

[51] (a) J. Vites, C. E. Housecroft, C. Eigenbrot, M. L. Buhl, G. J. Long, and T. P. Fehlner, *J. Am. Chem. Soc.*, 1986, **108**, 3304; (b) C. E. Housecroft and T. P. Fehlner, *Inorg. Chem.*, 1986, **25**, 404; (c) *idem, J. Am. Chem. Soc.*, 1986, **108**, 4867; (d) *idem, Organometallics*, 1986, **5**, 379; (e) *ibid.*, p. 1279; (f) C. E. Housecroft, *Inorg. Chem.*, 1986, **25**, 3108; (g) C. E. Housecroft and A. L. Rheingold, *J. Am. Chem. Soc.*, 1986, **108**, 6420; (h) R. D. Barreto, T. P. Fehlner, L.-Y. Hsu, D.-Y. Jan, and S. G. Shore, *Inorg. Chem.*, 1986, **25**, 3572.

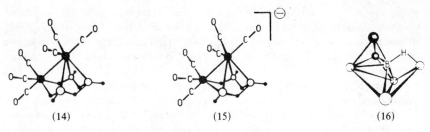

(14)   (15)   (16)

revealed in Figure 4. It was suggested that this structural feature might represent a trapped intermediate in the addition of a B—H bond across the formal Mo=P bond.[52]

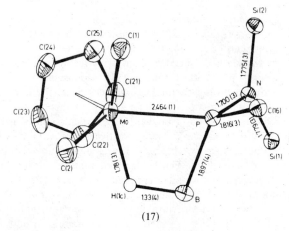

(17)

Figure 4 *View of the core atoms of* $CpMo(CO)_2\{P(Ph)[N(SiMe_3)_2](\mu\text{-}H\text{-}BH_2)\}$ (17)[52] (Reproduced by permission from *Organometallics*, 1986, **5**, 380)

In recent years the Leeds group has dominated the field of synthetic polyhedral metallaborane chemistry, and this year has been no exception. An immense amount of work has been reported, and it is only possible to mention here some of the main lines of approach, with selected examples. From a mechanistic point of view the reactions of the *arachno*-nonaborane derivative mentioned in the previous section have yielded interesting results. Thus, deprotonation of $1\text{-}Cl\text{-}4\text{-}(SMe_2)\text{-}B_9H_{12}$ (18a) followed by reaction with *cis*-$[PtCl_2(PMe_2Ph)_2]$ gave (19a), and similar treatment of $4\text{-}(SMe_2)\text{-}7\text{-}(OMe)\text{-}B_9H_{12}$ (18b) gave the corresponding compound (19b). The positions of the chloro and methoxy substituents in the nine-vertex platinaborane products of these reactions show that the metal has added to the B(4)B(5)B(6) edge of the substrates, and that the opposing B(8) vertex has been eliminated. Interestingly, the reaction appears not to proceed *via* attack at the B(6)B(7)B(8) position with elimination of the opposing B(4) vertex, even though the latter could be regarded as already partially sequestered by the Lewis base SMe_2.[34a]

[52] W. F. McNamara, E. N. Duesler, R. T. Paine, J. V. Ortiz, P. Kölle, and H. Nöth, *Organometallics*, 1986, **5**, 380.

(18) a; L = SMe₂, X = H, Y = Cl
b; L = SMe₂, X = OMe, Y = H

(19) a; X = H, Y = Cl
b; X = OMe, Y = H

Another interest of the group has been to examine comparative structural data and other behavioural trends across Periods or down Groups of the Periodic Table for particular structural classes. This process has generated many new compounds with the ten-vertex 6-MB$_9$ structure based on *nido*-B$_{10}$H$_{14}$, involving the metals tungsten, rhenium, ruthenium, osmium, iridium,[53a] rhodium,[53b] and molybdenum.[53c] The work has relied heavily on the use of single-crystal X-ray diffraction analysis and high-resolution n.m.r. spectroscopy to provide the detailed structural information which is a prerequisite for progress in this area, and in this respect it has been suggested that ^{1}H—^{1}H COSY n.m.r. spectroscopy may prove even more useful than ^{11}B—^{11}B COSY experimentation in providing connectivity data and related information.[53d] Continued success in the synthetic area depends on the identification of convenient and reactive metal-containing starting materials. The areneruthenium dimer [(η^6-C$_6$Me$_6$)RuCl$_2$]$_2$ has been shown to fulfil this role in that it reacts readily with polyhedral borane anions to yield a host of new ruthenaborane cluster compounds, each incorporating the {(η^6-C$_6$Me$_6$)Ru} moiety as a cluster vertex.[53e] A particularly interesting product of the reaction with *arachno*-[B$_6$H$_{11}$]⁻ is the novel *arachno*-nine-vertex ruthenaborane [(η^6-C$_6$Me$_6$)RuB$_8$H$_{14}$], which is unique for a metallaborane in having the same skeletal configuration as n-B$_9$H$_{15}$, rather than that of iso-B$_9$H$_{15}$.[53f] Also of interest are the novel *arachno* five-vertex species [(C$_6$Me$_6$)RuB$_4$H$_9$] [which has an *endo* hydrogen atom on the Ru(1)-position equivalent to the one on B(1) in B$_5$H$_{11}$ itself] and various 'isocloso' ten- and eleven-vertex species (*e.g.* [1-(C$_6$Me$_6$)RuB$_{10}$H$_{10}$]).[53e] The term 'isocloso' which has also been applied to the rhodaundecaboranes [(PMe$_2$Ph)$_2$RhHB$_{10}$H$_8$(OMe)$_2$] and [(PMe$_2$Ph)$_2$RhHB$_{10}$H$_8$Cl(OMe)],[53g] and the osmaundecaborane [2,5-(OEt)$_2$-1-(PPh$_3$-1-*o*-Ph$_2$PC$_6$H$_4$)-*closo*-1-OsB$_{10}$H$_7$-3],[53h] is intended to imply that the metal contributes four rather than three orbitals to the cluster.[54a] Others, however, have taken the view that such complexes contain two skeletal electrons *fewer* than their *closo* counterparts and are therefore best regarded as 'hyper-closo' metallaboranes,[54b] a view which has received some support from molecular-orbital calculations.[54c]

[53] (a) M. A. Beckett, N. N. Greenwood, J. D. Kennedy, and M. Thornton-Pett, *J. Chem. Soc., Dalton Trans.*, 1986, 795; (b) X. L. R. Fontaine, H. Fowkes, N. N. Greenwood, J. D. Kennedy, and M. Thornton-Pett, *ibid.*, p. 547; (c) N. N. Greenwood, J. D. Kennedy, I. Macpherson, and M. Thornton-Pett, *Z. Anorg. Allg. Chem.*, 1986, **540/541**, 45; (d) X. L. R. Fontaine and J. D. Kennedy, *J. Chem. Soc., Chem. Commun.*, 1986, 779; (e) M. Bown, N. N. Greenwood, and J. D. Kennedy, *J. Organomet. Chem.*, 1986, **309**, C67; (f) M. Bown, X. L. R. Fontaine, N. N. Greenwood, and J. D. Kennedy, and M. Thornton-Pett, *ibid.*, 1986, **315**, C1; (g) H. Fowkes, N. N. Greenwood, J. D. Kennedy, and M. Thornton-Pett, *J. Chem. Soc., Dalton Trans.*, 1986, 517; (h) M. Elrington, N. N. Greenwood, J. D. Kennedy, and M. Thornton-Pett, *ibid.*, p. 2277.

[54] (a) J. D. Kennedy, *Inorg. Chem.*, 1986, **25**, 111; (b) R. T. Baker, *ibid.*, p. 109; (c) R. L. Johnston and D. M. P. Mingos, *ibid.*, p. 3321.

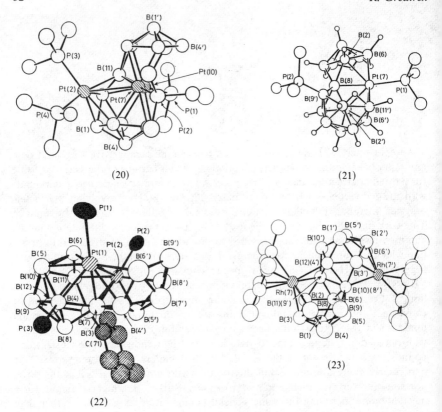

(20)   (21)

(22)   (23)

Figure 5 *Molecular   structures   of   the   new   macropolyhedral   metallaboranes*
[(PMe$_2$Ph)$_4$Pt$_3$B$_{14}$H$_{16}$]   (20),^{55a} [(PMe$_2$Ph)PtB$_{16}$H$_{18}$(PMe$_2$Ph)]   (21),^{55a}
[(PMe$_2$Ph)$_2$Pt$_2$B$_{16}$H$_{15}$(C$_6$H$_4$Me)(PMe$_2$Ph)]   (22),^{55b} and   [(C$_5$Me$_5$)$_2$Rh$_2$B$_{17}$H$_{19}$]
(23)^{55c}
[Reproduced from *J. Chem. Soc., Dalton Trans.*, 1986, 1879 (20) and (21); *J. Chem. Soc., Chem. Commun.*, 1986, 556 (22); *ibid.*, p. 1111 (23)]

Another area of activity within the Leeds group has been the synthesis of
macropolyhedral metallaboranes, and this year has witnessed reports of several new
variants in the thermolysis of the nine-vertex species [4-(PMe$_2$Ph)$_2$-*arachno*-4-
PtB$_8$H$_{12}$]. As well as the known yellow 14-vertex diplatinaborane
[(PMe$_2$Ph)$_2$Pt$_2$B$_{12}$H$_{18}$], two novel 17-vertex species are produced in boiling toluene,
the green triplatinaborane [(PMe$_2$Ph)$_4$Pt$_3$B$_{14}$H$_{16}$] (20) and the red monoplatinaborane
[(PMe$_2$Ph)PtB$_{16}$H$_{18}$(PMe$_2$Ph)] (21). The structures of both compounds have been
established by X-ray diffraction analysis, and are shown in Figure 5. Compound
(20) features a seven-membered $\overline{\text{PtB}_2\text{PtB}_3}$ open face, and can be interpreted in terms
of two fused subclusters: (i) an *arachno*-type nine-vertex 6',8'-Pt$_2$B$_7$ moiety and (ii)
a *nido*-type 11-vertex 2,7,10-Pt$_3$B$_8$ moiety. Compound (21) consists of two open
subclusters fused across their open faces and with Pt(7)B(8) as a common edge. The
larger one is a *nido* 11-vertex PtB$_{10}$ structure, and the smaller one an eight-vertex

PtB_7 structure analogous to that adopted by the binary boranes *nido*-B_8H_{12} and *arachno*-B_8H_{14}. The platinum centres in these complexes are considered to have available for bonding three orbitals in a T-shaped configuration, and it is suggested that this imposes 'planar' rather than 'conical' character on the metal-to-ligand bonding.[55a] Shorter reaction times favour the formation of yet another species in this reaction, namely the novel *conjuncto*-$[(PMe_2Ph)_2Pt_2B_{16}H_{15}(C_6H_4Me)(PMe_2Ph)]$ (22). This dark-orange air-stable compound is isolated in <*ca.* 1% yield, and consists of a *closo*-$\{Pt_2B_{10}\}$ subcluster and a *nido*-$\{Pt_2B_7\}$ subcluster fused *via* a common $\{Pt_2B\}$ closed triangular face (Figure 5); unexpectedly, a *p*-tolyl residue originating from the solvent is bound to the $\{Pt_2B_{10}\}$ subcluster, and this may have significance with regard to C—H bond activation under mild conditions.[55b] The other interesting structure shown in Figure 5 is that of the 19-vertex $[(C_5Me_5)_2Rh_2B_{17}H_{19}]$ (23) which is prepared *via* a degradative insertion from *anti*-$B_{18}H_{22}$ in the presence of $[(C_5Me_5)RhCl_2]$ and base. It comprises a 12-vertex *nido* $\{RhB_{11}\}$ subcluster and a 10-vertex *nido* $\{5-RhB_9\}$ subcluster conjoined *via* a common triangular $\{B_3\}$ face in *syn* orientation. It is suggested that the Rh(7′) has added to the skeleton of the *anti*-$B_{18}H_{22}$ (24) with elimination of an opposing vertex as in (25), to give the 18-vertex *syn*-$\{RhB_{17}\}$ (26), and that a movement of a boron vertex between these two sites in an 18-boron skeleton would provide a mechanism for *syn* ⇌ *anti* conversion.[55c]

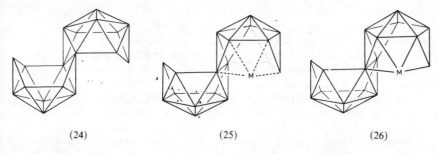

(24)   (25)   (26)

A final paper that should be mentioned in this section is a theoretical study which addresses the possibility that metal ions can exist within the icosahedral $B_{12}H_{12}^{2-}$ cage. On the basis of PRDDO calculational techniques it was shown that the centre of the cluster is a stable position of minimum energy for the Mg^{2+} ion.[56]

5 Carbaboranes

Some important papers relevant to this section are mentioned elsewhere (see refs. 29a, 29b, 31, 32, and 88). The literature covering the year 1984 on carbaboranes and their metal complexes has been reviewed.[57]

[55] (a) M. A. Beckett, J. E. Crook, N. N. Greenwood, and J. D. Kennedy, *J. Chem. Soc., Dalton Trans.*, 1986, 1879; (b) M. A. Beckett, N. N. Greenwood, J. D. Kennedy, P. A. Salter, and M. Thornton-Pett, *J. Chem. Soc., Chem. Commun.*, 1986, 556; (c) X. L. R. Fontaine, N. N. Greenwood, J. D. Kennedy, P. I. MacKinnon, and M. Thornton, Pett, *ibid.*, p. 1111.

[56] C. L. Beckel and I. A. Howard, *Chem. Phys. Lett.*, 1986, **130**, 254.

[57] T. R. Spalding, 'Carbaboranes, including their Metal Complexes' in 'Organometallic Chemistry', A Specialist Periodical Report, vol. 14, ed. E. W. Abel and F. G. A. Stone, The Royal Society of Chemistry, London, 1986, p. 40.

Ab initio calculations at the STO-3G level have been performed on 29 of the possible 52 *closo*-carbaborane isomers $C_2B_{n-2}H_n$ ($5 \leqslant n \leqslant 12$).[58a] Eighty two percent of the calculated B—B, C—B, C—H, and B—H distances were found to be within 4 pm of the experimentally determined values, and the calculated relative stabilities were in agreement both with experiment and with previous predictions[58b] based on topological charge stabilization considerations. A theoretical study[58c] of the trigonal-bipyramidal $C_2B_3H_5$ has shown that diamond-square-diamond (DSD) rearrangements are symmetry forbidden, which suggests that such processes have very high activation barriers. The results suggest that these species should be stereochemically rigid, which is consistent with the experimental observation of only one isomer ($1,5$-$C_2B_3H_5$). As discussed in an earlier section, mechanisms involving single DSD rearrangements are also forbidden for the carbaborane $C_2B_7H_9$, but double DSD processes are allowed. In addition it was suggested on the basis of MO calculations that it should be possible to isolate the 3,4- and 1,4-isomers of $C_2B_7H_9$, in addition to the known 4,5-isomer, provided rearrangement does in fact occur in this way.[32] It has been suggested that polyhedral rearrangements in cluster units may, in general, not proceed by commonly accepted routes, and that mechanisms involving single edge cleavage are preferred. The specific case of isomerization of icosahedral $1,2$-$C_2B_{10}H_{12}$ to $1,7$-$C_2B_{10}H_{12}$ was discussed as an example.[59]

Onak and co-workers have extended their careful n.m.r.-based rearrangement studies on *closo*-carbaboranes with a detailed comparison of the behaviour at 295 °C of the mixed disubstituted species 5-Me-6-Cl-$2,4$-$C_2B_5H_5$, with that of the monosubstituted 5-X-$2,4$-$C_2B_5H_6$ and disubstituted $5,6$-X$_2$-$2,4$-$C_2B_5H_5$ (X = Me, Cl). The results were considered to rule out a triangular-face rotation mechanism, in which cage-carbon atoms stay in the low coordination non-adjacent positions, but were consistent with a DSD mechanism.[60]

The versatility of the *C*-trimethylsilyl-derivative *nido*-$2,3$-$(Me_3Si)_2$-$2,3$-$C_2B_4H_6$ as a precursor in the synthesis of *C*-substituted carbaboranes has been further investigated. Reaction of the neat liquid with solid $NaHF_2$ in a pyrex reactor at 413 K was found to give, in quantitative yield, Me_3SiF and a mono-*C*-trimethylsilyl-substituted species shown by gas-phase electron diffraction to be *nido*-$(Me_3Si)C_2B_4H_7$ [(27), Figure 6].[61a] After being deprotonated to form the anion *nido*-$[2,3$-$(Me_3Si)_2$-$2,3$-$C_2B_4H_5]^-$, the same starting material has been used in a cage-expansion reaction with B_5H_9 to give the 10-vertex species *arachno*-6-(Me_3Si)-$6,9$-$C_2B_8H_{13}$ [(28, Figure 6] which is isoelectronic and similar in structure to $[B_{10}H_{14}]^{2-}$.[61b] A related compound, in which a silicon atom has become incorporated in the cluster, is more conveniently discussed in the following section (see ref. 69a).

Starting with the carbaborane *nido*-$2,3$-$Et_2C_2B_4H_6$ (29), new synthetic routes to the small *closo*-carbaboranes $2,3$-$R_2C_2B_5H_5$ (30) and $2,4$-$R_2C_2B_5H_5$ (31),[62a] and the

[58] (a) J. J. Ott and B. M. Gimarc, *J. Comput. Chem.*, 1986, **7**, 673; (b) *idem*, *J. Am. Chem. Soc.*, 1986, **108**, 4303; (c) B. M. Gimarc and J. J. Ott, *Inorg. Chem.*, 1986, **25**, 83.

[59] B. F. G. Johnson, *J. Chem. Soc., Chem. Commun.*, 1986, 27.

[60] Z. J. Abdou, G. Abdou, T. Onak, and S. Lee, *Inorg. Chem.*, 1986, **25**, 2678.

[61] (a) N. R. Hosmane, N. N. Maldar, S. B. Potts, D. W. H. Rankin, and H. E. Robertson, *Inorg. Chem.*, 1986, **25**, 1561; (b) J. R. Wermer, N. S. Hosmane, J. J. Alexander, U. Siriwardane, and S. G. Shore, *ibid.*, p. 4351.

[62] (a) J. S. Beck, A. P. Kahn, and L. G. Sneddon, *Organometallics*, 1986, **5**, 2552; (b) M. G. L. Mirabelli and L. G. Sneddon, *ibid.*, p. 1510.

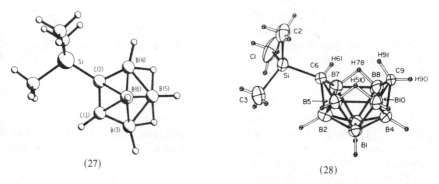

(27)

(28)

Figure 6 *Molecular structures of nido-*(Me$_3$Si)C$_2$B$_4$H$_7$ (27), *as determined by gas-phase electron diffraction,*[61a] *and arachno-*6-(Me$_3$Si)-6,9-C$_2$B$_8$H$_{13}$ (28), *as determined by X-ray diffraction*[61b]

[Reproduced by permission from *Inorg. Chem.*, 1986, **25**, 1561 (27); *Inorg. Chem.*, 1986, **25**, 4351 (28)]

four-carbon carbaborane *nido-*4,5,7,8-R$_4$C$_4$B$_4$H$_4$ (32)[62b] have been devised. Thus, the reaction of (29) with Et$_3$N.BH$_3$ at 140 °C results in the capping of the open face of the carbaborane, producing (30) in good yield; subsequent heating of (30) at 320 °C gives the 2,4-isomer (31).[62a] The reaction of (29) with 2-butyne in the presence of NaH and NiCl$_2$ produces the tetracarbon carbaborane (32) *via* a metal-promoted two-carbon insertion. The possibility that this type of reaction may be more generally useful is being further investigated.[62b]

(29) Et$_3$N·BH$_3$, 140 °C → (30) 320 °C → (31)

(29) + NaH + MeC ≡ CMe + NiCl$_2$ $\xrightarrow{\text{THF}}$ (32)

Insertion reactions involving Et$_3$N.BH$_3$ to provide an additional boron atom to the cluster have also been described by Czechoslovakian workers, who have synthesized the anion [1-H$_3$B-1,2-C$_2$B$_{10}$H$_{11}$]$^-$ from [7,8-C$_2$B$_9$H$_{12}$]$^-$, [63a] and 1-H$_3$N-1-

[63] (a) J. Plešek, T. Jelínek, S. Heřmánek, and B. Štíbr, *Collect. Czech. Chem. Commun.*, 1986, **51**, 81; (b) T. Jelínek, J. Plešek, S. Heřmánek, and B. Štíbr, *ibid.*, p. 819; (c) T. Jelínek, S. Heřmánek, B. Štíbr, and J. Plešek, *Polyhedron*, 1986, **5**, 1303; (d) T. Jelínek, B. Štíbr, J. Plešek, and S. Heřmánek, *J. Organomet. Chem.*, 1986, **307**, C13.

$CB_{11}H_{11}$ from $7\text{-}H_3N\text{-}7\text{-}CB_{10}H_{12}$;[63b] various substituted derivatives of these products were also described. In a separate study, substitution of $arachno\text{-}4,6\text{-}C_2B_7H_{11}$ has yielded a wide range of derivatives, including $3\text{-}X\text{-}4,6\text{-}C_2B_7H_{12}$ (X = Cl, Br, I, or HS), $5\text{-}X\text{-}4,6\text{-}C_2B_7H_{12}$ (X = Br, I, HS, or Bu), and $3,5\text{-}X_2\text{-}4,6\text{-}C_2B_7H_{11}$ (X = D, Cl, Br, or I), whose ^1H and ^{11}B n.m.r. spectra have been reported.[63c] An efficient route leading to the isolation of $nido\text{-}6\text{-}H_3N\text{-}6\text{-}CB_9H_{11}$, and the synthesis of the species $6\text{-}L\text{-}6\text{-}CB_9H_{11}$ (L = H, $Me_2C{=}NH$, or Me_2S) have also been reported.[63d]

Other topics of interest include a study of the electronic structure of the anion $[B_9C_2H_{11}]^{2-}$,[14] further work on the weakly coordinating anion $[B_{11}CH_{12}]^-$,[64] and the synthesis of an o-carbaborane in which framework boron and carbon atoms are linked through a simple carbon bridge.[65] Derivative chemistry of o-, m-, and p-carbaboranes, typified by the work of Zakharkin et $al.$[66] is not covered in this report.

6 Metallacarbaboranes

The literature from 1965 to 1983 on the synthesis and properties of polyhedral boranes, carbaboranes, and heteroboranes containing the $\eta^5\text{-}C_5H_5Fe$ fragment in the cluster has been reviewed.[67]

The reaction between $[(dppe)NiX_2]$ (X = Cl, Br) and $K[C_2B_4H_7]$ (33) is known to give the $closo$ species $[(dppe)NiC_2B_4H_6]$ (34). However, it has now been shown that for X = Cl an intermediate is present which is believed to be $nido\text{-}$ $[(dppe)Ni(Cl)C_2B_4H_7]$ (35), and it is suggested that the reaction proceeds according

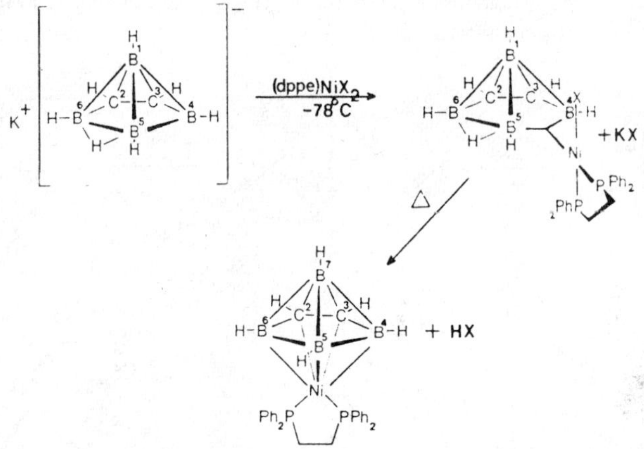

(34)

Scheme 2
(Reproduced by permission from $Inorg.$ $Chem.$, 1986, **25**, 91)

[64] K. Shelly, C. A. Reed, Y. J. Lee, and W. R. Scheidt, $J.$ $Am.$ $Chem.$ $Soc.$, 1986, **108**, 3117.
[65] S. Wu and M. Jones, $Inorg.$ $Chem.$, 1986, **25**, 482.
[66] L. A. Zakharkin, A. I. Kovredov, and V. A. Ol'shevskaya, $Izv.$ $Akad.$ $Nauk$ $SSSR$, $Ser.$ $Khim.$, 1986, 1388.
[67] L. I. Zakharkin, V. V. Kobak, and G. G. Zhigarëva, $Russ.$ $Chem.$ $Revs.$, 1986, **55**, 531 (Transl. from Uspekhi Khim., 1986, **55**, 974).

to Scheme 2, which may represent a general mechanism for insertion reactions leading to the formation of closed polyhedra.[68]

The reaction of $Na^+Li^+[(Me_3Si)_2C_2B_4H_4]^{2-}$ with $SiCl_4$ in THF has given (in 1% yield) $[(Me_3Si)_2C_2B_4H_4]Si^{II}$ (36), which represents the first reductive insertion of Si into a carbaborane cluster.[69a] In the same reaction, the silacarbaborane

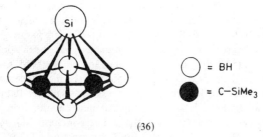

(36)

$[(Me_3Si)_2C_2B_4H_4]_2Si^{IV}$ (37) was produced in 18% yield. From an *X*-ray diffraction analysis, the silicon atom in (37) was shown to be η^5 bonded to each of the C_2B_3 faces (Figure 7), though detailed inspection of the bond distances [Si—C = 223(1)

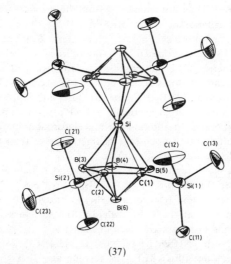

(37)

Figure 7 *ORTEP view of* $[(Me_3Si)_2C_2B_4H_4]_2Si^{IV}$ (37)[69]
(Reproduced from *J. Chem. Soc., Chem. Commun.*, 1986, 1421)

and 225(1) pm, Si—B = 210(1), 203(1), and 210(1) pm] reveals evidence for 'slippage' towards η^3. Similar structural features were observed for $[2,3-(Me_3Si)_2C_2B_4H_4]_2Ge^{IV}$ [69b] and the 2,2'-bipyridyl complex $(C_{10}H_8N_2)Sn(SiMe_3)_2C_2B_4H_4$.[69c] The former was isolated from the reaction between

[68] L. Barton and P. K. Rush, *Inorg. Chem.*, 1986, **25**, 91.
[69] (*a*) N. S. Hosmane, P. de Meester, U. Siriwardane, M. S. Islam, and S. S. C. Chu, *J. Chem. Soc., Chem. Commun.*, 1986, 1421; (*b*) *J. Am. Chem. Soc.*, 1986, **108**, 6050; (*c*) N. S. Hosmane, P. de Meester, N. N. Maldar, S. B. Potts, and S. S. C. Chu, *Organometallics*, 1986, **5**, 772.

Li$^+$[(Me$_3$Si)$_2$C$_2$B$_4$H$_5$]$^-$ and GeCl$_4$ in THF, and is thought to be the first example of a germanocene analogue. The latter was produced quantitatively in the reaction between the stannacarbaborane Sn(SiMe$_3$)$_2$C$_2$B$_4$H$_4$ and 2,2'-bipyridine in benzene, and can be regarded as an η^3-stannaborallyl complex. When the stannacarbaborane was allowed to react with THF, the air-sensitive complex (C$_4$H$_8$O)$_2$Sn(SiMe$_3$)$_2$C$_2$B$_4$H$_4$ was detected in small quantities, and it was suggested that this species is an intermediate in the formation of the η^3-stannaborallyl complex from SnCl$_2$ and Na$^+$[(Me$_3$Si)$_2$C$_2$B$_4$H$_5$]$^-$ in THF at 0 °C.[69c] A paramagnetic species closely related to these sandwich-type complexes has been prepared from tris(ethene)nickel and 2,3-dihydro-1,3-diborole, and is believed to be the first bis(tricarbahexaboranyl)metal complex analogous to the metallocenes.[70]

In continuation of their work on the metal atom synthesis of metallaboron clusters, Briguglio and Sneddon have isolated a range of new cobalt-, iron-, and nickel-containing clusters derived from 2,6-C$_2$B$_7$H$_{11}$.[71] These include 2-(η^5-C$_5$H$_5$)Co-1,4-C$_2$B$_7$H$_9$ and 4-(η^5-C$_5$H$_5$)Co-2,3-C$_2$B$_7$H$_{13}$ from the reaction of 2,6-C$_2$B$_7$H$_{11}$ with cobalt atoms and cyclopentadiene; the *closo* η^6-arene complexes 2-(η^6-MePh)Fe-6,9-C$_2$B$_7$H$_9$ and 2-(η^6-MePh)Fe-1,6-C$_2$B$_7$H$_9$ from the reaction of iron atoms and toluene; both *closo*, 2-(η^6-Me$_3$C$_6$H$_3$)Fe-1,6-C$_2$B$_7$H$_9$ (38), and *nido*, 6-(η^6-Me$_3$C$_6$H$_3$)Fe-9,10-C$_2$B$_7$H$_{11}$ (39), complexes from the reaction with mesitylene; and the unique metallacarbaborane cluster 5,7,8-Me$_3$-11,7,8,10-(η^3-C$_4$Me$_4$H)NiC$_3$B$_7$H$_7$ (40) from the reaction with nickel atoms, toluene, and 2-butyne. The structures of (38), (39), and (40) were determined by X-ray diffraction analysis and are shown in Figure 8. Compound (38) is composed of a *closo* bicapped square antiprism in which the iron and carbon atoms occupy adjacent five-coordinate positions; (39) is composed of a ten-vertex *nido*-B$_{10}$H$_{14}$-type structure; and (40) has a sandwich-type structure in which a nickel atom is bonded to both a η^3-cyclobutenyl group and a three-carbon carbaborane. In addition the nickelacarbaborane cage in (40) has a slip distortion resulting in an open-cage geometry rather than the *closo* structure predicted by simple electron-counting rules.[71]

Russian workers have reported the synthesis and structure of the interesting basket-like molecule (Ph$_3$P)$_2$Pt-6,9-C$_2$B$_8$H$_{10}$ (41). It is described as having a *nido*-(pseudo-*closo*) structure and features a (PPh$_3$)$_2$Pt-bridge between the two carbon atoms in the open face of the C$_2$B$_8$ skeleton. The structure of (41), and some important interatomic distances are given in Figure 9.[72]

Reaction of the salt [N(PPh$_3$)$_2$][W($\equiv$CC$_6$H$_4$Me-4)(CO)$_2$(η^5-1,2-C$_2$B$_9$H$_9$Me$_2$)] with [*trans*-PtH(acetone)(PEt$_3$)$_2$][BF$_4$] in acetone at −30 °C has been shown to afford [PtW(CO)$_2$(PEt$_3$)$_2${η^6-C$_2$B$_9$H$_8$(CH$_2$C$_6$H$_4$Me-4)Me$_2$}] [(42), Figure 10], in which there has been an unprecedented rearrangement of the 11-vertex carbaborane ligand. The B(4) atom has acquired a CH$_2$C$_6$H$_4$Me-4 group, the C(1)-C(2) separation (288 pm) is non-bonding, and the tungsten atom is ligated by the six atoms C(2), B(3), B(4), B(5), C(1), and (B6). Reaction of (42) with PMe$_3$ gave the complex [PtW(μ-H) {μ-σ,η^5-C$_2$B$_9$H$_7$(CH$_2$R)Me$_2$}(CO)$_2$(PMe$_3$)(PEt$_3$)$_2$] [(43), Figure 10], in which the PMe$_3$ group has added to the tungsten centre, the B(3) atom has formed

[70] J. Zwecker, H. Pritzkow, U. Zenneck, and W. Siebert, *Angew. Chem., Int. Ed. Engl.*, 1986, **25**, 1099.

[71] J. J. Briguglio and L. G. Sneddon, *Organometallics*, 1986, **5**, 327.

[72] G. A. Kukina, M. A. Porai-Koshits, and V. S. Sergienko, *Koord. Khim.*, 1986, **12**, 561.

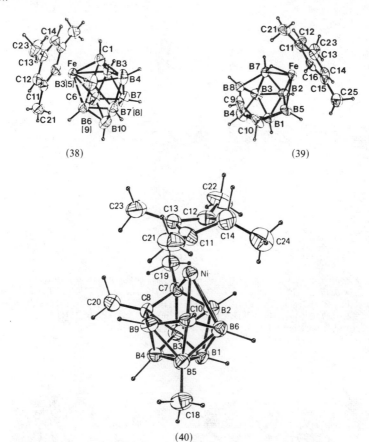

Figure 8 *ORTEP drawings of* 2-(η^6-Me$_3$C$_6$H$_3$)Fe-1,6-C$_2$B$_7$H$_9$ (38), 6-(η^6-Me$_3$C$_6$H$_3$)Fe-9,10-C$_2$B$_7$H$_{11}$ (39), *and* 5,7,8-Me$_3$-11,7,8,10-(η^3-C$_4$Me$_4$H)NiC$_3$B$_7$H$_7$ (40)[71]
(Reproduced by permission from *Organometallics*, 1986, **5**, 327)

an exopolyhedral B(3)—Pt σ bond, and the C(1)—C(2) interatomic distance (162 pm) has become bonding. The conversion of (42) into (43) by addition of an electron pair donor at tungsten involves a polyhedral rearrangment which is relevant to the *hypercloso/isocloso* controversy discussed earlier (see p. 51). The process may be viewed as a *hypercloso* to *closo* conversion corresponding to a two-electron addition to the cluster, or an *isocloso* to *closo* conversion in which a tungsten orbital is removed from the cluster bonding.[73]

The syntheses and crystal structures of two novel bis (η^5-dicarbollide) complexes have been reported, in which the planar bonding faces of the two dicarbollide ligands are nearly parallel and sandwich a single metal atom. The first was produced

[73] M. J. Attfield, J. A. K. Howard, A. N. de M. Jelfs, C. M. Nunn, and F. G. A. Stone, *J. Chem. Soc., Chem. Commun.*, 1986, 918.

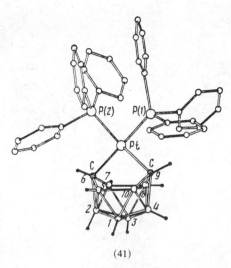

(41)

Figure 9 *Molecular structure of* $(Ph_3P)_2Pt$-6,9-$C_2B_8H_{10}$ (41). *Interatomic distances associated with the* Pt-*cluster linkage are* Pt—C(6) 214.1(9), Pt—C(9) 215.4(8), Pt—B(5) 265(1), Pt—B(7) 258(1), Pt—B(8) 265(1), Pt—B(10) 266(1), C(6)—B(5) 156(1), C(6)—B(7) 157(1), C(9)—B(4) 167(1), C(9)—B(8) 160(1), C(9)—B(10) 161(1)[72]
(Reproduced from *Koord. Khim.*, 1986, **12**, 561)

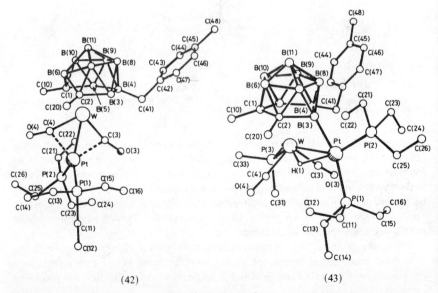

(42)   (43)

Figure 10 *Molecular structure of* $[PtW(CO)_2(PEt_3)\{\eta^6\text{-}C_2B_9H_8\text{-}(CH_2C_6H_4Me\text{-}4)Me_2\}]$ (42), *and* $[PtW(\mu\text{-}H)\{\mu\text{-}\sigma,\eta^5\text{-}C_2B_9H_7(CH_2C_6H_4Me\text{-}4)Me_2\}(CO)_2(PMe_3)(PEt_3)_2]$ (43)[73]
(Reproduced from *J. Chem. Soc., Chem. Commun.*, 1986, 918)

in a CO-catalysed 93% conversion of *closo*-3-Et-3,1,2,-AlC$_2$B$_9$H$_{11}$ in benzene, and was shown to be the bi-aluminacarbaborane *commo*-3,3'-Al[{*exo*-8,9-(μ-H)$_2$AlEt$_2$-3,1,2-AlC$_2$B$_9$H$_9$}(3',1',2'-AlC$_2$B$_9$H$_{11}$)]. It features a diethylaluminium moiety, bound to one of the dicarbollide clusters *via* two B—H—Al bridges, and can be thought of as a zwitterion composed of an anionic [Al(η^5-C$_2$B$_9$H$_{11}$)$_2$]$^-$ sandwich complexed with a [AlEt$_2$]$^+$ cation.[74a] The second is the isoelectronic silicon complex *commo*-3,3'-Si(3,1,2-SiC$_2$B$_9$H$_{11}$)$_2$, produced in 78% yield from the reaction between Li$_2$[*nido*-7,8-C$_2$B$_9$H$_{11}$] and SiCl$_4$ in benzene, which can be specifically deuterated at carbon to give [Si(C$_2$B$_9$H$_{10}$D)$_2$].[74b]

The electronically induced translational slip of a metal fragment across an η-bonded metallacarbaborane face continues to be a subject of keen interest, and one that has led to the determination of a series of crystal structures of compounds containing the [C$_2$B$_9$H$_{11}$]$^{2-}$ ligand or its derivatives. These include (i) caesium salts of the anions [Co(C$_2$B$_9$H$_8$Br$_3$)$_2$]$^{2-}$,[75a] [Co(C$_2$B$_9$H$_{10}$I)$_2$]$^{2-}$,[75b] and [Co(C$_2$B$_9$H$_{11}$)-(C$_2$B$_9$H$_{10}$I)]$^{2-}$;[75c] (ii) the complex 8,8'-MeOCOCH$_2$S-3-Co(1,2-C$_2$B$_9$H$_{10}$)$_2$, in which the dicarbaborane ligands are linked together by a monoatomic sulphur bridge to which a methoxycarbonylmethyl group is bound,[75d] (iii) the new indenylmetallacarbaborane 3-(η^5-C$_9$H$_7$)-3,1,2-CoC$_2$B$_9$H$_{11}$ (in two crystalline forms, one of which was studied at low temperature to give very accurate data) and the related compound 3-(η^5-C$_5$H$_5$)-3,1,2-CoC$_2$B$_9$H$_{11}$;[76a] and (iv) the cyclooctadiene species [3-(η^2,η^2,C$_8$H$_{12}$-3,1,2-PdC$_2$B$_9$H$_{11}$], for which the slipping distortion parameters cast doubt on the applicability of the previously suggested correlation with B(8) chemical shift data.[76b] A related compound of interest is the orthometallated phosphinocarbarhodaborane obtained in the reaction of [*closo*-3,3-{PPh$_3$}$_2$-3-H-3,1,2-RhC$_2$B$_9$H$_{11}$] with either isopropenyl acetate or vinyl acetate.[77]

7 Other Boron–Carbon Compounds

The role of organoboron compounds in organic synthesis is not covered in this article, though readers may wish to note two useful reviews on this topic.[78,79] Several other areas of work on the periphery of inorganic boron chemistry include, (i) the reaction of ethylmagnesium bromide with triethylborane, which exhibits an unusual solvent effect: in ethyl ether the reaction leads to triethylborane in 100% yield, whereas in THF the complex Et$_4$BMgBr.nTHF is the sole product;[80] (ii) the crystal structure for the crown ether salt [Li(12-crown-4)$_2$]$^+$[BMeS]$^{\cdot-}$ (Mes = 2,4,6-Me$_3$C$_6$H$_2$—), which constitutes the first such study on a boron-centred radical;[81] (iii) the interesting bis-ylides (44), which may be useful as complex

[74] (*a*) W. S. Rees, D. M. Schubert, C. B. Knobler, and M. F. Hawthorne, *J. Am. Chem. Soc.*, 1986, **108**, 5367; (*b*) W. S. Rees, D. M. Schubert, C. B. Knobler, and M. F. Hawthorne, *ibid.*, p. 5369.
[75] (*a*) P. Sivý, A Preisinger, O. Baumgartner, F. Valach, B. Koreň, and L. Mätel, *Acta Crystallogr.*, Sect. C, 1986, **42**, 24; (*b*) *ibid.*, p. 28; (*c*) *ibid.*, p. 30; (*d*) I. Císařová and V. Petříček, *ibid.*, p. 663.
[76] (*a*) D. E. Smith and A. J. Welch, *Organometallics*, 1986, **5**, 760; *erratum op. cit.*, *ibid.*, p. 2398; (*b*) *idem.*, *Acta Crystallogr.*, Sect. C, 1986, , 1717.
[77] C. B. Knobler, R. E. King, and M. F. Hawthorne, *Acta Crystallogr.*, Sect. C, 1986, **42**, 159.
[78] A. Suzuki, *Pure Appl. Chem.*, 1986, **58**, 629.
[79] G. W. Kabalka, *J. Organomet. Chem.*, 1986, **298**, 1.
[80] H. C. Brown and U. S. Racheria, *Organometallics*, 1986, **5**, 391.
[81] M. M. Olmstead and P. P. Power, *J. Am. Chem. Soc.*, 1986, **108**, 4235.

$$\begin{array}{c}
\text{R}^1 \\
/ \\
\overset{+}{\text{Ph}_3\text{P}}\!-\!\overset{..}{\text{C}} \\
\quad\quad\quad\searrow \\
\quad\quad\quad\quad \text{B}\!-\!\text{R}^2 \\
\quad\quad\quad\nearrow \\
\overset{+}{\text{Ph}_3\text{P}}\!-\!\overset{..}{\text{C}} \\
\quad\quad\searrow \\
\quad\quad\quad\quad \text{R}^3
\end{array}$$

(44) R^1 = H, Me, or Et; R^2 = c-C_6H_{11}, thexyl, or 2,2,6,6-tetramethylpiperidino (Tmp); R^3 = H or Me (not all combinations)

ligands;[82] and (iv) the work of Wrackmeyer and co-workers on the organoboration of alkylstannanes[83a-c] and *cis*-diethynylplatinum(II) complexes.[83d]

An excimer laser has been used as a heat source to evaporate bulk boron, and reactions of the resulting B atoms studied on a low temperature matrix. In pure argon both B and B_2 were detected, but with a matrix containing 10% methane, reaction occurred to generate what was thought to be a B—CH_3 species.[84] A novel graphite-like material of composition BC_3 has been synthesized, as a lustrous film of metallic appearance, from the interaction of BCl_3 with benzene at 800 °C. The material was shown by electron micrography to consist of sheets with a separation of 300–400 pm. Reaction with both strongly oxidizing and strongly reducing species led to intercalation compounds; sodium naphthalide in THF, for example, gave a bronze first-stage material with an interlayer spacing (I_c) of 430 pm, whereas oxidation with $(SO_3F)_2$ gave a blue first-stage material of composition $(BC_3)_2SO_3F$ with I_c = 810 pm. Reaction with liquid bromine gave a deep blue solid of composition $(B_{0.25}C_{0.75})_{15}Br$.[85]

The question of whether neutral helium compounds might be stable in their ground state has been addressed theoretically, and it was shown that HeBCH is predicted to be unstable towards helium dissociation by −25.1 kJ mol^{-1}.[86] Optimum geometries and relative energies of the five most probable isomers of formula C_2H_5B have been calculated by use of *ab initio* SCF theory, and it was shown that the closed-ring form, borirane, is a stable minimum on the potential energy surface.[87] In an important semiempirical and *ab initio* MO study it has been shown that a pathway of very low activation energy (46 kJ mol^{-1}), involving a highly non-planar four-centre transition state, is available for the dimerization of borirene (45) to 1,4-diboracyclohexa-2,5-diene (46). Boron acceptor orbitals were shown to play a crucial role, and it was suggested that borirenes should dimerize easily unless π-donor groups are present at boron. The dimerization could also be hindered by the presence of bulky groups, because of the close approach of two boron and two carbon atoms in the transition state. No low-energy pathway was found for the dimerization of (45) to 2,3,4,5-tetracarba-*nido*-hexaborane(6) (47), despite the fact

---

[82] H. J. Bestmann and T. Arenz, *Angew. Chem., Int. Ed. Engl.*, 1986, **25**, 559.

[83] (a) B. Wrackmeyer, *J. Chem. Soc., Chem. Commun.*, 1986, 397; (b) A. Sebald, P. Seiberlich, and B. Wrackmeyer, *J. Organomet. Chem.*, 1986, **303**, 73; (c) A. Sebald and B. Wrackmeyer, *ibid.*, 1986, **307**, 157; (d) *ibid.*, 1986, **304**, 271.

[84] G. Jeong and K. J. Klabunde, *J. Am. Chem. Soc.*, 1986, **108**, 7103.

[85] J. Kouvetakis, R. B. Kaner, M. L. Sattler, and N. Bartlett, *J. Chem. Soc., Chem. Commun.*, 1986, 758.

[86] W. Koch, J. R. Collins, and G. Frenking, *Chem. Phys. Lett.*, 1986, **132**, 330.

[87] C. A. Taylor, M. C. Zerner, and B. Ramsey, *J. Organomet. Chem.*, 1986, **317**, 1.

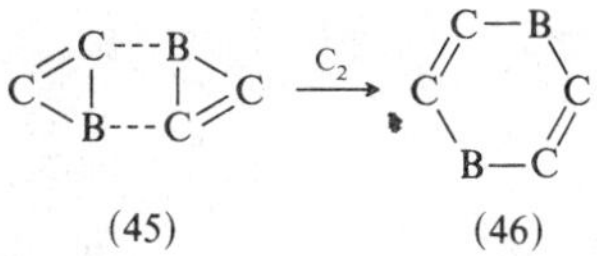

(45)   (46) 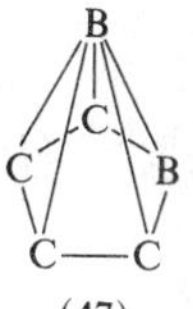   (47) 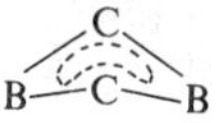   (48)

that energetically this was the most favourable reaction. In contrast, the other possible reaction of two borirenes, disproportionation to give 1,3-diboretene (48) and acetylene was predicted to be nearly thermoneutral and to have a very low activation energy. Several other $(CH)_4(BH)_2$ isomers were also examined, and their relative energies established.[88]

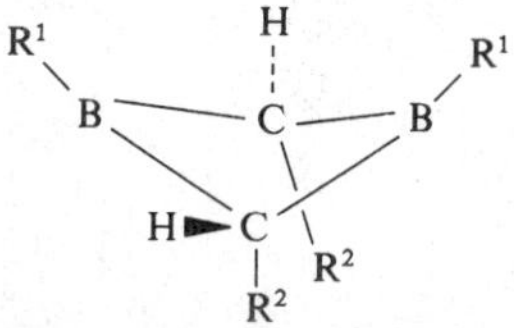

(49) (a) $R^1 = NMe_2$, $R^2 = Bu^t$; (b) $R^1 = NPr_2^i$, $R^2 = SiMe_3$ (50)

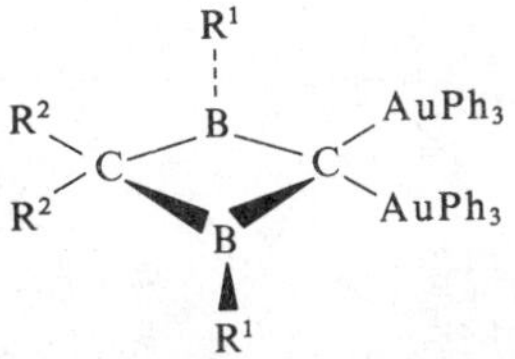

The first structurally confirmed 1,2 diboretanes (49) and (50) have been reported. Compound (49) has a puckered structure with a puckering angle of 32.2°, in very good agreement with the value of 36.2° derived previously from calculations of Schleyer and co-workers. However, the puckering angle for (50) is only 3°, which indicates that a planar structure is also possible.[89a] The simplest known organometallic cyclobutadiene, 1,3-diboracyclobutadiene, has been isolated as the lithium salt, and shown to have ^{11}B and ^{13}C n.m.r. parameters consistent with the resonance structures (51). An X-ray diffraction analysis indicated that (51) crystallizes as a dimer in which a Li_4 layer bridges two four-membered rings.[89b]

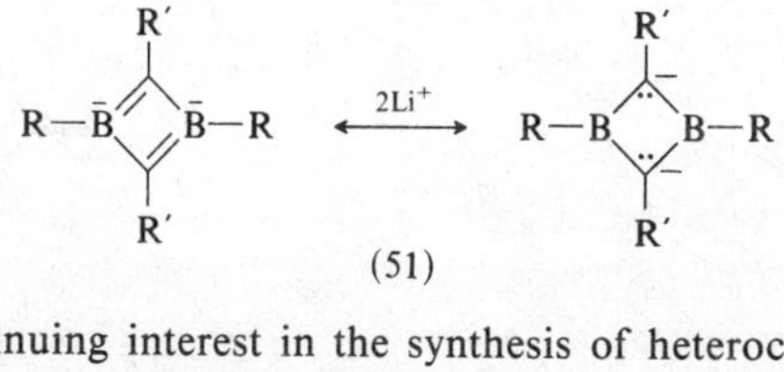

(51)

There is a continuing interest in the synthesis of heterocyclic ring compounds containing boron–carbon bonds, and in particular of metal derivatives, including sandwich complexes, in which these species act as η-bonded ligands. These include 2-, and 3-borolenes $[C_4H_6BR$ (R = Me, Ph, *etc.*)][90a] and their complexes with

[88] P. H. M. Budzelaar, S. M. van der Kerk, K. Krogh-Jespersen, and P. von Rague Schleyer, *J. Am. Chem. Soc.*, 1986, **108**, 3960.

[89] (*a*) P. Hornbach, M. Hildebrand, H. Pritzkow, and W. Siebert, *Angew. Chem., Int. Ed. Engl.*, 1986, **25**, 1112; (*b*) G. Schmidt, G. Baum, W. Massa, and A. Berndt, *ibid.*, p. 1111.

[90] (*a*)H. E. Herberich, W. Boveleth, B. Hessner, M. Hostalek, D. P. J. Köffer, H. Ohst, and D. Söhnen, *Chem. Ber.*, 1986, **119**, 420; (*b*) G. E. Herberich, W. Boveleth, B. Hessner, D. P. J. Köffer, M. Negele, and R. Saive, *J. Organomet. Chem.*, 1986, **308**, 153; (*c*) G. E. Herberich, U. Büschges, B. Hessner, and H. Lüthe, *ibid.*, 1986, **312**, 13.

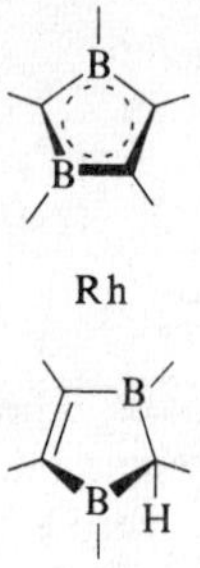

Rh

(52)

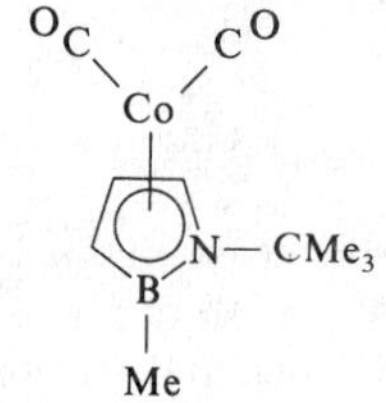

OC CO
Co
N—CMe₃
Me

(53)

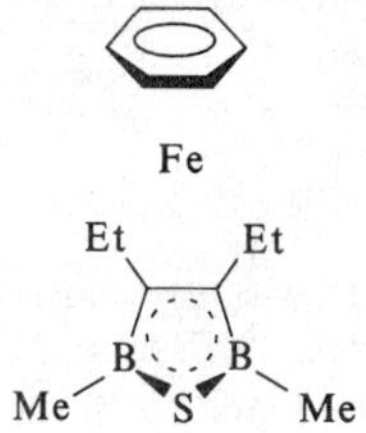

Fe
Et   Et
B   B
Me   S   Me

(54)

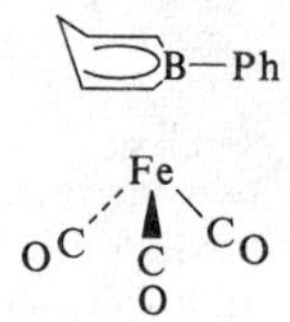

B—Ph
Fe
OC   C   CO
O

(55)

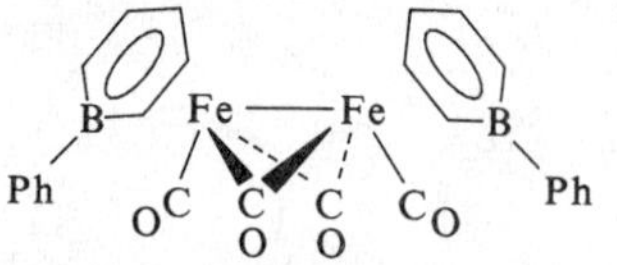

Ph B
Fe —— Fe
OC   C   C   CO
O   O
B Ph

(56)

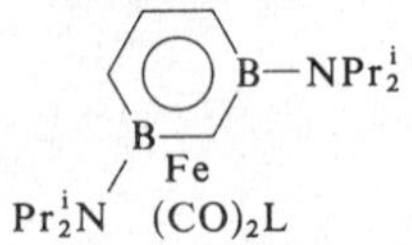

B—NPr$_2^i$
B
Pr$_2^i$N   Fe
(CO)₂L

(57) L = CO, PMe₃

Li₂
B   NMe₂
B   NMe₂

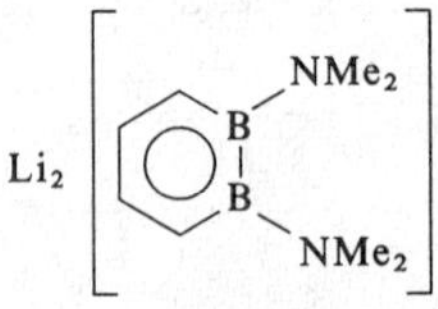

(58)

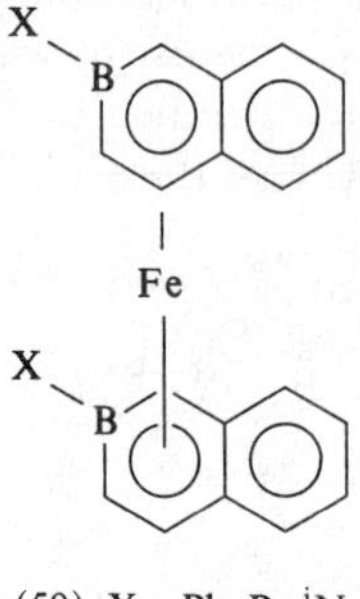

X—B
Fe
X—B

(59) X = Ph, Pr₂^{i}N

Ni
—B   B—
Ni
—B   B—
Ni
—B   B—
Ni

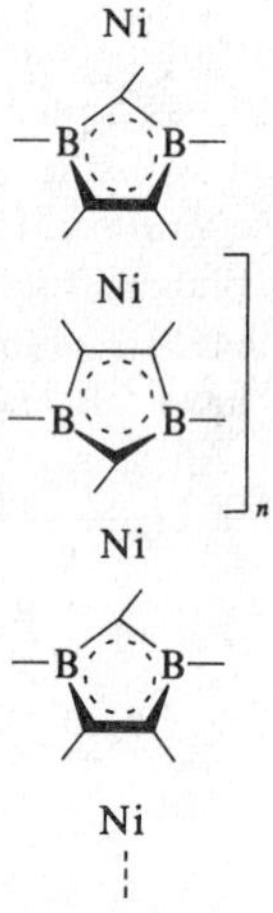

(60)

manganese, iron, and cobalt carbonyls;[90b] various η^5-borolene[90c] and diborolene[91] complexes of rhodium [e.g.(52)]; azaborolyl cobalt complexes such as (53);[92a] and the thiaborolenes (54).[92b] Phenylbora-2,5-cyclohexadiene has also proved to be a versatile ligand precursor in reactions with Fe(CO)₅, CO₂(CO)₈, Ni(CO)₄, and Ni(η^4-1,5-COD)₂; for example the photochemical reaction with Fe(CO)₅ gives (55)

[91] K. Geilich and W. Siebert, Z. Naturforsch., Teil B, 1986, 41, 671.
[92] (a)G. Schmid and F. Schmidt, Chem. Ber., 1986, 119, 1766; (b) U. Zenneck, L. Suber, H. Pritzkow, and W. Siebert, ibid., p. 971.

which under irradiation is slowly converted into the known borabenzene derivative (56).[93a] The first 1,3-diborabenzene complexes (57) have also been described.[93b] Other interesting species include the 1,2-diboratabenzene lithium salt (58), the anion of which has favourable ligating properties;[93c] iron and rhodium derivatives of the hexahapto-bonded ligand 2-boranaphthalene [*e.g.* (59)];[94] anionic triple-decker complexes having closed 30-electron valence shells, derived from [(η-C_5H_5)Fe(C_4H_4BPh)]$^-$ and the fragment Cr(CO)$_3$;[95a] and the first polymeric *poly-decker* sandwich compound (60), mentioned in last year's report, but now described in detail. The black film-like polymer (60) is a highly air-sensitive, though thermally stable, semi-conductor.[95b]

8 Boron–Nitrogen, –Phosphorus, and –Arsenic Compounds

Reactions of BH_3.THF with (dialkylamino)dimethylarsines, Me_2AsNR_2 (R = Me, Et, Pr^n, or Pr^i), have been carried out over a range of temperature to establish the initial co-ordination site of boron and to determine the nature of any As—N, As—B, and N—B bond dissociation–formation processes in solution.[96] The nature of the N—B bond in a series of aromatic amine boranes has been studied by ^{11}B and ^{1}H n.m.r. spectroscopy, but no simple relationship was established between the spectroscopic data and the nature of the bonding.[97] In contrast, ^{11}B quadrupole-perturbed n.m.r. spectra of the compounds $F_3BNH_xMe_{3-x}$ have provided a clear picture of their electronic structures.[98] The molecular structures of pyridine complexes of BF_3, BCl_3, and BBr_3 have been determined by gas-phase electron diffraction measurements; free rotation about the B—N bond was established in all cases.[99] The barrier to rotation about the B—N bond in methylaminodiphenylborane has been studied by n.m.r. techniques.[100]

A laser flash-photolysis-e.s.r. technique has been used to measure absolute rate coefficients and activation parameters for β-scission of the amine-boryl radicals $R_3N \rightarrow \dot{B}H_2$ ($R_3N = Pr_2^iNEt$, Bu^{t}NMe, or azetidine; *e.g.* equation 2). The process was shown to be much more favourable thermodynamically than the corresponding cleavage of their isoelectronic alkyl-radical counterparts.[101]

$$Bu^tMe_2N \rightarrow \dot{B}H_2 \rightarrow \dot{B}u^t + Me_2N \rightleftharpoons BH_2 \qquad (2)$$

Two members of a new class of substituted amine-boranes, $Me_3N.BH_2CRR'CN$ (R = H, R' = Ph; R = R' = Me), have been synthesized. The compounds are referred to as amine-(α-cyanoorgano)boranes and are closely related to

[93] (a) G. E. Herberich and E. Raabe, *J. Organomet. Chem.*, 1986, **309**, 143; (b) G. E. Herberich and H. Ohst, *ibid.*, 1986, **307**, C16; (c) G. E. Herberich, B. Hessner, and M. Hostalek, *Angew. Chem., Int. Ed. Engl.*, 1986, **25**, 642.

[94] P. Paetzold, N. Finke, P. Wennek, G. Schmid, and R. Boese, *Z. Naturforsch., Teil B*, 1986, **41**, 167.

[95] (a)G. E. Herberich, B. Hessner, J. A. K. Howard, D. P. J. Köffer, and R. Saive, *Angew. Chem., Int. Ed. Engl.*, 1986, **25**, 165; (b) T. Kuhlmann, S. Roth, J. Rozière, and W. Siebert, *ibid.*, p. 105.

[96] R. V. Kanjolia, L. K. Krannich, and C. L. Watkins, *J. Chem. Soc., Dalton Trans.*, 1986, 2345.

[97] C. Camacho, M. A. Paz-Sandoval, and R. Contreras, *Polyhedron*, 1986, **5**, 1723.

[98] J. Olliges, A. Lötz, and J. Voitländer, *J. Mag. Reson.*, 1986, **69**, 302.

[99] K. Iijima, T. Noda, M. Maki, T. Sasase, and S. Shibata, *J. Mol. Struct.*, 1986, **144**, 169.

[100] B. Wrackmeyer, *Z. Naturforsch., Teil B*, 1986, **41**, 59.

[101] J. A. Baban, J. P. Goddard, and B. P. Roberts, *J. Chem. Res. (S)*, 1986, 30.

$Me_3N.BH_2CN$ which shows potent antiarthritic, antihypolipidemic, and antitumour activity on rodents, and is the precursor to $Me_3N.BH_2CO_2H$, the (protonated) boron analogue of the dipolar amino acid betaine, $Me_3N.CH_2CO_2^-$.[102a] The synthesis of $(PPh_3)_2Cu(NC)_2BH_2$ from the reaction between $Na(NC)_2BH_2.0.65(dioxane)$ and $(PPh_3)_2CuCl$, and its X-ray crystal structure have also been reported; the CN groups are co-ordinated to different Cu^I atoms, with the result that the molecule exists as a linear polymer of an unprecedented type in cyanohydroborate chemistry.[102b] The synthesis and characterization of Cr^{III} and Fe^{III} complexes of $Me_3N.BH_2CO_2H$, both of which contain the cation $[M_3O(Me_3N.BH_2CO_2)_6R_3]^+$ (R = H_2O, MeOH), have also been reported.[102c]

Ab initio calculations at the 6-31G and 6-31G** levels have been performed on the ground state of CH_2NBH_2 (iminoborane) and its derivatives $(CH_2N)_2BH$ and $(CH_2N)_3B$. In all cases the perpendicular arrangement of the amino group with respect to the CNB-framework was found to be more stable than the coplanar arrangement, but with increased imino substitution the energy difference between the two configurations was found to decrease. Protonation of iminoborane was predicted to occur at N and hydride ion addition at either B or C; association of CH_2NBH_2 to form dimers and trimers was found to be favourable.[103a] The boron–nitrogen bond in aminoborane ($BH_2.NH_2$), diaminoborane $[BH(NH_2)_2]$, and aminodifluoroborane ($BF_2.NH_2$) has also been studied theoretically, and an appreciable amount of double-bond character proposed;[103b] the He(I) photoelectron spectrum has also been discussed.[103c]

Nöth and co-workers have continued their extensive investigations in the field of boron–nitrogen chemistry. Diisopropyl- and di-t-butylboranes R_2BX (X = NHR' and NR'$_2$, R' = H, Me, Et, CMe_3, etc.) have been synthesized,[104a] and their geometries probed by ^1H, ^{11}B, ^{13}C, and ^{14}N n.m.r. and He(I) photoelectron spectroscopy. Only the di-t-butyl(diorganylamino)boranes were found to deviate from a planar C_2BNC_2 conformation, the extent of the deviation increasing with the bulkiness of the R'$_2$N group.[104b] N-metallation of $RBNH_2$ or R_2BNMeH has afforded N-lithioaminoboranes $R_2BNR'Li$ with sterically demanding R groups (R = Bu^t, Pr^i), and these have been used in the synthesis of symmetrically and unsymmetrically substituted diborylamines.[104c] Crystal structures have been determined for $Ph_2BN(CMe_3)SiMe_3$ and five diborylamines to assess the influence of steric factors on the detailed conformations adopted.[104d] The synthesis of diborylamines of the type $R_3SnN(Br'R'')_2$ (R, R', and R'' = Me, Et, etc.) from $N(SnMe_3)_3$ or $N(SnEt_3)_3$ and boron halides R'R''BX has also been described.[104e]

[102] (a) J. L. Peters, V. M. Norwood, and K. W. Morse, Inorg. Chem., 1986, 25, 3713; (b) K. W. Morse, D. G. Holah, and M. Shimoi, ibid., p. 3113; (c) V. M. Norwood and K. W. Morse, ibid., p. 3690.

[103] (a) D. R. Armstrong and G. T. Walker, J. Mol. Struct. (Theochem.), 1986, 32, 47; (b) T.-K. Ha, ibid., 1986, 136, 165; (c) N. P. C. Westwood and N. H. Werstiuk, J. Am. Chem. Soc., 1986, 108, 891.

[104] (a) U. Höbel, H. Nöth, and H. Prigge, Chem. Ber., 1986, 119, 325; (b) H. Nöth and H. Prigge, ibid., p. 338; (c) H. Nöth, H. Prigge, and A.-R. Rotsch, ibid., p. 310; (d) D. Männig, H. Nöth, H. Prigge, A.-R. Rotsch, S. Gopinathan, and J. W. Wilson, J. Organomet. Chem., 1986, 310, 1; (e) H. Nöth, P. Otto, and W. Storch, Chem. Ber., 1986, 119, 2517; (f) E. Hanecker, H. Nöth, and U. Wietelmann, ibid., p. 1904; (g) B. Glaser and H. Nöth, ibid., p. 3253; (h) ibid., p. 3856; (i) P. Kölle and H. Nöth, ibid., p. 313; (j) ibid., p. 3849; (k) H. Nöth and B. Rasthofer, ibid., p. 2075; (l) T. Franz, E. Hanecker, H. Nöth, W. Stöcker, W. Storch, and G. Winter, ibid., p. 900.

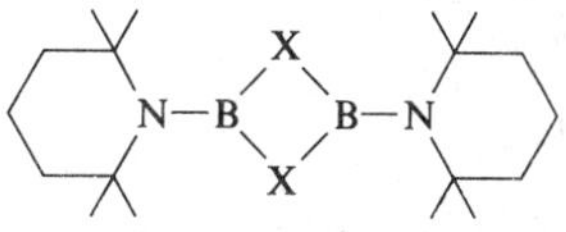

(61) X = O, Se 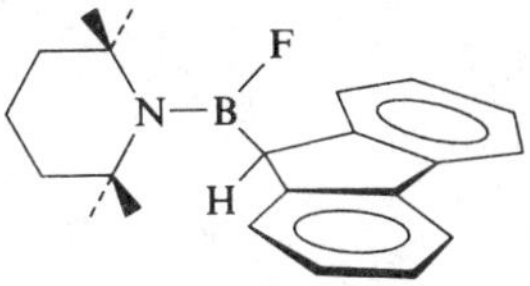   (62) 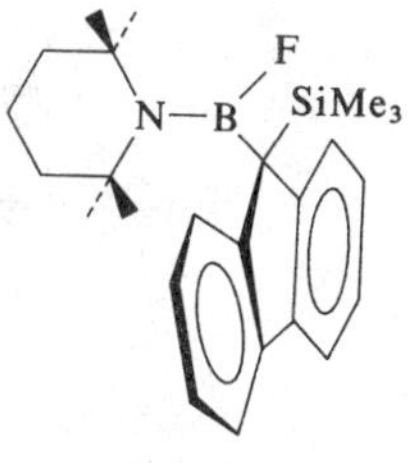   (63)

Crystal and molecular structures have been determined for the tetramethyl-piperidino (tmp) diboretanes (61) and compared with that of the related dithiadiboretane. Planar four-membered ring systems were found in all three molecules, and there were no marked differences in B—N bond length or internal ring angles. It was therefore suggested that the boron atoms were electronically saturated *via* the B—N bonds.[104f] The replacement of halogen atoms by metal organyls in (tmp)boron dihalides has been shown to depend primarily on steric factors. For example, single F atoms can easily be replaced by an isopropyl or a (trimethylsilyl)methyl group to yield boranes of the type tmpB(F)R, but Cl or Br atoms cannot be substituted. Similarly, $LiCHPh_2$ reacts with $tmpBX_2$ to produce $tmpB(X)CHPh_2$, but $LiC(SiMe_3)Ph_2$ can substitute only one F atom in $tmpBF_2$. Particularly interesting results were obtained in reactions of $tmpBF_2$ with 9-lithio-fluorene and 9-lithio-9-(trimethylsilyl)fluorene, respectively; these yielded (62) in which the F-atom and 9H-atom of the fluorene unit are seen to adopt an *anti* orientation, and (63) in which the trimethylsilyl moiety is *syn* oriented.[104g] Thermolysis of fluoro(tmp)[(trimethylsilyl)phenyl]boranes in solution at 285 °C has yielded Me_3SiF, and, when benzophenone was included, olefins were obtained, apparently *via* a Wittig-type reaction of the intermediate methylene borane (64) with the ketone.[104h]

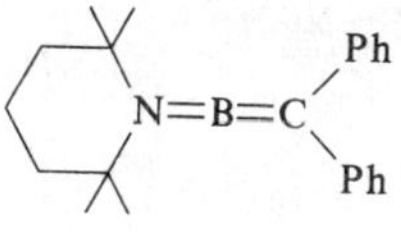

(64) 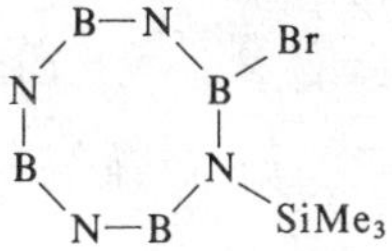   (65)

Bis(amino)boron salts have been obtained by halide abstraction from $(RR'N)_2BX$ (RR'NH = benzyl-t-butylamine; X = F, Cl) with $AlCl_3$ or BBr_3. The structure of $[(RR'N)_2B][AlCl_4]$ was determined by X-ray crystallography and found to possess an allene-type arrangement with a linear N=B=N skeleton and rather short BN bonds (133.2 pm). The barrier to rotation about the BN bond was estimated to be $\Delta G^* = 82 \pm 3 \text{ kJ mol}^{-1}$.[104i] Similar linear N=B=N structures have been found in (trimethyl)aminoborinium[104j] and bis(diisopropyl)aminoborinium[104k] cations. The former are unstable and at room temperature $[(Me_3Si)_2N=B=N(SiMe_3)_2][BBr_4]$ decomposes into Me_3SiBr and the eight-membered ring compound $(Me_3SiN—BBr)_4$ (65).[104j] The related compound $[ClB=NCMe_3]_4$, which is formed by dimerization of the metastable diazadiboretidine $[ClB=NCMe_3]_2$ at 20 °C, also possesses an eight-membered ring with alternating single and double BN bonds.[104l]

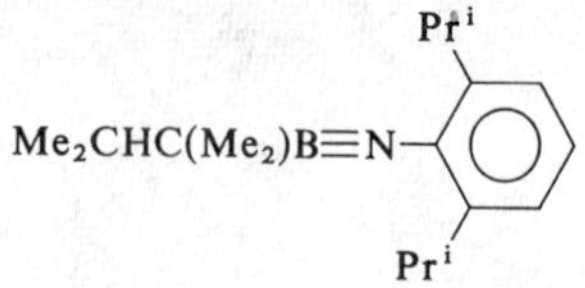

Me₂CHC(Me₂)B≡N—⟨benzene ring with Pr^i above and Pr^i below⟩

(66)

Me₃CMe₂Si   SiMe₂CMe₃
        N—N
          B
          N
Me₃Si   R

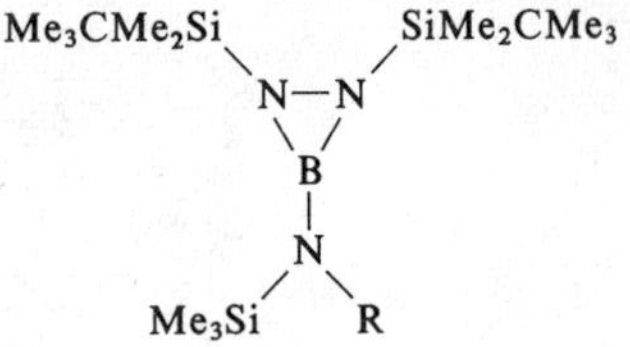

(67) R = CMe₃, SiMe₃

The iminoborane (66) has been synthesized from dibromothexylborane and 2,6-diisopropylaniline, and can be stored for long periods.[105] X-Ray diffraction analysis of the iminoborane $(Me_3Si)_3Si—B≡N—CMe_3$ has revealed the shortest B—N bond length found so far (122.1 pm).[106]

The three-membered diazaboracyclopropanes (67) have been obtained in reactions of N,N'-bis(silyl)-N-fluoroboryl-hydrazines with Bu^{t}Li, and shown to contain planar NBN_2 units.[107] Further information has appeared on the three-membered ring systems obtained by dehalogenation of diborylamines $[Me_3CN(BXNR_2)]_2$ $[NR_2 = tmp, X = Cl; NR_2 = N(CHMe_2)_2, X = Br]$.[108] Detailed *ab initio* MO calculations have also been carried out on these diazaboridine and azadiborane three-membered ring systems.[109]

The synthesis, crystal structure, and gas-phase photoelectron spectroscopic study of the highly sterically hindered compound (68) have been reported. The two halves of the molecule are linked by a B—B bond (172.1 pm) and the five-membered ring BN_2C_2 planes form a dihedral angle of 80.7°. A comparison of the photoelectron spectrum of (68) with that of $B_2(NMe_2)_4$ suggests that both compounds adopt a staggered configuration in the gas phase.[110] Borylated carbodiimides such as (69) have also been reported.[111]

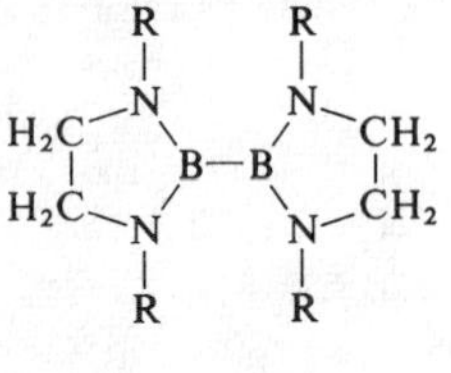

(68) R = cyclohexyl 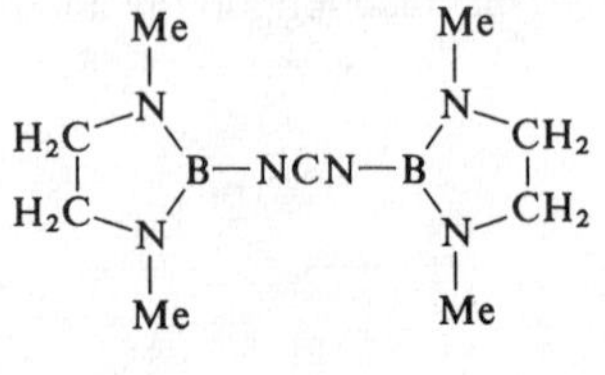   (69)

A novel type of pyrazabole (70), in which the two pyrazabole boron atoms are also bridged by an O—BR—O group, has been obtained in the reaction of tri-B-organylboroxins, $(—BR—O—)_3$ (R = Et, Ph), with pyrazole, Hpz.[112a,b] Detailed

[105] M. Armbrecht and A. Meller, *J. Organomet. Chem.*, 1986, **311**, 1.
[106] M. Haase, U. Klingebiel, R. Boese, and M. Polk, *Chem. Ber.*, 1986, **119**, 1117.
[107] R. Boese and U. Klingebiel, *J. Organomet. Chem.*, 1986, **306**, 295.
[108] F. Dirschl, E. Hanecker, H. Nöth, W. Rattay, and W. Wagner, *Z. Naturforsch., Teil B*, 1986, **41**, 32.
[109] P. H. M. Budzelaar and P. von R. Schleyer, *J. Am. Chem. Soc.*, 1986, **108**, 3967.
[110] G. Ferguson, M. Parvez, R. P. Brint, C. M. Power, T. R. Spalding, and D. R. Lloyd, *J. Chem. Soc., Dalton Trans.*, 1986, 2283.
[111] W. Einholz and W. Haubold, *Z. Naturforsch., Teil B*, 1986, **41**, 1367.
[112] (a) J. Bielawski and K. Niedenzu, *Inorg. Chem.*, 1986, **25**, 85; (b) *ibid.*, p. 1771; (c) J. Bielawski, M. K. Das, E. Hanecker, K. Niedenzu, and H. Nöth, *ibid.*, p. 4623; (d) J. Bielawski, T. G. Hodgkins, W. J. Layton, K. Niedenzu, P. M. Niedenzu, and S. Trofimenko, *ibid.*, p. 87.

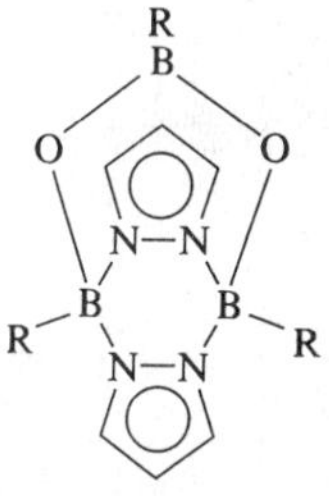

(70) 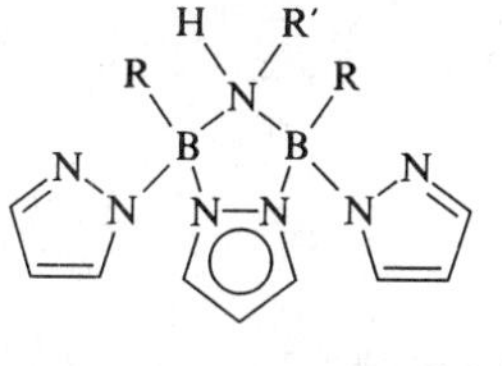 (71) R = Et, Ph; R' = H, Me 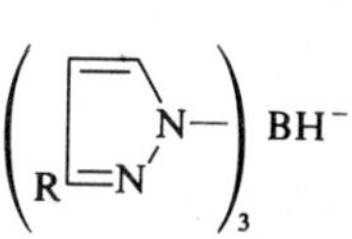 (72) R = Bu^t, Ph

investigations of the interaction of pyrazole with 1,3-diboroxanes, $(R_2B)_2O$ (R = Et, Ph),[112b] and borazines[112c] have also been reported. In the latter study, four compounds of the type $R(pz)B(\mu\text{-}pz)(\mu\text{-}NHR')BR(pz)$ (71) containing a central B_2N_3 ring system were obtained. Chain-type polynuclear pyrazolyl-bridged spiro species containing boron and metal centres have also been synthesized in reactions involving the anion $[B(pz)_4]^-$ and metal halide species LMX (L = non-reactive ligand; M = metal, *e.g.* Pd; X = halogen).[112d] Many other studies exploiting the useful ligating properties of the pyrazolylborate anions have been reported,[113] but the most noteworthy from the point of view of this article describes a new class of ligand (72), in which a bulky group is included in the 3-position of the pyrazole ring to prevent bis-chelate formation and dimerization. Crystal structures were presented for the cobalt complexes $Co\{HB(3\text{-}Bu^tpz)_3\}(SCN)$ and $Co\{HB(3\text{-}Phpz)_3\}(SCN)\text{-}$ (THF).[113h]

Current interest in the use of simple boron-nitrogen compounds as precursors to polymers and solid-state materials has prompted the report of a convenient synthesis and *X*-ray crystal structure determination of two isomers of 1,3,5-trimethylcycloborazane $[MeN(H)BH_2]_3$. The isomers were found to possess similar six-membered ring structures with alternating boron and nitrogen atoms in chair conformations, but differed in the disposition of the methyl groups on the nitrogen atoms; in one isomer these were all equatorial, whereas in the other isomer one of the methyl groups was in an axial position.[114]

From a comparison of the spectral properties of borazine and benzene, it has been concluded that the two molecules have similar delocalized π-electron systems.[115] Other boron-nitrogen ring systems reported during the year include various polycyclic borazines;[116] N,N',N''-trialkoxy- and N,N',N''-tris(amino)borazines;[117a] new mono- and bicyclic BNS systems such as (73);[117b]

[113] (a) S. Lincoln and S. A. Koch, *Inorg. Chem.*, 1986, **25**, 1594; (b) C. G. Young, S. A. Roberts, and J. Enemark, *ibid.*, p. 3667; (c) H. P. Kim, S. Kim, R. A. Jacobson, and R. J. Angelici, *Organometallics*, 1986, **5**, 2481; (d) H. P. Kim and R. J. Angelici, *ibid.*, p. 2489; (e) M. Green, J. A. K. Howard, A. P. James, C. M. Nunn, and F. G. A. Stone, *J. Chem. Soc., Dalton Trans.*, 1986, 187; (f) F. Mani, *Inorg. Chim. Acta*, 1986, **117**, L1; (g) S. K. Lee and B. K. Nicholson, *J. Organomet. Chem.*, 1986, **309**, 257; (h) J. C. Calabrese, S. Trofimenko, and J. S. Thompson, *J. Chem. Soc., Chem. Commun.*, 1986, 1122.

[114] C. K. Narula, J. F. Janik, E. N. Duesler, R. T. Paine, and R. Schaeffer, *Inorg. Chem.*, 1986, **25**, 3346.

[115] J. P. Doering, A. Gedanken, A. P. Hitchcock, P. Fischer, J. Moore, J. K. Olthoff, J. Tossell, K. Raghavachari, and M. B. Robin, *J. Am. Chem. Soc.*, 1986, **108**, 3602.

[116] R. H. Cragg and M. Nazery, *J. Organomet. Chem.*, 1986, **303**, 329.

[117] (a) A. Meller and M. Armbrecht, *Chem. Ber.*, 1986, **119**, 1; (b) D. Fest, C. Habben, and A. Meller, *ibid.*, p. 3121; (c) C. Habben and A. Meller, *ibid.*, p. 9; (d) C. Habben, A. Meller, M. Noltemeyer, and G. M. Sheldrick, *Angew. Chem., Int. Ed. Engl.*, 1986, **25**, 741.

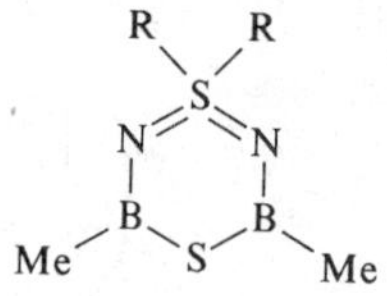

(73) R = Me, Et, Pr^i

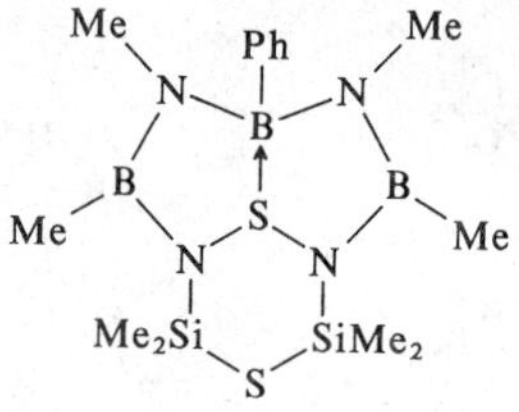

(74)

novel S- and Se-containing boron heterocycles;[117c] the new tricyclic B$_3$N$_4$S$_2$Si$_2$ system (74);[117d] and the eight-membered molecule Ph$_2$P—NPh—BH$_2$—NPh—PPh$_2$—NPh—BH$_2$—NPh.[118]

In comparison with the amount of work reported on boron-nitrogen compounds, there has been very little on boron-phosphorus or boron-arsenic systems. Nevertheless, there have been some interesting developments. The first diphosphadiboretanes, (75a),[119a] (75b),[119b] and (75c),[120] have been synthesized, and shown by X-ray

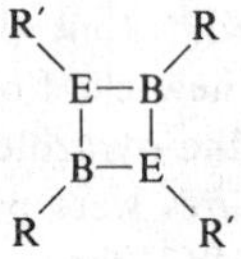

a; E = P, R = tmp, R' = 2,4,6-Me$_3$C$_6$H$_2$ (mes) (ref. 119a)
b; E = P, R = tmp, R' = 2,4,6-Bu$^t_3$C$_6$H$_2$ (ref. 119b)
c; E = P, R = tmp, R' = CEt$_3$ (ref. 120)
d; E = As, R = tmp, R' = mes (ref.119a)

(75)

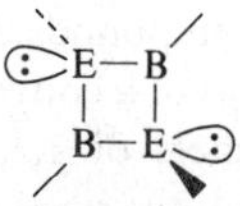

(76)

(77)

crystallography to feature planar P$_2$B$_2$ rings and pyramidal geometry at phosphorus. The arsenic analogue (75d) and the monomeric compound MesP(H)B(Cl)(tmp), which contains a P—B single bond (~195 pm), were also reported.[119a] The P—B bond lengths in (75a) and (75c) are similar (ca. 192 pm), and are indicative of a bond order of unity. On this basis it was suggested that the structure of (75a) (Figure 11) corresponds to canonical form (76), rather than the alternative form (77). The latter is, however, believed to be the appropriate description for the analogous diazaboretidines, which exhibit trigonal planar geometries at both N and B, and N—B bond orders >1.[119a] The crystal structure analysis of the sterically crowded diphosphadiboretane (75b) reveals considerable strain in the molecule, as evidenced by (i) the increase in the P—B bond length to 196 pm, (ii) the rhombic distortion in the P$_2$B$_2$ moiety, (iii) the 'bending' of the P atom out of the aryl plane, and (iv) the flattening of the pyramidal geometry at the phosphorus atom. Thermolysis of

[118] G. Süss-Fink, Chem. Ber., 1986, 119, 2393.
[119] (a) A. M. Arif, A. H. Cowley, M. Pakulski, and J. M. Power, J. Chem. Soc., Chem. Commun., 1986, 889; (b) A. M. Arif, J. E. Boggs, A. H. Cowley, J.-G. Lee, M. Pakulski, and J. M. Power, J. Am. Chem. Soc., 1986, 108, 6083.
[120] P. Kolle, H. Nöth, and R. T. Paine, Chem. Ber., 1986, 119, 2681.

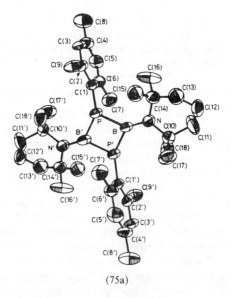

(75a)

Figure 11 *ORTEP view of* [MesPB(tmp)]$_2$ (75a)[119a]
(Reproduced from *J. Chem. Soc., Chem. Commun.*, 1986, 889)

(75b) at 250 °C was found to give the first boraphosphene (tmp)B=P(Ar) (Ar = 2,4,6-t-Bu$_3$C$_6$H$_2$) in the vapour phase.[119b] Phosphorus–boron multiple bonding has also been established in the novel monomeric complexes [Li(Et$_2$O)$_2$RPBMes$_2$] (Mes = 2,4,6-Me$_3$C$_6$H$_2$; R = Ph, C$_5$H$_5$, or Mes) and the ion pairs [Li(12-crown-4)$_2$][RPBMes$_2$].THF (R = Ph, C$_6$H$_{11}$, or Mes), all of which have relatively short P—B distances (~183 pm). The presence of the bulky mesityl group is crucial in stabilizing the P=B multiple bond, by preventing further attack at the boron centre.[121]

The products of the reaction between borane (BH$_3$) and bis(diphenylphos-phino)methane have been found to depend upon the stoicheiometry, and, if excess phosphine is used, a monoborane adduct is formed. In contrast, the phosphines (Ph$_2$P)$_2$(CH$_2$)$_n$ (n = 2—4) afford only the bis(borane) adducts. With BH$_2$I, the boronium iodide [(Ph$_2$P)$_2$CH$_2$BH$_2$]$^+$I$^-$ is formed.[122] Other investigations involving compounds with B—P bonds include the microwave study of MePH$_2$BH$_3$ and MePH$_2$BD$_3$,[123] and the pieces of work already discussed in refs. 34b and 52.

Attempts to isolate *stable* adducts containing As—B bonds, in reactions of BH$_3$.THF with (dialkylamino)dimethylarsines Me$_2$AsNR$_2$ (R = Me, Et, Pr^n, or Pr^i), have proved unsuccessful, though such species are present in solution at low temperature. When R = Me, Et, or Pr^n, formation of the N—B bonded adduct is thermodynamically preferred, and only when R=Pr^i is an As—B bonded adduct formed as the exclusive product at room temperature.[96] ^{1}H n.m.r. chemical shifts

[121] R. A. Bartlett, X. Feng, and P. P. Power, *J. Am. Chem. Soc.*, 1986, **108**, 6817.
[122] D. R. Martin, C. M. Merkel, and J. P. Ruiz, *Inorg. Chim. Acta*, 1986, **115**, L29
[123] W. Kasten, H. Dreizler, R. L. Kuczkowski, and M. Soltis La Barge, *Z. Naturforsch., Teil A*, 1986, **41**, 835.

for the adduct series $Me_nAsH_{3-n}BX_3$ and the ^{11}B n.m.r. chemical shifts for the adduct series $Me_3AsH_{3-n}BX_3$, $Me_nAsH_{3-n}BX_2Y$, and $Me_nAsH_{3-n}BXYZ$ (where $n = 1,2,3$; $X \neq Y \neq Z = Cl$, Br, or I) have been discussed in terms of their use as indicators of acid–base strength.[124]

9 Boron–Oxygen, –Sulphur, and –Selenium Compounds

An *ab initio* SCF-RHF and CI calculation on BO_2 has yielded results that are consistent with a linear molecular model for the compound in its ground and low-lying excited states.[125] The transient species HBS reported in 1967 as the first known molecule containing a B=S double bond, has also been studied by *ab initio* CI techniques.[126a] The BO_2 radical has been studied by diode laser spectrometry.[126b]

The structure and bonding of the lithium metaborate molecule has been the subject of an *ab initio* study. In the gas phase $LiBO_2$ was shown to have a linear $(C_{\infty h})$ conformation, but comparison between computed and observed vibrational frequencies suggests that a lowering of the molecular symmetry to C_s may take place under matrix conditions. The linear molecule is predicted to be very stable with regard to fragmentation into Li^+ and BO_2^-, but the migration of lithium from one oxygen atom to another has a relatively low energy barrier (34.3 kJ mol^{-1}). A knowledge of the geometry and fundamental vibrational frequencies of these molecules is of interest because of the important role played by alkali-metal metaborates in high temperature vapour transport and hot corrosion flame processes.[127]

Crystal structures have been determined for the new air-sensitive alkali-metal borates $CsLi_5[BO_3]_2$[128a] and $Na_4Li_5[BO_3]_3$,[128b] both of which were shown to contain boron in trigonal planar coordination. A product of the reaction between hexamethylenetetramine and boric acid has also been studied by X-ray diffraction analysis, and found to have the composition $[(CH_2)_6N_4H]^+[B_5O_6(OH)_4]^-.0.5H_2O$. The anion has the usual pentaborate structure, in which the central BO_4 tetrahedron shares vertices with four BO_3 triangles.[129] The role of ^{11}B magic angle sample spinning (MASS) n.m.r. spectroscopy as a technique for quantitatively estimating trigonal/tetrahedral ratios in natural and synthetic boron-containing materials such as glasses, minerals, and zeolites, has been explored.[130]

Several potentially important industrial applications of boron-containing oxide materials have been reported this year. Because of their possible application as ionic (Ag^+) conductors, interest has centred on glasses, in particular the AgI–Ag_2O–B_2O_3 system, in which B_2O_3 is the network-forming oxide.[131] The system Re_2O_7–$Al_2O_3.B_2O_3$, on the other hand, has proved useful as a highly active and selective catalyst for the metathesis of functionalized alkenes.[132] In the same vein, it has been

[124] J. M. Chehayber and J. E. Drake, *Inorg. Chim. Acta*, 1986, **112**, 209.
[125] P. Császár, W. Kosmus, and Y. Panchenko, *Chem. Phys. Lett.*, 1986, **129**, 282.
[126] (a) F. Grein, *J. Mol. Spectrosc.*, 1986, **115**, 47; (b) K. Kawaguchi and E. Hirota, *ibid.*, 1986, **116**, 450.
[127] M. T. Nguyen, *J. Mol. Struc. (Theochem)*, 1986, **29**, 371.
[128] (a) M. Miessen and R. Hoppe, *Z. Anorg. Allg. Chem.*, 1986, **536**, 92 (b) *ibid.*, p. 101.
[129] A. S. Batsanov, O. V. Petrova, Yu T. Struchov, V. M. Akimov, A. K. Molodkin, and V. G. Skvortsov, *Russ. J. Inorg. Chem.*, 1986, **31**, 637.
[130] G. L. Turner, K. A. Smith, R. J. Kirkpatrick, and E. Oldfield, *J. Magn. Reson.*, 1986, **67**, 544.
[131] G. Licheri, A. Musinu, G. Paschina, G. Piccaluga, G. Pinna, and A. Magistris, *J. Chem. Phys.*, 1986, **85**, 500.
[132] X. Xiaoding, C. Boelhouwer, J. I. Benecke, D. Vonk, and J. C. Mol, *J. Chem. Soc., Faraday Trans. 1*, 1986, **82**, 1945.

found that the catalytic properties of zeolites can be modified by the incorporation of boron into the lattice.[133]

Twelve new β-ketoamine complexes of boron have been synthesized and assigned tetrahedral geometry on the basis of a wide range of spectroscopic techniques. An X-ray diffraction study of $OC_6H_4OBOC(R)CHC(R')NR''$ (R = p-ClC$_6$H$_4$, R' = Ph, R'' = Me) showed that each boron is bonded to one catechol and one β-ketoamine moiety, with the tetrahedral environment around the boron atom being made up of two oxygen atoms from the catechol and an oxygen and a nitrogen atom from the β-ketoamine ligand.[134]

Organic derivatives of boron which contain one or more B—O bonds are not discussed in detail in this report, but it is perhaps appropriate to indicate the types of compound being studied. These include the mono- and bi-cyclic boroxazolidones (78) and (79);[135a] the monocyclic boronate (80);[135b] 9-bora-10-oxa-phenanthrenes and related compounds;[135c] 2-substituted 1,3,2-dioxaboroles;[135d] novel organoboron–oxygen–halogeno–aluminium compounds (81);[135e,f] chelated organodiboroxanes and organoboroxins (82);[135g] and alkoxypiperidinophenylboranes.[135h]

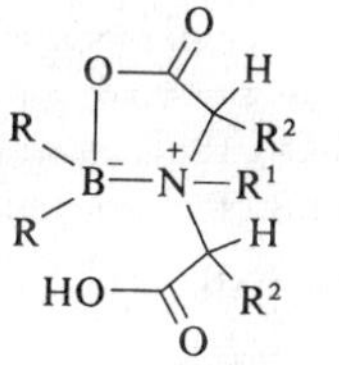

(78) 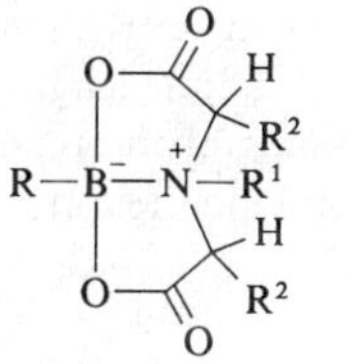   (79) 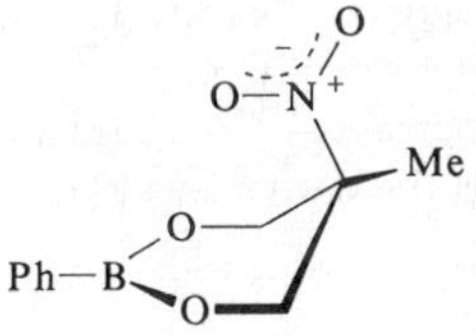   (80)

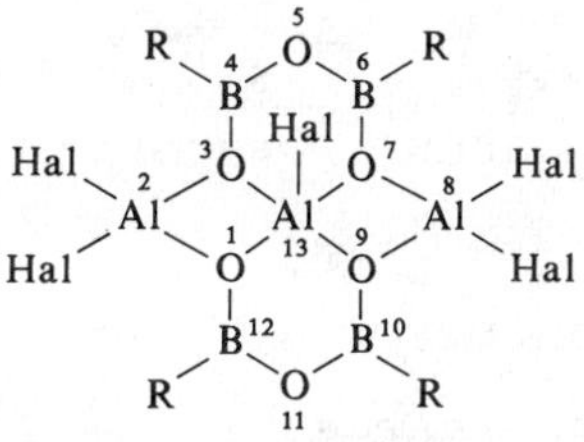

(81) R = Me, Et, Bu, Ph 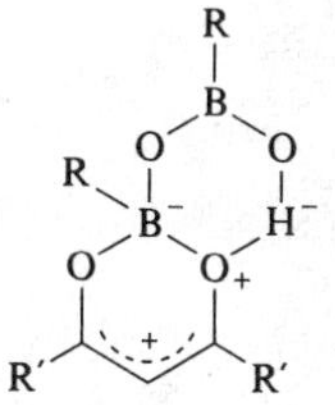   (82) R = R' = Me; R = Et, R' = Me; R = Et, R' = Ph

[133] (a) G. Coudurier and J. C. Védrine, *Pure Appl. Chem.*, 1986, **58**, 1389; (b) M. B. Sayed and J. C. Védrine, *J. Catal.*, 1986, **101**, 43; (c) P. Ratnasamy, S. G. Hegde, and A. J. Chandwadkar, *ibid.*, 1986, **102**, 467; (d) M. L. Occelli and T. P. Debies, *ibid.*, 1986, **97**, 357; (e) A. Thijs, S. Peeters, E. F. Vansant, G. Peeters, and I. Verhaert, *J. Chem. Soc., Faraday Trans. 1*, 1986, **82**, 963.

[134] Y. P. Singh, P. Rupani, A. Singh, A. K. Rai, R. C. Mehrotra, R. D. Rogers, and J. L Atwood, *Inorg. Chem.*, 1986, **25**, 3076.

[135] (a) B. Garrigues and M. Mulliez, *J. Organomet. Chem.*, 1986, **314**, 19; (b) W. Kliegel, L. Preu, S. J. Rettig, and J. Trotter, *Canad. J. Chem.*, 1986, **64**, 1855; (c) W. Maringgele, A. Meller, M. Noltemeyer, and G. M. Sheldrick, *Z. Anorg. Allg. Chem.*, 1986, **536**, 24; (d) G. Wulff and A. Hansen, *Angew. Chem., Int. Ed. Engl.*, 1986, **25**, 560; (e) R. Köster, Y.-H. Tsay, C. Krüger, and J. Serwatowski, *Chem. Ber.*, 1986, **119**, 1174; (f) R. Köster, K. Angermund, J. Serwatowski, and A. Sporzyński, *ibid.*, p. 1301; (g) R. Köster, K. Angermund, A. Sporzyński, and J. Serwatowski, *ibid.*, p. 1931; (h) R. H. Cragg, T. J. Miller, and D. O'N. Smith, *J. Organomet. Chem.*, 1986, **302**, 19.

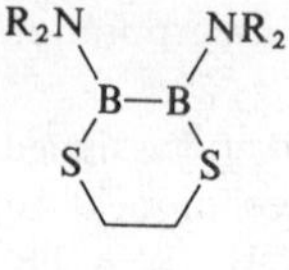

(83) 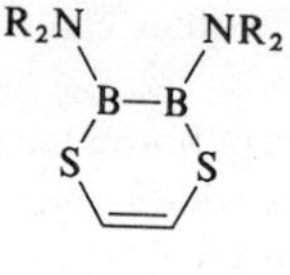   (84) 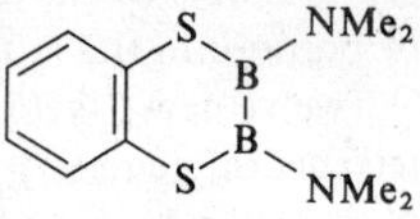   (85)

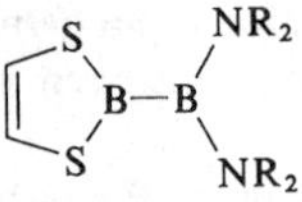

(86) 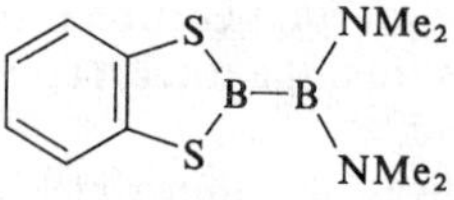   (87)

Apart from the *ab initio* study on HBS referred to earlier,[126a] work on compounds containing B—S bonds has been restricted to heterocyclic ring systems such as the dithiadiborines (83), dithiadiborinanes (84), and benzodithiadiborines (85). These compounds, which share the common feature of a six-membered S—C—C—S—B—B ring skeleton, isomerize on heating to give five-membered S—C—C—S—B ring systems such as (86) and (87), in which the B—B bond is retained. A redox disproportionation competes with the isomerization, and becomes increasingly important as the C—C bond length of the six-membered ring increases.[136] A wide range of compounds of the type $XBS_2Fe_2(CO)_6$ (88), as well as the derivatives (89) and (90), all of which feature metal-containing heterocyclic

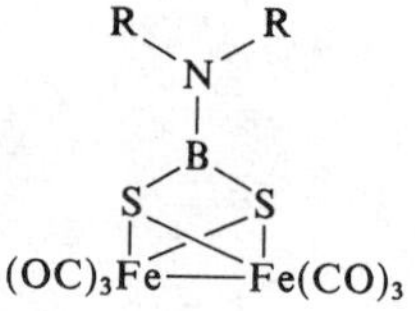

(88) $R_2N = Me_2N$, Pr_2^iN, tmp 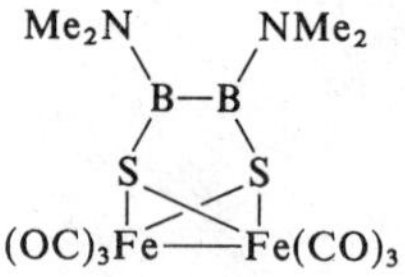   (89) 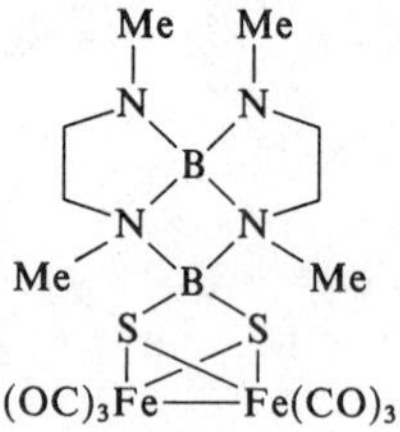   (90)

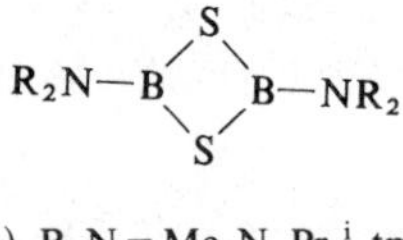

(91) $R_2N = Me_2N$, Pr_2^i, tmp 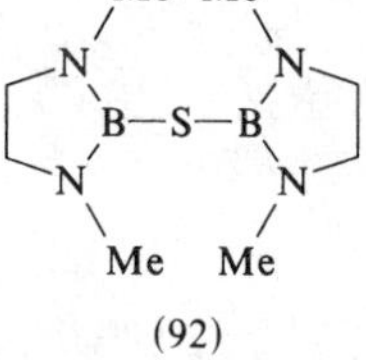   (92)

rings, have also been reported.[137a] Thermal decomposition of compounds (88) was found to give the dithiadiboretanes $(R_2NBS)_2$ (91). These B_2S_2 compounds are thought to be the kinetically-controlled products, because $(Me_2NBS)_2$ obtained in this way reacts with itself in solution to give $(Me_2NBS)_3$, which is thought to

[136] H. Nöth and H. Pommerening, *Chem. Ber.*, 1986, **119**, 2261.
[137] (a) H. Nöth and W. Rattay, *J. Organomet. Chem.*, 1986, **308**, 131; (b) *ibid.*, 1986, **312**, 139.

represent the thermodynamically-controlled product. Thermolysis of (89) and (90) was found to give $(Me_2NBS)_2$ and (92), respectively.[137b]

The reaction of the diselenaboroles (93) with n- and i-alkyl isocyanates (Scheme 3) to form the selenazaborin-4-ones (94) has been reported. The corresponding sulphur compounds can be obtained by reaction of compounds (94) with elementary sulphur.[138]

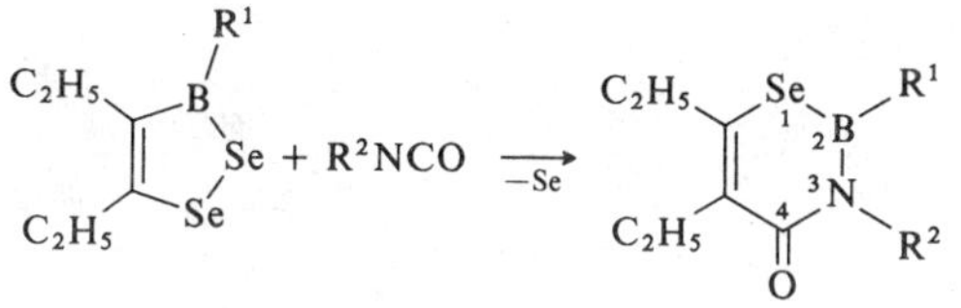

	R^1	R^2
	Me	Me
	Me	Et
	Me	Pr^n
	Me	Pr^i
	Me	Bu^n
	Bu^n	Me

(93)   (94)

Scheme 3

10 Halides

Spectroscopic properties of HBF^+ have been calculated theoretically as part of a systematic investigation of linear triatomic cations.[139a] The infrared absorption spectrum of the ν_2 band of BF_3 has been remeasured with a tunable diode laser, in order to study the vibration–rotation spectrum of this molecule in detail; the equilibrium B—F distance was found to be 130.70(1) pm.[139b] The high-resolution infrared spectrum of HBF_2,[139c] and the microwave spectrum of $MeBCl_2$ have also been studied.[139d]

He(I) photoelectron spectra have been obtained for B_4Cl_4, B_8Cl_8, and B_9Cl_9 (the three boron monohalides which have been structurally characterized) and interpreted on the basis of *ab initio* [GAUSSIAN 80(STO-36)] and semi-empirical MO calculations. The stability of the neutral boron chlorides relative to that of the (unknown) analogous boron hydrides was found to be related to differences in the make-up of the HOMO orbitals of the two sets of compounds.[140]

The structure of the benzaldehyde–boron trifluoride adduct has been determined by X-ray crystallography in the solid state, and by n.m.r. (NOE) spectroscopy in solution. In each case the Lewis acid BF_3 was found to be complexed in an *anti* disposition relative to the phenyl group in benzaldehyde. MNDO calculations of the related acetaldehyde–BF_3 adduct suggested that the *anti* configuration does indeed represent the lowest energy situation, though the *syn* adduct lies only 7.5 kJ mol^{-1} higher in energy. The results were discussed in terms of their relevance to a wide variety of important C—C bond-forming reactions of carbonyl compounds, that are mediated by Lewis acids.[141]

[138] C. Habben and A. Meller, *Chem. Ber.*, 1986, **119**, 1189.
[139] (a) P. Botschwina, *J. Mol. Spectrosc.*, 1986, **118**, 76; (b) S. Yamamoto, R. Kuwabara, M. Takami, and K. Kuchitsu, *ibid.*, 1986, **115**, 333; (c) M. C. L. Gerry, W. Lewis-Bevan, D. J. Maclennan, A. J. Merer, and N. P. C. Westwood, *ibid.*, 1986, **98**, 143; (d) S. D. Hubbard and A. P. Cox, *ibid.*, 1986, **115**, 188.
[140] P. R. LeBreton, S. Urano, M. Shahbaz, S. L. Emery, and J. A. Morrison, *J. Am. Chem. Soc.*, 1986, **108**, 3937.
[141] M. T. Reetz, M. Hüllmann, W. Massa, S. Berger, P. Rademacher, and P. Heymanns, *J. Am. Chem. Soc.*, 1986, **108**, 2405.

Topics of marginal interest include the electrochemical behaviour of BF_2-capped organocobalt chelates which serve as models for vitamin B_{12} group coenzymes,[142] the synthesis of arylhalogenoboranes,[143] and the spectral-luminescent properties of BF_2 β-diketonates.[144]

[142] G. Costa, A. Puxeddu, C. Tavagnacco, G. Balducci, and R. Kumar, *Gazz. Chim. Ital.*, 1986, **116**, 735.
[143] W. Haubold, J. Herdtle, W. Gollinger, and W. Einholz, *J. Organomet. Chem.*, 1986, **315**, 1.
[144] V E. Karasev and O. A. Korotkikh, *Russ. J. Inorg. Chem.*, 1986, **31**, 493.

4 Al, Ga, In, Tl

By S. M. GRIMES

Department of Chemistry, The City University, Northampton Square, London EC1V 0HB

This chapter follows the same format as in previous years reviewing the developments in the chemistry, in particular the inorganic chemistry, of Al, Ga, In, and Tl that have appeared in the literature over the past year.

A book published[1] in 1985 reviews the physics and chemistry of III–V compound semiconductor interfaces and highlights the value of such materials in applications in semiconductor technology.

1 Aluminium

Steady-state porous anodization of a high-purity aluminium electrode in sulphuric and propanedioic acid solutions has been studied[2] by the potentiostatic technique and in the 1.00—10.0 V *vs.* SCE potential range. The experimental results were found to be in agreement with the predicted linear function of the steady-state current density at low potentials. In another paper[3] the mechanism of plating on anodized aluminium has been investigated. Porous anodic films were formed on (1100) aluminium and the electrochemical behaviour of a number of test reactions $\{Ag^+/Ag, Cu^{2+}/Cu, [Fe(CN)_6]^{4-}/[Fe(CN)_6]^{3-}\}$ was determined.

Wilkinson and co-workers have published a number of papers[4-7] on the chemistry of transition-metal aluminohydride complexes. In their review Wilkinson and Barron[4] report on the recent advances in the field of covalent transition-metal aluminohydrides, and compare their structure and chemistry, where relevant, to that of analogous borohydride complexes. Aluminohydride complexes of chromium,[5] molybdenum,[5] tungsten,[5,6] rhenium,[6] ruthenium, and osmium have been characterized using a combination of techniques, *viz.* i.r. and n.m.r. spectroscopy and X-ray crystallography. The complexes $[(dmpe)_2HCr(\mu\text{-}H)_2AlH(\mu\text{-}H)]_2$ [dmpe = 1,2-bis(dimethylphosphino)ethane], $[(Me_3P)_4HM(\mu\text{-}H)_2AlH(\mu\text{-}H)]_2$ (M = Mo or W), and $[(Me_3P)_4ClW(\mu\text{-}H)_2AlH(\mu\text{-}H)]_2$ were synthesized[5] by reaction of the

[1] 'Physics and Chemistry of III–V Compound Semiconductor Interfaces', ed. C. W. Wilmsen, Plenum Press, New York, 1985, 465 pp.
[2] P. Ll. Cabot and E. Perez, *Electrochim. Acta*, 1986, **31**, 319.
[3] J. Gruberger and E. Gileadi, *Electrochim. Acta*, 1986, **31**, 1531.
[4] A. R. Barron and G. Wilkinson, *Polyhedron*, 1986, **5**, 1897.
[5] A. R. Barron, J. E. Salt, and G. Wilkinson, *J. Chem. Soc. Dalton Trans.*, 1986, 1329.
[6] A. R. Barron, D. Lyons, G. Wilkinson, M. Motevali, A. J. Howes, and M. B. Hursthouse, *J. Chem. Soc., Dalton Trans.*, 1986, 279.

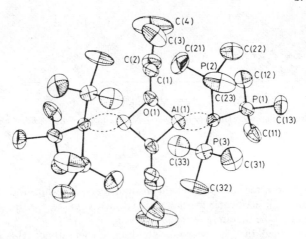

Figure 1 *The molecular structure of* $(Me_3P)_3H_3W(\mu\text{-}H)_2Al(H)(\mu\text{-}OBu^n)_2Al(H)(\mu\text{-}H)_2WH_3(PMe_3)_3$
(Reproduced from *J. Chem. Soc., Dalton Trans.*, 1986, 279)

corresponding metal halide complex with $LiAlH_4$ in ether. The broad singlet ^{27}Al-$\{^1H\}$ n.m.r. spectra [δ(range) 49—54 p.p.m., $w_{1/2}$(range) 5400—5590 Hz] for the chromium, molybdenum, and two tungsten aluminohydrides confirmed the geometry around the aluminium to be five-co-ordinate. The reaction of $[WCl_4(PMe_3)_3]$ with $LiAlH_4$ in refluxing THF leads to the aluminopolyhydride,[6] $(Me_3P)_3H_3W(\mu\text{-}H)_2Al(H)(\mu\text{-}OBu^n)_2Al(H)(\mu\text{-}H)_2H_3(PMe_3)_3$ (1); in diethyl ether the product is $(Me_3P)_3H_3W(\mu\text{-}H)_2Al(H)(\mu\text{-}H)_2Al(H)(\mu\text{-}H)_2WH_3(PMe_3)_3$ (2), and addition of N,N,N',N'-tetramethylethylenediamine to a toluene solution of (2) results in the formation of $(Me_3P)_3H_3W(\mu\text{-}H)_2Al(H)(\mu\text{-}H)_2WH_3(PMe_3)_3$ (3). The rhenium analogues[6] of (1) and (2), namely $(Me_2PhP)_3H_2Re(\mu\text{-}H)_2Al(H)(\mu\text{-}OBu^n)_2Al(H)(\mu\text{-}H)_2ReH_2(PMe_2Ph)$ (4) and $(PMe_2PhP)_3H_2Re(\mu\text{-}H)_2Al(H)(\mu\text{-}H)_2Al(H)(\mu\text{-}H)_2ReH_2(PMe_2Ph)_3$ have been obtained and the molecular structure of the butoxo compound (1) shown in Figure 1 has been determined. The aluminium atom is bridged to tungsten by two hydrogens and to a third terminal hydrogen, giving, together with the two oxygen atoms, five-co-ordination. The ready conversion of complex (2) into (1) on reflux for a few hours in THF implies that (2) is an intermediate in the formation of (1) in THF solution. Reaction of (2) with N,N,N',N'-tetramethylethylenediamine in toluene yielded complex (3), the molecular structure of which also showed a five-co-ordinate aluminium environment (Figure 2). The co-ordination geometry at the Al atom approximates closely to a trigonal bipyramid with average Al—H distances of 1.836 Å. The i.r. and ^{27}Al n.m.r. data for the tungsten derivatives and corresponding rhenium analogues are consistent with terminal and bridging hydrides.

The synthesis and characterization of a series of complexes with the general formula $L_3HM(\mu\text{-}H)_2AlH(\mu\text{-}H)_2\text{-}AlH(\mu\text{-}H)_2MHL_3$ (M = Ru, L = PMe_3, PEtPh_2, or PPh_3; M = Os, L = PMe_3 or PPh_3) have been reported.[7] Interaction of tetraethyl-

[7] A. R. Barron and G. Wilkinson, *J. Chem. Soc., Dalton Trans.*, 1986, 287.

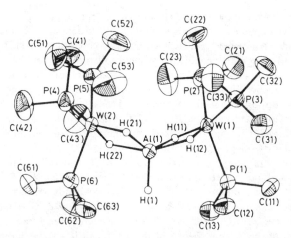

Figure 2 *The molecular structure of* $(Me_3P)_3H_3W(\mu\text{-}H)_2Al(H)(\mu\text{-}H)_2WH_3(PMe_3)_3$
(Reproduced from *J. Chem. Soc., Dalton Trans.*, 1986, 279)

enediamine with $[(Me_3P)_3HRuAlH_4]_2$ gives the complex $[(Me_3P)_3HRu]_2AlH_5$. The ^{27}Al n.m.r. spectroscopic features of all the complexes are similar and show that the M–H and $M(\mu\text{-}H)_2Al$ hydride-exchange process is very rapid and proceeds well below 180 K. The e.s.r. spectrum[8] of the bis(trimethylamine)alanyl radical both in fluid solution and trapped in polycrystalline $(Me_3N)_2AlH_3$ has been found to be in accordance with a quasi-trigonal-bipyramidal structure.

An extensive review of the structural, vibrational and ^{27}Al nmr spectroscopic properties of co-ordination derivatives of aluminium halides has been published.[9] Three-co-ordinate, mononuclear and polynuclear four-co-ordinate, trigonal-bipyramidal and square-pyramidal five-co-ordinate and six-co-ordinate aluminium halide derivatives have been investigated.

The preparation, Raman spectra, and crystal structure of $(TeCl_3)(AlCl_4)$ have been reported.[10] The tetrachloroaluminate salt of $[TeCl_3]^+$ crystallizes in the triclinic space group $P\bar{1}$ and is not monoclinic as previously reported; the cell data are as follows: $a = 6.554(1)$ Å, $b = 16.691(4)$ Å, $c = 8.391(1)$ Å, $\alpha = 92.79(1)°$, $\beta = 97.31(1)°$, $\gamma = 96.11(1)°$, $V = 1065.8$ Å^3 and $D_c = 2.51$ for $Z = 4$. Polarization phenomena at β''-alumina interfaces have been investigated[11] in the rechargeable cell $Na\beta''$-alumina/S^{IV} molten $AlCl_3/NaCl$, at ~ 200 °C, using cells equipped with β''-alumina disks. It was found that most of the polarization occurred at the β''-alumina molten salt interface and was dependent on the initial melt composition. The polarization could be reduced significantly by a predischarge of the cell containing an equimolar $AlCl_3$–$NaCl$ melt and elemental sulphur in the positive electrode compartment.

[8] R. L. Hudson and B. P. Roberts, *J. Chem. Soc., Chem. Commun.*, 1986, 1194.
[9] M. Dalibart and J. Derouault, *Coord. Chem. Rev.*, 1986, **74**, 1.
[10] B. H. Christian, M. J. Collins, R. J. Gillespie, and J. F. Sawyer, *Inorg. Chem.*, 1986, **25**, 777.
[11] A. Katagiri, J. Hvistendahl, K. Shimakage, and G. Mamatov, *J. Electrochem. Soc.*, 1986, **133**, 1281.

For the first time, the sequence distribution concepts applied for understanding organic copolymer microstructures have been used[12] to determine the microstructure of an inorganic copolymer, β-$(TiCl_3)(AlCl_3)_{0.33}$, consisting of $TiCl_3$ and $AlCl_3$ sequences. Whereas organic copolymer sequence distributions are invariably determined from interpretation of ^{13}C n.m.r. spectra, the authors have determined the $TiCl_3$, $AlCl_3$ sequence distributions in this organic copolymer by magnetic methods (e.s.r. spectroscopy and magnetic susceptibility) sensitive to unpaired electrons. The magnetic behaviour of this material is consistent with a periodic sequence distribution in which Al^{3+} and Ti^{3+} ions have fixed, alternating runs of one and three units, respectively. This sequence distribution gives rise to 0.33 spin/Ti.

Raman spectra of vapour mixtures of $ZrCl_4$-$AlCl_3$ and $HfCl_4$-$AlCl_3$ have been obtained[13] in the temperature range 300–500 °C. The spectra consist of superposition of bands due to the component gases plus a few new bands which were attributed to vapour complexation. For the MCl_4-$AlCl_3$ systems (M = Zr or Hf), two possible vapour complexes, MAl_2Cl_{10} and $MAlCl_7$, are considered. Temperature dependence measurements of the Raman spectra of the $ZrCl_4$-$AlCl_3$ systems indicate that the predominant vapour complex is probably 1:1 and permit the estimation of the enthalpy of the reaction:

$$ZrCl_4(g) + Al_2Cl_6(g) = ZrAlCl_7(g); \Delta H° = -34.9 \pm 2.5 \text{ kJ mol}^{-1}$$

The strongest Raman bands for the Zr–Cl–Al and the Hf–Cl–Al complexes are at 401 and 391 cm^{-1}, respectively, and are assigned to M—Cl (terminal) stretching.

Raman and i.r. spectra were obtained[14] of molten $AlCl_3NH_3$, $AlCl_3ND_3$, and $AlBr_3NH_3$ in addition to mixtures of $AlCl_3NH_3$ with $MAlCl_4$ (M = Li, Na, K, or Cs). The main spectral features of the gaseous $AlCl_3NH_3$ molecules which have C_{3v} symmetry are retained in the molten and glassy states. Some additional bands were observed both in the pure liquid and in mixtures with chloroaluminate melts. The spectra indicated that the dissociation reaction $2AlCl_3NH_3 = AlCl_4^- + [AlCl_2(NH_3)_2]^+$ occurs with $K \simeq 3 \times 10^{-3}$. Frequency shifts found on liquefaction demonstrate that extensive hydrogen bonding takes place between the $AlCl_3NH_3$ molecules. As hydrogen bonding weakens the N—H bonds, Al—N is strengthened; in turn, Al—Cl becomes weaker, the hydrogen bond stronger, and the N—H bond even weaker, resulting in an additional contribution to the Al—N bond strength. The authors confirmed that $Al_2Cl_6NH_3$ also exists in the melt and that the relative concentration of $Al_2Cl_6NH_3$ is less than the concentration of $[Al_2Cl_7]^-$ in the alkali chloride–$AlCl_3$ melts, possibly owing to factors such as hydrogen bonding, counterion influence, and the basicity of Cl^- vs. NH_3.

Densities of aqueous solutions of $AlCl_3$ have been measured at 10, 25, 40, and 55 °C with results that have led to defined apparent molar volumes.[15] The Pitzer ion interaction model has been used as the basis for analysing the apparent molar volumes to obtain standard-state partial molar volumes of $AlCl_3$(aq) at each temperature. Similar use of apparent molar heat capacities of aqueous solution of $AlCl_3$-HCl and $Al(NO_3)_3$-HNO_3 has also been made to obtain standard-state partial

[12] P. Brant and E. G. M. Tornqvist, Inorg. Chem., 1986, 25, 3776.
[13] S. Boghosian, G. N. Papathedorou, R. W. Berg, and N. J. Bjerrum, Polyhedon, 1986, 5, 1393.
[14] T. Ostvold, E. Rytter, and G. N. Papathedorou, Polyhedron, 1986, 5, 821.
[15] L. Barta and L. G. Hepler, Can. J. Chem., 1986, 64, 353.

molar heat capacities for $AlCl_3(aq)$ and $Al(NO_3)_3(aq)$. Finally, the standard-state partial molar volumes and heat capacities have been used with the Helgeson–Kirkham semi-theoretical equation of state for aqueous ions to provide a basis for estimating the thermodynamic properties of $Al^{3+}(aq)$ at high temperatures and pressures.

A number of papers have appeared[16-20] describing studies of the aluminium choride–butylpyridinium chloride (BuPyCl) and related systems. One paper reports[16] the anion–cation interactions that take place in several room-temperature halogen-aluminate molten salts investigated using n.m.r. spectroscopy. The systems studied were $AlCl_3$-1-(1-butyl)pyridinium chloride, $AlCl_3$-1-methyl-3-ethylimidazolium chloride, and $AlBr_3$-1-methyl-3-ethylimidazolium bromide. The ion–ion associations in these systems were examined by studying the ^1H and ^{13}C chemical shifts of the nuclei on the organic cations as a function of melt composition. Investigations conducted in basic melts, *i.e.* those melts with an aluminium halide mole fraction less than 0.5 and where the anion–cation interactions have been found to be most extensive, suggested that, in contrast to previous work, ion–ion interactions in these solvents involve more than just simple ion-pair formation. A model is proposed which considers that each cation may interact with two or more anions and *vice versa* to form oligomeric chains of alternating cations and anions. The latter model is shown to provide a more realistic interpretation of the experimental chemical shift data than a model based on simple ion-pair formation.

Osteryoung *et al.*[17] continue to report on the N-1-butylpyridinium chloride- and 1-methyl-3-ethylimidazole chloride (ImCl)–aluminium chloride systems. They report on the application of a method, developed by them earlier, to the study of the reaction of proton donor, hydroquinone (H_2Q), with BuPyCl–$AlCl_3$ and ImCl–$AlCl_3$ ambient-temperature ionic liquids. Amperometric titrations of free chloride ion indicate that H_2Q reacts with $2Cl^-$, suggesting that HCl is undissociated in the molten salt. Comparison of the $D\eta/T$ values where D, η, and T are the diffusion coefficient, viscosity, and temperature respectively, suggest that both water and H_2Q give rise to two HCl molecules.

Three basic organometallic compounds, $Fe(phen)_2(CN)_2$ (5), $Cp_4Fe_4(CO)_4$ (6), and $CpFe(CO)_2CN$ (7), have been studied[18] in the $AlCl_3$-n-butylpyridinium chloride (BPC) molten salt system. I.r. and u.v.-visible spectroscopies show that the ligands of all three complexes interact with Lewis acids contained in the melts, *i.e.* FeCN–Al and Fe$_3$CO–Al. Melts with molar ratios of $AlCl_3$ to BPC of less than one, previously referred to as 'basic', behave as weak Lewis acids relative to compound (5), whereas melts with ratios greater than one behave as strong Lewis acids toward all three compounds. A comparison of spectral shifts for these compounds in the various melts with shifts previously reported for isolated Lewis acid adducts of the same compounds provides insight into the relative electron pair acceptor ability of the various melts. Cyclic voltammetry of (5) in the molten salts demonstrates a very large shift in the oxidation potential whereas the oxidation potential of (6) is

[16] J. S. Wilkes, C. L. Hussey, and J. R. Sanders, *Polyhedron*, 1986, **5**, 1567.
[17] M. Lipsztajn, S. Sahami, and R. A. Osteryoung, *Inorg. Chem.*, 1986, **25**, 549.
[18] C. Woodcock and D. F. Shriver, *Inorg. Chem.*, 1986, **25**, 2137.
[19] D. L. Ward, R. R. Rhinebarger, and A. I. Popov, *Inorg. Chem.*, 1986, **25**, 2825.
[20] R. R. Rhinebarger, J. W. Rovang, and A. I. Popov, *Inorg. Chem.*, 1986, **25**, 4430.

unchanged. The authors have inferred that the HOMO of (5) is strongly perturbed by interaction with the Lewis acids in the molten salt whereas for (6) the HOMO is unaffected – an interpretation which is consistent with simple MO calculations.

The crystal and molecular structures of a lithium–C222B cryptate complex isolated from n-butylpyridinium chloride–aluminium chloride molten salt medium have been determined.[19] The complex, $Li(C_{22}H_{36}N_2O_6)AlCl_4$, crystallizes in the monoclinic space group $P2_1/n$ with unit cell data of $a = 11.099(2)$ Å, $b = 11.796(2)$ Å, $c = 22.345(5)$ Å, $\beta = 96.15(2)$ Å, and $z = 4$. The $[AlCl_4]^-$ ion is a nearly ideal tetrahedron, with an average Al—Cl distance of 2.127 Å and an average Cl-Al-Cl angle of 109.47°, but is far removed from the lithium–C222B cryptate group, with Al and Cl distances from Li exceeding 5.8 Å. In another paper by Popov and co-workers[20] ^{7}Li n.m.r. and potentiometric measurements have been used to identify and characterize lithium chloro-complexes in the $BuPyCl$–$AlCl_3$ ambient-temperature molten salt media. ^{7}Li Chemical shift data have been fitted to a two-site fast-exchange model that assumes an equilibrium between the $[LiCl_2]^-$ monomer and the $[Li_2Cl_4]^{2-}$ dimer species. Potentiometric measurements on LiCl and $LiClO_4$ solutions in basic melt (45 mol% $AlCl_3$) confirmed that two chloride ions were associated with each lithium ion.

Intensely coloured charge-transfer complexes have been formed[21] from $NOMX_4$ (M = B, Al, Ga, or Tl; X = Cl or F) and benzene or methyl-substitued benzenes, and their magnetic resonance and optical spectroscopic properties recorded. Solid adsorbents for ethylene have been prepared[22] by stirring copper(I) chloride, aluminium chloride, and a macroreticular type polystyrene resin in carbon disulphide, followed by removal of the liquid phase. The adsorbents effectively adsorb ethylene at 20 °C under 1 atm, and exhibit no measurable adsorptions of nitrogen, hydrogen, or methane.

Extraction of aluminium(III) and indium(III) with decanoic acid in octan-1-ol was carried out[23] at 25 °C and an aqueous ionic strength of 0.1 mol dm^{-3} ($NaClO_4$). Monomeric and tetrameric aluminium(III) decanoates and monomeric indium(III) decanoate are responsible for the extraction. In another study[24] the high-pressure kinetics of the complex formation of Al^{III}, Ga^{III}, and In^{III}, ions in Me$_2$SO have been investigated in order to understand more thoroughly the effect of the steric bulkiness of reactants and solvents on the reaction mechanism.

A method that allows the calculation of the profile of the whole X-ray diffraction pattern of a powder, taking into account simultaneously a size-effect broadening and the influence of a stacking fault in a layered structure, has been described by Grebille and Berar.[25] An application of the method has been given in the case of boehmite, AlOOH.

Proton n.m.r. spectra have been obtained[26] for a number of hydrolysed Al^{III} perchlorate salt solutions dissolved in deuterioacetone, with which they exchange deuterium to become partially deuteriated. Bound H_2O and OH entities can be

[21] S. Brownstein, A. Morrison, and L. K. Tan, *Can. J. Chem.*, 1986, **64**, 265.
[22] H. Hirai, M. Nakamura, S. Hara, and M. Komiyama, *Bull. Chem. Soc. Jpn.*, 1986, **59**, 3655.
[23] H. Yomada, H. Hayashi, Y. Fujii, and M. Mizuta, *Bull. Chem. Soc. Jpn.*, 1986, **59**, 789.
[24] K. Ishihara, S. Funahashi, and M. Takaka, *Inorg. Chem.*, 1986, **25**, 2898.
[25] D. Grebille and J. F. Berar, *J. Appl. Cryst.*, 1986, **19**, 249.
[26] J. W. Akitt, J. M. Elders, and X. L. R. Fontaine, *J. Chem. Soc., Chem. Commun.*, 1986, 1047.

distinguished since the former appear as HOD/H_2O doublets although the isotope shift is negative, in contrast to the normal positive effect seen with free water. The resonances also contain a fine isotope shift structure which appears to permit the number of protons around a given Al to be counted and so provide structural information about the polymeric species present.

Three papers describe studies involving the use of solid-state ^{27}Al n.m.r. spectroscopy to investigate Al–O-containing materials. ^{27}Al N.m.r. isotropic chemical shifts and quadrupole coupling data for AlO_4 tetrahedra in polycrystalline aluminates and aluminate hydrates with various degrees of anion condensation and with different cations have been studied.[27] It is shown that the various degrees of condensed AlO_4 tetrahedra differ in size and symmetry of the electric field gradient tensor at the Al site, but not in the ^{27}Al chemical shift. A slight variation in the chemical shift values of entirely connected AlO_4 tetrahedra around 80 p.p.m. indicates weak cation effects and other structural influences. A markedly increased Al shielding, however, was observed for tetrahedra with lengthened Al—O distances which occur in AlO_4 tetrahedra with neighbouring AlO_6 octahedra. The authors, Muller *et al.*, have published[28] in more detail results from the $CaO–Al_2O_3$ system. They have extended their systematic studies of ^{27}Al chemical shifts and electric field gradients at the Al sites to a series of calcium aluminates with varying $CaO:Al_2O_3$ ratios. Magic angle spinning n.m.r. spectra and spectra of the static samples were recorded at different magnetic fields up to 11.7 T. The study was aimed to show to what extent the different Al sites could be resolved in the ^{27}Al n.m.r. spectra. The results have shown that, apart from the known influence of the co-ordination state of aluminium, other structural factors such as interatomic distances and bond angles cause only weak changes in the ^{27}Al chemical shift values in the calcium aluminates studied. So, for example, only slight variations in the chemical shift values are observed between the various AlO_4 tetrahedra with different numbers of bridging oxygen atoms. Structural effects are reflected more sensitively in the quadrupole splitting data and, therefore, resolved n.m.r. spectra were only obtained in cases where the crystallographically different Al sites also have clearly differing electric field gradients. Solid-state MAS n.m.r. spectroscopy on naturally occurring montmorillonite has revealed[29] that the Fe^{III}, contained in the octahedral layer, exerts a greater influence on octahedral Al nuclei than on tetrahedral ones. The same trend is observed in a more pronounced manner with the material intercalated with $[AlO_4Al_{12}(OH)_{24}(OH_2)_{12}]^{7+}$ cations.

When it has not been possible to use crystallographic methods to determine co-ordination details, spectroscopic methods such as magic angle spinning solid-state n.m.r. have been able to distinguish between tetrahedral and octahedral co-ordination in aluminates. Dent Glasser and co-workers[30] have now shown that the technique can be applied to five-co-ordinate aluminates for which ^{27}Al n.m.r. shifts of approximately 35 p.p.m. are observed.

[27] D. Muller, W. Gessner, A. Samoson, E. Lippmaa, and G. Scheler, *J. Chem. Soc., Chem. Commun.*, 1986, 1277.

[28] D. Muller, W. Gessner, A. Samoson, E. Lippmaa, and G. Scheler, *Polyhedron*, 1986, **5**, 779.

[29] M. Matsumoto, M. Suzuki, H. Takahashi, and Y. Saito, *Bull. Chem. Soc. Jpn.*, 1986, **59**, 303.

[30] M. C. Cruickshank, L. S. Dent Glasser, S. A. I. Barri, and I. J. F. Poplett, *J. Chem. Soc., Chem. Commun.*, 1986, 23.

Ligand isotopic exchange kinetics of tris(acetylacetonato)aluminium(III) in acetylacetone under atmospheric and elevated pressures have been investigated.[31] The rate of exchange was expressed by $R = (k_1 + k_2[H_2O])[Al(acac)_3]$, where $[Al(acac)_3] = 0.0012$—0.0057 mol dm^3 and $[H_2O] = 0.022$—0.162 mol dm^3 at 15—40 °C under atmospheric pressure and at 25 °C under pressures up to 237 MPa. The rate constants k_1 and k_2 decreased with an increase in pressure on deuteration of protolytic hydrogens of water and acetylacetone, and on addition of acetonitrile.

Derivatives of the type $ZrAl(OPr^i)_{(7-x)}L_x$ (L = acetylacetone, benzoylacetone, methyl acetoacetate or ethyl acetoacetate; $x = 1$ or 2) have been prepared and analysed using i.r. and n.m.r. (^{27}Al, ^1H, and ^{13}C) spectroscopies.[32] The ^{27}Al chemical shift data confirm the tetrahedral nature of the aluminium in $ZrAl(OPr^i)_7$ and suggest five-co-ordination for Al in $[ZrAl(OPr^i)_6L]$ and six-co-ordination for Al in $[ZrAl(OPr^i)_5L_2]$.

Growth defects and the incommensurate phase have been observed in synthetic berlinite (AlPO$_4$) samples following high-temperature (298—853 K) X-ray topographic studies.[33] The crystal structures of quartz type $M^{III}X^VO_4$ (M = Al or Ga; X = P or As) were refined at 173, 293, and 373 K in the $P3_121$ space group.[34] In this temperature range, the packing shows very high rigidity. Distortions of tetrahedral sites (XO$_4$ and MO$_4$) have been evaluated and decrease when the temperature increases. The comparison of the packings shows increasing distortions with the size of the M^{III} and X^V elements.

Two papers by Pluth and Smith[35,36] describe the crystal structures of aluminophosphates. Aluminium phosphate hydrate, AlPO$_4$. 1.5H$_2$O crystallizes in the orthorhombic space group *Pbca* with cell dimensions $a = 19.3525(13)$ Å, $b = 9.7272(7)$ Å, and $c = 9.7621(8)$ Å. A four-connected framework contains PO$_4$ tetrahedra interposed between AlO$_4$ tetrahedra and AlO$_4$(H$_2$O)$_2$ octahedra at the nodes of cross-linked alternate 6^3 and 4.8^2 nets. A two-dimensional channel system, limited by 8-rings, lies between adjacent 6^3 nets. One H$_2$O of each octahedron lies in a 6-ring, and the other forms a continuous chain with a third H$_2$O which is held in place only by hydrogen bonds. An aluminophosphate gel, produced from hydrated alumina and phosphoric acid, was digested with piperidine for one week at 473 K to form AlPO$_4$-17, aluminium phosphate–piperidine hydrate.[36] This complex, Al$_{18}$P$_{18}$O$_{72}$. 4(C$_5$H$_{11}$H · H$_2$O) crystallizes in the hexagonal space group $P6_3/m$ with $a = 13.2371(9)$ Å and $c = 14.7708(10)$ Å. The Al and P atoms alternate near the tetrahedral nodes of the erionite framework. An extra-framework oxygen species in a partially occupied site bridges between a pair of Al atoms in the opposing 6-rings of each cancrinite cage. Each of the two ellipsoidal cavities contains two piperidine species.

Kneip[37] has presented a summary of the chemistry, in terms of composition, structure, and conditions of formation, of aluminium orthophosphates. The three-

[31] A. Nagasawa, H. Kido, T. M. Hattori, and K. Saito, *Inorg. Chem.*, 1986, **25**, 4330.
[32] R. Jain, A. K. Rai, and R. C. Mehrotra, *Polyhedron*, 1986, **5**, 1017.
[33] A. Zarka, B. Capelle, E. Philippot, and J.-C. Jumas, *J. Appl. Cryst.*, 1986, **19**, 477.
[34] A. Goiffon, J.-C. Jumas, M. Maurin, and E. Philippot, *J. Solid State Chem.*, 1986, **61**, 384.
[35] J. J. Pluth and J. V. Smith, *Acta Crystallogr., Sect. C*, 1986, **42**, 1118.
[36] J. J. Pluth, J. V. Smith and J. M. Bennett, *Acta Crystallogr., Sect. C*, 1986, **42**, 283.
[37] R. Kneip, *Angew. Chem., Int. Ed. Engl.*, 1986, **25**, 525.

component system $Al_2O_3 . P_2O_5 . H_2O$ encompasses several intermediate orthophosphates, the variety of which stems from the many ways of combining the cationic co-ordination polyhedra of aluminium with anionic phosphate species.

As in previous years a number of papers have appeared reporting on the chemistry of mixed-metal aluminates and alumina-based materials. The crystal structure and cation distribution of highly non-stoicheiometric potassium β-alumina $K_{1.50}Al_{11.0}O_{17.25}$ have been studied[38] by single-crystal X-ray diffraction. The single crystal was obtained by ion exchange from barium β-alumina ($Ba_{0.75}Al_{11.0}O_{17.25}$) and the least-squares refinement was accomplished with a final R value of 0.029. The occupation of K^+ ion in the mirror plane ($z = 0.25$) was explained by assuming three types of cell: singlet cell ($KAl_{11}O_{17}$), K–K–K triplet cell ($K_3Al_{11}O_{17}$), and K–K–O$_i$ triplet cell ($K_2O . Al_{11}O_{17}$), where O$_i$ stands for interstitial oxygen. This model was shown to be applicable to the ordinary potassium β-alumina ($K_{1.30}Al_{11.0}O_{17.15}$).

Black needles of $La_3Mo_{4.33}Al_{0.67}O_{14}$ have been[39] prepared by electrolytic reduction at 1100 °C of a melt containing Na_2MoO_4, MoO_3, and La_2O_3 in an alumina crucible. The compound is isomorphous with $La_3Mo_4SiO_{14}$, crystallizing in space group *Pnma* with $a = 17.750(3)$ Å, $b = 5.6600(9)$ Å, $c = 11.070(2)$ Å. As in $La_3Mo_4SiO_{14}$, the structure contains Mo_3O_{13} clusters and chains and edge-sharing MoO_6 octahedra with alternatively short (2.535 Å) and long (3.167 Å) Mo—Mo distances, but, unlike $La_3Mo_4SiO_{14}$, the tetrahedral silicon is replaced by a random distribution of 1/3 Mo and 2/3 Al. Electrical conductivity measurements show that $La_3Mo_4SiO_{14}$ and $La_3Mo_4Al_{2/3}Mo_{1/3}O_{14}$ are both highly anisotropic semiconductors with the easy direction of conduction being parallel to the chain cluster units.

Single crystals of $LaNiAl_{11}O_{19}$ have been grown[40] from the melt using either the Verneuil flame fusion process or the floating zone method. The compound is of the distorted magnetoplumbite type and the Ni is shared between tetrahedral and octahedral sites of the spinel blocks. This compound has also been prepared in powdered form by annealing coprecipitates of amorphous oxides at medium temperature. The thermal evolution of the coprecipitates leading to the magnetoplumbite phase has been studied simultaneously by X-ray diffraction and diffuse reflectance spectroscopy. All the main transitions of the electronic spectra of $LaNiAl_{11}O_{19}$ have been assigned. The existence of a certain amount of octahedral Ni in this material makes it a possible candidate as an i.r. tunable laser.

The structure of a mixed sodium–neodymium aluminate has also been investigated;[41] it was found to crystallize in the hexagonal space group $P6m2$. Its unit cell ($a = 5.57$ Å, $c = 22.25$ Å) is described as an alternate stacking of half β-alumina unit and half magnetoplumbite unit cells leading to an ordering between the layers of large ions Na^+ and Nd^{3+} in the c direction. Iyi *et al.* report on the use[42] of convergent beam electron diffraction and high-resolution electron microscopy to

[38] N. Iyi, Z. Inoue, and S. Kimura, *J. Solid State Chem.*, 1986, **61**, 81.
[39] W. H. McCarroll, K. Podejko, A. K. Cheetham, D. M. Thomas, and F. J. DiSalvo, *J. Solid State Chem.*, 1986, **62**, 241.
[40] F. Laville, M. Perrin, A. M. Lejus, M. Gasperin, R. Moncorge, and D. Vivien, *J. Solid State Chem.*, 1986, **65**, 301.
[41] A. Khan and J. Thery, *J. Solid State Chem.*, 1986, **64**, 102.
[42] N. Iyi, Y. Bando, S. Takekawa, Y. Kitami, and S. Kimura, *J. Solid State Chem.*, 1986, **64**, 220.

examine the superstructure of $BaPb^{II}$-alumina and in a second paper report[43] the crystal structure of Mg-doped potassium β-alumina $K_{1.875}Ba_{0.022}Mg_{0.010}Al_{10.081}O_{17.0}$ ($K_{1.875}$ Mg β-alumina). The material was obtained by ion-exchange from $Ba_{0.955}$ Mg β-alumina), the structure of which was also refined. The $K_{1.875}$ Mg β-alumina contains a large number of K ions while retaining the fundamental β-alumina structure and space group symmetry $P6_3/mmc$. It has been reported[44] that the series of interstitial compounds $(Na_2O)_{0.33}Na[AlSiO_4]$ exhibits the cubic symmetry of the cristabolite type $Na[AlSiO_4]$-carnegeite host lattice as concentrations of 0.01— 0.33 Na_2O are intercalated at elevated temperatures under closed system conditions (950—1100 K). A total of $6\frac{2}{3}$ sodium atoms and $1\frac{1}{3}$ oxygens statistically occupying the cage-like 12-fold oxygen-co-ordinate position of space group $F\bar{4}3m$ form a diamond-type sublattice in the high-cristobalite-type 'host lattice' $[AlSiO_4]^{1-}_{3d\infty}$. Superstructure characteristics on powder diffraction data indicate complex ordering mechanisms following long-term annealing (200 h, 370—970 K). At temperatures >1350 K in an open system, intercalated Na_2O is released with the subsequent collapse of the $Na[AlSiO_4]$ host lattice from cubic to triclinic, the regular symmetry of $Na[AlSiO_4]$-carnegeite at temperatures <960 K.

Detailed i.r. studies have revealed[45] that the adsorption of benzene on two faujasite-type zeolites with different cation contents viz. $Na_{56}(AlO_2)_{27}(SiO_2)_{136}$ and dealuminated $H_2Na_{25}(AlO_2)_{27}(SiO_2)_{165}$ shows the existence of three possible states of adsorbed benzene, interaction with S_{11} cations, interaction with oxygen atoms, and liquid benzene filling the cavity. ^{29}Si Magic angle spinning n.m.r. and i.r. spectroscopy and X-ray powder diffractometry have established[46] that the amount of framework aluminium increases dramatically (by nearly 48%) upon treatment of ultrastable zeolite Y with an aqueous solution of KOH. The authors have found a convenient method of modifying the siting of Si and Al in the framework sites of zeolitic catalysts.

High-temperature phase transformations of zeolite A with various degrees of exchange of Na^+ with Li^+ ions have been investigated.[47] An increase in the number of Li^+ ions per unit cell accelerates the thermal transformation of the zeolite framework to the amorphous state. Above 730 °C, four phases were identified. Of these, only γ- and β-eucryptite phases were obtained from pure LiA zeolite. γ-Eucryptite is a new metastable polymorph in the system $Li_2O-Al_2O_3-SiO_2$ and is transformed to β-eucryptite. Szostak and Thomas[48] have used i.r. spectroscopy to examine the gels containing iron and gallium in place of aluminium in order to determine the conditions necessary for successful incorporation of the elements before further attempting to crystallize a specific molecular sieve structure.

The results of a Mössbauer study on the effect of substitution of Al^{3+} and Ga^{3+} in ferromagnetic $FeBO_3$ close to T_c have been presented.[49] The anomalous spectra have been found to appear 1 K below the T_c for $FeBO_3$ (T_c = 352 K) but 15 K

[43] N. Iyi, Z. Inoue, and S. Kimura, J. Solid State Chem., 1986, 61, 236.

[44] R. Klingenberg and J. Felsche, J. Solid State Chem., 1986, 61, 40.

[45] A. de Mallmann and D. Barthomeuf, J. Chem. Soc., Chem. Commun., 1986, 476.

[46] X. Liu, J. Klinowski, and J. M. Thomas, J. Chem. Soc., Chem. Commun., 1986, 582.

[47] V. Dondour and R. Dimitrijevic, J. Solid State Chem., 1986, 63, 46.

[48] R. Szostak and T. L. Thomas, Inorg. Chem., 1986, 25, 4311.

[49] M. Vithal and R. Jagannathan, J. Solid State Chem., 1986, 63, 16.

below the T_c for $Fe_{0.9}Al_{0.1}BO_3$ (T_c = 317 K) and $Fe_{0.9}Ga_{0.1}BO_3$ (T_c = 319 K), and have been explained in terms of superparamagnetic relaxation effects.

The anodic degradation of β/β''-alumina of different compositions has been investigated by voltammetric techniques in cells of the type Cu/Rodar alloy/carbon fibres/graphite layer/solid electrolyte/$NaNO_3$/Pt/Cu and Cu/Rodar alloy/Hg/β''-alumina/$NaNO_3$/Pt/Cu at 350 °C.[50,51] Surface and bulk analyses were applied to specimens at the end of the electrochemical runs making it possible to monitor the progress of anodic degradation with time and to establish the value of the voltammetric technique as a potential diagnostic tool for the prediction of failure of β/β''-alumina electrolyte in sodium/sulphur cells. The preparation, stability, conductivity, and structure of selected hydronium and ammonium/hydronium β''-aluminas have recently been investigated.[52] The ammonium/hydronium and hydronium β''-aluminas which have room temperature conductivities of about 10^{-4} and $10^{-6}\ \Omega^{-1}$ cm, respectively do not appear to be distinct and independent materials, but only general compositions in a continuum of protonic β''-aluminas which contain various concentrations of water and/or ammonia. The results of thermal analyses, crystal structure studies, and ionic conductivity measurements indicate that it should be possible to convert one protonic β''-alumina composition into another by careful thermal treatment. Methods[53] used to synthesize multivalent β''-aluminas and the preparation of a number of specific transition-metal and lanthanide forms have been described. β''-Aluminas containing multivalent cations can be synthesized by ion-exchange reactions starting with the sodium form of the compound. These facile reactions make it possible to prepare a wide variety of β''-alumina compositions, many of which are metastable and inaccessible by normal synthetic means.

The results of a study of the interaction of β''-alumina ceramic with molten sodium and sodium vapour[54] have been explained on the basis of De Jonghe's mechanism for the interaction between β''-alumina ceramic and sodium, involving the formation of F centres and Mott and Gurney's model for the influence of an electric field on the drift rate of F centres in alkali halides. Single crystals of Co^{2+}-stabilized β''-alumina have been synthesized[55] by a flux growth technique using Bi_2O_3 as the flux, and crystallographic studies have shown that Co^{2+} ions substitute only at the Al(2) sites. The temperature-dependent ionic conductivities of Co^{2+}-stabilized β''-alumina, Co^{2+}-doped β-alumina, and undoped β-alumina have been measured and compared between 25 and 45 °C, the conductivity being highest for the first system and lowest for the last.

Enhancement of electrical conductivity was found[56] in the mixed $SrCl_2$–Al_2O_3 system. The enhancement of Cl-conductivity depended on the composition, the grain size of Al_2O_3, and the preparation methods. From these results, it was considered that the interface between the $SrCl_2$ matrix and Al_2O_3 particles played a

[50] N. S. Choudhury, M. W. Breiter, and E. L. Hall, *Electrochim. Acta*, 1986, **31**, 771.

[51] M. W. Breiter, N. S. Choudhury, and E. L. Hall, *J. Electrochem. Soc.*, 1986, **133**, 2064.

[52] K. G. Frase, J. O. Thomas, A. R. McGhie, and G. C. Farrington, *J. Solid State Chem.*, 1986, **62**, 297.

[53] S. Sattar, B. Ghosal, M. L. Underwood, H. Merywoy, M. A. Saltzberg, W. S. Frydrych, G. S. Rohrer, and G. C. Farrington, *J. Solid State Chem.*, 1986, **65**, 231.

[54] P. D. Yankulov, G. Staikov, A. Yanakiev, R. Kvachkoc, P. V. Angelov, and E. Budevski, *J. Solid State Chem.*, 1986, **63**, 1.

[55] S. Chen, D. R. White, H. Sato, J. B. Lewis, and W. R. Robinson, *J. Solid State Chem.*, 1986, **62**, 26.

[56] S. Fujitsu, H. Kobayashi, K. Koumoto, and H. Yanagida, *J. Electrochem. Soc.*, 1986, **133**, 1497.

role in giving rise to high ionic conductivity. The effective thickness and the conductivity of the interface were estimated using a simple mixing model. The thickness decreased from 0.6 to 0.15 μm, and conductivity increased with increase of temperature. The conductivity was about two orders of magnitude larger than that of pure $SrCl_2$.

The use of solid adsorbents such as alumina is not uncommon and indeed in many catalytic reactions alumina has played the role of an inert support. Hanson *et al.*[57] have studied the formation of $[HFe_3(CO)_{11}]^-$ on hydroxylated γ-Al_2O_3. They have found that, once formed, on hydroxylated γ-alumina the adsorbed anion appears to be susceptible to decarbonylation to yield subcarbonyls that may be represented as $[HFe(CO)_{11-x}]^-$(ads). Selective monoesterification of the longer carbon chain in a mixture of dicarboxylic acids, [dodecanedioic acid (C_{12}), glutaric acid (C_5), adipic acid (C_6), pimelic acid (C_7), suberic acid (C_8), or sebacic acid (C_{10})] has been achieved[58] by adsorbing and aligning the acids on alumina. From *in situ* Fourier transform i.r. studies[59] of gas-phase carbonylation experiments at high pressure, the cocatalytic function of the potassium oxide promoter in $Ru/K/Al_2O_3$ catalysts in stabilizing formyl intermediates during synthesis gas (CO + H_2) conversions has been established.

In addition to the esterification and carbonylation reactions described, alumina-based catalysts have found use in methanation,[60] hydrogenation,[61,62] hydrogenolysis,[63] and hydrodesulphurization[64-67] reactions. The catalytic activity and selectivity toward methanation of CO_2 using several alumina-supported ruthenium-cluster-derived catalysts have been studied[60] over the temperature range 180–250 °C. The catalysts were supported by impregnation over alumina and activated in hydrogen at 200 °C. The activity of catalysts derived from supported neutral species was observed to increase as the number of ruthenium atoms present in the precursor complex increased, *i.e.* $Ru(CO)_5 < Ru_3(CO)_{12} < H_4Ru_4(CO)_{12} < Ru_6C(CO)_{17}$. Catalysts derived from supported anionic ruthenium cluster derivatives were less active than their neutral counterparts, displaying a great deal of sensitivity to the nature of the accompanying cation.

A comparative study[61] of the effect of supports of hydrocracking and hydrogenation activities of molybdenum catalysts used for upgrading coal-derived liquids has shown the Al_2O_3 support to have the highest hydrogenation activity compared with TiO_2. However, both hydrocracking and hydrogenation activities of the catalysts increased when a double oxide of TiO_2–Al_2O_3 was used. The changes in the activity of the reduced *vs.* sulphided molebdena–alumina catalysts in the hydrogenation of simple olefins and diolefins have been entirely attributed[62] to changes in the active-

[57] B. E. Hanson, J. L. Bergmeister, J. T. Petty, and M. C. Connaway, *Inorg. Chem.*, 1986, **25**, 3089.
[58] M. Ogawa, T. Chihara, S. Teratani, and K. Taya, *J. Chem. Soc., Chem. Commun.*, 1986, 1336.
[59] I. L. C. Freriks, P. C. de Jong-Versloot, A. G. T. G. Kortbeek, and J. P. van den Berg, *J. Chem. Soc., Chem. Commun.*, 1986, 253.
[60] D. J. Darensbourg and C. Ovalles, *Inorg. Chem.*, 1986, **25**, 1603.
[61] A. Nishijima, H. Shimada, T. Sato, Y. Yoshimura, and J. Hiraishi, *Polyhedron*, 1986, **5**, 243.
[62] W. S. Millman, K.-I. Segawa, D. Smrz, and W. K. Hall, *Polyhedron*, 1986, **5**, 169.
[63] M. K. Huuska, *Polyhedron*, 1986, **5**, 233.
[64] S. Kasztelan, E. Payen, H. Toulhout, J. Grimblot, and J. P. Bonnelle, *Polyhedron*, 1986, **5**, 157.
[65] J. Brito and J. Laine, *Polyhedron*, 1986, **5**, 179.
[66] R. Prada Silvy, F. Delannay, P. Grange, and B. Delmon, *Polyhedron*, 1986, **5**, 195.
[67] P. C. H. Mitchell and C. E. Scott, *Polyhedron*, 1986, **5**, 237.

site concentration rather than to the nature of the catalytic sites. The hydrogenolysis of anisole was studied in a fixed-bed reactor at 30 °C and 5 MPa hydrogen pressure on pure carriers and different catalysts containing Mo or Ni. The Mo-γ-Al$_2$O$_3$ catalyst gave high conversion of anisole both in the sulphided and unsulphided forms. The Ni-γ-Al$_2$O$_3$ catalyst on the other hand gave the highest conversion of anisole with a low loading of Ni. The catalytic performances (in terms of activity, selectivity) of hydrotreating catalysts composed of supported Mo associated with a promoter (Co or Ni) and working in a sulphided state have been shown to be greatly dependent on the sequences and conditions of the preparation process.[64] In another[65] paper the effects of several variables such as type of support, calcination temperature, and amount of Mo and promoters upon the reduction characteristics of supported MoO$_3$ have been investigated using temperature-programmed reduction. It was found that the incorporation of Ni into MoO$_3$-Al$_2$O$_3$ catalysts activated hydrogen more easily than both cobalt-promoted and unpromoted catalysts.

A Co-Mo-γ-Al$_2$O$_3$ hydrodesulphurization[66] catalyst was activated under an H$_2$S–H$_2$ mixture at temperatures ranging from 673 to 1073 K using two different procedures. The results suggest that the decrease in activity with increasing activation temperature could be due to the sintering of MoS$_2$, the incomplete sulphuration of molybdenum, and the increase in the amount of sulphided cobalt. Mitchell and Scott[67] have studied the interaction of vanadium and nickel porphyrins and metal-free porphyrins with the Co–Mo–alumina hydrodesulphurization catalyst, to identify any metal poisoning and, as a result, catalyst deactivation. An LaAlO$_3$ catalyst prepared by the mist decomposition method has shown[68] a high activity (1.2 mol s^{-1} m^{-2}) and a high selectivity (48%) for C$_2$ hydrocarbon formation at 983 K in an atmospheric flow reactor. Facile cyclodehydration of several non-aromatic diols to the corresponding cyclic ethers has been achieved[69] using aluminium(III)-exchanged montmorillonite as a solid Brønsted acid catalyst.

The i.r. spectrum of Al(CO)$_2$, observed[70] in deposits of Al atoms and CO in adamantane at 77 K, has a symmetric CO stretching mode at 1985 cm^{-1} and an antisymmetric CO stretching mode at 1903.8 cm^{-1}; there is i.r. evidence for the formation of Al$_2$CO and Al$_2$(CO)$_4$ in this system.

The aluminium–phosphorus compound of simplest formula Me$_2$AlPMePh has been prepared[71] by an elimination–condensation reaction between Me$_2$AlH and PMePhH, characterized by elemental analyses and cryoscopic molecular weight measurements as well as i.r., ^{1}H n.m.r., and ^{31}P n.m.r. spectral data. The product exists in benzene solution as a trimer, Me$_2$AlP(MePh)$_3$. However the complexity of the spectral data suggest that (Me$_2$AlPMePh)$_3$ exists in benzene solution as a mixture of isomers due to different conformations of the ring and/or different orientations of the methyl and phenyl groups. The nature of the elimination reaction between Me$_2$AlH and PMePhH has also been investigated by following the rate of formation of H$_2$ at 22 °C in isomeric xylenes solution. The kinetic data are consistent with a second-order rate law that is complicated by equilibria. The initial intermediate from the elimination reaction is a monomeric Me$_2$AlPMePh species, which reacts

[68] H. Imai and T. Tagawa, *J. Chem. Soc., Chem. Commun.*, 1986, 52.
[69] D. Kotkar and P. K. Ghosh, *J. Chem. Soc., Chem. Commun.*, 1986, 650.
[70] J. H. B. Chenier, C. A. Hampson, J. A. Howard, and B. Mile, *J. Chem. Soc., Chem. Commun.*, 1986, 730.
[71] O. T. Beachley and L. Victoriano, *Inorg. Chem.*, 1986, **25**, 1948.

in turn with Me₂AlH and/or Me₂AlPMePh eventually to form the trimer. The equilibrium constant for the formation of the adduct HMe₂Al.PMePhH from monomeric alkane and phosphine, K_a, evaluated from the kinetic data has a value of 2.60 l mol⁻¹ at 22 °C. I.r. spectra of the monomers Me₂AlCl and MeAlCl₂ have been obtained[72] by thermal dissociation of the corresponding dimers followed by their isolation in argon matrices. The Al—C bond is found to be of similar stability for all monomers $Me_{3-n}AlCl_n$ (n = 0—2) whereas the strength of the Al—Cl bond decreases with higher alkyl contents. Several chlorine-bridged dimers ($Me_{6-n}Al_2Cl_n$ (n = 2—6) were identified in studied of dimeric dimethylaluminium chloride and methylaluminium dichloride. The experiments included elevated Knudsen cell temperatures, causing the following decompositions to occur: $Me_4Al_2Cl_2 \rightarrow Me_3AlCl_3 \rightarrow$ *trans*-$Me_2Al_2Cl_4$ (350–450 °C) and $Me_2Al_2Cl_4 \rightarrow$ $MeAl_2Cl_5 \rightarrow Al_2Cl_6$ (550 °C). These cracking reactions are more prominent than the mere dissociations.

Irradiation of the AlEtCl₂ complex of *endo*-tricyclo[5.2.1.0²·⁶]deca-4,8-dien-3-one has yielded[73] the AlEtCl₂ complex of *exo*-tricyclo-[5.3.0.0²·⁶]deca-4,8-dien-3-one, the structure of which was established by its reduction to give the known *exo*-tricyclo[5.3.0.0²·⁶]decan-3-one.

For the first time a compound containing aluminium–olefin bonds has been prepared[74] and its structure determined. The investigation of the structure of 1,4-dichloro-2,3,5,6-tetramethyl-1,4-dialuminacyclohexa-2,5-diene at −130 °C shows two non-planar 1,4-dialumina-cyclohexadiene moieties, twisted through 90° with respect to each other, coupled *via* four aluminium–olefin bonds to give the dimer shown in Figure 3. Carbonyl olefination has been found[75] to occur on reaction of

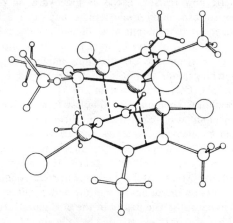

Figure 3 *The molecular structure of dimeric 1,4-dichloro-2,3,5,6-tetramethyl-1,4-dialuminacyclohexa-2,5-diene*
(Reproduced by permission from *Angew. Chem., Int. Ed. Engl.*, 1986, **25**, 921)

⁷² E. Rytter and S. Kvisle, *Inorg. Chem.*, 1986, **25**, 3796.
⁷³ R. F. Childs, B. M. Duffey, and M. Mahendran, *Can. J. Chem.*, 1986, **64**, 1220.
⁷⁴ H. Schnockel, M. Leinkuhler, R. Lotz, and R. Mattes, *Angew. Chem., Int. Ed. Engl.*, 1986, **25**, 921.
⁷⁵ T. Kauffmann, M. Enk, W. Kaschube, E. Toliopoulos, and D. Wingbermuhle, *Angew. Chem., Int. Ed. Engl.*, 1986, **25**, 910.

two molar equivalents each of trimethylaluminium with MoO_2Cl_2 and with $MoOCl_3.(THF)_2$ in THF.

The reaction of bis(triethylaluminium), Al_2Et_6, with a variety of tetradentate Schiff-base ligands in benzene–acetonitrile solution has yielded ethane and the monoethylaluminium Schiff-base complexes.[76] The ligands used include the 2:1 condensation products of salicylaldehyde with ethylenediamine, *o*-phenylenediamine, and propylenediamine (salen, salophen, and salpn, respectively) and of acetylacetone with ethylenediamine (acen). These complexes are stable in the solid state in the absence of moisture but react with moisture and anhydrous HCl in solution to yield ethane and the corresponding hydroxoaluminium and chloroaluminium complexes, respectively. The susceptibility of these complexes to moisture in solution is dependent to a remarkable degree on the flexibility of the ligand. It is proposed that the remarkable resistance to hydrolysis, as compared with non-chelated aluminium alkyl species, is related to the high-energy state of the square-planar four-co-ordinate Al^{3+} species remaining on loss of the alkyl group. The X-ray crystal structure of $C_{18}H_{19}N_2O_2Al$, reveals a five-co-ordinate, square-pyramidal aluminium environment with an Al—C bond distance of 1.966(7) Å. The average Al—O distance, 1.815(15) Å, is much shorter than the Al—N distance, 2.020(7) Å. The aluminium is displaced 0.540 Å from the least-squares plane of the ligating nitrogen and oxygen atoms, toward the ethyl group. The salen moiety adopts the usual 'inverted umbrella' structure observed for monomeric transition-metal species. The results of kinetic studies of reactions of the Schiff bases (SB) formed from pyridoxal-5′-phosphate (PLP) and 2-amino-3-phosphonopropionic acid (APP) and of the 1:2 Al^{III}-SB system have been reported.[77]

Improvements of anode properties of aluminium alloys were attempted[78] for application to a primary dry cell. Heat treatments were investigated chiefly for the alloy Al–3.0 Zn–0.13 Sn–0.01 Ga–0.07 Bi (%w/w) derived from high-purity aluminium (99.999%). The alloy heat-treated at 400 °C and combined with agar ($2g\,l^{-1}$) as a corrosion inhibitor gave the best electrochemical properties, with the Faradaic efficiency 99.8% and the average rate of corrosion 0.008 mg cm^{-2} h^{-1} (short term) and 0.005 mg cm^{-2} h^{-1} (long term). The lattice expansion of four Ag–Al alloys containing 5.8, 9.2, 15.8, and 18.2 at% Al has been determined[79] by X-ray measurements. The alloys were investigated in the temperature range 303–928 K. The lattice parameters increase non-linearly with temperature and from least-squares fits of the experimental data the average coefficients of linear thermal expansion (α_{av}) have been calculated for the four compositions as 20.41, 19.69, 19.93, and $20.37 \times 10^{-6}\,K^{-1}$.

The structural changes associated with the precipitation of metastable and equilibrium phases during decomposition of three supersaturated Al-rich Al–Ag–Mg alloys at 393, 456, and 508 K have also been established[80] by analysing the diffraction of monochromatic X-radiation by single-crystal alloy plates.

[76] S. J. Dzugan and V. L. Goedken, *Inorg. Chem.*, 1986, **25**, 2858.
[77] B. Szpoganicz and A. E. Martell, *Inorg. Chem.*, 1986, **25**, 327.
[78] Y. Hori, J. Takao, and H. Shomon, *Electrochim. Acta*, 1986, **31**, 555.
[79] S. K. Pradhan and M. De, *J. Appl. Cryst.*, 1986, **19**, 484.
[80] S. McK. Cousland, and G. R. Tate, *J. Appl. Crystallogr.*, 1986, **19**, 174.

2 Gallium

Macroreticular styrene–divinylbenzene copolymers functionalized with imino-diacetic acid and diethylenetriamine-N,N,N'',N'''-tetra-acetic acid have been examined[81] in order to apply them to the selective recovery of gallium(III) and/or indium(III) from acidic solutions. A detailed study of the equilibrium absorption capacities, distribution coefficients, and metal-adsorption rates of the polymer resins for these ions and other common metal ions has been made. It is reported[82] that a porous membrane impregnated with a long-chain alkylated cupferron [N-(alkylphenyl)-N-nitrosohydroxylamine ammonium salt] kerosene solution almost completely concentrated Ga^{3+} from Ga^{3+}-Al^{3+} binary solution against the concentration gradient, the counter flow of H^+ providing the driving force.

The recent finding[83] that the minimum value of the differential capacity of Ga at negative charges is systematically higher than that of Hg by about 20% has caused the relation between the structure of the interface at Hg and that at Ga to be looked at in a different light.

Temperature dependences of ^{69}Ga and ^{81}Br nuclear quadrupole resonance frequencies have shown[84] that phase transitions take place at 98 and 137 K in $(Me_4N)_2Ga_2Br_6$ and that the C_3 orientation of the $GaBr_3$ group occurs in $(Et_4N)_2Ga_2Br_6$ around room temperature.

Chlorogallium(III) porphyrins, Ga(Por)Cl, have been obtained[85] in good yield by treating the corresponding free bases with gallium trichloride. The action of HX (X = F) or I) on the chloro-derivatives Ga(Por)Cl has led to the corresponding halogeno-complexes Ga(Por)X, which derivatives have been characterized by mass spectrometry, ^1H n.m.r., and u.v.–visible data and the crystal structure of chloro-(5,10,15,20-tetraphenylporphyrinato)gallium(III), $C_{44}N_4H_{28}GaCl$ [Ga(TPP)Cl], has been determined. The material crystallizes in the tetragonal space group $I4/m$ and is isomorphous with Fe(TPP)Cl, the Ga atom lying 0.317 Å above the perfect porphyrin plane. The co-ordination polyhedron of the metal atom is a square pyramid with Ga—N and Ga—Cl distances of 2.021(1) and 2.196(2) Å respectively. The azido- and thiocyanato-gallium(III) porphyrin complexes have been prepared[86] from Ga(Por)Cl, and the stereochemistry of the metal has been established by i.r. and ^1H n.m.r. measurements and confirmed by the crystal structure determination of the azido-derivative, azido(2,3,7,8,12,13,17,18-octaethyl porphyrinato)gallium(III). The crystals, of monoclinic space group $P2_1/c$ with $a = 13.469(1)$ Å, $b = 13.487(1)$ Å, $c = 19.050(2)$ Å, and $\beta = 106.28(1)°$, contain a gallium atom which is five-co-ordinated by the four nitrogen atoms (N_p) of the porphyrin plane (average Ga—N_p = 2.034 Å) and one nitrogen atom (N) of the azido-group, (Ga—N = 1.955 Å).

Hydrous gallium(III) oxide was produced[87] as monodispersed spherical particles by forced hydrolysis in the presence of sulphate ions, though they partially dissolved and recrystallized. The rate of the reaction was empirically expressed under the

[81] T. M. Suzuki, T. Yokayama, H. Matsunaga, and T. Kimura, *Bull. Chem. Soc. Jpn.*, 1986, **59**, 865.
[82] T. Shimidzu and H. Okushita, *J. Chem. Soc., Chem. Commun.*, 1986, 1173.
[83] L. Doubova, A. De Battisti, and S. Trasatti, *Electrochim. Acta*, 1986, **31**, 881.
[84] H. Ishihara, K. Yamada, and T. Okuda, *Bull. Chem. Soc. Jpn.*, 1986, **59**, 3969.
[85] A. Coutsolelos, R. Guilard, D. Bayeul, and C. Lecomte, *Polyhedron*, 1986, **5**, 1157.
[86] A. Coutsolelos, R. Guilard, A. Boukhris, and C. Lecomte, *J. Chem. Soc., Dalton Trans.*, 1986, 1779.
[87] S. Hamada, K. Bando, and Y. Kudo, *Bull. Chem. Soc. Jpn.*, 1986, **59**, 2063.

given conditions as $d\alpha/dt = 1.6^{2/3}(1 - \alpha)^{0.874}$, where α is the degree of reaction. Polymeric hydroxo-complexes were said to act as precursors of the monodispersed spherical particles, judging from the fractional changes of the monomeric and polymeric species during the hydrolysis.

An e.p.r. study of Cr^{3+} in tris(acetylacetonato)gallium(III) single crystals has confirmed[88] the insensitivity of the e.p.r. parameters of Cr^{3+} to changes in the size of the host ion and temperature. Crystals containing between 0.5 and 1 mol% Cr were studied by e.p.r. at X-band between 298 and 20 K. Two crystallographically inequivalent centres for Cr^{3+} were observed whose spin-Hamiltonian parameters differed very little from each other and from those for isomorphous aluminium and cobalt(III) compounds. The crystal structure of tetragallium tetrakis(orthophosphate) hydrate–isopropylamine, $Ga_4(PO_4)_4OH.C_3H_{10}N.1.09H_2O$, which crystallizes in the triclinic space group $P\bar{1}$, has been found[89] to be novel with a powder pattern resembling that of $AlPO_4$-14. The structure of $GaPO_4$-14 consists of a framework $[Ga_4(PO_4)OH]$ enclosing pores containing protonated isopropylamine $(C_3H_{10}N)^+$. Within the framework there are PO_4 and GaO_4 tetrahedra cross-linking tetramers of Ga-centred polyhedra. These tetrahedra consist of edge- and corner-linked $GaO_4(OH_2)$ octahedra and $GaO_4(OH)$ trigonal bipyramids. Removal of the hydroxyl corner linkage leads to a continuous four-connected network with alternating GaO_4 and PO_4 polyhedra in four rings outlining a two-dimensional system of channels in (001). Parise[90] also reports the structure determination of $GaPO_4$-21, trigallium tris(orthophosphate) hydrate–isopropylamine, which is essentially as described for the aluminophosphate molecular sieve precursor $AlPO_4$-21, with the exception of the position of the attempted template isopropylamine, which is statistically disordered over two sites within the pores of the structure. The framework is formed by cross linking corrugated sheets composed of ribbons of edge-shared three- and five-membered rings containing five-co-ordinate gallium polyhedra and phosphorus-centred tetrahedra, with crankshaft-shaped chains of alternating GaO_4 and PO_4 tetrahedra. Upon calcination, loss of the hydroxyl group, which forms the bridge between three- and five-membered rings, converts $GaPO_4$-21 into the open-framework structure of $GaPO_4$-25.

Novel Pt ion-exchanged Ga-silicate[91] bifunctional catalysts have been developed which produce aromatics from propane more actively and selectively than previous catalysts such as Ga ion-exchanged and Pt ion-exchanged H-ZSM-5.

The structure of several gallium germanates has been established[92] or confirmed (Ga_4GeO_8 and forms, Ga_2GeO_5, $Ga_4Ge_3O_{12}$). Among them, α-Ga_4GeO_8 exhibits a new type of tunnel structure resulting from an intergrowth of GeO_2 rutile elements in the β-gallia network. Large hexagonal tunnels are formed at the junction of both lattices. This compound is the first of a series having as general formula $Ga_4M_{2n-1}O_{4n+4}$, although other compounds with M = Ge (n = 1 or 2), Ti (n = 2, 3, 11, *etc.*), or Sn (n = 1) belong to the same family. In a structure determination[93]

[88] G. Elbers, S. Remme, and G. Lehmann, *Inorg. Chem.*, 1986, **25**, 896.
[89] J. B. Parise, *Acta Crystallogr.*, Sect. C, 1986, **42**, 670.
[90] J. B. Parise, *Acta Crystallogr.*, Sect. C, 1986, **42**, 144.
[91] T. Inui, T. Makino, F. Okazumi, and A. Miyamoto, *J. Chem. Soc., Chem. Commun.*, 1986, 571.
[92] A. Kahn, V. Agafonov, D. Michel, and M. Perez y Jorba, *J. Solid State Chem.*, 1986, **65**, 377.
[93] S. Jaulmes, M. Julien-Pouzol, J. Dugue, P. Laurelle, and M. Guittard, *Acta Crystallogr.*, Sect C, 1986 **42**, 1111.

of $Ga_9Tl_3O_2S_{13}$ the Ga atoms were found to be in $[S_3O]$ or $[S_4]$ tetrahedra whilst two of the Tl atoms are nine-co-ordinate in tricapped trigonal prisms (TlS_9) and the third Tl atom lies in an elevenfold dodecahedral co-ordination (TlS_{11}).

Several gallium sulphide thiolate compounds that are structural analogues of the well-known $Fe^{III}-S^{2-}-RS^-$ complexes have been prepared and structurally characterized.[94] Crystalline $[Ga(SR)_4]^-$ complexes (SR = SMe, SEt, SPr^i, SPh, S-2,3,5,6-Me_4C_6H, or S-2,4,6-Pr$^i_3C_6H_2$) have been prepared by reaction of either $GaCl_3$ or $[GaCl_4]^-$ with five equivalents of LiSR, and structures of $[Pr^n_4N][Ga(SEt)_4]$, $[Et_4N]$-$[Ga(SPh)_4]$, and $[Et_4N]_2[Ga_2S_2(SPh)_4]$ have been determined.

The compound $[\{(Me_3Si)_3Si\}_2Ga(\mu Cl)_2Li(THF)_2]$ has been prepared[94a] from $GaCl_3$ and $Li[SiMe_3]$.3THF in Et_2O solution at $-78\,°C$ and its crystal structure determined. The structure is a double chloride bridged complex of $[(Me_3Si)_3Si]_2GaCl$ and solvated LiCl, with Si, Li, and Ga adopting distorted tetrahedral geometries and Ga having bonds to Si of the order 2.439 Å.

The reaction of $GaCl_3$ with three equivalents of Bu^t_2PLi, Bu^t_2AsLi, or ArP(H)Li (Ar = 2,4,6-$Bu^t_3C_6H_2$) affords $Ga(PBu^t_2)_3$, $Ga(AsBu^t_2)_3$, and $Ga[P(H)Ar]_3$, respectively[95] and the reaction of $GaCl_3$ with one equivalent of Bu^t_2ELi and two equivalents of RLi results in dimeric phosphido- or arsenido-bridged compounds of the type $[Ga(\mu\text{-}EBu^t_2)R_2]_2$ (E = P or As; R = Me or Bu^n). The successful syntheses of sterically hindered arsinogallanes using a lithium arsinide and the structure determination of tris(dimesitylarsino)gallane (Figure 4) have been reported.[96] This structure

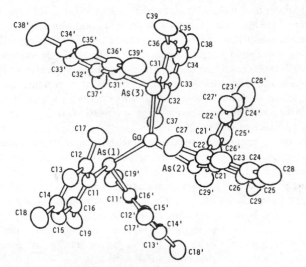

Figure 4 *ORTEP View of tris(dimesitylarsino)gallane*
(Reproduced by permission from *Inorg. Chem.*, 1986, **25**, 2483)

[94] L. E. Maelia and S. A. Koch, *Inorg. Chem.*, 1986, **25**, 1896.
[94a] A. M. Arif, A. H. Cowley, T. M. Elkins, and R. A. Jones, *J. Chem. Soc., Chem. Commun.*, 1986, 1776.
[95] A. M. Arif, B. L. Benac, A. H. Cowley, R. Gerets, R. A. Jones, K. B. Kidd, J. M. Power, and S. T. Schwab, *J. Chem. Soc., Chem. Commun.*, 1986, 1543.
[96] C. G. Pitt, K. T. Higa, A. T. McPhail, and R. L. Wells, *Inorg. Chem.*, 1986, **25**, 2483.

with GaAs bonds of 2.498, 2.508, and 2.470 Å is the first example of a monomeric three-co-ordinate Group IIIA–VA compound of the second and higher row elements.

$[(PhAsH)(R_2Ga)(PhAs)_6(RGa)_4]$ (R = Me_3SiCH_2) has been isolated[97] as a product of the reaction of $PhAsH_2$ with $(Me_3SiCH_2)_3Ga$ and determined by *X*-ray crystallographic analysis to be a cluster containing an As_7Ga_5 core. Dimethylgallane, best synthesized by the reaction between trimethylgallane and sodium tetrahydrogallate, has been characterized by its spectroscopic and chemical properties, and electron diffraction has established[98] the structure of the dimer $Me_2Ga(\mu\text{-H})_2GaMe_2$ (Figure 5).

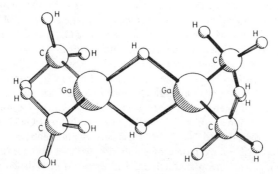

Figure 5 *Perspective view of the molecule* $Me_2Ga(\mu\text{-H})_2GaMe_2$
(Reproduced from *J. Chem. Soc., Chem. Commun.*, 1986, 805)

The reactions of the $[LMo(CO)_3]^-$ ions [L = MeGa(pz), HB(pz)$_3$, Me$_2$Ga(pz)-(OCH$_2$CH$_2$NMe$_2$); pz = pyrazolyl] with $[Cu(PPh_3)Cl]_4$ and $Rh(PPh_3)_3Cl$ have yielded[99] complexes with Mo—Cu and Mo—Rh bonds. The materials $[MeGa(pz)_3]Mo(CO)Cu(PPh_3)$ and $[MeGa(pz)_3]Mo(CO)_3Rh(PPh_3)_2$ have been found to crystallise in the monoclinic $P2_1/n$ and triclinic $P\bar{1}$ space groups respectively. In the kinetic studies[77] of the reaction of the Schiff bases (SB) formed from pyridoxal-5′-phosphate (PLP) and 2-amino-3-phosphonopropionic acid (APP) the transamination and dephosphonylation reactions were found to proceed more rapidly for the 1:2 Ga^{III}-SB complexes than other related metal complexes. The results show that the most active species are those in which the carboxylate group of the amino-acid moiety of the SB ligand is co-ordinated to the metal ion and the phosphonate is not co-ordinated.

The photoreduction of oxygen at p-GaAs gives rise to current doubling since electron capture by O_2 is followed by hole injection from $HO_2\cdot$. The rate constant for hole injection by $HO_2\cdot$ has been measured[100] directly using the intensity modulated a.c. photocurrent method and found to be 2.5×10^4 s^{-1} in good agreement with the value estimated from the intensity dependence of the steady-state quantum yield. When (100) wafers of undoped semi-insulating LEC GaAs are etched in the

[97] R. L. Wells, A. P. Purdy, A. T. McPhail, and C. G. Pitt, *J. Chem. Soc., Chem. Commun.*, 1986, 487.
[98] P. L. Baxter, A. J. Downs, M. J. Goode, D. W. H. Rankin, and H. E. Robertson, *J. Chem. Soc., Chem. Commun.*, 1986, 805.
[99] G. A. Banta, B. M. Louie, E. Onyriuka, S. J. Rettig, and A. Storr, *Can. J. Chem.*, 1986, **64**, 373.
[100] R. Peat and L. M. Peter, *Electrochim. Acta*, 1986, **31**, 731.

AB etch, both grooved etch features and ridges are produced at dislocations. Both these features have been studied[101] by a correlation of etching, scanning electron microscopy in the wavelength dispersive cathodoluminescence mode (SEM-WDCL), and transmission electron microscopy (TEM). It has been shown that a groove corresponds to a dislocation surrounded by matrix with a relatively lower etch rate whereas a ridge corresponds to a dislocation surrounded by matrix with a higher etch rate. These areas of different etch rate are shown by SEM-WDCL to correspond to areas with slightly different Fermi levels (E_F). The AB etch rate dependence on E_F is confirmed by determining the etch rate for samples with widely separated Fermi levels. An alternative explanation which ascribes grooved features to heavy decoration with As precipitates is shown to be inconsistent with TEM observations. Reaction mechanisms[102] of GaAs epitaxial growth using GaAs–AsCl$_3$–H$_2$ and GaAs–AsCl$_3$–N$_2$ systems have been investigated by means of i.r. spectroscopy; probable reaction models are discussed. Over the whole temperature range 400—840 °C, the dominant gallium chlorides observed by the sampling method are Ga$_2$Cl$_6$ and an unidentified gallium compound with chlorine that has an absorption band at 1600 cm^{-1} in the hydrogen carrier system, but only Ga$_2$Cl$_6$ in the nitrogen carrier system. Analyses by a mechanical balance have suggested that GaCl$_3$ samples from a reactor should dimerize to Ga$_2$Cl$_6$ in a gas cell. A photoenhancement of AsCl$_3$ reduction by H$_2$ has been examined by irradiation with an excimer laser. A photo-excitation of GaCl$_3$ has been ascertained, and epitaxial growth with a single flat temperature profile has been realized at low temperatures below 600 °C by irradiation with a 249 nm laser.

A survey of conventional etchants for GaAs has shown[103] them to be inappropriate for GaSb, so a novel etch for GaSb has been developed. It consists of HCl–H$_2$O–NaK (tartrate) solutions and gives controllable etch rates from 0.1 to 2.0 μm min^{-1}. Smooth surfaces free from etch pits and surface films have been obtained. The Ca$_{28}$Ga$_{11}$ structure, determined by Fornasini and Pani[104] is characterized by three types of calcium polyhedra surrounding the gallium atoms: tricapped trigonal prisms, distorted cubes, and distorted cubicosahedra. The linkage of these polyhedra gives rise to columns or slabs which alternate along the b axis of the cell. The co-ordinations of the Ga atoms in Ca$_{28}$Ga$_{11}$ are not unusual but are fairly typical of Ga-containing intermetallic phases. The structures of the metastable phase Mn$_2$Ga$_{0.5}$As$_{0.5}$(m) and the stable phases Fe$_3$GaAs and Co$_2$Ga$_{0.5}$As$_{0.5}$ isotypic with Ni$_2$In have been reported.[105] An important factor in terms of the stability of these phases is the valence-electron concentration. X-Ray topographs have been obtained[106] of as-grown GaAs crystals doped with 10^{20} atoms cm^{-3} of In, where the usual extinction criterion g, $b = 0$ leads to this type of defect. However, for several g satisfying the condition g, $b = 0$ with $b = a[111]$, the images of these dislocations were still clearly visible. Comparison between experimental and computer-simulated X-ray

[101] G. T. Brown and C. A. Warwick, *J. Electrochem. Soc.*, 1986, **133**, 2576.

[102] J. Nishizawa, H. Shimawaki, and Y. Sakuma, *J. Electrochem. Soc.*, 1986, **133**, 2567.

[103] J. G. Buglass, T. D. McLean, and D. G. Parker, *J. Electrochem. Soc.*, 1986, **133**, 2565.

[104] M. L. Fornasini and M. Pani, *Acta Crystallogr.*, Sect. C, 1986, **42**, 394.

[105] M. Ellner and M. Et-Boragy, *J. Appl. Crystallogr.*, 1986, **19**, 80.

[106] N. Burle-dubec, B. Pichaud, and F. Minari, *J. Appl. Crystallogr.*, 1986, **19**, 140.

topographic sections of these defects confirms the existence of Burgers vectors along ⟨111⟩.

The final two papers[107,108] in this section describe the crystal structures of two gallium(III)–dirhenium complexes. The first paper reports[107] the structure of tetra-carbonyldi - μ - iodo - μ - {[tetracarbonyl(triphenylphosphine)rhenio(0)]gallio(III)} - bis(triphenylphosphine)dirhenium(0)(*Re–Re*), which crystallized in the triclinic space group $P\bar{1}$. The molecular structure contains an inner GaI$_2$Re$_2$ molecular fragment with an Re—Re bond. This transition metal–transition metal bond is bridged by two iodine atoms and one trivalent Ga atom possessing a terminal Re(CO)$_4$PPh$_3$ group. The two Re atoms are central atoms from a pair of distorted octahedra sharing a face with the three bridge atoms. Each of these Re atoms has two CO and one PPh$_3$ as terminal ligands. Their two CO ligands are *trans* to the iodine atoms and the two phosphines *trans* to the Ga atom. The Re atom of the terminal Re(CO)$_4$PPh$_3$ group has a distorted octahedral ligand arrangement with the Ga and P atoms in axial and four CO ligands in equatorial positions.

In the second paper[108] the structure of octacarbonyl-bis-{μ-[pentacar-bonylrhenio(−I)]gallio(III)}-dirhenium(−II)(*Re–Re*) is described. The central frag-ment of the molecule consists of a planar Ga$_2$Re$_2$ rhombus with an Re—Re bond of 3.139(2) Å. The Re atoms of the two Re(CO)$_5$ ligands have a *trans* configuration with respect to the plane of the Ga$_2$Re$_2$ ring. The mean value of the Ga—Re bond length is 2.589(5) Å.

3 Indium

The structure of trichlorobis(trimethylamine)indium(III) has been determined[109] and is made up of discrete molecules of InCl$_3$(NMe$_3$)$_2$ with crystallographically imposed C_s symmetry. The indium atom has a five-co-ordinate undistorted trigonal-bipyramidal environment, being bonded to three chlorine atoms in the equatorial plane and two trimethylamine ligands in axial positions.

High-resolution He-I and He-II photoelectron spectra of the gas-phase indium(III) trihalides have been recorded.[110] The absence of any resolvable ligand field splitting the In 4d spectra of indium tri-iodide shows that this compound is a dimer with pseudotetrahedral geometry about each In, rather than having the planar monomeric structure as postulated in earlier gas-phase He–I photoelectron studies of the valence band. X_α-SW calculations on indium tri-iodide are also consistent with a dimeric structure. The structure of indium tribromide and indium trichloride cannot be obtained unambiguously from the In 4d valence-band spectra or X_α-SW calcula-tions; however, the overall evidence is that the tribromide is a dimer whereas the trichloride is probably mainly a monomer.

Structural investigations of the compound 1,1,4,4-tetramethylpiperazonium aquo-pentabromoindate monohydrate [C$_8$H$_{20}$N$_2$] [InBr$_5$(H$_2$O)].H$_2$O and characterization of the [InBr$_5$(OH$_2$)]$^{2-}$ anion complete[111] the series of six-co-ordinate indium halide complex anions [InX$_n$(OH$_2$)$_{6-n}$]$^{3-n}$ (X = Cl or Br; n = 4–6) for which

[107] U. Florke, P. Balsaa, and H.-J. Haupt, *Acta Crystallogr., Sect. C*, 1986, **42**, 275.
[108] H.-J. Haupt, U. Florke, and H. Preut, *Acta Crystallogr., Sect. C*, 1986, **42**, 665.
[109] R. Karia, G. R. Willey, and M. G. B. Drew, *Acta Crystallogr., Sect. C*, 1986, **42**, 558.
[110] J.-A. E. Bice, G. M. Bancroft, and L. L. Coatsworth, *Inorg. Chem.*, 1986, **25**, 2181.
[111] G. R. Clark, C. E. F. Rickard, and M. J. Taylor, *Can. J. Chem.*, 1986, **64**, 1697.

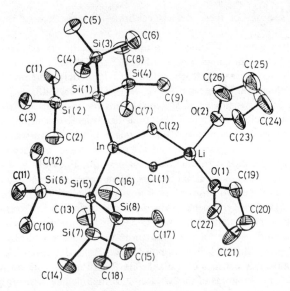

Figure 6 *ORTEP view of* $\{(Me_3Si)_3Si\}_2In(\mu\text{-}Cl)_2Li(THF)_2$
(Reproduced from *J. Chem. Soc., Chem. Commun.*, 1986, 1776)

crystallographic and spectroscopic data are available. Treatment of $InCl_3$ with three equivalents of $Li[Si(SiMe_3)_3].3THF$ in Et_2O solution at $-78\,°C$ has yielded[94a] a double chloride bridged complex $[\{(Me_3Si)_3\}_2In(\mu\text{-}Cl)_2Li(THF)_2]$ (Figure 6) in which the In—Si bond lengths of 2.591(7) Å are the first reported for indium–silicon bonds. Reaction of $[Li(THF)_3(\mu\text{-}Cl)InC(Me_3Si)_3Cl_2]$ and an excess of $LiAlH_4$ in THF at $-40\,°C$ yielded[112] a hydride of $(Me_3Si)_3CIn((H)(\mu\text{-}H)Li(THF)_2(\mu\text{-}H)In(\mu\text{-}H)(H)C(SiMe_3)_3$ in which there is crystallographic and spectroscopic evidence for the existence of hydrogen bridges between In and Li and between In and In.   Eaborn *et al.*[113] also report the structure of the compound $[O\{(Me_3Si)_3CIn\}_4(OH)_6]$, which is based on four-membered In_2O_2 rings. The central oxygen atom in the cage structure is surrounded by four indium atoms at the corners of a tetrahedron, each of which is linked to three bridging hydroxide ligands along the edges of the tetrahedron and to the large trimethylsilyl group on the periphery of the molecule, thus creating a five-co-ordinate environment about In.

Indium(III) chloride reacted[114] at $60\,°C$ with the macrocyclic ligands (1 : 1) 1,4,7-triazacyclonnoane (L^1) 1,4,7-trimethyl-1,4,7-triazacyclonnoane (L^2), and 1,4,7-trithiacyclonane (L^3) in chloroform (acetronitrile) has yielded the complexes L^1InCl_3, L^2InCl_3, and L^3InCl_3, respectively. L^1InBr_3 was prepared analogously in aqueous solution. Hydrolysis of L^1InBr_3 in alkaline aqueous solution afforded the

[112] A. G. Avent, C. Eaborn, P. B. Hitchcock, J. D. Smith, and A. C. Sullivan, *J. Chem. Soc., Chem. Commun.*, 1986, 988.

[113] S. S. Al-Juaid, N. H. Buttrus, C. Eaborn, P. B. Hitchcock, A. T. L. Roberts, J. D. Smith, and A. C. Sullivan, *J. Chem. Soc., Chem. Commun.*, 1986, 908.

[114] K. Wieghardt, M. Kleine-Boymann, B. Nuber, and J. Weiss, *Inorg. Chem.*, 1986, **25**, 1654.

colourless, tetrameric cation $[L_4^1In_4(\mu\text{-}OH)_6]^{6+}$, the first well characterized μ-hydroxo-complex of indium(III), which has been isolated as the perchlorate, dithionate, iodide, and hexafluorophosphate salts. $[L_4In_4(\mu\text{-}OH)_6](S_2O_6)_3.4H_2O$, which crystallizes in the orthorhombic space group $P2_12_12_1$, consists of discrete tetrameric cations and dithionate anions. The $In_4(OH)_6$ core has an adamantane-like skeleton with an average In—O bond distance of 2.14(1) Å. Hydrolysis of $LInBr_3$ in 2M sodium acetate at 60 °C and addition of sodium perchlorate yielded colourless crystals of $[L_2In_2(MeCo_2)_4(\mu\text{-}O)].2NaClO_4$, which crystallizes in the triclinic space group $P1$ and consists of the neutral μ-oxo-bridged dimer $[L_2In_2(MeCO_2)_4(\mu\text{-}O)]$ and sodium perchlorate. The average In—O bond distance of the oxo-bridge is 2.114(4) Å.

Three papers by Kadish *et al.* describe[115-117] studies of indium(III) porphyrins. The synthesis and spectroscopic properties of a new series of σ-bonded indium porphyrins are reported.[115] Perfluoroaryl groups (C_6F_4H and C_6F_5) in some derivatives are σ-bonded to indium(III) complexes containing four different types of porphyrin ring. All of the complexes can be oxidized or reduced by multiple single-electron transfers. Two reversible reductions were observed in non-aqueous media and corresponded to ring-centred reactions. The first of these generates a stable $[(P)In(C_6F_4X)]^{\cdot-}$ radical anion. The compounds could also be oxidized by two one-electron abstractions generating stable radical cations.

The reactions of *N*-methylimidazole (*N*-MeIm) and pyridine with (P)In(X), where P is either the dianion of octaethylporphyrin (OEP) or the dianion of tetraphenylporphyrin (TPP) and X is Cl^-, AcO^-, $PhSO_3^-$, or $MeSO_3^-$, were monitored by ^1H n.m.r., electronic absorption spectroscopy, and conductivity measurements.[116] All three methods were self-consistent in demonstrating the stepwise formation of six-co-ordinate monomeric In-porphyrin species of the types (P)InX(L) and $[(P)In(L)_2]^+$, where L = pyridine or *N*-MeIm. This is the first time that monomeric six-co-ordinate In^{III} porphyrins have been reported. Detailed electron transfer and ligand exchange reactions[117] of In^{III} porphyrins that occur in neat pyridine or in CH_2Cl containing pyridine or *N*-methylimidazole have been reported. Eight porphyrin complexes were selected for investigation, *viz.* (P)InSO_3Me, (P)InSO_3Ph, (P)InCl, and (P)InOAc. The authors have shown that the electrochemical behaviour of (P)InX in the presence of pyridine or *N*-MeIm is very different from the electrochemical behaviour in CH_2Cl_2, the singly and doubly reduced porphyrins readily binding with pyridine or *N*-MeIm.

$In_{11}Mo_{40}O_{62}$ has been prepared from a stoicheiometric mixture of In, Mo, and MoO_2 at 1100 °C and its crystal structure determined.[118] Chains of four and five Mo_6 octahedra condensed *via* opposite edges occur that are surrounded by O atoms above all free octahedra edges. The channels between these cluster units are filled with chains of five and six In atoms with metal–metal-bonded In_5^{7+} and In_6^{8+} ions (In—In distances between 2.62 and 2.68 Å). The authors point out in another paper[119] on the structures with oligomeric condensed clusters in indium

[115] A. Tabard, R. Guilard, and K. M. Kadish, *Inorg. Chem.*, 1986, **25**, 4277.
[116] J. L. Cornillon, J. E. Anderson, and K. M. Kadish, *Inorg. Chem.*, 1986, **25**, 991.
[117] J. L. Cornillon, J. E. Anderson, and K. M. Kadish, *Inorg. Chem.*, 1986, **25**, 2611.
[118] H. Mattausch, A. Simon, and E. M. Peters, *Inorg. Chem.*, 1986, **25**, 3428.
[119] A. Simon, W. Meetin, H. Mattausch, and R. Gruehn, *Angew. Chem., Int. Ed. Engl.*, 1986, **25**, 845.

oxomolybdates that $In_{11}Mo_{40}O_{62}$ is the first compound where structural disorder occurs at regular intervals with formation of defined fragments of the Mo_4O_6 chain.

A continuous solid solution of $In_{1-x}Li_{3x}VO_4$ ($0 < x < 0.4$) has been observed[120] in the $InVO_4$–Li_3VO_4 system. The solid solution is of two types (a) with interstitial Li^+ ions in vacant tetrahedral sites of $InVO_4$ and (b) Li^+ ions in octahedral sites vacated by In^{3+}.

The structure determinations of two lead indium bismuth chalcogenides have been carried out. $Pb_4In_3Bi_7S_{18}$ consists of In–S octahedra [In—S 2.583(12)— 2.853(8) Å], in which In^{3+} is partially substituted by Bi^{3+}, mono- and bi-capped trigonal Pb–S prisms, and Bi–S octahedra as well as $Bi–S_5$ pyramids with two additional S atoms below the basal plane.[121] The In–S chains are mutually linked by edge-sharing octahedra forming a sixfold band parallel to $(\bar{2}03)$. The second structure, $Pb_4In_2Bi_4S_{13}$, consists of In–S tetrahedra and octahedra and mono- and bi-capped trigonal prisms of Pb–S and Bi–S octahedra, all polyhedra being distorted and forming chains parallel to [010].[122] In this structure also there is evidence of partial substitution of the octahedral In^{3+} atoms by Bi^{3+}. Sodium ion motion in $Na_{1+x}Zr_{2-x}In_x(PO_4)_3$ Nasicon-related materials ($0 < x < 1.85$) has been studied by n.m.r. at resonance frequencies from 170 to 450 K[123] At 8 MHz the resonance lines due to M1- and M2-type sites are well resolved. They are split by a strong second-order quadrupolar effect ($v_Q = 1.5$ Megacycles (Mc) s^{-1}). For $NaZr_2(PO_4)_3$ only the M1 site is occupied at room temperature, but a new site M5 appears at higher temperatures (410 K). At 21 Mc s^{-1} the line narrowing observed at increasing temperature illustrates the Na^+ ion mobility within the skeleton.

Time-resolved photoelectrochemical measurements following illumination with a pulsed laser source have been performed on surface-modified n- and p-InP photo-electrodes.[124] The rate of the initial fast transient photopotential decay was shown to be sensitive to the nature of the surface modification used. The chemisorption of Ru ions onto the n-InP surface reduced the rapid initial decay of the transient photopotential response observed on the unmodified n-InP surface. For p-InP either unmodified or modified by Co, Rh, Pt, Pb, or Rh/Pt, the transient photopotential decay exhibited a two-segment response. The fast initial decay was found to be dependent on the previous surface treatment of the In-P electrode. The second, slower decay, however, was found to be independent of previous surface treatments. The similar decay rate values obtained with metal surface modifications possessing significantly different work functions (Co and Pt) suggested that the barrier height was pinned at the interface region. The origin of this pinned barrier height was associated with the proposed formation of intrinsic surface states at 0.9 and 1.2 eV (with respect to the valence band maximum) after p-InP surface modification. Since the barrier height appears to be pinned, these results suggest that the major role of metallic surface modifications in promoting photoelectrochemical hydrogen evolution may, in part, be that of a conventional electrocatalyst.

[120] M. Touboul and A. Popot, *J. Solid State Chem.*, 1986, **65**, 287.
[121] V. Kramer and I. Reis, *Acta Crystallogr., Sect. C*, 1986, **42**, 249.
[122] V. Kramer, *Acta Crystallogr., Sect. C*, 1986, **42**, 1089.
[123] F. Cherkaoul, G. Villeneauve, C. Delmas, and P. Hagenmuller, *J. Solid State Chem.*, 1986, **65**, 293.
[124] R. L. Cook, P. F. Dempsey, and A. F. Sammells, *J. Electrochem. Soc.*, 1986, **133**, 2287.

Polycrystalline samples of n-type $AgIn_5Se_8$ have been prepared[125] from the constituent elements and used as electrodes in photoelectrochemical cells. These semiconductor electrodes were characterized by spectral response, Mott–Schottky plots, current–voltage measurements, and complex impedance analyses. The polycrystalline n-$AgIn_5Se_8$ electrodes show a very encouraging photoelectrochemical behaviour, capable of offering solar to electrical conversion efficiencies of the order of 3%. However, stability tests, run both under illumination and in the dark, revealed that this interesting photoelectrochemical performance worsened upon time of illumination. This was interpreted in terms of photocorrosion, which leads to the formation of a selenium film on the surface of the semiconductor electrode.

High-resolution electron microscopic observations have shown[126] that $CuIn_5Te_8$ has a somewhat disordered thiogallate structure. The domain structure, which is due to different arrangements of the cations within a common face-centred cubic tellurium sublattice, is consistent with the symmetry of the thiogallate structure.

The preparation of polycrystalline samples of $Cd_{2x}(CuIn)_yMn_{2z}Te_2$ ($x + y + z = 1$) alloys prepared by a melt and anneal technique has been reported.[127] By X-ray powder photography it was found that, in addition to the zinc blende and chalcopyrite structures, a partially ordered cubic structure was obtained in the alloys plus a two-phase field higher z values.

4 Thallium

Detailed investigations on the mono- and multi-layer formation of thallium on to a glassy carbon electrode have been reported.[128] The monolayer region shows two peaks: one obeying Kolb's correlation which follows absorption–desorption kinetics and the other occurring with a large underpotential shift which follows progressive nucleation growth kinetics. The multilayer formation of thallium occurs by 3D nucleation-growth kinetics at higher concentrations of thallium and is independent of the presence or absence of underpotentially deposited thallium or the nature of the substrate.

The relatively abundant structural data for the halogen compounds ABX_3, A_4BX_6, and A_2BX_4 ($X = Cl$, Br, or I; A = bulky alkali metal or pseudo-alkali metal; B = $+2$ transition-metal ion) have enabled[129] the determination of the Tl^I effective radius for a given co-ordination number and ionicity f_{A-X} and the establishment of an ionicity–effective radius diagram for this cation. The diagram forms a coherent set which enables the calculation *a priori* of $Tl-X$ interatomic distances in different compounds by the application of the rule of addition of effective radii stated by Shannon and Prewitt for oxides and fluorides.

Jouini[130] has investigated the correlation between the covalency and the ^{205}Tl n.m.r. chemical shift data in oxides and halides of Tl^I. A correlation was established between chemical shift and stereochemical activity of the lone pair of Tl^I

[125] G. Razzini, L. Peraldo Bicelli, M. Arfelli, and B. Scrosaati, *Electrochim. Acta*, 1986, **31**, 1293.
[126] N. Frangis, G. Van Tendeloo, J. Van Landuyt, and S. Amelinckx, *J. Solid State Chem.*, 1986, **61**, 369.
[127] M. Quintero, L. Dierker, and J. C. Woolley, *J. Solid State Chem.*, 1986, **63**, 110.
[128] S. Jaya, T. P. Rao, and G. P. Rao, *Electrochim. Acta*, 1986, **31**, 1601.
[129] N. Jouini, *J. Solid State Chem.*, 1986, **63**, 431.
[130] N. Jouini, *J. Solid State Chem.*, 1986, **63**, 439.

for Tl^I halides the shift data increased with covalency. The structure of tribromobis(pyridine-*N*-oxide)thallium(III) [TlBr$_3$(C$_5$H$_5$NO)$_2$] consists[131] of neutral monomeric complexes, in which the Tl atom shows a fivefold co-ordination in the form of a distorted trigonal bipyramid.

In trithallium(I) phosphate the thallium(I) ion is co-ordinated[132] to oxygen from three different phosphate groups at distances of 2.529(7), 2.553(5), and 2.555(8) Å, creating a three-dimensional network of thallous and phosphate ions. Each Tl atom is nearest neighbour to two Tl atoms in the layer above it and to two Tl atoms in the layer below it at distances of 3.690 Å. The structure of the mixed-valence thallium compound Tl$_3$[Tl$_{0.5}$(H$_3$O)$_{0.5}$]H$_{14}$(PO$_4$)$_8$.4H$_2$O has been redetermined in the monoclinic space group *C*2/*c* by Marsh[133] and found to be isostructural with (H$_3$O)-[Al$_3$(H$_2$PO$_4$)$_6$(HPO$_4$)$_2$]4H$_2$O. A study[134] of the structure of (TlOTeF$_5$(1,3,5-Me$_3$C$_6$H$_3$)$_2$]$_2$ has shown that the co-ordination sphere about each Tl^I atom is pseudo-tetrahedral being made up of two oxygen atoms from two bridging (OTeF$_5$)$^-$ groups and two η^6-mesitylene molecules. The compound provides an example with which to gauge the variability or constancy of two recently discovered structural features, Tl^I–arene co-ordination and bridging OTeF$_5$ groups.

Reaction of the macrocycles 1,4,7-triazacyclonane (L^1) and *N,N',N''*-trimethyl-1,4,7-triazacyclononane (L^2) with TlCl$_3$.4H$_2$O(1:1) have afforded L^{1}TlCl$_3$ and L^{2}TlCl$_3$ respectively, with thallium(III) nitrate and L^1 (1:4) afforded [TlL$^1_2$](NO$_3$)$_3$, and with thallium(I) nitrate and L^2 and NaPF$_6$ or NaClO$_4$.H$_2$O yielded L^{2}Tl(PF$_6$) and L^{2}Tl(ClO$_4$) respectively.[135] The crystal structure of L^{2}Tl(PF$_6$), which is the first example of a thallium(I) complex with a saturated N-donor ligand, contains L^{2}Tl$^+$ cations and non-co-ordinated [PF$_6$]$^-$ anions.

The synthesis and spectroscopic and structural studies[136] of dimethylthallium(III) complexes containing heterocyclic N-donor ligands, including 'TlC$_2$N$_3$O$_2$' and 'TlC$_2$O$_6$' geometries have been reported. The crystal structures of [{TlMe[(py)$_2$CH$_2$]-(NO$_3$)}$_2$] (py = pyridine), [TlMe$_2$(terpy)(H$_2$O)]NO$_3$ (terpy = 2,2':6',2''-terpyridyl), and [(TlMe$_2$)$_3$(Et$_3$terpy)$_2$(NO$_3$)$_3$] have 'TlC$_2$N$_2$O$_3$', 'TlC$_2$N$_3$O$_2$', and 'TlC$_2$N$_3$O$_2$ and TlC$_2$O$_6$' co-ordination environments about thallium(III) respectively.

Complexes of quasi-cylindrical crown ethers with dimethyl- and diethyl-thallium(III) perchlorates have been prepared[137] and their spin–spin coupling constants between the thallium nucleus and protons measured. As part of the same work, the authors determined the crystal structure of [TlMe$_2$(L^1)]ClO$_4$ {L^1 = 1,4,7,18,21,24-hexaoxaperhydro[7,7](41,8a)naphthalenophane} and in another paper[138] reported the structure of [TlMe$_2$(L)]ClO$_4$ {L = 1,4,15,18,29,32-hexaoxaperhydro[4,4,4]-(4a,8a)naphthalenophane}. In both complexes the Tl atom is situated in the O$_6$ plane of the crown ether with the dimethylthallium unit perpendicular to this plane,

[131] W. Huller, M. E. Garcia-Fernandez, M. R. Bermejo, and M. V. Castano, *Acta Crystallogr., Sect. C*, 1986, **42**, 60.

[132] A. Zalkin, D. H. Templeton, D. Eimerl, and S. P. Velsko, *Acta Crystallogr., Sect. C*, 1986, **42**, 1686.

[133] R. E. Marsh, *Acta Crystallogr., Sect. C*, 1986, **42**, 511.

[134] S. H. Strauss, M. D. Noirot, and O. P. Anderson, *Inorg. Chem.*, 1986, **25**, 3850.

[135] K. Wieghardt, M. Kleine-Boymann, B. Nuber, and J. Weiss, *Inorg. Chem.*, 1986, **25**, 1309.

[136] A. J. Canty, K. Mills, B. W. Skelton, and A. H. White, *J. Chem. Soc., Dalton Trans.*, 1986, 939.

[137] K. Kobiro, S. Takada, Y. Odaira, and Y. Kawasaki, *J. Chem. Soc., Dalton Trans.*, 1986, 1767.

[138] K. Kobiro, M. Takahashi, Y. Odaira, Y. Kawasaki, Y. Kai, and N. Kasai, *J. Chem. Soc., Dalton Trans.*, 1986, 2613.

and the decalin moieties creating a cylinder wall. A new regioselective α-monochlorination of simple aliphatic ketones using aqueous $TlCl_3$ which proceeds *via* a mono-organothallium(III) intermediate followed by ^1H, ^{13}C, and ^{205}Tl n.m.r. spectroscopy has been reported.[139]

Six thallium(I) molybdates with allotropic modifications have been found[140] in the binary system Tl_2O-MoO_3. $Tl_4Mo_2O_5$, $Tl_2Mo_2O_7$, $Tl_8Mo_{19}O_{34}$ (three forms), and $Tl_2Mo_7O_{22}$ (two forms) melt incongruently; Tl_2MoO_4 and $Tl_2Mo_4O_{13}$,which are trimorphic, melt congruently.

The structures of tetrathallium(I) tetrathiosilicate(IV) and tetrathallium(I) tetraselenosilicate(IV) have been reported.[141] $Tl_4[SiS_4]$ is isostructural with $Tl_4[GeS_4]$ and contains $[SiS_4]^{4-}$ anions held together by Tl^+ cations in irregular sixfold co-ordination. $Tl_4[SiSe_4]$ is built up from slabs parallel to (001) which consist of almost planar 3^6 nets of Tl^+ cations enclosed between two parallel layers of $[SiSe_4]^{4-}$ tetrahedra. Further Tl^+ cations are located in voids between the $[SiSe_4]^{4-}$ tetrahedra of each layer and between the slabs.

Mössbauer studies have been carried out[142] on another thallium–selenium intermetallic, $TlFe_{2-x}Se_2$ over a temperature range 100—460 K. From the study the magnetic properties of the material were investigated and a magnetic transition was observed at ~450 K. Multinuclear n.m.r. characterization[143] of anionic clusters of the elements Ge, Sn, Sb, Tl, Pb, and Bi in liquid ethylenediamine has identified, for the first time, the clusters $[Sn_{8-x}Pb_xTl]^{5-}$ ($x = 1$—4) and $[Sn_2Bi_2]^{2-}$.

Finally, the structure and low temperature behaviour of $Tl_{0.82}V_5S_{4.36}Se_{3.64}$ have been reported.[144] The structure is built of distorted vanadium chalcogen octahedra connected by face and edge sharing. Owing to the three-dimensional framework, large quasi-rectangular channels are built in which the Tl atoms are inserted. Magnetic measurements were performed on a crystalline powder sample in the temperature range 74–290 K. The measured susceptibility obeys a Curie–Weiss law with a term for the temperature-independent paramagnetism. No ordering of the magnetic moments was observed down to the lowest temperatures. A slight discontinuity appears in the susceptibility *vs.* temperature curve at *ca.* 170 K which may be due to a phase transition.

[139] J. Glaser and I. Toth, *J. Chem. Soc., Chem. Commun.*, 1986, 1336.
[140] M. Touboul, P. Toledano, C. Idoura, and M.-M. Bolze, *J. Solid State Chem.*, 1986, **61**, 354.
[141] G. Eulenberger, *Acta Crystallogr., Sect. C*, 1986, **42**, 528.
[142] L. Haggstrom, H. R. Vewrma, S. Bjarman, R. Wappling, and R. Berer, *J. Solid State Chem.*, 1986, **63**, 401.
[143] W. L. Wilson, R. W. Rudolph, L. L. Lor, R. C. Taylor, and P. Pyykko, *Inorg. Chem.*, 1986, **25**, 1535.
[144] W. Bensch, J. Abart, E. Amberger, H. W. Schmalle, and J. Kopf, *Acta Crystallogr., Sect. C*, 1986, **42**, 6.

5 C, Si, Ge, Sn, Pb; N, P, As, Sb, Bi

By P. G. HARRISON

Department of Chemistry, University of Nottingham, University Park, Nottingham NG7 2RD

1 Carbon

Ab initio quantum mechanical calculations using triple-ζ-plus double polarization quality basis sets in conjunction with self-consistent-field, two-configuration SCF, and configuration interaction methods show that the three halogenocarbenes, CHF, CHCl, and CHBr, all have singlet ground states with singlet–triplet separations predicted to be 13.2, 5.4, and 4.1 kcal mol^{-1}, respectively.[1]

Two methods are available for the generation of carbenes: the decomposition of the appropriate diazo or diazirine derivatives. The diazirine method has been employed for the generation of phenylchlorocarbene,[2] methoxycarbene,[3] and phenoxycarbene[4] in argon or nitrogen matrices at 10 K. Warming phenylchlorocarbene in an argon matrix containing oxygen to 35 K resulted in the formation of the corresponding yellow–green carbonyl oxide, photolysis of which with visible light gave the corresponding dioxirane, benzoyl chloride, and ozone (Scheme 1). Irradiation of the dioxirane gave mainly phenyl chloroformate together with a small

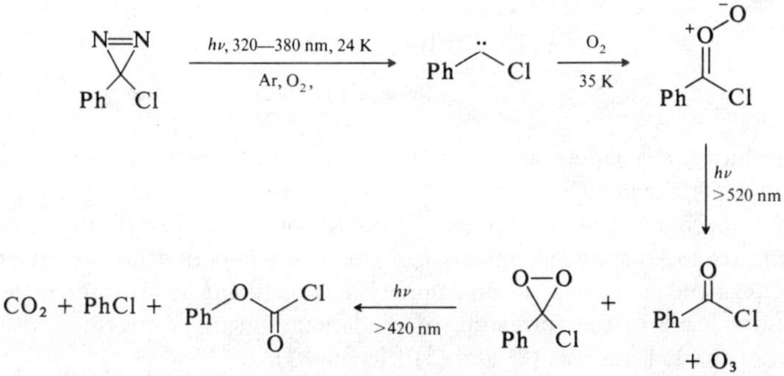

Scheme 1

amount of chlorobenzene and CO_2. Deuterium and ^{18}O labelling indicate significant C—O double bond character in methoxychlorocarbene [due to the charge-separated form (1)], which exhibits the intense i.r. absorption at *ca.* 1300 cm^{-1}. The data are

[1] G. E. Scuseria, M. Durán, R. G. A. R. Maclagan, and H. F. Schaeffer, *J. Am. Chem. Soc.*, 1986 **108**, 3248.
[2] G. A. Ganzer, R. S. Sheridan, and M. T. H. Liu, *J. Am. Chem. Soc.*, 1986, **108**, 1517.
[3] M. A. Kesselmayer and R. S. Sheridan, *J. Am. Chem. Soc.*, 1986, **108**, 99.
[4] M. A. Kesselmayer and R. S. Sheridan, *J. Am. Chem. Soc.*, 1986, **108**, 844.

$$R \diagdown \!\! \underset{\displaystyle }{\overset{+}{O}} \!\! \diagup R$$

(1)

indicative of two geometric isomers for this carbene, the *cis*-carbene showing a significantly lower C—Cl stretching frequency than the *trans*, consistent with an anomeric interaction. Irradiation of the carbene in argon matrices gives acetyl chloride, ketene, and HCl, and in nitrogen small amounts of CO and methyl chloride are also observed. Similar results are also obtained from phenoxycarbene.

Photolysis of diphenyldiazomethane in frozen alcoholic matrices affords ground-state triplet diphenylcarbene, which at 77 K reacts primarily with alcohols by OH insertion to give ethers.[5] Photolysis produces an excited carbene which reacts with the matrix by hydrogen-atom abstraction to give ultimately alcohol-type products (Scheme 2). Raney nickel causes the elimination of N_2 from diazopropane quantitatively at 373 K yielding not only propene (*cf.* the decomposition on quartz wool,

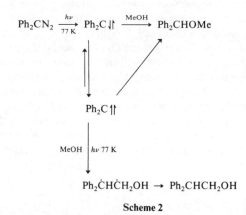

Scheme 2

which requires a temperature of 523 K and affords only propene) but also tetramethylethylene in 40% yield, an observation which lends support to the surface methylene mechanism of the Fischer–Tropsch synthesis.[6] The diazo method has also been used to generate diadamantylcarbene[7] and 1-naphthylcarbene.[8] Irradiation of dry, degassed solutions of diadamantyldiazomethane in *cis*-butene at room temperature leads to the formation of diadamantylmethane together with small amounts of the 1:1 adducts (2) and (3) (Scheme 3).

The reactions of 1-naphthylcarbene have been examined by flash photolysis techniques. Generation of the carbene in hydrocarbon solvents leads to the formation of 1-naphthylmethyl radicals. However, in cyclohexane and $[^2H_{12}]$cyclohexane, the main reaction pathway is carbene insertion into the C—H bond rather than hydrogen

[5] E. Leyva, R. L. Barcus, and M. S. Platz, *J. Am. Chem. Soc.*, 1986, **108**, 7786.
[6] H. Boch, G. Tschmutowa, and H. P. Wolf, *J. Chem. Soc., Chem. Commun.*, 1986, 1068.
[7] D. R. Myers, V. P. Senthilnathan, M. S. Platz, and M. Jones, *J. Am. Chem. Soc.*, 1986, **108**, 4232.
[8] H. Barcus, L. M. Hadel, L. J. Johnston, M. S. Platz, T. G. Savino, and J. C. Scaiano, *J. Am. Chem. Soc.*, 1986, **108**, 3928.

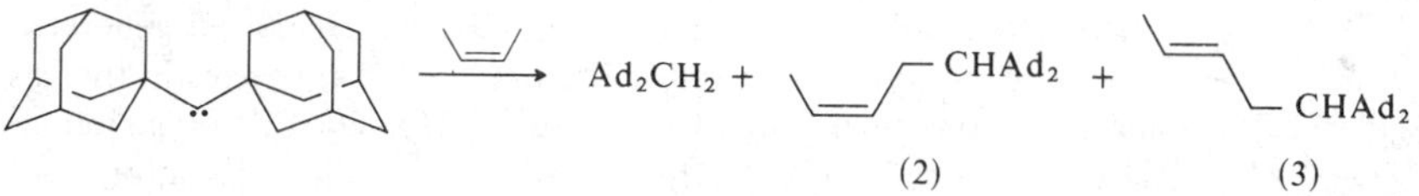

Scheme 3

$$Ad_2CH_2 + \quad CHAd_2 \quad + \quad CHAd_2$$

$$(2) \hspace{5cm} (3)$$

abstraction. The carbene reacts readily with nitriles to yield nitrile ylides, which can also be generated from the corresponding azirine. Reaction with oxygen affords the carbonyl oxide.

Ab initio calculations have shown that, in contrast to ordinary carbenes, unsaturated carbenes such as $H_2C=C$ ($^1A_1$) undergo insertion into the O—H bond of water with a non-zero energy barrier (15.3 kcal mol^{-1} for $H_2C=C$ ($^1A_1$). The correlation effect is also important in determining the barrier height.[9] The univalent species CF appears to be formed as an intermediate in the reaction of arc-generated carbon atoms with CF_4 at 77 K. When the reactions are carried out in the presence of alkenes as trapping reagents the observed products are fluorocyclopropanes, postulated to arise by addition of CF to the double bond generating cyclopropyl radicals which then abstract hydrogen (Scheme 4a), and 1,1-difluoroalkanes. These latter products are postulated to form by reaction of triplet CF_2 (Scheme 4b). The addition of CF to alkenes is stereospecific and gives both *cis*- and *trans*-fluorocyclopropanes with the latter generally predominating.[10]

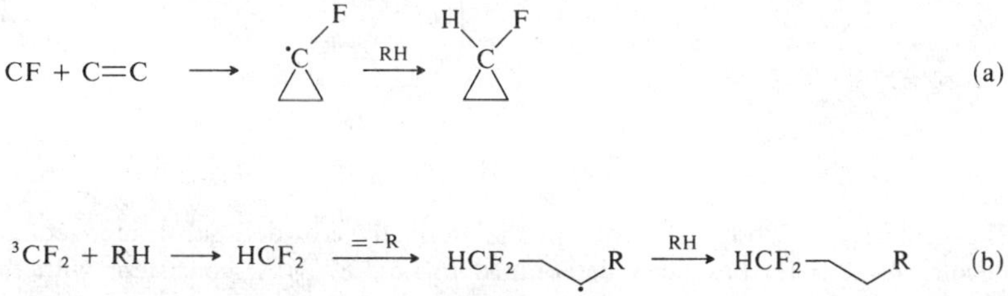

$$CF + C=C \quad \longrightarrow \quad \overset{F}{\underset{\triangle}{\dot{C}}} \quad \overset{RH}{\longrightarrow} \quad \overset{H \quad F}{\underset{\triangle}{C}} \hspace{3cm} (a)$$

$$^3CF_2 + RH \quad \longrightarrow \quad HCF_2 \quad \overset{=-R}{\longrightarrow} \quad HCF_2{\text-}R \quad \overset{RH}{\longrightarrow} \quad HCF_2{\text-}R \hspace{1.5cm} (b)$$

Scheme 4

The kinetics and mechanism of the reactions of HCO radicals with unsaturated hydrocarbons have been studied experimentally by the flash photolysis–laser resonance absorption technique in the temperature range 350—510 K as well as theoretically in the case of ethylene by *ab initio* SCF MO calculations with double-ζ basis sets. The radicals were generated by photolysis of acetaldehyde, and reaction with alkenes and butadiene essentially proceeds *via* an addition mechanism rather than a hydrogen-atom transfer from HCO to the double bond, which is also energetically favourable. The experimental results are supported by the calculated potential energy surfaces involved in the two possible reaction channels, and in particular that the energy barrier is much higher for the hydrogen atom transfer.[11]

[9] M. T. Nguyen and A. F. Hegarty, *J. Chem. Soc., Chem. Commun.*, 1986, 773.
[10] M. Rahmann, M. L. McKee, and P. B. Shevlin, *J. Am. Chem. Soc.*, 1986, **108**, 6296.
[11] R. Lesclaux, P. Roussel, B. Veyret, and C. Pouchan, *J. Am. Chem. Soc.*, 1986, **108**, 3872.

The oxygen-methylated carbon monoxide cation, CH_3OC^+, has been generated in mass spectrometric experiments as a stable (lifetime $>10^{-5}$ s) species in the gas phase. The data were consistent with the structure $CH_3-\overset{+}{O}=C$.[12] Reaction of $[Fe(C_5Me_5)_2]^+[BF_4]^-$ with $K[C(CN)_3]$ leads to the isolation of $[Fe(C_5Me_5)_2]^+[C(CN)_3]^-$ as green needle crystals. The crystals contain independent cations and anions, but there is no evidence for a structure with C_{2v} symmetry corresponding to the $[(NC)_2C=C=N]^-$ resonance form. I.r. and Raman data suggest a D_{3h} structure, which was confirmed by *ab initio* calculations.[13]

In agreement with experimental data, MNDO calculations show that excited state $[S(^1D)]$ atomic sulphur inserts into the CH bonds of alkanes such as methane and ethane whereas ground-state $[S(^3P)]$ atomic sulphur does not. Similarly, MNDO calculations offer a reasonable rationalization for the observed differences in the addition of ground- and excited-state atomic sulphur to ethylene. Here, the stereoselectivity of the ground state is postulated to result from rapid intersystem crossing rather than from a high methylene rotational barrier in the triplet biradical. Ethenethiol is predicted to result from isomerization of hot thiirene rather than from sulphur insertion into the C—H bond. In addition to acetylene, ground-state atomic sulphur is predicted to give thioketocarbene in contrast to thiirene, thioketene, and ethynethiol which are predicted to result from reaction of excited-state atomic sulphur.[14]

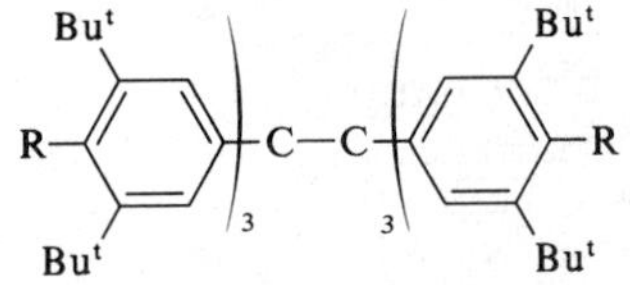

(4) R = H or Ph

The central C—C bond distance in the sterically crowded substituted ethane molecule (4; R = H) has been determined to be 1.67(3) Å, consistent with the calculated (MM2) value of 1.65 Å, and in sharp contrast to the extremely short value previously determined for the similar molecule (4; R = Ph).[15] The structure of ethylene oxide has been determined at 150 K. In the crystal the molecule is within experimental error an equilateral triangle, the C—O bond distance being very similar to those in other cyclic ethers whereas the C—C distance is somewhat shorter than that found by other methods. MP2/6-31G* calculations predict the same C—O distance, but a longer C—C bond distance. In the crystal lattice, adjacent molecules are connected by short C—H···O contacts.[16]

The geometries of fluorinated ethylenes have been optimized at the SCF level with a double-ζ-plus polarization function on carbon ($DZ + D_c$) basis set. The C=C and C—F bond distances for C_2H_4, C_2H_3F, CH_2CF_2, *cis*- and *trans*-CHFCHF, and C_2F_4 have been optimized at the configuration–interaction level, including all

12 B. van Barr, P. C. Burgers, J. K. Terlouw, and H. Schwarz, *J. Chem. Soc., Chem. Commun.*, 1986, 1607.
13 D. A. Dixon, J. C. Calabrese, and J. S. Miller, *J. Am. Chem. Soc.*, 1986, **108**, 2582.
14 M. L. McKee, *J. Am. Chem. Soc.*, 1986, **108**, 5059.
15 B. Kahr, D. Van Engen, and K. Mislow, *J. Am. Chem. Soc.*, 1986, **108**, 8305.
16 P. Luger, C. Zaki, J. Buschmann, and R. Ruders, *Angew. Chem., Int. Ed. Engl.*, 1986 **25**, 276.

single and double excitations. Good agreement with electron diffraction data was found. The values of $r(C=C)$ and $r(C—F)$ decrease with increasing fluorine substitution. Calculated values for the heats of formation of *cis*- and *trans*-CHFCHF are in excellent agreement with experimental data.[17]

(5a)   (5b)

Microwave spectra, electric dipole moments, and molecular structures have been reported for *cis*- and *trans*-1,2-difluorocyclopropane,[18,19] *cis,trans*-1,2,3-trifluoro-cyclopropane,[20] and methylenecyclopropene (5a).[21] All ring bonds in *trans*-1,2-difluorocyclopropane are shorter than the C—C bonds in cyclopropane, but those in the *cis*-isomer are longer than those in the *trans*-isomer. The orientation of the HCF group with respect to the ring plane is almost the same for both isomers and does not differ significantly from the CH$_2$ orientation in cyclopropane. The ring bonds in *cis,trans*-1,2,3-trifluorocyclopropane are also shortened compared with cyclopropane, with a greater reduction occurring in the two equivalent *trans* ring bonds. The C—F bonds are inequivalent, with longer C—F bonds observed in the HCF moiety exclusively *trans* to neighbouring HCF groups. Analysis of experimental and calculated (MP2/6-31G*) data for (5a) indicates that the dipolar resonance form (5b) constitutes about 20% of the ground-state character, but it only contributes a π-delocalization energy comparable to that of buta-1,3-diene. (5a) is concluded to be non-aromatic.

The molecular structure of the free allyl radical, produced with 75% relative abundance by vacuum pyrolysis of hexa-1,5-diene at 960 °C, has been determined by high-temperature electron diffraction coupled with mass spectrometry. The data are consistent with a planar symmetric geometry with a C—C distance of 1.428(13) Å and a C—C—C angle of 124.6(34)°.[22] Isomeric allyl, prop-2-enyl, prop-1-enyl, and cyclopropenyl anions and the vinyl anion have been generated in the gas phase by collision-induced dissociation of the corresponding carboxylate anions using an FT-mass spectrometer. Interconversion of isomers does not occur under the conditions of the experiments. Each ion produces a unique set of products in the bimolecular reaction with N$_2$O which is characteristic of its structure. The vinylic isomer and cyclopropyl anion exhibit acid–base behaviour which is consistent with their expected high basicities. *Ab initio* MO calculations indicate that propene is more acidic than ethylene at the 2-position but less so at C-1. The vinyl anions and cyclopropyl anion exhibit bent and pyramidal structures, respectively, with relatively high barriers to inversion. The effect of methyl substitution on both vinyl anion

[17] D. A. Dixon, T. Fukumaga, and B. E. Smart, *J. Am. Chem. Soc.*, 1986, **108**, 1585.
[18] S. K. Sengupta, H. Justnes, C. W. Gillies, and N. C. Craig, *J. Am. Chem. Soc.*, 1986, **108**, 876.
[19] H. Justnes, J. Zozom, C. W. Gillies, and S. K. Sengupta, and N. C. Craig, *J. Am. Chem. Soc.*, 1986, **108**, 881.
[20] R. N. Beauchamp, J. W. Agopovich, and C. W. Gillies, *J. Am. Chem. Soc.*, 1986, **108**, 2552.
[21] T. D. Norden, S. W. Staley, W. H. Taylor, and M. D. Harmony, *J. Am. Chem. Soc.*, 1986, **108**, 7912.
[22] E. Vajda, J. Tremmel, B. Rozsondai, L. Hargittai, A. K. Maltsev, N. D. Kagramanov, and O. M. Nefedov, *J. Am. Chem. Soc.*, 1986, **108**, 4352.

basicity and inversion barriers can be rationalized in terms of charge polarization and hyperconjugative interactions.[23]

The lowest-energy form of the $C_4H_2^{2+}$ dication has been calculated (at the MP3/6-31G* //HF/6-31G* level) to have a linear ($D_{\infty h}$) structure. The alternative four-membered ring (D_{2h}) structure is 13.3 kcal mol^{-1} less stable. However, at the MP4SDQ/6-31G* //3-21G level, the linear C_4H^+ cation ($C_{\infty h}$) is only 3.6 kcal mol^{-1} more stable than its four-membered ring (C_{2v}) isomer. Despite the very high estimated heat of formation (733 kcal mol^{-1}) for the dication, all the modes of dissociation explored are calculated to be exothermic, the most favourable being dissociation into $C_3H^+ + CH^+$ and into $C_3H_2^+ + C^+$. All the four-membered ring structures show σ-deficient character for the bridging carbon atoms, the HOMOs being σ-orbitals of non-bonding nature with the significant stabilization resulting from four-centre, two-electron aromatic π-bonding.[24]

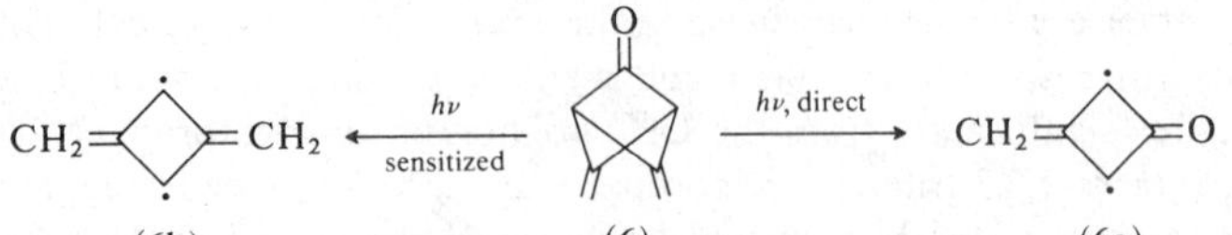

Scheme 5

The direct irradiation of the ketone (6) at 10 K affords the 3-methylenecyclobutanone diradical (6a), whereas a different product, dimethylenecyclobutadiene (6b), is obtained by photolysis in the presence of acetophenone as a sensitizer (Scheme 5).[25] The mono- and di-cations of both hydroxyacetylene[26] and amino acetylene[27] have been characterized by mass spectrometry. The acetylene (7) can be transformed into many other halogeno- and metallo-substituted derivatives (Scheme 6).U.v. photolysis of (7) produces the cyclic trimer (8), and reaction with $[Co_2(CO)_8]$ affords the complexes (9)—(13). The crystal structure of (11) shows it to contain a helical six-membered carbon chain as a ligand.[28] Cyclo-C_3I_4 (14) has been synthesized by two different ways (Scheme 7) and can be isolated as a yellow–brown crystalline powder which decomposes violently at 80 °C. The properties of (14), *e.g.* the formation of (15) and (16) by reaction with an excess of MeSSMe or Me$_3$SiNMe$_2$, and spectroscopic data, indicate that it is best formulated as tri-iodocyclopropenium iodide (17).[29]

Codeposition of lithium atoms and carbon monoxide molecules in a krypton matrix at 12 K leads to the spontaneous formation of numerous products: (i) the mononuclear species Li(CO)$_n$ with $n = 1, 2, 3$, or ≥ 4, (ii) species with several lithium atoms and one or two CO molecules in which the carbonyl groups are only weakly coupled, and (iii) species identified by stretching modes of either CO single bonds or strongly coupled double bonds and therefore species in which true chemical

[23]. S. W. Froelicher, B. S. Freiser, and R. R. Squires, *J. Am. Chem. Soc.*, 1986, **108**, 2853.

[24] K. Lammertsma, J. A. Pople, and P. von R. Schleyer, *J. Am. Chem. Soc.*, 1986, **108**, 7.

[25] P. Dowd and Y. H. Paik, *J. Am. Chem. Soc.*, 1986, **108**, 2788.

[26] B. van Baar, T. Weiske, J. K. Terlouw, and H. Schwarz, *Angew. Chem., Int. Ed. Engl.*, 1986, **25**, 282.

[27] B. van Baar, W. Koch, C. Lebrilla, J. K. Terlouw, T. Weiske, and H. Schwarz, *Angew. Chem. Int. Ed. Engl.*, 1986, **25**, 827.

[28] J. Wessel, H. Hard, and K. Seppelt, *Chem. Ber.*, 1986, **119**, 453.

[29] R. Weiss, G. E. Miess, A. Haller, and W. Reinhardt, *Angew. Chem., Int. Ed. Engl.*, 1986, **25**, 103.

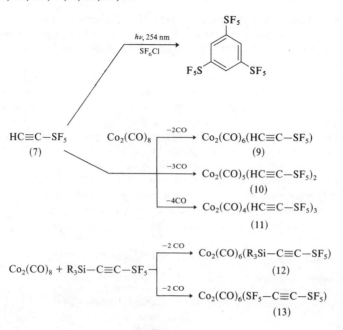

Scheme 6

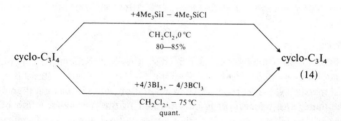

Scheme 7

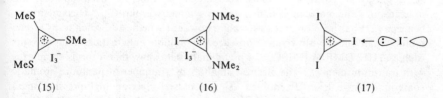

bonds are formed between carbonyls.[30] *Ab initio* MO calculations have been carried out on the low-stoicheiometry complexes and show that the 1:1 Li—CO complex corresponds to a linear $^2\Pi$ state in which the lithium atom faces the carbon end of the carbonyl group. An analogous $^2\Pi$ structure is also predicted for the linear 1:2 (OC—Li—CO) complex, while for the 2:1 (Li—CO—Li) complex two inequivalent

---

[30] O. Ayed, A. Loutellier, L. Manceron, and J. P. Perchard, *J. Am. Chem. Soc.*, 1986, **108**, 8138.

linear geometries are found corresponding to $^1\Pi$ and $^3\Sigma$ states. In all these complexes large electron transfer toward oxygen occurs, leading to large dipole moments for Li—CO and Li—CO—Li. The bonding in these complexes is described as charge transfer between Li^+ and CO^-.[31]

The gas-phase hydration of carbon dioxide has also been studied by MO calculations (PRDDO and 4-31G SCF). In the most favourable pathway, H and O of water approach respectively O and C of carbon dioxide, and after the transition state is passed the new OH bond is formed followed by formation of the new CO bond. Deformation energies of CO_2 and H_2O contribute most to the energy barrier; exchange repulsive interactions are also important. No barrier is found for the reaction of HO^- with CO_2. In this case the deformation energy for CO_2 is smaller than the charge transfer and electrostatic interactions as CO_2 and HO^- react, owing to the extra negative charge. In the reverse reaction, the dehydration of H_2CO_3, the barrier arises primarily from the loss of interaction energies which may be described as charge transfer and electronic interactions.[32]

$$RCO_2Et + MeCOMe \xrightarrow[65\%]{i} RCOCH_2COCH_2COR \xrightarrow[65\%]{ii} RCOCN_2COCN_2COR$$

$$\Big\downarrow iii$$

$$RCO)_5R \xleftarrow{iv} R(CO)_2C(OH)_2(CO)_2R$$

$$R = Bu^t \text{ or } Ph$$

Reagents: i, NaH monoglyme; ii, $p\text{-MeC}_6H_4SO_2N_3$–Et_3N. MeCN; iii, Bu^tOCl/HCO_2H; iv, P_3O_5, $CHCl_3$

Scheme 8

The first vicinal pentaketones have been synthesized (Scheme 8).[33] Diphenylcarbene undergoes oxidation to benzophenone O-oxide (18) by molecular oxygen in an argon matrix. (18) is more sensitive to irradiation with visible light than diphenyldiazomethane, and hence (18) is only formed in the thermal reaction of oxygen with the carbene. Photolysis of (18) yields diphenyldioxirane (19) and some benzophenone. The final product is (20) (Scheme 9).[34] The microwave spectrum of benzyne, generated by the pyrolysis of benzocyclobutene-1,2-dione, has been reported, and indicates that the molecule is planar and of symmetry group C_{2v}.[35] Hexamethynylbenzene (21) has been obtained as a white powder (Scheme 10) which turns brown rapidly in air but only slowly in its absence. It is soluble only in more polar solvents such as THF, DME or DMSO, and preliminary data show that it forms complexes with transition metals.[36] The electron affinities of a number of perfluoro-aromatic compounds have been determined using a pulsed electron high-pressure mass spectrometer. Compounds studied include C_6F_6 (0.52 eV), C_6F_5CN (1.1 eV),

[31] B. Silvi, O. Ayed, and W. B. Person, *J. Am. Chem. Soc.*, 1986, **108**, 8148.
[32] . J.-Y. Liang and W. N. Lipscomb, *J. Am. Chem. Soc.*, 1986, **108**, 5051.
[33] R. Gleiter, G. Krennrich, and M. Langer, *Angew. Chem., Int. Ed. Engl.*, 1986, **25**, 999.
[34] W. Sander, *Angew. Chem., Int. Ed. Engl.*, 1986, **25**, 255.
[35] R. D. Brown, P. D. Godfrey, and M. Rodler, *J. Am. Chem. Soc.*, 1986, **108**, 1296.
[36] R. Diercks, J. C. Armstrong, R. Boese, and K. P. C. Vollhardt, *Angew. Chem., Int. Ed. Engl.*, 1986, **25**, 268.

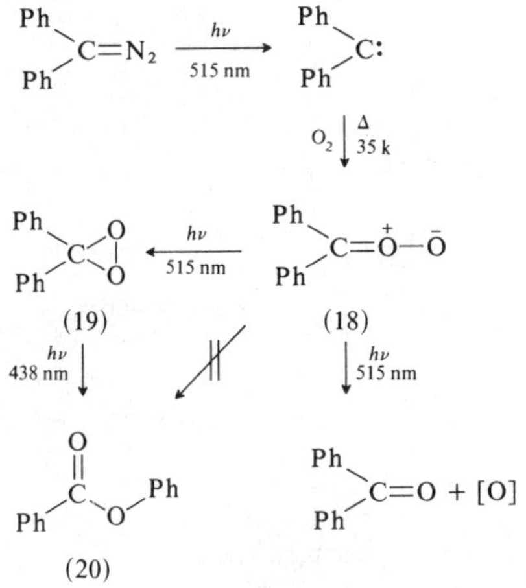

Scheme 9

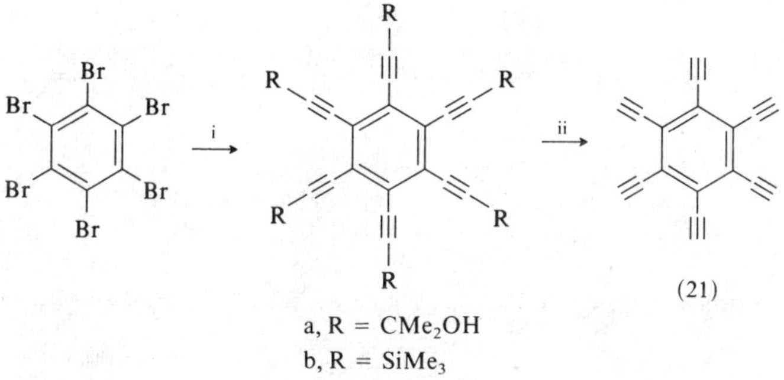

a, R = CMe$_2$OH

b, R = SiMe$_3$

Reagents: i, [PdCl$_2$(PPh$_3$), CuI. PPh$_3$ (for a only), Et$_3$N. RC≡CH (a: 34%; b: 28%): ii, from a: KOBu^t. Bu^{t}OH (undetermined yield); from b: KF.2H$_2$O,[18]crown-6, DME, 10 min (99%)

Scheme 10

C$_6$F$_5$—C$_6$F$_5$ (0.91 eV), C$_6$F$_5$CF$_3$ (0.94 eV), C$_6$F$_5$COCH$_3$ (0.94 eV), (C$_6$F$_5$)$_2$CO (1.61 eV), and 1,4-(CN)$_2$C$_6$F$_4$.[37]

The salt/molecule reaction technique coupled with matrix isolation has been used to demonstrate the formation of two isomers, *cis* and *trans*, of the C$_2$F$_3$O$_2^-$ anion ion-paired with a Cs$^+$ cation. I.r. data are suggestive of a fluorine-bridged structure for each isomer. On warming, the bridged form rearranges to the more stable CF$_3$CO$_2^-$ form.[38] Trifluoromethyl hypochlorite reacts with substituted alkenes to give mixtures of isomeric ethers. Typical reactions and further transformations of the products are summarized in Schemes 11-13.[39]

Perfluoroalkyl fluorosulphates, R$_f$OSO$_2$F [R$_f$ = CF$_3$CHMe, CF$_3$CMe$_2$, or (CF$_3$)$_2$CH], have been synthesized by the reaction of polyfluoro-alcohols with

[37] S. Chowdhury, E. P. Grimsrud, T. Heinis, and P. Kebarle, *J. Am. Chem. Soc.*, 1986, **108**, 3630.
[38] B. S. Ault, *Inorg. Chem.*, 1986, **25**, 1013.
[39] W. A. Kamil, F. Haspel-Hentrich, and J. M. Shreeve, *Inorg. Chem.*, 1986, **25**, 376.

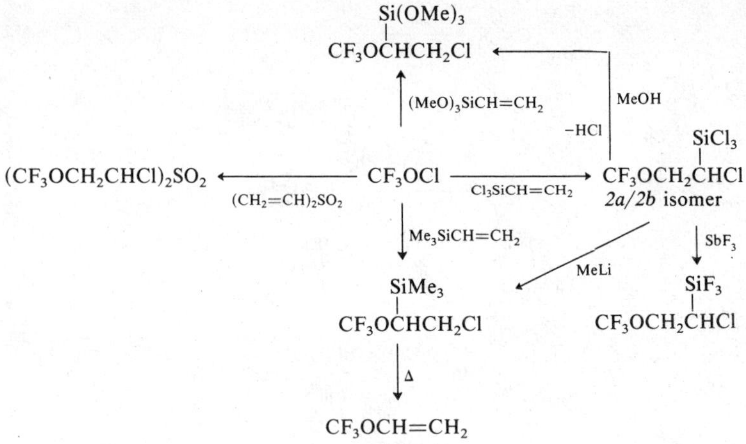

$$Si(OMe)_3$$
$$CF_3OCHCH_2Cl \longleftarrow$$

$$\uparrow (MeO)_3SiCH=CH_2 \qquad MeOH$$

$$-HCl$$
$$SiCl_3$$

$$(CF_3OCH_2CHCl)_2SO_2 \longleftarrow \quad CF_3OCl \quad \xrightarrow{Cl_3SiCH=CH_2} \quad CF_3OCH_2CHCl$$
$$\qquad\qquad (CH_2=CH)_2SO_2 \qquad\qquad\qquad\qquad\qquad 2a/2b \text{ isomer}$$

$$\downarrow Me_3SiCH=CH_2 \qquad\qquad\qquad \downarrow SbF_3$$

$$SiMe_3 \qquad\qquad\qquad MeLi \qquad\qquad SiF_3$$
$$CF_3OCHCH_2Cl \qquad\qquad\qquad\qquad CF_3OCH_2CHCl$$

$$\downarrow \Delta$$

$$CF_3OCH=CH_2$$

Scheme 11

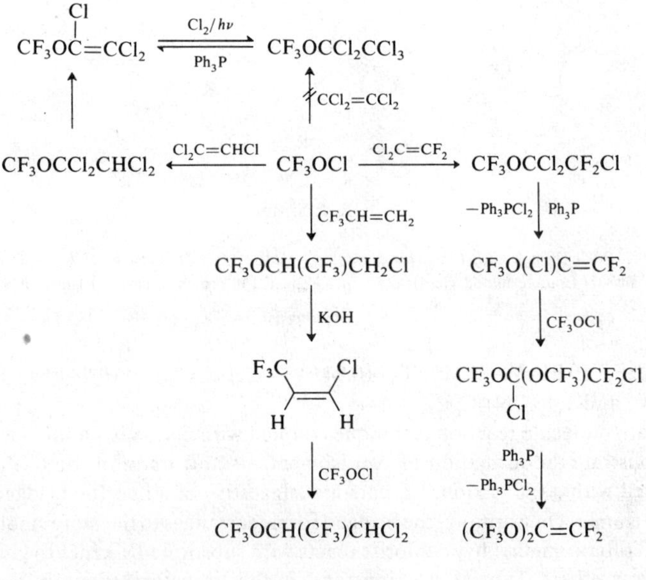

$$Cl$$
$$CF_3OC=CCl_2 \quad \underset{Ph_3P}{\overset{Cl_2/h\nu}{\rightleftharpoons}} \quad CF_3OCCl_2CCl_3$$

$$\uparrow \qquad\qquad\qquad\qquad \not\uparrow CCl_2=CCl_2$$

$$CF_3OCCl_2CHCl_2 \quad \xleftarrow{Cl_2C=CHCl} \quad CF_3OCl \quad \xrightarrow{Cl_2C=CF_2} \quad CF_3OCCl_2CF_2Cl$$

$$\downarrow CF_3CH=CH_2 \qquad\qquad -Ph_3PCl_2 \downarrow Ph_3P$$

$$CF_3OCH(CF_3)CH_2Cl \qquad\qquad CF_3O(Cl)C=CF_2$$

$$\downarrow KOH \qquad\qquad\qquad\qquad \downarrow CF_3OCl$$

$$F_3C \qquad Cl \qquad\qquad\qquad CF_3OC(OCF_3)CF_2Cl$$
$$\qquad\qquad\qquad\qquad\qquad\qquad\qquad |$$
$$H \qquad H \qquad\qquad\qquad\qquad Cl$$

$$\downarrow CF_3OCl \qquad\qquad \underset{-Ph_3PCl_2}{\overset{Ph_3P}{\downarrow}}$$

$$CF_3OCH(CF_3)CHCl_2 \qquad\qquad (CF_3O)_2C=CF_2$$

Scheme 12

sulphuryl fluoride or sulphuryl chloride fluoride, and react with nucleophiles such as amines, polyfluoro-alcohols, polyfluoro-alkoxides, and bromide ion to afford sulphamates, dialkyl sulphates, and polyfluoroalkyl bromides, respectively. In the reaction of $CF_3MeCOSO_2F$ with bromide ion, the polyfluoroalkyl bromide loses

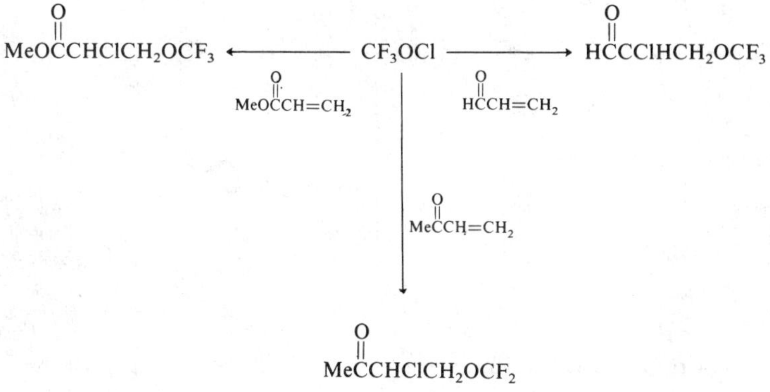

Scheme 13

hydrogen bromide to give 2-(trifluoromethyl)propene in high yield.[40] Reaction of 2,2,4,4-tetrafluoro-1,3-dithietane and arsenic or antimony pentafluoride affords the stable 2,4,4-trifluoro-1,3-dithietan-2-ylium salts (22) and (23), which can add chloride, bromide, or iodide to give the corresponding 2,2,4-trifluoro-4-halogeno-1,3-dithietanes (24).[41]

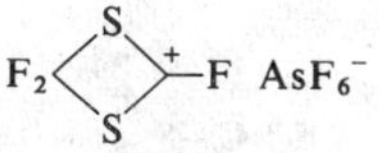

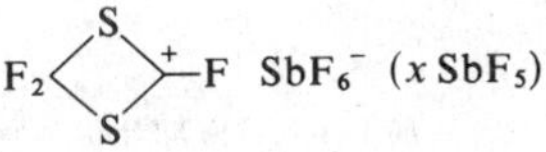

The structures of two dithietanes have been determined: (22) by crystallography[42] and $(SF_4CF_2)_2$ by electron diffraction.[43] That of bis(pentafluorothio)difluoromethane, $(SF_5)_2CF_2$, has also been determined by electron diffraction.[43] Contrary to other sulphines, Bis(trifluoromethyl)sulphine (25) reacts with amines, alcohols, and hydrogen chloride to yield derivatives of the corresponding sulphinic acid (Scheme 14).[44] The molecular structure and electronic properties of $CF_3C{\equiv}SF_3$ have been calculated using a double-ζ basis set augmented by sets of polarization functions on both carbon and sulphur. The lowest-energy structure has a short (1.412 Å), polar $C{\equiv}S$ bond and a linear $C-C{\equiv}S$ skeleton, somewhat different from the experimental data which indicate an angle of 171.5°. However, a bent structure with an angle of 171.6° is only 210 cal mol^{-1} higher in energy.[45]

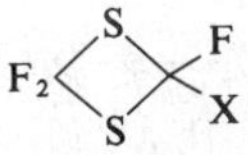

A more convenient synthesis of $O{=}C{=}C{=}C{=}S$, by the pyrolysis of (26) (Scheme 15) has been described. The photoelectron spectrum of $O{=}C{=}C{=}C{=}S$ was also

[40] T. Huang and J. M. Shreeve, *Inorg. Chem.*, 1986, **25**, 496.
[41] A. Waterfeld and R. Mews, *Chem. Ber.*, 1986, **118**, 4997.
[42] J. Antel, K. Harms, P. G. Jones, R. Mews, G. M. Sheldrick, and A. Waterfeld, *Chem. Ber.*, 1986, **118**, 5006.
[43] K. D. Gupta, R. Mews, A. Waterfeld, J. M. Shreeve, and H. Oberhammer, *Inorg. Chem.*, 1986, **25**, 275.
[44] M. Schwab and W. Sundermeyer, *Chem. Ber.*, 1986, **118**, 2458.
[45] D. A. Dixon and B. E. Smart, *J. Am. Chem. Soc.*, 1986, **108**, 2688.

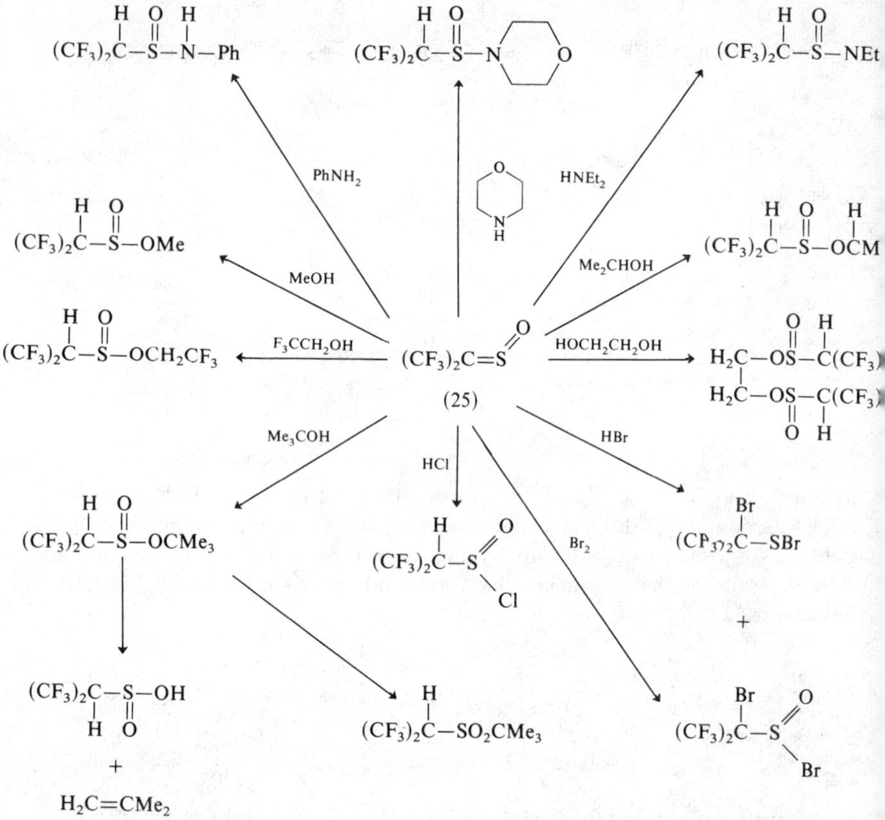

$$(CF_3)_2C\overset{H}{\underset{}{-}}S\overset{O}{\underset{}{-}}N\overset{H}{\underset{}{-}}Ph \qquad (CF_3)_2C\overset{H}{\underset{}{-}}S\overset{O}{\underset{}{-}}N\diagdown O \qquad (CF_3)_2C\overset{H}{\underset{}{-}}S\overset{O}{\underset{}{-}}N\overset{H}{\underset{}{-}}Et$$

$$(CF_3)_2C\overset{H}{\underset{}{-}}S\overset{O}{\underset{}{-}}OMe \qquad\qquad (CF_3)_2C\overset{H}{\underset{}{-}}S\overset{O}{\underset{}{-}}OCM$$

$$(CF_3)_2C\overset{H}{\underset{}{-}}S\overset{O}{\underset{}{-}}OCH_2CF_3 \qquad (CF_3)_2C{=}S{=}O \quad (25)$$

$$H_2C{-}OS\overset{O}{\underset{}{-}}C(CF_3)$$
$$H_2C{-}OS\overset{}{\underset{O}{-}}C(CF_3)\overset{}{\underset{H}{}}$$

Reagents: PhNH₂, morpholine, HNEt₂, Me₂CHOH, MeOH, F₃CCH₂OH, HOCH₂CH₂OH, Me₃COH, HCl, HBr, Br₂

$$(CF_3)_2C\overset{H}{\underset{}{-}}S\overset{O}{\underset{}{-}}OCMe_3$$

$$(CF_3)_2C\overset{H}{\underset{}{-}}S\overset{O}{\underset{Cl}{-}}$$

$$(CF_3)_2C\overset{Br}{\underset{}{-}}SBr$$

$$(CF_3)_2C\overset{H}{\underset{O}{-}}S{-}OH$$
$$+$$
$$H_2C{=}CMe_2$$

$$(CF_3)_2C\overset{H}{\underset{}{-}}SO_2CMe_3$$

$$(CF_3)_2C\overset{Br}{\underset{}{-}}S\overset{O}{\underset{Br}{-}}$$

Scheme 14

reported.[46] Microwave spectra show that butatrienone, $H_2C{=}C{=}C{=}C{=}O$, is not kinked in its equilibrium configuration.[47] The reaction of $H_2N{-}CMe{=}NH$ with CS_2 at $-15\,°C$ yields the acetamidinium salt of N-acetimidoyldithiocarbamic acid, $[(H_2N)_2CMe][S_2C{-}N{=}O(Me)NH_2]$, which reacts with metal hydroxides to form the corresponding metal derivatives (M = Na, K, Rb, Cs, Tl, $\frac{1}{2}$Pb, or $\frac{1}{2}$Cd).[48] The structures of these salts have been investigated spectroscopically,[49] and that of $[(H_2N)_2CMe][S_2CN{=}O(Me)NH_2]$ has been determined by X-ray crystallography.[50] In the crystal, the cation is associated with one anion, which is not planar, by $S{\cdots}H{-}N$ and $N{\cdots}H{-}N$ hydrogen bridges forming an eight-membered ring as

[46] H. Bock, R. Dammal, and D. Jaculi, *J. Am. Chem. Soc.*, 1986, **108**, 7844.
[47] R. D. Brown, P. D. Godfrey, M. J. Ball, S. Godfrey, D. McNaughton, M. Rodler, B. Kleibömer, and R. Champion, *J. Am. Chem. Soc.*, 1986, **108**, 6534.
[48] W. Eul and G. Gattow, *Z. Anorg. Allg. Chem.*, 1986, **537**, 189.
[49] W. Eul and G. Gattow, *Z. Anorg. Allg. Chem.*, 1986, **538**, 151.
[50] W. Eul, G. Kiel, and G. Gattow, *Z. Anorg. Allg. Chem.*, 1986, **542**, 182.

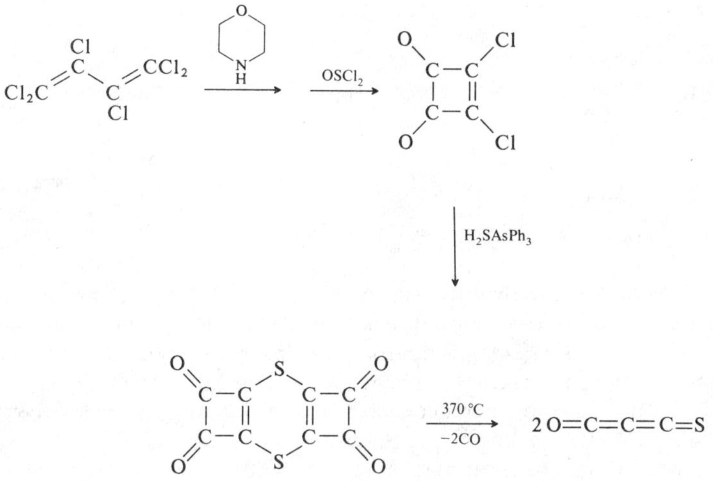

Scheme 15 structures:

$Cl_2C=C(-CCl_2)-C(Cl)=$ with morpholine (O-containing ring, NH), then $OSCl_2$ gives a four-membered ring with O, Cl substituents; then H_2SAsPh_3 gives (26); then $\xrightarrow[-2CO]{370\,°C}$ $2\,O=C=C=C=S$

(26)

Scheme 15

in (27). Orange N-acetimidoyldithiocarbamic acid has been prepared by reaction of $[(H_2N)_2CMe][S_2CN=C(Me)NH_2]$ in aqueous solution with hydrochloric acid at 0 °C. The acid, from spectroscopic data, exists in the zwitterionic form $H_2N^+=CMe-NH-CS_2^-$.[51] Spectroscopic and thermogravimetric data have also been reported for metal N,N'-diphenyl-N-formimidoyl dithiocarbamate solvates, $M[S_2C-NPh-CH=NPh].xL$ (L = water, acetonitrile, dioxane, DME, or acetone; M = Na, K, Rb, Cs, Tl, $\frac{1}{2}$Ba, or $\frac{1}{2}$Pb).[52] In crystals of the potassium salt–dioxane solvate the potassium cation is surrounded by one oxygen, one nitrogen, and three sulphur atoms to form a distorted trigonal bipyramid. The $[S_2CNCN]$ framework of the anion is planar with the E,E conformation.[53] The potassium salt also reacts with alkyl halides to form the esters $PhN=CH-NPh-CSSR$ (R = Me, Et, or CH_2Ph) and $(PhN=CH-NPh-CSS)_2CH_2$.[54] Reaction of $K_2[S_2C-NH-NH-CS_2]$ or $K_2[S_2C=N-NH-CSSMe]$ with H_3CI affords 2,5-dimethylthio-1,3,4-thiadiazole (28; $R^1 = R^2 = Me$) as the major product along with a small amount of bis(methylthio)ketazine (29). Reaction of hydrazine with ClCSSEt gives exclusively (28; $R^1 = R^2 = Et$). With benzyl bromide, $K_2[S_2C=N-NH-CSSMe]$ affords a mixture of dibenzyl sulphide, (28; $R^1 = R^2 = CH_2Ph$), and (28; $R^1 = Me$, $R^2 = CH_2Ph$).[55] Hydrazine reacts with carbonyl sulphide in the presence of sodium methoxide to give $Na_2[SOCNH-NHCOS]$, which forms the corresponding methyl ester with methyl iodide.[56] The anion of the N-

[51] W. Eul and G. Gattow, Z. Anorg. Allg. Chem., 1986, **543**, 81.
[52] W. Eul and G. Gattow, Z. Anorg. Allg. Chem., 1986, **535**, 159.
[53] W. Eul, G. Kiel, and G. Gattow, Z. Anorg. Allg. Chem., 1986, **535**, 167.
[54] W. Eul, and G. Gattow, Z. Anorg. Allg. Chem., 1986, **536**, 119.
[55] G. Gattow and S. Lotz, Z. Anorg. Allg. Chem., 1986, **533**, 99.
[56] G. Gattow and S. Lotz, Z. Anorg. Allg. Chem., 1986, **533**, 109.

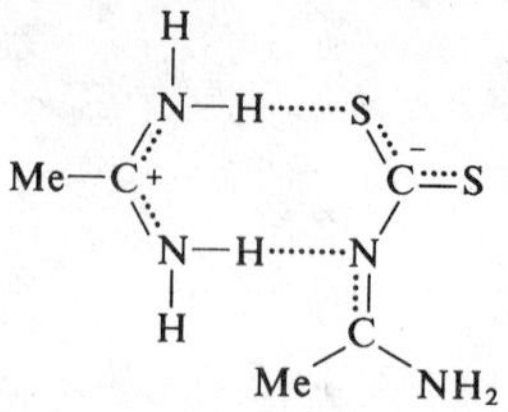

(27) 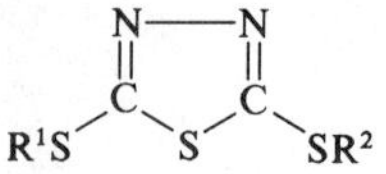   (28) 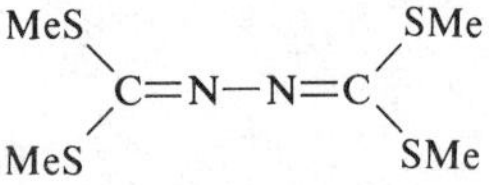   (29)

methyl-N-thioformylcarbamate salt, $[NBu_4^n][S_2C-N(Me)CSH]$, has an essentially planar skeleton.[57] Tetramethylethylenediamine-complexed lithium monothiobenzoate, $(PhCOSLi.TMEDA)_2$, is dimeric, with an eight-membered ring composed of coplanar carbon, oxygen, and sulphur atoms and the lithium atoms lying above and below this plane in order to reduce excessive angles at oxygen and to accommodate the bulky TMEDA ligands.[58]

Extended-Hückel band calculations for patterns of ordered overlayers of hydrogen atoms on unreconstructed graphite (11% coverage) show energy differences as large as 15 kcal mol^{-1}, ascribed primarily to differences in interactions between the hydrogen atoms and graphite rather than direct interactions between hydrogen atoms.[59] The formation of graphite fluoride, $(C_2F)_n$, from artificial graphite has been investigated. Elemental fluorine is occluded into the particle of fluorinated graphite and distributed in the interface between unreacted graphite and already formed $(C_2F)_n$. Reaction of fluorine with graphite then occurs in two steps: (i) where both $(CF)_n$ and $(C_2F)_n$ coexist with unreacted graphite, and (ii) when the fluorine content of the product slightly increases in spite of the absence of unreacted graphite.[60]

A novel graphite-like material of composition BC_3 has been prepared by the reaction of benzene with boron trichloride at 800 °C. The probable structure of this material is shown in Figure 1. A similar nitrogen–carbon graphite analogue was obtained from chlorine and pyridine at the same temperature.[61] Intercalation of graphite by $SbCl_4F$ affords various stage products, which are stable in air. Aqueous HCl or KOH removes only pentavalent and no trivalent antimony.[62] Other metal halides which have been intercalated into graphite are $SbCl_5$, $SbCl_2F_3$, SbF_5, $AsCl_3$, $SbCl_3$, $BiCl_3$, $AlCl_3$, $GaCl_3$, and $FeCl_3$.[63,64]

2 Silicon, Germanium, Tin, and Lead

Transient Intermediates and their Stable Analogues.—The gas-phase pyrolysis of hexafluorodisilane is a convenient method for the generation of monomeric

[57] B. Gerner and G. Kiel, Z. Anorg. Allg. Chem., 1986, **532**, 99.
[58] D. R. Armstrong, A. J. Bannister, W. Clegg, and W. R. Gibb, J. Chem. Soc., Chem. Commun., 1986, 1672.
[59] J. P. LaFemina and J. P. Lowe, J. Am. Chem. Soc., 1986, **108**, 2527.
[60] S. Koyama, T. Maeda, and K. Okamura, Z. Anorg. Allg. Chem., 1986, **540/541**, 117.
[61] J. Kouvetakis, R. B. Kaner, M. L. Sattler, and N. Bartlett, J. Chem. Soc., Chem. Commun., 1986, 1758.
[62] H. Preiss, M. Goerlich, and H. Sprenger, Z. Anorg. Allg. Chem., 1986, **533**, 37.
[63] H. Preiss, U. Nissel, and H. Sprenger, Z. Anorg. Allg. Chem., 1986, **543**, 133.
[64] H. Preiss, A. Lehmann, and H. Sprenger, Z. Anorg. Allg. Chem., 1986, **543**, 143.

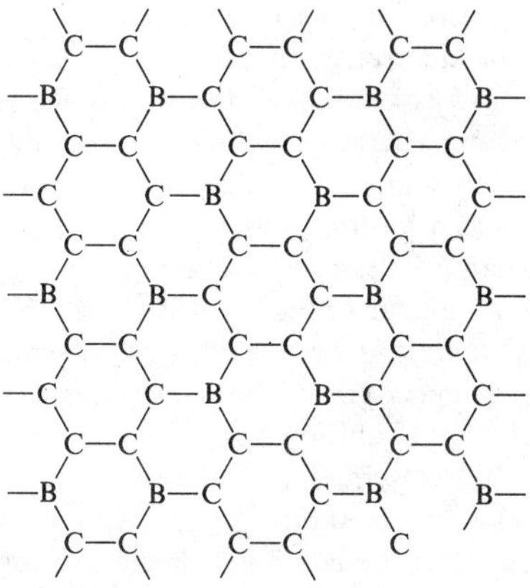

Figure 1. *The probable atomic arrangement in a layer of the boron-carbon hybrid* BC_3

difluorosilylene. Thus, fast-flow pyrolysis at 670–720 °C in a mixture with an excess of but-1,3-diene affords high yields of the addition product 1,1-difluoro-1-silacyclopent-3-ene (30). A lower limit of $k_2 = 10^6 \, l \, mol^{-1} \, s^{-1}$ was estimated for the addition reaction.[65] Several new organosilicon compounds (31)–(35) have been obtained

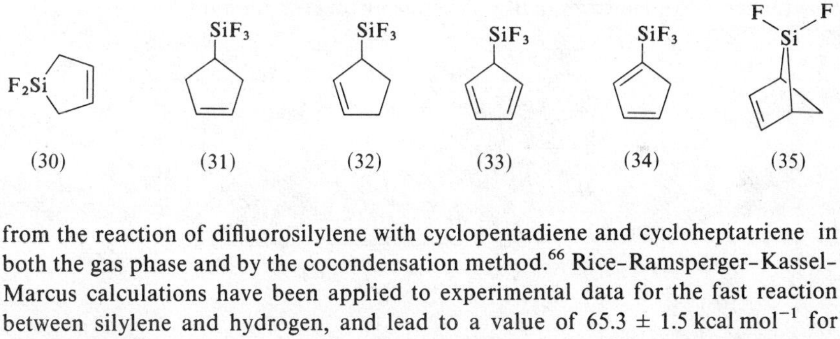

(30)   (31)   (32)   (33)   (34)   (35)

from the reaction of difluorosilylene with cyclopentadiene and cycloheptatriene in both the gas phase and by the cocondensation method.[66] Rice–Ramsperger–Kassel–Marcus calculations have been applied to experimental data for the fast reaction between silylene and hydrogen, and lead to a value of $65.3 \pm 1.5 \, kcal \, mol^{-1}$ for $\Delta H°_f(SiH_2)$.[67]

An ion-beam apparatus has been used to investigate the reactions of organosilanes with transition-metal ions in the gas phase. Co^+ and Ni^+ react with silane to yield metal silylenes. Reaction with methylsilanes leads to the formation of metal silylenes as the major reaction channels, along with several other processes including hydride abstraction, dehydrogenation, and methane loss. Reaction with hexamethyldisilane proceeds mostly by Si—Si bond cleavage. The metal ion–silylene bond energy, $D°(M^+—SiH_2)$ (M = Co or Ni), has been estimated to be $67 \pm 6 \, kcal \, mol^{-1}$. In these complexes, back-donation of paired $3d$ electrons from the metal into the

[65] S. Konieczny, P. P. Gaspar, and J. Wormhoudt, *J. Organomet. Chem.*, 1986, **307**, 151.
[66] W. L. Lee, C. F. Shieh, and C. S. Liu, *J. Organomet. Chem.*, 1986, **302**, 23.
[67] H. M. Frey, R. Walsh, and I. M. Watts, *J. Chem. Soc., Chem. Commun.*, 1986, 1189.

vacant $3p$ orbitals is suggested to supplement the donation of silicon lone-pair electrons to a $4s$ orbital on the transition metal.[68]

The ion–molecule chemistry of dimethylsilylene with fluoride and amide ions has been explored by the flowing afterglow technique. Fluoride adds to the silylene to give an adduct anion which was characterized by reaction with nitrous oxide. In contrast, amide ion abstracts a proton from dimethylsilylene giving an anion which was characterized by reaction with carbon disulphide and various acids. The gas-phase acidity of this anion is near to that of methanol. The reaction with fluoride ion followed by nitrous oxide has been employed to study the isomerization of dimethylsilylene to methylsilene. At the highest temperatures, the latter predominates slightly over the silylene.[69] The equilibrium geometries, vibrational frequencies, and i.r. intensities of the three lowest-lying states of dimethylsilylene have been predicted by *a priori* quantum mechanical methods.[70]

Irradiation of matrix-isolated dimethyldiazidosilane affords dimethylsilylene as the major product. Further irradiation with polarized 488 nm light converts the silylene into 1-methylsilene, and the reverse process is accomplished by irradiation of 1-methylsilene with polarized 248 nm light. The resulting map of i.r. transition moment directions together with other data allow little uncertainty as to the correctness of the vibrational assignments in both molecules.[71] Pyrolysis of dimethyl-*cis*-1-propenylvinylsilane leads to the extrusion of dimethylsilylene and the formation of a mixture of *cis*- and *trans*-piperylenes *via* a sigmatropic hydrogen shift which gives rise to a *cis*-1,1,2-trimethyl-3-vinylsilirane intermediate.[72] Thermally generated dimethylsilylene reacts with phenylated alkynes to give 1,4-disilacyclohexadienes (36) (Scheme 16). However, bulky substituents prevent addition.

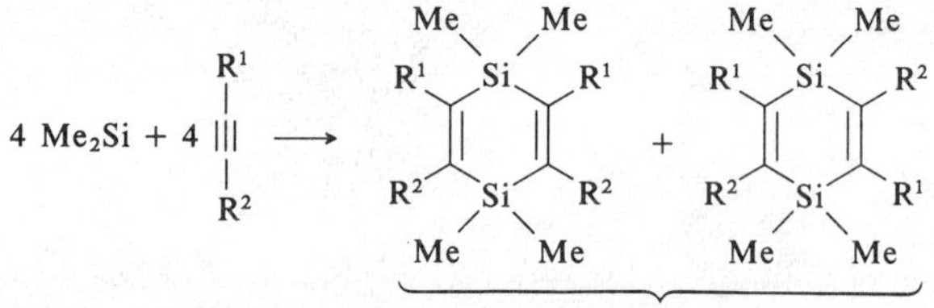

(36)

Scheme 16

The strained cycloalkyne (37) affords the silirene (38) (Scheme 17) whose transformation into the 1,4-disilacyclohexadiene is also prevented by steric effects. The reaction with thermally stable 1,3-dienes gives either 1-silacyclopent-2-enes or -3-enes depending on the substituents (Scheme 18).[73] The thermal generation of (allyloxy)methylsilylene by flash vacuum pyrolysis of 1,1-bis(allyloxy)tetramethyldisilane

[68] H. Kang, D. B. Jacobson, S. K. Shin, J. L. Beauchamp, and M. T. Bowers, *J. Am. Chem. Soc.*, 1986, **108**, 5668.

[69] R. Damrauer, C. H. DePuy, I. M. T. Davidson, and K. Hughes, *Organometallics*, 1986, **5**, 2054.

[70] R. S. Grev and H. F. Schaeffer, *J. Am. Chem. Soc.*, 1986, **108**, 5804.

[71] G. Raabe, H. Vancik, R. West, and J. Michl, *J. Am. Chem. Soc.*, 1986, **108**, 671.

[72] P. P. Gaspar and D. Lei, *Organometallics*, 1986, **5**, 1276.

[73] H. Appler and W. P. Neumann, *J. Organomet. Chem.*, 1986, **314**, 261.

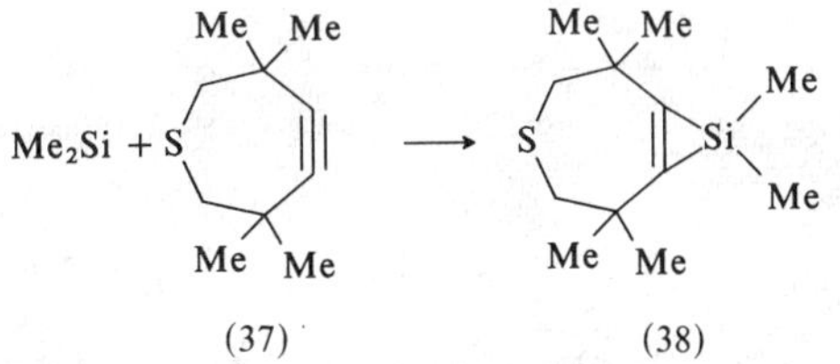

(37)   (38)

Scheme 17

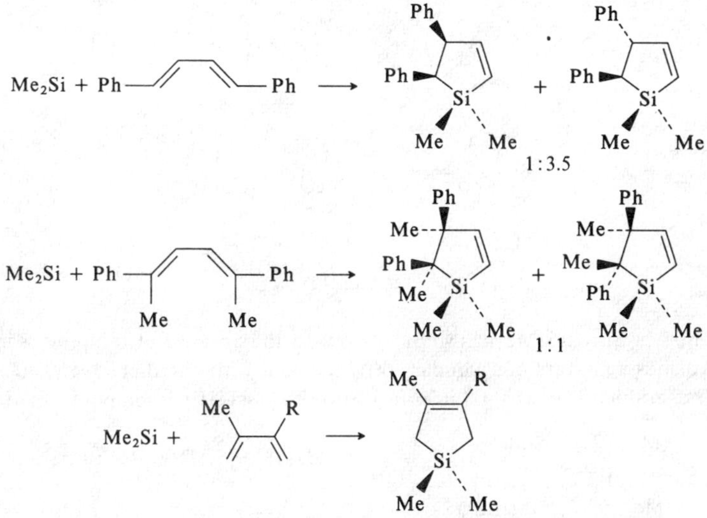

R = H or Me

Scheme 18

affords unexpected products which indicate the formation of intermediate allyl-methylsilanone, $C_3H_5MeSi=O$ (Scheme 19).[74] Evidence has been found for the rearrangement of Me_3SiSiH to $Me_2HSiSiMe$ and for the intermediacy of the silene $Me_2Si=SiHMe$.[75]

In an interesting article, Brook[76] has related how his research led to the synthesis of stable silaethylenes. Wiberg[77-81] has published more data on the synthesis and reactions of stable sila- and germa-ethenes. On gentle heating the sterically crowded trisilylmethane $Bu^t_2SiF=CLi(SiMe_3)_2$ rearranges into the compound $Me_2SiF-CLi(SiMeBu^t_2)$, which in turn decomposes at 100 °C into LiF and the

[74] L. Linder, A. Revis, and T. J. Barton, *J. Am. Chem. Soc.*, 1986, **108**, 2742.
[75] B. H. Boo and P. P. Gaspar, *Organometallics*, 1986, **5**, 698.
[76] A. G. Brook, *J. Organomet. Chem.*, 1986, **300**, 21.
[77] N. Wiberg and G. Wagner, *Chem. Ber.*, 1986, **119**, 1455.
[78] N. Wiberg and G. Wagner, *Chem. Ber.*, 1986, **119**, 1467.
[79] N. Wiberg and H. Köpf, *J. Organomet. Chem.*, 1986, **315**, 9.
[80] N. Wiberg and C.-K. Kim, *Chem. Ber.*, 1986, **119**, 2966.
[81] N. Wiberg and C.-K. Kim, *Chem. Ber.*, 1986, **119**, 2980.

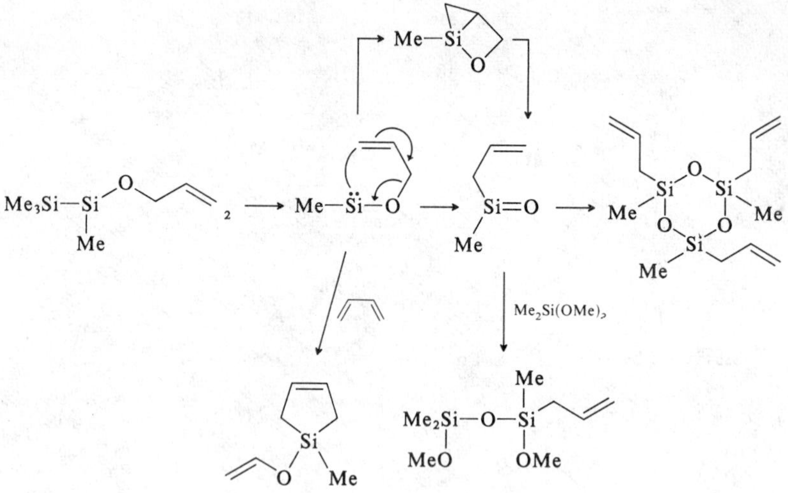

Scheme 19

silaethene $Me_2Si=C(SiMe_3)(SiMeBu^t_2)$ (39). In the absence of trapping reagents, (39) furnishes a mixture of secondary products, but with butadiene reacts to afford (40). The adduct $Me_2SiF-CLi(SiMe_3)(SiMeBu^t_2).4THF$ decomposes in diethyl

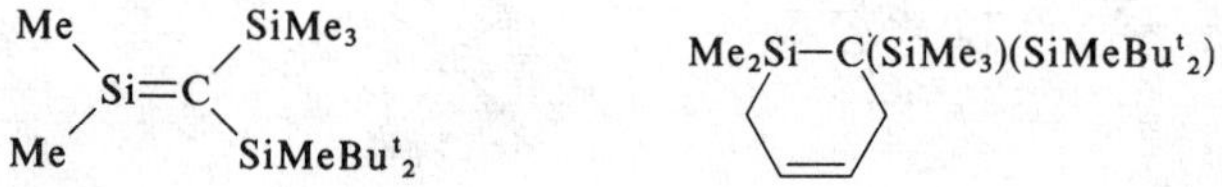

ether in the presence of Me_3SiCl at room temperature into the monotetrahydrofuran adduct of $Me_2Si=C(SiMe_3)(SiMeBu^t_2)$ (Scheme 20). The unsolvated silaethene, which may be obtained by removal of the THF by azeotropic distillation, is kinetically stable at ambient temperatures but decomposes slowly at 60 °C. In solution the

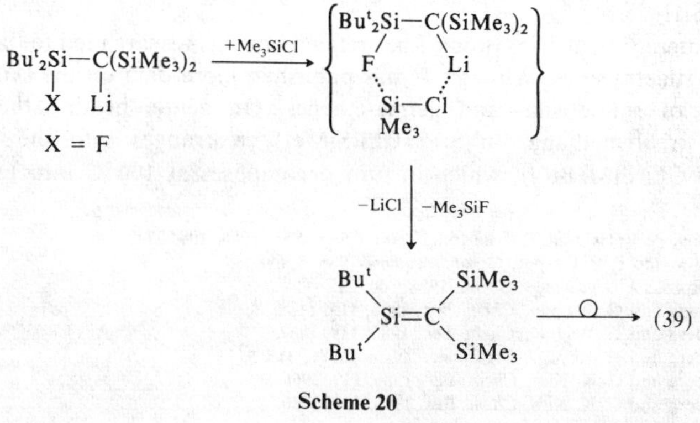

Scheme 20

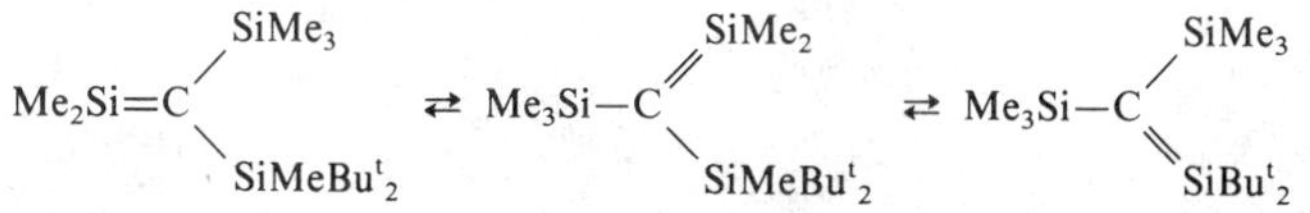

Scheme 21

silaethene undergoes rapid intramolecular methyl group exchange (Scheme 21). Reactions are summarized in Scheme 22. $Me_2Si=C(SiMe_3)_2$ is unstable at $-100\ ^\circ C$

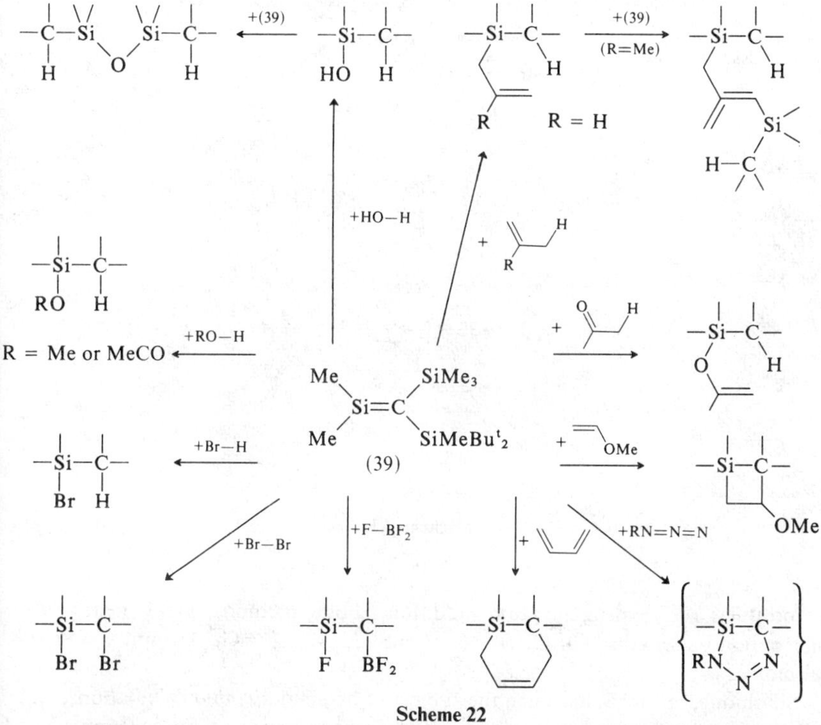

Scheme 22

with respect to dimerization, but forms the adduct $Me_2Si=C(SiMe_3)_2$ (41) which is metastable at $0\ ^\circ C$. Further heating results in dissociation to its components, and hence the adduct is a useful source of the free silaethene. Typical reactions of (41) are shown in Scheme 23. $Me_2Si=C(SiMe_3)_2$ also forms 1:1 adducts with other donors (F^-, NMe_3, NEt_3, Br^-, or THF), whose tendency towards thermal decomposition increases as the Lewis basicity of the donor increases. The analogous germaethene $Me_2Ge=C(SiMe_3)$ may be generated by similar methods, *e.g.* thermal elimination of LiX from $Me_2XGe-CLi(SiMe_3)_2$ or the thermal cycloreversion from the adduct (42), and undergoes similar types of reactions to the silaethene analogues.

π-Bond strengths in the alkene analogues $H_2X=CH_2$ ($X = C$, Si, Ge, or Sn) have been calculated from the energy differences between planar (π-bonded) and perpendicular (diradical) structures and from the energies of disproportionation of

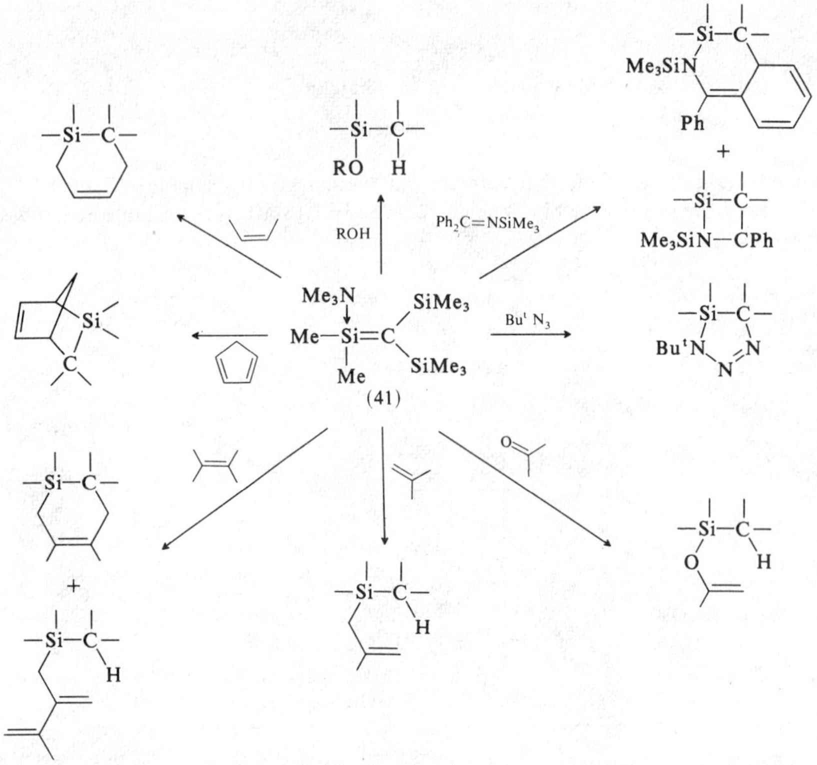

Scheme 23

the products of hydrogen atom addition. Both methods yield nearly the same π bond strengths: C=C$_3$ 64—68, C=Si 35—36, C=Ge 31, and C=Sn 19 kcal mol^{-1}.[82]

Silathione, H$_2$Si=S, has been the subject of detailed *ab initio* calculations, and has been found to be kinetically stable with respect to unimolecular decomposition reactions, *e.g.* to H$_2$ + SiS, H + HSiH, and HSiSH (*cf.* H$_2$Si=0 and H$_2$C=O), and is more thermodynamically stable than H$_2$Si=O.[83] Methylsilanone and dimethyl-silanone have been generated in an argon matrix by the cocondensation of the appropriate silane and ozone and the positions of the ν(Si=O) vibration identified.[84] *Ab initio* calculations with a 6-31G* basis set indicate that silaimine, H$_2$Si=NH, is bent at nitrogen (126.6°), but has a low barrier to linearization (6.0 kcal mol^{-1}). hence, in accord with experimental results, substitution of an electropositive SiH$_3$ group produces a nearly linear skeleton (175.6°). The Si=N double bond is 54.1 kcal-mol^{-1} weaker than two Si—N single bonds (*cf.* carbon-nitrogen bonds where

[82] K. D. Dobbs and W. J. Hehre, *Organometallics*, 1986, **5**, 2057.
[83] T. Kudo and S. Nagase, *Organometallics*, 1986, **5**, 1207.
[84] R. Withnall and L. Andrews, *J. Am. Chem. Soc.*, 1986, **108**, 8118.

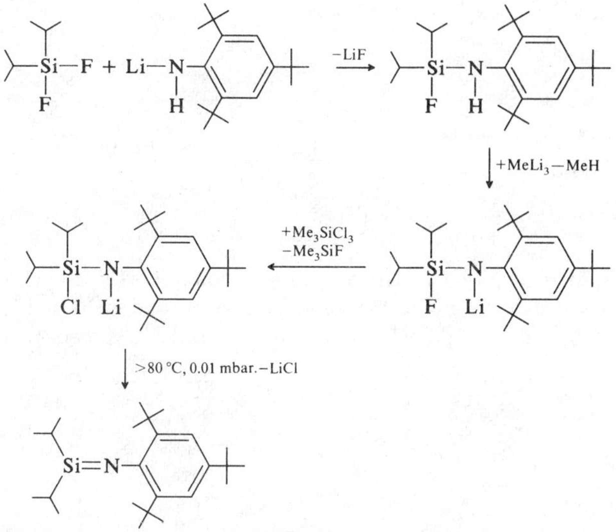

(43)

Scheme 24

the analogous differences is only 1.6 kcal mol^{-1}).[85] The syntheses of two stable silaimines have been reported. Di-isopropyl(2,4,6-tri-t-butylphenylimino)silane (43) has been obtained by the route shown in Scheme 24. (43) is a sublimable orange crystalline solid which melts without decomposition to a deep red liquid and is very oxygen- and moisture-sensitive.[86] The silaketimine (44), from the reaction between azido-di-t-butylchlorosilane and tri-t-butylsilylsodium in dibutyl ether at −78 °C, forms pale yellow needles. The structure of (44) shows it to have an essentially linear skeleton (cf. the ab initio calculations above) with a Si=N bond distance of 1.568(3) Å.[87]

Tetra-aryldisilenes undergo a facile intramolecular rearrangement involving the exchange of two aryl substituents between the silicon atoms of the Si=Si double bond.[88] Intermediates of the type (45) are indicated from n.m.r. data in the reaction of disilenes with mercury(II) trifluoroacetate.[89] (46) is a synthon for dimethyldisilyne, MeSi≡SiMe.[90]

Low-valent Compounds.—The most significant advance in this area has been the synthesis and characterization by Jutzi of the first bivalent silicon compound which is stable under ordinary conditions. Decamethylsilicocene [bis(pentamethylcyclopentadienyl)silicon(II)] is obtained by reduction of dichloro[bis(pentamethylcyclopentadienyl)silicon(IV)] by naphthalene-lithium, -sodium, or -potassium in THF

[85] P. von R. Schleyer and P. D. Stout, J. Chem. Soc., Chem. Commun., 1986, 1373.
[86] M. Hesse and U. Klingebiel, Angew. Chem., Int. Ed. Engl., 1986, **25**, 649.
[87] N. Wiberg, K. Schurz, G. Reber, and G. Müller, J. Chem. Soc., Chem. Commun., 1986, 591.
[88] H. B. Yokelson, J. Maxka, D. A. Siegel, and R. West, J. Am. Chem. Soc., 1986, **108**, 4239.
[89] C. Zybill and R. West, J. Chem. Soc., Chem. Commun., 1986, 857.
[90] A. Sekiguchi, S. S. Zigler, and R. West, J. Am. Chem. Soc., 1986, **108**, 4241.

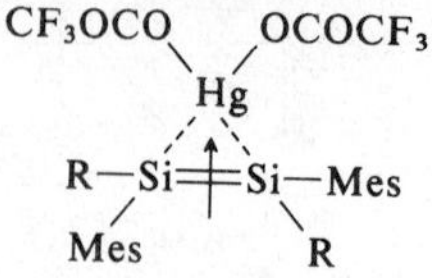

(45) 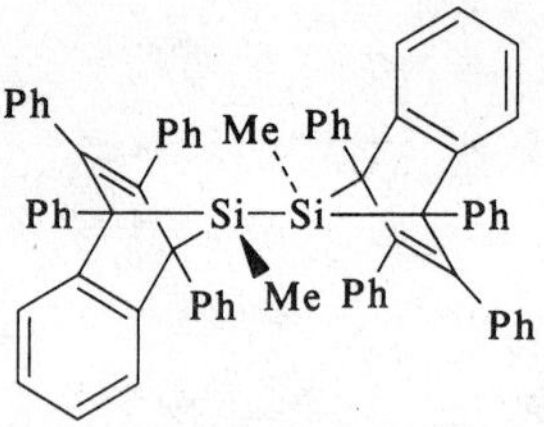   (46)

(Scheme 25; M = Si, X = Cl). The compound is colourless (*cf.* the yellow or orange colours of the heavier homologues), sublimes readily, and melts at 171 °C without decomposition, but is extremely air-sensitive. Surprisingly, two conformers are

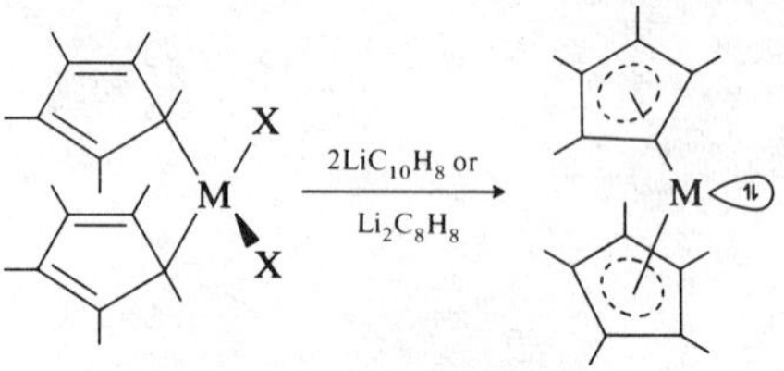

M = Si, Ge, or Sn

Scheme 25

present in the crystal in the ratio 1:2. Both have sandwich structures, but whereas in one the two cyclopentadienyl rings are parallel and staggered, in the other the rings form an interplanar angle of 25.3°, most probably due to intermolecular interactions and crystal packing effects.[91]

Decamethylgermanocene and decamethylstannocene have also been prepared similarly (Scheme 25; M = Ge, Sn).[92] These compounds undergo a wide variety of reactions. Typical reactions of decamethylgermanocene and $Me_5C_5GeCH(SiMe_3)_2$ with electrophiles are shown in Scheme 26 and Scheme 27, respectively, whereas

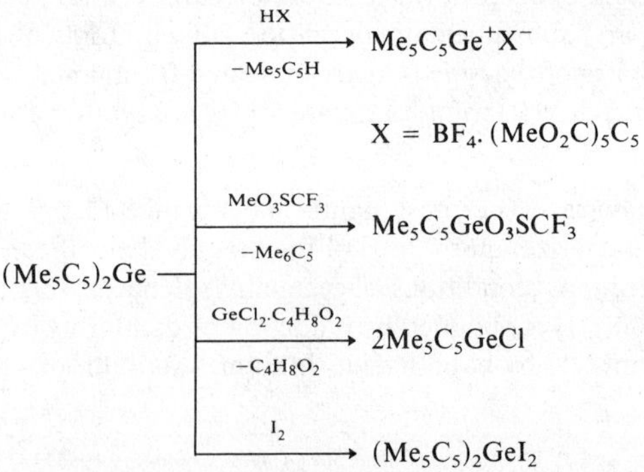

Scheme 26

[91] P. Jutzi, D. Kanne, and C. Krüger, *Angew. Chem., Int. Ed. Engl.*, 1986, **25**, 164.
[92] P. Jutzi and B. Hielscher, *Organometallics*, 1986, **5**, 1201.

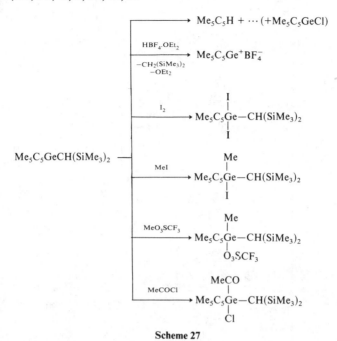

Scheme 27

reactions of Me_5C_5GeCl are shown in Scheme 28.[93,94] The unsymmetrically substituted germylenes exhibit no tendency to rearrange into the symmetrical compounds. X-Ray analysis of $Me_5C_5GeCH(SiMe_3)_2$ shows it to be monomeric with the germanium atom bonded in a *dihapto*-manner to the C_5 ring as in (47). Alkyllithiums react with decamethylstannocene by nucleophilic substitution at the tin atom and the displacement of a C_5Me_5 group. Thus reaction with $(Me_3Si)_2CHLi$ yields Me_5C_5Li and $[(Me_3Si)_2CH]_2Sn$, but reaction with MeLi produces Me_5C_5Li and a mixture of oligomeric stannylenes $[Me_2Sn]_n$. However, trapping experiments demonstrated the initial formation of a short-lived $(\eta^1-Me_5C_5)_2Sn(Me)Li$ intermediate.[95]

Fenske–Hall MO calculations have been reported for stannocene and decaphenylstannocene and predict that the tin lone pair resides in the HOMO, a tin 5s-like orbital, in the latter compound, and is not delocalized onto the ring system of the ligands. Computer graphics show that the experimental structure (parallel rings) minimizes steric repulsion by placing the phenyl rings on the cyclopentadienyl ligands in a nearly perpendicular orientation.[96] Reaction of $GeCl_2$.dioxane with mono-, bis-, and tris-(trimethylsilyl)cyclopentadienyl-lithium yields bis-, tetrakis-, and hexakis-(trimethylsilyl)germanocene, respectively. In the

[93] P. Jutzi and B. Hampel, *Organometallics*, 1986, **5**, 730.
[94] P. Jutzi, B. Hampel, M. B. Hursthouse, and A. J. Howes, *Organometallics*, 1986, **5**, 1944.
[95] P. Jutzi and B. Hielscher, *Organometallics*, 1986, **5**, 2511.
[96] R. L. Williamson and M. B. Hall, *Organometallics*, 1986, **5**, 2142.

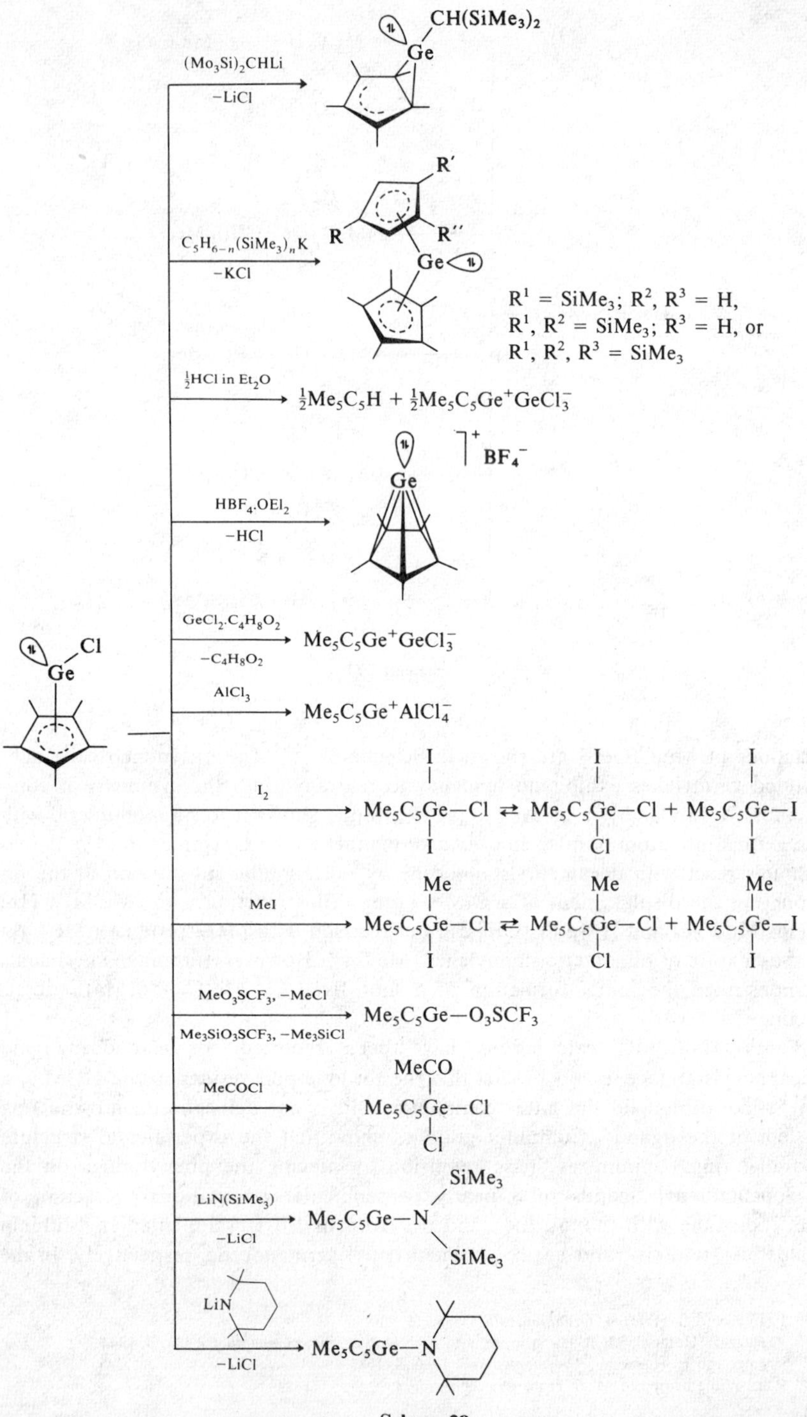

$CH(SiMe_3)_2$

$(Mo_3Si)_2CHLi$
$-LiCl$

$C_5H_{6-n}(SiMe_3)_nK$
$-KCl$

$R^1 = SiMe_3;\ R^2, R^3 = H,$
$R^1, R^2 = SiMe_3;\ R^3 = H,$ or
$R^1, R^2, R^3 = SiMe_3$

$\frac{1}{2}HCl$ in Et_2O

$\frac{1}{2}Me_5C_5H + \frac{1}{2}Me_5C_5Ge^+GeCl_3^-$

$HBF_4 \cdot OEl_2$
$-HCl$

$[\ ]^+ BF_4^-$

$GeCl_2 \cdot C_4H_8O_2$
$-C_4H_8O_2$

$Me_5C_5Ge^+GeCl_3^-$

$AlCl_3$

$Me_5C_5Ge^+AlCl_4^-$

I_2

$$Me_5C_5\overset{\overset{I}{|}}{\underset{\underset{I}{|}}{Ge}}-Cl \rightleftarrows Me_5C_5\overset{\overset{I}{|}}{\underset{\underset{Cl}{|}}{Ge}}-Cl + Me_5C_5\overset{\overset{I}{|}}{\underset{\underset{I}{|}}{Ge}}-I$$

MeI

$$Me_5C_5\overset{\overset{Me}{|}}{\underset{\underset{I}{|}}{Ge}}-Cl \rightleftarrows Me_5C_5\overset{\overset{Me}{|}}{\underset{\underset{Cl}{|}}{Ge}}-Cl + Me_5C_5\overset{\overset{Me}{|}}{\underset{\underset{I}{|}}{Ge}}-I$$

$MeO_3SCF_3, -MeCl$
$Me_3SiO_3SCF_3, -Me_3SiCl$

$Me_5C_5Ge-O_3SCF_3$

$MeCOCl$

$$Me_5C_5\overset{\overset{MeCO}{|}}{\underset{\underset{Cl}{|}}{Ge}}-Cl$$

$LiN(SiMe_3)$
$-LiCl$

$$Me_5C_5Ge-N\overset{\diagup SiMe_3}{\diagdown SiMe_3}$$

$LiN\langle\text{(piperidine)}\rangle$
$-LiCl$

$Me_5C_5Ge-N\langle\ \rangle$

Scheme 28

latter, the two C_5 rings are nearly parallel and have an eclipsed conformation.[97] Decabenzyl-germanocene, -stannocene, and -plumbocene have been synthesized by substitution from GeI_2, $SnCl_2$, or $Pb(O_2CMe)_2$ and pentabenzyl-lithium. All are air-stable and exhibit interplanar angles of between 31° and 36°.[98]

The compound $(\eta^2\text{-Me}_4C_5H)Me_2Si(\eta^5\text{-Me}_4C_5)Ge + GeCl^-$ (48) exhibits the extremely novel feature of an alkene–main Group metal π-type interaction. Colourless crystals of (48) are obtained according to Scheme 29, and the co-ordination sphere of the germanium contains a *pentahapto*-tetramethylcyclopentadienyl ring, the two carbon atoms of the second C_5 ring, and two chlorine atoms of the counterion. An additional contact from the third chlorine atom results in the formation of centrosymmetric dimers. N.m.r. data indicate that the alkene co-ordination is preserved in solution.[99]

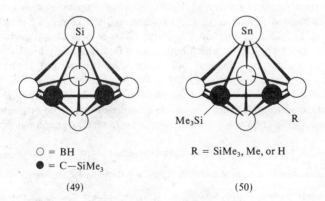

$B = C_4H_8O_2$ (48)

Scheme 29

Several interesting analogous carbaborane compounds have also been described, and again the most intriguing is the bivalent silicon compound $[(Me_3Si)_2C_2B_4H_4]Si$ (49) which is formed together with $[(Me_3Si)_2C_2B_4H_4]_2Si$ in the reaction of $SiCl_4$ with $Na^+Li^+[(Me_3Si)_2C_2B_4H_4]^{2-}$ in THF.[100] Similarly, the *closo*-germarcarbaborane

$\bigcirc$ = BH $R = SiMe_3$, Me, or H
$\bullet$ = $C-SiMe_3$

(49)   (50)

[97] P. Jutzi, E. Schlüter, M. B. Hursthouse, A. M. Arif, and R. L. Short, *J. Organomet. Chem.* 1986, **299**, 285.

[98] H. Schumann, C. Janiak, E. Hahn, C. Kolax, J. Loebel, M. D. Rausch, J. J. Zuckerman, and M. J. Heeg, *Chem. Ber.*, 1986, **119**, 2656.

[99] F. Kohl, R. Dickbreder, P. Jutzi, G. Müller, and B. Huber, *J. Organomet. Chem.*, 1986, **309**, C43.

[100] N. S. Hosmane, P. de Meester, U. Siriwardane, M. S. Islam, and S. S. C. Chu, *J. Chem. Soc., Chem. Commun.*, 1986, 1421.

$[(Me_3Si)_2B_4H_4]Ge,^{101}$ and the stannacarbaboranes $[(Me_3Si)_2C_2B_4H_4]Sn$, $[(Me_3Si)CH_3C_2B_4H_4]Sn$, and $[(Me_3Si)C_2B_4H_5]Sn$ (50) have been obtained. The three stannacarbaboranes form 1 : 1 donor–acceptor complexes with 2,2′-bipyridine, as well as weaker 1 : 2 complexes with THF. The crystal structures of (50) and the bipyridine complex $[(Me_3Si)_2C_2B_4H_4]Sn$.bipy have been determined, which confirm the *closo* structures, and show that the bipyridine group co-ordinates to the tin atom opposite to the carbaborane framework. Unusually, the co-ordination of bipyridine has only a small effect on the Mössbauer parameters, but all the spectroscopic data are consistent with a distorted pentagonal-bipyramidal framework with the tin atom occupying an apical position and bonded exclusively to the three boron atoms of the carbaborane ring.[102]

Lappert[103,104] has published full details of the synthesis, structural determinations, and MO calculations on the compounds $M[CH(SiMe_3)_2]_2$ (M = Ge or Sn). The germanium compound is conveniently prepared by the reaction of $GeCl_2$.dioxane (from the improved synthesis from $GeCl_4$ and Bu_3SnH) and magnesium alkyls. The tin compound is obtained from $SnCl_4$ by successive reaction with two moles of the lithium alkyl and dilithium cyclo-octatetraenide. Electron diffraction shows that both compounds are V-shaped monomers in the vapour with CMC angles of 107(2)° (Ge) and 97(2)° (Sn). In the solid both have the dimetallene centrosymmetric, *trans*-folded M_2R_2 framework, with fold angles of 32° (Ge) and 41° (Sn). The M—M distances in each are slightly shorter than those found in the tetrahedral elements. Double-ζ *ab initio* calculations predict somewhat shorter M—M distances, but show the *trans*-folded structure to be more stable than planar and M—M bond dissociation energies to be about half those found in H_3GeGeH_3 or Me_3MMMe_3.

The preparation and properties of the compounds $M[C(PR_2)_2X)]_2$(M = Ge, Sn, or Pb; X = H, PR_2, or $SiMe_3$) have been reported.[105—107] $Li[CH(PPh_2)_2]$ reacts with $GeCl_2.L$ (L = dioxane or PPh_3), $SnCl_2$, or $PbCl_2$ in THF to afford $M[CH(PPh_2)_2]_2$ (M = Ge, Sn or Pb) which are three-co-ordinate in the solid with one group functioning as a chelating diphosphido-ligand and the other as a uniden-tate carbon-bonded ligand as in (51). In contrast, the complex $Sn[C(PMe_2)_3]_2$ synthesized by similar methods has the four-co-ordinate pseudo-trigonal-bipyramidal geometry (52) in which both groups function as diphosphido-ligands. All the compounds are fluxional in solution. The tin(II) and lead(II) bis-(t-butyl)phosphido 'ate' complexes $Li(THF)[M(PBu^t_2)_3]$ (M = Sn or Pb) (53) have been obtained from $Li (PBu^t_2)$ and the metal(II) chloride in THF. In the crystal both metals are pyramidal with two of the phosphorus atoms co-ordinated to the lithium atom.[108] Reaction of *o*-(diphenylphosphino)phenylmagnesium bromide with

[101] N. S. Hosmane, P. de Meester, U. Siriwardane, M. S. Islam, and S. S. C. Chu, *J. Am. Chem. Soc.*, 1986, **108**, 6050.

[102] N. S. Hosmane, P. de Meester, N. N. Maldar, S. B. Potts, S. S. C. Chu, and R. H. Herber, *Organometallics*, 1986, **5**, 772.

[103] T. Fjeldberg, A. Haaland, B. E. R. Schilling, M. F. Lappert, and A. J. Thorne, *J. Chem. Soc., Dalton Trans.*, 1986, 1551.

[104] D. E. Goldberg, P. B. Hitchcock, M. F. Lappert, K. M. Thomas, A. J. Thorne, T. Fjeldberg, A. Haaland, and B. E. R. Schilling, *J. Chem. Soc., Dalton Trans.*, 1986, 2387.

[105] A. L. Balch and D. E. Oram, *Organometallics*, 1986, **5**, 2159.

[106] H. H. Karsch, A. Appelt, and G. Hanika, *J. Organomet. Chem.*, 1986, **312**, C1.

[107] H. H. Karsch, A. Appelt, and G. Müller, *Organometallics*, 1986, **5**, 1664.

[108] A. M. Arif, A. H. Cowley, R. A. Jones, and J. M. Power, *J. Chem. Soc., Chem. Commun.*, 1986, 1446.

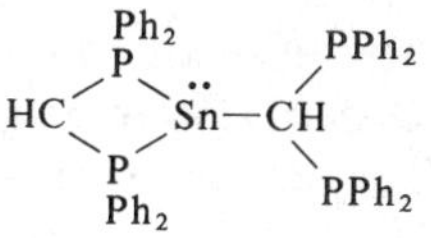

(51) 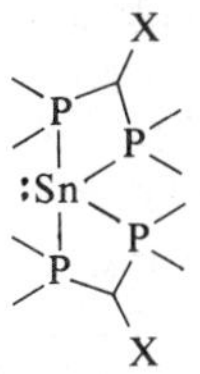   (52) 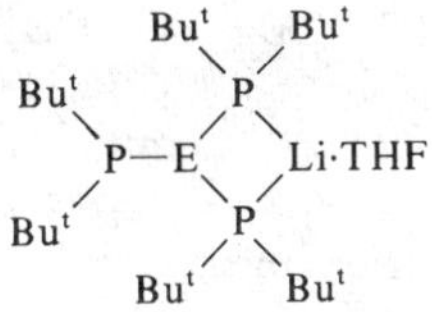   E = Sn or Pb (53)

$SnCl_2$ leads to the formation of the mixed-valence compound (54) (Scheme 30). Treatment of the complex $Cl_2Sn-W(CO)_5$ with the same Grignard reagent produces

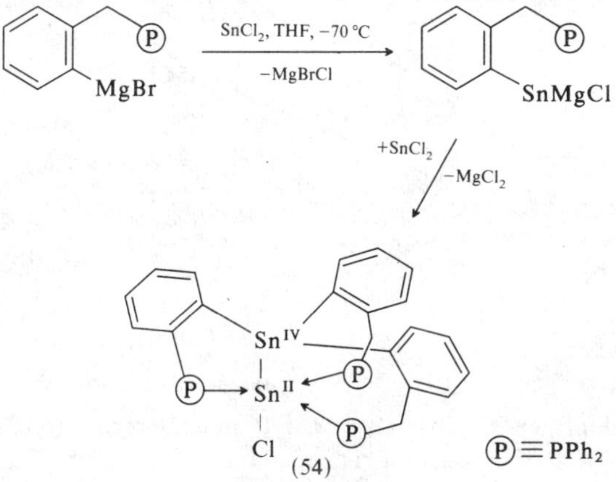

Scheme 30

(55).[109] Several other complexes with transition-metal carbonyl fragments have been described including the germylene complexes $Me_5C_5(R)Ge \rightarrow W(CO)_5$ [R = Cl, Me, $(Me_3Si)_2N$, or $(Me_3Si)_2CH$][110,111] and $L.X_2Ge \rightarrow M(CO)_5$ (M = Cr or W; X = F or Cl; L = nitrone)[112] the stannylene complexes $(CO)_5M \leftarrow$

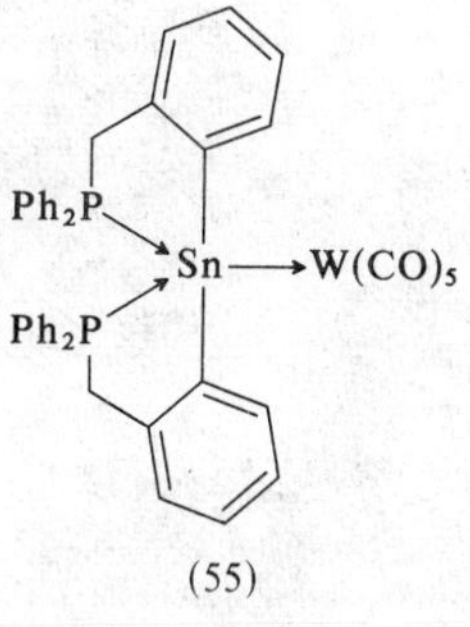

(55)

[109] K. Jurkschat, H.-P. Abicht, A. Tzschach, and R. Mahieu, *J. Organomet. Chem.*, 1986, **309**, C47.
[110] P. Jutzi and B. Hampel, *J. Organomet. Chem.*, 1986, **301**, 283.
[111] P. Jutzi, B. Hampel, M. B. Hursthouse, and A. J. Howes, *J. Organomet. Chem.*, 1986, **299**, 19.
[112] A. Castel, P. Riviere, J. Satge, M. Ahbala, and J. Jaud, *J. Organomet. Chem.*, 1986, **307**, 205.

Sn(XCH$_2$CH$_2$)$_2$E (M = Cr, Mo, or W; X = O or S; E = NR, PPh, O, or S) and (CO)$_5$W ← Sn(XCH$_2$CH$_2$)$_2$E.py (E = NMe, O, or S)[113,114], [(C$_5$H$_5$)-(CO)$_3$]Sn[(R$_2$P)$_2$CX] (M = W, R = Ph, X = PPh; M = Mo, R = Me, X = SiMe$_3$),[115] and the carbonyl-iron complexes *eq*-[Fe(CO)$_4${M(OAr)$_2$}] (M = Ge or Sn; Ar = C$_6$H$_2$Bu$^t_2$-2,6-Me-4).[116] The crystal structures of several have been determined.

Surprisingly, the germanium(II) pseudohalides Ge(CN)$_2$, Ge(NCO)$_2$, and Ge(NCS)$_2$ were unknown prior to their synthesis by the Toulouse group from the corresponding halides and either potassium or silver salts. They are stable in THF and acetone solution but such solutions are extremely sensitive to moisture, and undergo typical cycloaddition, insertion, and Lewis acid–base reactions characteristic of highly reactive germylenes (Scheme 31).[117] The germylenes X$_2$Ge, RGeX, and

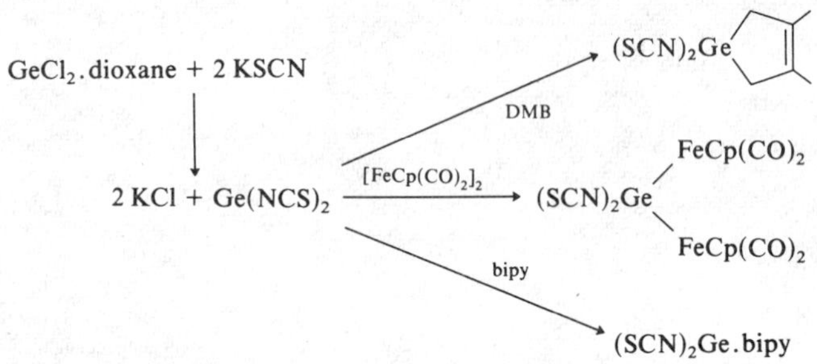

Scheme 31

R$_z$Ge (X = halogen or OR; R = alkyl or aryl) undergo regioselective cycloaddition to 3,5-di-t-butylorthoquinone at room temperature to afford substituted 2-germa-1,3-dioxolanes in good yield (*e.g.* Scheme 32).[118] E.s.r. evidence has been presented for

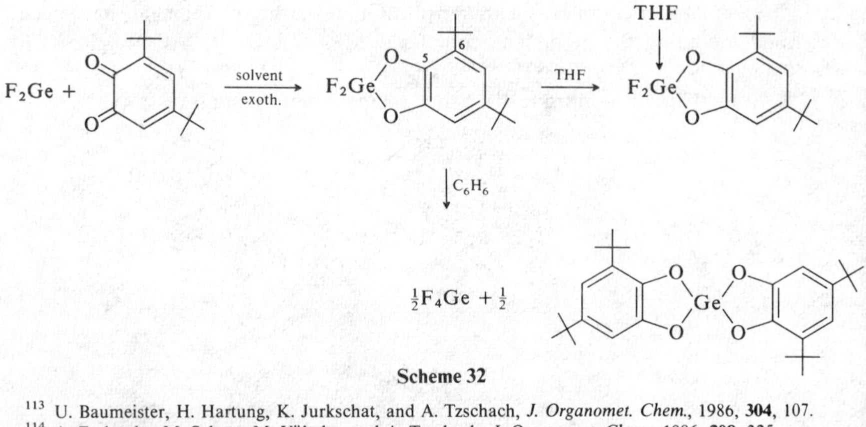

Scheme 32

[113] U. Baumeister, H. Hartung, K. Jurkschat, and A. Tzschach, *J. Organomet. Chem.*, 1986, **304**, 107.
[114] A. Zschunke, M. Scheer, M. Völtzke, and A. Tzschach, *J. Organomet. Chem.*, 1986, **308**, 325.
[115] H. H. Karsch and A. Appelt, *J. Organomet. Chem.*, 1986, **312**, C6.
[116] P. B. Hitchcock, M. F. Lappert, S. A. Thomas, A. J. Thorne, A. J. Carty, and N. J. Taylor, *J. Organomet. Chem.*, 1986, **315**, 37.
[117] M. Onyszchuk, A. Castel, P. Rivere, and J. Satge, *J. Organomet. Chem.*, 1986, **317**, C35.
[118] P. Riviere, A. Castel, J. Satge, and D. Guyot, *J. Organomet. Chem.*, 1986, **315**, 157.

the existence of a radical species in the vapour of tin(II) fluoride.[119] Treatment of the iridium complex $[Ir(n-C_8H_{14})_2(\mu-Cl)_2]$ with $Ge[N(SiMe_3)_2]_2$ in n-hexane at 20 °C in the absence or presence of CO leads to the formation of the complexes (56) and (57), respectively, whose structures were also determined.[120]

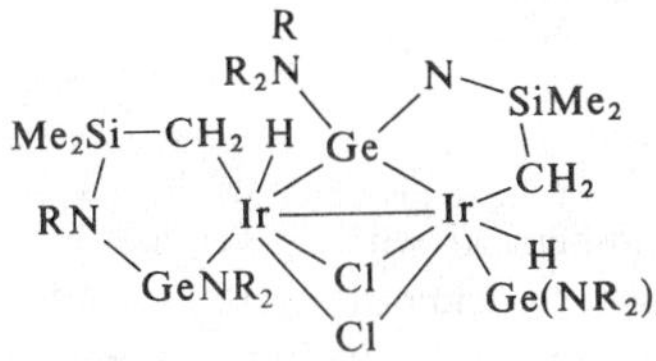

(56)

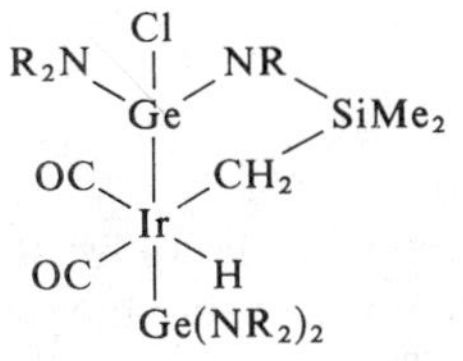

(57)

Monomeric 1,4,2,3,5-λ^2-diazadisila-stannolidines and -plumbolidines (58) have been obtained according to Scheme 33 as thermochromic oily red liquids or orange

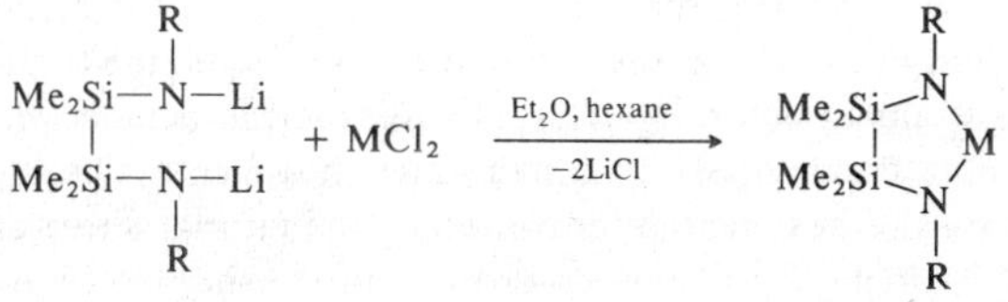

(58)

Scheme 33

solids.[121] Heating the acid–base adduct (59a) to 120 °C in toluene leads to the formation of the pentacyclic compound (59b), which has a centrosymmetric structure with a central $[(CH)_2Sn_2]$ ring. The decomposition follows first-order kinetics, and proceeds *via* the intermediate (60), which, although possessing the same composition as (59a), exhibits a completely different structure and is characterized by an intramolecular hydrogen bond between an *o*-carbon atom of a phenyl group and a nitrogen atom of the four-membered $[SiN_2Sn]$ ring which may be considered responsible for a hydrogen atom transfer.[122]

Strontium and barium bis(t-butoxides) react with tin(II) t-butoxide to afford the mixed alkoxides $[Sn(OBu^t)_3M(OBu^t)_3Sn]$ (M = Sr or Ba) (61). The structure of the

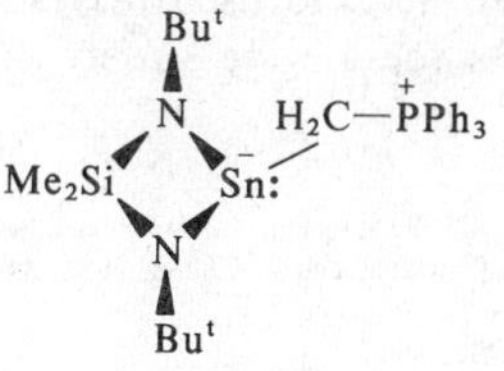

(59a)

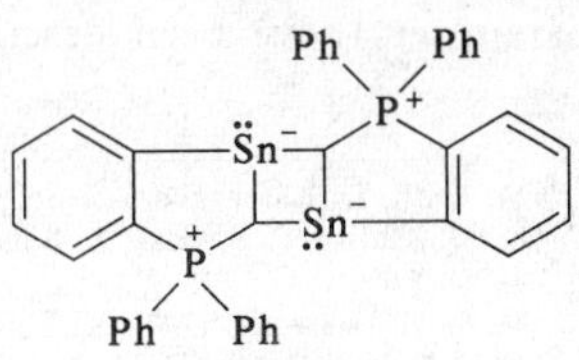

(59b)

[119] D. L. Perry, J. L. Margrave, R. H. Hauge, and P. F. Meier, *Inorg. Chim. Acta*, 1986, **116**, L17.

[120] S. M. Hawkins, P. B. Hitchcock, M. F. Lappert, and A. K. Rai, *J. Chem. Soc., Chem. Commun.*, 1986, 1689.

[121] K. Horchler, C. Stader, and B. Wrackmeyer, *Inorg. Chim. Acta*, 1986, **117**, L39.

[122] M. Veith and V. Huch, *J. Organomet. Chem.*, 1986, **308**, 263.

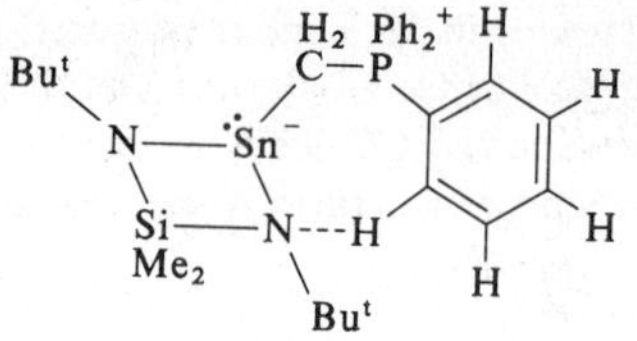

(60)

strontium derivative (61; M = Sr) is highly symmetrical with S_6 symmetry, and comprises a strontium atom sandwiched between two terdentate $[Sn(OBu^t)_3]$ units.[123]

The germanium(II) and tin(II) porphyrin-tetracarbonyliron complexes $(P)MFe(CO)_4$ (M = Ge or Sn) have been synthesized by the reaction of $Na_2Fe(CO)_4$ and the corresponding $(P)M^{IV}Cl_2$ complexes in THF. The complexes were extremely stable after electroreduction, and two reversible one-electron-transfer steps are observed without cleavage of the metal–metal bond. Whereas these reductions occur at the π ring system, irreversible oxidation takes place at the metal centre leading to $[(P)M^{IV}]^{2+}$ and $Fe_3(CO)_{12}$ as the main products.[124] Both the trichlorostannate(II) and perchlorate salts of chloro(1,4,7,10,13,16-hexacyclo-octadecane)tin(II) contain the discrete $[Sn(18$-crown-$6)Cl]^+$ cation in which the tin enjoys hexagonal-pyramidal co-ordination by the hexadentate crown ether (equatorial sites) and the chlorine (axial site).[125] A similar pyramidal co-ordination is exhibited by the lead complex $[PbCl(DAPSC)]^+NO_3$ – (DAPSC = 2,6-diacetylpyridinedisemicarbazone) in which again a chlorine occupies the axial site and the five donor sites of the DAPSC ligand in the equatorial sites.[126] Lead is ten-co-ordinate in the complex $Pb(PHENSC)(NO_3)_2$ (PHENSC = 2,9,-diformyl-1,10-phenanthrolinedisemicarba-zone),[127] and has highly irregular co-ordination polyhedra in the 1,4,7-trizacyclo-nonane (L) complexes $LPb(ClO_4)_2$ and $LPb(NO_3)_2$.[128] Lead(II) tri-t-butoxy-silanethiolate is dimeric in the crystal with a central puckered four-membered $(Pb_2S_2]$ ring. The lead atoms are three-co-ordinate.[129]

The 'inert-pair' effect has a substantial effect on electronegativity. Thus, whereas the electronegativities of germanium(IV), tin(IV) and lead(IV) are 2.62, 2.30, and 2.29, respectively, those of germanium(II), tin(II), and lead(II) are only 0.56, 1.49, and 1.92, respectively.[130]

Molecular Tetravalent Compounds.—This area has presented its usual plethora of data and only a relatively small fraction can be reported here. Revised MNDO parameters have been described for silicon with results for a wide variety of

[123] M. Veith, D. Käfer, and V. Huch, *Angew. Chem., Int. Ed. Engl.*, 1986, **25**, 375.
[124] K. M. Kadish, C. Swistak, B. Boisselier-Cocolios, J. M. Barbe, and R. Guilard, *Inorg. Chem.*, 1986, **25**, 4336.
[125] M. G. B. Drew and D. G. Nicholson, *J. Chem. Soc., Dalton Trans.*, 1986, 1543.
[126] A. E. Koziol, R. C. Palenik, and G. J. Palenik, *Inorg. Chim. Acta*, 1986, **116**, L51.
[127] H. Aghabozorg, R. C. Palenik, and G. J. Palenik, *Inorg. Chim. Acta*, 1986, **111**, L53.
[128] K. Wieghardt, M. Kleine-Boymann, B. Nuber, J. Weiss, L. Zsolnai, and G. Huttner, *Inorg. Chem.*, 1986, **25**, 1647.
[129] W. Wojnowski, M. Wojnowski, K. Peters, E. M. Peters, and H. G. von Schnering, *Z. Anorg. Allg. Chem.*, 1986, **535**, 56.
[130] R. T. Sanderson, *Inorg. Chem.*, 1986, **25**, 1856.

silicon-containing compounds in much better agreement with experiment.[131] MNDO methods have been used to estimate skeletal bending frequencies in linear and quasi-linear silyl compounds.[132] Semi-empirical methods have been employed to examine the ability of carbon, silicon, and germanium to form square-planar geometries. In all cases such planar geometries were found to be less stable than tetrahedral structures.[133]

The primary thermal decomposition processes for both silane and disilane have been calculated using extended basis sets.[134,135] The transition state for the molecular elimination from silane is predicted to be 56.9 kcal mol^{-1} above silane, whereas for the reverse reaction it lies only 2 kcal mol^{-1} over the two fragments, silylene and H_2. Of the four competing unimolecular decomposition pathways, the 1,1- and 1,2-eliminations of H_2 and the elimination of silylene to form silane all have high endothermicities, but the very high activation energy for the 1,2-elimination excludes this process as a significant contributor at low energies. The most likely source of disilene in the thermal decomposition is the rearrangement of its high-energy isomer silylsilylene. *Ab initio* and MNDO calculations have been carried out to evaluate the gas-phase acidity of silane and the affinities of silane and the SiH_3^- anion. Only a marginally stable charge–dipole complex is predicted for SiH_5^-;[136] however, this anion has been synthesized and characterized along with several of its simple alkyl derivatives in a flowing afterglow apparatus at 298 K.[137] Both the SiH_3^- anion and the SiH_3 radical are pyramidal, with inversion barriers of 9000 ± 2000 cm^{-1} and 1900 ± 300 cm^{-1} and HSiH bond angles of 94.5° and 112.5°, respectively. The Si—H bond dissociation energy in silane was estimated to be 90.3 ± 2.4 kcal mol^{-1}.[138] Because of the more favourable electrostatic interactions in the ion pair SiH_3^- Li$^+$, the inverted C_{3v} geometry (62a) is calculated to be 2.4 kcal mol^{-1} more stable than the 'tetrahedral' isomer (62b).[139] All the silicon hydrides, SiH$_n$ ($n = 1$—4), and the

(62a)   (62b)

silyl compounds H$_3$SiX (X = Li, BeH, BH$_2$, CH$_3$, NH$_2$, OH, F, Na, MgH, AlH$_2$ SiH, Ph$_2$, SH, or Cl) have been investigated by *ab initio* methods and compared with the corresponding methyl compounds. In most cases the equilibrium geometries of the methyl and silyl molecules are similar. The most notable exception is in silylamine, where a planar geometry is found at nitrogen. Addition of d-orbital functions to the second-row atoms leads to a decrease in the bond lengths. The relative H$_3$Si-X and H$_3$C—X bond energies depend principally on the electronegativ-

[131] M. J. S. Dewar, J. Friedheim, G. Grady, E. F. Healy, and J. J. P. Stewart, *Organometallics*, 1986, **5**, 375.
[132] C. Glidewell, *Inorg. Chim. Acta*, 1986, **112**, 14.
[133] G. Schultz, R. Lück, and L. Kolditz, *Z. Anorg. Allg. Chem.*, 1986, **532**, 57.
[134] M. S. Gordon, D. R. Gano, J. S. Binkley, and M. J. Frisch, *J. Am. Chem. Soc.*, 1986, **108**, 2191.
[135] M. S. Gordon, T. N. Truong, and E. K. Bonderson, *J. Am. Chem. Soc.*, 1986, **108**, 1421.
[136] M. S. Gordon, L. P. Davis, L. W. Burggraf, and R. Damrauer, *J. Am. Chem. Soc.*, 1986, **108**, 7890.
[137] D. J. Hajdasz and R. R. Squires, *J. Am. Chem. Soc.*, 1986, **108**, 3139.
[138] M. R. Nimlos and G. B. Ellison, *J. Am. Chem. Soc.*, 1986, **108**, 6522.
[139] P. von R. Schleyer and T. Clark, *J. Chem. Soc., Chem. Commun.*, 1986, 1371.

ity of the group X. Since SiH_3 has a higher electron affinity and a lower ionization potential than CH_3, groups which are very electronegative or very electropositive have stronger bonds to silicon than to carbon.[140]

Hexa-t-butyldisilane, formed by reaction of nitrosyl cations with tri-t-butylsilyl-sodium or -potassium, shows an unusually long Si—Si distance of 269.7 pm.[141] The structures of several other catenated silicon, germanium, or tin compounds have also been determined including the α,ω-substituted permethylpolysilanes $R(SiMe_2)_n R$ ($R = (Bu^tO)_3SiS$; $n = 2,3$, or 6),[142] hexa-t-butyl-1,3-di-iodotrisilane,[143] the tetrasilabicyclo[1,1,0]butane (63),[144] the alkali metal silicides K_3LiSi_4 and $K_7Li(Si_4)_3$, which contain tetrahedral $[Si_4]$ structural units,[145] the large-ring cyclo-polysilanes $(Me_2Si)_n$ ($n = 13$ or 16),[146] the linear phenylpolygermanes Ge_3Ph_8, Ge_4Ph_{10} (as a bis-benzene solvate), Ge_5Ph_{12} and $Cl(GePh_2)_nCl$ ($n = 2,3$, or 4),[147—149] and the cyclotetrastannanes, $(Bu_2^tSn)_4$ (planar) and $(t\text{-amyl}_2Sn)_4$ (puckered).[150]

Fluorination of chlorophenyldisilanes with zinc fluoride with silver powder as catalyst yields the corresponding fluorophenyldisilanes. 1,2-Difluorotetraphenyl-disilane can be obtained by u.v. irradiation of bis-(fluorodiphenylsilyl)mercury.[151]

$$R_2Si \qquad SiR_2$$
$$Si \!\!-\!\! Si$$
$$Bu^t \qquad\qquad Bu^t$$

$$R = 2,6\text{-}Et_2C_6H_3 \ (63)$$

The electronic structure of polysilane molecules has been investigated both theoretically and experimentally.[152—154] In poly(di-n-hexyl)silane the polysilane chain is effectively separated into a series of chromophores, which appear to be all-*trans* segments separated by *gauche* links, communicating by rapid energy transfer. Energy band structures have been calculated for polysilane models, $-(SiXY)_n-$ (X, Y = H, Me, Et, Pr, or Ph). Polysilane has a directly allowed type band structure, and a σ'-σ^* interband optical transition is allowed. Band-edge states are formed mainly of skeleton Si atomic orbitals, which results in a skeleton band

[140] B. T. Luke, J. A. Pople, M. B. Krogh-Jespersen, V. Apeloig, J. Chandrasekhar, and P. von R. Schleyer, *J. Am. Chem. Soc.*, 1986, **108**, 260.

[141] N. Wiberg, H. Schuster, A. Simon, and K. Peters, *Angew. Chem., Int. Ed. Engl.*, 1986, **23**, 79.

[142] W. Wojnowski, B. Dreczewski, K. Peters, and H. G. von Schnering, *Z. Anorg. Allg. Chem.*, 1986, **540/541**, 271.

[143] M. Weidenbruch, B. Blintjer, K. Peters, and H. G. von Schnering, *Angew. Chem., Int. Ed. Engl.*, 1986, **23**, 1129.

[144] R. Jones, D. J. Williams, Y. Kabe, and S. Masamune, *Angew. Chem., Int. Ed. Engl.*, 1986, **23**, 173.

[145] H. G. von Schnering, M. Schwarz, and R. Nesper, *Angew. Chem., Int. Ed. Engl.*, 1986, **23**, 566

[146] F. Shafiee, K. J. Haller, and R. West, *J. Am. Chem. Soc.*, 1986, **108**, 5478.

[147] S. Roller, D. Simon, and M. Dräger, *J. Organomet. Chem.*, 1986, **301**, 27.

[148] S. Roller and M. Dräger, *J. Organomet. Chem.*, 1986, **316**, 57.

[149] K. Häberle and M. Dräger, *J. Organomet. Chem.*, 1986, **312**, 155.

[150] H. Puff, C. Bach, W. Schuh, and R. Zimmer, *J. Organomet. Chem.*, 1986, **312**, 313.

[151] E. Hengge and F. Schrank, *J. Organomet. Chem.*, 1986, **299**, 1.

[152] R. W. Bigelow, *Organometallics*, 1986, **5**, 1502.

[153] K. A. Klingensmith, J. W. Downing, R. D. Miller, and J. Michl, *J. Am. Chem. Soc.*, 1986, **108**, 7438.

gap. This gap tends to be reduced when larger alkyl groups are substituted for side chains. The σ' and σ^* band-edge states are well delocalized on the skeleton axis. Poly(phenylsilane) exhibits a characteristic band-edge structure due to σ'-π mixing between the Si skeleton and the phenyl-side chains. Variable-temperature ^{29}Si and ^{13}C CPMAS n.m.r. spectra confirm that poly(di-n-hexyl)silane is in a rigid form at temperatures below 310 K. The observed thermochromism of the u.v. spectrum is due to backbone disordering above the transition temperature resulting from increasing contributions from *gauche* conformations at elevated temperatures.[155]

The novel persilylcyclotrisilane (64) has been obtained by the reaction of 2,2-dibromohexamethyltrisilane with sodium. U.v. irradiation of (64) in the presence of methanol affords (66), presumably *via* the disilene (65) (Scheme 34).[156] Permethylcyclosilanes rearrange in the presence of an Al(Fe)Cl$_3$ catalyst to form isomeric branched cyclopentasilanes or cyclohexasilanes.[157] Calculations show that thermodynamically the preferred isomer is that with the lower steric energy.[158] The reaction of Ph$_4$Ge$_2$Cl$_2$ with Bu$_2^t$Ge(OH)$_2$ leads to the germanium–oxygen heterocycle (67), which has an almost planar [Ge$_3$O$_2$] ring.[159]

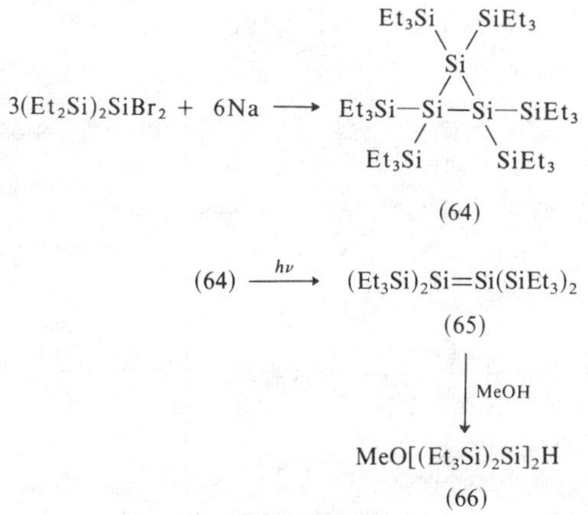

$$3(\text{Et}_2\text{Si})_2\text{SiBr}_2 + 6\text{Na} \longrightarrow$$

$$
\begin{array}{c}
\text{Et}_3\text{Si} \quad \text{SiEt}_3 \\
\diagdown \; / \\
\text{Si} \\
/ \\
\text{Et}_3\text{Si}-\text{Si}-\text{Si}-\text{SiEt}_3 \\
/ \qquad \diagdown \\
\text{Et}_3\text{Si} \qquad \text{SiEt}_3
\end{array}
$$

(64)

$$(64) \xrightarrow{h\nu} (\text{Et}_3\text{Si})_2\text{Si}{=}\text{Si}(\text{SiEt}_3)_2$$

(65)

$$\Big\downarrow \text{MeOH}$$

$$\text{MeO}[(\text{Et}_3\text{Si})_2\text{Si}]_2\text{H}$$

(66)

Scheme 34

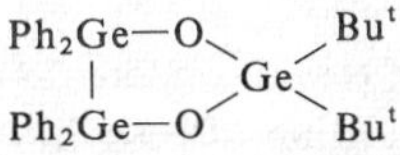

$$
\begin{array}{c}
\text{Ph}_2\text{Ge}-\text{O} \qquad \text{Bu}^t \\
\quad | \qquad\qquad \diagdown\;\diagup \\
\quad\qquad\qquad \text{Ge} \\
\quad | \qquad\qquad \diagup\;\diagdown \\
\text{Ph}_2\text{Ge}-\text{O} \qquad \text{Bu}^t
\end{array}
$$

(67)

The triphenylsilyl cation has been obtained by hydride transfer from the silane to the trityl cation.[160] The structures of two silylmethyl-lithium derivatives have been determined. That of trimethylsilylmethyl-lithium comprises hexameric

[154] K. Takeda, H. Teramae, and N. Matsumoto, *J. Am. Chem. Soc.*, 1986, **108**, 8186.
[155] G. C. Gobbi, W. W. Fleming, R. Sooriyakumaran, and R. D. Miller, *J. Am. Chem. Soc.*, 1986, **108**, 3624.
[156] H. Matsumoto, A. Sakamoto, and Y. Nagai, *J. Chem. Soc., Chem. Commun.*, 1986, 1768.
[157] T. A. Blinka and R. West, *Organometallics*, 1986, **5**, 128.
[158] T. A. Blinka and R. West, *Organometallics*, 1986, **5**, 133.
[159] H. Püff, H. Heisig, W. Schuh, and W. Schwab, *J. Organomet. Chem.*, 1986, **303**, 343.
[160] J. B. Lambert, J. A. McConnell, and W. J. Schulz, *J. Am. Chem. Soc.*, 1986, **108**, 2482.

(LiCH$_2$SiMe$_3$)$_6$ units with two distinct Li—Li distances.[161] The organolithium reagent derived from (MeOMe$_2$Si)$_3$CCl crystallizes as a dimer [{LiC(SiMe$_2$OMe)$_3$}$_2$], in which the Li—C bonds are unusually long and the C—SiMe$_2$ bonds unusually short.[162] Metal vapours of cadmium and zinc react with trifluorosilyl radicals to give bis(trifluorosilyl)-cadmium and -zinc, which were isolated at low temperatures. Both compounds are unstable at room temperature.[163] Treatment of gallium(III) and indium(III) chloride with three equivalents of Li[Si(SiMe$_3$)$_3$].3THF affords the compounds [{(Me$_3$Si)$_3$Si}$_2$M(μ-Cl)$_2$Li(THF)$_2$] (M = Ga or In). The structures of the compounds can be regarded as a double-bridged complex of [{(Me$_3$Si)$_3$Si}$_2$MCl] and solvated LiCl.[164]

The germyl-lithium compound (Me$_3$Si)$_3$GeLi has been prepared from (Me$_2$Si)$_4$Ge, and its reactions are summarized in Scheme 35.[165] The novel bis(η^5-dicarbollide) (68) has been obtained by substitution from SiCl$_4$ and the lithium salt of the carbollide. It is stable in dry air and soluble in most organic solvents. The silicon atom resides on a crystallographic centre of symmetry, being equidistant from the planar parallel faces of the two ligands.[166]

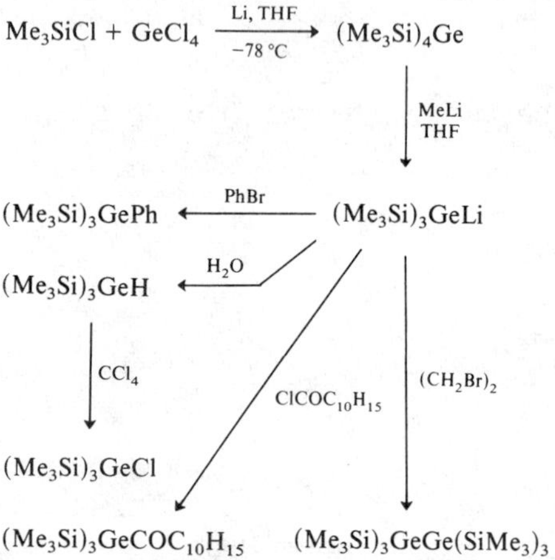

$$\text{Me}_3\text{SiCl} + \text{GeCl}_4 \xrightarrow[-78\,°C]{\text{Li, THF}} (\text{Me}_3\text{Si})_4\text{Ge}$$

Scheme 35

Flash vacuum pyrolysis of (69) with cyclotetra- or cyclopenta-siloxanes leads to products resulting from the insertion of [O=Si=O] (or equivalent synthon) into the Si—O bonds of the siloxanes.[167,168] The synthesis and crystal structures of the cyclodisiloxanes tetramesitylcyclodisiloxane, trans-1,3-dimesityl-1,3-di-t-butylcyclo-disiloxane, and cis-1,3-bis[bis(trimethylsilyl)amino]-1,3-dimesitylcyclodisiloxane

[161] B. Tecle, A. F. M. M. Rahman, and J. P. Oliver, J. Organomet. Chem., 1986, **317**, 267.
[162] N. H. Buttrus, C. Eaborn, S. H. Gupta, P. B. Hitchcock, J. D. Smith, and A. C. Sullivan, J. Chem. Soc., Chem. Commun., 1986, 1043.
[163] M. A. Guerra, T. R. Bierschenk, and R. J. Lagow, J. Am. Chem. Soc., 1986, **108**, 4103.
[164] A. M. Arif, A. H. Cowley, T. M. Elkins, and R. A. Jones, J. Chem. Soc., Chem. Commun., 1986, 1776.
[165] A. G. Brook, F. Abdesaken, and H. Söllradl, J. Organomet. Chem., 1986, **299**, 9.
[166] W. S. Rees, D. M. Schubert, C. B. Knobler, and M. F. Hawthorn, J. Am. Chem. Soc., 1986, **108**, 5369.
[167] C. D. Juengst, W. P. Weber, and G. Manuel, S. Roller, J. Organomet. Chem., 1986, **308**, 187.
[168] G. K. Henry, D. R. Dowd, R. Bau, G. Manuel, and W. P. Weber, Organometallics, 1986, **5**, 1818.

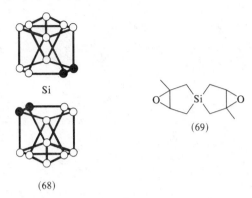

(68)

(69)

have been reported. Whereas the [Si_2O_2] unit in the former and latter compounds is slightly puckered, that in *trans*-1,3-dimesityl-1,3-di-t-butylcylodisiloxane is planar. The Si—Si non-bonding distances in all three compounds are short and of the order of normal Si—Si bond distances, and were considered to arise from antibonding interactions between the oxygen atoms.[169]

Tetraphenyldisiloxane crystallizes at 298 K in the monoclinic $P2_1/n$ space group but undergoes a second-order phase transition at 200 K to a triclinic phase with an almost unchanged structure. At 298 K the Si—O—Si bridge is bent with an angle of 160° with static or dynamic disorder of the bridging oxygen atom.[170] The Si—O—Si fragment in the *R,S*-diastereoisomer of [(C_5H_5)(CO)$_2$FeSiCH$_2$F]$_2$O is linear even at 120 K, but the data indicate severe disorder.[171] Hexa(tri-t-butoxy)disiloxane also has a linear Si—O—Si skeleton, but the corresponding disilthiane is bent.[172] The product of the reaction of Me_2SiCl_2 with 1,2-dihydroxybenzene, bis(*o*-phenyl-enedioxy)dimethylsilane, is dimeric with a ten-membered [$Si_2C_4O_4$]ring.[173] Siloxy cage compounds such as (70) have been synthesized and characterized as models for silica-supported transition-metal catalysts.[174]

The question as to why cyclodisilazane rings are more stable than cyclodisiloxanes has been examined by simple MO theory. In the latter, high silicon $3p_z$ orbital contribution to the siloxane HOMO prevents any strengthening of the Si—O bond by silicon $3d$ orbitals. In contrast, considerable Si($3d_\pi$)-N($2p_\pi$) bonding may occur in the disilazanes, which is responsible for their relative stability.[175] The structure of chlorosilyl-*N,N*-dimethylamine has been determined in both the gaseous and solid phases. In the gas phase the molecule is monomeric with four-co-ordinate silicon, but crystals comprise well separated dimers.[176] Disilazanes have been

[169] M. J. Michalczyk, M. J. Fink, K. J. Haller, R. West, and J. Michl, *Organometallics*, 1986, **5**, 531.
[170] W. Wojnowski, K. Peters, E. M. Peters, T. Meyer, and H. G. von Schnering, *Z. Anorg. Allg. Chem.*, 1986, **537**, 31.
[171] W. Ries, T. Albright, J. Silvestre, I. Bernal, W. Malisch, and C. Burschka, *Inorg. Chim. Acta*, 1986, **111**, 119.
[172] W. Wojnowski, W. Bochenska, K. Peters, E. M. Peters, and H. G. von Schnering, *Z. Anorg. Allg. Chem.*, 1986, **533**, 165.
[173] J. A. Hawari, H. J. Gabe, F. L. Lee, M. Lesage, and D. Griller, *J. Organomet. Chem.*, 1986, **299**, 279.
[174] F. J. Feher, *J. Am. Chem. Soc.*, 1986, **108**, 3850.
[175] I. Silaghi-Dumitrescu and I. Haiduc, *Inorg. Chim. Acta*, 1986, **112**, 159.
[176] D. G. Anderson, A. J. Blake, S. Cradock, E. A. V. Ebsworth, D. W. H. Rankin, and A. J. Welch, *Angew. Chem., Int. Ed. Engl.*, 1986, **23**, 107.

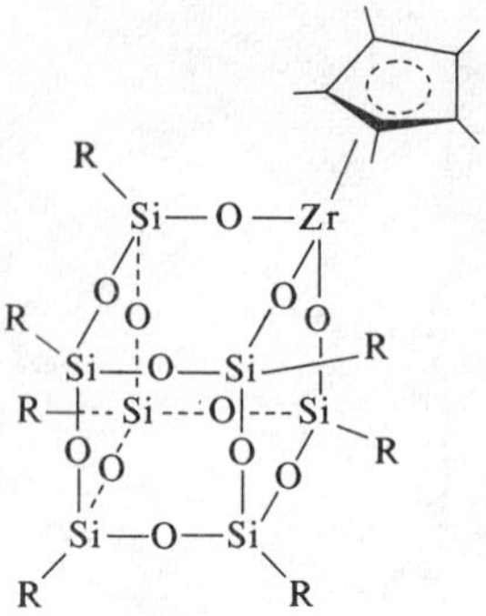

(70)

obtained by deprotonation of dimethylhydroaminosilanes by a mechanism thought to involve an intermediate silaimine species.[177] The [Si_2N_2] ring in ($Ph_2SiNC_6F_5$)$_2$ is planar but not square. The pentafluorophenyl groups are twisted by *ca.* 160° from the plane of the four-membered ring.[178] Cyclosilazane cations have been obtained by treatment of the disilazane with aluminium(III) chloride (Scheme 36).[179] A variety of homogeneous and heterogeneous catalysts promote the ring-opening oligomerization of octamethylcyclotetrasilazane. Low pressures of hydrogen (1 atm) also enhance transition-metal catalysed ring opening by up to two orders of magnitude. The metal hydrides which are thus formed are also active catalysts in the absence of hydrogen.[180]

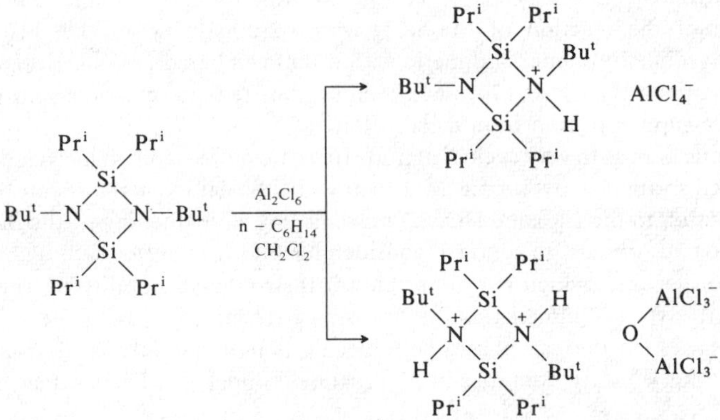

Scheme 36

N.m.r. continues to prove an exceptionally useful tool for structural investigations, especially with the advent of solid-state techniques. The combined technique of

[177] G. H. Wiseman, D. R. Wheeler, and D. Seyferth, *Organometallics*, 1986, **5**, 146.
[178] P. Clare, D. B. Sowerby, and I. Haiduc, *J. Organomet. Chem.*, 1986, **310**, 161.
[179] U. Kliebisch, U. Klingebiel, D. Stalke, and G. M. Sheldrick, *Angew. Chem., Int. Ed. Engl.*, 1986, **23**, 915.
[180] Y. Blum and R. M. Laine, *Organometallics*, 1986, **5**, 2081.

high-power decoupling, cross-polarization, and magic-angle rotation have been applied to ^{207}Pb n.m.r. of solid organolead compounds.[181] ^{13}C MAS n.m.r. has been used to study organotin compounds, and is extremely useful for structural elucidation of powders[182] and the study of polymorphism.[183] The structures of dimethyltin diacetate and its hydrolysis product [Me$_2${Sn(OAc)}$_2$O]$_2$ have been studied by both solid-state and solution n.m.r. Coupling constant data suggest a value of *ca.* 135° for the C—Sn—C bond angle in Me$_2$Sn(OAc)$_2$, and ^{13}C spectra indicate the occurrence of a second crystalline modification. The distannoxane forms dimeric units with a central [Sn$_2$O$_2$] ring, with an overall polymeric structure formed by an additional bonding interaction between an exocyclic tin atom and an oxygen atom of a carboxylate group of adjacent dimers. Unlike other similar molecules, no centre of symmetry is present. Solution n.m.r. data indicate the presence of two types of tin atom, both of which are six-co-ordinated and adopt distorted *trans*-dimethyl-octahedral conformations (C—Sn—C angles *ca.* 141° and 145°).[184] The two-bond $^2J(^{119}$Sn-^{1}H) coupling constant in methyltin compounds has been shown to be related (by a smooth curve) to the C—Sn—C bond angle;[185] similarly the dependence of the one-bond $^1J(^{119}$Sn-^{13}C) coupling constant to the C—Sn—C bond angle in butyltin compounds has been demonstrated.[186]

Several other solution n.m.r. studies are worthy of mention. A one-bond ^{117,119}Sn-^{14}N coupling has been observed for the first time (in tricyclohexyltin isothiocyanate).[187] Tetra-alkoxygermanes have been investigated by ^{13}C, ^{17}O, and ^{73}Ge n.m.r. The ^{73}Ge signals are more sensitive to structure variations than those of ^{29}Si in isostructural alkoxysilanes.[188] The ^{119}Sn chemical shift and one-bond $^1J(^{119}$Sn-^{13}C) coupling constant can be used semi-quantitatively to describe the shape of the co-ordination polyhedra about tin in dibutyltin compounds and their complexes. Values of the chemical shift define regions with different co-ordination numbers, so that four-co-ordinate compounds have $\delta(^{119}$Sn) ranging from +200 to −60 p.p.m, five-co-ordinate compounds −90 to −190 p.p.m., and six-co-ordinate compounds −210 to −400 p.p.m. Magnitudes of the coupling constants can be used to estimate the C—Sn—C angle.[189] Similar conclusions have also been arrived at by French workers in studies of bisalkoxytin derivatives of the type R$_3$SnO—A—OSnR$_3$.[190] The combination of ^{119}Sn n.m.r. and ^{119}Sn Mössbauer spectroscopies can also be powerful,[191-192] and the application of multinuclear (^{119}Sn, ^{205}Tl, and ^{207}Pb) n.m.r. has enabled the characterization of the anionic clusters formed by dissolving alkali metal alloys of Ge, Sn, Sb, Tl, Pb, and Bi in liquid ethylenediamine.[193] Mass

[181] J. R. Ascenso, R. K. Harris, and P. Granger, *J. Organomet. Chem.*, 1986, **301**, C23.

[182] T. Lockhart and W. F. Manders, *Inorg. Chem.*, 1986, **25**, 1068.

[183] T. Lockhart and W. F. Manders, *Inorg. Chem.*, 1986, **25**, 583.

[184] T. Lockhart, W. F. Manders, and E. M. Holt, *J. Am. Chem. Soc.*, 1986, **108**, 6611.

[185] T. Lockhart and W. F. Manders, *Inorg. Chem.*, 1986, **25**, 892.

[186] J. Holocek and A. Lycka, *Inorg. Chim. Acta*, 1986, **118**, L15.

[187] K. C. Molloy, K. Quill, S. J. Blunden, and R. Hill, *J. Chem. Soc., Dalton Trans.*, 1986, 875.

[188] E. Liepins, I. Zicmane, and E. Lukevics, *J. Organomet. Chem.*, 1986, **306**, 327.

[189] J. Holocek, M. Nadvornik, K. Handlik, and A. Lycka, *J. Organomet. Chem.*, 1986, **315**, 299.

[190] C. Picard, P. Tisnes, and L. Cazaux, *J. Organomet. Chem.*, 1986, **315**, 277.

[191] S. J. Blunden, D. Searle, and P. J. Smith, *Inorg. Chim. Acta*, 1986, **119**, L31.

[192] B. King, H. Eckert, D. Z. Kenney, and R. H. Herber, *Inorg. Chim. Acta*, 1986, **122**, 45.

[193] W. L. Wilson, R. W. Rudolph, L. L. Lohr, R. C. Taylor, and P. Pyykko, *Inorg. Chem.*, 1986, **25**, 1535.

spectroscopy has been employed to characterize neutral analogues of these anionic clusters in the gas phase such as Pb_3Sb_2, Pb_5Sb_4, and Sn_5Bi_4.[194]

Crystals of dicyclohexyltin dichloride and dibromide (isomorphous)[195] and methyltin tri-iodide[196] comprise discrete tetrahedral molecules. Theoretical MO calculations indicate a small back-donation from the iodine atoms to tin. Lewis acidities of triorganotin halides R_3SnX (R = Me, Et, Pr, Bu, or Ph; X = Cl, Br, or I)[197] organotin iodides SnI_4, R_2SnI_2 (R = Me, Ph, or $PhCH_2$), and R_3SnI (R = Ph or $PhCH_2$), and di- and tri-benzyltin chlorides[198] have been estimated by calorimetric and n.m.r. methods. Acidities decrease slightly as the size of the alkyl group increases and increase as the size of the halogen increases. ^1H T_1 measurements have been made with dimethyltin dichloride and its complex with bipyridyl in mixed solvents of dichloromethane and weak bases to elucidate the role of the solvent in the dynamic behaviour of organotin compounds. In the lineshape analysis of the exchange reaction between Me_2SnCl_2 and its bipyridyl complex, the dissociation rate constant of the complex supports an exchange mechanism of dissociation followed by a recombination step. Data indicate more extensive solvation in the activated state than in the ground state of the complex.[199]

Crystallography continues to play a very important role in the elucidation of the more subtle structural features of tin compounds. The large number reported in the past year precludes more than a brief mention of most. In some cases crystallographic data are used to aid the interpretation of spectroscopic data. Unique is the trinuclear methylchlorotin anion, $[Sn_3Me_6Cl_8]^{2-}$, present in crystals of the salt $[DBTTF]_3[Sn_3Me_6Cl_8]^{2-} \cdot PhCN$ (DBTTF = dibenzotetrathiafulvalene), in which each tin atom is six-co-ordinate and chlorine-bridged.[200] 1,2-Dichlorotetramethyldistannane, $ClMe_2SnSnMe_2Cl$, forms a tin–tin connected double helical structure.[201] Five co-ordination at tin has been confirmed in $Cl_3SnCH_2CH_2CO_2Pr^i$,[202] $(NCS)Ph_3Sn(2-O_2CC_5H_4N) \cdot H_2$,[203] 2-$Me_2NC_6H_4CH(SiMe_3)SnMePhBr$,[204] ($p$-tolyl)$_3SnBr \cdot$(quinoline N-oxide),[205] {3-[t-butyl(phenyl)phosphinol]propyl}dimethyltin chloride,[206] and $Me_2ClSn[(1-pyrazolyl)_2BH_2]$,[207] and six-co-ordination is present in $SnCl_2Br_2(DMSO)_2$,[208] cis-(p-tolyl)$SnCl_2 \cdot 2,2'$-bipyridyl,[209] and (for the

[194] R. G. Wheeler, K. Laihing, W. J. Wilson, J. D. Allen, R. B. King, and M. A. Duncan, J. Am. Chem. Soc., 1986, 108, 8101.
[195] P. Ganis, G. Valle, D. Furlani, and G. Tagliavini, J. Organomet. Chem., 1986, 302, 165.
[196] J. S. Tse, M. J. Collins, F. L. Lee, and E. J. Gabe, J. Organomet. Chem., 1986, 310, 169.
[197] J. N. Spencer, R. B. Beiser, S. R. Moyer, R. E. Haines, M. A. DiStravalo, and C. H. Yoder, Organometallics, 1986, 5, 118.
[198] Y. Farhangi and D. P. Graddon, J. Organomet. Chem., 1986, 317, 153.
[199] F. Sakai, H. Fujiwara, and Y. Sasaki, J. Organomet. Chem., 1986, 310, 293.
[200] K. Shimizu, G. E. Matsubayashi, and T. Tanaka, Inorg. Chim. Acta, 1986, 122, 37.
[201] S. Adams, M. Dräger, and B. Kathiasch, Z. Anorg. Allg. Chem., 1986, 532, 81.
[202] R. A. Howie, E. S. Paterson, J. L. Wardell, and J. W. Burley, J. Organomet. Chem., 1986, 304, 301.
[203] E. J. Gabe, F. L. Lee, L. E. Khoo, and F. E. Smith, Inorg. Chim. Acta. 1986, 112, 41.
[204] J. T. B. H. Jastrzebski, G. van Koten, C. T. Knaap, A. M. M. Schreurs, J. Kroon, and A. L. Spek, Organometallics, 1986, 5, 1551.
[205] V. G. Kumar Das, Y. C. Keong, N. S. Weng, C. Wei, and T. C. W. Mak, J. Organomet. Chem., 1986, 311, 289.
[206] H. Weichmann, J. Meunier-Piret, and M. van Meerssche, J. Organomet. Chem., 1986, 309, 267.
[207] S. K. Lee and B. K. Nicholson, J. Organomet. Chem., 1986, 309, 257.
[208] D. Tudela, V. Fernandez, J. D. Tornero, and A. Vegas, Z. Anorg. Allg. Chem., 1986, 532, 215.
[209] V. G. Kumar Das, C. Wei, Y. C. Keong, and T. C. W. Mak, J. Organomet. Chem., 1986, 299, 41.

Sn^{IV} atoms) Sn$^{II}_2$SN$^{IV}_2$F$_4$(O$_2$CCF$_3$)$_8$.2CF$_3$CO$_2$H (the Sn^{II} sites have square-pyramidal geometries.[210]

The synthesis and structures of several organotin carboxylates have been reported Di-t-butyltin bis(carboxylates) have been synthesized from di-t-butyltin oxide trimer and the appropriate acid, and are monomeric and five-co-ordinate except for the picolinate which is six-co-ordinate.[211] Tin is tetrahedrally co-ordinated in tricyclohexylstannyl 3-indolylacetate, but an S-shaped polymer is present in the crystal, propagated by intermolecular NH···OC hydrogen bonds.[212] Of the two chlorobenzoates, Ph$_3$Sn(2-O$_2$CC$_6$H$_4$Cl) is a five-co-ordinate monomer in the crystal whereas Ph$_3$Sn(4-O$_2$CC$_6$H$_4$Cl) is a one-dimensional polymer,[213] as are 2-Me$_3$SnO$_2$CC$_6$H$_4$OMe and 2-Me$_3$SnO$_2$CC$_6$H$_4$OH.[214] Bis(trimethyltin) carbonate has a polymeric structure resulting from the terdentate mode of co-ordination of the carbonate dianion.[215]

Other structures which have been described include 2,2-dibutyl-1,3,2-dioxastannolane (associated into an infinite ribbon structure).[216] [SnBu$_3$(OH$_2$)$_2$]-[C$_5$(CO$_2$Me)$_5$],[217] Bu$_2$Sn[O$_2$C$_3$HPh$_2$]$_2$,[218] {SnPh$_3$NO$_3$).dpaoe [dpaoe = 1,2-bis(diphenylarsonyl)ethane] and SnPh$_3$(NO$_3$)$_2$.dppom [dppom = bis(diphenylphosphoryl)methane],[219] Ph$_3$PbO$_2$CCO$_2$Me,[220] Ph$_3$Sn(S$_2$COPr^i),[221] [Bu$^t_2$Sn(ebdtc)]$_2$.4THF [ebdtc = ethylenebis(dithiocarbamate)],[222] [Me$_2$SnS$_2$P(Ph)CH$_2$CH$_2$].HMPT,[223] (C$_5$H$_5$)$_2$USnPh$_3$ (the first example of a uranium–tin bond),[224] (C$_5$H$_5$)$_2$TaH$_2$MeCl$_2$,[225] the osmium–tin clusters Os$_3$(μ-H)$_2$(CO)$_{10}$(SnMe$_3$)$_2$[226] and Os$_3$SnCl$_2$(CO)$_{11}$(μ-CH$_2$),[227] and [Et$_4$N]-[Pb{Fe(CO)$_4$}$_2${Fe$_2$(CO)$_8$}].[228]

3 Nitrogen

Pyrolysis of phenyl azide at 500 °C affords phenylnitrene (71), also obtained on photolysis of matrix-isolated phenyl azide, and identified by e.s.r. Phenylnitrene

210 T. Birchall and V. Manivannan, *J. Chem. Soc., Chem. Commun.*, 1986, 1441.
211 S. Dietzel, K. Jurschat, A. Tzschach, and A. Zschunke, *Z. Anorg. Allg., Chem.*, 1986, **537**, 163.
212 K. C. Molloy, T. G. Purcell, E. Hahn, H. Schumann, and J. J. Zuckerman, *Organometallics*, 1986, **5**, 85.
213 R. R. Holmes, R. O. Day, V. Chandrasekhar, J. F. Vollano, and J. M. Holmes, *Inorg. Chem.*, 1986, **25**, 2490.
214 P. J. Smith, R. O. Day, V. Chandrasekhar, J. M. Holmes, and R. R. Holmes, *Inorg. Chim. Acta*, 1986, **118**, 2495.
215 E. R. T. Tiekink, *J. Organomet. Chem.*, 1986, **302**, C1.
216 A. G. Davies, A. J. Price, H. M. Dawes, and M. B. Hursthouse, *J. Chem. Soc., Dalton Trans.*, 1986, 297.
217 A. G. Davies, J. P. Goddard, M. B. Hursthouse, and N. P. C. Walker, *J. Chem. Soc., Dalton Trans.*, 1986, 1873.
218 G. Poll, C. J. Cheer, and W. H. Nelso, *J. Organomet. Chem.*, 1986, **306**, 347.
219 S. Dondi, M. Nardelli, C. Pelizzi, G. Pelizzi, and G. Predieri, *J. Organomet. Chem.*, 1986 **308**, 195.
220 A. Glowacki, F. Huber, and H. Preut, *J. Organomet. Chem.*, 1986, **306**, 9.
221 E. R. T. Tiekink and G. Winter, *J. Organomet. Chem.*, 1986, **314**, 85.
222 O. S. Jung, Y. S. Sohn, and J. A. Ibers, *Inorg. Chem.* 1986, **25**, 2273.
223 H. Weichmann, J. Meunier-Piret, and M. van Meerssche, *J. Organomet. Chem.*, 1986, **309**, 273.
224 M. Porchia, U. Casellato, F. Ossola, G. Rossetto, P. Zanella, and R. Graziana, *J. Chem. Soc., Chem. Commun.*, 1986, 1034.
225 T. M. Arkhireeva, B. M. Bulychev, A. N. Protsky, G. L. Soloveichik, and V. K. Belsky, *J. Organomet. Chem.*, 1986, **317**, 33.
226 F. W. B. Einstein, R. K. Pomeroy, and A. C. Willis, *J. Organomet. Chem.*, 1986, **311**, 257.
227 N. Viswanathan, E. D. Morrison, G. L. Geoffroy, S. J. Geib, and A. L. Rheingold, *Inorg. Chem.*, 1986, **25**, 3100.
228 C. B. Lagrone, K. H. Whitmire, M. R. Churchill, and J. C. Fettinger, *Inorg. Chem.*, 1986, **25**, 1080.

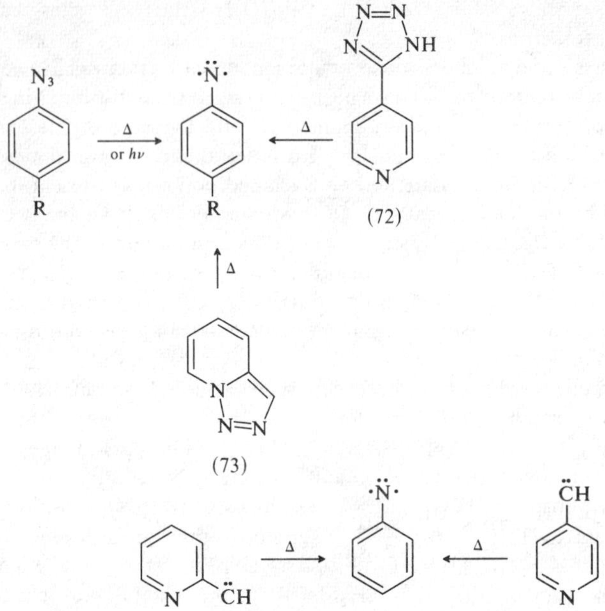

Scheme 37

spectra were also observed on pyrolysis of the carbene precursors (72) and (73), thereby verifying the carbene → nitrene rearrangements (72) → (71) and (73) → (71) (Scheme 37).[229] The prevailing reaction pathway in the photolysis of phenyl azide at ambient temperature is different from that at 77 K, owing to a temperature-dependent branching ratio from the primary photoproduct, singlet phenylnitrene. The photoproduct observed by absorption spectroscopy at 77 K is predominantly triplet phenylnitrene.[230]

Methyl nitrile ylide (74) can be generated by both photochemical and desilylation routes, and undergoes smooth cycloaddition with a variety of dipolarophiles in the presence of silver fluoride (Scheme 38).[231] Nitrogen dichloride reacts with trialkylphosphines to afford trialkyldichlorophosphoranes, R_3PCl_2, and with PCl_3 to yield PCl_5.[232] Cleavage of aminostannanes $Me_3SnN(SO_2R)_2$ with sulphonyl chlorides gives trisulphonyl amines, $N(O_2SR)(O_2SR')_2$. Both $N(O_2SEt)_3$ and $HN(O_2SEt)_2$ are planar at nitrogen.[233] N-Alkylbis(difluorophosphoryl)amides, $RN(POF_2)_2$, react with silylated nucleophiles such as Me_3SiOMe and Me_3NEt_2 with cleavage of the PNP bridge to form derivatives of di- and mono-fluorophosphoric acid.[234] The structures of both N,N'-diformyl- and N,N'-dithioformylanilines (obtained by the reaction of N,N'-dichloromethylaniline with SiS_2) have been

[229] M. Kuzaj, H. Lüessen, and C. Wentrup, *Angew. Chem., Int. Ed. Engl.*, 1986, **25**, 480.
[230] E. Leyva, M. S. Platz, G. Persey, and J. Wirz, *J. Am. Chem. Soc.*, 1986, **108**, 3783.
[231] A. Padwa, J. R. Gasdaska, M. Tomas, N. J. Turro, Y. Cha, and I. A. Gould, *J. Am. Chem. Soc.*, 1986, **108**, 6739.
[232] R. M. Kren and H. H. Sisler, *Inorg. Chim. Acta*, 1986, **111**, 31.
[233] A Blaschette, E. Wieland, D. Schomburg, and M. Adelhelm, *Z. Anorg. Allg. Chem.*, 1986, **533**, 7.
[234] L. Riesel and D. Sturm, *Z. Anorg. Allg. Chem.*, 1986, **539**, 187.

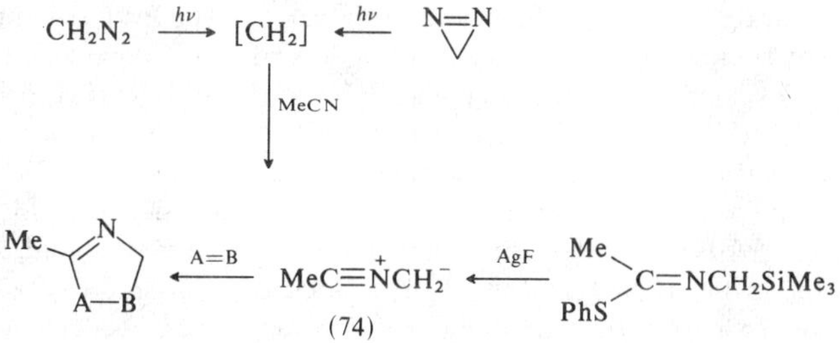

$$CH_2N_2 \xrightarrow{h\nu} [CH_2] \xleftarrow{h\nu} N{=}N$$

MeCN

$$\text{Me–C(N)=CH–A—B} \xleftarrow{A=B} MeC{\equiv}\overset{+}{N}CH_2^{-} \xleftarrow{AgF} \underset{PhS}{\overset{Me}{>}}C{=}NCH_2SiMe_3$$

(74)

Scheme 38

described. In both, the plane of the aromatic ring is not coplanar with the rest of the molecule.[235,236]

Some of the previously reported chemistry of tetramethylammonium superoxide has been found to be erroneous. The purported dimer is actually a peroxide adduct of acetonitrile [$MeC(OO^-){=}NH$], which hydrolyses to the base adduct of acetamide [$MeC(O^-)(OH)NH_2$]. The previously reported synthetic procedure can also yield substantial amounts of Me_4NOH and Me_4NOOH, and a better procedure for the preparation which gives a purer product is by the reaction of $Me_4NOH.H_2O$ or $(Me_4N)_2CO_3$ with KO_2.[237] The fluoromethylammonium cations have been obtained by the quaternization of the corresponding amines. In the cations $Me_3NCHF_2^+$ and $Me_3NCF_3^+$ the $N{-}CH_3$ distances are longer than the $N{-}CF$ distances.[238]

The kinetics of the reaction of dissolved nitric oxide with sulphite and bisulphite ions has been studied over a pH range of 4—10 by monitoring the reaction product, N-nitrosohydroxylamine N-sulphate.[239] The rate-determining step in the reduction of nitric oxide by hydroxylamine and the substituted hydroxylamines MeNHOH, NHOMe and MeNHOMe is the abstraction of a nitrogen-bound hydrogen by NO to form HNO. Trace amounts of oxygen catalyse the H_2NON reaction. In the case of the N,N-dialkyl compounds R_2NOH (R = Me or Et), Abstraction occurs at the α-carbon. Reaction is inhibited by O-methylation since no reactive anion can form.[240,241]

4. Phosphorus and Arsenic

Both $O{=}P{-}Cl$ and $O{=}P{-}F$ have been characterized.[242,243] The former is formed by the pyrolysis of allyl dichlorophosphite at 1150 K, and the latter is obtained by the reaction between $P(O)FBr_2$ and silver. Matrix i.r. data and MO calculations have been reported for $O{=}P{-}F$[243] and PO_2Cl[244] with usually good agreement. Two

[235] E. Allenstein, F. J. Hofmann, and H. Riffel, *Z. Anorg. Allg. Chem.*, 1986, **534**, 7.
[236] E. Allenstein, F. J. Hofmann, and H. Riffel, *Z. Anorg. Allg. Chem.*, 1986, **534**, 13.
[237] K. Yamaguchi, T. S. Calderwood, and D. T. Sawyer, *Inorg. Chem.*, 1986, **25**, 1289.
[238] D. J. Brauer, H. Bürger, M. Grunwald, G. Pawelke, and J. Wilke, *Z. Anorg. Allg. Chem.*, 1986, **537**, 63.
[239] D. Littlejohn, K. Y. Hu, and S. G. Chang, *Inorg. Chem.*, 1986, **25**, 3131.
[240] F. T. Bonner and N. Y. Wang, *Inorg. Chem.*, 1986, **25**, 1858.
[241] N. Y. Wang and F. T. Bonner, *Inorg. Chem.*, 1986, **25**, 1863.
[242] M. Binnewies, H. Schnöckel, R. Gereke, and R. Schmutzler, *Z. Anorg. Allg. Chem.*, 1986, **534**, 143.
[243] R. Ahlrichs, R. Becherer, M. Binnewies, H. Borrmann, M. Lakenbrink, S. Schunck, and H. Schnöckel, *J. Am. Chem. Soc.*, 1986, **108**, 7905.
[244] R. Ahlrichs, C. Ehrhardt, M. Lakenbrink, S. Schunck, and H. Schnöckel, *J. Am. Chem. Soc.*, 1986, **108**, 3596.

theoretical studies of H_2P_2 isomers have appeared.[245,246] The $P=P$ π-bond strength determined by computing the rotational barrier (36 kcal mol^{-1}) compares with values of 63.5 and 47 kcal mol^{-1} for the analogues $HN=NH$ and $HN=PH$, respectively. The ground state in the H_2P_2 system is *trans*-HP=PH, which lies 3.5 kcal mol^{-1} below *cis*-HP=PH. The $H_2P=P$ isomer lies at a relative energy of 28 kcal mol^{-1} above the ground state. For the ground state, the highest occupied MO is n_x, a symmetric lone-pair combination; the second HOMO is the P—P π-bond. In $H_2P=P$ a lone pair n_y is the HOMO, and the P—P π-bond is the second HOMO. In both molecules, the LUMO is the antibonding P—P π^*-orbital. SCF calculations on the molecules $HPCH_2$, *trans*-HPNH, HPO, $HP(NH)_2$, and HPO_2 show essential differences. HPX_2 structures are heteroallyl systems with strongly dipolaric character rather than with multiple bonding at phosphorus. The bis(imino)phosphoranes prefer a *cis* conformation of the hydrogens (at the nitrogens) owing to the smaller dipole moment in comparison with its *trans-isomer*. The *cis-trans* conformational equilibrium depends on the substituents.[247] The electronic nature of white phosphorus, cyclotriphosphane and tetraphosphabicyclobutane has been investigated by *ab initio* calculations at a double-ζ level with the inclusion of electron correlation, as well as the p.e. spectrum of tetraphosphabicyclobutane. In this compound the central bond has almost the same *p* character as the peripheral bonds. In contrast to bicyclobutane, tetraphosphabicyclobutane and cyclotriphosphane are almost free of ring strain.[248] The detailed molecular structure of conformers of methyl phosphinate, the simplest ester of the four-co-ordinate phosphorus oxyacids, has been determined by *ab initio* methods.[249] Calculations on the model cyclophosphazenes $(H_2PN)_2$ and $(H_2PN)_3$ satisfactorily reproduce known experimental data, including the X-ray structure of $[(Pr^i_2N)_2PN]_2$. The cyclodimerization of two H_2PN units results in an energy benefit of 80 kcal mol^{-1}, and the cyclotrimerization process is beneficial by 210 kcal mol^{-1}.[250] Local minima on the singlet and triplet $[CH_3P]$ potential energy surfaces correspond to methylphosphinidene, phosphaethene, and phosphinocarbene. The global minimum is singlet phosphaethene.[251]

The geometric structures of mixed chlorofluorophosphoranes PCl_nF_{6-n} ($n = 1$—4) have been determined by gas-phase electron diffraction. Both P—F and P—Cl bond distances increase with increasing fluorine substitution, the axial bond being generally affected more. Angular distortions are small.[252] Pure $AsCl_4F$ has been obtained and is much more stable than $AsCl_2F_3$ or $AsClF_4$.[253] Tetrahydrofuran undergoes ring opening with phosphorus trihalides and $POCl_3$ in the presence of mercury(II) halides as catalysts.[254] The solvolysis of methylphosphorus(V) compounds proceeds faster in 100% H_2SO_4 than in 25 oleum, and for P—Cl than for P—Br bonds. The supposed 'compounds' $PMeCl_2Br_2$ and PMe_2ClBr_2 actually

[245] M. W. Schmidt and M. S. Gordon, *Inorg. Chem.*, 1986, **25**, 248.
[246] T. L. Allen, A. C. Scheiner, Y. Yamaguchi, and H. F. Schaeffer, *J. Am. Chem. Soc.*, 1986, **108**, 7579.
[247] W. W. Schoeller and C. Lerch, *Inorg. Chem.*, 1986, **25**, 576.
[248] W. W. Schoeller, V. Staemmier, P. Rademacher, and E. Nieke, *Inorg. Chem.*, 1986, **25**, 4382.
[249] J. R. Van Wazer and C. S. Ewig, *J. Am. Chem. Soc.*, 1986, **108**, 4354.
[250] G. Trinquier, *J. Am. Chem. Soc.*, 1986, **108**, 568.
[251] M. T. Nguyen, M. A. McGinn, and A. F. Hegarty, *Inorg. Chem.*, 1986, **25**, 2185.
[252] C. Macho, R. Minkwitz, J. Rohmann, B. Steger, V. Wölfel, and H. Oberhammer, *Inorg. Chem.*, 1986, **25**, 2828.
[253] R. Minkwitz and H. Prenzel, *Z. Anorg. Allg. Chem.*, 1986, **534**, 150.
[254] P. Peringer and P. P. Winkler, *Inorg. Chim. Acta*, 1986, **118**, L1.

comprise mixtures of the corresponding halogenophosphonium cations $[PMeCl_nBr_{3-n}]^+$ ($n = 0$—3) and $[PMe_2Cl_nBr_{2-n}]^+$ ($n = 0$—2) in these solvents.[255] Tetrabasic 2,2,3,3-tetrafluoro butane-1,4-diyl bis(phosphate) and 2,2,3,3,4,4,5,5-octafluorohexane-1,6-diyl bis(phosphate) and monobasic 5,5,6,6-tetrafluoro-2-hydroxy-1,3,2-dioxaphosphane 2-oxide have been obtained by the controlled hydrolysis of the appropriate bis(phosphorodichloridates) and chloro-5,5,6,6-tetrafluoro-1,3,2-dioxaphosphane 2-oxide.[256] Six-co-ordinate phosphorus is present in $F(CF_3)_3PO_2CNMe_2$. The unique fluorine atom lies in the molecular plane containing that fluorine, one CF_3 group, the phosphorus centre, and the planar carbamate ligand.[257]

Several novel complexes incorporating naked phosphorus or arsenic atoms or small clusters of atoms have been described. The complex anion $[\{Fe_3(CO)_{10}\}P\{Fe(CO)_4\}]$, in which the phosphorus atom caps a $Fe_3(CO)_{10}$ group and is also bonded to a $Fe(CO)_4$ fragment, is obtained by the reaction of PCl_3 with the $[Fe_4(CO)_{13}]$ anion.[258] A more general route to complexes of this type are obtained by the reaction of phosphorus(III) halides co-ordinated to transition-metal centres with dinuclear carbonyls such as $Co_2(CO)_8$ and $Fe_2(CO)_9$. Using this method, complexes such as (75)—(77) have been isolated. An alternative approach is by

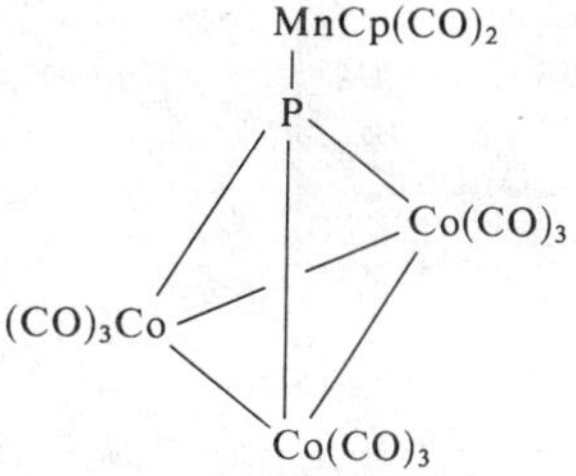

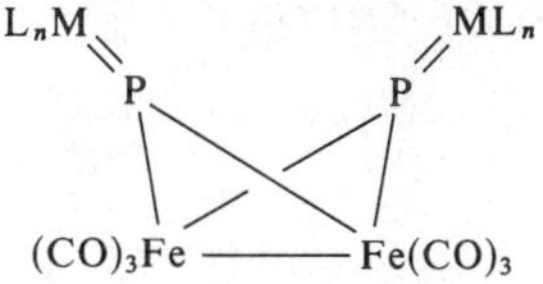

substitution at phosphorus or arsenic with a carbonyl anion by which complexes such as (78)—(84) were obtained. X-Ray structures of several of the complexes have been determined.[259,260] A redox reaction occurs when $[FeCl_2(PPh_3)_2]$ reacts

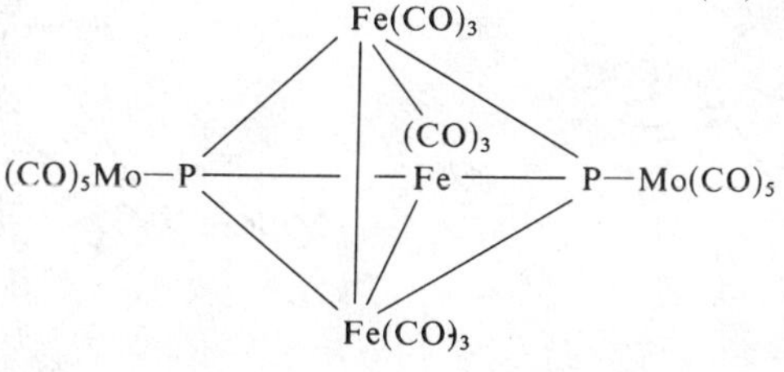

[255] R. M. K. Deng and K. B. Dillon, *J. Chem. Soc., Dalton Trans.*, 1986, 1443.
[256] T. Mahmood and J. M. Shreeve, *Inorg. Chem.*, 1986, **25**, 4081.
[257] R. G. Cavell and L. V. Griend, *Inorg. Chem.*, 1986, **25**, 4699.
[258] A. Gourdon and Y. Jeannin, *J. Organomet. Chem.*, 1986, **304**, C1.
[259] H. Lang, G. Hüttner, B. Sigwarth, I. Jibril, L. Zsolnai, and O. Orama, *J. Organomet. Chem.*, 1986, **304**, 137.
[260] H. Lang, G. Hüttner, B. Sigwarth, L. Zsolnai, G. Mohr, B. Sigwarth, U. Weber, O. Orama, and I. Jibril, *J. Organomet. Chem.*, 1986, **304**, 157.

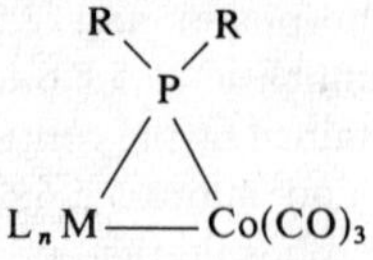

$L_n M = (C_5H_5)(CO)_2Mn$ or
$(MeC_5H_4)(CO)_2Mn$

(78)

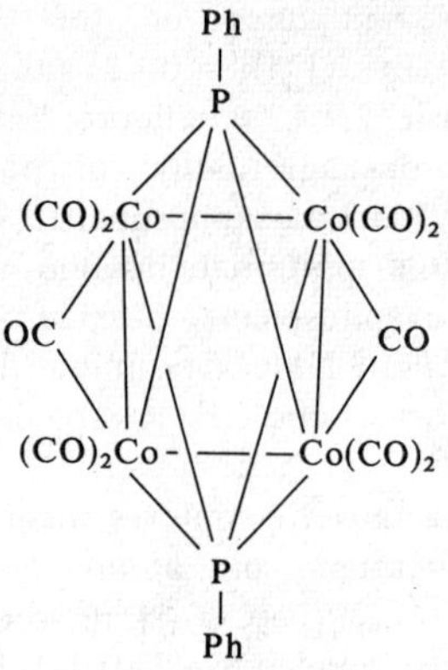

(79)

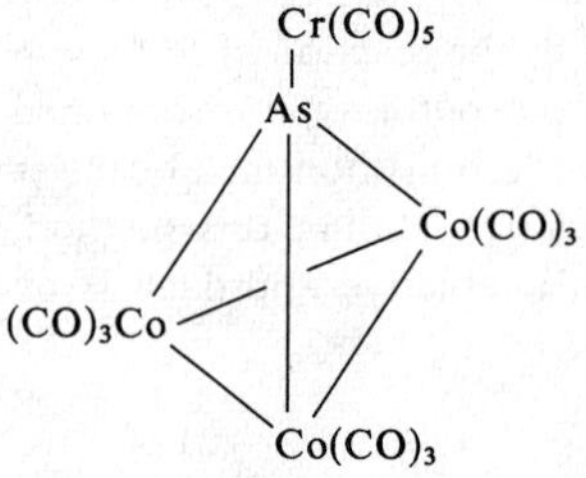

(80)

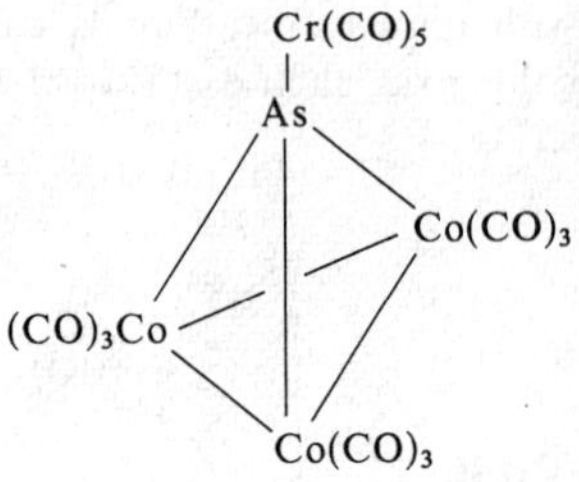

(81)

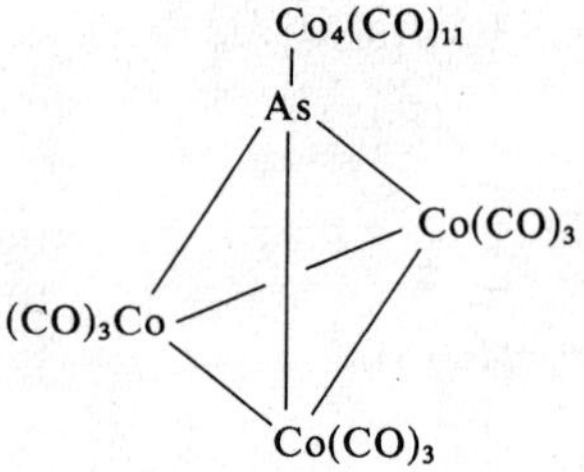

(82)

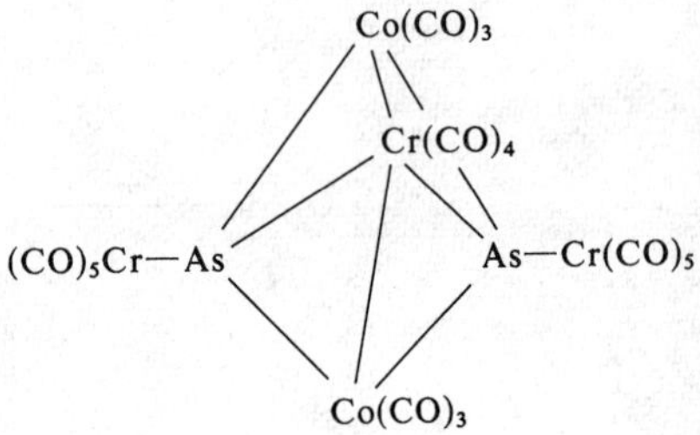

(83)

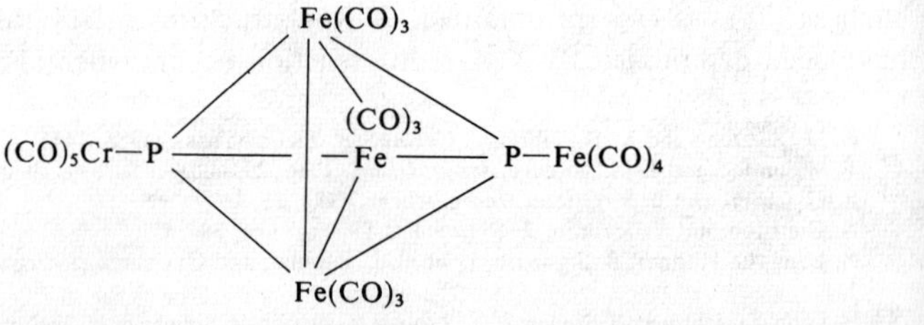

(84)

with PhAs(SiMe$_3$)$_2$, affording the complex [(Co$_4$(μ_3-As)$_3$(μ_3,η^3-AS$_3$)(PPh$_3$)$_4$], which contains two types of [As$_3$] ligand.[261] Direct reaction of elemental P$_4$ or As$_4$ with Cr≡Cr or Mo≡Mo triply bonded complexes also yields complexes of naked Group V atom clusters. Treatment of the complex (85) with white phosphorus at 140 °C in xylene affords the triple-decker complex (86) besides the [η^3-P$_3$] complex (87) (Scheme 39).[262] Similar reaction of the Mo≡Mo complex (88) with As$_4$ under the same conditions gives three products (Scheme 40).[263] A linear η-phosphabutadiene-cobalt complex [Co(Ph$_2$PCH$_2$P(Ph)$_2$P$_4$P(Ph)$_2$CH$_2$PPh$_2$)]BF$_4$ results from treatment of cobalt(II) tetrafluoroborate hexahydrate with white phosphorus in the presence of bis(diphenylphosphino)methane.[264] Co-ordinated (η^3-P$_3$) ligands can be methylated by trimethyloxonium tetrafluoroborate or methyl trifluoromethanesulphonate.[265]

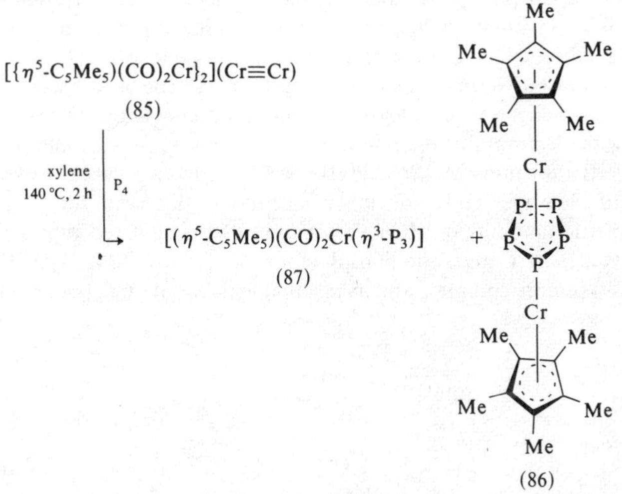

[{η^5-C$_5$Me$_5$)(CO)$_2$Cr}$_2$](Cr≡Cr)

(85)

xylene
140 °C, 2 h   P$_4$

[(η^5-C$_5$Me$_5$)(CO)$_2$Cr(η^3-P$_3$)]

(87)

(86)

Scheme 39

[Cp*(CO)$_4$Mo$_2$](Mo≡Mo)
(Cp* = η^5-C$_5$Me$_5$)

As$_4$, xylene
ca. 140 °C C, ca. 30h

→ [Cp*(CO)$_2$Mo(η^3-As$_3$)]

→ [Cp$_2^*$(CO)$_4$Mo$_2$(μ,η^2-As$_2$)]

→ cis-[Cp*(CO)Mo(μ,η^2-As$_2$)]$_2$

Scheme 40

[261] D. Fenske and J. Hachgenei, *Angew. Chem., Int. Ed. Engl.,* 1986, **25**, 175.
[262] O. J. Scherer, J. Schwalb, G. Wolmershäuser, W. Kaim, and R. Gross, *Angew. Chem., Int. Ed. Engl.,* 1986 **25**, 363.
[263] O. J. Scherer, H. Sitzmann, and G. Wolmershäuser, *J. Organomet. Chem.* 1986, **309**, 77.
[264] F. Cecconi, C. A. Ghilardi, S. Midollini, and A. Orlandini, *Inorg. Chem.,* 1986, **25**, 1766.
[265] G. Capozzi, L. Chiti, M. Di Vaira, M. Peruzzini, and P. Stoppioni, *J. Chem. Soc., Chem. Commun.,* 1986, 1799.

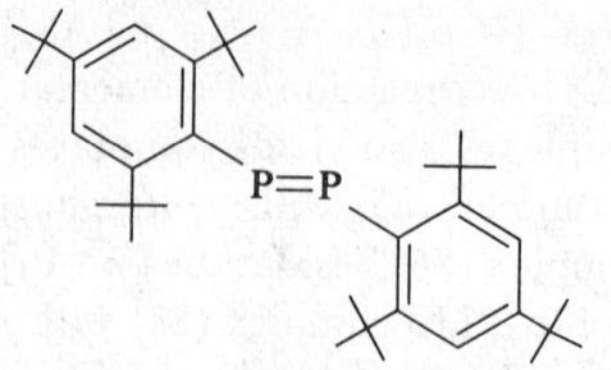

(89)

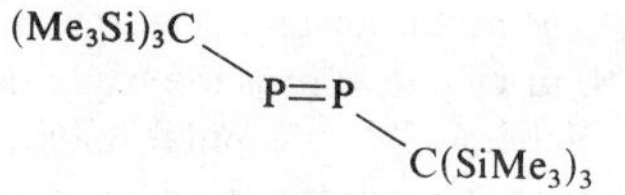

(90)

Interest in the chemistry of diphosphenes continues unabated. A new general route for their preparation *via* the dehydrohalogenation of the germyl phosphines $RP(H)GeCl_3$, using an excess of 1,5-diazabicyclo[5.4.0]undec-5-ene] (DBU), has been devised. By this method, stable diphosphenes such as bis(2,4,6-tri-t-butylphenyl)diphosphene (89) and bis[bis(trimethylsilyl)methyl]diphosphene (90) are isolable in excellent yields, whilst unstable dienophilic diphosphenes can be characterized by cycloaddition with 1,3-dienes. The cycloadduct (91) formed by the reaction of di-t-butyldiphosphene with cyclopentadiene is a clean precursor of di-t-butyldiphosphene.[266] Cowley[267] has published details of the reactivity of (89) and (90) with electrophilic and nucleophilic reagents. A stoicheiometric quantity of HCl converts (90) into $(Me_3Si)_3CP(H)-P(Cl)(SiMe_3)_3$, whereas excess results in P—P bond cleavage. Only the latter reaction is observed for (89) even with stoicheiometric amounts of HCl. Protonation of the silylated diphosphene using $HBF_4.Et_2O$ at −78 °C gives the phosphonium salt $[PH_3\{C(SiMe_3)_3\}][BF_4]$. Protonation of (89) is more complex, and its reaction with $Ag[SO_3CF_3]$ or $[Au(PEt_3)][PF_6]$

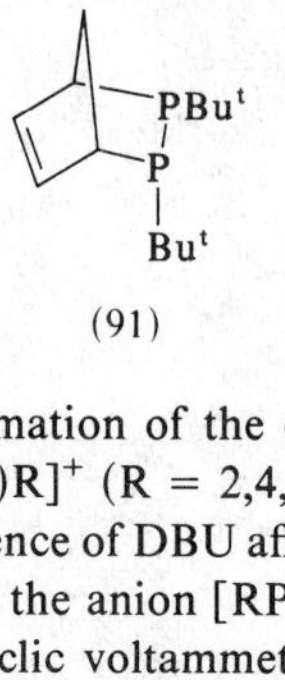

(91)

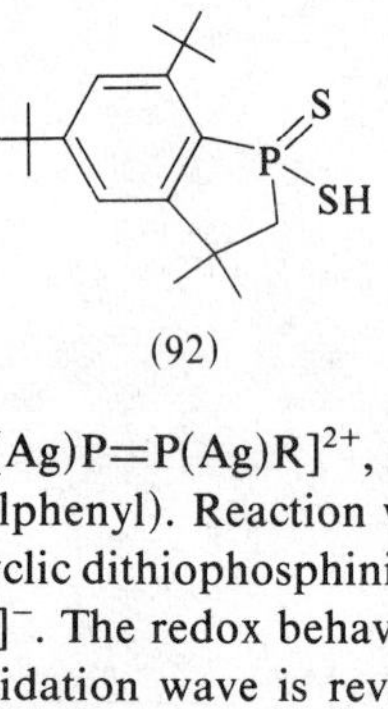

(92)

leads to the formation of the cations $[R(Ag)P=P(Ag)R]^{2+}$, $[R(Ag)P=PR]^+$, and $[RP=P(AuPEt_3)R]^+$ (R = 2,4,6-tri-t-butylphenyl). Reaction with an excess of sulphur in the presence of DBU affords the cyclic dithiophosphinic acid (92); treatment with MeLi gives the anion $[RP-P(Me)R]^-$. The redox behaviour of (90) has been examined by cyclic voltammetry. The oxidation wave is reversible even at a fast sweep wave and no cation radical could be detected. In contrast, a mono-radical anion, characterized by e.s.r., was observed from a rapid reversible one-electron reduction.[268] Chromium, molybdenum, and tungsten carbonyl complexes of less stable diphosphenes may be obtained by the reaction of the decacarbonyl metallates $Na_2M_2(CO)_{10}$ (M = Cr, Mo, or W) with dichlorophosphines. The P=P double bond in the resulting trinuclear complexes undergoes addition reaction with acetic acid or methanol and [2 + 1] cycloaddition reactions to yield complexes with cyclic, three-membered $[P_3]$, $[P_2N]$, or $[P_2S]$ ligands. *X*-Ray structures of several of these

[266] C. Couret, J. Escudie, H. Ranaivonjatovo, and J. Satge, *Organometallics*, 1986, **5**, 112.
[267] A. H. Cowley, J. E. Kilduff, N. C. Norman, and M. Pakulski, *J. Chem. Soc., Dalton Trans.*, 1986, 1801.
[268] M. Culcasi, G. Gronchi, J. Escudie, C. Couret, L. Pujol, and P. Tordo, *J. Am. Chem. Soc.*, 1986, **108**, 3130.

complexes have been described.[269,270] X-Ray crystallography has been used to identify unequivocally the stages of the iron carbonyl-induced carbonylation of (89) to afford the P-metallated diphosphinomethanone (93) (Scheme 41).[271] Dimethylarsine reacts irreversibly with Me_2AsNMe_2 and its borane complex to afford high yields of $Me_2AsAsMe_2$.[272] Tetramethyldiarsine undergoes exchange with tetramethyldiphosphine to afford the unsymmetrical phosphinoarsine, Me_2AsPMe_2. Several other unsymmetrical derivatives including $Me_2AsSbMe_2$, $Me_2SbBiMe_2$, and the mixed Group V–Group VI compounds $Me_2M^1M^2Me$) (M^1 = P, As, Sb, or Bi; M^2 = S, Se, or Te) can be obtained similarly.[273,274]

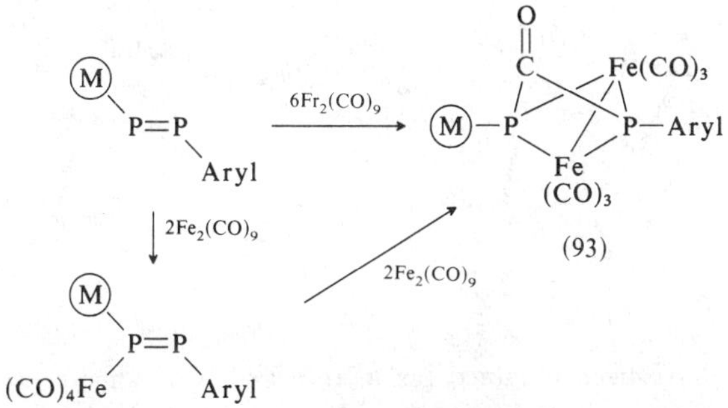

Aryl = $2,4,6\text{-}Bu^tC_6H_2$; M = $C_5Me_5(CO)_2Fe$ or $C_5Me_5(CO)_2Ru$

Scheme 41

Linear silylated triphosphanes have been obtained by first reacting PCl_3 with the trimethylsilylated phosphine $P(SiMe_3)R$ to afford $R(Me_3)PPCl_2$, which is then treated with the lithiated phosphine $LiP(SiMe_3)R^1$.[275] Reaction of the fully silylated triphosphane, $(Me_3Si)_2PP(SiMe_3)P(SiMe_3)_2$, with Bu^tPCl_2 affords *cis*-$P_4(SiMe_3)Bu^t$.[276] The cyclotetraphosphanes $P_4Bu^t_3(SiMe_3)$ and *trans*-$P_4Bu^t_2(SiMe_3)_2$ react with MeLi and Bu^nLi in THF *via* cleavage of a trimethylsilyl group leaving the [P_4] ring intact. *cis*-$P_4Bu^t_2(SiMe_3)_2$ and P_4 rings with a higher degree of silylation, however, react differently, undergoing P—P bond cleavage to produce primary n-tetraphosphides which rearrange even at low temperature in THF to form the corresponding secondary n-tetraphosphides.[277]

The question as to the possible stability of hexaphosphabenzene has been addressed by *ab initio* calculations. Cyclic P_6 is predicted to be 6 kcal mol^{-1} above the free $3P_2$ fragments and to decompose to the latter with an energy barrier of 13 kcal mol^{-1}. It was suggested that the molecule may be stabilized by low-temperature matrix isolation techniques.[278] The first polycyclophosphane oxides

[269] G. Hüttner, J. Borm and L. Zsolnai, *J. Organomet. Chem.*, 1986, **304**, 309.
[270] J. Borm, G. Hüttner, and O. Orama, *J. Organomet. Chem.*, 1986, **306**, 29.
[271] L. Weber, K. Reizig, and R. Boese, *Angew. Chem., Int. Ed. Engl.*, 1986, **25**, 755.
[272] V. K. Gupta, L. K. Krannich, and C. L. Watkins, *Inorg. Chem.*, 1986, **25**, 2553.
[273] A. J. Ashe and E. G. Ludwig, *J. Organomet. Chem.*, 1986, **303**, 197.
[274] A. J. Ashe and E. G. Ludwig, *J. Organomet. Chem.*, 1986, **308**, 289.
[275] G. Fritz and K. Stoll, *Z. Anorg. Allg. Chem.*, 1986, **538**, 78.
[276] G. Fritz and K. Stoll, *Z. Anorg. Allg. Chem.*, 1986, **538**, 115.
[277] G. Fritz and K. Stoll, *Z. Anorg. Allg. Chem.*, 1986, **539**, 65.
[278] M. T. Nguyen and A. F. Hegarty, *J. Chem. Soc., Chem. Commun.*, 1986, 383.

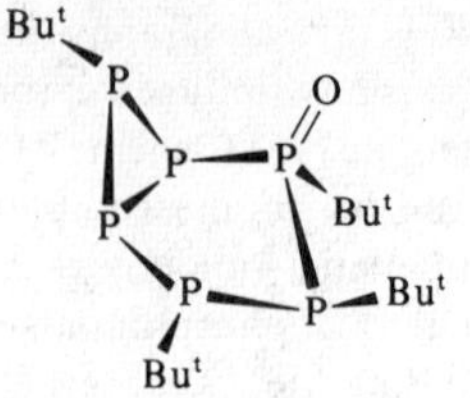

(94a)

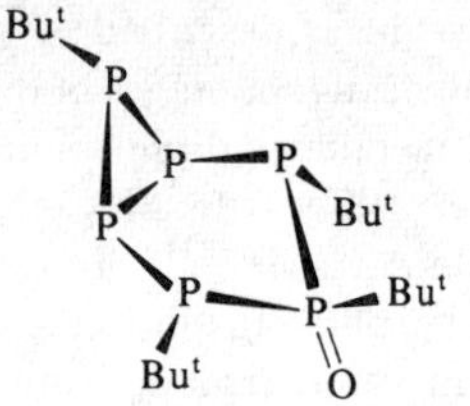

(94b)

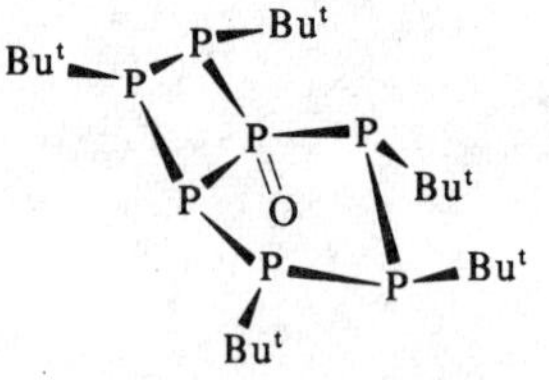

(95a)

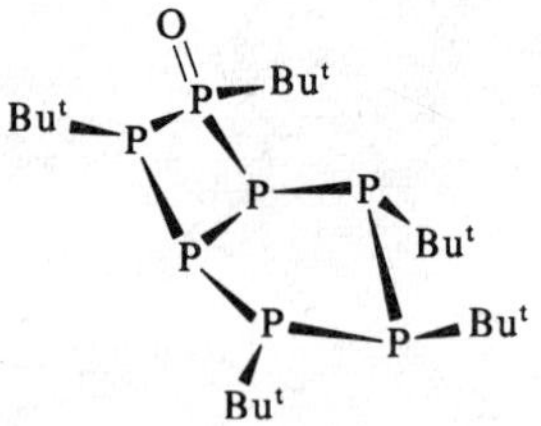

(95b)

(94) and (95) have been obtained [as mixtures of two isomers (a) and (b)] by reaction of the appropriate phosphanes with aerobic oxygen.[279] Crystals of Li_3P_7 comprise $[P_7]$ units with the nortricyclene structure which are interconnected by lithium cations.[280] Solutions of the mixed salt $Rb_3[P_{7-x}As_x]$ ($x = ca.$ 3) in ethyl-enediamine exhibit six singlet signals in the ^{31}P n.m.r. spectrum, corresponding to the various possible $[P_{7-x}As_x]$ species which interconvert by valence tautomerism.[281]

Several novel unsaturated organophosphorus compounds have been reported in the past year. The simplest of these is 2-phosphapropene, $MeP=PCH_2$, obtained by the pyrolysis of dimethylchlorophosphine at 770 K.[282] Others include 1-phospha-allenes, $ArP=C=CR^1R^2$,[283] the first example of a triphosphabutadiene (96) (synthe-sized according to Scheme 42),[284] 2,4-diphosphabut-1-enes (97),[285] and 1-phos-phabuta-1,2,3-trienes (98).[286]

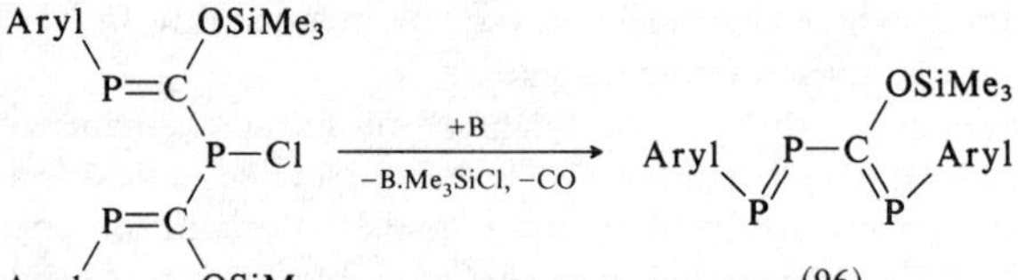

Scheme 42

[279] M. Naudler, M. Michels, M. Pieroth, and J. Hahn, Angew. Chem., Int. Ed. Engl., 1986, 25, 471.
[280] V. Manriquez, W. Hönle, and H. G. von Schnering, Z. Anorg. Allg. Chem., 1986, 539, 95.
[281] W. Hönle and H. G. von Schnering, Angew. Chem., Int. Ed. Engl., 1986, 25, 352.
[282] H. Bock and M. Bankmann, Angew. Chem., Int. Ed. Engl., 1986, 25, 265.
[283] R. Appel, V. Winkhaus, and F. Knoch, Chem. Ber., 1986, 119, 2466.
[284] R. Appel, B. Niemann, W. Schuhn, and F. Knoch, Angew. Chem., Int. Ed. Engl., 1986, 25, 932.
[285] R. Appel, C. Casser, and F. Knoch, Chem. Ber., 1986, 119, 2609.
[286] G. Märkl, H. Sejpka, S. Dietl, B. Nuber, and M. L. Ziegler, Angew. Chem., Int. Ed. Engl., 1986, 1003.

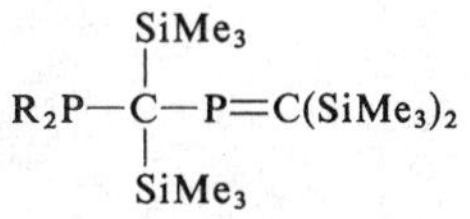

$$R_2P-\underset{\underset{SiMe_3}{|}}{\overset{\overset{SiMe_3}{|}}{C}}-P{=}C(SiMe_3)_2$$

(97)

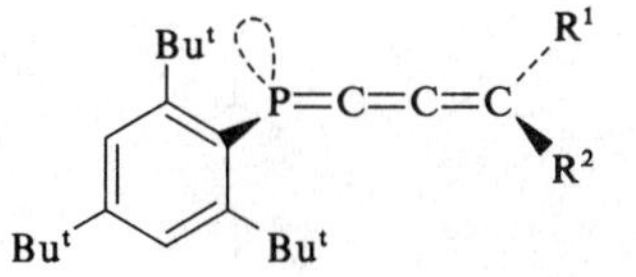

(98)

Although phospha- and arsa-alkenes and phospha-alkynes are now well documented, a stable arsa-alkyne (99) has only recently been synthesized. The procedure is analogous to one employed for phospha-alkynes and involves the spontaneous elimination of hexamethyldisiloxane from the acylarsenic intermediate (100) (Scheme 43).[287] The cycloaddition of phospha-alkynes to cyclobutadienes affords

$$-(Me_3Si)_2O$$

(100)

(99)

Scheme 43

examples of the previously unknown 2-Dewar phosphinines (101) as a mixture of isomers (Scheme 44). The predominant (or even exclusive) isomer is that (101a) in

$$+ \ P{\equiv}C-R^2 \xrightarrow{\ [4+2]\ }$$

(a)

(b)

(101)

Scheme 44

which the ester function on the cyclobutadiene and the organic group on the phospha-alkyne are in neighbouring positions.[288] Similar cycloaddition to 1,3-diphosphabutadienes affords the 1,3,5-triphosphabenzene (102).[289] A formally for-

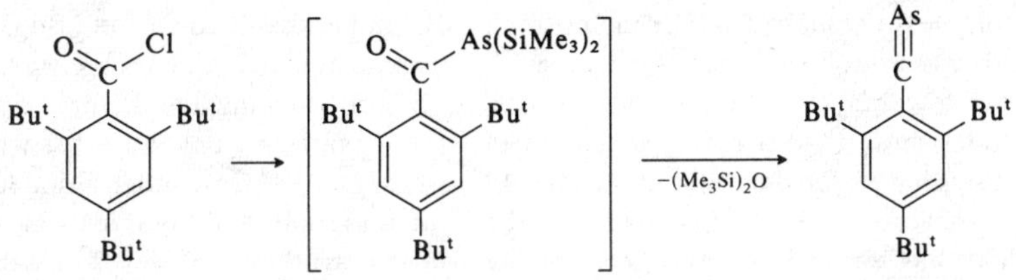

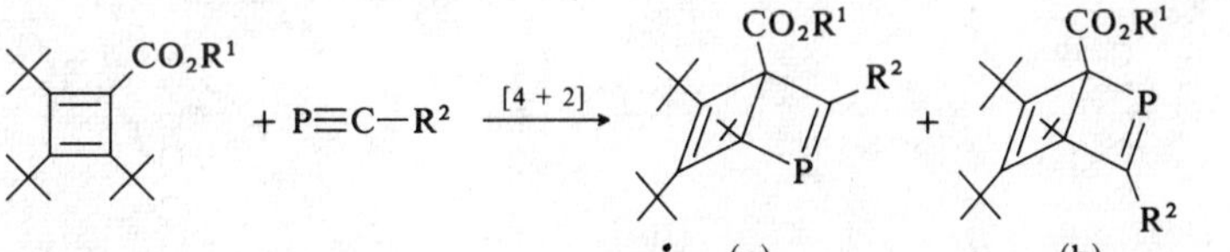

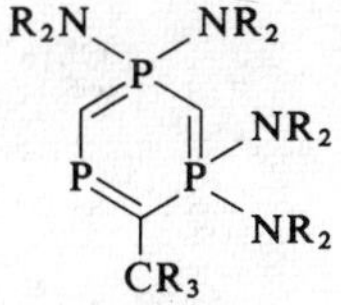

(102)

[287] G. Märkle and H. Sejpka, *Angew. Chem., Int. Ed. Engl.*, 1986, **25**, 264.
[288] J. Fink, W. Rösch, U. J. Vogelbacher, and M. Regitz, *Angew. Chem., Int. Ed. Engl.*, 1986, **25**, 280.
[289] E. Fluck, G. Becker, B. Neumüller, R. Knebl, G. Heckmann, and H. Riffel, *Angew. Chem., Int. Ed. Engl.*, 1986, **25**, 1002.

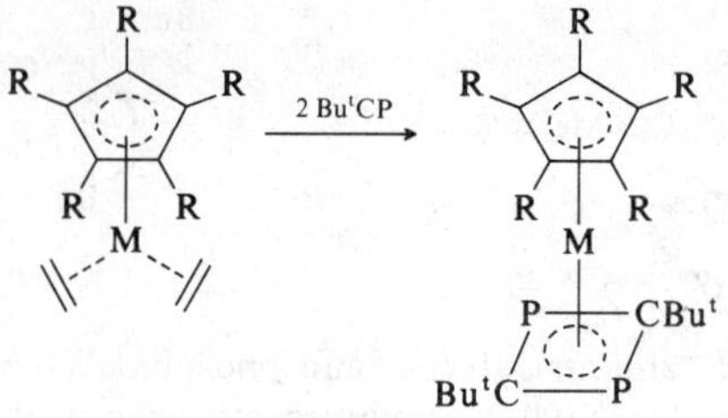

Scheme 45

bidden [2 + 2] cyclodimerization of two $Bu^tC{\equiv}P$ molecules occurs on displacement of ethylene ligands from cobalt, rhodium, and iridium complexes (Scheme 45). That the complexes contain the four-membered $[C_2P_2]$ rather than two discrete $Bu^tC{\equiv}P$ ligands was established by X-ray studies.[290] Phospha- and arsa-alkenes are versatile ligands towards transition metals as has been well documented before. Space precludes more than a cursory mention of novel complexes, but those for which structures have been determined include (103)—(107).[291—294] Co-condensation techniques have been used little in the synthesis of these types of complex, but this method has proved an excellent means for the synthesis of bis(η^6-arsabenzene)chromium(0) (108).[295]

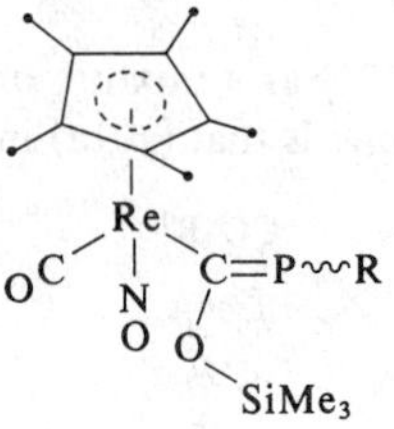

(103)

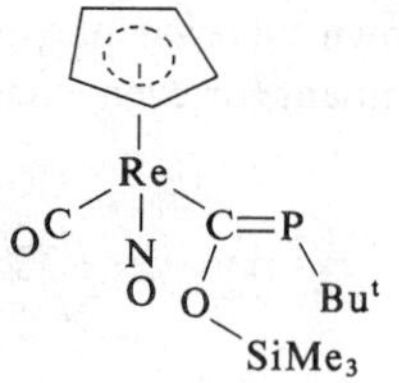

(104)

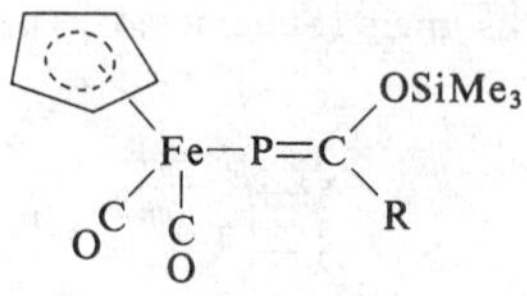

(105)

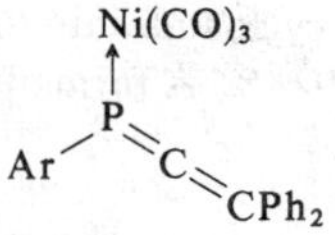

(106)

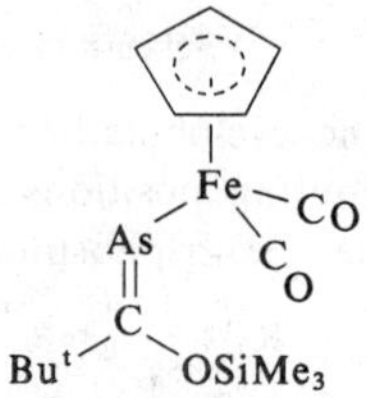

(107)

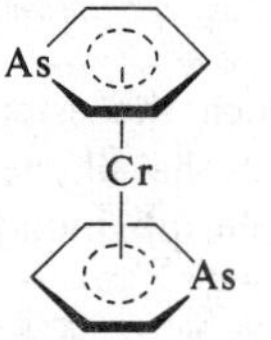

(108)

[290] P. B. Hitchcock, M. J. Maah, and J. F. Nixon, *J. Chem. Soc., Chem. Commun.*, 1986, 737.
[291] L. Weber, K. Reizig, R. Boese, and M. Polk, *Organometallics*, 1986, **5**, 1098.
[292] L. Weber, R. Reizig, M. Frebel, R. Boese, and M. Polk, *J. Organomet. Chem.*, 1986, **306**, 105.
[293] R. Appel, F. Knoch, and V. Winkhaus, *J. Organomet. Chem.*, 1986, **307**, 93.
[294] L. Weber, G. Meine, and R. Boese, *Angew. Chem., Int. Ed. Engl.*, 1986, **25**, 469.
[295] C. Eischenbroich, J. Kroker, W. Massa, M. Wünsch, and A. J. Ashe, *Angew. Chem., Int. Ed. Engl.*, 1986, **25**, 571.

The gas-phase structures of $CF_3P=CF_2$ and its cyclic dimer (which is stongly puckered) have been determined by electron diffraction,[296] and the ability of the arsenic analogue $CF_3As=CF_2$ to undergo [2 + 4] cycloaddition reactions with 1,3-dienes has been amply illustrated.[297] The silylene–sila-alkene rearrangement has been the subject of much investigation in recent years, and perhaps not surprisingly the phosphinocarbene–phospha-alkene rearrangement has now been studied. Photolysis of bis(phosphino)diazomethane in benzene leads to the phospha-alkene (109) in near quantitative yield. Methanol trapping studies clearly demonstrated the intermediacy of the phosphinocarbene (110) in the reaction.[298]

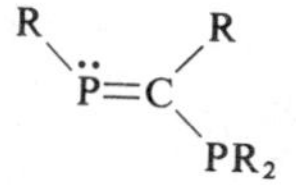

(109) 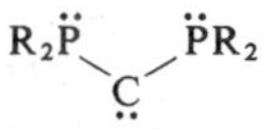   (110)

Treatment of PCl_3 or $AsCl_3$ with the sterically hindered lithium amide Li(NHAr) (Ar = 2,4,6-tri-t-butylphenyl), afford the two-co-ordinate imides, $ArN=E—NArH$ (E = P or As), whose structures have been determined. Both are angular [NEN = 103.8(5)° (P), 98.8(3)° (As).[299] In its chemistry, the phosphorus compound behaves as a source of its conjugate base, or as a neutral hydrido-P^V ligand as in the complex $[NiCl_2\{(ArN)_2PH\}]$. The arsenic analogue undergoes facile As—NHAr bond scission with $(AlMe_3)_2$.[300] Phosphonitriles $R^1R^2P\equiv N$, from the photolysis or thermolysis of the azides $R^1R^2PN_3$, are versatile synthetic intermediates for the synthesis of a variety of phosphazene derivatives (Scheme 46).[301] Other notable syntheses are

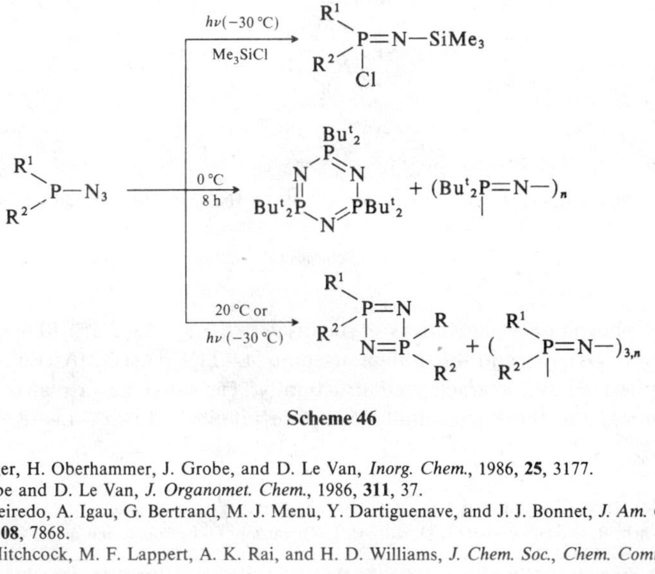

Scheme 46

[296] B. Steger, H. Oberhammer, J. Grobe, and D. Le Van, *Inorg. Chem.*, 1986, **25**, 3177.
[297] J. Grobe and D. Le Van, *J. Organomet. Chem.*, 1986, **311**, 37.
[298] A. Baceiredo, A. Igau, G. Bertrand, M. J. Menu, Y. Dartiguenave, and J. J. Bonnet, *J. Am. Chem. Soc.*, 1986, **108**, 7868.
[299] P. B. Hitchcock, M. F. Lappert, A. K. Rai, and H. D. Williams, *J. Chem. Soc., Chem. Commun.*, 1986, 1633.
[300] P. B. Hitchcock, H. A. Jasim, M. F. Lappert, and H. D. Williams, *J. Chem. Soc., Chem. Commun.*, 1986, 1634.
[301] J. Böske, E. Niecke, E. Ocando-Mavarez, J. P. Majoral, and G. Bertrand, *Inorg. Chem.*, 1986, **25**, 2695.

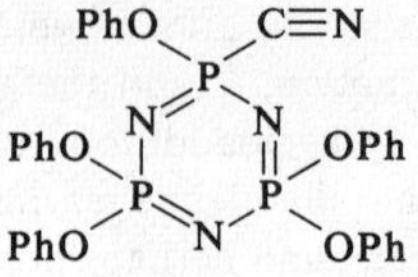

(111) 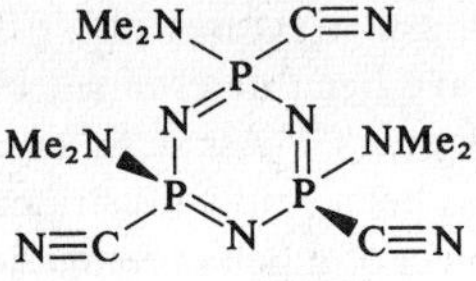 (112)

those of stable cyanocyclophosphazenes such as (111) and (112),[302] linear phos-
phazenes (Schemes 47 and 48),[303,304] and diphospha- and diarsa-diboretenes.[305] The

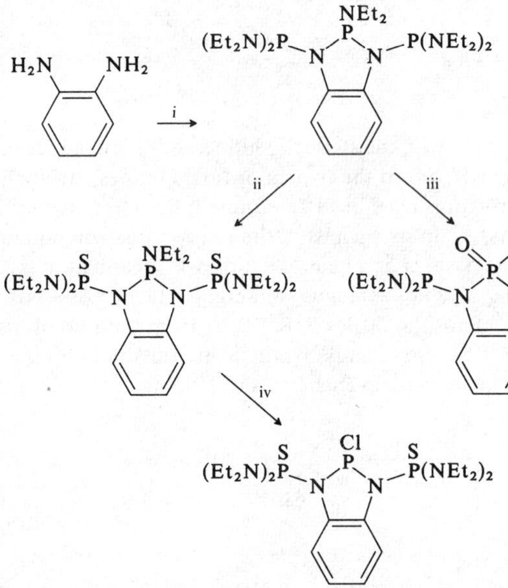

Reagents: i, 3P(NEt$_2$)$_3$, −4EtNH; ii, 1:4 S$_8$-toluene; iii, H$_2$O−CH$_2$Cl$_2$; iv, 2HCl(anhydrous), −Et$_2$NH$_2$Cl-
toluene

Scheme 47

lithium phosphide derivatives {Li(Et$_2$O)PPh$_2$}$_n$, Li(THF)$_2$PPh$_2$}$_n$, and
{Li(THF)P(C$_6$H$_{11}$)$_2$}$_n$ and the lithium arsenide [Li(THF){As(Bu^t)As(Bu^t)$_2$}]$_2$ have
been synthesized and characterized structurally. The latter has a planar [Li$_2$As$_2$]
core, whereas the three phosphides comprise infinite −Li−P−Li−P−Li−P−
chains.[306,307]

[302] J. S. Rutt, M. Parvez, and H. R. Allcock, *J. Am. Chem. Soc.*, 1986, **108**, 6089.
[303] H. J. Chen, R. C. Haltiwanger, T. O. Hill, M. L. Thompson, D. E. Coons, and A. D. Norman, *Inorg. Chem.*, 1986, **25**, 4323.
[304] J. M. Barendt, R. C. Haltiwanger, and A. D. Norman, *J. Am. Chem. Soc.*, 1986, **108**, 3127.
[305] A. M. Arif, A. H. Cowley, M. Pakulski, and J. M. Power, *J. Chem. Soc., Chem. Commun.*, 1986, 889.
[306] R. A. Bartlett, M. M. Olmstead, and P. P. Power, *Inorg. Chem.*, 1986, **25**, 1243.
[307] A. M. Arif, R. A. Jones, and K. B. Kidd, *J. Chem. Soc., Chem. Commun.*, 1986, 1440.

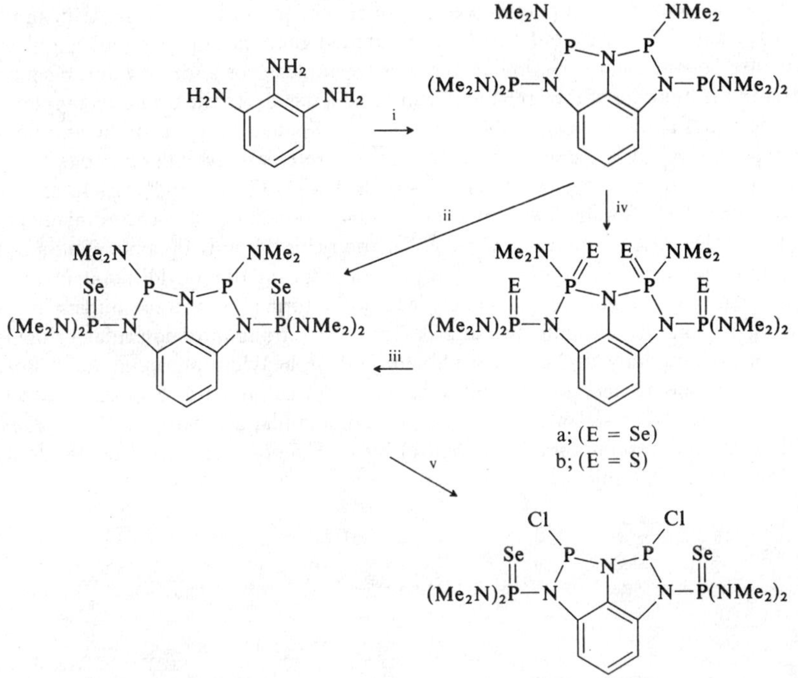

Reagents: i, $(Me_2N)_3P-Et_2NH$, 70 °C, 36 h; ii, excess Se, CH_2Cl_2, 25 °C, 20 h; iii, excess Se, CH_2Cl_2, 40 °C, 3 days; iv, S_8, CH_2Cl_2, 25 °C, 12 h; v, HCl(g)-Et_2NH_2Cl, −196 °C, toluene, 1 h

Scheme 48

MAS [31]P n.m.r. has been applied to a number of phosphates.[308—310] Many factors, including *inter alia* the nature of the counterion, hydration state, and crystalline form, as well as inequivalence within the molecule, affect the observed chemical shifts.

5 Antimony and Bismuth

As with tin and lead, crystallographic studies dominate the publications relating to antimony and bismuth. The most interesting of these are the structures of three arene–metal complexes, one of antimony and two of bismuth. The co-ordination polyhedron of the metal in $(SbCl_3)_2.(C_6H_6)$ resembles a deformed pentagonal bipyramid in which the benzene molecule occupies an axial site.[311] That in [1,3,5-$Me_3C_6H_3$][$BiCl_3$] is somewhat similar, but the bismuth atom enjoys distorted octahedral six-co-ordination with the arene molecule occupying one co-ordination site. In crystals of the hexamethylbenzene complex, $(C_6Me_6)(BiCl_3)_2$, the centres of the

[308] A. K. Cheetham, N. J. Clayden, and C. M. Dobson, *J. Chem. Soc., Chem. Commun.*, 1986, 195.
[309] U. Haubenreisser, G. Scheler, and A. R. Grimmer, *Z. Anorg. Allg. Chem.*, 1986, **532**, 157.
[310] L. Griffiths, A. Root, R. K. Harris, K. J. Packer, A. M. Chippendale, and F. H. Tromans, *J. Chem. Soc., Dalton Trans.*, 1986, 2247.
[311] D. Mootz and V. Händler, *Z. Anorg. Allg. Chem.*, 1986, **533**, 23.

arene ring form crystallographic centres of inversion, so that the same bismuth chloride unit is found on either side of the arene ligand, *i.e.* an 'inverted' sandwich structure.[312] Although very similar, the arene complexes of antimony and bismuth differ in the bonding of the arene molecules to the metals. In the bismuth complexes the metal lies above the centre of the ring, *i.e.* η^6-bonded, whereas in the antimony complex the metal is acentric and is associated preferentially with one edge of the ring. The complexes, [Li(12-crown-4)$_2$][SbPh$_2$].1/3THF and [Li(12-crown-4)$_2$[Sb$_3$Ph$_4$].THF, are the first fully characterized examples of two-co-ordinate antimony.[313] Distorted square-pyramidal geometry is found in phenylantimony bis(monothioacetate),[314] diphenyldithiophosphinate, and diphenyldithioarsinate,[315] but in each case longer Sb···S contacts lead to the formation of weak dimers in the crystal. In contrast, a chain structure is formed by strong intermolecular bridging in diphenylantimony diphenylphosphinate and diphenylthiophosphinate.[316] Both triphenylantimony and triphenylbismuth bis(phenylsulphonate) comprise discrete monomers in the crystal with usual trigonal-bipyramidal geometry.[317] The unusual [BiFe$_4$(CO)$_{16}$]$^{3-}$ anion contains a central bismuth atom surrounded tetrahedrally by four [Fe(CO)$_4$] units.[318]

[312] A. Schier, J. M. Wallis, G. Müller, and H. Schmidbaur, *Angew. Chem., Int. Ed. Engl.*, 1986, **25**, 757.
[313] R. A. Bartlett, H. V. Rasika Dias, H. Hope, D. Murray, M. M. Olmstead, and P. P. Power, *J. Am. Chem. Soc.*, 1986, **108**, 6921.
[314] M. Hall, D. B. Sowerby, and C. P. Falshaw, *J. Organomet. Chem.*, 1986, **315**, 321.
[315] C. Silvestru, L. Silaghi-Dumitrescu, I. Haiduc, M. J. Begley, M. Nunn, and D. B. Sowerby, *J. Chem. Soc., Dalton Trans.*, 1986, 1031.
[316] M. J. Begley, D. B. Sowerby, D. Wesolek, C. Silvestru, and I. Haiduc, *J. Organomet. Chem.* 1986, **316**, 281.
[317] R. Rüther, F. Huber, and H. Preut, *Z. Anorg. Allg. Chem.*, 1986, **539**, 110
[318] M. R. Churchill, J. C. Fettinger, K. H. Whitmire, and C. B. Lagrone, *J. Organomet. Chem.*, 1986, **303**, 99.

6 O, S, Se, Te

By F. J. BERRY

Department of Chemistry, University of Birmingham, P.O. Box 363, Birmingham B15 2TT

1 Introduction

Several studies of compounds which differ only in the nature of the Group VI element have been reported and will be cited in appropriate sections of this chapter. However, it is relevant to note here a theoretical study[1] of the geometries and binding energies of CX and X_2 groups (X = O, S, Se, Te) and of H_2CX and CX_2 groups (X = O, S) when complexed to $Ru(CO)_4$ which were calculated by the Hartree–Fock–Slater transition-state method. The increase in the donor and acceptor abilities of CX along the series X = O, S, Se, Te was explained in terms of the decrease in electronegativity from oxygen to tellurium. The X—X distances in $Ru(CO)_4X_2$ were found to be 0.2 Å longer than in free X_2 and similar to those of X_2^-, whilst complexed H_2CX in the η^2 conformation contained hydrogen atoms which were bent back by 30° and C—X bonds elongated by 0.1 Å. The CX_2 ligands were found to prefer the η^2-CX functionality as opposed to coordination through carbon.

2 Oxygen

The reactions of dioxygen with transition-metal ions in solution have attracted much attention both because of their fundamental importance and because of their relevance to chemical and biological processes. For example, the reaction of hexaaquamolybdenum(III) with dioxygen in *p*-toluenesulphonic acid solutions has been shown[2] to give the di-μ-oxomolybdenum(V) ion $[Mo_2O_4(H_2O)_6]^{2+}$. Reactions involving a large excess of Mo^{3+} gave an intermediate formulated as MoO_2Mo^{6+}. The intermediate was not formed from reactions in which dioxygen exceeded the concentration of Mo^{3+} and, from the kinetic data, a route for the conversion of MoO_2^{3+} to $Mo_2O_4^{2+}$ was identified. Further kinetic studies of the fundamental reactions of the hexaaquamolybdenum(III) ion with dioxygen have subsequently been reported.[3]

The oxidation of some cobalt(II) pentaamine complexes by molecular oxygen has been found[4] to take place through the formation of μ-peroxo-bridged cobalt(III) dioxygen complexes and to give ultimately either oxidative dehydrogenation of the coordinated ligand or simple metal-centred oxidation and the formation of

[1] T. Ziegler, *Inorg. Chem.*, 1986, **25**, 2721.
[2] E. F. Hills, P. R. Norman, T. Ramasami, D. T. Richens, and A. G. Sykes, *J. Chem. Soc., Dalton Trans.*, 1986, 157.
[3] E. F. Hills and A. G. Sykes, *Polyhedron*, 1986, **5**, 511.
[4] C. J. Rayleigh and A. E. Martell, *Inorg. Chem.*, 1986, **25**, 1190.

cobalt(III) complexes of the unchanged ligand and hydrogen peroxide. In other studies of the reactions of dioxygen complexes the mechanism and kinetics of the oxidative dehydrogenation of 2-(aminomethyl)pyridine through cobalt(II) dioxygen complex formation has been examined.[5] A number of representative cobalt(II) complexes of the imines of L-amino acids derived from salicylaldehyde and pyridoxal have been prepared and characterized as high-spin five- or six-coordinate structures.[6] The binding of dioxygen to some cobalt(II) 1-naphthyl porphyrins substituted with amido, carboxy, or hydroxymethyl groups at the 8-naphthyl position have been found[7] to display significant enhancements in affinity for dioxygen which correlate well with intramolecular hydrogen-bonding effects. Studies such as these are of interest in connection with current interests in oxygen binding in myoglobin and haemoglobin as well as oxygen activation and reduction processes in other haemoproteins. The synthesis of some cobalt(II) complexes with a series of potentially tetradentate Schiff-base ligands have been reported[8] and it is especially interesting to note that their subsequent reactions with dioxygen appear to be significantly different from the reactions of dioxygen with the well studied [Co(salen)] system. Also of note is the four-electron reduction of oxygen to water which occurs on a planar dinuclear cobalt chelate adsorbed onto pyrolytic graphite in alkaline solution.[9] A study of the kinetics and mechanism of the cobalt-catalysed reaction of oxygen and sulphite at low concentrations has been reported.[10]

Studies of the binding of dioxygen with other transition-metal-containing compounds have also been reported *e.g.* the dioxygen-binding behaviour of an iron(II) lacunar complex has been examined[11] and the axial ligand-dependence of the dioxygen affinity described in terms of a multiple-equilibria model comprising four separate, but mutually related, equilibria. The *cis*- and *trans*-isomers of the osmium(IV) dioxo complex $[(bpy)_2Os(O)_2](ClO_4)_2$ (bpy = 2,2'-bipyridine) have been prepared and characterized and their redox properties in aqueous solution investigated by electrochemical techniques.[12]

Interest in the reactions of the superoxide ion has continued. For example, the mechanism of the reactions of the superoxide ion, O_2^-, with molybdenum(V) tetraphenylporphyrins in aprotic solvents under anaerobic conditions has been stoicheiometrically elucidated[13] and has led to the synthesis and characterization of a new dioxygen complex of molybdenum tetraphenylporphyrin. *Ab initio* calculations have been performed on the Cu^{2+}-O_2^- system[14] by considering the effects of an ammonia ligand on copper and of an ammonium ion interacting with superoxide. The results of the calculations of the Cu^{2+}-O_2^- interaction were used as a model

[5] A. K. Basak and A. E. Martell, *Inorg. Chem.*, 1986, **25**, 1182.

[6] L. Casella and M. Gullotti, *Inorg. Chem.*, 1986, **25**, 1293.

[7] C. K. Chang, and M. P. Kondylis, *J. Chem. Soc., Chem. Commun.*, 1986, 316.

[8] F. M. Ashmawy, R. M. Issa, S. A. Amer, C. A. McAuliffe, and R. V. Parish, *J. Chem. Soc., Dalton Trans.*, 1986, 421.

[9] A. van de Patten, A. Elzing, W. Visscher, and E. Barendrecht, *J. Chem. Soc., Chem. Commun.*, 1986, 477.

[10] D. B. Hobson, P. J. Richardson, P. J. Robinson, E. A. Hewitt, and I. Smith, *J. Chem. Soc., Faraday Trans. 1*, 1986, **82**, 869.

[11] K. A. Goldsby, B. D. Beato, and D. H. Busch, *Inorg. Chem.*, 1986, **25**, 2342.

[12] J. C. Dobson, K. J. Takeuchi, D. W. Pipes, D. A. Geselowitz, and T. J. Meyer, *Inorg. Chem.*, 1986, **25**, 2357.

[13] K. Hasegawa, T. Imamusa, and M. Fujimoto, *Inorg. Chem.*, 1986, **25**, 2154.

[14] M. Rosi, A. Sgamellotti, F. Tarantelli, I. Bertini, and C. Luchinat, *Inorg. Chem.*, 1986, **25**, 1005.

for the mechanism of copper/zinc superoxide dismutase. An improved synthesis of tetramethylammonium superoxide, $[(Me_4N)O_2]$, has been described[15] and the purported dimer of $(Me_4N)O_2$ shown to be a peroxide adduct of acetonitrile.

Although uranium is the most useful of the actinide metals and is known to form simple peroxides, its heteroligand peroxy chemistry seems to have been practically overlooked in the past until the publication of a study of complex peroxyuranates[16] which involved the synthesis and structural characterization of the alkali-metal and ammonium dioxoperoxy (sulphate) aquouranates(VI), $A_2[UO_2(O_2)SO_4(H_2O)]$ (A = NH_4, Na) and the alkali-metal and ammonium dioxoperoxy(oxalato) uranate(VI) hydrates, $A_2[UO_2(O_2)C_2O_4]\cdot H_2O$. The preparation of some oxygen-donor complexes of acetylacetonatotrichlorothorium(IV), bis(acetylacetonato)-dichloro- and di-N-thiocyanato-thorium(IV) and uranium(IV) have also been described[17] and their stoicheiometries discussed in terms of the steric crowding about the metal atom.

Various structural studies of oxygen-containing solids have been reported and a few citations are included here to illustrate the types of materials currently attracting interest and the techniques which are being used. For example, the topotactic dehydration of the solid solution $Mo_{1-x}W_xO_3\cdot H_2O(O \leqslant \chi \leqslant 1.0)$ has been shown[18] to give rise to monophasic oxides with ReO_3-type structures which transform to the layered structure in the case of pure MoO_3. The slightly reduced mixed oxides contain disordered {102} shear planes which are different from those found in the pure oxides. It is also interesting to note the isolation[19] and structural characterization of the first mixed oxo-thio-bridged tritungsten triangular cluster complex containing the $[W_3(\mu_2-O)_3(\mu_3-S)(NCS)_9]^{5-}$ unit. In continued studies of transmetallation of tetranuclear copper complexes the spectral evidence for the substoicheiometric transmetallation of $(\mu_4-O)[(DENC)Cu]_4X_6$ complexes, (DENC = N,N-diethylnicotinamide; X = Cl or Br), by a $Ni(NS)_2$ transmetallating agent has been reported.[20] An improved synthesis of rubidium ozonide involving ozonization of the hyperoxide at room temperature and extraction with liquid ammonia has produced, for the first time, pure rubidium ozonide in preparative amounts and as single crystals.[21] The crystal structure of calcium 1,1-dicyanoethylene-2,2-dithiolate pentahydrate has shown[22] the calcium atom to be located in a distorted bicapped trigonal-prismatic coordination of six oxygen and two nitrogen atoms with two oxygen atoms being attached to two different Ca^{2+} ions and forming dimeric units.

Other types of investigations involving oxygen have included studies of the use of ozone-enriched oxygen[23] in the development of a novel procedure for the removal of carbonaceous deposits from zeolites, which is also applicable to current commercial interests in the regeneration of zeolite catalysts. Matrix-isolation techniques have also been used to examine the reactivity of oxygen. For example, the oxidation

[15] K. Yamaguchi, T. S. Calderwood, and D. T. Sawyer, *Inorg. Chem.*, 1986, **25**, 1289.

[16] M. Bhattacharjee, M. K. Chandhuri, and R. N. D. Purkayastha, *Inorg. Chem.*, 1986, **25**, 2354.

[17] A. G. M. Al-Daher and K. W. Bagnall, *J. Chem. Soc., Dalton Trans.*, 1986, 355.

[18] L. Ganapathi, A. Ramanan, J. Gopalkrishnan, and C. N. R. Rao, *J. Chem. Soc., Chem. Commun.*, 1986, 62.

[19] Z. Dori, F. A. Cotton, R. Llusar, and W. Schwotzer, *Polyhedron*, 1986, **5**, 907.

[20] G. Davies, M. A. El-Sayed, and A. El-Toukhy, *Inorg. Chem.*, 1986, **25**, 2269.

[21] W. Schnick and M. Jansen, *Z. Anorg. Allg. Chem.*, 1986, **532**, 37.

[22] C. Wolf and H. U. Hummel, *J. Chem. Soc., Dalton Trans.*, 1986, 43.

[23] R. G. Copperthwaite, G. J. Hutchings, P. Johnston, and S. W. Orchard, *J. Chem. Soc., Faraday Trans. 1*, 1986, **82**, 1007.

of diphenylcarbene to benzophenone *O*-oxide by molecular oxygen has been elucidated by matrix isolation methods.[24] In another study[25] the identification of peroxo
and dioxo metal carbonyl intermediates was achieved by monitoring the pyrolysis
of $M(CO)_6$ molecules (M = Cr, Mo, W), which had been isolated in oxygen-doped
argon or methane matrices at 10–20 K, by infrared, Raman, and u.v.-visible spectroscopy. Oxygen-17 n.m.r. has been used[26] to examine pressure effects in scalar coupling
in $Co(H_2O)_6^{2+}$ and high-spin–low-spin equilibria in (1,4,8,11-tetrazaundecane)
nickel(II). A $^{31}P(^{18}O)$ positional isotope investigation[27] has shown that hypophos-
phate is oxidized quantitatively by bromine in ^{18}O-labelled water to [^{18}O]pyrophos-
phate in which the ^{18}O label is located exclusively in the non-bridging position. The
development of the chemistry of polyoxoanions has frequently been frustrated by
difficulties in characterization and it is therefore especially interesting to record the
first use of fast atom bombardment mass spectroscopy (FABMS) to examine very
large, highly charged non-volatile polyoxoanions.[28] Also of note is the synthesis
and study of mono-nuclear ruthenium(II) complexes of sterically hindering diimine
chelates[29] since the results have implications for current interests in the catalytic
oxidation of water to molecular oxygen.

3 Sulphur

A review[30] of the generation, growth, and properties of homo- and hetero-nuclear
clusters including some containing sulphur has appeared. Given the substantial
interest which has developed in aqueous polysulphide solutions as primary electrolytes in photoelectrochemical solar cells, it is pertinent to record the application[31]
of a computer iterative technique to calculate the distribution of species such as
S^{2-}, S_2^{2-}, S_3^{2-}, S_4^{2-}, S_5^{2-}, together with OH^-, H^+, H_2S, HS^-, water and alkali-metal
cations, in aqueous alkali-metal polysulphide solutions. From the calculated distribu-
tions of the species in various polysulphide electrolytes the rate of disproportionation
of the electrolytes was deduced and the electrolyte instability shown to influence
the gradual degradation of polysulphide-based photoelectrochemical cells. A qualitative analysis[32] of hypervalency in sulphur in terms of valence bond theory has
been published. In a different area of sulphur chemistry, the reaction of a perrhenate
solution reduced by $NH_2OH \cdot HCl$ with an aqueous polysulphide solution gives an
ammonium salt of the cluster anion $[Re_4S_{22}]^{4-}$ in which the highly symmetrical
metal–sulphur cluster contains six S_3^{2-} ligands.[33]

The chemistry of sulphur–nitrogen compounds continues to attract significant
attention for several reasons. The compounds provide useful intermediates in various
areas of sulphur–nitrogen chemistry: they can stabilize otherwise unobservable
sulphur–nitrogen anions, they can provide interesting comparisons with isoelectronic

[24] W. Sander, *Angew. Chem., Int. Ed. Engl.*, 1986, **25**, 255.
[25] M. J. Almond, J. A. Crayston, A. J. Downs, M. Poliakoff, and J. J. Turner, *Inorg. Chem.*, 1986, **25**, 19.
[26] R. M. Nielson, H. W. Dodgen, J. P. Hunt, and S. E. Wherland, *Inorg. Chem.*, 1986, **25**, 582.
[27] B. V. L. Potter, *J. Chem. Soc., Chem. Commun.*, 1986, 21.
[28] K. S. Suslick, J. C. Cook, B. Rapko, M. W. Droege, and R. G. Finke, *Inorg. Chem.*, 1986, **25**, 241.
[29] J. P. Collin and J. P. Sauvage, *Inorg. Chem.*, 1986, **25**, 135.
[30] T. P. Martin, *Angew. Chem., Int. Ed. Engl.*, 1986, **25**, 197.
[31] S. Licht, G. Hodes, and J. Manassen, *Inorg. Chem.*, 1986, **25**, 2486.
[32] F. Volatron, and O. Eisenstein, *J. Chem. Soc., Chem. Commun.*, 1986, 301.
[33] A. Muller, E. Krickmeyer, and H. Bogge, *Angew. Chem., Int. Ed. Engl.*, 1986, **25**, 272.

sulphur oxygen complexes, and they have the potential to form one-dimensional materials with interesting electrical properties. An excellent review[34] is available of the preparation and structural properties of complexes containing simple sulphur-nitrogen ligands. The review gives a timely outline of the diversity of structural types encountered in this area of chemistry.

Some novel complexes of copper(I) with the S_3N^- chelate ligand have been described[35] and the structures of the $[Ph_4As][Cu(S_3N)(CN)]$, $[(Ph_3P)_2N]$-$[Cu(S_3N)(S_7N)]$, and $[Ph_4As]_2[(S_3N)Cu(S_2O_3)Cu(S_3N)]$ materials shown to involve the trigonal-planar coordination of copper(I) in which the S_3N^- chelate group coordinates the copper by two sulphur atoms. The reaction of the six membered rings $(R_2NCN)(NSCl)_2$ (R = Me, Et, Pr^i) with $Me_3SiNSNSiMe_3$ or Me_3SiNSO has been shown[36] to give the bicyclic compounds $R_2NCS_3N_5$. The X-ray crystal structure of the folded eight-membered ring $1,5$-$Me_2NC(NSN)_2SCl$ was also described. In a subsequent study[37] of the synthesis and electronic structure of the $R_2NCS_2N_3$ ring, the X-ray crystal structure of the bicyclic compound $Pr^i_2NCS_3N_5$ was described and the preparation of $R_2NCS_2N_3 \cdot C_7H_8$ (R = Me, Et, Pr^i), $Et_2NCS_2N_2^+Cl^-$, and some salts of the $(R_2NCN)(NSCl)(NS)^+$ cation were reported.

The reaction of cis-$PtCl_2(PMe_3)_2$ with $[Me_2Sn(S_2N_2)]_2$ and $[NH_4][PF_6]$ in dichloromethane has been shown[38] to give the compound $[Pt(S_2N_2H)(PMe_3)_2]$-$[PF_6]$. Another platinum compound of composition $[Pt(S_2N_2H)(PMe_2Ph)_2][BF_4]$ was formed by protonation of $[Pt(S_2N_2)(PMe_2Ph)_2]$. The two new compounds are the first examples of a novel class of platinum stacking compounds involving the $S_2N_2H^-$ ligand. The crystal structures, illustrated in Figures 1 and 2, show that the platinum adopts square-planar coordination with the $S_2N_2H^-$ ligands having two short and one long SN distances, comparable with those in the compound $Pt(S_2N_2)$-$(PPh_3)_2$. Of particular interest is the stacking arrangement in the compounds $[Pt(S_2N_2H)(PMe_3)_2][PF_6]$ and $[Pt(S_2N_2H)(PMe_2Ph)_2][BF_4]$ where, in both cases, the (PtS_2N_2) rings lie almost directly on top of each other with alternate cations reversed so as to accommodate the relatively bulky phosphine ligands. This results in structures with channels of cations and anions.

The preparative routes to $1,3,2,4$-benzodithiadiazine, $C_6H_4S_2N_2$, $1,3,5,2,4$-benzotrithiadiazepine, $C_6H_4S_3N_2$, and their respective norbornadiene adducts $C_6H_4S_2N_2 \cdot C_7H_8$ and $C_6H_4S_3N_2 \cdot C_7H_8$ have been described.[39] The molecular structures of $C_6H_4S_2N_2$ and $C_6H_4S_3N_2$ were correlated with MNDO molecular orbital calculations on their ground-state electronic structures and the aromatic-antiaromatic characters of the two compounds were discussed in relation to their electronic and ^1H n.m.r. spectra, their electrochemical behaviour, and the ease of dissociation of their norbornadiene adducts. In an investigation of three cyclic sulphur-nitrogen compounds with similar electronic structures, namely 4-methyl-$1,2,3,5$-dithiadiazolyl, 4-phenyl-$1,2,3,5$-dithiadiazolyl, and $1,2,4,3,5$-trithiadiazolylium the

---

[34] P. F. Kelly, and J. D. Woollins, *Polyhedron*, 1986, **5**, 607.

[35] J. Weiss, *Z. Anorg. Allg. Chem.*, 1986, **532**, 184.

[36] T. Chivers, J. F. Richardson, and N. R. M. Smith, *Inorg. Chem.*, 1986, **25**, 272.

[37] T. Chivers, F. Edelmann, J. F. Richardson, N. R. M. Smith, O. Treu, and M. Trsic, *Inorg. Chem.*, 1986, **25**, 2119.

[38] R. Jones, P. F. Kelly, C. P. Warrens, and J. D. Woollins, *J. Chem. Soc., Chem. Commun.*, 1986, 711.

[39] A. W. Cordes, M. Hojo, H. Koenig, M. C. Noble, R. T. Oakley, and W. T. Pennington, *Inorg. Chem.*, 1986, **25**, 1137.

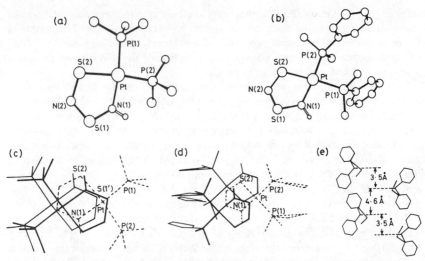

Figure 1 *(a) The molecular structure of the cation in* $[Pt(S_2N_2H)(PMe_3)_2][PF_6]$, *selected bond distances* (Å) *and angles* (°): Pt—P(1) 2.263(1), Pt—P(2) 2.249(1), Pt—N(1) 2.038(12), Pt—S(2) 2.287(4), N(1)—S(1) 1.663(13), S(1)—N(2) 1.583(18), N(2)—S(2) 1.672(17); P(1)—Pt—P(2) 96.4(1), N(1)—Pt—S(2) 83.7(4), Pt—N(1)—S(1) 117.6(7), N(1)—S(1)—N(2) 101.9(8), S(1)—N(2)—S(2) 119.9(10), N(2)—S(2)—Pt 104.3(6). *(b)* $[Pt(S_2N_2H)(PMe_2Ph)_2][BF_4]$ *selected bond lengths* (Å) *and angles* (°): Pt—P(1) 2.303(2), Pt—P(2) 2.261(2), Pt—N(1) 2.015(7), Pt—S(2) 2.283(2), N(1)—S(1) 1.595(7), S(1)—N(2) 1.538(9), N(2)—S(2) 1.649(6); P(1)—Pt—P(2) 94.3(1), N(1)—Pt—S(2) 84.6(2), Pt—N(1)—S(1) 121.4(4), N(1)—S(1)—N(2) 108.1(4), S(1)—N(2)—S(2) 121.6(5), N(2)—S(2)—Pt 104.4(3). *(c) and (d) show overlap of the cations in* $[Pt(S_2N_2H)(PMe_3)_2][PF_6]$ *and* $[Pt(S_2N_2H)-$ $(PMe_2Ph)_2][BF_4]$ *respectively. The uppermost cation is shown by a thick line, the next is shown by a broken line and the lowermost by a thin line. (e) Side view of the stacking of the cations in* $[Pt(S_2N_2H)(PMe_2Ph)_2][BF_4]$ *showing the alternative stacking sequence* (Reproduced from *J. Chem. Soc., Chem. Commun.*, 1986, 711)

isotropic and anisotropic e.s.r. spectra of the $Me\overset{\frown}{C}NSS\overset{\frown}{N}$, $Ph\overset{\frown}{C}NSS\overset{\frown}{N}$, and $\overset{\frown}{S}NSS\overset{\frown}{N}^{\cdot+}$ free radicals were recorded.[40] The minimum-energy structures of the radicals were calculated and the hyperfine coupling constants determined. All the radicals were found to dimerize in solution at low temperatures to give crystalline solids containing detectable amounts of the monomeric free-radical.

The reaction of $Ph_4P_2N_3SCl$ with triphenylantimony, which has been reported to give a twelve-membered $P_4N_6S_2$ macrocycle, has been shown[41] to yield a persistent radical species which, on the basis of molecular orbital calculations and e.s.r. spectroscopic evidence, was described as the 1,2,4,6,3,5-thiatriazadiphosphininyl radical $[Ph_4P_2N_3S]^{\cdot}$. In a study of the redox chemistry of 1,2,4,6-thiatriazinyls[42] the oxidation of the 3,5-diphenyl-1,2,4,6-thiatriazinyl dimer with the nitrosonium salts

[40] S. A. Fairhurst, K. M. Johnson, L. H. Sutcliffe, K. F. Preston, A. J. Banister, Z. V. Hauptmann, and J. Passmore, *J. Chem. Soc., Dalton Trans.*, 1986, 1465.

[41] R. T. Oakley, *J. Chem. Soc., Chem. Commun.*, 1986, 596.

[42] R. T. Boere, A. W. Cordes, P. J. Hayes, R. T. Oakley, R. W. Reed, and W. T. Pennington, *Inorg. Chem.*, 1986, **25**, 2445.

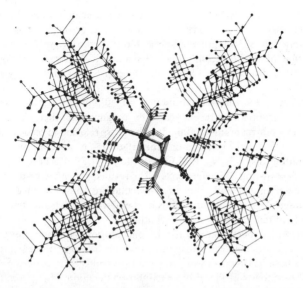

Figure 2 *Perspective view, down the crystallographic* a-*axis, of the infinite stacking of anions and cations in* $[Pt(S_2N_2H)(PMe_3)_2][PF_6]$
(Reproduced from *J. Chem. Soc., Chem. Commun.*, 1986, 711)

$[NO]^+X^-$ ($X^- = BF_4^-$, PF_6^-) has been shown to yield the corresponding salts of the $[Ph_2C_2N_3S]^+$ cation. Treatment of the same compound with sodium in liquid ammonia gave, following acidification, the reduced system 3,5-diphenyl-4-hydro-1,2,4,6-thiatriazine, $Ph_2C_2N_3SH$. The structures of $[Ph_2C_2N_3S]^+[PF_6]^-$ and $Ph_2C_2N_3SH \cdot 0.5CH_2Cl_2$ were determined by X-ray diffraction.

The reaction of dialkylcyanamides with NSCl units generated from $(NSCl)_3$ in CCl_4 at *ca.* 60 °C has been shown[43] to produce either the six-membered mixed cyanuric-thiazyl ring $(R_2NCN)(NSCl)_2$ (R = Me, Et, Pr^i) or, in the presence of Me_2NCN, the eight-membered mixed cyanuric-thiazyl ring 1,3-$(Me_2NCN)_2$ $(NSCl)_2$. The crystal structure of $(Et_2NCN)(NSCl)_2$, which was determined by X-ray crystallography, showed that the exocyclic chlorine substituents adopt a *cis*-configuration with respect to the almost planar CN_3S_2 ring.

The linear dithianitronium cation S_2N^+, which can be conveniently prepared as S_2NAsF_6, has been found[44] to undergo general quantitative cycloaddition reactions with olefins to give 1:1 and 1:2 stoicheiometric cationic products. The generality of the reaction leading to products such as 1,3,2-dithiazolium and 1,4-dithia-7-azanorbornylium cations was rationalized and the X-ray crystal structure of the 1,4-dithia-7-azanorbornylium hexafluoroarsenate(V) described.

In a study[45] of the structure and properties of the sulphur dinitride dianion the properties of N_2S^{2-} have been calculated using Hartree–Fock–Slater and MNDO

[43] T. Chivers, J. F. Richardson, and N. R. M. Smith, *Inorg. Chem.*, 1986, **25**, 47.
[44] N. Burford, J. P. Johnson, J. Passmore, M. J. Schriver, and P. S. White, *J. Chem. Soc., Chem. Commun.*, 1986, 966.
[45] M. Conti, M. Trsic, and W. G. Laidlaw, *Inorg. Chem.*, 1986, **25**, 254.

methods and compared with those of N_2S and experimental data. A locally stable NSN^{2-} species with a NSN angle of $135°$ was characterized and a decomposition channel to N_2 and S^{2-} identified. The stability of the symmetric N_2S^{2-} species relative to that of the apparently elusive symmetric N_2S reaction fragment was associated with a stronger $N{-}S$ bond and a higher barrier to decomposition in the former species.

Finally, despite the substantial interest in sulphur–nitrogen heterocycles as reported in this chapter over recent years, the assignment of their electronic excited states has been a subject of uncertainty. It is therefore relevant to record the electronic absorption and magnetic circular dichroism study[46] of several cyclic π-electron systems including S_2N_2, S_4N_2, $S_3N_3O^-$, $1,5$-$Bu^t_2C_2N_4S_2$, $1,5$-$Ph_2C_2N_4S_2$, $S_4N_4O_2$, and two condensed-ring SN heterocycles. The results, which were analysed in terms of the perimeter model and π-electron calculations, have enabled the energies, absorption intensities, and polarizations of the first few electronic transitions to be accounted for in terms of $\pi^* \to \pi^*$ excitations.

Interest has also been maintained in phosphorous–sulphur compounds. For example, the reaction between phosphorus trichloride and hydrogen sulphide in the presence of a base has been shown[47] to yield P_4S_3 and $[PS_2Cl_2]^-$ besides some other products. The synthesis of the trithiometalphosphate anion, PS_3^-, from the reaction of P_4S_{10} with KCN and H_2S and its isolation as the Ph_4As salt has been reported.[48] The characterization of the anion, which was described in terms of a planar gaseous species with D_{3h} symmetry or a monomeric pyramidal solid structure, may be significant in terms of the quest for the hitherto unreported monomeric metaphosphate anion PO_3^- in the solid state.

In a study of mixed crystals from A_4B_3 molecules (A = P, As; B = S, Se) the P_4S_3-P_4Se_3-As_4S_3-As_4Se_3 system has been investigated by thermal and X-ray diffraction methods.[49] Five regions of solid solubility with different crystal structures were identified in the system at room temperature and the changes in the phases which are induced by thermal treatment were described. The existence of all molecules of formulation $P_nAs_{4-n}S_mSe_{3-m}$ ($n = 0$-4, $m = 0$-3) and $As_4S_mSe_{4-m}$ ($m = 1$-3) was reported.

The compounds $[Ph_2P(O)]_n[Ph_2P(S)]_{3-n}CH$ ($n = 1$, 2, or 3) have been synthesized by the oxidation of $[Ph_2P]_n[Ph_2P(S)]_{3-n}$ CH with hydrogen peroxide.[50] The corresponding anions $\{[Ph_2P(O)]_n[Ph_2P(S)]_{3-n}C\}^-$ were prepared from the neutral precursors by proton abstraction with LiOMe. Several new compounds of the type $[Ph_2P(X)][Ph_2P(Y)][Ph_2P(Z)]CH$, where X, Y and Z are various combinations of O, S, Se and electron pairs, were also synthesized. The proton chemical shifts of the methine protons in the compounds were found to be linearly related to the sum of substituent constants for the respective chalcogens, X, Y, and Z.

The stabilization has been reported[51] of an almost complete arsenic–sulphur cage molecule as a complex ligand, $[(C_5Me_5)_2Co_2As_2S_3]$, by the photochemical reaction

[46] H.-P. Klein, R. T. Oakley, and J. Michl, *Inorg. Chem.*, 1986, **25**, 3194.

[47] E. Fluck and B. Neumuller, *Z. Anorg. Allg. Chem.*, 1986, **534**, 27.

[48] H. W. Roesky, R. Ahlrichs, and S. Brode, *Angew. Chem., Int. Ed. Engl.*, 1986, **25**, 82.

[49] R. Blachnik, Th. Weber, and U. Wickel, *Z. Anorg. Allg. Chem.*, 1986, **532**, 90.

[50] S. O. Grim, S. A. Sangokoya, I. J. Colquhoun, W. McFarlane, and R. K. Khanna, *Inorg. Chem.*, 1986, **25**, 2699.

[51] H. Brunner, H. Kauermann, B. Nuber, J. Wachter, and M. L. Ziegler, *Angew. Chem., Int. Ed. Engl.*, 1986, **25**, 557.

of $[C_5Me_5Co(CO)_2]$ with As_4S_4. The ligand was found to consist of a basket formed by the cobalt, arsenic, and sulphur atoms.

Several studies of the fundamental properties of main-group sulphides are worthy of note. The low frequency (<30 cm^{-1}) Raman scattering lineshapes recorded from the chalcogenide glasses As_2S_3 and As_2Se_3 have been interpreted[52] in terms of a model which explicitly includes acoustic and optic vibrational modes. In another study[53] involving compounds containing group V and VI elements the preparation, infrared, and Raman vibrational spectra, and crystal structures of the penta-bromothio-diarsenate and -diantimonate species $PPh_4[As_2SBr_5]$ and $PPh_4[Sb_2SBr_5]$ were described. The increasing use of electron microscopy to study the microstructures of non-stoichiometric compounds has revealed that point defects are not the only structural way of accommodating changes in non-metal to metal ratios. This is well illustrated by a study[54] of non-stoicheiometric forms of Sb_2S_3 occurring in the $PbSb_2S_4$–Sb_2S_3 region of the Pb–Sb–S phase diagram.

Some diphenylantimony(III) dialkyldithio- and diaryldithio-, -phosphinates and -arsinates of composition $Ph_2SbS_2MR_2$ (M = P, R = Me, Et, Pr, or Ph; M = As, R = Me or Ph) have been synthesized and characterized by i.r. and ^{1}H n.m.r. spectroscopy.[55] The crystal structures of $Ph_2SbS_2MPh_2$ (M = P or As), which were determined by X-ray crystallography, are depicted in Figure 3. In each compound

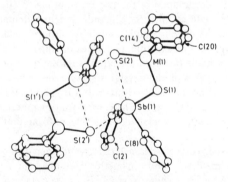

Figure 3 *Molecular structure and atom-numbering scheme for the dimers formed by* $Ph_2SbS_2MPh_2$ *[M = P or As]*
(Reproduced from *J. Chem. Soc., Dalton Trans.*, 1986, 1031)

there is one short Sb—S contact at 2.490(1) Å for M = P and at 2.486(2) Å for M = As, but there are also two longer Sb$\cdots$S contacts (3.440 and 3.474 Å for M = P and 3.590 and 3.369 Å for M = As) which lead to the formation of dimers and eight-membered $Sb_2S_4M_2$ rings with transanular Sb$\cdots$S interactions. The crystal structures of the hexachloroantimonate salts $[Me_2SSMe]SbCl_6$, $[(MeS)_3]SbCl_6$ and $[(MeSe)_3]SbCl_6$ have also been determined by X-ray crystallography[56] and each cation found to involve a three coordinated chalcogen atom. The nature of the

[52] U. Strom and J. A. Freitas, *Solid State Commun.*, 1986, **59**, 565.
[53] U. Muller and A. T. Mohammed, *Z. Anorg. Allg. Chem.*, 1986, **533**, 65.
[54] R. J. D. Tilley and A. C. Wright, *J. Solid State Chem.*, 1986, **64**, 1.
[55] C. Silvstru, L. Silaghi-Dumitrescu, I. Haiden, M. J. Begley, M. Nunn, and D. B. Sowerby, *J. Chem. Soc., Dalton Trans.*, 1986, 1031.
[56] R. Laitinen, R. Steudel, and R. Weiss, *J. Chem. Soc., Dalton Trans.*, 1986, 1095.

packing in the compounds containing the $(MeS)_3^+$ and $(MeSe)_3^+$ ions was found to be disimilar and to be associated with the different colours of the compounds.

An important feature of the crystal structures of salts containing MX_3^+ cations (M = S, Se, Te; X = F, Cl, Br, I) is the strong secondary interaction between the cations and the accompanying anions; it is interesting to note therefore the reported[57] synthesis, Raman spectra, and crystal structures of the compounds $(SCl_3)(SbCl_6)$, $(SeCl_3)(SbCl_6)$, $(SCl_{1.2}Br_{1.8})(SbCl_6)$, $(TeCl_3)(AlCl_4)$, $(TeCl_3)(SbF_6)$, $(TeCl_3)(AsF_6)$, and $(TeF_3)_2(SO_4)$. All the MX_3^+ cations in the compounds were found to be involved in significant anion–cation secondary bonding interactions of varying strengths and geometries. The trends in the overall MX_3^+ geometry and the strength of the anion–cation interactions with changing M and/or X were discussed in terms of *trans*-relationships between the lengths of these interactions and primary bond lengths. The $SBr_{1.2}Cl_{1.8}^+$ cation in the hexachloroantimonate salt was shown to consist of a disordered mixture of $SBr_xCl_{3-x}^+$ cations. The tetrachloroaluminate salt of $TeCl_3^+$ was shown to be a triclinic modification with a different crystal packing from that previously reported.

The reaction of bismuth(III) chloride with a polysulphide solution in acetonitrile has been found to give the complex anion $[Bi_2S_{34}]^{4-}$ in which the bismuth atoms are coupled by an S_6^{2-} ligand and in which two bidentate S_7^{2-} ligands complete the square-pyramidal coordination of the two metal atoms.[58] The report is the first example of a complex with coupled S_6^{2-} ligands and/or bidentate S_7^{2-} ligands.

Interest in the chemistry of compounds containing Group IV elements and sulphur has also been maintained. For example, in studies of silicon–sulphur compounds[59] hexa(tri-t-butoxy)disilthiane has been prepared from the reaction of R_3SiSNa with R_3SiCl (R = tri-t-butoxy) and the crystal structure described. In a subsequent investigation[60] the compound lead(II)-bis-tri-butoxysilanethiolate was prepared from $(Bu^tO)_3SiSH$ and lead(II) oxide. The central four-membered Pb_2S_2 ring of the dimer was found to adopt a puckered butterfly conformation with the folding occurring at the lead atoms.

The synthesis and spectroscopic characterization of seven dithiocarbamate derivatives of methyldimethyl-, and trimethyl-germane of general formula $Me_nGeX_{4-m-n}[S_2CNMe_2]_m$ (where X = Cl, Br, I, n = 1-3, m = 1,2) has been described.[61] The crystal structure of dimethylbis(N,N-dimethyldithiocarbamato)germanium showed the germanium atom to be at the centre of a distorted octahedron. Investigations of metal complexes of thiocholinate, $^-SCH_2CH_2NMe_3^+$, and the preparation and crystal structure of pentakis(thiocholinato)di-lead(II) hexafluorophosphate have been reported.[62]

Many studies have been reported of transition-metal sulphides and of these considerable interest continues in catalytically active compounds containing molybdenum and sulphur. For example, the optical and catalytic properties of molybdenum

[57] B. M. Christian, M. J. Collins, R. J. Gillespie, and J. F. Sawyer, *Inorg. Chem.*, 1986, **25**, 777.

[58] A. Muller, M. Zimmermann, and H. Bogge, *Angew. Chem., Int. Ed. Engl.*, 1986, **25**, 273.

[59] W. Wojnowski, W. Bochenska, K. Peters, E.-M. Peters, and H. G. von Schnering, *Z. Anorg. Allg. Chem.*, 1986, **533**, 165.

[60] W. Wojnowski, M. Wojnowski, K. Peters, E.-M. Peters, and H. G. von Schnering, *Z. Anorg. Allg. Chem.*, 1986, **535**, 56.

[61] R. K. Chadha, J. E. Drake, and A. B. Sarkar, *Inorg. Chem.*, 1986, **25**, 2201.

[62] I. G. Dance, P. J. Guerney, A. D. Rae, M. L. Scudder, and A. T. Baker, *Aust. J. Chem.*, 1986, **39**, 383.

sulphide surfaces have been studied in single crystals and disordered powders[63] and the i.r. optical absorption due to edge planes in the single crystal platelets found to be proportional to the catalytic activity for hydrodesulphurization. The results enabled a correlation to be made between the electronic structure of these materials, as measured by optical absorption, and their catalytic performance. In a different type of study, nuclear quadrupole interactions as measured by time differential perturbed angular correlations on ^{99}Tc were used to characterize molybdenum sulphides and an alumina-supported molybdenum catalyst.[64] The technique was shown to be successful for the detection of anionic vacancies and hydrogen uptake in highly dispersed unsupported molybdenum disulphide.

A review[65] has appeared of the coordination chemistry of molybdenum- and tungsten–sulphur compounds, including aspects relevant to their properties as hydro-desulphurization catalysts. Recent work on ligand- and induced-internal redox processes in molybdenum- and tungsten–sulphur systems has also been well summarized[66] and the ability of S^{2-} ligands to act as donors and of molybdenum and tungsten to act as acceptors of electron density shown to be exploitable in the synthesis of new heteronuclear 'thiocubane' clusters. It is also interesting to note the low-temperature synthesis of mixed nickel–molybdenum sulphides together with an assessment of their structural, textural, and catalytic properties[67] and also a study[68] by X-ray photoelectron spectroscopy of electronic effects in mixed nickel-tungsten sulphide catalysts.

Studies of phase relationships in the vanadium–molybdenum–sulphur system formed in sealed silica tubes at 1373 K have proved[69] the existence of three solid solution series: cubic metallic alloy $V_xMo_{1-x}(0 \leqslant x \leqslant 1)$, monoclinic $(V_xMo_{1-x})_{2.06}S_3$ $(0 \leqslant x \leqslant 0.06)$, and monoclinic $V_{3-x}Mo_xS_4$ $(0 \leqslant x \leqslant 2)$. The structures of VMo_2S_4 and V_2MoS_4 were determined from X-ray powder diffraction data. It is also interesting to note the studies[70] of di- and tri-nuclear complexes of $(WS_4)^{2-}$ with copper(I) chloride, including the structural characterization of $(PPh_4)_2Cu_2Cl_2WS_4$, and the preparation[71] of the incomplete cubane-type triangular tungsten(IV) aqua ion $W_3S_4^{4+}$ and the X-ray crystal structure of a derivative complex $(bpyH)_5[W_3S_4(NCS)_9] \cdot 3H_2O$. The preparation and characterization of a trimeric complex containing the incomplete cubane-type sulphur capped $Mo_3O_2S_2^{4+}$ aqua ion has been described[72] and the X-ray structure analysis of the derivative complex $(pyH)_5[Mo_3O_2S_2(NCS)_9] \cdot 2H_2O$ reported. A novel[73] conversion of $[NH_4]_2 [W^{VI}OS_3]$ into $[W^V_2O_2(\mu\text{-}S)_2(S_2)_2]^{2-}$ by its treatment with elemental iodine in aqueous solution and the transformation of the product back to the starting material by treatment with S_x^{2-} has been described.

[63] C. B. Roxlo, M. Daage, A. F. Ruppert, and R. R. Chianelli, *J. Catalysis*, 1986, **100**, 176.
[64] C. Vogt, T. Butz, A. Lerf, and H. Knozinger, *Polyhedron*, 1986, **5**, 95.
[65] A. Muller, *Polyhedron*, 1986, **5**, 323.
[66] M. A. Harmer, T. R. Halbert, W.-H. Pan, C. L. Coyle, S. A. Cohen, and E. I. Stiefel, *Polyhedron*, 1986, **5**, 341.
[67] F. B. Garreau, H. Toulhoat, S. Kasztelan, and R. Paulus, *Polyhedron*, 1986, **5**, 211.
[68] L. Blanchard, J. Grimblot, and J. P. Bonnelle, *J. Catalysis*, 1986, **98**, 229.
[69] H. Wada, M. Onoda, H. Nozaki, and I. Kowada, *J. Solid State Chem.*, 1986, **63**, 369.
[70] F. Secheresse, M. Salis, C. Potvin, and J. M. Manoli, *Inorg. Chim. Acta*, 1986, **114**, L19.
[71] T. Shibahara, K. Kohda, A. Ohtsuji, K. Yasuda, and H. Kuroya, *J. Am. Chem. Soc.*, 1986, **108**, 2757.
[72] T. Shibahara, T. Yamada, and H. Kuroya, *Inorg. Chem. Acta*, 1986, **113**, L19.
[73] S. Sarkar and M. A. Ansari, *J. Chem. Soc., Chem. Commun.*, 1986, 324.

Significant interest has also been maintained in the metal molybdenum chalco-genides, particularly sulphides, which are known as the Chevral phases. A review[74] has been published describing the impact which these new superconducting materials have made on preparative solid-state chemistry and of their novel pseudo-molecular structures in terms of cluster coupling. The review also assesses the performance of these materials as superconductors with high critical temperatures and critical magnetic fields and their ability to exhibit both superconductivity and long-range magnetic order. Some interesting prospects for the future including magnetic-field induced superconductivity, the development of superconducting wires, the electron donor–acceptor role of the Mo_6 cluster, and the cation donor–acceptor role of the channels in the rigid Mo_6X_8 units which give the materials potential as cationic electrodes for secondary batteries were also evaluated. The tendency of the M_6X_8 clusters to form *exo* bonds to donor ligands has been explained[75] in terms of the electronic structure of the cluster core and the σ-acceptor character of the LUMO orbitals of the cluster. The phase field of the Chevrel phase $PbMo_6S_8$ has been investigated by X-ray diffractometry[76] and, at 900 °C, the $PbMo_6S_8$ phase shown to be surrounded by three two-phase regions: $PbMo_6S_8$ and Pb, $PbMo_6S_8$ and Mo, and $PbMo_6S_8$ and MoS_2. The superconducting transition-temperatures of various $Pb_xMo_6S_{8-y}$ samples were measured inductively and the critical temperature T_c was found to be higher in materials with the higher sulphur concentrations. Studies by X-ray analysis of single crystals of $Eu_xMo_6S_8$ have shown[77] an upper stoicheiometry of $x = 1.0$. The europium atoms were found to be divalent and a well-defined anomaly in the electrical resistivity at 107 K was shown to separate the metallic high-temperature rhombohedral phase from the low temperature non-metallic (tri-clinic) phase. The large residual resisitivity ratios obtained from stoicheiometric crystals were interpreted in terms of the semiconducting nature of the low tem-perature $Eu_{1.0}Mo_6S_8$ phase. Finally, an interesting study[78] of the condensation of molybdenum–sulphur clusters has been published describing the synthesis of a new series of ternary molybdenum sulphide phases of composition $M_{3.4}Mo_{15}S_{19}$ (M = vacancy, Li, Na, K, Zn, Cd, Sn, or Tl). The materials were prepared by removing indium from the $In_{3.4}Mo_{15}S_{19}$ compound by oxidation with hydrogen chloride to give the new binary phase $Mo_{15}S_{19}$, from which the other ternary compounds were synthesized.

Considerable interest has also been maintained in cadmium sulphide particles. For example, additional boundary conditions for excitonic polaritons in CdS have been experimentally determined[79] and, representative of a different area of interest, the photodecomposition of hydrogen sulphide by visible light (>400 nm) has been examined[80] in alkaline aqueous media in the presence of CdS dispersions and in mixtures composed of combinations of CdS and alumina or ruthenium dioxide. The fivefold increase in the hydrogen evolution rate obtained with the latter system was

[74] R. Chevrel, M. Hirrien, and M. Sergent, *Polyhedron*, 1986, **5**, 87.

[75] T. Hughbanks, *Inorg. Chem.*, 1986, **25**, 1492.

[76] H. Yamasaki and Y. Kimura, *Mat. Res. Bull.*, 1986, **21**, 125.

[77] O. Pena, R. Horyn, C. Geantet, P. Gougeon, J. Padiou, and M. Sergent, *J. Solid State Chem.*, 1986, **63**, 62.

[78] J. M. Tarascon and G. W. Hull, *Mat. Res. Bull.*, 1986, **21**, 859.

[79] I. Broser, K.-H. Pantke, and N. Rosenzweig, *Solid State Commun.*, 1986, **58**, 441.

[80] E. Borgarello, N. Serpone, M. Gratzel, and E. Pelizzetti, *Inorg. Chim. Acta*, 1986, **112**, 197.

associated with an inter-particle electron-transfer pathway in which the conduction band electrons of the excited CdS were transferred to the redox catalyst in competition with electron–hole pair recombination. In a study[81] of a number of powdered semiconductors the photoassisted reduction of nitrogen oxyanions to ammonia in alkaline aqueous sulphide solutions was achieved by illumination with visible light in the presence of suspended CdS as CdS–ZnS. It has been demonstrated[82] that CdS semiconductor particles can induce the photodecomposition of formate and the recent report[83] of photocatalysed transformation of cyanide to thiocyanate by rhodium-loaded cadmium sulphide in alkaline aqueous sulphide media has environmental implications.

X-ray-induced luminescence measurements of barium sulphide phosphors have shown[84] that the luminescence may be associated with native defects and those induced by impurities. The mechanism of the $^4T_1 \rightarrow {}^6A_1$ transition which gives rise to the orange emission of Mn^{2+} in the cubic crystal field of CaS has been investigated[85] in powders of composition $CaS:Mn_x$ where $x = 0.1, 0.3, 0.8, 1.3$ atom %. The substitution of europium and sulphur for calcium in calcium sulphoapatite has been found[86] to give the continuous series $Ca_{10-x}Eu_x(PO_4)_6S_{1-x/2}$ in which Eu^{3+} substitutes at only certain calcium sites. It is also interesting to note the study[87] of the kinetics and mechanism of the thermal decomposition of sodium sulphide pentahydrate to an essentially amorphous form in the controlled water vapour pressure range 3–20 Torr.

Transition-metal disulphides have also continued to attract substantial interest. Titanium disulphide has received significant attention, for example highly stoicheiometric TiS_2 has been compositionally characterized and studied chemically, structurally, magnetically, and thermogravimetrically.[88] The solid solutions $(Zr_xTi_{1-x})_{1+y}S_2$ have been prepared and characterized;[89] when $x < 0.5$ the reaction mixture contained both non-stoicheiometric $M_{1+y}S_2$ and a small amount of MS_3 whilst when $x > 0.5$ only stoicheiometric MS_2 was formed. X-ray photoelectron spectroscopy and conductivity measurements showed that the metal-to-insulator transition which occurs near an anomaly in the lattice parameters at $x = 0.5$ may be associated with non-stoicheiometry and disorder phenomena.

Molybdenum disulphide has also attracted interest. For example, MoS_2 which had been exfoliated into monolayers by intercalation with lithium and reaction with water was shown[90] by X-ray diffraction to exist as one-molecule-thick sheets whilst the dried films were found to be randomly stacked. In another study[91] the photoelectrochemical properties of MoS_2 single crystals grown in the presence of cobalt were compared with those of pure MoS_2 and the results associated with the segregation of cobalt to the surface edges of the doped crystals. The transport properties of

[81] M. Halmann and K. Zuckerman, *J. Chem. Soc., Chem. Commun.*, 1986, 455.
[82] I. Willner and Z. Goren, *J. Chem. Soc., Chem. Commun.*, 1986, 172.
[83] E. Borgarella, R. Terzian, N. Serpone, E. Pelizzetti, and H. Barbeni, *Inorg. Chem.*, 1986, **25**, 2135.
[84] R. P. Rao, *Mat. Res. Bull.*, 1986, **21**, 249.
[85] T. Yamase, *Inorg. Chim. Acta*, 1986, **114**, L35.
[86] P. R. Suitch, A. Taitai, J. L. Lacout, and R. A. Young, *J. Solid State Chem.*, 1986, **63**, 267.
[87] J. Y. Andersson and M.Azoulay, *J. Chem. Soc., Dalton Trans.*, 1986, 469.
[88] M. J. McKelvy and W. S. Glaunsinger, *Mat. Res. Bull.*, 1986, **21**, 835.
[89] D. T. Hodul and A. M. Stacy, *J. Solid State Chem.*, 1986, **62**, 328.
[90] P. Joensen, R. F. Frindt, and S. R. Morrison, *Mat. Res. Bull.*, 1986, **21**, 457.
[91] D. M. D'Ambra, R. Kershaw, J. Baglio, K. Dwight, and A. Wold, *J. Solid State Chem.*, 1986, **64**, 108.

layered-type molybdenum sulphoselenide MoS_xSe_{2-x} $(0 < x < 2)$ single crystals grown by direct vapour transport techniques have also been examined.[92]

It is interesting to report the new intercalation compound $[Fe_6S_8(PEt_3)_3]_{0.05}TaS_2$ which has been synthesized[93] by flocculation of TaS_2^- layers dispersed in water-N-methylformamide. The orientation of the intercalated cluster was determined from the intensities of the 001 X-ray diffraction peaks. The projection of the electron density along the c-axis (Figure 4a), which was obtained from the X-ray data, was used to derive the schematic illustration of the material depicted in Figure 4b.

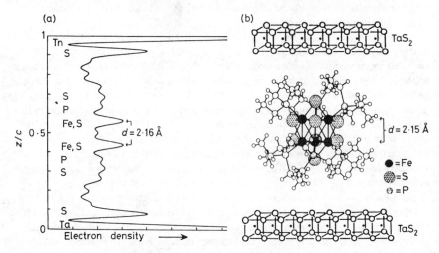

Figure 4 *(a) Projection of the electron density of $[Fe_6S_8(PEt_3)_3]_{0.05}TaS_2$ along the c-axis. (b) Schematic illustration showing the orientation of the iron cluster between the TaS_2 layers as deduced from the interlayer electron density*
(Reproduced from *J. Chem. Soc., Chem. Commun.*, 1986, 570)

The susceptibility of the iron-intercalated zirconium disulphide $Fe_{0.25}ZrS_2$ has been shown[94] to depend strongly on the applied field at lower temperatures, and the cusp at lower fields has been attributed to spin-glass properties below 15 K. The magnetic, but unresolved, Mössbauer spectrum at 4 K showed a high degree of disorder of spin orientations which underwent relaxation at higher temperatures. New polymorphs of the rare-earth polysulphides PrS_2 and NdS_2 have been synthesized under high temperature/pressure conditions.[95]

The triniobium tetrachalcogenides Nb_3X_4 (X = S, Se, Te), which were structurally characterized over twenty years ago, continue to attract interest. Tight-binding band calculations have been performed[96] on Nb_3S_4 and model Nb_2S_6 chains to examine the relationship between the electronic structure of Nb_3S_4 and its crystal structure. Various metal trisulphides have also been examined and it is interesting to note the

[92] M. K. Agarwal and L. T. Talele, *Solid State Commun.*, 1986, **59**, 549.
[93] L. F. Nazar and A. J. Jacobson, *J. Chem. Commun.*, 1986, 570.
[94] M. A. Buhannic, P. Colombet, and M. Danot, *Solid State Commun.*, 1986, **59**, 77.
[95] Y. Yanagisawa and S. Kume, *Mat. Res. Bull.*, 1986, **21**, 379.
[96] E. Canadell and M.-M. Whangbo, *Inorg. Chem.*, 1986, **25**, 1488.

use of EXAFS to investigate[97] the structural changes which occur during the lithiation of the amorphous materials MoS_3, WS_3, and WSe_3 and also the studies[98] of the conductivity and thermoelectric power of NbS_3-type compounds. The low-temperature specific heats of the semiconducting compounds ZrS_3 and $ZrSe_3$ have been measured[99] in the temperature range between 8 and 200 K.

The layered metal phosphorous trichalcogenides have continued to attrct attention. The structure of $ZnPS_3$ has been determined[100] from single crystal X-ray diffraction data and has been shown to adopt the expected $CdCl_2$-type arrangement. The S K and P K absorption spectra of the compounds MPS_3 (M = Mg, Mn, Fe, Ni, Zn, Cd, Sn) have been recorded[101] and discussed in terms of the electronic structures of the materials.

The compound Na_5FeS_4 has been synthesized[102] from a stoicheiometric mixture of Na_2S, Fe, and S at 970 K and characterized as the first thioferrate(III) compound with discrete tetrahedral anions. The phase diagrams of the quaternary systems MS-Cr_2S_3-In_2S_3 (M = Mn, Fe, Ni) have been studied[103] by powder X-ray diffraction and immiscibility domes found to exist below 800 and 850 °C in the quasibinary sections $MnIn_2S_4$-$MnCr_2S_4$ and $FeIn_2S_4$-$FeCrS_4$ respectively. The quaternary system $NiCr_{2-2x}Ga_{2x}S_4$ has also been studied[104] by powder X-ray diffraction and the crystal structure of the ternary $NiGa_2S_4$ determined from single crystal data.

The recent interest in the zinc–indium–sulphur system which emanates from its interesting semiconductive, photoconductive, and luminescent properties has been reflected in several studies. For example, seven different phases have been identified by X-ray diffraction methods in the indium-rich side of the quaternary system $ZnIn_2S_4$-$ZnIn_2Se_4$-In_2Se_3-In_2S_3 when quenched from 800 °C to room temperature.[105] The optical and electrical properties of the layered compound $Zn_3In_2S_6$, which is the third member of the system $mZnS$-In_2S_3 (m = 3), have also been described.[106]

The vibrational spectra recorded by infrared and Raman spectroscopy from polycrystalline samples of two-dimensional P_2NbS_8, PNb_2S_{10}, and $P_2Nb_4S_{21}$ at 300 K have been[107] compared with those from NbS_2Cl_2 and various other phosphorus/sulphide anion-containing salts, and interpreted in terms of $\{Nb_2(S_2)_2\}$ cluster motions and the structural relationships within the niobium–phosphorus–sulphur system. The u.v.-visible spectra confirmed the broadband semiconductor properties of the materials.

Some lanthanide–sulphur systems have also attracted attention. For example, the extent of the glass forming region in the La_2S_3-La_2O_3-Ga_2O_3-Ga_2S_3 system has been

[97] R. A. Scott, A. J. Jacobson, R. R. Chianelli, W.-H. Pan, E. I. Stiefel, K. O. Hodgson and S. P. Cramer, *Inorg. Chem.*, 1986, **25**, 1461.
[98] B. Fisher and M. Fibich, *Solid State Commun.*, 1986, **59**, 187.
[99] R. Provencher, C. Ayache, S. Jandl, and J.-P. Jay-Gerim, *Solid State Commun.*, 1986, **59**, 553.
[100] E. Prouzet, G. Ouvrard, and R. Brec, *Mat. Res. Bull.*, 1986, **21**, 195.
[101] Y. Ohno and K. Hirama, *J. Solid State Chem.*, 1986, **63**, 258.
[102] K. O. Klepp and W. Bronger, *Z. Anorg. Allg. Chem.*, 1986, **532**, 23.
[103] H. D. Lutz, M. Jung, and K. Wussow, *Mat. Res. Bull.*, 1986, **21**, 161.
[104] H. D. Lutz, W. Buchmeier, and H. Siwert, *Z. Anorg. Allg. Chem.*, 1986, **533**, 118.
[105] H. Haeuseler and M. Himmrich, *Z. Anorg. Allg. Chem.*, 1986, **535**, 13.
[106] A. Anagnostopoulos, K. Kambas, and J. Spyridelis, *Mat. Res. Bull.*, 1986, **21**, 407.
[107] M. Queignec, M. Evain, R. Brec, and C. Sourisseau, *J. Solid State Chem.*, 1986, **64**, 89.

described[108] as a function of the composition and the cooling speed and three new phases of formula $(EuO)_2Ga_4S_7$, $EuGaOS_2$, and $(EuO)_4Ga_2S_5$ have been identified in the Ga_2S_3–Eu_2O_2S system.[109] In a study of high pressure polymorphism in rare-earth sulphideiodides[110] several compounds from GdSI to DySI with the CdSI-type structure were shown to transform to the SmSI-type within the pressure range up to 40 kbar.

The structure of a new synthetic sulphosalt $KHgSbS_3$ has been shown[111] to consist of trigonal pyramids of SbS_3 units which are linked together by mercury atoms to form a two-dimensional sheet-like network containing weakly bonded potassium atoms. In a study of complex formation in pyrosulphate melts[112] the calorimetric properties of the systems V_2O_5–$K_2S_2O_7$, $V_2O_5 \cdot K_2S_2O_7$–K_2SO_4, $V_2O_5 \cdot 2K_2S_2O_7$–K_2SO_4, and $V_2O_5 \cdot 3K_2S_2O_7$–K_2SO_4 at 430 °C were described and discussed in terms of their implications for the catalytic oxidation of SO_2 to SO_3. In a subsequent study[113] the crystal structure, i.r., and Raman spectra of $KV(SO_4)_2$, which was synthesized by the dissolution of V_2O_5 in a $KHSO_4$ melt at 450 °C under sulphur dioxide, were reported and interpreted. The aqueous chemistry of oxovanadium(v) complexes has been extensively studied whilst oxosulphidovanadium(v) chemistry has received little attention and so it is interesting to note that the three known monomeric sulphido- and oxosulphido-vanadium(v) species $[VS_4]^{3-}$, $[VOS_3]^{3-}$, $[VO_2S_2]^{3-}$ have been identified[114] by high field vanadium-51 n.m.r. spectroscopy together with the previously unobserved ions $[VO_3S]^{3-}$, $[V_2S_7]^{4-}$, $[O_3VSVO_3]^{4-}$, $[SO_2VSVO_2S]^{4-}$ and the unprotonated monomers.

Sulphur dioxide has been found[115] to reduce a number of first-row transition-metal compounds in molten ternary sulphate eutectic. The results are relevant to the catalytic oxidation processes mentioned above and to current interests in the corrosion of high-temperature boilers and heat exchangers. Electron spin resonance has been used to identify[116] a sulphoperoxy radical on titanium dioxide following ultraviolet irradiation in the presence of sulphur dioxide and oxygen. In another study[117] the net reduction of peroxotitanium(IV) by sulphur(IV) in acidic solution was shown to proceed by a rate-determining dissociation of peroxide from TiO_2^{2+} followed by the rapid reaction of H_2O_2 and HSO_3^-. Separate peaks in the oxygen-17 n.m.r. spectra of sodium bisulphite solutions have given evidence[118] for the existence in solution of two isomers of the bisulphite ion: one with the proton bonded to the sulphur atom (HSO_3^-) and the other with the proton bonded to an oxygen atom (SO_3H^-). The kinetics of the reactions of dissolved nitric oxide with sulphite and

[108] M. Guittard, A. M. Loireau-Lozach, F. Berguer, S. Barnier, and J. Flahaut, *J. Solid State Chem.*, 1986, **62**, 191.
[109] I. B. Bakhtiyarov, P. G. Rustamov, S. M. Nakhmetov, and V. A. Gasymov, *Z. Anorg. Allg. Chem.*, 1986, **533**, 186.
[110] H. P. Beck and C. Strobel, *Z. Anorg. Allg. Chem.*, 1986, **535**, 229.
[111] M. Imafuka, I. Nakai, and K. Nagashima, *Mat. Res. Bull.*, 1986, **21**, 493.
[112] R. Fehrmann, M. Gaune-Escard, and N. J. Bjerrum, *Inorg. Chem.*, 1986, **25**, 1132.
[113] R. Fehrmann, B. Krebs, G. N. Papatheodorou, R. W. Berg, and N. J. Bjerrum, *Inorg. Chem.*, 1986, **25**, 1517.
[114] A. T. Harrison and O. W. Hawarth, *J. Chem. Soc., Dalton Trans.*, 1986, 1405.
[115] R. I. Dearnaley, D. H. Kerridge, and D. J. Rogers, *Inorg. Chem.*, 1986, **25**, 1721.
[116] A. R. Gonzalez-Elipe and J. Soria, *J. Chem. Soc., Faraday Trans. 1*, 1986, **82**, 739.
[117] R. C. Thompson, *Inorg. Chem.*, 1986, **25**, 184.
[118] D. A. Horner and R. E. Connick, *Inorg. Chem.*, 1986, **25**, 2414.

bisulphite ions over the pH range of 4–10 have been described.[119] The sulphur K-edge X-ray absorption spectra of SO_4^{2-}, SO_3^{2-}, $S_2O_3^{2-}$, and $S_2O_x^{2-}$ ($x = 5$–8) have been measured by synchrotron radiation[120] and the strong similar absorption bands were found to be associated with bound-state transitions to antibonding $3p$-type orbitals whilst the in shape resonances were related to $3d$-type orbitals.

The nature of the bonding and the extent of reactivity of carbon disulphide with iron has been investigated[121] by matrix isolation experiments in which either pure CS_2 or CS_2-Ar and CS_2-Kr mixtures were co-condensed with iron atoms and examined by i.r. and Mössbauer spectroscopy. The rate of reduction of plutonium(VI) and neptunium(VI) by bisulphide ion in neutral and mildly alkaline solutions has been studied[122] and the mechanism discussed in terms of an intramolecular reaction involving an unusual actinide(VI)–hydroxide–bisulphide complex.

Several new multi-sulphur 1,2-dithiolene complexes have been prepared and characterized,[123] and the syntheses and reactions of several polyfluorosulphates have been examined.[124] The stereochemistry of intramolecular homolytic substitution at the sulphur atom of a chiral sulphoxide has been described[125] and three hexacyclen complexes of the sulphate anion have been reported.[126] Hydrazine has been shown[127] to react with COS in the presence of sodium methoxide to produce the reactive product $Na_2[SOC-NH-NH-COS]$. The cycloaddition of dithiocyanogen and trithiocyanogen with hexafluoroacetone has been reported and the cleavage of the sulphur–sulphur bond in the dithiocyan-hexafluoroacetone adduct by elemental chlorine has been described.[128]

The coordination chemistry of thioethers has come under increasing scrutiny as a result of their potential to parallel the behaviour of phosphines as well as their occurrence in the blue copper proteins. In this connection a study[129] has been reported of the electronic consequences of thioether coordination which has demonstrated the ability of hexathia-24-crown-6 to wrap around nickel(II) to yield an octahedral cation with meridional stereochemistry as the first reported complex of this ligand. In an investigation[130] of palladium(II) and platinum(II) complexes of 1,4,7,10,13,16-hexathiacyclooctadecane the palladium or platinum atoms were found to be coordinated to four sulphur atoms in a square-planar fashion with only long-range weak axial interactions being observed between the two remaining sulphur atoms of the macrocycle.

The structural chemistry of metal thiolate complexes has been reviewed[131] with special attention being given to complexes of monofunctional thiolate ligands. Within

[119] D. Littlejohn, K. Y. Hu, and S. G. Chang, *Inorg. Chem.*, 1986, **25**, 3131.

[120] H. Sekiyama, N. Kosugi, H. Kuroda, and T. Ohta, *Bull. Chem. Soc. Jpn.*, 1986, **59**, 575.

[121] M. M. Doeff, *Inorg. Chem.*, 1986, **25**, 2474.

[122] K. L. Nash, J. M. Cleveland, J. C. Sullivan, and M. Woods, *Inorg. Chem.*, 1986, **25**, 1169.

[123] R. Kato, M. Kobayashi, A. Kobayashi, and Y. Sasaki, *Bull. Chem. Soc. Jpn.*, 1986, **59**, 627.

[124] T.-J. Huang and J. M. Shreeve, *Inorg. Chem.*, 1986, **25**, 496.

[125] A. L. J. Beckwith and D. R. Boate, *J. Chem. Soc., Chem. Commun.*, 1986, 189.

[126] R. I. Gelb, L. M. Schwartz, and L. J. Zompa, *Inorg. Chem.*, 1986, **25**, 1527.

[127] G. Gattow and S. Lotz, *Z. Anorg. Allg. Chem.*, 1986, **533**, 109.

[128] H. W. Roesky, N. K. Homsy, and H. G. Schmidt, *Z. Anorg. Allg. Chem.*, 1986, **532**, 131.

[129] S. C. Rawle, J. A. R. Hartman, D. J. Watkin, and S. R. Cooper, *J. Chem. Soc., Chem. Commun.*, 1986, 1083.

[130] A. J. Blake, R. O. Gould, A. J. Lavery, and M. Schroder, *Angew. Chem., Int. Ed. Engl.*, 1986, **25**, 274.

[131] I. G. Dance, *Polyhedron*, 1986, **5**, 1037.

the field of metal–thiolate chemistry significant interest has been maintained in the nature of molybdenum–thiolate complexes because of their general relevance to the study of molybdenum-containing enzymes and to nitrogenase in particular. This is well illustrated by a study[132] of mononuclear thiolate complexes of molybdenum(II) and molybdenum(III) which included a crystal-structure determination of trans-[Mo(SBu^n)$_2$(Ph$_2$PCH$_2$CH$_2$PPh$_2$)$_2$], and also by a different investigation[133] of the synthesis and structural properties of molybdenum complexes with sterically hindered thiolato-ligands. In another study[134] of a number of five-coordinate nitrosyl thiolato-complexes of molybdenum and tungsten the crystal structures of [NHEt$_3$]-[Mo(SPh)$_4$(NO)], [PPh$_4$][WCl(SPh)$_3$(NO)], and [Mo(SC$_6$H$_2$Pr$^i_3$-2,4,6)$_3$(NH$_3$)-(NO)] were found to be characterized by the trigonal-bipyramidal geometries of the anions. The synthesis, crystal structure, spectroscopic, and electrochemical properties of tris(quinoxaline-2,3-dithiolato)molybdate(IV), [Mo(qdt)$_3$]$^{2-}$, have been reported[135] and the complex shown to undergo reversible protonation on the ligand nitrogen atoms with the triply protonated species [Mo(Hqdt)$_3$]$^+$ being isolated and characterized as its chloride salt. The synthesis, spectral properties, and substitution reactions of a series of seven-coordinate M(CO)$_3$(S$_2$CNC$_4$H$_4$)$_2$ complexes, where M = Mo, W, each containing the electronically unique pyrrole-N-carbodithioate ligand have been reported.[136] The electrochemical half-wave potentials, charge-transfer absorption properties, e.s.r. spectra, and magnetic properties of a series of seven- and eight-coordinated complexes of molybdenum(V) and (VI) with 2-amino-1-carbodithioato ligands have been described.[137] In an investigation of the unique reactivity characteristics of molybdenum-coordinated S$_2^{2-}$ and S$_4^{2-}$ ligands the synthesis and properties of the new dithiolene complexes [{(MeOOC)$_2$C$_2$S$_2$}$_2$Mo(μ_2-S)]$_2^-$ and [OMo{S$_2$C$_2$(COOMe)$_2$}$_2$]$^{2-}$ have been reported.[138] The molecular structure of the dimethylacetylenedicarboxylate adduct of bis(diethyldithiocarbamato) (p-tolylimido)molybdenum(IV), Mo(Ntol)(DMAC)(S$_2$CNEt$_2$)$_2$ has been described.[139]

Whilst attention to biological nitrogenase and industrial hydrodesulphurization processes has undoubtedly spurred interest in molybdenum–sulphur chemistry in recent years, it must also be acknowledged that the variety of structural types and the diversity of reactions observed in molybdenum–sulphur systems has also contributed to continued interest in this area of chemistry. Studies[140] of reactions at dimolybdenum(V) sulphur bridges and the formation of disulphide(S$_2^{2-}$), monosubstituted organic disulphide (RSS$^-$), and tetrasulphide (S$_4^{2-}$) bridges. The redox properties of nickel(II) and palladium(II) complexes containing tetrathiomolybdate, tetrathiotungstate, and dialkyldithiocarbamate ligands, [M(M'S$_4$)$_n$(R$_2$NCS$_2$)$_{2-n}$]$^{n-}$ where n = 0—2; M = Ni or Pd; M' = Mo or W, have been studied by e.s.r. and by d.c. and a.c. cyclic voltammetry at a platinum electrode in dichloromethane

[132] D. C. Povey, *Polyhedron*, 1986, **5**, 369.
[133] P. T. Bishop, P. J. Blower, J. R. Dilworth, and J. A. Zubieta, *Polyhedron*, 1986, **5**, 363.
[134] P. T. Bishop, J. R. Dilworth, J. Hutchinson, and J. Zubieta, *J. Chem. Soc., Dalton Trans.*, 1986, 967.
[135] S. Boyde, C. D. Garner, J. H. Enemark, and R. B. Ortega, *Polyhedron*, 1986, **5**, 377.
[136] R. S. Herrick and J. L. Templeton, *Inorg. Chem.*, 1986, **25**, 1270.
[137] M. Chaudhury, *Polyhedron*, 1986, **5**, 387.
[138] D. Coucouvanis, A. Hadjikyriacou, M. Draganjac, M. G. Kanatzidis, and O. Ileperuma, *Polyhedron*, 1986, **5**, 349.
[139] D. D. Devore, E. A. Maatta, and F. Takusagawa, *Inorg. Chim. Acta*, 1986, **112**, 87.
[140] M. E. Noble, *Inorg. Chem.*, 1986, **25**, 3311.

solution.[141] The series of complexes showed initial reversible or quasi-reversible one-electron reduction and gave species containing a single unpaired electron. The electronic properties of a series of mixed-valence tetrasulphur-bridged Mo^{III}/Mo^{IV} dimers of general formula $[(CpMo)_2(S)_x(SR)_{4-x}]^n$ $(Cp = C_5H_5, n = +1, 0, -1)$ have been characterized.[142] The sodium salts of the anionic dimers were, on the basis of e.s.r. spectroscopy, characterized as tight ion pairs and discussed in terms of an extended Hückel molecular orbital model. The equilateral trinuclear molybdenum(IV) ion of incomplete cubane-type structure $Mo_3S_4^{4+}$ has been prepared and characterized and the X-ray crystal structure of $Ca[Mo_3S_4\{HN(CH_2CO_2)_3\}_3] \cdot 11.5H_2O$ has been described.[143] Some complexes of molybdenum(VI) with lactic, thiolactic, and thiomalic acids have been studied[144] by ^1H and ^{13}C n.m.r.

Recent interest in the preparation of synthetic models for the Fe—Mo cofactor of nitrogenase has been reflected in continued attention to the iron–molybdenum–sulphur systems. For example, a new synthetic route to Mo—Fe—S clusters related to nitrogenase has been developed[145] in which the bridging sulphides originate with the iron reagent, rather than with MoS_4^{2-} as is the case for other known Mo—Fe—S units, to produce clusters containing the novel $(Mo(\mu_3$-S$)_2Fe_2$ structural unit. The redox potentials of $[Fe_4S_4(SR)_4]^{2-}$ and $[Mo_2Fe_6S_8X_3(SR)_6]^{3-}$, where R = C_6H_4-p-n-$C_{18}H_{17}$, X = SEt, OMe, in aqueous micellar solutions have, under various conditions, been found[146] to resemble the behaviour of some ferredoxins in water. In a subsequent study[147] the multi-electron reductions of alkyl azides with $[Mo_2Fe_6S_8(SPh)_9]^{3-}$ and $[Mo_2Fe_6S_8(\mu$-SEt$)_3(SCH_2CH_2OH)_6]^{3-}$ in homogeneous systems and with a $(Bu^n_4N)[Mo_2Fe_6S_8(SPh)_9]$ modified glassy-carbon electrode were examined. More recently,[148,149] the catalytic reduction of NO_3^- to NH_3 via NO_2^- and NH_2OH has been achieved for the first time using a $(Bu^n_4N)_3[Mo_2Fe_6S_8(SPh)_9]$ modified glassy-carbon electrode.

Substantial interest has been maintained in tungsten–sulphur systems and in this respect it is especially interesting to note the synthesis[150] of the octathiotritungstate anion $W_3S_8^{2-}$, which is the first example of square-planar coordination about a tungsten atom. The structure, which is shown in Figure 5, involves tetrahedral coordination about the terminal tungsten atoms but with the central tungsten atom being located at the centre of a square plane of atoms with a very slight tetrahedral distortion. The isolation and structural characterization of the first mixed oxo thio tritungsten species $[W_3(\mu_2$-O$)_3(\mu_3$-S$)(NCS)_9]^{5-}$ is also to be noted.[19]

Hexanuclear iron–sulphur clusters have attracted attention because of their novel structures and their relevance to current interest in proteins, especially the iron–molybdenum cofactor of nitrogenase. The synthesis and properties[151] have been

[141] G. A. Bowmaker, P. D. W. Boyd, G. K. Campbell, and M. Zvagulis, *J. Chem. Soc., Dalton Trans.*, 1986, 1065.
[142] C. J. Casewit and M. Rakowski DuBois, *Inorg. Chem.*, 1986, **25**, 74.
[143] T. Shibahara and H. Kuroya, *Polyhedron*, 1986, **5**, 357.
[144] M. M. Caldeira and V. M. S. Gil, *Polyhedron*, 1986, **5**, 381.
[145] K. S. Bose, P. E. Lamberty, J. E. Kovacs, E. Sinn, and B. A. Averill, *Polyhedron*, 1986, **5**, 393.
[146] K. Tanaka, M. Moriya, and T. Tanako, *Inorg. Chem.*, 1986, **25**, 835.
[147] S. Kuwabata, K. Tanaka, and T. Tanaka, *Inorg. Chem.*, 1986, **25**, 1691.
[148] S. Kuwabata, S. Uezumi, K. Tanaka, and T. Tanaka, *J. Chem. Soc., Chem. Commun.*, 1986, 135.
[149] S. Kuwabata, S. Uezumi, K. Tanaka, and T. Tanaka, *Inorg. Chem.*, 1986, **25**, 3018.
[150] S. Bhaduri and J. A. Ibers, *Inorg. Chem.*, 1986, **25**, 3.
[151] G. L. Lilley, E. Sinn, and B. A. Averill, *Inorg. Chem.*, 1986, **25**, 1073.

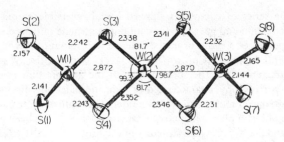

Figure 5 *Structure of the* $W_3S_8^{2-}$ *ion. the labelling scheme and some bond distances and bond angles are shown. The estimated standard deviations in the* W—W *and* W—S *bond distances and the* S—W—S *bond angles are 0.001 Å, 0.004 Å and 0.1° respectively*
(Reproduced by permission from *Inorg. Chem.*, 1986, **25**, 3)

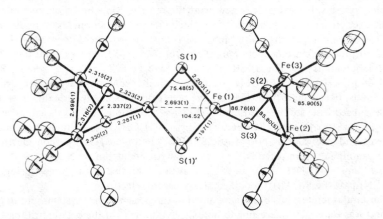

Figure 6 *Structure of the* $[Fe_6S_6(CO)_{12}]^{2-}$ *ion showing the atomic labelling scheme and selected distances and angles*
(Reproduced by permission from *Inorg. Chem.*, 1986, **25**, 1073)

reported of a novel hexanuclear iron–sulphur–carbonyl cluster, the $[Fe_6S_6(CO)_{12}]^{2-}$ dianion which consists of a central $[Fe_2S_2]^{2+}$ core ligated by a $[Fe_2S_2(CO)_6]^{2-}$ unit on each iron as shown in Figure 6. Noteworthy features of the structure include the acute S(2)—Fe(1)—S(3) angle of 86.76(6)°, the relatively large S(2)—Fe(2,3)—S(3) angle of *ca.* 85.95°, and the short Fe(2)—Fe(3) distance of 2.499(1) Å which is consistent with an iron–iron bonding interaction. In a study of the chemistry of $[Fe_6S_6]^{3+}$ prismatic cages the synthesis, structural characterization, and electronic structures of the $[Et_4N]_3[Fe_6S_6L_6]$ clusters, where L = p-MeOC$_6$H$_4$O$^-$, Br$^-$ were examined.[152] In a subsequent report,[153] the synthesis and spectroscopic characterization of the new heteropolymetallic $[Et_4N]_3[Fe_6S_6(p\text{-MeC}_6H_4O)_6\{W(CO)_3\}_2]$ cluster was described and the structure shown to contain an anion core consisting of a W_2Fe_6 cube with the two tungsten atoms located in opposite corners of the elongated body diagonal as illustrated in Figure 7.

[152] M. G. Kanatzidis, A. Salifoglou, and D. Coucouvanis, *Inorg. Chem.*, 1986, **25**, 2460.
[153] A. Salifoglou, M. G. Kanatzidis, and D. Coucouvanis, *J. Chem. Soc., Chem. Commun.*, 1986, 559.

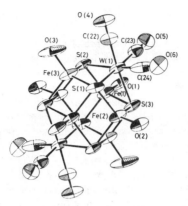

Figure 7 *Structure of the* $[Fe_6S_6(p\text{-}MeC_6H_4O)_6\{W(CO)_3\}_2]^{3-}$ *anion with the phenyl rings omitted for clarity*
(Reproduced from *J. Chem. Soc., Chem. Commun.*, 1986, 559)

A Mössbauer and e.p.r. study of the compound $(NBu_4)[Fe_4S_4(S\text{-}2,4,6\text{-}Pr^i_3\text{-}C_6H_2)_4]$ which is a synthetic analogue for the $[Fe_4S_4]^{3+}$ cluster observed in oxidized high potential iron proteins has been reported[154] and the iron–sulphur cluster $(Bu^n_4N)_2Fe_4S_4(SPh)_4$ shown[155] to act as an electron-transfer agent in the reduction of S-phenyl thiobenzoate and phenyl benzoate with n-butyl-lithium and phenyl-lithium to give benzil and benzoin.

The series $[Fe_4(\mu_3\text{-}S)_3(\mu_3\text{-}S_2)Cp_4]^{n+}$, where $n = 0$, 1, and 2, has been investigated[156] to examine the influence of variations in overall oxidation states on the core geometries, magnetic properties, and electronic distributions. In a study[157] of the resonance Raman spectra of rubredoxin new assignments and vibrational coupling mechanisms have been determined from iron-54/iron-56 isotope shifts and variable wavelength excitation data. The materials $BaFe_2S_4$ and $Ba_{13}(Fe_2S_4)_{12}$ have been imaged at high resolution in an electron microscope and the structures discussed in terms of solid solutions of moderately ordered structures.[158]

Some iron–sulphur dimers with benzimidazolate–thiolate, benzimidazolate–phenolate, or bis(benzimidazolate) terminal chelating ligands have been characterized by electronic and n.m.r. spectroscopy and shown to undergo one-electron reduction to the corresponding trianions.[159] The e.s.r. spectra showed a range of g-tensor values which were compared with those of $[2Fe-2S]^+$ proteins of the Rieske type. Several tetrapeptide 2Fe–2S complexes have been synthesized by a ligand-exchange reaction from $[Et_4N]_2[Fe_2S_2(SBu^t)_4]$ and peptide models of 2Fe ferredoxin have been characterized by electrochemistry and spectroscopy.[160] Iron and selenium EXAFS data have been used[161] to derive structural evidence on the

[154] V. Papaefthymiou, M. M. Millar, and E. Münck, *Inorg. Chem.*, 1986, **25**, 3010.
[155] H. Inoue and T. Nagata, *J. Chem. Soc., Chem. Commun.*, 1986, 1171.
[156] N. Dupré, P. Ausic, H. M. J. Hendriks, and J. Jordanov, *Inorg. Chem.*, 1986, **25**, 1391.
[157] R. S. Czernuszewicz, J. LeGall, I. Moura, and T. G. Spiro, *Inorg. Chem.*, 1986, **25**, 696.
[158] A. C. Holladay and L. Eyring, *J. Solid State Chem.*, 1986, **64**, 113.
[159] P. Beardwood and J. F. Gibson, *J. Chem. Soc., Chem. Commun.*, 1986, 490.
[160] S. Ueno, N. Ueyama, A. Nakamura, and T. Tukihara, *Inorg. Chem.*, 1986, **25**, 1000.
[161] T. D. Weatherill, T. B. Rauchfuss, and R. A. Scott, *Inorg. Chem.*, 1986, **25**, 1466.

frontier orbitals in $[Fe_2E_2(CO)_6]^{2-}$ redox-active dichalcogen ligands, where E = S or Se.

The dithiocarbene complex $Cp(CO)(MeCN)Fe=CC(SMe)_2^+$ has been shown[162] to react with $Fe(CO)_3(NO)^-$ to give the dinuclear complex $Cp(CO)Fe(\mu$-CO)$[\mu$-C-$(SMe)_2]Fe(CO)(NO)$ in which the $C(SMe)_2$ ligand bridges the two metal atoms. The synthesis, properties, and structure of tetraethylammonium tris(pyridine-2-thiolato) ferrate(II) has been described[163] and the structural, magnetic, and spectroscopic properties of some tetrakis(pyridine) complexes of iron(III) sulphonates have been reported.[164] Several gallium–sulphide–thiolate compounds that are structural analogues of the well-known Fe^{III}-S^{2-}-RS^- complexes have been prepared and structurally characterized.[165] The compound $[Fe\{S_2P(OPr^i)_2\}_3]$ has been prepared and shown by X-ray diffraction to adopt distorted octahedral coordination.[166] The rate of the reaction between sulphite and hexacyanoferrate(III) ions has been measured in several concentrated salt solutions and several isodielectric water-solvent mixtures.[167] The mode of axial thiocyanate ligand bonding in four low-spin iron macrocyclic complexes has been studied[168] and the crystal structure of $[Fe(TIM)(SCN)_2]PF_6$, where TIM is 2,3,9,10-tetramethyl-1,4,8,11-tetraazacyclotetradeca-1,3,8,10-tetraene, has been shown to involve iron–sulphur bonding. A novel series of fluxional homodinuclear iron(II) complexes with two different coordination spheres around the magnetic centres has been identified during investigations of the reactions of iron(III) halogenobisdithiocarbamates with halogens.[169] The reduction with sodium-amalgam of the cations $[Fe(\eta^2$-$CS_2Me)(CO)_2L_2]PF_6$ which contain the basic phosphines, L = PMe_3, PMe_2Ph, and PBu_3^n, has been shown[170] to give the low-valent thiocarbonyl–iron complexes of composition $Fe(CS)(CO)_2L_2$.

Substantial interest has also been maintained in copper–sulphur systems. For example, the electrochemical oxidation of anodic copper in an acetonitrile solution of RSH (R = Ph, $C_{10}H_7$, CPh_3, C_6F_4H, o-MeC_6H_4, CMe_2Et) has been shown[171] to give CuSR which, when synthesized as $CuSCMe_2Et$ and treated with carbon disulphide, yields a product of composition $Cu_8(SC_5H_{11})_4(S_2CSC_5H_{11})_4$. The structure of the compound is based on an unusual complex cage of eight copper atoms linked through thiolato (RS) and alkylthiocarbonato ($RSCS_2$) ligands. The overall structure is shown in Figure 8 and the Cu_8S_{12} cage is outlined in Figure 9. There are a number of interesting features to this unusual structure. The eight copper atoms form a series of distorted tetrahedra joined along the appropriate edges giving a structure of lower symmetry than those observed in polymeric organocopper compounds and the Cu_8 framework is capped by sulphur atoms which may be

[162] J. R. Matachek and R. J. Angelici, *Inorg. Chem.*, 1986, **25**, 2877.
[163] S. G. Rosenfield, S. A. Swedberg, S. K. Arora, and P. K. Mascharak, *Inorg. Chem.*, 1986, **25**, 2109.
[164] J. S. Haynes, S. J. Rettig, J. R. Sams, R. C. Thompson, and J. Trotter, *Can. J. Chem.*, 1986, **64**, 429.
[165] L. E. Maelia and S. A. Koch, *Inorg. Chem.*, 1986, **25**, 1896.
[166] M. G. B. Drew, W. A. Hopkins, P. E. Mitchell, and T. Colclough, *J. Chem. Soc., Dalton Trans.*, 1986, 351.
[167] A. Rodriguez, S. Lopez, M. Carmen Carmona-Guzman, and F. Sanchez, *J. Chem. Soc., Dalton Trans.*, 1986, 1265.
[168] M. J. Maroney, E. O. Fey, D. A. Baldwin, R. E. Stenkamp, L. H. Jensen, and N. J. Rose, *Inorg. Chem.*, 1986, **25**, 1409.
[169] F. D. Vakoulis, G. A. Katsoulos, and C. A. Tsipis, *Inorg. Chim. Acta*, 1986, **112**, 139.
[170] D. Touchard, C. Lelay, J.-L. Fillaut, and P. H. Dixneuf, *J. Chem. Soc., Chem. Commun.*, 1986, 37.
[171] R. Chadha, R. Kumar, and D. G. Tuck, *J. Chem. Soc., Chem. Commun.*, 1986, 188.

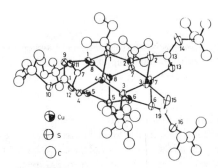

Figure 8 *The structure of* $Cu_8(SC_5H_{11})_4(S_2CSC_5H_{11})_4$. *Copper and sulphur atoms are drawn with 50% probability elipsoids*
(Reproduced from *J. Chem. Soc., Chem. Commun.*, 1986, 188)

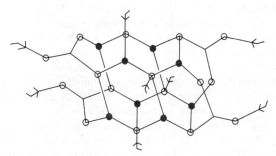

Figure 9 *The outline of the* Cu_8S_{12} *cage*
(Reproduced from *J. Chem. Soc., Chem. Commun.*, 1986, 188)

mono-, bi-, or tri-dentate. It is also interesting to note the presence in the Cu_8S_{12} cage of a series of six-membered rings, some of which are in the boat form.

The conditional stability constants for complexes formed between copper(II) and several cyclic polythiaether ligands in the presence of perchlorate ion have been determined spectrophotometrically[172] and the structure of dicopper thiobis(ethylenenitrilo)tetraacetate pentahydrate has been shown[173] to consist of dinuclear units containing two independent copper(II) ions, each in a tetragonally distorted octahedral environment, bridged by a thioether ligand. The synthesis and structural properties of a number of ethane-1,2-dithiolato complexes have been described[174] and the thermodynamics of formation of dinuclear complexes of some sulphur-containing dipeptides with copper(II) have been reported.[175]

In a series of studies of transmetallation of tetranuclear copper complexes, the structural implications of the kinetics of direct transmetallation of tetranuclear copper(II) complexes by $Ni(NS)_2$ reagents have been assessed,[176] the transmetallation of copper(I) complexes and the stoicheiometry and kinetics of oxidation of

[172] I. R. Young, L. A. Ochrymowycz, and D. B. Rorabacher, *Inorg. Chem.*, 1986, **25**, 2576.
[173] J. M. Berg and K. O. Hodgson, *Inorg. Chem.*, 1986, **25**, 1800.
[174] C. P. Rao, J. R. Dorfman, and R. H. Holm, *Inorg. Chem.*, 1986, **25**, 428.
[175] L. D. Pettit and A. Q. Lyons, *J. Chem. Soc., Dalton Trans.*, 1986, 499.
[176] G. Davies, M. A. El-Sayed, and A. El-Toukhy, *Inorg. Chem.*, 1986, **25**, 1925.

neutral tetranuclear complexes by dioxygen in aprotic solvents have been described,[177] and the transmetallation of $L_4Cu_4O_2$ complexes, where L = 6-methyl-2-hydroxypyridinate, by $M(NS)_2$ reagents has been reported.[178]

A variety of other transition-metal complexes with sulphur-containing compounds have been investigated. For example, the direct reaction of nickel powders with several sodium dithiocarbamates in organic solvents to produce the corresponding metal complexes has been described[179] and the reactions of tetraalkylthiuram sulphides and sodium dialkyldithiocarbamates with metals such as nickel, chromium, iron, and copper in the ion source of a mass spectrometer have been shown to give rise to the formation of chelate-type complexes.[180] The reaction of Li_2S_2 with $(MeC_5H_4)TiCl_3$ followed by exposure to oxygen has been found[181] to give the oxygenated titanium sulphide clusters formulated as $(MeC_5H_4)_4Ti_4S_8O$ and $(MeC_5H_4)_4Ti_4S_8O_2$ in which the four faces of the tetrahedral Ti_4O core are capped by μ_3-S_2 ligands. The compound $(MeC_5H_4)_2V_2S_4$ has been used[182] as an organometallic ligand in reactions with iron, cobalt, nickel, and iridium complexes and has resulted in the preparation of a series of new mixed-metal clusters. It is also interesting to note the reaction[183] of the tetrathiovanadate(V) species by an induced internal electron transfer process to give the vanadium(IV) dimer $V_2(\mu$-$S_2)_2)_2(Bu_2^iNCS_2)_4$.

The photoinduced insertion of sulphur ligands into the $Cr{\equiv}S{\equiv}Cr$ multiple bond of $(\eta^5$-$C_5H_5)_2Cr_2(CO)_4S$ has been shown[184] to provide a novel route to the new complex $(\eta^5$-$C_5H_5)_2Cr_2S_4$ which X-ray crystallography has identified as containing one bidentate μ-S_2 and two μ-S bridges which are symmetrically placed between the $(\eta^5$-$C_5H_5)Cr$ moieties. The dinuclear complexes $(\eta^6$-$C_6H_5R)(CO)_2CrCSCr$-$(CO)_5$ and $(\eta^6$-$C_6H_3Me_3)(CO)_2CrCSCr(CO)_5$, where R is OMe or Me, which contain thiocarbonyl bridges have been synthesized from the π-arene thiocarbonyl complexes and have been structurally characterized.[185] The reduction of carboxylato-bound chromium(V) with bisulphite in aqueous solutions has been examined[186] and the synthesis and structural characterization of new salts of the dimeric chromium(III) cation μ-sulphato-di-μ-hydroxobis-[{N,N-bis(2-pyridylmethyl)-amine} chromium(III)] has been described.[187]

Sulphur compounds of cobalt have also received attention. During studies of the kinetics and mechanism of the reactions of sulphito complexes in aqueous solutions the formation, aquation, and intramolecular electron-transfer reactions of the *trans*-aquo(sulphito-O)bis(ethylenediamine)cobalt(III) ion have been investigated[188] and,

[177] G. Davies, M. A. El-Sayed, A. El-Toukhy, T. R. Gilbert, and K. Nabih, *Inorg. Chem.*, 1986, **25**, 1929.
[178] G.-Z. Cai, G. Davies, M. A. El-Sayed, A. El-Toukhy, T. R. Gilbert, and K. D. Onan, *Inorg. Chem.*, 1986, **25**, 1935.
[179] T. Tetsumi, M. Sumi, M. Tanaka, and T. Shono, *Polyhedron*, 1986, **5**, 707.
[180] T. Tetsumi, M. Sumi, M. Tanaka, and T. Shono, *Polyhedron*, 1986, **5**, 703.
[181] G. A. Zank, C. A. Jones, T. B. Rauchfuss, and A. L. Rheingold, *Inorg. Chem.*, 1986, **25**, 1886.
[182] C. M. Bolinger, T. M. Weatherill, T. B. Rauchfuss, A. L. Rheingold, C. S. Day, and S. R. Wilson, *Inorg. Chem.*, 1986, **25**, 634.
[183] T. R. Halbert, L. L. Hutchings, R. Rhodes, and E. I. Stiefel, *J. Am. Chem. Soc.*, 1986, **108**, 6437.
[184] L. Y. Goh and T. C. W. Mak, *J. Chem. Soc., Chem. Commun.*, 1986, 1474.
[185] S. Lotz, R. R. Pille, and P. H. VanRooyen, *Inorg. Chem.*, 1986, **25**, 3053.
[186] R. N. Bosc, N. Rajasekar, D. M. Thompson, and E. S. Gould, *Inorg. Chem.*, 1986, **25**, 3349.
[187] S. Larsen, K. Michelsen, and E. Pedersen, *Acta Chem. Scand., Ser. A*, 1986, **40**, 63.
[188] A. A. El-Awady and G. M. Harris, *Inorg. Chem.*, 1986, **25**, 1323.

in other studies,[189] the formation and redox reactions of sulphur-bonded aqua(sulphito)- and bis(sulphito)[bis(phenanthroline) or -bis(bipyridine)]cobalt(III) complex ions have been reported. The reaction of NO with $Co(NCS)_2(PMe_3)$ has been found[190] to occur with disporportionation and to give rise to the (trimethylphosphine) cobalt(III) complex $Co(NCS)_3(PMe_3)_3$. A mixture of isomers of bis(2-aminoethyl-3-aminopropyl sulphide) cobalt(III) chloride has been prepared[191] and, from the mixture, four geometric isomers have been isolated on a preparative scale. It is interesting to note that exchange between the two aryl groups in the aryl(arylthiolato)nickel complexes with tertiary phosphine ligands of composition *trans*-Ni(Ar)-$(SAr')(PEt_3)_2$ and *cis*-Ni(Ar)(SAr')(dmpe), where dmpe is 1,2-bis(dimethylphosphino)ethane, has been shown[192] to give a mixture of $Ni(SAr)(Ar')L_2$ and Ni(Ar)-$(SAr')L_2$ complexes by the reversible reductive elimination and oxidative addition of diaryl sulphides *via* a mechanism involving C—S bond cleavage and reformation at the nickel centre.

Reactions of the Group IIB metals with sulphur-containing compounds have continued to receive attention. For example, the synthesis and crystal structures of the compounds $[Zn\{S_2P(OPr^i)_2\}_2]\cdot H_2NCH_2CH_2NH_2$ and $[Zn\{S_2P(OPr^i)_2\}_2]\cdot NC_5H_5$ have been described.[193] The synthesis and cadmium-113 n.m.r. spectra of mixed-ligand tetranuclear clusters of the type $[Cd_4(EPh)_x(E'R)_{10-x}]^{2-}$, where E,E' are S, Se, or Te, and of the mixed-metal clusters $[Cd_xZn_{4-x}(SPh)_{10}]^{2-}$ have been reported[194] and studies of the extraction of cadmium(II) with sulphur ligands have been described.[195] The compounds [MgR(L)], where R is Me or Ph and L is pyridine-2-thiolate, have been isolated by reaction of HL with methylmercury(II) hydroxide and phenylmercury(II) acetate respectively, and the crystal structure of methyl(pyridine-2-thiolato)mercury(II) has been described.[196] The synthesis of a series of complexes of mercury(II) halides with dimethyl- and diethyl-disulphide has been attempted[197] but, of all the products formed, only $HgCl_2\cdot Me_2S_2$ was sufficiently stable to be characterized by X-ray crystallography.

The stereochemistry and metal-centred rearrangements of some eight-coordinate niobium(V) and tantalum(V) dithiocarbamates and monothiocarbamates have been described.[198] The tantatum complex $(SMe_2)Cl_2Ta(\mu-Cl)_2(\mu-SMe_2)TaCl_2(SMe_2)$ possessing a $\sigma^2\pi^2$ Ta=Ta double bond has been reported[199] to react readily with diphenyl disulphide by a mechanism involving the oxidative addition of the S—S single bond to the Ta=Ta double bond, and giving rise to the compound formulated as $(SMe_2)Cl_3Ta(\mu-SPh)_2TaCl_3(SMe_2)$. In this reaction the bridging SMe_2 ligand is

189 V. K. Joshi, R. van Eldik, and G. M. Harris, *Inorg. Chem.*, 1986, **25**, 2229.
190 O. Alnaji, Y. Peres, F. Dahan, M. Dartiguenave, and Y. Dartiguenave, *Inorg. Chem.*, 1986, **25**, 1383.
191 M. J. Bjerrum, T. Laier, and E. Larsen, *Inorg. Chem.*, 1986, **25**, 816.
192 K. Osakada, M. Maeda, Y. Nakamura, T. Yamamoto, and A. Yamamoto, *J. Chem. Soc., Chem. Commun.*, 1986, 442.
193 M. G. B. Drew, M. Hasan, R. J. Hobson, and D. A. Rice, *J. Chem. Soc., Dalton Trans.*, 1986, 1161.
194 P. A. W. Dean and J. J. Vittal, *Inorg. Chem.*, 1986, **25**, 514.
195 R. Lahiri and G. N. Rao, *Bull. Chem. Soc. Jpn.*, 1986, **59**, 1567.
196 A. Castineiras, W. Hiller, J. Strahle, J. Bravo, J. S. Casas, M. Gayoso, and J. Sordo, *J. Chem. Soc., Dalton Trans.*, 1986, 1945.
197 P. M. Boorman, C. L. Merrit, W. A. Shantha Nandane, and J. F. Richardson, *J. Chem. Soc., Dalton Trans.*, 1986, 1251.
198 J. R. Weir and R. C. Fay, *Inorg. Chem.*, 1986, **25**, 2969.
199 G. C. Campbell, J. A. M. Canich, F. A. Cotton, S. A. Duraj, and J. F. Haw, *Inorg. Chem.*, 1986, **25**, 287.

lost from the starting material and two chloride bridges are broken with the subsequent formation of only two new SPh bridges in a product which is the first example of a d^1-d^1 ditantalum thiolate-bridged dimer.

In a study of rhenium complexes with di- and tri-thiolate ligands[200] the synthesis and X-ray crystal structures of square-pyramidal [PPh$_4$][ReO(SCH$_2$CH$_2$S)$_2$] and [NMe$_4$][ReS(SCH$_2$CH$_2$S)$_2$], and of trigonal prismatic [PPh$_4$][Re{(SCH$_2$)$_3$CCH$_3$}$_2$], have been described. A thionitrosyl-chlorothionitrene complex of rhenium of composition [ReCl$_2$(NS)(NSCl)(pyridine)$_2$] has also been reported.[201]

In a series of studies of ruthenium(II)-catalysed thioether oxidation reactions the synthesis of mixed-sulphide and sulphoxide complexes has been described,[202] the synthesis and crystal structures of two ruthenium(II) complexes with the new linear tridentate ligand 3-(ethylthio)-1-{(3-(ethylthio)propyl}sulphinyl propane were reported,[203] and the synthesis and structures of model complexes with another new linear tridentate ligand bis{3-(ethylsulphinyl)propyl}sulphide have been discussed.[204] The synthesis, reactions and structural properties of some ruthenium complexes of N,S- and C,N,S-coordinating azo ligands have been described[205] and the crystal structure of the ruthenium thionitrosyl complex (PPh$_4$)$_2$[{RuBr$_4$(NS)}$_2$-(μ-N$_2$S$_2$)] has been reported.[206]

The reactions of the sulphidometal carbonyl clusters Os$_3$(CO)$_9$(μ-S)$_2$ with dimethylamine have been found[207] to differ significantly from those of analogous iron-containing clusters. The cluster compound Os$_4$(CO)$_{12}$(μ_3-S) has been reported[208] to undergo facile addition reactions with the Lewis donors Me$_2$NH and CO to form the 1:1 adducts Os$_4$(CO)$_{12}$(NHMe$_2$)(μ_3-S) and Os$_4$(CO)$_{13}$(μ_3-S). Cleavages of carbon–sulphur and carbon–hydrogen bonds during the reactions of phenyl vinyl sulphide with triosmium clusters have been discussed[209] and the crystal structure of [Os$_3$(μ-SPh)(μ-CH=CH$_2$)(CO)$_{10}$] has been described. The reaction of [Os$_3$(CO)$_{10}$(MeCN)$_2$] with heterocyclic thioamides has been found[210] to give products formulated as [Os$_3$H(CO)$_{10}$L], where L is the deprotonated thioamide coordinated via the exocyclic sulphur atom, and the structure of [Os$_3$(μ-H)(CO)$_{10}$-(μ-SC=NCH$_2$CH$_2$S)] has been described. The reaction of Os$_3$(CO)$_{10}$(μ_3-S) with trimethylamine N-oxide dihydrate has been reported[211] to lead to the formation of the compound Os(CO)$_8$(NMe$_3$)(μ-OH)(μ_3-S)(μ-H) which consists of an open cluster of three osmium atoms with bridging hydroxyl and hydrido ligands, a triply bridging sulphido ligand, and one trimethylamine ligand. The compound reacts with Os(CO)$_{10}$(μ_3-S) to give the hexanuclear cluster Os$_6$(CO)$_{18}$(μ-OH)(μ_4-S)-(μ_3-S)(μ-H). The role of sulphido ligands in the synthesis of heteronuclear clusters

[200] P. J. Blower, J. R. Dilworth, J. P. Hutchinson, T. Nicholson, and J. Zubieta, J. Chem. Soc., Dalton Trans., 1986, 1339.

[201] H. G. Hanck, W. Willing, U. Muller, and K. Dehnicke, Z. Anorg. Allg. Chem., 1986, **534**, 77.

[202] D. P. Riley and J. D. Oliver, Inorg. Chem., 1986, **25**, 1814.

[203] D. P. Riley and J. D. Oliver, Inorg. Chem., 1986, **25**, 1821.

[204] D. P. Riley and J. D. Oliver, Inorg. Chem., 1986, **25**, 1825.

[205] A. K. Mahapatra, S. Datta, S. Goswami, M. Mukherjee, A. K. Mukherjee, and A. Chakravorty, Inorg. Chem., 1986, **25**, 1715.

[206] U. Demant, W. Willing, U. Muller, and K. Dehnicke, Z. Anorg. Allg. Chem., 1986, **532**, 175.

[207] R. D. Adams and J. E. Babin, Inorg. Chem., 1986, **25**, 3418.

[208] R. D. Adams and S. Wang, Inorg. Chem., 1986, **25**, 2534.

[209] E. Boyar, A. J. Deeming, K. Henrick, M. McPartlin, and A. Scott, J. Chem. Soc., Dalton Trans., 1986, 1431.

[210] A. M. Brodie, H. D. Holden, J. Lewis, and M. J. Taylor, J. Chem. Soc., Dalton Trans., 1986, 633.

[211] R. D. Adams, J. E. Babin, and H. S. Kim, Inorg. Chem., 1986, **25**, 1122.

has been investigated during an investigation[212] of the synthesis and structural properties of the mixed-metal cluster compounds $PtOs_4(CO)_{11}(PMe_2Ph)_2(\mu_3-S)_2$ and $PtOs_3(CO)_9(PMe_2Ph)_2(\mu_3-S)_2$. In a subsequent study[213] the reaction of $Os_5(CO)_{15}(\mu_4-S)$ with $Pt(PPh_3)_2C_2H_4$ was found to give $Os_5(CO)_{14}(PPh_3)(\mu_4-S)$ and several new platinum–osmium clusters containing quadruply bridging sulphido ligands. The X-ray crystal structure of trans-$[Os^{IV}(salen)(SPh)_2]$ has been determined[214] during a study of high-valent Schiff-base complexes of osmium and, in another study[215] the structure of $(p\text{-}MeC_6H_4CHMe_2)Os(Me_2SO)Cl_2$ has been described in terms of a complex with a π-arene ligand and a S-coordinated Me_2SO group.

The compound $[Rh_2\{\mu_2\text{-}SC_6H_2Pr^i_2\text{-}4,6;\ 2\text{-}(\eta^2\text{-}MeC{=}CH_2)\}_2\{SC_6H_2Pr^i_3\text{-}2,4,6\}_2\text{-}(NCMe)]$ has been synthesized[216] by the reaction of rhodium trichloride with 2,4,6-tri-isopropylbenzenethiol in MeCN and shown to be an asymmetric dinuclear complex of rhodium(II) with two bridging thiolato-ligands chelated via sulphur and an alkene substituent. The crystal and molecular structure of the compound $Rh_2(acam)_4(Me_2SO)_2\cdot2H_2O$, where acam is $MeCONH^-$, has been determined during an investigation of the nature of the Rh-axial ligand bonds in the complex.[217].

The palladium complex $(PPh_4)_2(NH_2Me_2)[Pd_2(NH_4)(S_7)_4]$ has been shown[218] to contain a novel anion of composition $[Pd_2S_{28}]^{4-}$ which acts as a thirty-membered cage containing an entrapped cation. The cage consists of two PdS_4 planes, which form an almost square antiprism, and are coupled by four polysulphide chains. The formation and decomposition of the α-methoxyalkyl species $\{PdCl[CH(OMe)\text{-}CMe_2CH_2SMe]\}_2$ has been studied[219] and, in an investigation of optical resolutions involving metal complexation,[220] the synthesis, resolution, and structure of a palladium complex involving $(\pm)$-(2-mercaptoethyl)methylphenylarsine has been described. The preparation of bis(perhalogenophenyl) derivatives of palladium(II) and platinum(II) which contain $S_2CP(C_6H_{11})_3$ as a mono- or bi-dentate ligand have been described[221] and the cis-$[Pd(C_6F_5)_2\{S_2CP(C_6H_{11})_3\}]$ and cis-$[Pt(C_6F_5)_2\text{-}\{SC(S)P(C_6H_{11})_3\}(CO)]$ derivatives shown to be the first compounds in which $S_2CP(C_6H_{11})_3$ is chelate-bonded to palladium or platinum. A general synthetic route for the preparation of platinum sulphur dioxide cluster compounds has been reported[222] and the X-ray structural characterization of $[Pt_5(CO)(\mu\text{-}CO)_2\text{-}(\mu\text{-}SO_2)_3(PPh_3)_4]\cdot2CH_2Cl_2\cdot Me_2CH(OH)$ has been described. Variable-temperature ^{195}Pt n.m.r. spectroscopy has been demonstrated[223] as being a potentially powerful

[212] R. D. Adams, I. T. Horvath, and S. Wang, Inorg. Chem., 1986, **25**, 1617.

[213] R. D. Adams, J. E. Babin, R. Mathab, and S. Wang, Inorg. Chem., 1986, **25**, 1623.

[214] C. M. Che, W. K. Chang, and T. C. W. Mak, Inorg. Chem., 1986, **25**, 703.

[215] J. A. Cabeza, H. Adams, and A. J. Smith, Inorg. Chim. Acta, 1986, **114**, L17.

[216] P. T. Bishop, J. R. Dilworth, T. Nicholson, and J. A. Zubieta, J. Chem. Soc., Chem. Commun., 1986, 1123.

[217] M. Y. Chaven, X. Q. Lin, M. Q. Ahsan, I. Bernal, J. L. Bear, and K. M. Kadish, Inorg. Chem., 1986, **25**, 1281.

[218] A. Muller, K. Schmitz, E. Krickemeyer, M. Penk, and H. Bogge, Angew. Chem., Int. Ed. Engl., 1986, **25**, 453.

[219] R. McCrindle, D. K. Stephenson, A. J. McAlees, and J. M. Wilson, J. Chem. Soc., Dalton Trans., 1986, 641.

[220] P. H. Leung, G. M. McLaughlin, J. W. L. Martin, and S. B. Wild, Inorg. Chem., 1986, **25**, 3392.

[221] R. Uson, J. Fornies, M. A. Uson, J. F. Yague, P. G. Jones, and K. Meyer-Base, J. Chem. Soc., Dalton Trans., 1986, 947.

[222] C. E. Briant, D. G. Evans, and D. M. P. Mingos, J. Chem. Soc., Dalton Trans., 1986, 1535.

[223] D. D. Gummin, E. M. A. Ratilla, and N. M. Kostic, Inorg. Chem., 1986, **25**, 2429.

new technique for the study of stereodynamic phenomena in relatively complex plantinum-containing biomolecules and processes causing subtle changes in molecular structure. In another study, ^{31}P and ^{195}Pt n.m.r. spectroscopy and X-ray crystallography were used[224] to characterize the compound $[PtCl(PEt_3)\{C(PPh_2S)_3\}]$ as an S,S-bonded chelate in which the dynamic stereochemistry is controlled by one labile and one inert platinum–sulphur bond. Some platinum(IV) complexes of the type $[(PtXMe_3)_2(ECH_2CMe_2CH_2E)]$, where E is S or Se; X is Cl, Br, I, have been synthesized[225] and shown to possess dinuclear structures with highly strained Pt_2X_2 moieties. Variable-temperature ^{1}H n.m.r. studies identified pyramidal inversion of the chalcogen atom pairs and scrambling of the Pt-methyl groups.

The acidic properties of (dimethyl suphoxide)(1,5-diamino-3-azapentane) platinum(II) perchlorate and the kinetics of the displacement of dimethyl sulphoxide from the conjugate base have been reported.[226] The replacement of the labile water molecules in the diplatinum(III) complex $Pt_2(\mu\text{-}PO_4H)_4(H_2O)_2{}^{2-}$ by halide, amine, thioether, and thiolato ligands has been investigated.[227] The disulphur dioxide species S_2O_2, which is the unstable initial product of SO decomposition in the gas phase has been trapped for the first time by complexation with bis(triphenylphosphane)platinum.[228]

The dinuclear platinum(III) dithiocarboxylato-derivatives $[Pt_2(RCS_2)_4I_2]$, where $R = PhCH_2$ or Me_2CH, have been synthesized, characterized, and structurally described.[229] The triplatinum compound $[Pt_3(2,4,6\text{-}Bu_3^tC_6H_2NC)_6]$ has been shown[230] to react with sulphur or carbon disulphide to give diplatinum complexes $[Pt_2(2,4,6\text{-}Bu_3^tC_6H_2NC)_4S]$ and $[Pt_2(2,4,6\text{-}Bu_3^tC_6H_2NC)_4(CS_2)]$ which contain Pt_2S and Pt_2CS cores. The X-ray crystal structure determination of $[Pt_2(2,4,6\text{-}Bu_3^tC_6H_2NC)_4S]$ showed the compound to contain a triangular Pt_2S core containing terminal isocyanide groups as depicted in Figure 10. The compound tris(tetrathiafulvalenium)hexachloroplatinate has been prepared[231] by the diffusion of tetrathiafulvalene (TTF) and $[NBu_4^n]_2[PtCl_6]$ in acetonitrile. The compound was shown by X-ray crystallography to be composed of TTF-trimer units which are arranged perpendicular to each other to form a two-dimensional layer with significantly close sulphur–sulphur contacts existing amongst the trimers. It is also interesting to note the synthesis and structural characterization of some tetrahedral platinum–gold cluster compounds containing the sulphur dioxide ligand.[232]

The polymeric silver complexes $[Ag\{S_n(CN)_2\}_2][AsF_6]$ ($n = 3$ or 4) have been formed[233] as examples of stable coordination compounds of dicyanotri- and dicyanotetra-sulphanes. The sulphur-rich compound $[(PPh_3)_2N][Ag(S_9)]\cdot S_8$ has been prepared[234] by the reaction of a $S_x{}^{2-}$ solution with silver nitrate and shown to possess an $[Ag(S_9)]^-$ anion in the form of a ten-membered ring system.

[224] J. Browning, K. A. Beveridge, G. W. Bushnell, and K. R. Dixon, *Inorg. Chem.*, 1986, **25**, 1987.

[225] E. W. Abel, P. K. Mittal, K. G. Orrell, and V. Sik, *J. Chem. Soc., Dalton Trans.*, 1986, 961.

[226] R. Romeo, D. Minniti, G. Alibrandi, L. deCola, and M. L. Tobe, *Inorg. Chem.*, 1986, **25**, 1944.

[227] R. El-Mehdawi, F. R. Fronczek, and D. M. Roundhill, *Inorg. Chem.*, 1986, **25**, 1155.

[228] I.-P. Lorenz and J. Kull, *Angew. Chem., Int. Ed. Engl.*, 1986, **25**, 261.

[229] C. Bellitto, M. Bonamico, G. Dessy, V. Fares, and A. Flamini, *J. Chem. Soc., Dalton Trans.*, 1986, 595.

[230] Y. Yamamoto and H. Yamazaki, *J. Chem. Soc., Dalton Trans.*, 1986, 677.

[231] K. Ueyama, G. Matsubayashi, and T. Tanaka, *Inorg. Chim. Acta*, 1986, **112**, 135.

[232] D. M.P. Mingos and R. W. M. Wardle, *J. Chem. Soc., Dalton Trans.*, 1986, 73.

[233] H. W. Roesky, T. Gries, J. Schimkowiak, and P. G. Jones, *Angew. Chem., Int. Ed. Engl.*, 1986, **25**, 84.

[234] A. Muller, M. Romer, H. Bogge, E. Krickemeyer, and H. Zimmermann, *Z. Anorg. Allg. Chem.*, 1986, **534**, 69.

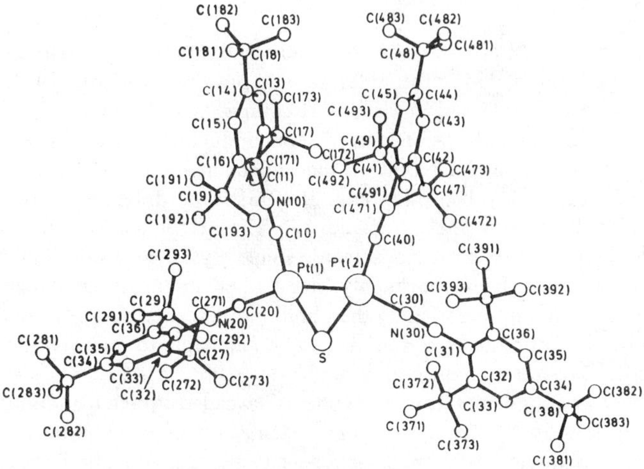

Figure 10 *Structure of* [Pt$_2$(2,4,6-Bu$^t_3$C$_6$H$_2$NC)$_4$S]
(Reproduced from *J. Chem. Soc., Dalton Trans.*, 1986, 677)

It is interesting to note the synthesis and structural characteristics of two novel hydrocarbon-soluble gold(II) compounds, [Au(CH$_2$)$_2$P(C$_6$H$_5$)$_2$]$_2$S$_8$ and [Au(CH$_2$)$_2$P(C$_6$H$_5$)$_2$]$_2$S$_9$, which contain 12- and 13-atom catenated gold and sulphur rings and which are the largest metal–sulphur rings prepared to date.[235] It is also interesting to record the preparation and characterization[236] of (ET)$_2$AuBr$_2$ (ET is bis(ethylenedithio)tetrathiafulvalene), which contains a new type of structure for the organic radical cation salts (ET)$_2$AuX$_2$. The material was formed as a part of a search for new organic metals and superconductors and contains a novel arrangement of the S$\cdots$S network, which is decisive for the electrical properties, involving the ET radical cation forming stacks with S$\cdots$S separations of <3.8 Å and linked together to form two-dimensional layers.

The structure of the lanthanum(III) complex La{(EtO)$_2$PS$_2$}$_3${(PhCH$_2$)$_2$SO}$_2$ has been described[237] and the oxygen atoms of dibenzyl sulphoxide molecules have been shown to be ligated at both ends of the base of a trapezia. The trivalent uranium metallocenes (MeC$_5$H$_4$)$_3$U·THF and (Me$_3$SiC$_5$H$_4$)$_3$U have been shown[238] to react with carbon disulphide to form the dinuclear uranium(IV) complexes [(RC$_5$H$_4$)$_3$U]$_2$[μ-η^1, η^2-CS$_2$] (R is Me or SiMe$_3$), and the geometry about the carbon disulphide ligands has been found to be consistent with two full one-electron transfers into the carbon disulphide. In a subsequent investigation[239] the (MeC$_5$H$_4$)$_3$U·THF compound was shown to react with COS, SPPh$_3$, SePPh$_3$, and TeP(Bu^n)$_3$ to form the bridging chalcogenide complexes [(MeC$_5$H$_4$)$_3$U]$_2$E (E is S, Se, or Te). Some substituted urea complexes of thorium(IV) and uranium(IV) N-thiocyanates have also been reported.[240]

235 J. P. Fackler and L. C. Porter, *J. Am. Chem. Soc.*, 1986, **108**, 2750.
236 E. Amberger, K. Polborn, and H. Fuchs, *Angew. Chem., Int. Ed. Engl.*, 1986, **25**, 727.
237 T. Imai, M. Shimoi, and A. Ouchi, *Bull. Chem. Soc. Jpn.*, 1986, **59**, 669.
238 J. G. Brennan, R. A. Andersen, and A. Zalkin, *Inorg. Chem.*, 1986, **25**, 1756.
239 J. G. Brennan, R. A. Andersen, and A. Zalkin, *Inorg. Chem.*, 1986, **25**, 1761.
240 A. G. M. Al-Daher and K. W. Bagnall, *J. Chem. Soc., Dalton Trans.*, 1986, 843.

4 Selenium

Deuteron and proton n.m.r. studies of potassium and rubidium deuterium (hydrogen) selenites have been related to the structural and physical properties of the solids.[241] Interest has been maintained in Group III compounds of selenium. For example, the Raman scattering spectra of layered semiconducting ε-GaSe have been interpreted in terms of a model based on the molecular nature of the compounds[242] whilst the shape of the absorption edge in $TlGaSe_2$ at different temperatures has been associated with new phase-transitions.[243] The elecrochemical behaviour of van der Waals forces in p-type iridium selenide single crystal electrodes in 0.5M H_2SO_4 and 1M NaOH have been thoroughly investigated under different conditions of illumination.[244] In a different study the photoluminescence data recorded from the layered In_2Se_3 compound were found[245] to be amenable to interpretation in terms of pure α-In_2Se_3 and a mixture of α- and γ-In_2Se_3.

Selenium compounds of Group IV elements have also attracted attention. Silicon-selenium glasses in the Si_xSe_{1-x} alloy system, where $0 < x < 0.40$, have been prepared and characterized by bulk glass density and glass transition measurements. The results were used in the construction of the first equilibrium phase-diagram for the selenium-rich region of this technologically and theoretically interesting Si–Se alloy system.[246] In a very different type of investigation[247] some asymetrically substituted selenides of the type Me_3MCH_2SeR (M is Si, Ge, Sn; R is Me or Ph) have been prepared by the reaction of the appropriate Group IV halides with lithium organoselenolates and characterized by i.r., Raman, ^1H, ^{13}C, and ^{77}Se n.m.r. spectroscopy. It is interesting to note the steady-state and transient photoconductivity measurements of amorphous thin films of $Ge_{22}Se_{78-x}Bi_x$ ($x = 0$, 2, or 10) which have been related to defect states produced by the incorporation of bismuth within the Ge—Se systems.[248] Some studies of phosphorus-containing compounds with selenium have been cited earlier.[49,50] Other interests in these types of materials has emanated from the growing importance of chalcogenide glasses in current technology. For example, Raman and resonance Raman studies of tetraphosphorus triselenide have been interpreted in terms of the layered lattice which confers semiconducting properties on the helical chains of crystalline P_4Se_3 and which gives rise to anisotropic resonant transitions which cause an enhancement of the Raman-active phonons.[249] Some low-frequency Raman spectral studies of As_2Se_3 glasses have been mentioned in a previous section.[52] The alkali selenoarsenates(III), $KAsSe_3 \cdot H_2O$, $RbAsSe_3 \cdot 1/2\ H_2O$, and $CsAsSe_3 \cdot 1/2\ H_2O$ have been prepared by hydrothermal reaction of the respective alkali carbonate with As_2Se_3 at 135 °C and the compounds shown to contain Ψ-$AsSe_3$ tetrahedra linked through Se—Se bonds

[241] I. Vinogradova and S. I. Vasiljeva, *J. Solid State Chem.*, 1986, **62**, 138.
[242] K. R. Allakhverdiev, E. Yu. Salaev, M. M. Tagyeu, S. S. Babaev, and L. Genzel, *Solid State Commun.*, 1986, **59**, 133.
[243] K. R. Allakhverdiev, M. A. Aldzanov, T. G. Mamedov, and E. Yu. Salaev, *Solid State Commun.*, 1986, **58**, 295.
[244] K. Uosaki, S. Kaneko, H. Kita, and H. Chevy, *Bull. Chem. Soc. Jpn.*, 1986, **59**, 599.
[245] M. Balkanski, C. Julien, and A. Chevy, *Solid State Commun.*, 1986, **59**, 423.
[246] R. W. Johnson, S. Susman, J. McMillan, and K. J. Volin, *Mat. Res. Bull.*, 1986, **21**, 41.
[247] J. M. Chehayber and J. E. Drake, *Inorg. Chim. Acta*, 1986, **111**, 51.
[248] R. Mathur and A. Kumar, *Solid State Commun.*, 1986, **59**, 163.
[249] G. R. Burns, R. R. Rollo, and R. J. H. Clark, *Inorg. Chem.*, 1986, **25**, 1145.

into infinite single chains.[250] The compounds $[AsPh_4]_2Se_2Cl_{10}$ and $[AsPh_4]Se_2Cl_9$ have been prepared[251] and characterized as the first dinuclear halogenoselenates(IV). Both compounds contain two distorted octahedral $SeCl_6$ groups connected through a common edge. The structural properties of the hexachloroantimonate salts of $(MeSe)_3^+$ have been reported[56] and the preperation, Raman spectra, and crystal structures of $(SeCl_3)(SbCl_6)$ have been described.[57] The compound $CsSb_2Se_4$ has been prepared[252] by the hydrothermal reaction of Cs_2CO_3 with Sb_2Se_3 at 115 °C and shown to contain polyselenoantimonate(III) anions $(Sb_2Se_4^-)_n$ which exhibit both $(Sb)Se-Sb$ and $(Sb)Se-Se(Sb)$ bridges. The molecular addition compounds $K_2SeO_4(SbF_3)_2$, $Rb_2SeO_4(SbF_3)_2$, $K_2SeO_4(SbF_3)_2 \cdot H_2O$ and the double decomposition compounds $MSbF_2SeO_4$ have been isolated from the $SbF_3-M_2SeO_4$ (M = K, Rb, Cs) systems in selenic aqueous solution and the structural properties studied.[253]

The selenium-rich six- and seven-membered chalcogen rings Se_5S, Se_5S_2, and Se_7 have been prepared[254] from titanocene pentaselenide and disulphur dichloride. The compound Se_5S_2 crystallizes as a mixture of two isomers in a ratio which is temperature-dependent. In another study ^{77}Se n.m.r. spectroscopy has been shown[255] to have attractive possibilities for the characterization of selenium and selenium sulphide ring molecules. ^{77}Se n.m.r. and Raman spectroscopy have also been used to characterize a series of square-planar $S_xSe_{4-x}^{2+}$ cations, where x = 0—3, and the crystal structure of $(S_{3.0}Se_{1.0})_2(Sb_4F_{17})(SbF_6)_3$ shows that the material contains a disordered mixture of $S_xSe_{4-x}^{2+}$ cations.[256] The synthesis and structural characterization of a tellurium(IV) and selenium(IV) mixed oxide of composition Te_3SeO_8 which was prepared by the solid-state reaction of tellurium trioxide and selenium dioxide has been reported.[257] It is pertinent to note the identification[258] by ^{77}Se n.m.r spectroscopy of the novel $Se_4I_4^{2+}$ cation in solution and the study of its equilibrium with $Se_6I_2^{2+}$ and SeI_3^+.

Interest has continued in zirconium triselenide which crystallizes with quasi one-dimensional character.[99] New ternary chalcogenides of composition $InTi_5Se_{7.8}$, $KCr_5Se_{7.8}$, $InCr_5Se_{7.8}$, $KTi_6Se_{7.8}$, and $Tl_{1.4}V_6(Se_{6.8}P_{1.0})$ have been reported and examined by magnetic susceptibility measurements.[259] A study has been made of the reactivity of chromium–selenium triple bonds[260] in which the first example of a planar Cr_3Se multiply bonded system was generated by a selective reduction mechanism.

[250] W. S. Sheldrick and J. Kaub, *Z. Anorg. Allg. Chem.*, 1986, **535**, 179.
[251] B. Krebs, N. Rieskamp, and A. Schaffer, *Z. Anorg. Allg. Chem.*, 1986, **532**, 118.
[252] W. S. Sheldrick and J. Kaub, *Z. Anorg. Allg. Chem.*, 1986, **536**, 114.
[253] D. Mascherpa-Corral, B. Ducourant, R. Fourcade, G. Mascherpa, and S. Alberola, *J. Solid State Chem.*, 1986, **63**, 52.
[254] R. Steudal, M. Papavassiliou, E.-M. Strauss, and R. Laitinen, *Angew. Chem., Int. Ed. Engl.*, 1986, **25**, 99.
[255] R. S. Laitinen and T. A. Pakkaneu, *J. Chem. Soc., Chem. Commun.*, 1986, 1381.
[256] M. J. Collins, R. J. Gillespie, J. F. Sawyer, and G. J. Schrobilgen, *Inorg. Chem.*, 1986, **25**, 2053.
[257] C. Pico, A. Castro, M. L. Veiga, E. Gutierrez-Pueblo, M. A. Monge, and C. Ruiz-Valero, *J. Solid State Chem.*, 1986, **63**, 172.
[258] M. M. Carnell, F. Grein, M. Murchie, J. Passmore, and C.-M. Wong, *J. Chem. Soc., Chem. Commun.*, 1986, 225.
[259] T. Ohtani and S. Onoue, *Mat. Res. Bull.*, 1986, **21**, 69.
[260] W. A. Herrmann, J. Rohrmann, E. Herdtweck, H. Bock, and H. Veltmann, *J. Am. Chem. Soc.*, 1986, **108**, 3134.

Molybdenum–selenium compounds have continued to attract interest and some examples[74,75,92] have been cited in earlier sections. Tetrafluoro(seleno)tungsten(VI), $WSeF_4$, has been shown[261] by gas-phase electron diffraction to adopt a square-pyramidal structure. Lithiation of crystalline $NbSe_3$ has been shown[97] to result in a slight contraction of the average Nb—Se bond length to give a material with significant Se—Se interactions and which contrasts with the structural properties of other amorphous transition-metal trichalcogenides.

X-ray diffraction and ^{57}Fe Mössbauer spectroscopy have been used[262] to investigate the structural properties and magnetic transition in the antiferromagnetic layered compound $TlFe_{2-x}Se_2$. Some new ternary and quarternary transition-metal chalcogenides, for example $Ta_2Pt_3S_8$ and $Co_2Nb_4PdSe_{12}$, have been synthesized by the substitution of known phases and characterized by single crystal X-ray techniques and electrical measurements.[263]

An *in situ* gas chromatographic and mass spectrometric investigation has been made of the role of substrate temperature in the production of zinc selenide from dimethylzinc and hydrogen selenide *via* metal–organic vapour phase epitaxy.[264] The spin glass properties of $Zn_{1-x}Mn_xSe$ and $Zn_{1-x}Mn_xTe$ have been investigated.[265] Studies of the quarternary Zn-In-S-Se system have been mentioned earlier.[105] Alkaline selenide and polyselenide photoelectrochemical cells, sometimes involving n-cadmium telluride, have been the subject of investigation[266,267] and the far i.r. reflection spectra of $Cd_xHg_{1-x}Se$ have been interpreted[268] in terms of lattice vibrations and free carrier effects. Mercury selenide stoicheiometry and phase relationships in the condensed mercury–selenium system have also been examined.[269]

Studies of the structural and bonding properties of the Fe_2Se_2 cores in several compounds have been mentioned in an earlier section.[161] The reactivity of multiple bonds between chromium and selenium has been investigated[270] and the chemistry of selenoaldehydes and selenoketones in the coordination sphere of tungsten have been examined.[271] The reactions of $Se_4(AsF_6)_2$, $Se_8(AsF_6)_2$, and $Se_{10}(SbF_6)_2$ with $Mo(CO)_9$ and $W(CO)_6$ liquid sulphur dioxide have been shown[272] to give the diamagnetic products $[M_2(CO)_{10}Se_4][EF_6]_2$, where M = Mo, W and E = As, Sb. The crystal structure of $[W_2(CO)_{10}Se_4][AsF_6]$ was found to contain the shortest solid-state Se—Se bond of 2.213 Å yet reported. In another study,[273] the reaction of $W(CO)_8$ and $Fe_2(CO)_9$ with Se_4^{2+} in sulphur dioxide has been shown to give the mixed cationic cluster $[FeW(CO)_8Se_2]^{2+}$ which contains a distorted tetrahedral Se_2FeW core.

[261] K. Hagen, D. A. Rice, J. H. Holloway, and V. Kancic, *J. Chem. Soc., Dalton Trans.*, 1986, 1821.

[262] L. Haggstrom, H. R. Verma, S. Bjarman, R. Wappling, and R. Berger, *J. Solid State Chem.*, 1986, **63**, 401.

[263] P. J. Squattrito, S. A. Sunshine, and J. A. Ibers, *J. Solid State Chem.*, 1986, **64**, 261.

[264] J. I. Davies, G. Fan, M. J. Parrott, and J. O. Williams, *J. Chem. Soc., Chem. Commun.*, 1986, 68.

[265] A. Twardowski, C. J. M. Denissen, W. J. M. de Jonge, A. T. de Waele, M. Demianiuk, and R. Triboulet, *Solid State Commun.*, 1986, **59**, 199.

[266] L. E. Lyons and T. L. Young, *Aust. J. Chem.*, 1986, **39**, 347.

[267] L. E. Lyons and T. L. Young, *Aust. J. Chem.*, 1986, **39**, 511.

[268] K. Kumazaki, N. Nishiguchi, and M. Cardona, *Solid State Commun.*, 1986, **58**, 425.

[269] N. Z. Boctor and G. Kullerud, *J. Solid State Chem.*, 1986, **62**, 177.

[270] W. A. Herrmann, *Angew. Chem., Int. Ed. Engl.*, 1986, **25**, 56.

[271] H. Fischer, U. Gerbing, J. Riede, and R. Benn, *Angew. Chem., Int. Ed. Engl.*, 1986, **24**, 78.

[272] M. J. Collins, R. J. Gillespie, J. W. Kolis, and J. F. Sawyer, *Inorg. Chem.*, 1986, **25**, 2057.

[273] D. J. Jones, T. Makani, and J. Roziere, *J. Chem. Soc., Chem. Commun.*, 1986, 1275.

The synthesis of $(Me_4N)_2[Cd_4(SePh)_{10}]$ has been described[194] and the compound has been shown by ^{113}Cd n.m.r. to undergo ligand redistribution with the sulphur analogue to give the series $[Cd_4(SPh)_x(SePh)_{10-x}]^{2-}$ ($x = 0$—10). The linkage isomers of the pentaamine(selenocyanato)ruthenium(III) cation have been prepared and characterized[274] and the crystal and molecular structure of $[AsPh_4][Ru(DL-MeSeCH_2CH_2SeMe)Cl_4]$ has been described.[275]

It is important to note the study of some single crystals of $[TMTSF]_2[M\{Se_2C_2(CF_3)_2\}_2]$, where TMTSF is tetramethyltetraselenafulvalene and M is Ni or Pt, which were shown[276] to have room temperature conductivities of 20—65 Ω^{-1} cm^{-1} which, on cooling, underwent an unprecedented phase transition to give a low-temperature phase showing enhanced electrical conductivity at *ca.* 275 K for the nickel derivative and at *ca.* 245 K for the platinum compound. These new molecular conductors are some of the first metal bis-diselenoalkene complexes to have been reported and are pertinent to current interests in molecular metals. Six Pd^{II} and Pt^{II} complexes of general formula $(Me_3MCH_2SeR)_2M'Cl_2$ (M is Si, or Ge, R is Me or Ph, and M' is Pd or Pt) have been prepared and shown to contain selenium in pyramidal coordination.[277]

Several bidentate selenoether complexes of the tetravalent platinum metals have been reported[278] and characterized by i.r., u.v., visible, ^1H, ^{77}Se, and ^{195}Pt n.m.r. spectroscopy. The structure of $[Pt\{o\text{-}C_6H_4(SeMe)_2\}Cl_4]$ was determined by single crystal X-ray diffraction and shown (Figure 11) to consist of discrete six-coordinate

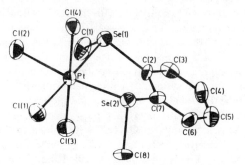

Figure 11 *The $[Pt\{o\text{-}C_6H_4(SeMe)_2\}Cl_4]$ molecular structure*
(Reproduced from *J. Chem. Soc., Dalton Trans.,* 1986, 1003)

platinum with the selenium ligand adopting the *meso*-conformation. Some studies[225] by variable temperature n.m.r. of configurational non-rigidity about a selenium atom in dinuclear platinum(IV) complexes have been mentioned in an earlier section; similar investigations[279] have been made into the novel series of complexes of composition $[PdX_2\{Se(CH_2)_n\}_2]$ ($n = 4$—6 and X is Cl, Br or I) and also of the series $[PdX_2\{SeCH_2CMe_2CH_2\}_2]$. Although conformational changes of the various-

[274] V. Palaniappan and U. C. Agarwala, *Inorg. Chem.,* 1986, **25**, 4064.
[275] E. G. Hope, H. C. Jewiss, W. Levason, and M. Webster, *J. Chem. Soc., Dalton Trans.,* 1986, 1479.
[276] W. B. Heuer, and B. M. Hoffman, *J. Chem. Soc., Chem. Commun.,* 1986, 174.
[277] R. K. Chadha, J. M. Chehayber, and J. E. Drake, *Inorg. Chem.,* 1986, **25**, 611.
[278] E. G. Hope, W. Levason, M. Webster, and S. G. Murray, *J. Chem. Soc., Dalton Trans.,* 1986, 1003.
[279] E. W. Abel, T. E. Mackenzie, K. G. Orrell, and V. Sik, *J. Chem. Soc., Dalton Trans.,* 1986, 205.

sized ligand rings were always fast on the n.m.r. chemical-shift time-scale, the rates of inversion of the coordinated selenium atoms were much slower and were measured directly by bandshape fitting methods. The ΔG^+ values for selenium inversions, which were calculated as being in the range 66–78 KJ mol^{-1}, were observed to increase consistently with decreasing ligand ring-size and were associated with the extent of angle strain required to achieve the transition-state structure of the inversion process. The synthesis and characterization of the trivalent uranium metallocene $[(MeC_5H_4)_3U]_2Se$ has been mentioned earlier.[239]

The compound $(\eta^5\text{-Bu}^tC_5H_4)_2ZrSe_2C_6H_4\text{-}o$ has been shown[280] to undergo electrophilic cleavage to give new Se,Se'-disubstituted derivatives of benzene-1,2-diselenol. The reaction of $Co(BF_4)_2(H_2O)$ with the arsenic selenides and tellurides As_4Se_3, As_2Se_3, and As_2Te_3 in the presence of 1,1,1-tris-(diphenylphosphinomethyl)ethane (triphos) has been found[281] to give the complexes $[(triphos)Co(As_2Se)](BF_4)(EtOH)$ and $[(triphos)Co(As_2Te)](BF_4)(C_6H_6)_{0.5}$.

5 Tellurium

The lattice of elemental tellurium has been known for many years to consist of helical chains, which spiral around the axis parallel to the crystallographic c-axis, and in which the bonding between the individual chains is much weaker than the bonding in the chain. Resistivity measurements[282] of vapour-grown tellurium single crystals which were performed from 298 to 4.2 K showed the material to have a metallic character above 40 K consistent with the carriers flowing along the chain and with the interchain interaction being weak. The resistivity measurements below 40 K were indicative of the carriers being scattered by the magnetic centres.

Significant interest has been maintained in tellurium compounds containing Group III elements, for example a new metastable crystalline compound has been identified in the germanium–tellurium system.[283] A Bridgman synthesis[284] of single crystals of some mixed valence tellurohalides of gallium and indium of composition E_3Te_3X (E is Ga or In; X is Cl, Br, or I) has been reported. The one-dimensional macromolecular structural units in Ga_3Te_3I were shown to consist of $GaTe_{3/5}I$ tetrahedra with gallium(III) centres and $Ga_2Te_{6/3}$ units with gallium(II) centres. The crystal structure of $LiInTe_2$ has been described[285] and compared with that of analogous copper and silver compounds.

Studies of the preparation and structural properties of the compounds $(TeCl_3)$-$(AlCl_4)$, $(TeCl_3)(SbF_6)$, and $(TeCl_3)_2(SO_4)$ have been reported;[57] tetrachloroaluminate salt of $TeCl_3^+$ has been shown to adopt a triclinic modification with a crystal packing which is significantly different from that previously reported. The synthesis and structure of the compound Te_3SeO_8 has been mentioned previously.[257]

Interest in tellurium–oxygen compounds has been sustained. It is relevant to note the growth of large single crystals of Te_2O_5 by a reactive hydrothermal method[286]

280   P. Meunier, B. Gautheron, and A. Mazouz, *J. Chem. Soc., Chem. Commun.*, 1986, 424.
281   M. Di Vaira, M. Peruzzini, and P. Stoppioni, *Polyhedron*, 1986, **5**, 945.
282   T. Ikari, H. Berger, and F. Levy, *Mat. Res. Bull.*, 1986, **21**, 99.
283   S. Asokan, G. Parthasarathy, and E. S. R. Gopal, *Mat. Res. Bull.*, 1986, **21**, 217.
284   S. Paashaus and R. Kniep, *Angew. Chem., Int. Ed. Engl.*, 1986, **25**, 752.
285   W. Honle, G. Kohn, and H. Neumann, *Z. Anorg. Allg. Chem.*, 1986, **532**, 150.
286   J. Alaoui, J. C. Peuzin, M. Couchaud, and J. G. Gay, *Mat. Res. Bull.*, 1986, **21**, 421.

starting from an aqueous solution of $Te(OH)_6$. The compound $[N(Bu^n)_4^+]$-$[H(OTeF_5)_2^-]$ has been prepared from $HOTeF_5$ and $[N(Bu^n)_4^+][OTeF_5^-]$ and structurally characterized.[287] Subsequent studies have shown[288] that the penta-fluoroorthotellurate anion $OTeF_5^-$ has coordinating properties which are very different from that of other anionic ligands including ClO_4^- and $CF_3SO_3^-$. The structure of $[TlOTeF_5(1,3,5-Me_3C_6H_3)_2]_2$, which is a thallium(I) complex with neutral arene ligands, has also been described.[288] Polycrystalline samples of the ammonium fluorotellurates NH_4TeF_5 and $(NH_4)_2TeF_6$ have been investigated by Raman spectroscopy and new features, which can be associated with internal and external modes of the NH_4^+ and TeF_5^- ions, have been identified.[289] It is relevant to note the isolation of a new class of infrared transmitting glasses in the Te—Cl and Te—Cl—S systems.[290] Four new oxides of composition $Pb_2[M_{1.5}Te_{0.5}]O_{6.5}$ and which adopt a pyrochlore-type structure have been obtained[291] from mixtures of PbO, MO_2, (where M is Ti, Zr, or Hf), SnO, and TeO_2.

The substantial interest in cadmium tellurides has been cited in an earlier section.[266] It is also relevant to note the improved optoelectronic properties of thin-film CdTe which has been achieved by treating the surface with potassium cyanide.[292] The improvement has been associated with the removal of free tellurium from the surface of the CdTe film. The interaction of hydrogen with impurities and defects in semiconductors is currently of interest *e.g.* a recent study of hydrogen-acceptor interactions in CdTe and ZnTe by photoluminescence techniques.[293] Another area of topical interest involves the so called diluted magnetic semiconductors *e.g.* the study[294] of the magnetic-field-dependence of the energy levels in CdTe—Cd(Mn)Te double quantum-well structures. In another investigation[295] the electrical properties of antimony-doped p-type $Hg_{0.78}Cd_{0.22}Te$ liquid-phase-epitaxy films have been examined.

Mercury manganese tellurides are known to be semimagnetic semiconductors and to have analogous band structures to HgCdTe and a study[296] of the photolumines-cence of the semimagnetic semiconductor $Hg_{1-x}Mn_xTe$ and[297] of the ground state of the shallow acceptor in highly doped p-type $Hg_{1-x}Mn_xTe$ has been carried out. The investigations of the spin glass properties of $Zn_{1-x}Mn_xTe$ semiconductors are also to be noted.[265]

The compound $TlCr_5Te_{7.8}$ has been found[259] to be ferromagnetic with a Currie temperature of *ca.* 150 K. It is interesting to record the use[298] of electron microscopy to identify superlattice reflections from charge-density wave-domains in the quasi-one-dimensional conductor Nb_3Te_4. The structure of the new ferrimagnetic semicon-

[287] S. H. Strauss, K. D. Abney, and O. P. Anderson, *Inorg. Chem.*, 1986, **25**, 2806.
[288] S. H. Strauss, M. D. Noiret, and O. P. Andersen, *Inorg. Chem.*, 1986, **25**, 3850.
[289] J.-C. Bureau, S. Bendaoud, H. Eddaondi, and G. Perachon, *Mat. Res. Bull.*, 1986, **21**, 1986.
[290] J. Lucas and Zhang Hua, *Mat. Res. Bull.*, 1986, **21**, 871.
[291] J. A. Alonso, C. Cascales, and I. Rasines, *Z. Anorg. Allg. Chem.*, 1986, **537**, 213.
[292] T. K. Bandyopadhyay and S. R. Chaudhuri, *Mat. Res. Bull.*, 1986, **21**, 265.
[293] L. Svob and Y. Marfaing, *Solid State Commun.*, 1986, **58**, 343.
[294] J. A. Brum, G. Bastard, and M. Voos, *Solid State Commun.*, 1986, **59**, 561.
[295] M. C. Chen and J. A. Dodge, *Solid State Commun.*, 1986, **59**, 449.
[296] R. R. Galazka, B. L. Gelmont, W. I. Ivanov-Omskii, U. T. Postolaki, and W. A. Smirnov, *Solid State Commun.*, 1986, **58**, 311.
[297] R. Buezko, *Solid State Commun.*, 1986, **59**, 495.
[298] K. Suzuki, M. Ichihara, and I. Nakada, *Solid State Commun.*, 1986, **59**, 291.

ducting silicon manganese telluride of composition $Mn_3Si_2Te_6$ has been described in terms of a layer-type compound.[299] Polycrystalline samples of $Cd_{2x}(CuIn)_yMn_{2z}Te_2$ alloys (where $x + y + z = 1$) have been prepared by melting and annealing techniques and the structural properties have been examined.[300] The infrared reflection spectra of the CdI_2-type layer compounds $HfS_{2-x}Te_x$ have been measured between 30 and 4000 cm^{-1} and related to the electrical properties of the compounds.[301] Platinum-metal dichalcogenides such as PtS_2 and RuS are currently under investigation as electrode materials for photoelectrochemical cells; $RuTe_2$ crystals have been grown which crystallize with a pyrite structure with n-type semiconductor properties.[302] The optical bandgap and magneto-optical effects in (Pb, Eu)Te thin single crystal films grown by molecular beam epitaxy have been measured by luminescence and absorption techniques.[303]

Mixed oxides containing tellurium have continued to attract attention. For example, iron selenium tellurates of composition $Fe_aSe_bTe_cO_x$ have been prepared and found to be suitable selective and active low-temperature ammoxidation catalysts for the conversion of propylene, ammonia, and air into acrylonitrile.[304] The atomic arrangement of a zinc-tellurite glass of composition 80% TeO_2, 20% ZnO has been studied by neutron diffraction.[305] Luminescence studies of tellurium-rich and lanthanide-rich anti-glass tellurites of La, Gd, and Y have been reported[306] and the first observed Te^{4+} luminescence in such materials has been identified.

The reaction between NaTeR (where R is C_6H_4OEt-p) and $Br(CH_2)_nBr$ (where $n = 1,5,6,7,9$ or 10) has been shown[307] to give the bis(tellurides) $(RTe)_2(CH_2)_n$. In the case of $n = 5$, a telluronium salt of composition $[\overline{R}Te(\overline{CH_2})_5]Br$ was also isolated under some experimental conditions. Triphenyltellurium(VI) trifluoride, Ph_3TeF_3, has been prepared[308] by the reaction of triphenyltellurium(IV) fluoride with xenon difluoride and the crystal structure of mer-Ph_3TeF_3 shown to contain tellurium in a slightly distorted octahedral environment. Stereoselective intermolecular fluorine exchange was found to occur on addition of $Ph_3TeF_2{}^+$ to mer-Ph_3TeF_3.

The synthesis of the (2-phenylazophenyl-C,N') tellurium(II) dithiocarbamates, $Te(C_6H_4N_2Ph)(dtc)$, where dtc is dimethyl-, diethyl-, or dibenzyl-dithiocarbamate, together with the corresponding series of tris compounds $Te(C_6H_4N_2Ph)(dtc)_3$ have been reported.[309] The ^{125}Te n.m.r. data indicated that the tris compounds dissociate to tellurium(II) compounds in solution and ^{125}Te Mössbauer data suggested that the tris compounds are better formulated as loose charge-transfer compounds. The structure of dimethyldithiocarbamato(2-phenylazophenyl-C,N') tellurium(II) was shown to contain tellurium in a distorted planar environment with no significant

[299] H. Vincent, D. Leroux, D. Bijaoui, R. Rimet, and C. Schlenker, *J. Solid State Chem.*, 1986, **63**, 349.

[300] M. Quintero, L. Dierker, and J. C. Woolley, *J. Solid State Chem.*, 1986, **63**, 110.

[301] G. Kliche, *Solid State Commun.*, 1986, **59**, 587.

[302] J. W. Foise, H. Ezzaouia, and O. Gorochov, *Mat. Res. Bull.*, 1986, **21**, 7.

[303] W. C. Goltsos, A. V. Nurmikko, and D. L. Partin, *Solid State Commun.*, 1986, **59** 183.

[304] R. K. Grasselli, J. F. Brazdil, and J. D. Burrington, *Appl. Catal.* 1986, **25**, 335.

[305] V. Kozhukharov, H. Bruger, S. Neov, and B. Sidzhimov, *Polyhedron*, 1986, **5**, 771.

[306] G. Blasse, G. J. Dirksen, E. W. J. L. Oomen, and M. Tromel, *J. Solid State Chem.*, 1986, **63**, 148.

[307] H. M. K. K. Pathirana and W. R. McWhinnie, *J. Chem. Soc., Dalton Trans.*, 1986, 2003.

[308] A. S. Secco, K. Alam, B. J. Blackburn, and A. F. Janzen, *Inorg. Chem.*, 1986, **25**, 2125.

[309] M. A. K. Ahmed, A. E. McCarthy, W. R. McWhinnie, and F. J. Berry, *J. Chem. Soc., Dalton Trans.*, 1986, 771.

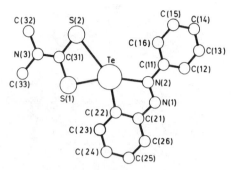

Figure 12 *Molecular structure of* Te(C$_6$H$_4$N$_2$Ph)(dmdtc)
(Reproduced from *J. Chem. Soc., Dalton Trans.*, 1986, 771)

intermolecular contacts as shown in Figure 12. Other new tellurium dithiocarbamate complexes have been reported;[310] these have included [Te^{IV}(S$_2$CNR$_2$)$_4$] compounds, where R is Bu^i, Ph, or CH$_2$Ph. Evidence has been found from ^{125}Te n.m.r. spectroscopy of a ligand exchange reaction involving [Te^{IV}(S$_2$CNR$_2$)$_4$], where R is Et or CH$_2$Ph, which has a rate comparable with the time scale of the n.m.r. experiment. The preparation, ^{125}Te n.m.r., and mass spectra of some dithiotellurides, Te(RS)$_2$, where R is Ph, 2-PhCOOH, CPh$_3$, CH$_2$Ph, 3-PrCOOH, Bu^n, Pr^i, and Bu^t, have been discussed.[311]

The crystal structures of trimethylammonium hexaiodotellurate(IV) and heptaiodotellurate(IV) have been described.[312] The reaction of Co(BF$_4$)$_2$(H$_2$O)$_6$ with the arsenic telluride As$_2$Te$_3$ in the presence of triphos to give the compound [(triphos)Co(As$_2$Te)](BF$_4$)(C$_6$H$_6$)$_{0.5}$ has been reported.[281]

The synthesis of acetylene adducts of an iron–molybdenum cluster (C$_5$H$_5$)$_2$Mo$_2$FeTe$_2$(CO)$_x$ has been described and the conversion of *arachno*-(C$_5$H$_5$)$_2$Mo$_2$FeTe$_2$(CO)$_7$ to *closo*-(C$_5$H$_5$)$_2$Mo$_2$FeTe$_2$(CO)$_3$(RC$_2$H) has been reported.[313] The reaction of a ditelluride solution with M(ClO$_4$)$_2$·6H$_2$O (M is Co or Ni) in the presence of triphos, where triphos is 1,1,1-tris(diphenylphosphinomethyl)ethane, has been shown[314] to give the compounds [{(triphos)-M}$_2$(μ-η^2-Te$_2$)]·2C$_4$H$_8$O. The Te$_2$ unit in these compounds was shown by single crystal X-ray diffraction to bridge side-on between two M(triphos) moieties as shown in Figure 13. Such in-plane bridging by a Te$_2$ unit between two metal moieties was previously unknown.

The compounds Cd$_4$[SPh]$_x$(TePh)$_{10-x}$]$^{2-}$ (x is 8—10) and Cd$_4$(SePh)$_x$(TePh)$_{10-x}$]^2 (x = 7—10) have been described.[194] The structure of 1,6-bis-2-butyltellurophenyl-2,5-diazahexa-1,5-diene has been shown[315] to involve a weak tellurium–nitrogen interaction which gives the tellurium atom an effective coordination number of three. The compound forms a monomeric complex with

[310] M. A. K. Ahmed, W. R. McWhinnie, and P. Granger, *Polyhedron*, 1986, **5**, 859.
[311] W. Mazurek, A. G. Moritz, and M. J. O'Connor, *Inorg. Chim. Acta*, 1986, **113**, 143.
[312] H. Kiriyama, Y. Mizuhashi, and J. Ootani, *Bull. Chem. Soc. Jpn.*, 1986, **59**, 581.
[313] L. E. Bogan, G. R. Clark, and T. B. Rauchfuss, *Inorg. Chem.*, 1986, **25**, 4050.
[314] M. DiVaira, M. Peruzzini, and P. Stopioni, *J. Chem. Soc., Chem. Commun.*, 1986, 374.
[315] N. Al-Salim, T. A. Hamor, and W. R. McWhinnie, *J. Chem. Soc., Chem. Commun.*, 1986, 453.

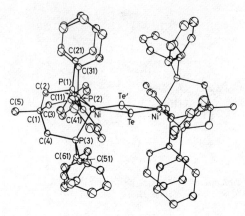

Figure 13 *View of the* [{(triphos)Ni}$_2$(μ-η^2-Te$_2$)] *molecule*
(Reproduced from *J. Chem. Soc., Chem. Commun.*, 1986, 374)

HgCl$_2$ which contains a 13-membered chelate ring. The compound [(MeC$_5$H$_4$)$_3$U]$_2$Te has been prepared[239] from the trivalent uranium metallocene by reaction with TeP(Bu^n)$_3$.

7   F, Cl, Br, I, At, and Noble Gases

By M. J. K. THOMAS

Chemical Laboratories, University of London Goldsmiths' College, London SE14 6NW

1 Introduction

This chapter follows the format of previous years in reviewing developments in the chemistry of the halogens and the noble gases that have appeared in the literature of the past year.

A number of reviews appeared in 1986 to commemorate the isolation of F_2 by Henri Moissan on 26th June 1886 and these can be found in reference 1.

2 Interhalogens and Related Ions

The dynamics of the fluorescence decay of the halogens and interhalogens have been reviewed.[2] The proceedings of the International Symposium on Radiohalogens held in Banff, Canada, in September 1985 are reported in reference 3.

A chemical synthesis of elemental fluorine from the displacement reaction shown in equation 1 is the first such synthesis since both starting materials can be readily prepared without using fluorine.[4]

$$K_2MnF_6 + 2\,SbF_5 \rightarrow 2\,KSbF_6 + MnF_3 + 1/2\,F_2 \qquad (1)$$

The reaction of alkanes with the reagent F_2–$MeCN$–H_2O gives good yields of epoxides rather than fluorinated compounds.[5] Alkanes with a terminal double bond and enones react smoothly. A matrix-isolation study of the sulphurones SCl_3F, SCl_2F_2, and $SClF_3$ which are formed in the gas phase during the co-condensation of SCl_2 with ClF and F_2, respectively, reveals that the most stable species is SCl_2F_2, which has C_{2v} symmetry.[6]

Combinations of ClF_3 with several hydrocarbons or halogenocarbons can be made to explode by sudden mixing at temperatures from 25 °C downwards.[7] All of the fuels initiate in less than 1 ms and an ionic mechanism is likely. All of the bonds, including the C—H bonds, are labilized. The suggestion that AX_6E systems should distort from octahedral symmetry in the gas phase because of the presence of a stereochemically active lone-pair of electrons has been strengthened by examination

[1] *J. Fluorine Chem.*, 1986, **33**.
[2] M. C. Heaven, *Chem. Soc. Rev.*, 1986, **15**, 405.
[3] Proceedings of the International Symposium of Radiohalogens, *Applied Radiation and Isotopes, Int. J. Appl. Radiat. Isot. Part A*, 1986, **37**.
[4] K. O. Christe, *Inorg. Chem.*, 1986, **25**, 3721.
[5] S. Rozen and M. Brand, *Angew. Chem., Int. Ed. Engl.*, 1986, **25**, 554.
[6] R. Minkwitz, U. Nass, and J. Sawatzki, *J. Fluorine Chem.*, 1986, **31**, 175.
[7] K. R. Brower, *J. Fluorine Chem.*, 1986, **31**, 333.

of the calculated electronic charge distribution of ClF_6^-.[8] The reaction products formed by the codeposition of cyclopropane and its derivatives with Cl_2, ClF, and Br_2 in an argon matrix have been investigated by infrared spectroscopy.[9] In each case a 1:1 complex is formed in which there is an interaction with the midpoint of one of the C—C bonds. This behaviour is comparable to the analogous HX complexes (see 1985 report).

Microwave spectroscopy has been used to get a partial structure of BrNCO in the gas phase.[10] The molecule is planar with a Br—N bond distance of 185 pm. $NF_4^+BrF_4^-$ and $NF_4^+BrF_4O^-$ have been prepared by low-temperature metathetical reactions.[11] Both compounds are unstable, decomposing according to equations 2 and 3.

$$NF_4BrF_4 \rightarrow NF_3 + BrF_5 \tag{2}$$

$$NF_4BrF_4O \rightarrow NF_3 + F_2 + BrF_3O \tag{3}$$

Bromine(I) fluorosulphate has been used for the preparation of new N-bromosulphonimides.[12] The use of $BrOSO_2F$ is general for the preparation of electropositive N-bromo-compounds, for example in the formation of the new compound $(FSO_2)_2NBr$ from $Hg[N(SO_2F)_2]_2$. The first example of a completely symmetrical Br_3^- ion has been identified in $[VBr_2(MeCN)_4]Br_3$.[13]

The kinetics of the hydrolysis of I_2 to form HOI and I^- have been studied by temperature-jump spectrophotometry.[14] In contrast to Cl_2 and Br_2, the rate of hydrolysis of I_2 is controlled by the disproportionation of the conjugate base, I_2OH^-. I_2 is oxidized by MF_6 (M = Mo, U) in MeCN at ambient temperature to $[I(NCMe)_2]MF_6$.[15] The vibrational spectra are consistent with linear coordination of MeCN to I^+. In MeCN solution the solvated I^+ decomposes to give I_2. In SO_2 solution I_2 is oxidized by AsF_5 to $[I_5]^+[AsF_6]^-$.[16] The structure consists of planar centrosymmetric I_5^+ cations (Figure 1) and AsF_6^- anions with cation–anion interactions. The reaction of I_2Cl_6 with $AgNO_3$ occurs sequentially, the overall reaction is shown in equation 4.[17]

$$3 I_2Cl_6 + 24 AgNO_3 + 12 H_2O \rightarrow 18 AgCl + 2 AgI + 4 AgIO_3 + 24 HNO_3 \tag{4}$$

Extremely efficient cyclopentene annulations occur *via* sequential Michael-carbene insertion reactions of alkynyliodonium salts.[18] A variety of (per- and polyfluoroalkyl)phenyl and p-fluorophenyl iodonium triflates were synthesized by the oxidation of the corresponding iodofluoroalkanes with trifluoroperacetic acid followed by treatment with benzene or fluorobenzene and triflic acid.[19] Several

[8] P. J. MacDougall, *Inorg. Chem.*, 1986, **25**, 4400.
[9] B. S. Ault, *J. Phys. Chem.*, 1986, **90**, 2825.
[10] H. M. Jemson, W. Lewis-Bevan, N. P. C. Westwood, and M. C. L. Gerry, *J. Mol. Spectrosc.*, 1986, **118**, 481.
[11] K. O. Christe and W. W. Wilson, *Inorg. Chem.*, 1986, **25**, 1904.
[12] S. Singh and D. D. DesMarteau, *Inorg. Chem.*, 1986, **25**, 4596.
[13] F. A. Cotton, G. E. Lewis, and W. Schwotzer, *Inorg. Chem.*, 1986, **25**, 3528.
[14] D. A. Palmer and R. van Eldik, *Inorg. Chem.*, 1986, **25**, 928.
[15] G. M. Anderson and J. M. Winfield, *J. Chem. Soc., Dalton Trans.*, 1986, 337.
[16] A. Apblett, F. Grein, J. P. Johnson, J. Passmore, and P. S. White, *Inorg. Chem.*, 1986, **25**, 422.
[17] B. M. Derakhshan, A. Finch, and P. N. Gates, *Polyhedron*, 1986, **5**, 1543.
[18] M. Ochiai, M. Kunishima, Y. Nagao, K. Fuji, M. Shiro, and E. Fujita, *J. Am. Chem. Soc.*, 1986, **108**, 8281.
[19] T. Umemoto, Y. Kuriu, H. Shuyama, O. Miyamo, and S.-I. Nakayama, *J. Fluorine Chem.*, 1986, **31**, 37.

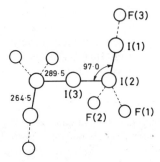

Figure 1 *Bond distances* (pm) *and angles* (°) *of the* I_5^+ *cation*

Figure 2 *The coordination around the iodine atom in diacetato*(m-*tolyl*)*iodine*(III)

crystal structures of organo-iodine(III) compounds have been reported. Diacetato(*m*-tolyl)iodine(III)[20a] (Figure 2) and μ-oxo-bis[trifluoroacetato(*m*-tolyl)iodine(III)][20b] have the characteristic T-shaped primary geometry around the iodine atoms which is found in most dsp^3 hybridized I^{III} compounds. The geometry about the iodine in $PhIC_6H_4$-o-$CO_2.H_2O$ consists of primary bonds to the two phenyl rings and a weak primary bond or a strong intramolecular secondary I—O bond to the carboxylate group in an approximately trigonal pyramidal AX_3E_2 arrangement.[21] In (2,2-biphenylylene)(1-pyrrolidinecarbodithioato)iodine(III) chloroform solvate the coordination around the iodine is planar tetragonal comprising two normal bonds to C atoms and two secondary bonds to S atoms of different ligands.[22]

The reactions of R_3E, R_2ECl, $RECl_2$, and R_2PF (E = P, As, Sb, Bi; R = alkyl or aryl) with IF_5 have been studied.[23] While R_3E (E = P, As, Sb) can be transformed into the corresponding phosphoranes, arsoranes, and stiboranes, the bismuthines give different types of product depending on the organic group: oxidation to difluorobismuthoranes; nucleophilic displacement of F by aryl groups; formation of fluorobismuthines by alkyl fluoride elimination from unstable difluorobis-muthoranes. R_nECl_{3-n} and IF_5 give the corresponding tri- and tetra-fluoroorganometallic compounds.

[20a] S. C. Kokkou and C. J. Cheer, *Acta Crystallogr. Section C*, 1986, **42**, 1159.
[20b] S. C. Kokkou and C. J. Cheer, *Acta Crystallogr. Section C*, 1986, **42**, 1748.
[21] R. J. Batchelor, T. Birchall, and J. F. Sawyer, *Inorg. Chem.*, 1986, **25**, 1415.
[22] A. P. Bozopoulos and P. J. Rentzeperis, *Acta Crystallogr. Section C*, 1986, **42**, 1014.
[23] H. J. Frohn and H. Maurer, *J. Fluorine Chem.*, 1986, **34**, 73; H. J. Frohn and H. Maurer, *J. Fluorine Chem.*, 1986, **34**, 129.

Far infrared and Raman spectra of Me_4NI_5, $(TMA.H_2O)_{10}HI_5$, $(phen)_2HI_5$, Me_4NBrI_4, Et_4NI_7, and Me_4NI_9 show that the first three compounds contain bent I_5 ions, linear I_5 ions, and $(I_3^-).(I_2)$ species respectively.[24] The last two compounds can be described as $(I_3^-).(2I_2)$ and $(I_5^-).(2I_2)$ respectively. The results on BrI_4^- are less conclusive but they do suggest a bent structure for this ion.

3 Noble Gas Compounds

XeF_2 reacts with CrO_2F_2 to give $CrOF_3$ in good yield.[25] PuF_6 is formed at ambient or lower temperatures by treatment of various solid substrates with KrF_2.[26] Formation of actinide hexafluorides has been confirmed for the reaction of KrF_2 in anhydrous HF with UO_2 and with uranium and neptunium fluorides. AmO_2 does not yield AmF_6 with KrF_2. CrF_4O can be prepared from CrO_2F_2 and KrF_2.[27] It is a strong Lewis acid and with KrF_2 it forms an unstable 1:1 adduct which has an essentially covalent structure containing a $Kr—F\cdots Cr$ bridge (1).

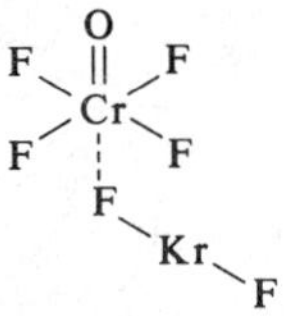

(1)

The $Xe—N$ bonded molecule $FXeN(SO_2F)_2$ exhibits F^- donor properties to give adducts which are predominantly ionic, *viz.*: $XeN(SO_2F)_2^+AsF_6^-$, $XeN(SO_2F)_2^+Sb_3F_{16}^-$, and $F[XeN(SO_2F)]_2^+AsF_6^-$.[28] In general the cations are thermally less stable than their fluorine analogues XeF^+ and $Xe_2F_3^+$. In $XeN(SO_2F)_2^+Sb_3F_{16}^-$ the anion is *cis*-fluorine-bridged and weakly covalently bonded by means of a fluorine bridge to the $XeN(SO_2F)_2$ cation. The cation is $Xe—N$ bonded and the $Xe—N$ bond is significantly shorter than that in $FXeN(SO_2F)_2$.

4 Hydrogen Halides

Ab initio calculations on F_2H^+ and the topological form FHF^+ show that the more stable species is the former, which is simply protonated F_2.[29]

The infrared spectra of HI and DI in argon matrices show the presence of monomers, dimers, and trimers of HI and DI as well as 1:1 and 2:1 complexes with water.[30] Binary complexes of HI with added HCl, HBr, and I_2 are also observed and there is evidence of a hydrogen-bonded complex between HI and atomic iodine. Matrix isolation studies of hydrogen-bonded complexes formed between HF and a range of molecules have continued. There are two primary 1:1 complexes formed

[24] E. M. Nour, L. H. Chen, and J. Laane, *J. Phys. Chem.*, 1986, **90**, 2841.
[25] M. McHughes, R. D. Willett, H. B. Davis, and G. L. Gard, *Inorg. Chem.*, 1986, **25**, 426.
[26] L. B. Asprey, P. G. Eller, and S. A. Kinkead, *Inorg. Chem.*, 1986, **25**, 670.
[27] K. O. Christe, W. W. Wilson, and R. A. Bougon, *Inorg. Chem.*, 1986, **25**, 2163.
[28] R. Faggiani, D. K. Kennepohl, C. J. L. Lock, and G. J. Schrobilgen, *Inorg. Chem.*, 1986, **25**, 563.
[29] R. L. DeKock, R. Dutler, A. Rauk, and R. D. Van Zee, *Inorg. Chem.*, 1986, **25**, 3329.
[30] A. Engdahl and B. Nelander, *J. Phys. Chem.*, 1986, **90**, 6118.

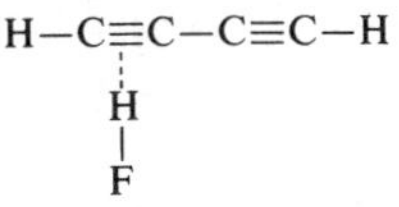

H—C≡C—C≡C—H
　　｜
　　H
　　｜
　　F

(2)

H—C≡C—C≡C—H···F
　　　　　　　　　＼
　　　　　　　　　　H

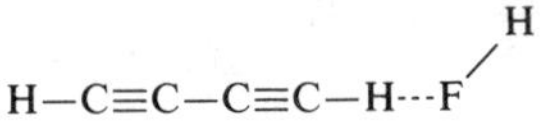

(3)

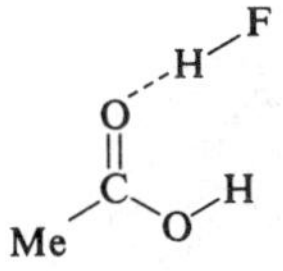

(4)

between HF and diacetylene, a π-complex (2) and a σ-complex (3).[31a] With acetic acid and methyl acetate the HF bonds to the carbonyl oxygen in the skeletal plane, presumably on the same side as the ester molecule (4).[31b] Methyl-substituted amines give 1:1 and 1:2 complexes with HF.[31c] The relative yield of the 1:2 complex increases with the methyl substitution. Hydrogen-bonded complexes of HF with halogenobenzenes are of two types.[31d] In the first the HF is bonded to the halogen atom attached to the ring (5) and the second involves hydrogen-bonding to the ring (6). The strength of the hydrogen-bond to the halogen increases with increasing

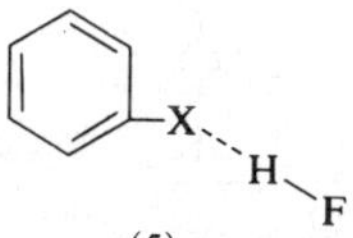

(5)

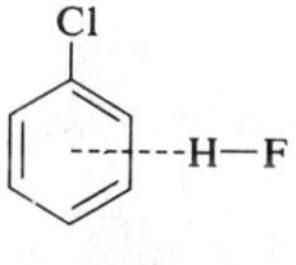

(6)

atomic number. The hydrogen-bond in $Me_2S\cdots HF$ is weaker than that in $Me_2O\cdots HF$ because, although the sulphide base has a higher proton affinity, it is too diffuse in electron density to bind HF as strongly as the smaller oxide base.[31e] Similar results were obtained for Et_2S, MeSH, and EtSH. A study of the complexes formed between HCl or HBr and cyclopropanes containing an electron-withdrawing substituent suggests that the hydrogen halide is bonded to the substituent rather than to the ring.[32]

A new selective radiofluorinating agent has been prepared from $H^{18}F$ and Sb_2O_3.[33] The nature of this new fluorinating agent is unknown but it is not SbF_3. The reaction of HCl(g), DCl(g), and HBr(g) with pyrazine-phosphorus(v) chloride is shown in Scheme 1 for HCl.[34] The reaction occurs *via* (a) rather than (b).

The compounds KF.2.5HF and KF.3HF have been examined by X-ray crystallography.[35] Both contain $H_nF_{n+1}^-$ anions of three types, formed by strong F—H···F hydrogen bonds (Figure 3). The structural formula of KF.2.5HF is $K_2[H_2F_3][H_3F_4]$. An isomeric $H_3F_4^-$ anion is present in the structure of KF.3HF.

[31a] K. O. Patten and L. Andrews, *J. Phys. Chem.*, 1986, **90**, 3910.

[31b] K. O. Patten and L. Andrews, *J. Phys. Chem.*, 1986, **90**, 1073.

[31c] L. Andrews, S. R. Davies, and G. L. Johnson, *J. Phys. Chem.*, 1986, **90**, 4273.

[31d] S. R. Davis and L. Andrews, *J. Phys. Chem.*, 1986, **90**, 2600.

[31e] L. Andrews, R. T. Arlinghaus, and R. D. Hunt, *Inorg. Chem.*, 1986, **25**, 3205.

[32] C. E. Truscott and B. S. Ault, *J. Phys. Chem.*, 1986, **90**, 2566.

[33] G. Angelini, M. Speranza, C.-Y. Shiue, and A. P. Wolf, *J. Chem. Soc., Chem. Commun.*, 1986, 924.

[34] H. C. Knachel, S. D. Owens, S. H. Lawrence, M. E. Dolan, M. C. Kerby, and T. A. Salupo, *Inorg. Chem.*, 1986, **25**, 4606.

[35] D. Motz and D. Boenigk, *J. Am. Chem. Soc.*, 1986, **108**, 6634.

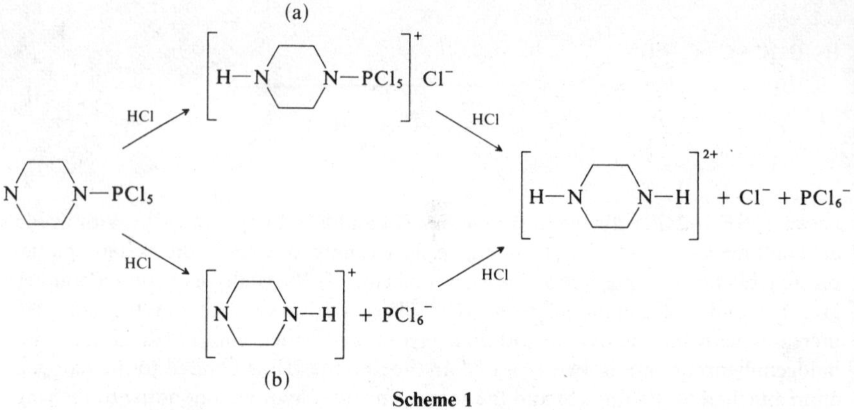

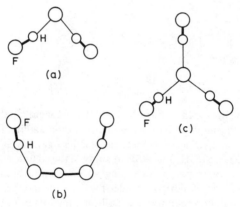

Scheme 1

Figure 3 *Structures of the* $H_nF_{n+1}^-$ *anions*: (*a*) $H_2F_3^-$; (*b*) $H_3F_4^-$ *of* KF.2.5HF; (*c*) $H_3F_4^-$ *of* KF.3HF

5 Oxo Compounds

The isomerization of HOCl to HClO has been studied by *ab initio* methods.[36] The reaction is endothermic with an activation energy of $311 \, \text{kJ mol}^{-1}$. Microwave spectroscopy has been used to determine the molecular structure of HOCl.[37]

The reaction between Cl^{III} and Br_2 results in the oxidation of Cl^{III} to ClO_2.[38a] The main products from the reaction of Cl^{III} and Br^- are Br_3^- and Cl^- when bromide is in excess and ClO_2 and Br_2 when ClO_2^- or $HClO_2$ is in excess.[38b] The latter reaction has a complex rate law which is explained by a 16-step mechanism including oxidation of Br^- to Br_2 by Cl^{III} and decomposition of chlorous acid.

[36] A. G. Turner, *Inorg. Chim. Acta*, 1986, **111**, 157.
[37] W. D. Anderson, M. C. L. Gerry, and R. W. Davis, *J. Mol. Spectrosc.*, 1986, **115**, 117.
[38a] O. Valdes-Aguilera, D. W. Boyd, I. R. Epstein, and K. Kustin, *J. Phys. Chem.*, 1986, **90**, 6696.
[38b] O. Valdes-Aguilera, D. W. Boyd, I. R. Epstein, and K. Kustin, *J. Phys. Chem.*, 1986, **90**, 6702.

The coordination structures of the alkali-metal chlorates have been studied.[39] The heavier alkali metals prefer a tridentate association in matrix isolated $M^+ClO_3^-$. The data for the Li^+ and Na^+ cations are conflicting. Deliberate hydration of the ion pairs is consistent with bidentate binding of ClO_3^- to Li and for Na, bidentate and tridentate coordination have comparable abundance.

Cl_2O_6 is present as ionic chloryl perchlorate $ClO_2^+ClO_4^-$ in the solid state.[40] The Cl—O distance in the cation (141 pm) is much shorter than in neutral ClO_2 in the gas phase. Each cation is coupled with two anions, and conversely, to give polymeric groups. The reaction of Cl_2O_6 with $TiCl_4$ gives $Ti(ClO_4)_4.xCl_2O_6$, which on warming gives $Ti(ClO_4)_4$.[41] The metal is strongly chelated by four perchlorato-ligands, the eight oxygen atoms adopting a slightly flattened dodecahedral arrangement. The interaction between ClO_4^- and Nd^{III}, Tb^{III}, Eu^{III}, and Er^{III} in MeCN has been studied by vibrational spectroscopy.[42] Vibrations due to unassociated, monodentate, and bidentate ClO_4^- were observed in solutions of the perchlorates. The number of unassociated ions depends on the cation but is always about two. The proportion of bidentate ClO_4^- is low and the amount of unidentate coordination depends on the cation.

The structure of $(NH_2)_2CO.HIO_3$ has been studied by n.q.r., infrared spectroscopy, and thermal analysis.[43] The structure contains $(NH_2)_2COH\cdots OIO_2$ groupings in which the hydrogen has transferred to the urea. Iodine is in a distorted octahedral environment in the structure of $Na_5HI_2O_{10}.14H_2O$.[44] Two IO_6 octahedra are connected by a common edge to form dimeric anions.

6 Structural Chemistry of Solid Complex Halides containing Main-group Elements

Two crystallographically independent ICl_4^- anions are found in the structure of $[PCl_4][ICl_4]$.[45] One of the iodine atoms is in almost ideal D_{4h} symmetry and the other anion is slightly distorted to a C_{2v} symmetry. The double salt $[PCl_4]_3[TiCl_6]$-$[PCl_6]$, prepared from PCl_5 and $TiCl_4$ in mole ratio 5:1, is isostructural with $[PCl_4]_3[SnCl_6][PCl_6]$ reported last year.[46] $PPh_4[As_2SBr_5]$ and $PPh_4[Sb_2SBr_5]$ can be obtained from PPh_4Br, HBr, and As_2S_3 or Sb_2S_3.[47] Both anions have the same general structure: including the lone-pair of electrons, the As and Sb atoms have distorted trigonal bipyramidal coordination. Two bipyramids share a common edge with S and Br bridging atoms. The $[As_2SBr_5]^-$ anions are associated into chains *via* As$\cdots$Br contacts and the $[Sb_2SBr_5]^-$ ions form pseudodimeric units by Sb$\cdots$S and Sb$\cdots$Br contacts (Figure 4). The compounds $(Hpy)_2[Sb_2OBr_nCl_{6-n}]$ ($n = 0, 2,$ and 4) are isostructural.[48] All contain a discrete $[Sb_2OX_6]^{2-}$ anion.

[39] G. Ritzhaupt and J. P. Devlin, *J. Phys. Chem.*, 1986, **90**, 6764.
[40] K. M. Tobias and M. Jansen, *Angew. Chem., Int. Ed. Engl.*, 1986, **25**, 993.
[41] M. Fourati, M. Chaabouri, C. H. Belin, M. Charbonnel, J.-L. Pascal, and J. Potier, *Inorg. Chem.*, 1986, **25**, 1386.
[42] J.-C. G. Bunzli and C. Mabillard, *Inorg. Chem.*, 1986, **25**, 2750.
[43] A. M. Peyrsyan, V. A. Shishkin, and V. B. Gavalyan, *Russ. J. Inorg. Chem. (Engl. Transl.)*, 1986, **31**, 32.
[44] K. M. Tobias and M. Jansen, *Z. Anorg. Allg. Chem.*, 1986, **538**, 159.
[45] J. Shamir, S. Schneider, A. Bino, and S. Cohen, *Inorg. Chim. Acta*, 1986, **114**, 35.
[46] J. Shamir, S. Schneider, A. Bino, and S. Cohen, *Inorg. Chim. Acta*, 1986, **111**, 141.
[47] U. Muller and A. T. Mohammed, *Z. Anorg. Allg. Chem.*, 1986, **533**, 65.
[48] M. J. Begley, M. Hall, M. Nunn, and D. B. Sowerby, *J. Chem. Soc., Dalton Trans.*, 1986, 1735.

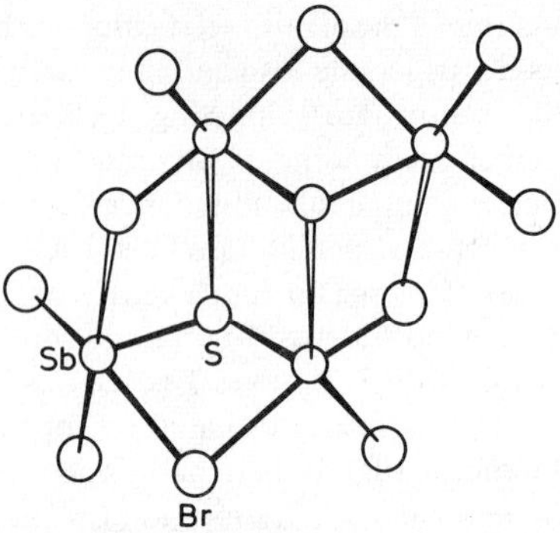

Figure 4 *The structure of the* [Sb$_2$SBr$_5$]$^-$ *anion*

The novel Se$_4$I$_4^{2+}$ cation has been identified in solution by its ^{77}Se n.m.r. spectrum.[49] The spectrum is consistent with an I$_2$Se$^+$SeSeSe$^+$I$_2$ structure for the cation (7). The cation is in equilibrium with lesser amounts of SeI$_3^+$ and Se$_6$I$_2^{2+}$. The first dinuclear halogenoselenate(IV) complexes were prepared from (SeCl$_4$)$_4$ and stoicheiometric amounts of Cl$^-$ in POCl$_3$ solution.[50] They contain the Se$_2$Cl$_{10}^{2-}$ and Se$_2$Cl$_9^-$ anions and were isolated as crystalline AsPh$_4^+$ salts. Each structure has two distorted SeCl$_6$ octahedra, connected through a common edge in the first complex and a common face in the second. The stereochemical activity of the Se^{IV} lone-pair of electrons

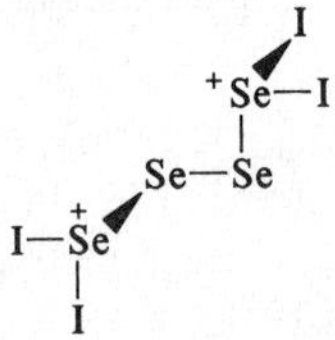

(7)

causes severe distortion of the central Se$_2$Cl$_2$ ring in the centrosymmetric Se$_2$Cl$_{10}^{2-}$ ion. The preparation and characterization of (SCl$_3$)(SbCl$_6$), (SeCl$_3$)(SbCl$_6$), (SBr$_{1.2}$Cl$_{1.8}$)(SbCl$_6$), (TeCl$_3$)(AlCl$_4$), (TeCl$_3$)(AsF$_6$), (TeCl$_3$)(SbCl$_6$), and (TeF$_3$)$_2$SO$_4$ have been reported.[51] All of the MX$_3^+$ cations are involved in significant anion–cation secondary bonding interactions of varying strengths and geometries. The [SBr$_{1.2}$Cl$_{1.8}$] cation consists of a disordered mixture of [SBr$_x$Cl$_{3-x}$]$^+$ cations. The coordination polyhedron around the TeCl$_3^+$ cation in [TeCl$_3$][AuCl$_4$], including secondary Te$\cdots$Cl interactions is square pyramidal.[52] These secondary interactions link the ions together to form centrosymmetric (TeCl$_3$AuCl$_4$)$_2$ dimers (Figure 5).

[49] M. M. Carnell, F. Grein, M. Murchie, J. Passmore, and C.-M. Wong, *J. Chem. Soc., Chem. Commun.*, 1986, 225.
[50] B. Krebs, N. Rieskamp, and A. Schaffer, *Z. Anorg. Allg. Chem.*, 1986, **532**, 118.
[51] B. H. Christian, M. J. Collins, R. J. Gillespie, and J. F. Sawyer, *Inorg. Chem.*, 1986, **25**, 777.
[52] P. G. Jones, D. Jentsch, and E. Schwarzmann, *Z. Naturforsch., Teil B*, 1986, **41**, 1483.

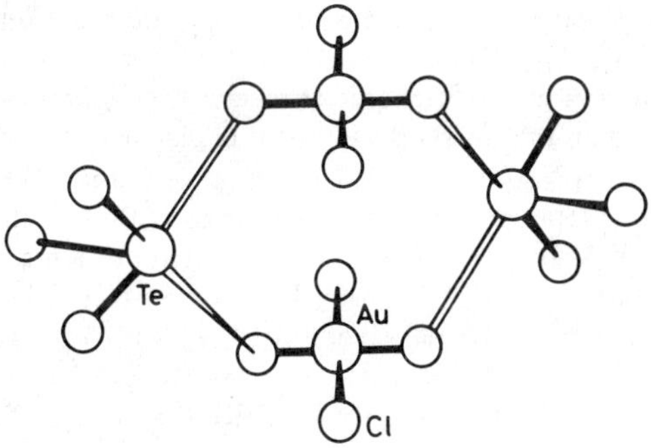

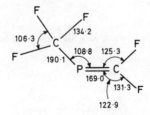

Figure 5 *The structure of* [TeCl₃][AuCl₄]

7 Group IV Halides

The gas-phase structures of $Me_{4-n}SiF_n$ with $n = 1$–3 have been determined by electron diffraction.[53] A steady decrease in the Si—F and Si—C bond lengths is observed with increasing fluorination, because of increasing polar contributions and contraction of the Si valence shell. *Ab initio* calculations on SiF_4 show that $(p-d)\pi$ bonding is negligible. Semi-empirical molecular orbital calculations were used to examine the stability of SiF_6^{2-}, SiF_4 in planar and tetrahedral forms, SiF_3^+ in planar and pyramidal forms, and SiF_2 as linear or bent.[54] The calculations show that planar structures are energetically possible in solid-state reactions. It was not possible to find an energetic difference between the linear and bent forms of SiF_2.

Rapid thermal decomposition of GeI_2 gives the metastable sub-iodide $Ge_{4.06}I$, which is a semiconductor with a clathrate structure.[55] The clathrate framework is made up of Ge and I^{3+}, with I^- ions in the cavities.

8 Fluoro- and Perfluoro-carbon Derivatives of Nitrogen and Phosphorus

The structures of $CF_3P{=}CF_2$ (Figure 6) and its cyclic dimer $(CF_3PCF_2)_2$ have been determined by gas-phase electron diffraction.[56] The four-membered PCPC ring of the dimer is strongly puckered. The electronic structures and energies of the model ylides H_3PCXY and their isomeric phosphines H_2PCHXY (X = Y = H, F, CF_3;

Figure 6 *Bond distances* (pm) *and angles* (°) *in* $CF_3P{=}CF_2$

[53] B. Rempfer, H. Oberhammer, and N. Auner, *J. Am. Chem. Soc.*, 1986, **108**, 3893.
[54] R. Luck, G. Scholz, and L. Kolditz, *Z. Anorg. Allg. Chem.*, 1986, **538**, 191.
[55] R. Nesper, J. Corda, and H.-G. von Schnering, *Angew. Chem., Int. Ed. Engl.*, 1986, **25**, 350.
[56] B. Steger, H. Oberhammer, J. Grobe, and D. LeVan, *Inorg. Chem.*, 1986, **25**, 3177.

and X = H, Y = F) have been determined by *ab initio* molecular orbital calculations.[57] The H_3PCH_2, H_3PCHF, and $H_3PC(CF_3)_2$ ylides exhibit varying degrees of P—C multiple bonding and zwitterionic character. The ylidic carbon in H_3PCHF is more pyramidal than that of H_3PCH_2, but it is planar in $H_3PC(CF_3)_2$.

The compounds $(F_2P)S(CH_2)_nS(PF_2)$ for $n = 2$—6 and $(F_2P)O(CH_2)_nO(PF_2)$ for $n = 3$—6 were prepared from the reaction of $S(PF_2)_2$ with the appropriate dithiols and diols.[58] Most are stable in solution at room temperature for several hours but $(F_2P)Y(CH_2)_2Y(PF_2)_2$ (Y = O, S) decompose by elimination of PF_3 to give the cyclic compound (8). Oxidation of $RN(PF_2)_2$ by N_2O gives the corresponding bis(difluorophosphoryl)amides, $RN(POF_2)_2$ (R = Me, Et).[59] Only one phosphorus atom is oxidized by phenylazide to give $[F_2PN(R)PF_2NPh]_2$ (9). Cleavage of the

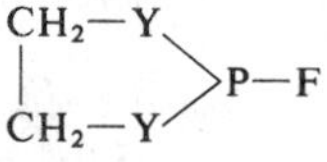

(8)

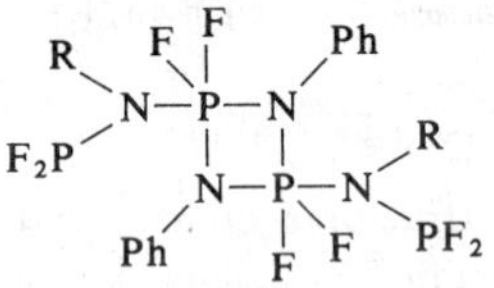

(9)

PNP bridge of $RN(POF_2)_2$ occurs in the reaction with Me_3SiOMe and Me_3SiNEt_2 to form derivatives of di- and mono-fluorophosphoric acid.[60] A large number of monophosphazenes and diphosphetidines have been prepared by the Staudinger reaction between R_nPF_{3-n} ($n = 1, 2$; R = $(CH_2)_5N$, $O(CH_2)_4N$, alkyl, aryl) and phenylazides, $XC_6H_4N_3$ (X = H, 4-Me, 4-Cl, 4-NO_2, and 3-NO_2).[61] PF_3 does not react with phenylazide.

Reaction of *N,O*-bis(trimethylsilyl) derivatives of *o*-aminophenol with RPF_4 (R = F, Me, Ph, 1-adamantyl) in 1:1 molar ratio gives monocyclic $C_6H_4(O)$-$(NH)PF_2R$ (equation 5).[62] A trigonal bipyramidal structure is suggested from the [19]F n.m.r. spectrum, with the $C_6H_4(O)(NH)$ group attached to one axial and one equatorial position.

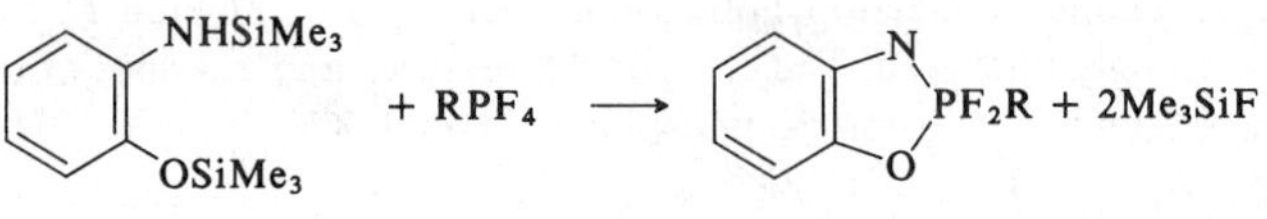

(5)

9 Halides of Phosphorus, Arsenic, and Antimony and their Derivatives

Reaction between $P(O)FBr_2$ and silver gives OPF.[63] The results of *ab initio* calculations show that the molecule is bent. In the presence of base PCl_3 reacts with H_2S

[57] D. A. Dixon and B. E. Smart, *J. Am. Chem. Soc.*, 1986, **108**, 7172.
[58] G. A. Bell and D. W. H. Rankin, *J. Chem. Soc., Dalton Trans.*, 1986, 1689.
[59] L. Reisel and D. Sturm, *Z. Anorg. Allg. Chem.*, 1986, **539**, 183.
[60] L. Reisel and D. Sturm, *Z. Anorg. Allg. Chem.*, 1986, **539**, 187.
[61] L. Reisel, D. Sturm, A. Nagel, S. Taudien, A. Beuster, and A. Karwatzki, *Z. Anorg. Allg. Chem.*, 1986, **542**, 157.
[62] R. Bartsch, J.-V. Weiss, and R. Schmutzler, *Z. Anorg. Allg. Chem.*, 1986, **537**, 53.
[63] R. Ahlrichs, R. Becherer, M. Binnewies, H. Borrmann, M. Lakenbrink, S. Schunk, and H. Schnockel, *J. Am. Chem. Soc.*, 1986, **108**, 7905.

to give P_4S_3 and $[PS_2Cl_2]^-$, besides other products.[64] The reaction scheme proposed is shown in Scheme 2. Tetraphenoxyphosphonium chloride and hexachlorophosphate ion are formed in the reaction of PCl_5 with phenol.[65] The geometric structures of PCl_nF_{5-n} for $n = 1$—4 have been determined by gas-phase electron-diffraction.[66] The lengths of the P—F and P—Cl bonds increase with increasing F/Cl substitution. In general the axial bonds are more strongly affected.

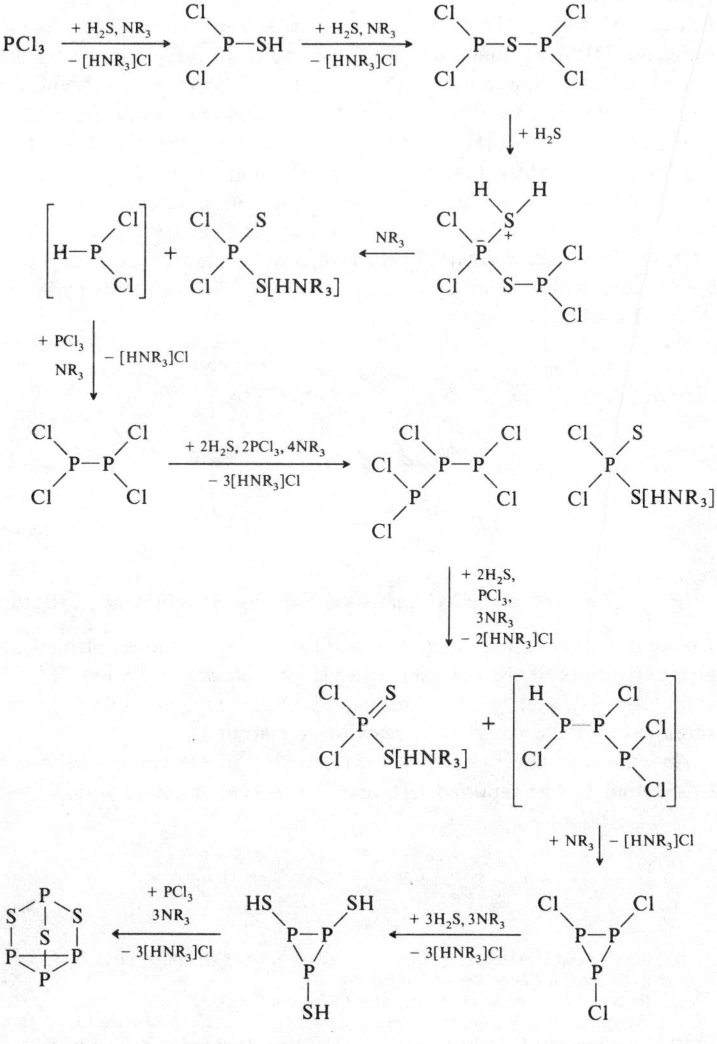

Scheme 2

[64] E. Fluck and B. Neumuller, *Z. Anorg. Allg. Chem.*, 1986, **534**, 27.
[65] J. Gloede and R. Waschke, *Phosphorus Sulfur*, 1986, **27**, 341.
[66] C. Macho, R. Minkwitz, J. Rohmann, B. Steger, V. Wolfel, and H. Oberhammer, *Inorg. Chem.*, 1986, **25**, 2828.

AsCl$_4$F is much more unstable than AsCl$_2$F$_3$ and AsClF$_4$.[67] It decomposes to give Cl$_2$ and AsCl$_3$ (equation 6).

$$6\,AsCl_4F \rightarrow AsCl_4 + AsF_6^- + 4\,Cl_2 + 4\,AsCl_3 \tag{6}$$

In the structure of C$_6$H$_6$.2SbCl$_3$ the SbCl$_3$ molecules form corrugated layers with the benzene molecule in between.[68] Both Sb atoms have a deformed pentagonal pyramidal coordination with the C$_6$H$_6$ molecule in an axial position. The electronic structure of SbF$_5^{2-}$ and the isoelectronic species TeF$_5^-$, IF$_5$, and XeF$_5^+$ were calculated using the DVM-X$_\alpha$ method.[69] The results of the calculations cast doubt on the validity of the assumption that the geometry of the anion is influenced by the lone pair of electrons. The structure of Mn(SbF$_4$)$_2$.2H$_2$O contains fluoroantimonate(III) layers linked by MnF$_4$(H$_2$O)$_2$ octahedra to form a three-dimensional network.[70] A large number of SbCl$_5$.L adducts have been studied by vibrational spectroscopy.[71] The extent of auto-ionization increases with the polarity of the solvent and the donor strength of the ligand, but it is always less than 3%.

Dissolution of BiCl$_3$ in mesitylene gives a 1:1 complex (10), which is characterized by η^6 coordination of mesitylene to the metal.[72] With hexamethylbenzene a novel inverted sandwich structure (11) is obtained.

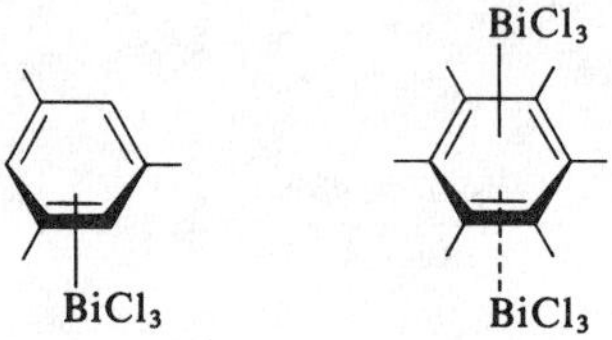

(10)   (11)

10 Perfluoroalkyl and Alkyl Sulphur, Selenium, and Tellurium Halides

Unlike other sulphines, (CF$_3$)$_2$C=S=O reacts with amines, alcohols, and HCl to give derivatives of the corresponding sulphinic acid (equation 7).[73] The reaction with HBr, however, gives *C*-brominated sulphenyl bromide (equation 8a) and sulphinyl bromide *via* a redox reaction (equation 8b).

Ab initio calculations on CF$_3$C≡SF$_3$ confirm that the molecule has a short, polar C≡S bond as was reported last year.[74] The bent structure with a C—C≡S bond

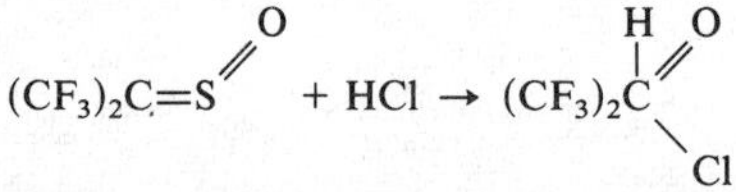

$$(CF_3)_2C=S\!\!\!\diagup^{O} + HCl \rightarrow (CF_3)_2\overset{H}{\underset{Cl}{C}}\diagdown\diagup^{O} \tag{7}$$

[67] R. Minkwitz and H. Prenzel, *Z. Anorg. Allg. Chem.*, 1986, **534**, 150; S. Elbel, G. Runger, H. Egsgaard, and L. Carlsen, *J. Chem. Res. (S)*, 1986, 294.
[68] D. Mootz and V. Handler, *Z. Anorg. Allg. Chem.*, 1986, **533**, 23.
[69] V. I. Sergienko, V. I. Kostin, L. N. Ignat'eva, and G. L. Gutsev, *J. Fluorine Chem.*, 1986, **32**, 367.
[70] T. M. Shchegoleva, L. D. Iskhakova, A. A. Shakhnazaryan, and V. K. Trunov, *Russ. J. Inorg. Chem.* (*Engl. Transl.*), 1986, **31**, 515.
[71] J. E. Kessler, C. T. G. Knight, and A. E. Merbach, *Inorg. Chim. Acta*, 1986, **115**, 75; J. E. Kessler, C. T. G. Knight, and A. E. Merbach, *Inorg. Chim. Acta*, 1986, **115**, 85.
[72] A. Schier, J. M. Wallis, G. Muller, and H. Schmidbauer, *Angew. Chem., Int. Ed. Engl.*, 1986, **25**, 757.
[73] M. Schwab and W. Sundermeyer, *Chem. Ber.*, 1986, **119**, 2458.
[74] D. A. Dixon and B. E. Smart, *J. Am. Chem. Soc.*, 1986, **108**, 2688.

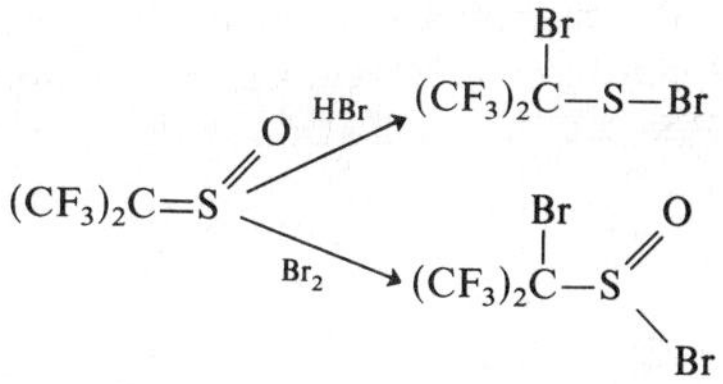

$$(CF_3)_2C{=}S{=}O \xrightarrow{\text{HBr}} \underset{\underset{\text{Br}}{|}}{(CF_3)_2C}{-}S{-}Br \qquad (8a)$$

$$(CF_3)_2C{=}S{=}O \xrightarrow{\text{Br}_2} (CF_3)_2\underset{\underset{\text{Br}}{|}}{C}{-}\underset{\underset{\text{Br}}{}}{S}\diagdown_{O} \qquad (8b)$$

angle of 171.6°, found by experiment, is *ca.* 900 kJ mol^{-1} higher in energy than a linear C—C≡S geometry. Very small crystal-packing forces may account for the observed non-linearity in the solid state. The structures of C$_4$F$_8$SF$_4$ (Figure 7), (SF$_5$)$_2$CF$_2$, and (SF$_4$CF$_2$)$_2$ have been determined by gas-phase electron-diffraction.[75,76] Very long S—C bonds and a large SCS bond angle were observed for (SF$_5$)$_2$CF$_2$ and the SCSC ring is planar in cyclic (SF$_4$CF$_2$)$_2$.

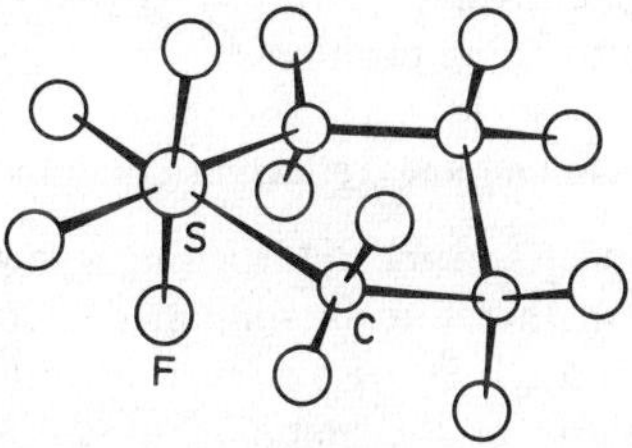

Figure 7 *The structure in the gas phase of* C$_4$F$_8$SF$_4$

The halogen ethynes XC≡CSF$_5$ (X = Cl, Br, I, or R$_3$Si) have been prepared from HC≡CSF$_5$ (equation 9).[77]

$$HC{\equiv}C{-}SF_5 \xrightarrow{\text{LiN(SiR}_3)_2} Li^+\bar{C}{\equiv}C{-}SF_5 \xrightarrow[\text{or R}_3\text{SiCl}]{X_2} X{-}C{\equiv}C{-}SF_5 \qquad (9)$$

The reaction of HC≡CSF$_5$ with Co$_2$(CO)$_8$ gives Co$_2$(CO)$_6$(HC≡CSF$_5$), Co$_2$(CO)$_5$(HC≡CSF$_5$)$_2$, and Co$_2$(CO)$_4$(HC≡CSF$_5$)$_3$. Decomposition of Co$_2$(CO)$_4$(HC≡CSF$_5$)$_3$ gives (12) and photolysis of HC≡CSF$_5$ gives (13).

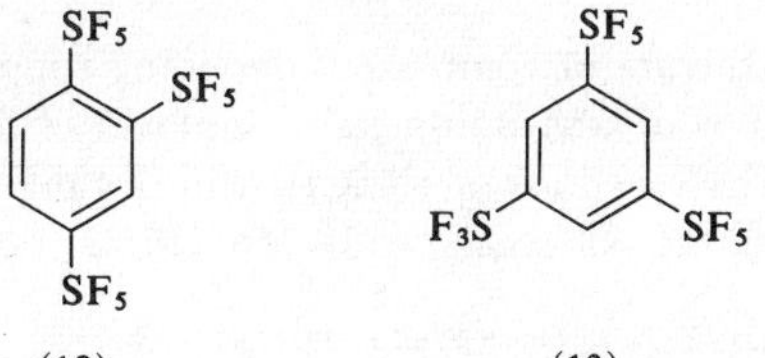

(12)

(13)

[75] K. D. Gupta, J. M. Schreeve, and H. Oberhammer, *J. Mol. Struct.*, 1986, **147**, 363.
[76] K. D. Gupta, R. Mews, A. Waterfeld, J. M. Schreeve, and H. Oberhammer, *Inorg. Chem.*, 1986, **25**, 275.
[77] J. Wessel, H. Hartl, and K. Seppelt, *Chem. Ber.*, 1986, **119**, 453.

Trifluoromethylselenyl thiocarbonyls react with $CF_{3-n}Cl_nSCl$ or CF_3SeCl to give the corresponding disulphones or selanesulphones (equation 10).[78] Oxidation of $CF_3SeC(Se)X$ (X = CF_3Se, CF_3S, Cl, Br) with $3—ClC_6H_4C(O)OOH$ gives $CF_3Se(X)C=S=O$.

$$CF_3SeC(S)X + CF_{3-n}Cl_nECl \rightarrow F_3C—Se—\overset{\overset{\displaystyle Cl}{|}}{\underset{\underset{\displaystyle X}{|}}{C}}—S—E—CF_{3-n}Cl_n \qquad (10)$$

mer-Ph_3TeF_3 has been prepared from Ph_3TeF and XeF_2.[79] The Te atom has a slightly distorted octahedral environment and the phenyl groups are rotated with respect to each other as a result of intramolecular steric repulsion. The molecules are packed in a chain-like pattern. The reaction of PF_5 with *mer*-Ph_3TeF_3 gives $Ph_3TeF_2 + PF_6^-$. Intermolecular fluorine exchange occurs on addition of $Ph_3TeF_2^+$ to Ph_3TeF_3. [19]F and [125]Te n.m.r. spectra indicate that only the axial fluorine exchanges, consistent with the fluorine-bridged intermediate $[Ph_3TeF_2—F—TeF_2Ph_3]^+$, in which only the axial fluorine occupies the bridging position.

11 Sulphur, Selenium, and Tellurium Oxofluorides and their Derivatives

The nature of the active species in the chlorinating system S_2Cl_2-$AlCl_3$-SO_2Cl_2 is $SCl_3^+AlCl_4^-$.[80] The system can, therefore, be simplified to comprise S_8-$AlCl_3$-SO_2Cl_2. The very unstable compound $CF_3S(O)I$ has been identified by correlating its u.v. visible absorptions with those of the homologous $CF_3S(O)X$ series and by the decomposition products.[81] The series of compounds R_FOSO_2F [R_F = CF_3CHMe, CF_3CMe_2, or $(CF_3)_2CH$] were synthesized from the reaction of polyfluoroalcohols with sulphuryl fluoride or sulphuryl chloridefluoride.[82] They react with nucleophilic reagents such as amines and polyfluoroalcohols to give sulphamates and dialkylsulphates.

The $H(OTeF_5)_2^-$ ion in $(NBu_4^n)^+[H(OTeF_5)_2]^-$ is unusual.[83] Data from i.r. and n.m.r. spectroscopy suggest that there is a very strong O—H—O hydrogen bond. Despite the fluorine-like behaviour of —$OTeF_5$, the O—H—O hydrogen bond in $H(OTeF_5)_2^-$ does not rival HF_2^- as the strongest, shortest hydrogen bond. The compound was prepared from $HOTeF_5$ and $(NBu_4^n)^+(OTeF_5)^-$. Bridging —$OTeF_5$ groups are found in the structure of $[TlOTeF_5(1,3,5-Me_3C_6H_3)_2]_2$.[84]

12 Sulphur–Nitrogen Halides and their Fluorocarbon Derivatives

The preparation and structure of complexes containing simple S—N ligands have been reviewed.[85] A review of transiton-metal complexes of S—N ligands has also appeared this year.[86] The reactions of $(NSCl)_3$ with a range of starting materials continue to be of interest. It reacts with Mo, MoO_3, or Na_2MoO_4 to give

[78] F. Fockenberg and A. Haas, *Z. Naturforsch., Teil B*, 1986, **41**, 413.
[79] A. S. Secco, K. Alam, B. J. Blackburn, and A. F. Janzen, *Inorg. Chem.*, 1986, **25**, 2125.
[80] C. Glidewell, *Inorg. Chim. Acta*, 1986, **117**, L7.
[81] R. Minkwitz and R. Lekies, *Z. Anorg. Allg. Chem.*, 1986, **537**, 169.
[82] T.-J. Huang and J. M. Shreeve, *Inorg. Chem.*, 1986, **25**, 496.
[83] S. H. Strauss, K. D. Abney, and O. P. Anderson, *Inorg. Chem.*, 1986, **25**, 2806.
[84] S. H. Strauss, M. D. Noirot, and O. P. Anderson, *Inorg. Chem.*, 1986, **25**, 3850.
[85] P. F. Kelly and J. D. Woollins, *Polyhedron*, 1986, **5**, 607.
[86] T. Chivers and F. Edelmann, *Polyhedron*, 1986, **5**, 1661.

$S_4N_3[MoCl_4(N_3S_2)]$[87a] and with $PPh_4[RuCl_4(NO)]_2$ the product is $PPh_4[RuCl_4(NO)(NSCl)]$, in which the NO and NSCl groups occupy *cis*-positions.[87b] Other complexes prepared from $(NSCl)_3$ are: $[SNBr_{0.4}]_x$ from $BrSiMe_3$;[88a] $[S_5N_5][SnCl_5(MeCN)]$ from $SnCl_2$;[88b] $MoCl_3(N_3S_2)$ from $MoCl_5$,[88c] and $[SN]^+[AsF_6]^-$ by reaction with excess $AgAsF_6$.[88d] The last compound provides a convenient small-scale route to NSF. The reaction of $(NSCl)_3$ with As_2O_3 gives $(S_5N_5)_4[As_8Cl_{28}].2S_4N_4$.[89] The structure of the $[As_8Cl_{28}]^{4-}$ anion can be described as an $[As_4Cl_{16}]^{4-}$ ion with a cubane-like structure to which four $AsCl_3$ molecules are attached *via* chloro-bridges. The corresponding reaction with Sb_2O_3 gives (S_5N_5)-$[SbCl_6]$ and with Bi_2O_3 the products are $[S_4N_5][BiCl_4]$ and $[S_4N_4Cl][BiCl_4]$.

Fluorination of FSO_2NH_2 at room temperature gives FSO_2NF_2 (Figure 8) in high yield.[90] The reaction of FSO_2NF_2 with Et_2NH gives FSO_2NEt_2 and HNF_2. Oxidative-addition reactions between ClF and $CF_3N{=}S(OR_F)_2$, $[R_F = CF_3CH_2, CF_3CF_2CH_2,$

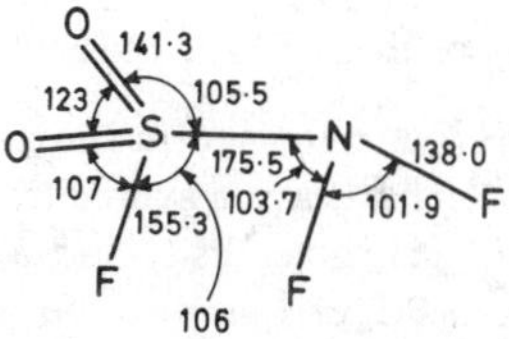

Figure 8 *Bond distances (pm) and angles (°) in* FSO_2NF_2

or $CF_3(CF_2)_2CH_2]$, result in the formation of both the *cis*- and *trans*-tetra-fluorobis(polyfluoroalkoxy)sulphur isomers, $F_4S(OR_F)_2$, with the *trans*-form predominating.[91] The reaction of ClF with $(CF_3CH_2O)_2S{=}NC(O)OCH_2CF_3$ also results in *cis*- and *trans*-$F_4S(OCH_2CF_3)_2$. In an attempt to prepare a compound containing the $F_4S{=}N$ moiety, the reaction between SF_5NCO and Me_3OMeSi was carried out.[92] The product which was isolated was *cis*-methoxytetrafluorosulphanyl isocyanate which was formed by an unusual migration of a methoxy group (Scheme 3). Friedel–Crafts reactions of $(NPCl_2)_2NSOCl$ with a series of monosubstituted benzenes give $(NPCl_2)_2NSOC_6H_4R$ in good yield.[93] Both *ortho*- and *para*-isomers are formed when R = Me or OMe, but only the *para*-isomer is obtained when R = Et, or halogen. The reactions between $NPCl_2(NSOX)_2 (X = Ph, Cl, F)$ and diaminoethane or 2-aminoethanol in 1:2 mole ratio give the spirocyclic compounds $NP[NH(CH_2)_2Y][NSOX]_2$ with Y = NH or O.[94]

[87a] H. Wadle, E. Conradi, U. Muller, and K. Dehnicke, *Z. Naturforsch., Teil B*, 1986, **41**, 429.
[87b] W. Willing, U. Muller, U. Demant, and K. Dehnicke, *Z. Naturforsch., Teil B*, 1986, **41**, 560.
[88a] U. Demant and K. Dehnicke, *Z. Naturforsch., Teil B*, 1986, **41**, 929.
[88b] U. Patt-Siebel, S. Ruangsuttinarupap, U. Muller, J. Pebler, and K. Dehnicke, *Z. Naturforsch., Teil B*, 1986, **41**, 1191.
[88c] K. Volp, W. Willing, U. Muller, and K. Dehnicke, *Z. Naturforsch., Teil B*, 1986, **41**, 1196.
[88d] A. Apblett, A. J. Bannister, D. Biron, A. G. Kendrick, J. Passmore, M. Schriver, and M. Stojanac, *Inorg. Chem.*, 1986, **25**, 4451.
[89] W. Willing, U. Muller, J. Eicher, and K. Dehnicke, *Z. Anorg. Allg. Chem.*, 1986, **537**, 145.
[90] H. W. Roesky, U. Otten, and H. Oberhammer, *Z. Anorg. Allg. Chem.*, 1986. **539**, 191.
[91] H. M. Marsden and J. M. Shreeve, *Inorg. Chem.*, 1986, **25**, 4021.
[92] J. S. Thrasher, J. L. Howell, M. Clark, and A. F. Clifford, *J. Am. Chem. Soc.*, 1986, **108**, 3526.
[93] F. J. Viersen, E. Bosma, B. DeRuiter, K. S. Dhathathreyan, J. C. Van De Grampel, and F. Van Bolhuis, *Phosphorus Sulfur*, 1986, **26**, 285.
[94] W. Hoeve, K. S. Dhathathreyan, J. C. Van De Grampel, and F. Van Bolhuis, *Phosphorus Sulfur*, 1986, **26**, 293.

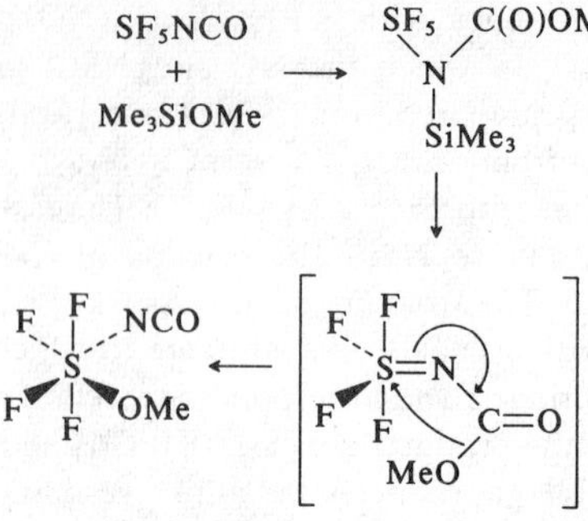

$$SF_5NCO + Me_3SiOMe \longrightarrow \begin{array}{c} SF_5 \quad C(O)OMe \\ \diagdown N \diagup \\ | \\ SiMe_3 \end{array}$$

$$\begin{array}{c} F \quad F \\ F \diagdown \overset{|}{\underset{|}{S}} \diagup NCO \\ F \diagup F \diagdown OMe \end{array} \longleftarrow \left[\begin{array}{c} F \quad F \\ F \diagdown \overset{|}{\underset{|}{S}} = N \\ F \diagup F \quad \diagdown C = O \\ MeO \diagup \end{array} \right]$$

Scheme 3

13 Binary Halides of the *d*-Block Elements

The nitrido complexes $M_2NX_{10}^{3-}$ are formed in the ammonolysis reaction of NH_4^+ with $NbBr_5$, WBr_5, and TaI_5.[95] They have a symmetrical $M=N=M$ nitrido bridge. The thermal decomposition of PtI_4 gives Pt_3I_8 and α-PtI_2, and β-PtI_2 is obtained by hydrothermal synthesis from PtI_4, KI, and I_2.[96] In β-PtI_2 two square-planar PtI_4 units are connected by a common edge to planar Pt_2I_6 groups which are linked by common corners to form puckered layers. Pt_3I_8 can be formulated as a mixed-valence $Pt^{II,IV}$ iodide, $PtI_4.2PtI_2$. Octahedral PtI_6 and square planar PtI_4 units are linked together in a three-dimensional network.

The gas-phase structures of $ZnCl_2$, $ZnBr_2$, and ZnI_2 have been determined by electron diffraction.[97] All the molecules are linear with $Zn—Cl = 207.2$, $Zn—Br = 220.4$, and $Zn—I = 240.1$ pm.

14 Halogenometallates of the *d*-Block Elements

The ammonium halides react with Y_2O_3 and Y_2S_3 in a molar ratio of 12:1 to give $(NH_4)_3YX_6$ (X = Cl, Br, I).[98] If smaller ratios than 12:1 are used the product is YOX.

The thermal decomposition of $N_2H_6MF_6$ (M = Zr, Hf) gives as a first step $N_2H_5MF_5$.[99] In the second step the M = Zr compound gives NH_4ZrF_5 as an intermediate, but an analogous Hf complex is not obtained in the pure state. The final products are the corresponding tetrafluorides. Vapour mixtures of MCl_4 (M = Zr, Hf) and $POCl_3$ or $AlCl_3$ in the temperature range 300–500 °C were studied by Raman spectroscopy.[100] The spectra indicate the presence of $MCl_4.POCl_3$ complexes, in which the complexation takes place through oxygen bridging. In the MCl_4-$AlCl_3$ system two complexes are possible: MAl_2Cl_{10} and $MAlCl_7$. The predominant species present is probably the 1:1 complex. A wide range of cluster halides of zirconium

[95] M. Horner, K.-P. Frank, and J. Strahle, *Z. Naturforsch., Teil B*, 1986, **41**, 423.
[96] G. Thiele, W. Weigl, and H. Wochner, *Z. Anorg. Allg. Chem.*, 1986, **539**, 141.
[97] M. Hargittai, J. Tremmel, and I. Hargittai, *Inorg. Chem.*, 1986, **25**, 3163.
[98] G. Meyer and T. Staffel, *Z. Anorg. Allg. Chem.*, 1986, **532**, 31.
[99] D. Gantar and A. Rahten, *J. Fluorine Chem.*, 1986, **34**, 63.
[100] S. Boghosian, G. N. Papatheodorou, R. W. Berg, and N. J. Bjerrum, *Polyhedron*, 1986, **5**, 1393.

and rare-earth metals have been synthesized and characterized in which a 3*d* transition metal is encapsulated within the cluster.[101]

The addition of CCl_4 to a solution of PPh_4Cl and excess VCl_4 in H_2Cl_2 gives $(PPh_4)_2[V_2Cl_9][VCl_5].CH_2Cl_2$.[102] The structure consists of PPh_4^+ cations, nearly ideal trigonal-pyramidal VCl_5^-, and $V_2Cl_9^-$ anions. $[VNCl_3]^-$ is isoelectronic with $VOCl_3$ and is formed in the reaction of Ph_3PNPPh_3Cl with Cl_3NSiMe_3.[103]

The reaction of SCl_2 with $Cr(CO)_5THF$ produces a solid which does not contain carbonyl groups.[104] Spectroscopic and chemical analysis suggest the formula $[Cr(H_2O)(THF)(SCl_2)Cl_3].2THF$. The crystal structure of $KCrF_4$ consists of infinite triangular anionic columns comprising approximately planar Cr_3F_9 units linked above and below by fluoride bridges.[105] Each Cr^{III} is octahedrally coordinated by six F atoms. In the structure of $K_2[Cr_3Cl_2F_6]$ there are two crystallographically independent Cr^{II} atoms which have distorted octahedral coordination by four equatorial F atoms and two axial Cl atoms. The Cr atoms are linked by single Cr—F—Cr bridges and by a novel chloridefluoride double bridge. The structure of $Cs_3[Cr_2Br_3Cl_6]$ is based on closest packing of $Cs(Cl, Br)_3$ layers with Cr ordered in octahedral interstices to form isolated pairs of face-shared octahedra.[106] The bromine atoms selectively occupy the terminal positions. Addition of X_2 (X = Br or I) to hydrocarbon solutions of $W_2(O_2CBu^t)_4.2L$ (L = THF or Bu^tCONMe_2) leads to facile carboxylate exchange.[107] Addition of I_2 allowed the isolation of the salt $[W_2(O_2CBu^t)_5.2L]^+[W_2I_4(O_2CBu^t)_2]^-$ (L = Bu^tCONMe_2) while addition of Br_2 gives $W_2Br_2(O_2CBu^t)_3.2L$ (L = THF). The reaction between $[W_6X_8]X_4$ and X_2 (X = Cl, Br) was investigated by changing the Cl_2/Br_2 ratio and temperature.[108] All of the $[W_6Br_{12-n}Cl_n]Cl_{6-m}Br_m$ compounds isolated are isotypic with W_6Cl_{18} and W_6Br_{18}.

The reaction of $[TcCl_6]^{2-}$ with aqueous HBr followed by rapid precipitation of $Cs_2[SnCl_6]$ from the solution gives nine of the ten stereoisomers of $[TcCl_xBr_{6-x}]^{2-}$ (x = 0—6).[109] Similar results were obtained starting with $[TcBr_6]^{2-}$ in HCl. There is no evidence for the dominance of a *trans*-effect in the substitution reaction.

The first carbonyl compound of Os^{IV}, $[OsCl_5(CO)]^-$, was prepared by chlorination of *trans*-$[OsX_4(CO)_2]^-$ (X = Br, I).[110] It is reduced immediately by Br^- and I^- to $[OsCl_5(CO)]^{2-}$. Oxidation of the pure fluorochloroosmates(IV) with BrF_3 gives $[OsF_5Cl]^-$, *cis*-$[OsF_4Cl_2]^-$, *fac*-$[OsF_3Cl_3]^-$, and *cis*-$[OsF_2Cl_4]^-$ without replacement of ligands.[111] *trans*-$[OsF_4Cl_2]^-$ and *trans*-$[OsF_2Cl_4]^-$ are formed from the corresponding Os^{IV} compounds by oxidation with PbO_2.

X-Ray photoelectron spectroscopy of several palladium fluorine compounds shows that shifts in the binding energies can be correlated with the Pd oxidation

[101] T. Hughbanks, G. Rosenthal, and J. D. Corbett, *J. Am. Chem. Soc.*, 1986, **108**, 8289.
[102] K.-D. Scherfise, W. Willing, U. Muller, and K. Dehnicke, *Z. Anorg. Allg. Chem.*, 1986, **534**, 85.
[103] K.-D. Scherfise and K. Dehnicke, *Z. Anorg. Allg. Chem.*, 1986, **538**, 119.
[104] C. Diaz, *Polyhedron*, 1986, **5**, 2001.
[105] J. C. Dewan, A. J. Edwards, and J. J. Guy, *J. Chem. Soc., Dalton Trans.*, 1986, 2623.
[106] I. E. Grey, I. C. Madsen, S. E. Butler, P. W. Smith, and R. Stranger, *Acta Crystallogr., Section C*, 1986, **42**, 769; S. E. Butler, P. W. Smith, R. Stranger, and I. E. Grey, *Inorg. Chem.*, 1986, **25**, 4375.
[107] M. H. Chisholm, H. T. Chin, and J. C. Huffman, *Polyhedron*, 1986, **5**, 1377.
[108] U. Lange and H. Schafer, *Z. Anorg. Allg. Chem.*, 1986, **542**, 207.
[109] C. D. Flint and P. F. Long, *J. Chem. Soc., Dalton Trans.*, 1986, 921.
[110] M. Bruns and W. Preetz, *Z. Naturforsch., Teil B*, 1986, **41**, 25.
[111] W. Preetz and Th. Groth, *Z. Naturforsch., Teil B*, 1986, **41**, 885.

state and covalency.[112] The spectra confirm the unusual Pd^{III} oxidation state in $NaPdF_4$, Na_3PdF_6, K_3PdF_6, and K_2NaPdF_6. The reaction of the NBu_4^+ salt of *trans*-$[Pt(C_6Cl_5)_2]^{2-}$ with $AgNO_3$ gives polymeric $[Pt(C_6Cl_5)_2(\mu\text{-}Cl)_2Ag]_x^{x-}$ as the NBu_4^+ salt.[113] The chain anion (Figure 9) consists of planar *trans*-$Pt(C_6Cl_5)_2Cl_2$ units linked by Ag atoms in which there are Ag—Cl bonds of two types: those which result in $Pt(\mu\text{-}Cl)Ag$ linkages and those in which *o*-Cl atoms of the aromatic groups make close approaches to the Ag atoms. Bonding between Pt and Ag is not sterically feasible and is unnecessary since the Ag atom is receiving electron density from six Cl atoms. This polymeric compound reacts with neutral ligands to give either anionic dinuclear complexes of the type $(NBu_4)[PtAgCl_2(C_6Cl_5)_2L]$ (L = PPh_3) in which there is a direct Pt—Ag bond, or neutral trinuclear compounds of the type $[PtAg_2Cl_2(C_6Cl_5)_2L_2]$ (L = $PMePh_2$).[114]

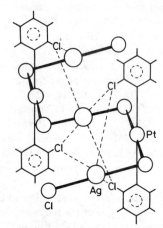

Figure 9 *Part of the polymeric* $[Pt(C_6Cl_5)_2(\mu\text{-}Cl)_2Ag]_x^{x-}$ *anion*

$Cs_2[CuF_4]$ has been prepared for the first time by the thermal decomposition of $Cs_2Rb[CuF_6]$.[115] It is isotypic to $K_2[NiF_4]$. The structures of the following copper-containing anions have been determined this year: $[Cu_2Cl_3]^-$,[116a] $[Cu_2Cl_5]^-$,[116b] $[Cu_4Cl_{12}]^-$,[116c] $[CuI_2]^-$,[116d] and $[Cu_3I_5]^{2-}$.[116e] $[NMe_4][Ag_2Br_3]$ is isostructural with $[NMe_4][Cu_2Cl_3]$.[117] The structure of $[NH_3(CH_2)_2NH_2(CH_2)_2NH_3]HgCl_8$ consists of diethylenetrammonium cations, Cl^-, and discrete $[HgCl_5]^{3-}$ anions.[118] The $HgCl_5^{3-}$ anion is a compressed trigonal-bipyramid, in which the axial bonds are

[112] A. Tressaud, S. Khairoun, H. Touhara, and N. Watanabe, *Z. Anorg. Allg. Chem.*, 1986, **540/541**, 291.
[113] R. Uson, J. Fornies, M. Tomas, J. M. Casas, F. A. Cotton, and L. R. Falvello, *Polyhedron*, 1986, **5**, 901.
[114] R. Uson, J. Fornies, M. Tomas, J. M. Casas, F. A. Cotton, and L. R. Falvello, *Inorg. Chem.*, 1986, **25**, 4519.
[115] D. Kissel and R. Hoppe, *Z. Anorg. Allg. Chem.*, 1986, **540/541**, 135.
[116a] S. Andersson and S. Jagner, *Acta Chem. Scand., Ser. A*, 1986, **40**, 177.
[116b] W.G. Haije, J. A. L. Dobbelaar, and W. J. A. Maaskant, *Acta Crystallogr., Section C*, 1986, **42**, 1485.
[116c] R. D. Willett and U. Geiser, *Inorg. Chem.*, 1986, **25**, 4558.
[116d] N. P. Rath and E. M. Holt, *J. Chem. Soc., Chem. Commun.*, 1986, 311.
[116e] H. Hartl and J. Fuchs, *Angew. Chem., Int. Ed. Engl.*, 1986, **25**, 569.
[117] S. Jagner, S. Olsen, and R. Stromberg, *Acta Chem. Scand., Ser. A*, 1986, **40**, 230.
[118] L. P. Battaglia, A. B. Corradi, L. Antolini, T. Manfredini, L. Menabue, G. C. Pellaconi, and G. Ponticelli, *J. Chem. Soc., Dalton Trans.*, 1986, 2529.

significantly shorter than the equatorial bonds. It can be considered as a linear $HgCl_2$ molecule perturbed by interaction of three weakly bonded Cl atoms.

15 Oxohalides, Chalcogen-halides, and Oxohalogenometallates of the *d*-Block Elements

The complexes $[VOCl_{4-n}F_n]^-$ ($n = 1$—4), $[VO(CF_3CO_2)_{4-n}F_n]^-$ ($n = 1$—3), $[VOF_3(NO_3)]^-$, $[VOF_3(MeSO_3)]^-$, $[VOClF]^-$, $[VOF_3(bipy)]$, $[VOF_3(phen)]$, $VOCl_2.n\,MeCN$, and $VOClF_2.n\,MeCN$ were characterized in organic solvents by ^{51}V and ^{19}F n.m.r. spectroscopy.[119] The magnitude of the V—F coupling depends upon the degree of covalency of the V—F bond. Two isomers of $[VOCl_2F_2]^-$ and $[VO(CF_3CO_2)_2F_2]^-$ were observed, suggesting that the geometry around the vanadium atom is square-pyramidal. All $[VOF_3X]^-$ (X = Cl, CF_3CO_2, NO_3, or $MeSO_3$) are fluxional. The structures of $Na_2[VF_5O]$ and $K_{2n}[(VF_3O_2)_n]$ have octahedral coordination around the vanadium atom.[120] The former contains disordered V atoms and the latter contains infinite chains of octahedra linked by *cis*-bridging fluorine atoms.

The molecular structures of $WSeF_4$ and $OsOCl_4$ were determined in the gas phase by electron diffraction.[121] The molecules are square-pyramidal in which the tungsten or osmium is above the plane of the four halogen atoms. In the $[WO_2Cl_3]_2^{2-}$ anion the tungsten atoms are linked *via* two oxo-bridges.[122] A distorted octahedral coordination is completed by three terminal chlorine atoms and one terminal oxygen atom.

$RuCl_3$ reacts with $(NSCl)_3$ to give *cis*-$RuCl_4(NS)_2$, which with Ph_3MePCl gives the complex anion $[RuCl_4(NS)_2Cl]^-$ in which a chloride ion is bonded between the sulphur atoms (14).[123]

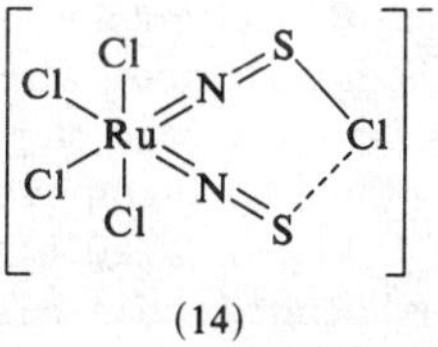

(14)

The quaternary chalcogen halides, $AgHgSI$, $AgHgSBr$, and $CuHgSI$ were obtained by annealing mixtures of Ag(Cu) halides and HgS.[124] In the AgI–HgS system the only stable compound is $AgHgSI$; Ag_2HgSI_2 is metastable. Substitution of Hg^{2+} by Cd^{2+} in CuHgSI gives $Cu_2Cd_3S_2I_4$.

16 Actinide Halides

Two stable intermediates, $UOCl_2.L$ and UO_2Cl, were isolated in the preparation of UX_4L_2 (X = Cl, Br; L = tetramethylurea).[125] The complexes are isomorphous, with

[119] R. C. Hibbert, *J. Chem. Soc., Dalton Trans.*, 1986, 751.
[120] R. Stomberg, *Acta Chem. Scand., Ser. A*, 1986, **40**, 425.
[121] K. Hagen, D. A. Rice, J. H. Holloway, and V. Kaucic, *J. Chem. Soc., Dalton Trans.*, 1986, 1821; K. Hagen, R. J. Hobson, C. J. Holwill, and D. A. Rice, *Inorg. Chem.*, 1986, **25**, 3659.
[122] W. Willing, F. Schmock, U. Muller, and K. Dehnicke, *Z. Anorg. Allg. Chem.*, 1986, **532**, 137.
[123] U. Demant, W. Willing, U. Muller, and K. Dehnicke, *Z. Anorg. Allg. Chem.*, 1986, **532**, 175.
[124] R. Blacknik and H. A. Dreisbach, *Monatsh. Chem.*, 1986, **117**, 305.
[125] J. G. H. du Preez, B. Zeelie, U. Casellato, and R. Graziani, *Inorg. Chim. Acta*, 1986, **122**, 119.

the uranium octahedrally coordinated. A thermally stable UI_4 starting material which can be readily prepared in high yield and which contains weakly coordinated donor ligands was prepared from uranium metal, I_2, and diphenyl ketone (dpk).[126] The product $UI_4(dpk)_2$ can be used to prepare other iodo-complexes according to equation (11).

$$UI_4(dpk)_2 \xrightarrow{2L} UI_4L_2(s) \qquad (11)$$

The crystal structures of $UBr_4.2ddu$, two polymorphic forms of $UCl_4.2ddu$, $UCl_4.2tprpo$, $UCl_4.2dibso$, and $UBr_4.2tpao$ (ddu = N,N'-dimethyl-N,N'-diphenylurea; tprpo = tris(pyrrolidinyl)phosphine oxide; dibso = diisobutyl sulphoxide; tpao = triphenylarsine oxide) have been determined.[127] The geometry around the uranium atom is approximately octahedral. It is found that the inverse relationship between the U—halogen and U—O bond length changes for complexes containing the UO_2X_4 chromophore (the axial–equatorial effect) is maintained in these complexes. The data lend support for the donor-strength sequence As=O > P=O > S=O > C=O, based on chemical evidence. The trinuclear U^{III} arene complex $[U_3(\mu^3\text{-}Cl)_2(\mu^2\text{-}Cl)_3(\mu^1\eta^2\text{-}AlCl_4)_3(\eta^6\text{-}C_6Me_6)_3][AlCl_4]$ was prepared from UCl_4, $AlCl_3$, Al, and C_6Me_6.[128] The three uranium atoms define approximately an equilateral triangle. Three chlorine atoms bridge the edges of the triangle while two additional chlorine atoms cap the triuranium fragment on each side. The coordination polyhedron of each uranium is a distorted pentagonal-bipyramid.

17 Graphite–Halide Intercalation Compounds and Graphite Fluorides

A study of the formation process with respect to some of the properties of graphite fluoride $(C_2F)_n$ prepared from artificial graphite shows that the process can be divided into two steps.[129] In the first step both $(CF)_n$ and $(C_2F)_n$ are formed and unreacted graphite still exists. In the second step the fluorine content increases slightly in spite of the absence of unreacted graphite. Mass spectrometry shows that the thermal decomposition of $C_4F_{1.88}Cl_{0.19}0.20N_2O_4$ takes place in two stages: in the first step intercalated N_2O_4 is liberated and in the second graphite chloridefluoride decomposes.[130]

The compounds formed by the gas-phase intercalation of graphite by $SbCl_4F$ are air stable.[131] In the presence of aqueous HCl or KOH only pentavalent antimony is removed. The intercalation of MF_6 into graphite proceeds according to equation 12, where x is the degree of charge transfer, which can be determined by magnetic measurements.[132] The degree of charge transfer varies approximately linearly with the electron affinity of MF_6.

$$nC + MF_6 \rightarrow C_n^{x+}(MF_6^-)_x(MF_6)_{1-x} \qquad (12)$$

[126] J. G. H. du Preez and B. Zeelie, *J. Chem. Soc., Chem. Commun.*, 1986, 743.
[127] J. F. de Wet and M. R. Caira, *J. Chem. Soc., Dalton Trans.*, 1986, 2035; J. F. de Wet and M. R. Caira, *J. Chem. Soc., Dalton Trans.*, 1986, 2043.
[128] F. A. Cotton, W. Schwotzer, and C. Q. Simpson, *Angew. Chem., Int. Ed. Engl.*, 1986, **25**, 637.
[129] S. Koyama, T. Maeda, and K. Okamura, *Z. Anorg. Allg. Chem.*, 1986, **540/541**, 117.
[130] A. F. Antimonov, V. M. Grankin, P. P. Semyannikov, A. S. Nazarov, and I. T. Yakovlev, *Russ. J. Inorg. Chem. (Engl. Transl.)*, 1986, **31**, 206.
[131] H. Preiss, M. Goerlich, and H. Sprenger, *Z. Anorg. Allg. Chem.*, 1986, **533**, 37.
[132] D. Vaknin, D. Davidov, and H. Selig, *J. Fluorine Chem.*, 1986, **32**, 345.

8 Ti, Zr, Hf; V, Nb, Ta; Cr, Mo, W; Mn, Tc, Re

By J. E. NEWBERY

Chemical Laboratories, University of London Goldsmiths' College, London SE14 6NW

1 Introduction

This section follows the style established over previous volumes of *Annual Reports* in that the chemistry of each group of elements is dealt with by starting with a note of any interesting review articles. Next structural and behavioural aspects of such species as oxides and halides are covered before beginning a detailed account of coordination compounds and organometallic species. Within each of these areas the emphasis has been placed on the structural complexity and donor type of the ligand rather than a consideration of the oxidation number of the metal. The overriding aim of the section is to present a representative sample of the type of work currently in progress rather than attempting to produce a comprehensive list of all articles published in 1986.

Amongst the papers of general significance to the early transition elements are two that can broadly be classified as dealing with organometallics. It has been noted that in cluster compounds containing open transition-metal polyhedra, X-ray structural methods have become the dominant tool[1] and that, with the exception of the isolobal model first employed by F. G. A. Stone in the construction of alkylidene and alkylidyne complexes, there are few methods of cluster assembly that are generally applicable. An interesting account has been made[2] of compounds containing multiple bonds between a transition element and a substituent-free main-group element. The first examples of this class of compounds had serendipitous origins, but with expansion of interest more patterns are becoming evident.

1986 saw the publication of three reviews on the chemistry of sulphur-containing ligands. Two of these[3,4] deal specifically with S—N donor ligands of mainly inorganic origin. The major emphasis is on the structural and synthetic chemistry of the resultant complexes. A similar position is taken in a massive (*ca.* 350 references) account of metal thiolate complexes.[5]

2 Titanium, Zirconium, and Hafnium

The layered structure of zirconium hydrogen phosphate has been well-characterized, but there is little information on any of the metal-exchanged materials since they

[1] M. O. Albers, D. J. Robinson, and N. J. Coville, *Coord. Chem. Rev.*, 1986, **69**, 127.
[2] W. A. Herrmann, *Angew. Chem., Int. Ed. Engl.*, 1986, **25**, 56.
[3] P. F. Kelly and J. D. Woollins, *Polyhedron*, 1986, **5**, 607.
[4] T. Chivers and F. Edelmann, *Polyhedron*, 1986, **5**, 1661.
[5] I. G. Dance, *Polyhedron*, 1986, **5**, 1037.

are usually powders. It has been reported[6] that it is possible to make $Zr(O_3PONa)_2$ in a crystalline form. The unit cell is more than twice the volume of the acid form. It has been demonstrated[7] that if the layered structure of α-zirconium phosphate is first swelled by the intercalation of ethanol, it becomes possible to take up the complex ion $Cu(1,10\text{-phen})_2^{2+}$. A maximum ratio of 0.2Cu : 1.0Zr is attained and the tetragonal form of the copper complex is retained. Removal of water causes the formation of a pillared structure with the copper complex becoming square-planar. A synthesis of $A_2[Ti(O_2)_2F_2]$, where A = Na, K, or NH_4, by the direct interaction of titanium dioxide and hydrogen peroxide in the presence of hydrofluoric acid has been described.[8] The structure of the related compound $A_3Ti(O_2)F_5$ has been determined by X-ray diffraction methods.[9] The titanium is essentially octahedral if the sideways-on peroxide ligand (O—O distance of 1.44 Å) can be regarded as occupying one coordination position. Reaction between KBH_4 and $TiBr_4$ in liquid ammonia has been shown[10] to produce the compound $[NH_3 \cdot NH_4^+]_2\text{-}$ $[Ti_4Br_4(NH_2)_{12}^{2-}]$. The formula was established by elemental analysis including Ce^{IV} titration to estimate the Ti^{III} content. The crystal structure of the complex $TiCl_3 \cdot 4MeCN$ has been investigated.[11] The titanium co-ordination was found to be a regular octahedron with three MeCN groups coordinated in a *mer-* fashion. The two *trans*-acetonitriles are closer to the metal (2.15 Å av.) than the third ligand (2.234 Å).

Iso-structural cluster compounds of the type $Zr_6I_{12}X$, where X is B, C, Al, or Si, have been described.[12,13] In the related $CsZr_6I_{14}C$ it was further demonstrated by [13]C n.m.r. spectroscopy that the carbon has axially-symmetric shielding in agreement with the structure determined by X-ray methods.

A unique sideways-bound NNPh ligand has been located[14] in the complex $Ti(NNPh)(Cl)_2Cp$ (1). The N—N bond length was found to be rather short at 1.219 Å and could be regarded as a triple bond that has been lengthened by olefinic-style binding. The ligand is readily removed with acid. It was also shown[15] that hydrazido-groups can be bound to titanium in this sideways fashion. The main objective of these two papers was to provide information on the mechanism of dinitrogen protonation in such reactions as shown[16] in Scheme 1. It seems that both

$$
\begin{array}{c}
Cp \\
| \\
Ph-N \overset{\displaystyle\diagup}{\underset{\displaystyle\diagdown}{}} Ti \diagdown Cl \\
N \quad Cl
\end{array}
$$

(1)

[6] D. P. Vliers, W. J. Mortier, and R. A. Schoonheydt, 1986, **5**, 1997.

[7] C. Ferragina, M. A. Massucci, P. Patrono, A. La Ginestra, and A. A. G. Tomlinson, *J. Chem. Soc., Dalton Trans.*, 1986, 265.

[8] M. K. Chaudhuri and B. Das, *Inorg. Chem.*, 1986, **25**, 168.

[9] R. Schmidt and G. Pausewang, *Z. Anorg. Allg. Chem.*, 1986, **537**, 175.

[10] L. Maya, *Inorg. Chem.*, 1986, **25**, 4213.

[11] S. I. Troyanov and A. I. Toorsina, *Koord. Khim.*, 1986, **12**, 1559.

[12] J. D. Smith and J. D. Corbett, *J. Am. Chem. Soc.*, 1986, **108**, 1927.

[13] C. G. Fry, J. D. Smith, B. C. Gerstein, and J. D. Corbett, *Inorg. Chem.*, 1986, **25**, 117.

[14] I. A. Latham, G. J. Leigh, G. Huttner, and I. Jibril *J. Chem. Soc., Dalton Trans.*, 1986, 377.

[15] D. L. Hughes, I. A. Latham, and G. J. Leigh, *J. Chem. Soc., Dalton Trans.*, 1986, 393.

[16] I. A. Latham and G. J. Leigh, *J. Chem. Soc., Dalton Trans.*, 1986, 399.

$$6TiCpCl_3 \overset{i}{\rightarrow} 6TiCpCl_2.thf + 2N_2 \; (or \; 2NH_3)$$

Reagents: i, 3 Me$_3$SiNHNHSiMe$_3$ or 3 LiNHNH$_2$ in thf

Scheme 1

NHNH^{2-} and NHNH$_2^-$ are unstable when bonded to Ti^{IV}, leading to the formation of a Ti^{II} species and dinitrogen.

Titanium is known to form complexes with tetra-aza ligands such as porphyrins and phthalocyanines, and has now been shown to have a number of interesting products with the dianionic ligand L (2). If the species Ti(L)Cl$_2$ is reacted with NaC$_5$H$_5$ in thf a Ti^{III} complex is produced, Ti(L)C$_5$H$_5$. The interesting point of this reaction proved to be that the product contains a π-bonded rather than a σ-bonded C$_5$H$_5$ ligand. Indeed the complex adopts almost a sandwich structure with the titanium displaced 0.900 Å from the N$_4$ donor plane, and with the two cyclic moieties almost parallel.[17]

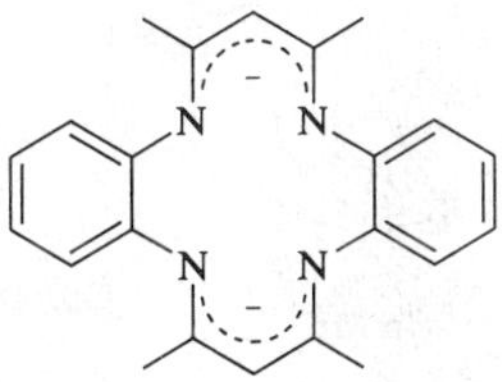

(2)

The cavities of the calixarenes are suitable sites for coordination and these compounds are studied as molecular transporters and as enzyme mimics. If hexamethoxycalix[6]arene is reacted with TiCl$_4$ in a toluene solution,[18] light red crystals are obtained with the formulation calix[6]arene[TiCl$_2$(μ—O)TiCl$_3$]$_2$. As can be seen from the structure (3), two of the titaniums are bound directly to the

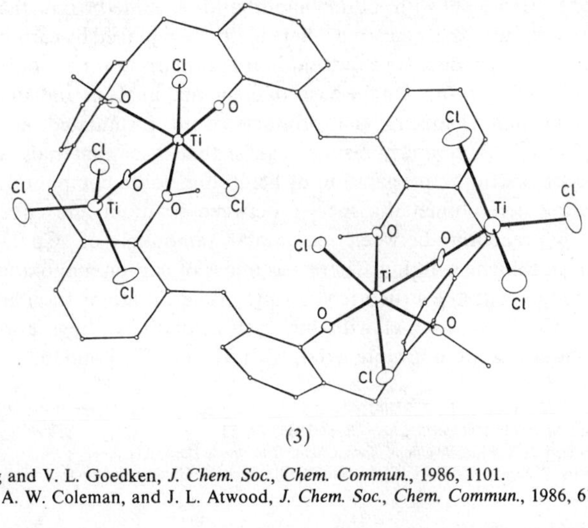

(3)

[17] C.-H. Yang and V. L. Goedken, *J. Chem. Soc., Chem. Commun.*, 1986, 1101.
[18] S. G. Bott, A. W. Coleman, and J. L. Atwood, *J. Chem. Soc., Chem. Commun.*, 1986, 610.

calix and have six-coordination from two chlorines, two methoxy groups, one $O-TiCl_3$, and $-O-$aryl made possible by the breaking of an $O-Me$ bond. The coordinative unsaturation of the tetrahedral titanium atoms is probably responsible for the $Ti-Cl$ distances $(2.17-2.22 \text{ Å})$ being much shorter than those of the octahedral centre $(2.25-2.29 \text{ Å})$.

Titanium clusters containing oxygen are quite common, but those featuring sulphur are much rarer. If Li_2S_2 is reacted[19] with $(MeCp)TiCl_3$ and a little air allowed to enter the vessel, it is possible to obtain crystals with the formula $(MeCp)_4Ti_4S_8O_x$, where $x = 1$ or 2. The structure of the latter was obtained from X-ray diffraction analysis (4), and there is also n.m.r. evidence to show that the mono-oxo complex can convert into the di-oxygen species without fragmentation of the cluster. It thus seems likely that the mono complex also contains the central μ_4 oxo ligand.

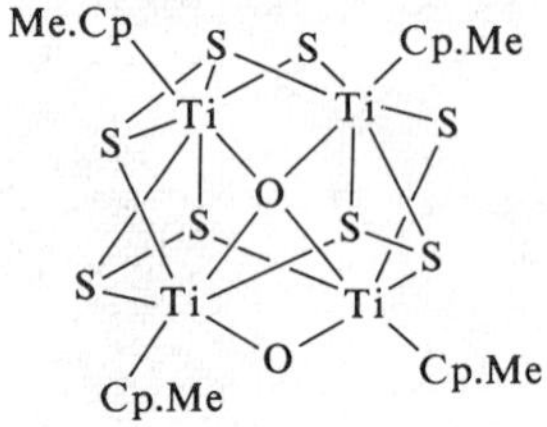

(4)

Over 300 titanium organometallic compounds are included in a useful review[20] that concentrates on the classification of recent X-ray crystallographic results. The tabulated information simplifies the observation of structural trends amongst a rich variety of compounds.

The synthesis has been reported of a new class of zero-valent Group 4 carbonyl derivatives.[21] Reductive carbonylation of $(C_5Me_5)MCl_3$, where M is Ti or Zr, is accomplished by treatment with sodium naphthalide in dimethoxymethane solution. After about 10 minutes the argon atmosphere is slowly replaced by carbon monoxide, and a dark coloured product (red for titanium and maroon for zirconium) isolated after a conventional work-up. On the basis of elemental analysis and an examination of the i.r. and n.m.r. spectra the products were formulated as the anions $[(C_5Me_5)M(CO)_4]^-$. Preliminary results suggest that these materials will prove to be useful precursors in the preparation of heterobimetallic compounds.

An interesting heterobimetallic species between titanium and cobalt has been produced[22] by reaction between equimolar amounts of $Cp_2Ti(CO)_2$ and $CpCo(C_2H_4)_2$ in toluene solution. There was a loss of carbon monoxide and ethene to give the product with the structure shown (5). The Co atoms form an equilateral triangle (Co—Co $2.38-2.40 \text{ Å}$) with the cyclopentadienyl rings almost perpendicular. The Ti atoms are not quite axial, with $Ti-\widehat{O}-C$ about 162°.

[19] G. A. Zank, C. A. Jones, T. B. Rauchfuss, and A. L. Rheingold, *Inorg. Chem.*, 1986, **25**, 1886.

[20] D. Cozak and M. Melnik, *Coord. Chem. Rev.*, 1986, **74**, 53.

[21] B. A. Kelsey and J. E. Ellis, *J. Chem. Soc., Chem. Commun.*, 1986, 331.

[22] S. Gambarotta, S. Stella, C. Floriani, A. Chiesi-Villa, and C. Guastini, *Angew. Chem., Int. Ed. Engl.*, 1986, **25**, 254.

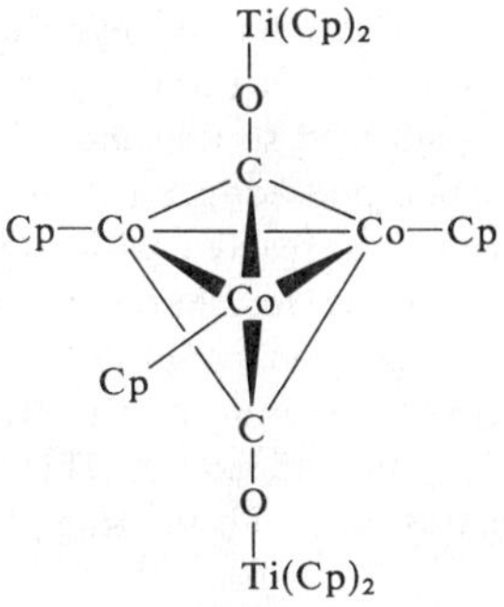

(5)

The reaction between adenine or purine and a number of titanium complexes has been studied.[23] When using $(Cp)_2TiCl$ or $(Cp)_2TiCl_2$ ligand-substitution occurs, but with $(Cp)_2Ti(CO)_2$ a first-order oxidative reaction occurs with the release of both carbon monoxide and hydrogen (4:1 ratio).

A facile route into Zr^{III} chemistry is provided[24] from $(Cp)_2ZrCl_2$ by the route shown in Scheme 2. The structure of the final product was confirmed by both ^1H

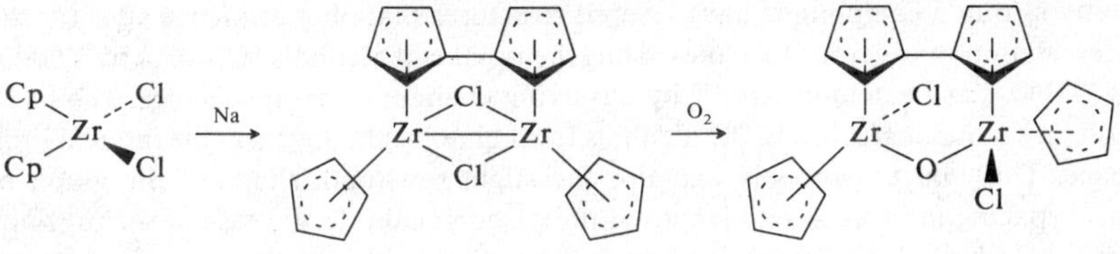

Scheme 2

n.m.r. spectroscopy and *X*-ray diffraction analysis. It should be noted that the ligands take up an eclipsed conformation about the central Zr—O—Zr structure $(Zr—\widehat{O}—Zr = 156°)$.

The structures of the three types of product shown in Scheme 3 were investigated[25] by i.r. and ^1H and ^{13}C n.m.r. spectroscopy and shown to be 18-electron species with

$$ZrCl_4$$
$$\downarrow \text{i}$$
$$\{Zr(\eta\text{-}C_5Me_4R)(\eta\text{-}C_8H_8)Cl\}$$
$$\swarrow \text{ii} \qquad \text{iii} \searrow$$
$$\{Zr(\eta\text{-}C_5Me_4R)(\eta\text{-}C_8H_8)H\} \qquad \{Zr(\eta\text{-}C_5Me_4R)(\eta\text{-}C_8H_8)R^1\}$$

Reagents: i, $K_2C_8H_8 + LiC_5Me_4R$(R = Me or Et); ii, $LiAlH_4$ or $Mg(Cl)CHMe_2$; iii, MR^1 (M = MgCl or Li; R′ = Me, Et, CH_2Ph, *etc.*)

Scheme 3

[23] D. Cozak, A. Mardhy, and A. Morneau, *Can. J. Chem.*, 1986, **64**, 751.
[24] T. A. Ashworth, T. C. Agreda, E. Herdtweck, and W. A. Herrmann, *Angew. Chem., Int. Ed. Engl.*, 1986, **25**, 289.
[25] W. J. Highcock, R. M. Mills, J. L. Spencer, and P. Woodward, *J. Chem. Soc., Dalton Trans.*, 1986, 821.

planar C_8H_8 rings. An *X*-ray examination of the hydride complex (R = Me) broadly confirms this analysis and shows that a sandwich-style structure with both rings tilted away from the centre is adopted so that the hydride can be accommodated. The C_8H_8 ring is not quite planar and adopts a shallow envelope configuration. If the chloro-complex in Scheme 3 is reacted with $MgCl(CH_2CH=CH_2)$ then[26] a substantial change occurs with the formation of the Zr^{II} species $Zr(C_5Me_5)$-$(C_8H_8)(C_3H_5)$. The C_8H_8 ring now is only a η^5 donor, and hence has a non-planar conformation. Heating a benzene solution of Cp_2ZrPh_2 with trimethylphosphine allows[27] a benzyne adduct (6) to be trapped out. This product is notable for having almost equal C—C bond lengths all round the ring (1.377—1.406 Å).

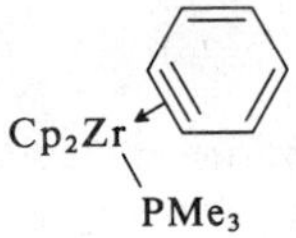

$$Cp_2Zr \diagdown \diagup$$
$$\big|$$
$$PMe_3$$

(6)

3 Vanadium, Niobium, and Tantalum

Many Group 5 compounds have layered structures that offer attractive sites for the intercalation of organic molecules. Thus the uptake of alcohols (C_1—C_8) by $VOSO_4$ or $VOPO_4$ has been monitored[28] by observing changes in basal spacings. The basic structure remains stable and the alcohols form bimolecular layers in the intercalation space. This bilayer probably explains the slight saw-tooth shape of the graph of basal spacing increment with length of the carbon chain. In the case of the protonic oxide $HNb_3O_8 \cdot H_2O$, structural evolution during the intercalation of alkyl-monoamines has been followed.[29] The amine chain has a transverse orientation with respect to the $Nb_3O_8{}^-$ layers.

The use of ^{51}V n.m.r. techniques has allowed a number of interesting vanadium anionic species to be identified.[30-32] Thus in the reaction of hydrogen sulphide with vanadate solutions the known ions $VS_4{}^{3-}$, $VOS_3{}^{3-}$, and $VO_2S_2{}^{3-}$ together with the new species VO_3S^{3-}, $V_2S_7{}^{4-}$, $O_3VSO_3{}^{4-}$, and $SO_2VSVO_2S^{4-}$ were identified.[30] The monomer species were present at high pH (*ca.* 13) and as the pH was lowered the dimers were identified by splittings of the ^{51}V resonance. Similar n.m.r. experiments were made[31] with $VOCl_3$ and $VO(NO_3)_3$. The two materials were subjected to controlled hydrolysis. The nitrate gave polymeric material, whereas the chloride produced mostly $VO_2Cl_2{}^-$. However, when nitrate was introduced, then polymeric material was again formed {Scheme 4}. This behavioural difference may be ascribed to the inability of chloride to chelate. The ^{51}V n.m.r. spectra of a wide range of fluorine-containing complexes dissolved in MeCN have been measured.[32] The most

[26] W. J. Highcock, R. M. Mills, J. L. Spencer, and P. Woodward, *J. Chem. Soc., Dalton Trans.*, 1986, 829.
[27] S. L. Buchwald, B. T. Watson, and J. C. Huffman, *J. Am. Chem. Soc.*, 1986, **108**, 7411.
[28] L. Beneš, J. Votinsky, J. Kalousová, and J. Klikorka, *Inorg. Chim. Acta*, 1986, **114**, 47.
[29] R. Nedjar, M. M. Borel, and B. Raveau, *Z. Anorg. Allg. Chem.*, 1986, **540/541**, 198.
[30] A. T. Harrison and O. W. Howarth, *J. Chem. Soc., Dalton Trans.*, 1986, 1405.
[31] R. C. Hibbert, N. Logan, and O. W. Howarth, *J. Chem. Soc., Dalton Trans.*, 1986, 369.
[32] R. C. Hibbert, *J. Chem. Soc., Dalton Trans.*, 1986, 751.

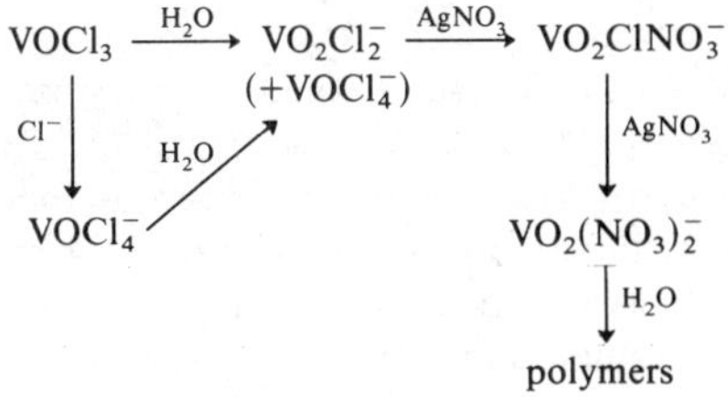
VOCl3
H2O
VO2Cl2−
(+VOCl4−)
AgNO3
VO2ClNO3−
Cl−
H2O
AgNO3
VOCl4−
VO2(NO3)2−
H2O
polymers

Scheme 4

notable point was the observation of V—F coupling which is usually absent in aqueous solution probably as a result of the presence of traces of hydrofluoric acid.

An interesting complex ion was identified[33] when carbon tetrachloride was added to a dichloromethane solution of PPh_4Cl and excess VCl_4. The product was formulated $(PPh_4)_2(V_2Cl_9)(VCl_5)$ and the structure was determined by *X*-ray techniques (7). The $V_2Cl_9^-$ ion contains two face-sharing octahedra where the shared chlorines are uniformally disposed (V—Cl = 2.43—2.46 Å) but significantly longer than the non-shared (V—Cl = 2.18 Å). The VCl_5^- species adopts a near perfect trigonal-bipyramidal structure.

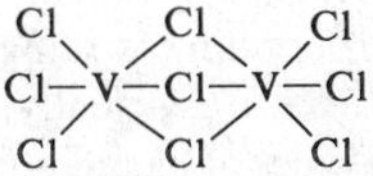

(7)

The last report to be considered in this introduction to Group 5 concerns the chance synthesis of an unusual niobium compound.[34] When attempting to make Rb_3As_7 by a reaction in a niobium tube, too much rubidium was added and after the addition of a solution of 2,2,2-crypt in ethylenediamine, dark red prismatic crystals were isolated which were found to be $[Rb(2,2,2\text{-crypt})]_2[Rb(NbAs_8)]$. The arsenic has been shown to have formed a monocyclic crown $(As_8)^-$ around the Nb^V.

Coordination Compounds—A wide range of donor complexes formed by coordination to $VOCl_3$ has been investigated by ^{51}V n.m.r. spectroscopy.[35] The spectra were obtained by either dissolving the $VOCl_3$ in the neat liquid or as CCl_4 or $CHCl_3$ solutions. It was found that the observations could be fitted into a simple pattern (Scheme 5) where some donors, such as NEt_3 and PhCN were mainly Class I

$$\mathrm{I} \rightleftharpoons \mathrm{II} \rightleftharpoons \mathrm{III}$$

$$\text{Class I: } \{VOCl_3.nL\}_x,\ n = 1, 2$$

$$\text{Class II: } \{VOCl_{3-m}.L\}Cl_m,\ m = 0, 1$$

$$\text{Class III: } \{VOCl_{3-m}.L_2\}.Cl_m,\ m = 0, 2$$

Scheme 5

33 K.-D. Scherfise, W. Willing, U. Müller, and K. Dehnicke, *Z. Anorg. Allg. Chem.*, 1986, **534**, 85.
34 H.-G. von Schnering, J. Wolf, D. Weber, R. Ramirez, and T. Meyer, *Angew. Chem., Int. Ed. Engl.*, 1986, **25**, 353.
35 C. Weidemann and D. Rehder, *Inorg. Chim. Acta*, 1986, **120**, 15.

(oligomers), others such as terpy and DMF were Class III (monomers) and a further group including py, MeCN, and uracil formed mixtures having appreciable quantities of all three Classes.

Sometimes it is more useful to use e.p.r. spectroscopy to follow vanadium complex formation, and a good example of this style of work is given by an investigation[36] into V^{IV} and V^V interactions with proteins in serum. Transferrin was found to form stronger complexes than albumin and for VO^{2+} it was possible to estimate that the ratio of the equilibrium constants was 6:1 in favour of transferrin.

Reaction of $LiS(Bu^t)$ with the iminovanadium(v) chloride, $Bu^tN=VCl_3$, has been used to produce[37] a range of species such as $Bu^tN=V(OBu^t)_n(SBu^t)_{3-n}$, where n is 0, 1, or 2. The use of single crystal X-ray diffraction techniques showed these to be essentially tetrahedral, and multinuclear n.m.r. spectroscopy was used to look at aspects of the hydrolysis behaviour.

The structures of an interesting set of saccharinate (8) complexes with V^{II} have been established.[38] Starting from $VSO_4.6H_2O$, treatment with sodium saccharinate gives $[V(sacc)_2(H_2O)_4].2H_2O$ where the vanadium retains an octahedral structure with the two saccharinate ligands being N-bound in the *trans*-configuration. Recrystallization from pyridine causes a replacement of the water to give $[V(sacc)_2(py)_4].2py$ but this process is accompanied by a switch to binding through the carbonyl group. This style of bonding is rather unusual for saccharinate complexes and rather surprisingly there are virtually no detectable differences in the structural parameters for the two types of ligand other than the change of attachment point. It is suggested that steric crowding around the metal is responsible for the change.

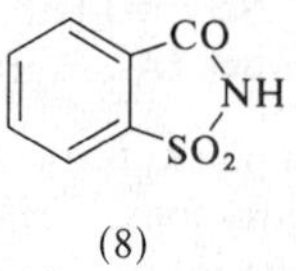

(8) 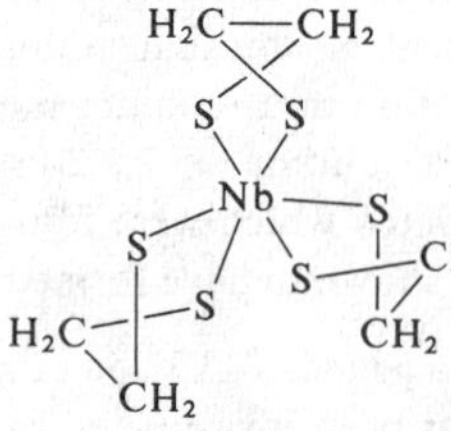   (9)

The reaction of $NbCl_5$ with $LiSCH_2CH_2SLi$ in acetonitrile solution produces,[39] in the presence of Et_4NCl, the material $[Et_4N][Nb(SCH_2CH_2S)_3]$. The structure of the product was determined by X-ray methods and niobium was found to have a coordination shell with a shape mid-way between that of a trigonal prism and an octahedron (9). The ligands adopt a propeller arrangement and are all twisted with torsion angles ranging from 29.4 to 31.4°. The S—Nb—S angles are about 82° within a ligand and about 157° for *trans*-atoms. Exposure to water, methanol, or phenol

[36] N. D. Chasteen, J. K. Grady, and C. E. Holloway, *Inorg. Chem.*, 1986, **25**, 2754.
[37] F. Preuss, H. Noichl, and J. Kaub, *Z. Naturforsch., Teil B*, 1986, **41**, 1085.
[38] F. A. Cotton, L. R. Falvello, R. Llusar, E. Libby, C. A. Murillo, and W. Schwotzer, *Inorg. Chem.*, 1986, **25**, 3423.
[39] K. Tatsumi, Y. Sekiguchi, A. Nakamura, R. E. Cramer, and J. J. Rupp, *Angew. Chem., Int. Ed. Engl.*, 1986, **25**, 86.

causes an isomerization to occur and results[40] in the formation of the ion $[NbS(SCH_2CH_2S)(SCH_2CH_2SCH_2CH_2S)]^-$. This ion (10) is notable for having thioether, thiolate, and terminal sulphide donors to a single metal centre. The central sulphur of the terdentate ligand is *trans* to the sulphide and makes the longest Nb—S bond (2.740 Å), the Nb=S is the shortest metal–sulphur bond (2.192 Å) and the others are around 2.46 Å.

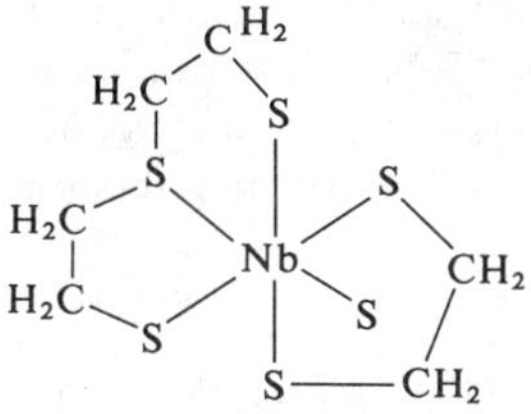

(10) 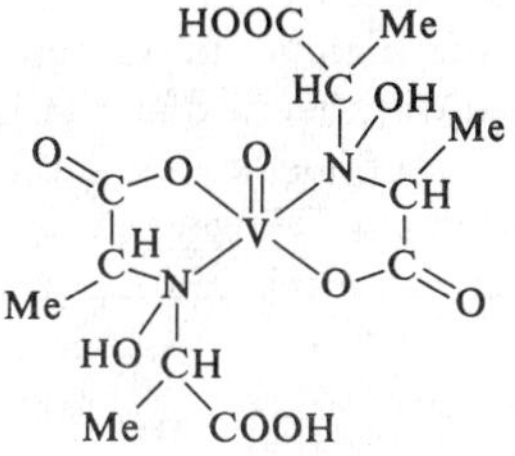   (11)

Although vanadium is known to have a number of biological effects there is little information on the structure of naturally occurring vanadium species, partly because of the low levels of the metal in most organisms. A number of living species do concentrate vanadium and the complex ('amavadin') shown in (11) was isolated[41] from the mushroom *Amanita muscaria*. Each ligand has two asymmetric carbon atoms and after an ingeneous synthetic exercise the stereochemistry of amavadin has been established. Circular dichroism was used to establish that the product was indeed identical to the naturally occurring material.

A detailed investigation into some reactions of the complex VO(L), where L is the ligand shown in (12), has been carried out with the aid of *X*-ray diffraction techniques.[42] If reacted with $TiCl_3(thf)_3$ the products are the octahedral complex V(L)Cl(thf) and $TiOCl_2$. If the reaction is repeated at a lower temperature (−70 °C)

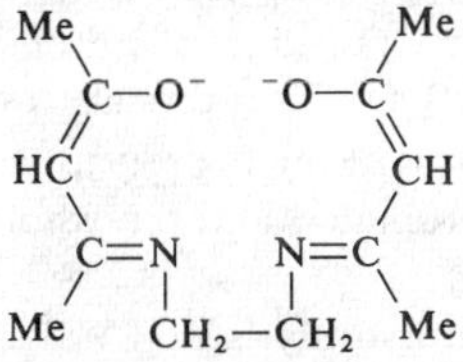

(12) 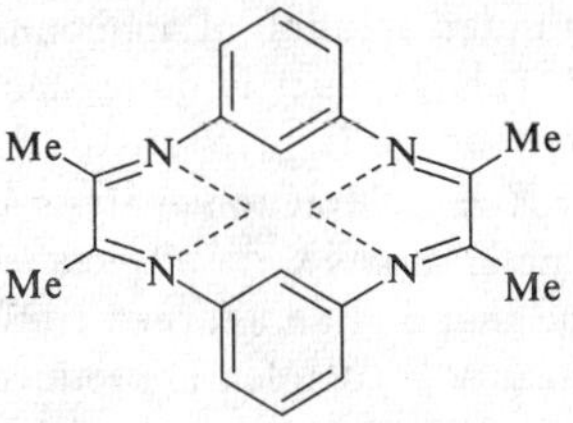   (13)

the product isolated has $-O-TiCl_3(thf)_2$ in the *trans*-position to the chlorine instead of the thf molecule found in the former complex. Perhaps the most surprising aspect of this pair of compounds is that the V—Cl distances are identical (2.352 Å).

The 14-member macrocycle shown in (13) will have two axial positions available if a complex is formed with a suitable metal. A number of coordination compounds

[40] K. Tatsumi, Y. Sekiguchi, A. Nakamura, R. E. Cramer, and J. J. Rupp, *J. Am. Chem. Soc.*, 1986, **108**, 1358.
[41] H. Kneifel and E. Bayer, *J. Am. Chem. Soc.*, 1986, **108**, 3075.
[42] M. Mazzanti, C. Floriani, A. Chiesi-Villa, and C. Guastini, *Inorg. Chem.*, 1986, **25**, 4158.

formulated as $[VO(L)X]^+X^-$, where X is Cl, Br, NO_3 or NCS, have been reported.[43] The materials were prepared by reacting *m*-phenylenediamine, 2,3-butanedione, and a VO^{2+} salt in a mixture of methanol and glacial acetic acid.

The remaining papers in this section on coordination chemistry are all concerned with species containing bridging groups or metal–metal bonds. Several authors draw parallels between the behaviour of Group 5 and their neighbours in Group 6. While it often seems that Group 6 offers more variety, there are a number of reactions that are quite unique to Group 5. For example if $TaCl_5$ is reacted with $Me_3SiNCNSiMe_3$ then a dimer $[Cl_4Ta(\mu\text{-}NCNSiMe_3)]_2$ is produced. The bridge structure was confirmed by i.r. and Raman spectroscopy and a C_{2h} planar environment suggested (14). However, when repeated with either tungsten or molybdenum chloride then polymers are formed.[44]

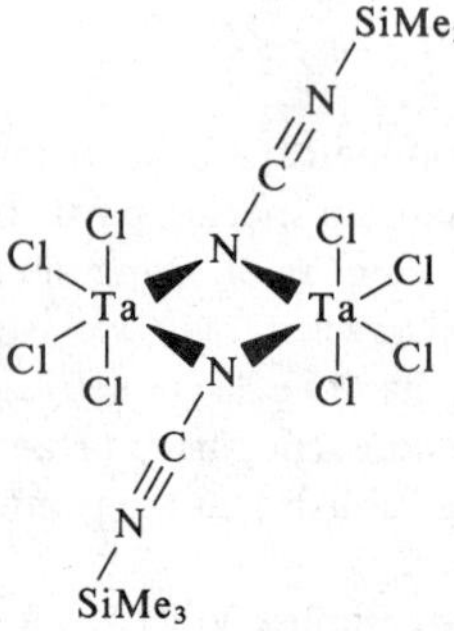

(14) 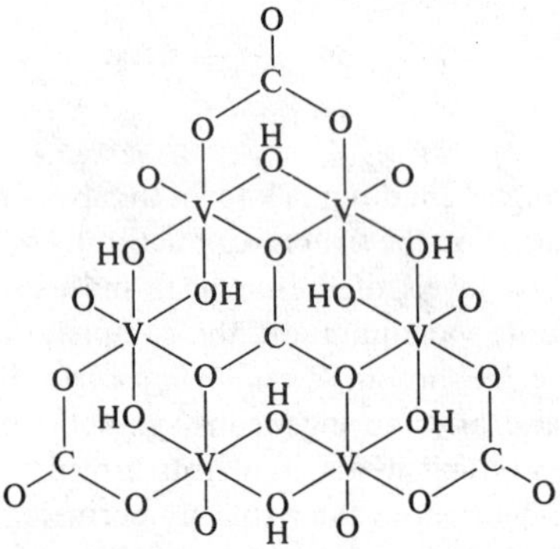   (15)

Addition of oxalic acid to V_2O_5 in boiling water followed by concentrated hydrochloric acid gives rise to $VOCl_2$. If this deep blue solution is now added to a carbon dioxide-saturated solution of ammonium bicarbonate then deep violet crystals are obtained.[45] This product is quite stable under a CO_2 atmosphere and has been characterized as $(NH_4)_5[(VO)_6(CO_3)_4(OH)_9].10H_2O$. The crystals proved very suitable for X-ray diffraction analysis and even most of the hydrogens were located (15). The metal atoms were all octahedrally coordinated and the whole structure takes on the shape of an adorned crown.

A dinuclear V^{III} complex is produced[46] by the hydrolysis of $LVCl_3.dmf$ in the presence of sodium acetate, where L is the cyclic triaza-ligand shown (16). The complex was found to be $[L_2V_2(\mu\text{-}O)(\mu\text{-}MeCO_2)_2]^{2+}$ and to have a rather distorted octahedral structure about the metal centres and with one oxo and two acetate bridges (17).

A niobium bridged by acetates can be produced[47] by reacting a 1:1 acetic acid–acetic anhydride mixture with $Nb_2Cl_6(tetrahydrothiophene)_3$. The product is

[43] V. K. Chauhan, S. K. Agarwal, S. P. Ratra, and V. B. Rana, *Acta Chim. Hung.*, 1986, **121**, 385.
[44] G. Rajca and J. Weidlein, *Z. Anog. Allg. Chem.*, 1986, **538**, 36.
[45] T. C. W. Mak, P. Li, C. Zheng, and K. Huang, *J. Chem. Soc., Chem. Commun.*, 1986, 1597.
[46] K. Wieghardt, M. Köppen, B. Nuber, and J. Weiss, *J. Chem. Soc., Chem. Commun.*, 1986, 1530.
[47] F. A. Cotton, M. P. Diebold, R. Llusar, and W. J. Roth, *J. Chem. Soc., Chem. Commun.*, 1986, 1276.

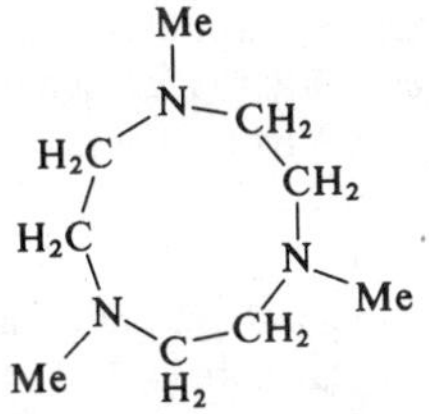

(16)

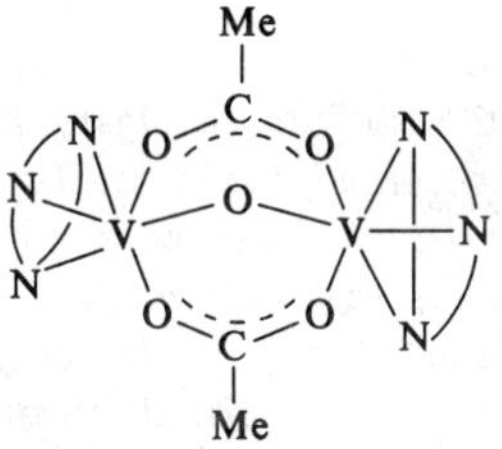

(17)

the trimeric species $[Nb_3(O)_2(O_2CMe)_6(thf)_3]^+$ which has a triangular core capped on each face by a μ_3-O and with two bridging acetates on each arm.

One of the developing features of Group 6 chemistry in recent years has been the rise of interest in the thiometallates, formed for example from MoS_4^{2-}. Using the diagonal relationship concept it might be expected that related species exist for VS_4^{3-}. However, to examine this idea it is first necessary to obtain a suitable solution. Attempts to dissolve the ammonium salt in DMF lead only to the precipitation of polymeric vanadium sulphides. However, a stable solution can be prepared[48] by stirring tricaprylmethylammonium chloride with $(NH_4)_3VS_4$ in toluene and filtering off the precipitated ammonium chloride. If this solution is then treated with $(Bu^i_2NCS_2)_2$ an internal redox reaction takes place with the formation of the dimeric species (18). The metals are bridged by two μ-η^2-S_2 ligands. There is approximate octahedral coordination about each metal if the S_2 ligands can be regarded as occupying one position. The V—V distance is 2.85 Å and since the compound is diamagnetic it is likely that a metal–metal bond is present. Using a slightly different strategy[49] it is possible to assemble a cubane-type cluster by reacting $(NH_4)_3(VS_4)$, Me_4NBr, and $FeCl_2$ in a DMF slurry. The product (19) is very similar to the analogous $MoFe_3S_4$ clusters.

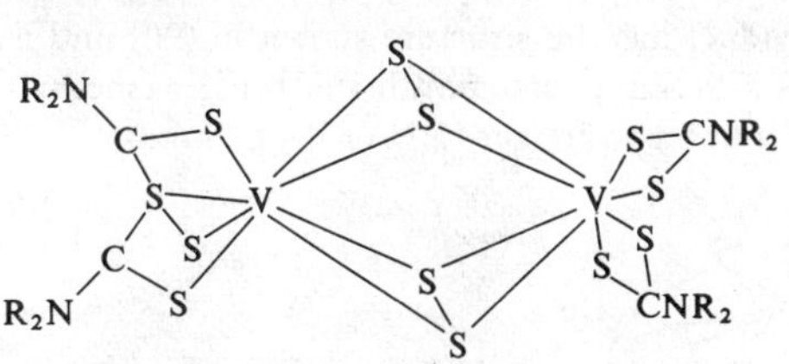

(18)

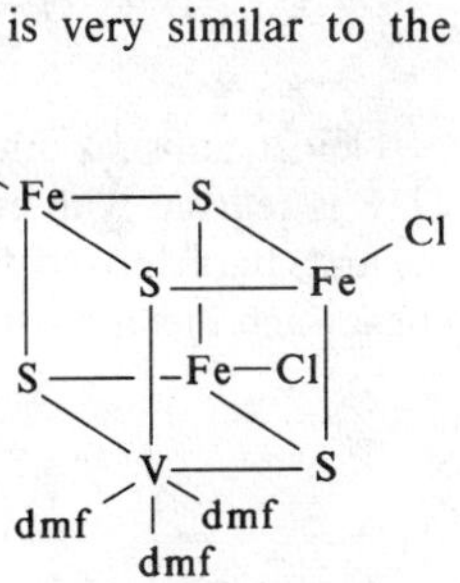

(19)

An oxidative-addition process can be accomplished[50] by reacting PhSSPh with $[(SMe_2)Cl_2Ta]_2(\mu\text{-}Cl)_2(\mu\text{-}SMe_2)$. The product was shown to be $[(SMe_2)Cl_3Ta]_2(\mu\text{-}SPh)_2$, the core of which is an edge-sharing bioctahedron. The use of magic-angle spinning ^{13}C n.m.r. spectroscopy indicated that the species was diamagnetic. It is likely that metal–metal bonding is present since as well as the diamagnetism there

[48] T. R. Halbert, L. L. Hutchings, R. Rhodes, and E. I. Stiefel, *J. Am. Chem. Soc.*, 1986, **108**, 6437.
[49] J. A. Kovacs and R. H. Holm, *J. Am. Chem. Soc.*, 1986, **108**, 340.
[50] G. C. Campbell, J. A. M. Canich, F. A. Cotton, S. A. Duraj, and J. F. Haw, *Inorg. Chem.*, 1986, **25**, 287.

is a Ta—Ta distance of 3.165 Å and non-square geometry for the Ta_2S_2 core. A tantalum(II) dimeric compound with three bridging sulphur groups can be produced[51] by the sodium amalgam reduction of $Ta_2Cl_6(L)_3$, where L is tetrahydrothiophene. The product is $[Ta_2Cl_6(\mu\text{-}L)_3]$ and has a confacial bioctahedral structure. The Ta—Ta distance is only 2.62 Å and it is probably safe to ascribe this as a formal triple Ta—Ta bond.

The vanadium entity shown in Scheme 6 makes a good template for the construction of a range of heterometallic clusters. Of particular note is the 2:1 reaction with the nickel complex which leaves space for further reaction with platinum species.[52]

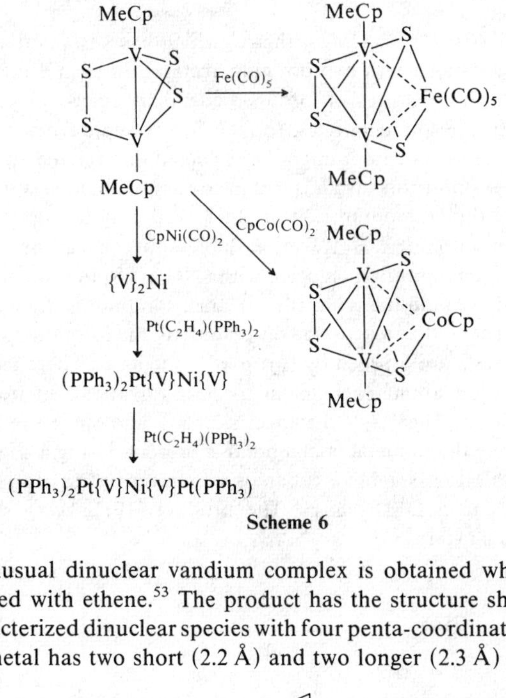

Scheme 6

A most unusual dinuclear vandium complex is obtained when $CpV(C_{10}H_8)$ in THF is reacted with ethene.[53] The product has the structure shown in (20) and is the first characterized dinuclear species with four penta-coordinated bridging methylenes. Each metal has two short (2.2 Å) and two longer (2.3 Å) V—C bonds.

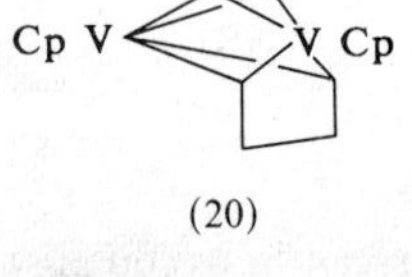

(20)

The vanadocene Cp_2VCl_2 is an anti-tumour agent and a detailed study has been made of the reactivity with the phosphate groups of a variety of nucleotides.[54] The main investigation method used was ^{31}P n.m.r. spectroscopy and in particular the

[51] F. A. Cotton, M. P. Diebold, and W. J. Roth, *J. Am. Chem. Soc.*, 1986, **108**, 3538.
[52] C. M. Bolinger, T. D. Weatherill, T. B. Rauchfuss, A. L. Rheingold, C. S. Day, and S. R. Wilson, *Inorg. Chem.*, 1986, **25**, 634.
[53] K. Jonas, W. Rüsseler, C. Krüger, and E. Raabe, *Angew. Chem., Int. Ed. Engl.*, 1986, **25**, 925.
[54] J. H. Toney, C. P. Brock, and T. J. Marks, *J. Am. Chem. Soc.*, 1986, **108**, 7263.

measurement of T_1 and T_2 relaxation rates. The complex interacts with the phosphate groups of d-AMP by the formation of an outer-sphere complex at a distance of *ca.* 5.5 Å. This is indicative of a very labile interaction and is quite distinct from that observed with cisplatin.

4 Chromium, Molybdenum, and Tungsten

The first part of this section will be concerned with general reviews and aspects of the behaviour of binary compounds and various ionic species of the Group 6 metals.

A well-documented discussion of the behaviour of molybdate as a corrosion inhibitor has been published.[55] Molybdates are broadly applicable as inhibitors due to their efficacy with both ferrous and non-ferrous metals and low toxicity. Stress is given to environmental aspects.

The interrelationship between processes occurring at active centres in molybdo-enzymes and at the surfaces of heterogeneous catalysts has drawn attention[56] to the common chemistry involved in many of the transformations. Several of the species are able to act interchangeably between these two classes.

An authoritative account of prospects and perspectives of superconducting Chevrel phases has been published.[57] The discussion centres around structural relationships.

An X-ray diffraction study has been made of $MoSe_2$, which was found[58] to be isotypic with hexagonal MoS_2. The authors comment that despite the importance of molybdenum sulphide there is a widespread mistake concerning the structural parameters so it is perhaps worthwhile to record in Table 1 details of both substances.

The structure of $WSeF_4$ in the gas-phase has been elucidated by electron diffraction procedures.[59] The tungsten was located above the plane of the four fluorines in a C_{4v} square-pyramidal structure with Se—W—F of 105°. The W=Se bond was found to be 2.226 Å, slightly longer than in the corresponding chloride.

Table 1 *Interatomic distances* (Å) *for hexagonal* MoX_2
($X = S$ *or* Se)

	S	Se
Within the MoX_2 *sandwiches:*		
Mo—6X	2.42	2.527
Mo—6Mo (or X—6X)	3.16	3.289
X—1X	3.17	3.335
Between the sandwiches:		
X—3X	3.49	3.660

An interesting matrix-isolation experiment has been reported[60] for the reaction between HBr and CrO_2Cl_2. An examination of the chromium isotopic pattern allowed an estimation of 107° to be made for the O—Cr—O in the product CrO_2Br_2. Some evidence was collected for the presence of CrO_2BrCl.

[55] M. S. Vukasovich and J. P. G. Farr, *Polyhedron*, 1986, **5**, 551.
[56] P. C. H. Mitchell, *J. Inorg. Biochem.*, 1986, **28**, 107.
[57] R. Chevrel, M. Hirrien, and M. Sergent, *Polyhedron*, 1986, **5**, 87.
[58] K. D. Bronsema, J. L. deBoer, and F. Jellinek, *Z. Anorg. Allg. Chem.*, 1986, **540/541**, 15.
[59] K. Hagen, D. A. Rice, J. H. Holloway, and V. Kaučič, *J. Chem. Soc., Dalton Trans.*, 1986, 1821.
[60] E. G. Hope, W. Levason, J. S. Ogden, and M. Tajik, *J. Chem. Soc., Dalton Trans.*, 1986, 1587.

The uptake of Cr^{III} ions by clay minerals has implications for any catalytic activity. For bentonite, which has a layered structure, the exact position of uptake depends on the preparation mode. If no treatment is given to the clay then ions are taken into the layers, but if pretreatment with an aluminium hydroxy polymer is given then pillars are formed between the layers and more positions become available. It has been shown[61] that the hydrocracking of n-decane is improved by using clays with a higher chromium content in the pillars. The interlayers can be swelled by small molecules as well as polymers, and even after heating to 200 °C up to two water molecules or eight molecules of dmso per chromium cation are retained.[62]

The heteropolyanion $SiW_{10}O_{36}{}^{8-}$ can be isolated[63] from a pH 9.1 solution of β-$SiW_{11}O_{39}{}^{8-}$. Good agreement was obtained between the structure deduced from an X-ray diffraction investigation and the resonance pattern obtained by ^{183}W n.m.r. spectroscopy. The stability of this ion was found to depend markedly on the nature of the counterion. With potassium, rubidium, or caesium the product is quite stable, but the lithium and sodium salts slowly revert to the original species. The acid was found to be stable for about a month, and use of a rapid titration procedure showed that seven of the protons were strongly held and that the remaining one was rather weak.

Reaction of sodium molybdate with hydroxylamine hydrochloride in acid solution gives rise to a red cystalline product that had sufficient stability in the mother liquor to form the subject of single-crystal X-ray structural analysis.[64] The formula was established as $[Mo_{36}O_{110}(NO)_4(H_2O)_{14}].52H_2O$. There are essentially two Mo_{18} sub-units containing 32, mostly edge-sharing, MoO_6 octahedra and 4 $Mo(NO)O_6$ pentagonal bipyramids which form a ring with a centre filled by six water molecules. As might be imagined such an edifice contains a wide variety of bonding modes and there are μ_4, μ_3, μ_2, and terminal oxo-ligands. The Mo—Mo separations are all greater than 3.2 Å and metal–metal bonding is thus probably not a feature of this molecule. The occurrence of both octahedal and pentagonal-bipyramidal molybdenum centres in one substance is also observed in smaller systems, such as[65] $K_6[Mo_7O_{22}(O_2)_2].10H_2O$.

Use of phenylhydrazine on the polyanion $[Mo_6O_{19}]^{2-}$ has been found[66] to lead to the replacement of a terminal oxo group by N_2Ph and the formation of the ion $[Mo_6O_{18}N_2Ph]^{3-}$. An investigation by X-ray diffraction was used to show that the framework of the original ion is retained after this replacement and also to note that the oxo group replaced had a shorter (2.16 Å) than average (2.35 Å) Mo—O distance. The Mo—N distance was found to be 1.76 Å, which implies a considerable contribution from a Mo—N triple bonded form. Similar substituted polyanions can[67] be produced with (NNR′R) ligands. Starting from the ion $[Mo_8O_{26}]^{4-}$, reaction with $H_2NNR′R$ in methanol–dichloromethane gives rise to a light red solution from which orange crystals of $[Mo_4O_{10}(OMe)_2(NNR′R)_2]^{2-}$ can be isolated. If these are

[61] K. A. Carrado, S. L. Suib, N. D. Skoularikis, and R. W. Coughlin, *Inorg. Chem.*, 1986, **25**, 4217.
[62] S. G. Garcia, J. M. Cardeso, S. R. Alonso, and G. D. Cancela, *Anal. Quim.*, 1986, **82**, 13.
[63] J. Canny, A. Tézé, R. Thouvenot, and G. Hervé, *Inorg. Chem.*, 1986, **25**, 2114.
[64] S. Zhang, D. Liao, M. Shao, and Y. Tang, *J. Chem. Soc., Chem. Commun.*, 1986, 835.
[65] I. Persdotter, L. Trysberg, and R. Stomberg, *Acta Chem. Scand.*, 1986, **A40**, 335.
[66] T.-C. Hsieh and J. A. Zúbieta, *Polyhedron*, 1986, **5**, 1655.
[67] S. N. Shaikh and J. A. Zubieta, *Inorg. Chem.*, 1986, **25**, 4613.

dissolved in wet alcohol about equal proportions of the starting material and of a new species $[Mo_8O_{16}(OMe)_6(NNR'R)_6]^{2-}$ are formed.

Much recent endeavour has been directed towards the application of n.m.r. methods to study such polyanions. This might involve the use of ^{183}W resonances such as in the determination[68] of the structure of $K_4H_2[P_2W_{21}O_{71}(OH_2)_3].28H_2O$. Inherent structural disorder gave problems in the interpretation of the X-ray diffraction results which were overcome by the application of tungsten n.m.r. However, the value of smaller nuclei should not be overlooked, and the formation of molybdophosphates during the acidification of a mixture of sodium phosphate and molybdate has been monitored[69] by an application of ^{31}P n.m.r. Several new lines were detected corresponding to the formation of the ions $P_2Mo_5O_{23}{}^{6-}$, $PMo_9O_{31}(OH)_3{}^{6-}$, and $PMo_{11}O_{39}{}^{7-}$. A fair correlation was noted between the chemical shift and the P—O vibration observed in the Raman spectrum, perhaps not an unexpected observation since both properties are affected by the electron density at the phosphorous nucleus.

Whilst much use has been made of ^{183}W n.m.r. in solution, the application of magic-angle spinning to obtain spectra from the solid-state is a useful advance. Chemical shifts ranging from +29 ppm for $CaWO_4$ to −3470 ppm for $W(CO)_6$ have been recorded.[70]

Typical of the application in solution is a paper[71] dealing with the location of the so-called 'blue' electrons in such ions as α-$[P_2Mo_3W_{15}O_{62}]^{8-}$ formed by reduction from α-$[P_2Mo_3W_{15}O_{62}]^{6-}$. The ions have the Wells-Dawson 18-metal centre structure formed from octahedra.

There is a central belt of twelve octahedra and two three-member caps. In this particular example one of the caps is formed from three molybdenums and the n.m.r. method was able to demonstrate quite convincingly that the 'blue' electrons of the reduced form are located around the tungsten belts.

In the case of slightly more complicated ions, 1,4-$PV_2W_{10}O_{40}{}^{5-}$ and 1,4,9-$PV_3W_9O_{40}{}^{6-}$, 1- and 2-dimension ^{183}W n.m.r. with ^{51}V decoupling was used[72] in the initial identification of the species. However, once a particular sample was identified, n.m.r. spectroscopy using the nucleii ^{51}V and ^{31}P was both more sensitive and quicker. It is necessary to be exceedingly careful in these type of studies since as well as the inherent complexity of the material, the splitting patterns include[73] four-bond vanadium to tungsten coupling constants.

Vibrational spectroscopy has a different role to fulfil in discussing the structure of such ions. In a normal coordinate analysis study of $NbW_5O_{19}{}^{3-}$ and $MoW_5O_{19}{}^{2-}$ where the Nb or Mo were regarded as perturbations to the tungsten framework it was found[74] that the most important feature was the charge on the ion, and hence the oxidation state of the metal, rather than a consideration of which perturbing metal was being discussed.

[68] C. M. Tourné, G. F. Tourné, and T. J. R. Weakley, *J. Chem. Soc., Dalton Trans.*, 1986, 2237.
[69] J. A. R. van Veen, O. Sudmeijer, C. A. Emeis, and H. de Wit, *J. Chem. Soc., Dalton Trans.*, 1986, 1825.
[70] C. T. G. Knight, G. L. Turner, R. J. Kirkpatrick, and E. Oldfield, *J. Am. Chem. Soc.*, 1986, **108**, 7426.
[71] M. Kozik, C. F. Hammer, and L. C. W. Baker, *J. Am. Chem. Soc.*, 1986, **108**, 2748.
[72] P. J. Domaille and G. Watunya, *Inorg. Chem.*, 1986, **25**, 1239.
[73] R. G. Finke, B. Rapko, R. J. Saxton, and P. J. Domaille, *J. Am. Chem. Soc.*, 1986, **108**, 2947.
[74] C. Rocchiccioli-Deltcheff, M. Fournier, R. Franck, and R. Thouvenot, *Spectrosc. Lett.*, 1986, **19**, 765.

The view has been propounded[75] that the heteropolyanions can be considered as quasi-metalloporphyrins, in the sense that framework is provided against which a metal can be studied in various oxidation states. In this context the species $[H_xSiW_{11}O_{39}V^{IV}O]^{(6-x)-}$, where $x = 0, 1$, or 2, has been shown to be reducible by a constant potential process to $[SiW_{11}O_{39}V^{III}(OH)]^{6-}$. If a suitable salt of this ion is isolated, and then dissolved in toluene a metastable V^{IV} species is obtained by O_2 or Br_2 oxidation, which in turn reverts to the monoprotonated form of the original material.

Coordination Compounds. —A good number of specialized review articles were published in 1986. Thus the kinetics and mechanisms of reductions at carboxylate-bound chromium(V) centres have been catalogued.[76] Over 70 references are listed dealing with reductants ranging from hydrazine to U^{IV}.

The use of ^{95}Mo n.m.r. in the study of di-, tri-, tetra-, homo-, and hetero-metallic complexes has been discussed.[77] It was noted that the method was particularly good for the identification of sulphido complexes and thus is of interest in the general field of catalysis. Aspects of shielding in ^{95}Mo n.m.r. are considered[78] in a review that deals with a wide range of styles of molybdenum bonding. Exceptions to the usual expectation that a decrease in shielding follows from increases in oxidation number are noted for many of the complexes discussed. It was also observed that ^{95}Mo shielding decreases smoothly along the series from Mo—Mo to the $(Mo)_2$ quadruple bond, whereas for ^{13}C the corresponding shielding series runs C—C < C≡C < C=C. It is said that this difference is indicative of the importance of nephalauxetic effects.

A systematic account of the coordination chemistry of molybdenum and tungsten sulphur compounds has been given.[79] The discussion centres on the structural types encountered, but extensive reference is made to their reactivity and to n.m.r. and other methods suitable for the characterization of these substances.

A structural comparison has been made[80] between the tungsten and molybdenum versions of $(Bu^t-O)_3M≡N$, which were prepared by reaction between $Cl_3M≡N$ and $LiO-Bu^t$ in thf solution. The main details are recorded in Table 2. The products

Table 2 *Structural details for* $(Bu^tO)_3M≡N$
  $(M = Mo$ or $W)$

	W	Mo
M≡N	1.740 Å	1.661 Å
M—N	2.661 Å	2.883 Å
M—O	1.872 Å	1.882 Å
M—O—C	136.6°	135.1°
O—M—O	116.1°	114.9°
N≡M—O	101.6°	103.3°

[75] S. P. Harmalker and M. T. Pope, *J. Inorg. Biochem.*, 1986, **28**, 85.

[76] E. S. Gould, *Acc. Chem. Res.*, 1986, **19**, 66.

[77] C. G. Young, M. Minelli, J. H. Enemark, G. Miessler, N. Janietz, H. Kauermann, and J. Wachter, *Polyhedron*, 1986, **5**, 407.

[78] R. A Grieves and J. Mason, *Polyhedron*, 1986, **5**, 415.

[79] A. Müller, *Polyhedron*, 1986, **5**, 323.

[80] D. M.-T. Chan, M. H. Chisholm, K. Folting, J. C. Huffman, and N. S. Marchant, *Inorg. Chem.*, 1986, **25**, 4170.

are polymeric and each metal has one short M—N distance (the triple bond) and a long M—N, *trans* to the triple bond, linking the monomers into an infinite linear chain. The main structural difference between the two products is that the tungsten compound has both the longer M≡N bond and the shorter M—N bond. A short review of the main advances in chemistry of dinitrogen complexes and related species published in the years 1983—85 has been reported.[81]

A series of reactions for the chromium dinitrogen complex *trans*-$Cr(N_2)_2(diphos)_2$, where diphos is dimethylphosphinoethane, has been established[82] and shown in Scheme 7. Most of these structural assignments are based on X-ray diffraction investigations, but others are supported by e.s.r. or n.m.r.

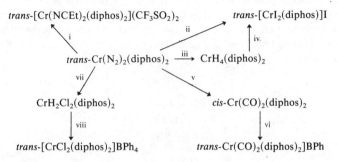

Reagents: i, CF_3SO_3H-EtCN; ii, I_2-MeOH; iii, $h\nu$-hexane-H_2; iv, I_2-CH_2Cl_2; v, $h\nu$-hexane-CO; vi, $AgCF_3SO_3$-MeCN-$NaBPh_4$; vii, HCl-Et_2O; viii, MeCN-$NaBPh_4$-MeOH

Scheme 7

A new synthesis has been devised[83] for $MoN_2(triphos)(PMe_2Ph)_2$, where triphos is the tripod ligand $PhP(CH_2CH_2PPh_2)_2$, by the reaction of $MoCl_3(triphos)$ with PR_3 under a nitrogen atmosphere in the presence of sodium amalgam. The product is notable for being the first mono dinitrogen molybdenum complex to give appreciable amounts of ammonia and hydrazine when decomposed by anhydrous HCl.

Treatment of *trans*-$MoN_2(diphos)_2$ with RSH causes displacement of the dinitrogen and the production[84] of *trans*-$Mo(SR)_2(diphos)_2$; a wide range of examples were produced, for R = various alkyl and aryl groups. As might be expected the electronic properties of these complexes were found to be dependent upon the nature of R and it seems that the main effect is transmitted by a σ-inductive mechanism. If the starting material, $Mo(N_2)_2(diphos)_2$ is irradiated[85] by a cobalt source it is possible to discern absorption bands from a radiation-induced ionic species $Mo(N_2)_2(diphos)_2^-$. When carried out in an oxygen atmosphere vibrations from the presence of dinitrogen were seen to diminish whilst those from P=O were seen to grow. This destruction of the complex was accompanied by oxidation of the molybdenum.

[81] G. J. Leigh, *Transition Met. Chem.*, 1986, **11**, 118.
[82] J. E. Salt, G. Wilkinson, M. Motevalli, and M. B. Hursthouse, *J. Chem. Soc., Dalton Trans.*, 1986, 1141.
[83] T. A. George and R. C. Tisdale, *Polyhedron*, 1986, **5**, 297.
[84] D. C. Povey, R. L. Richards, and C. Shortman, *Polyhedron*, 1986, **5**, 369.
[85] J. O. Dziegielewski, B. Jezowska-Trzebiatowska, R. Gil-Bortnowska, and R. Grzybek, *Polyhedron*, 1986, **5**, 833.

During the examination of the n.m.r. spectrum obtained from a sample of $trans$-$Mo(N_2)_2(Pr^n_2PhP)_4$ under a nitrogen atmosphere, it was noticed[86] that spectral changes were occurring, but not of a type associated with decomposition. The product of this process was obtained in a larger yield by simply stirring the complex in methanolic solution under nitrogen. This gave a colour change from red to yellow and allowed crystals of mer-$Mo(N_2)_3(Pr^n_2PhP)_3$ to separate out.

A mechanistic study of the reaction to form $trans$-$Mo(N_2)_2(diphos)_2$ from the action of dinitrogen, in the presence of a strong base, on $trans$-$[Mo(NNH_2)X(diphos)_2]^+$ has been made.[87] Stopped-flow kinetic techniques were used to establish the probable pathways, which are shown in Scheme 8. For X = F or Br the left-hand path is followed via species (II) and (III) to reach the product (IV). However for X = p-$MeC_6H_4SO_3$ the right-hand path is preferred. It will be seen that this route involves a considerable interaction with the solvent. Hence it is not entirely unexpected to find that general base catalysis and primary kinetic isotope effects are observed.

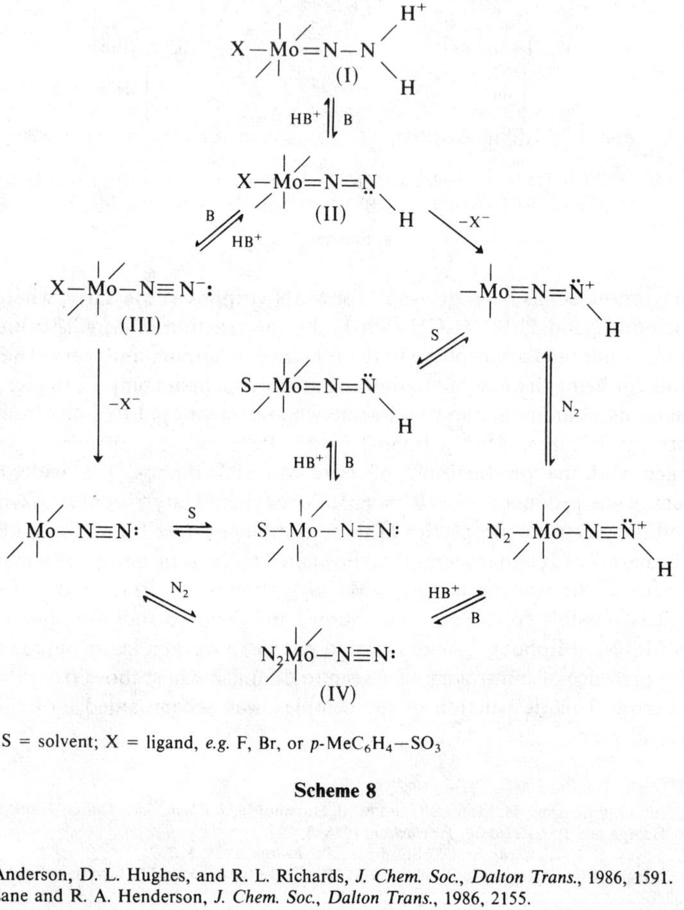

B = base; S = solvent; X = ligand, $e.g.$ F, Br, or p-MeC_6H_4—SO_3

Scheme 8

[86] S. N. Anderson, D. L. Hughes, and R. L. Richards, *J. Chem. Soc., Dalton Trans.*, 1986, 1591.
[87] J. D. Lane and R. A. Henderson, *J. Chem. Soc., Dalton Trans.*, 1986, 2155.

An account has been given[88] of the formation of a series of bis-dinitrogen complexes $Mo(N_2)_2(triphos)PMe_{3-n}Ph_n$, for n = 0, 1, or 2. The terdentate ligand is $PhP(CH_2CH_2CH_2PPh_2)_2$ and was found attached in a meridional fashion with the monodentate phosphine *trans* to the centre phosphorus of the tripod. This arrangement means that the dinitrogens have to be *trans* to each other and a slight deformation of octahedral symmetry is noted with $N—\hat{Mo}—N$ = 173°.

Reaction of *cis*-$Mo(N_2)_2(PMe_3)_4$ under a moderate pressure (4—5 atm) of carbon dioxide gives rise to the product *trans*-$Mo(CO_2)_2$. A number of interesting reactions of this carbon dioxide complex were investigated[89] and some of these are shown in Scheme 9. The structures of the products were determined by X-ray methods. One

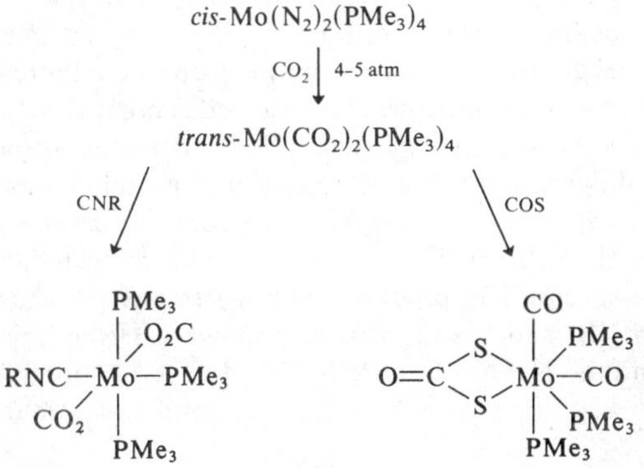

cis-$Mo(N_2)_2(PMe_3)_4$

CO_2 ↓ 4-5 atm

trans-$Mo(CO_2)_2(PMe_3)_4$

Scheme 9

of the most interesting results was obtained by the reaction with COS. A seven-coordinate pentagonal bipyramidal complex having a planar dithiocarbonate ligand and two carbonyl ligands have been formed by this reaction. With isocyanides one of the phosphine ligands is displaced to give *trans,mer*-$Mo(CO_2)_2(CNR)(PMe_3)_3$. The important feature of these species is the attachment mode of the carbon dioxides. The CO_2 ligands are dihapto bonded through one of the C=O bonds. This leads to a bent O=C=O with an angle *ca.* 130° that is dependent on the nature of R. The reaction between WCl_6, PR_2Ph, and dinitrogen in the solution in the presence of magnesium has been shown[90] to be a mixture of products (Scheme 10). The structure of each of the products was checked by single-crystal X-ray diffraction. Compound (I) has octahedral tungsten atoms with a near linear (W—N≡N—W) bridge and a phosphine *trans* to the bridge. The second compound is very nearly a regular octahedron, but the phosphines in the equatorial plane make a very shallow tetrahedron about the metal. The final substance, which is formed in greatest amounts when the nitrogen pressure is highest, takes a classic piano-stool arrangement of ligands with the (η^6-$C_6H_5PR_2$) as the 'seat'.

[88] L. Dahlenburg and B. Pietsch, *Z. Naturforsch., Teil B*, 1986, **41**, 70.
[89] R. Alvarez, E. Carmona, J. M. Marin, M. L. Poveda, E. Gutierrez-Puebla, and A. Monge, *J. Am. Chem. Soc.*, 1986, **108**, 2286.
[90] S. N. Anderson, R. L. Richards, and D. L. Hughes, *J. Chem. Soc., Dalton Trans.*, 1986, 245.

$$WCl_6$$

$$\downarrow \; {}_{Mg-PR_2Ph-N_2-thf}$$

I $\{W(N_2)_2(PR_2Ph)_3\}_2(\mu\text{-}N_2)$
    +
II $trans\text{-}W(N_2)_2(PR_2Ph)_4$
    +
III $W(\eta^6\text{-}C_6H_5PR_2)N_2(PR_2Ph)$

Scheme 10

A novel alkenyldiazenido complex, $mer\text{-}[W(acac)(NNCMeCHCOMe)\text{-}(PMe_2Ph)_3]$, has been prepared[91] from $cis\text{-}[W(N_2)_2(PMe_2Ph)_4]$ by the action of acetylacetone in methanol. The formation of this ligand may come from an initial protonation of the dinitrogen followed by nucleophilic addition of the resultant diazenido ligand to an acetylacetone molecule and subsequent elimination of water to give the product. Alternatively the diazenido species could form a hydrazido(-2) complex followed by condensation with acetylacetone and a final loss of the active methylene proton. Some support for this route comes from an examination of the structure of the alkenylhydrazido complex formed by reaction of HBr on the alkenyldiazenido species. This product is formulated $mer\text{-}[W(acac)(NNHCMeCH\text{-}COMe)(PMe_2Ph)_3]Br$ and has the structure shown (21) the linkage $W\equiv N-N$ is almost linear, and the $N-N-C$ angle is 120°. The $N-N$ bond length, 1.375 Å, is intermediate between that expected for a single bond and a double bond.

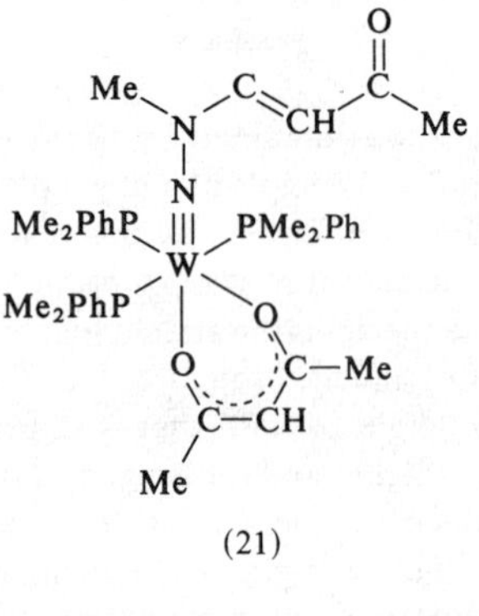

(21)

The mechanism of the reaction shown in Scheme 11 has been firmly established.[92] The kinetics were monitored by the application of peak current monitoring during cyclic voltammetry and found to conform to a classic S_N2 process, being first-order with respect to both the complex and to methyl iodide. There was a good correlation of results for the two metals and also between $\ln(k_2)$ and the Hammett σ parameter. The change of substituent caused roughly a thousand-fold change in the rate between the extremes of $-OMe$ and $-CF_3$.

[91] M. Hidai, S. Aramaki, K. Yoshida, T. Kodama, T. Takahashi, Y. Uchida, and Y. Mizobe, *J. Am. Chem. Soc.*, 1986, **108**, 1562.
[92] W. Hussain, G. J. Leigh, H. Modh-Ali, and C. J. Pickett, *J. Chem. Soc., Dalton Trans.*, 1986, 1473.

$$trans\text{-}[MBr(NNEt)(R_2PCH_2CH_2PR_2)_2] + MeI$$

$$\downarrow k_2$$

$$trans\text{-}[MBr(NNMeEt)(R_2PCH_2CH_2PR_2)_2] + I^-$$

$$R = p\text{-}XC_6H_4; \quad X = CF_3, Cl, H, Me, MeO; \quad M = Mo, W$$

Scheme 11

Reaction between WF_6 and $LSiMe_3$, where L is $Ph_3P=N$, produces[93] the products WF_5L and WF_4L_2. The latter has the L groups in a *cis*-octahedral arrangement with a considerable degree of distortion evident $(N-\hat{W}-N$ 96.3°$)$. The $W-N-P$ angle of 157.2° and the $W-N$ bond length of 1.825 Å are evidence for multiple-bond character in the ligand attachment. An improved synthesis of $MoCl_3(N_3S_2)$ from $MoCl_5$ and $(NSCl)_3$ has been developed.[94] Crystallization from the appropriate solvent gives the complex $MoCl_3(N_3S_2)L$, where L is py or thf. The X-ray diffraction results show that both of these are essentially octahedral with virtually no differences in bond lengths apparent between the two complexes. The six-membered ring $(-Mo-N-S-N-S-N-)$ was found to be almost planar. This type of ring arrangement was found[95] also in the complex ion $[WOCl_3(HN_3S_2)]^-$ which was isolated after the partial hydrolysis of $AsPh_4[WCl_4(N_3S_2)]$. The hydrogen is located on the nitrogen *trans* to the oxo group and is apparent by the difference in the $W-N$ bond length (2.26 Å) compared to that of the unprotonated (1.83 Å) nitrogen.

Despite the importance of chromium in a range of biologically important processes, there has been no example of a crystal structure for a Cr^{III}-peptide until the report[96] of structural investigations conducted with the species $Cr_2(\beta\text{-Ala-L-}H_{-1}His)_2(OH)(OMe)$, $Cr(Gly\text{-}H_{-1}Gly)_2^-$, and $Cr(L\text{-Pro-}H_{-1}Gly)_2^-$. The dipeptide ligands are bonded to the octahedral metal in a meridional style for each of the examples studied. The amide groups were found to be essentially coplanar and thus deprotonated. The dimensions of the two monomeric complexes were more precisely established than those of the dimer and have the $Cr-N(amino)$ bond length at 2.09 Å, significantly longer than the $Cr-N(peptide)$ bond (av. 1.96 Å), reflecting the difference between sp^3 and sp^2 hybridization. The dimer was obtained from dilute methanol solution and redissolution of the product in water led to suggestions of the formation of monomeric entities.

An air-stable high-spin chromium(II) complex, $trans\text{-}[Cr^{II}(nic)_2(H_2O)_4]$, was formed[97] as a red–green precipitate by mixing chromium(II) chloride with deoxygenated nicotinic acid. An X-ray structural analysis showed the ligands were N-bonded and also the presence of an extensive hydrogen bonding network, possibly accounting for the air-stability. The complex was found to form mixed crystals with a similarly formulated zinc compound. It is suggested that this ability accounts for

[93] H. W. Roesky, U. Seseke, M. Noltemeyer, P. G. Jones, and G. M Sheldrick, *J. Chem. Soc., Dalton Trans.*, 1986, 1309.

[94] K. Völp, W. Willing, U. Müller, and K. Dehnicke, *Z. Naturforsch., Teil B*, 1986, **41**, 1196.

[95] E. Conradi, H. Wadle, U. Müller, and K. Dehnicke, *Z. Naturforsch., Teil B*, 1986, **41**, 48.

[96] C. M. Murdoch, M. K. Cooper, T. W. Hambley, W. N. Hunter, and H. C. Freeman, *J. Chem. Soc., Chem. Commun.*, 1986, 1329.

[97] W. E. Broderick, M. R. Pressprich, U. Geiser, R. D. Willett, and J. I. Legg, *Inorg. Chem.*, 1986, **25**, 3372.

the significant differences from the previous structure quoted[98] for the chromium species.

The structure of the molybdenum amide complex $MoO_2Cl_2(MeCONH_2)_2$ has been established[99] by X-ray diffraction methods. This has an octahedral molybdenum with cis-oxo-groups. Although one of these is involved in hydrogen bonding with an NH_2 group (22), there is little difference in the two Mo—O bond lengths (1.692 and 1.688 Å).

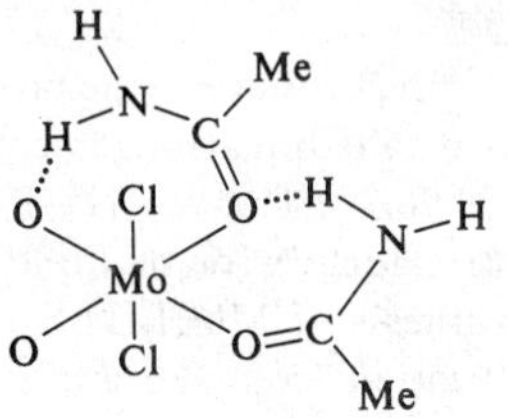

(22) 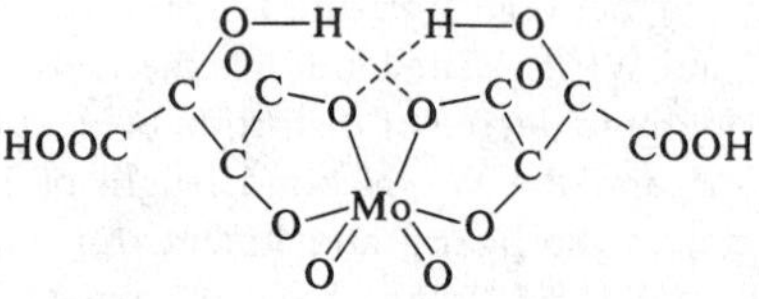   (23)

The same cis-oxo arrangement was found[100] in the complex formed between MoO_2^{2+} and tartrate (23). There are several interesting points about this substance, for example the inter-ligand hydrogen-bonding between the uncoordinated hydroxyls and the coordinated carboxyls. The H(O—C—C—COOH) fragments of the tartrates have near coplanarity and the C_4 skeleton has a torsion angle of ca. 180°. The other interesting aspect of the complex is that the tartrate functions as a bidentate ligand by employing one hydroxyl and an adjacent carboxyl oxygen. The same style of bonding has been observed[101] in the species $K_4[Mo_2O_2(O_2)_4(tart)].4H_2O$. This was prepared by reaction between K_2MoO_4, tartaric acid, and hydrogen perioxide and the complex ion takes the structure shown (24). Each molybdenum has essentially an octahedral-bipyramidal environment with

(24)

the deprotonated hydroxyls taking axial positions. The Raman scattering spectrum is quite similar in the solid to that of an aqueous solution and it seems likely that the spatial arrangement is indeed maintained in solution.

[98] J. A. Cooper, B. F. Anderson, P. D. Buckley, and L. F. Blackwell, Inorg. Chim. Acta, 1984, 91, 1.
[99] V. S. Sergienko, N. A. Ovchinnikova, M. A. Porai-Koshits, and M. A. Glushkova, Koord. Khim., 1986, 12, 1650.
[100] W. T. Robinson and C. J. Wilkins, Transition Met. Chem., 1986, 11, 86.
[101] A. C. Dengel, W. P. Griffith, R. D. Powell, and A. C. Skapski, J. Chem. Soc., Chem. Commun., 1986, 555.

An interesting series of Cr^{III} complexes involving a tetradentate and a bidentate ligand, $[Cr(L)(LL)]^{n-}$, has been prepared.[102] The tetra-species were nta or N,N-β-alaninediacetate and the bidentate ligands were malonate, oxalate, or acetylacetone. Both of the tetradentate ligands employ one nitrogen and three oxygen donors, but whereas nta is symmetrical the alanine derivative has one oxygen coordination arm shorter than the other two. This means that for the complex type studied here, two isomeric forms are possible differing in the disposition of their ligand attachments. Two products were separated chromatographically for the malonate and the acetylacetonate derivatives. There were some differences in their absorption spectra, but the major effect was found by examining the ^2D n.m.r. spectra. A large number of new products have been characterized[103] for the reaction of various reagents on the tungsten–phosphine–hydride complex shown in Scheme 12. The hydride ligand has

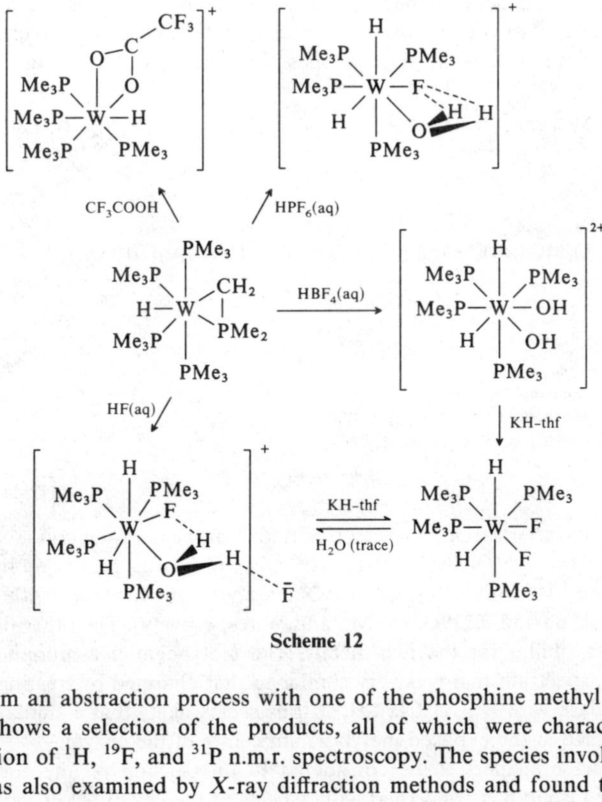

Scheme 12

arisen from an abstraction process with one of the phosphine methyl groups. The Scheme shows a selection of the products, all of which were characterized by a combination of ^1H, ^{19}F, and ^{31}P n.m.r. spectroscopy. The species involving a water ligand was also examined by X-ray diffraction methods and found to have both internal hydrogen-bonding to the fluorine ligand and external hydrogen-bonding to the fluoride counterion.

Moving on to sulphur-donor species, the nature of the ligand substitution reaction of chromium(VI) with thiols has been investigated.[104] The kinetics were followed

[102] N. Koine, R. J. Bianchini, and J. I. Legg, *Inorg. Chem.*, 1986, **25**, 2835.
[103] M. L. H. Green, G. Parkin, M. Chen, and K. Prout, *J. Chem. Soc., Dalton Trans.*, 1986, 2227.
[104] P. H. Connett and K. E. Wetterhahn, *J. Am. Chem. Soc.*, 1986, **108**, 1842.

by observing changes in λ_{max} for CrO_4^{2-} and fitted to the mixed second order rate equation, rate = $k[\text{thiol}][Cr^{VI}]$. This looks quite straightforward, but the complicating factor was the nature of the thiol species used which included such compounds as cysteine, cysteamine, cysteine ethyl ester, and thioglycolic acid. The process was also studied over a pH range of 2—10. Thus the simple rate-equation had to be rewritten as a linear combination of terms to represent the various protonated species present. The results indicate that ligand substitution at Cr^{VI} with thiols involves attack of chromate or hydrogen chromate by the protonated thiol with the proton transfer from the sulphydryl to the chromium as the rate-determining step. Thioglycolic acid gave a higher rate than other thiols, perhaps reflecting a chelating advantage.

A variety of trigonal bipyramidal thiolate complexes of Mo and W^{III} have been produced[105] by the reactions shown in Scheme 13. In all the cases examined the equatorial plane was occupied by thiolates. Little effect could be discerned on the nitrosyl by the presence of the various types of thiolate, and the linkage M—N—O was linear.

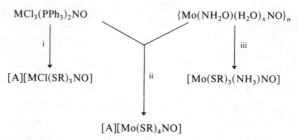

Reagents: i, SR^--methanol (R = Bu^t, M = Mo; R = Ph, M = W);
        ii, SR^--methanol (R = Ph, tol, p-Cl-Ph; Mo);
        iii, SR^--methanol (R = 2,4,6-$C_6H_2(Pr^i)_3$)

Scheme 13

Structural characterization has been reported[106] of the tungsten and molybdenum thiolate complexes $[MO(SCH_2CH_2S)_2][PPh_4]$, which was produced by thiolate exchange from $[MO(SPh)_4]^-$. The M=O sits above an S_4 plane (with the metal displaced by 0.760 and 0.739 Å for Mo and W respectively). The other dimensions are also quite similar for the two metals. The molybdenum compound gave an X-band e.s.r. spectrum that was very similar to that obtained by treating oxidized xanthine oxidase with $HSCH_2CH_2SH$. It thus seems likely that a similar complex has been formed at the exposed molybdenum centre of the enzyme.

The final set of papers to be considered in this section on the coordination complexes of Group 6 is concerned with aspects of the chemistry of mononuclear compounds involving macrocyclic ligands. Chromium(III) complexes with tetraza[107] and hexaza[108] macrocyclic ligands have been reported.

[105] P. T. Bishop, J. R. Dilworth, J. Hutchinson, and J. Zubieta, *J. Chem. Soc., Dalton Trans.*, 1986, 967.
[106] S. R. Ellis, D. Collison, C. D. Garner, and W. Clegg, *J. Chem. Soc., Chem. Commun.*, 1986, 1483.
[107] B. U. Nair, T. Ramasami, and D. Ramaswamy, *Inorg. Chem.*, 1986, **25**, 51.
[108] P. Comba, I. I. Creaser, L. R. Gahan, J. M. Harrowfield, G. A. Lawrance, L. L. Martin, A. W. H. Mau, A. M. Sargeson, W. H. F. Saisse, and M. R. Snow, *Inorg. Chem.*, 1986, **25**, 384.

An unexpected result was obtained from an experiment[109] to explore the use of ozonides as oxygen-transfer agents to Cr^{III} porphyrins. The reaction of styrene ozonide on (tpp)Cr^{III}Cl, where tpp is tetraphenyl porphyrin, produced neither a Cr^{IV} nor a Cr^{V} product, but instead led to the formation of an iso-porphryin.

The starting point for the molybdenum porphyrins is the reaction between $Mo(CO)_6$ and the appropriate porphyrin. The reaction solutions are then worked-up *via* chromatographic separations. It has now[110] been observed that the initial product from the reaction is a dimeric species with a Mo—Mo bond linking the two ring structures. An X-ray diffraction analysis of [Mo(tpp)]$_2$ shows that the Mo—Mo bond length is quite long (2.239 Å). Each ring is slightly dome-shaped with the metal displaced 0.458 Å from the N_4 plane. The two rings are mutually rotated by 18°. The structure appears to be a balance between the enhancement of the δ-bond contribution and a minimization of conflict from a purely eclipsed conformation. The reaction of superoxide ion in an aprotic solvent upon the porphyrin MoO(tpp)X, with X = Cl, Br, or NCS, is known to produce $Mo^{IV}O$(tpp). If the reaction is carried out at −80 °C instead of room temperature then it is possible to isolate[111] the intermediate (18-crown-6-K)$^+$[$Mo^{V}O$(tpp)$O_2{}^{2-}$]$^-$. This material was sufficiently robust for an analysis to be undertaken by X-ray diffraction methods, and the peroxide was shown to be coordinated in a sideways fashion. The photolysis in benzene of $Mo^{V}O$(X)(ttp), where X is a range of —OR groups or Cl, has been shown[112] to generate a radical species at a rate controlled by the nature of X.

Bridged Complexes.—The review continues with an account of the various types of bridged entities, ranging from simple dinuclear with a single μ-ligand up to the intricacies of heterobimetallics.

The crystal structure of the μ-hydroxy Cr^{III} dimer, [(NH$_3$)$_5$Cr]$_2$[μ-OH]Cl$_5$.H$_2$O, has been established for some time. An investigation of optical spectra (absorption, Zeeman, and site-selective luminescence) of single crystals from 1.5—90 K has revealed[113] much structural information that complements that produced by X-ray diffraction. In particular it is shown that the two sites in the dimer are non-equivalent, and demonstrates clear advantages for the selectivity of spectroscopic methods over other potential procedures such as inelastic neutron scattering. The related dimer, but having an NCO group acting as bridge instead of the hydroxyl, was isolated[114] from heating urea and CrCl$_3$.6H$_2$O at *ca.* 150 °C for 24 hours. It is insoluble in common organic solvents and decomposed by water. The main evidence in favour of the μ-NCO group comes from interpretation of the infrared spectra, but the magnetic susceptibility results indicate a weak antiferromagnetic coupling and predict little deviation from linearity in the bridge.

A range of high-spin Cr^{II} complexes, of empirical formula Cr(O$_2$PRR')$_2$.xH$_2$O, have been characterized by infrared and reflectance spectroscopy.[115] The products are all believed to be polymeric in nature (Scheme 14), with the linear type produced

[109] E. S. Schmidt, T. C. Bruice, R. S. Brown, and C. L. Wilkins, *Inorg. Chem.*, 1986, **25**, 4799.
[110] C.-H. Yang, S. J. Dzugan, and V. L. Goedken, *J. Chem. Soc., Chem. Commun.*, 1986, 1313.
[111] K. Hasegawa, T. Imamura, and M. Fujimoto, *Inorg. Chem.*, 1986, **25**, 2154.
[112] Y. Matsuda, T. Takaki, and Y. Murakami, *Bull. Chem. Soc. Jpn.*, 1986, **59**, 1839.
[113] H. Riesen and H. U. Güdel, *Inorg. Chem.*, 1986, **25**, 3566.
[114] T. Schönherr, *Inorg. Chem.*, 1986, **25**, 171.
[115] L. F. Larkworthy and K. A. R. Salib, *Transition Met. Chem.*, 1986, **11**, 121.

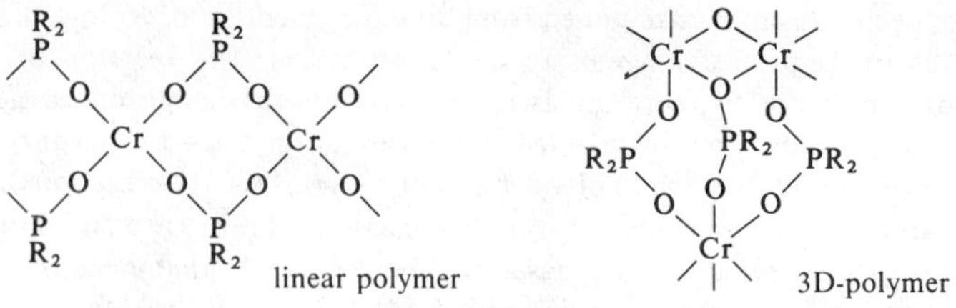

linear polymer   3D-polymer

Scheme 14

for R = R' = Ph or OPh and the three dimensional polymer formed for (R = H, R' = Ph) and (R = OH, R' = F).

A careful analysis of two examples of the $Mo_2Cl_9^{3-}$ ion has been reported.[116] The structure could be broadly described as that of two face-sharing octahedra. The bonds are all very uniform with the average Mo—Cl separation for the bridging positions (2.47 Å) being slightly longer than for the terminal positions (2.42 Å). The Mo—Mo distance was found to be 2.619 Å, which is shorter than in related singly-charged ions.

There are plenty of examples of single oxygens or halogens acting as bridging ligands with Group 6 metals, but the occurrence of a solitary sulphur is much rarer. If molybdenum hexacarbonyl is reacted with tetraethyl thiuram disulphide then it forms $Mo_4(\mu\text{-}S)_4(S_2CNEt_2)_6$, which has previously been shown to have a cubane-like core of alternate metal and sulphur atoms. If this material is aerated in a toluene solution then it is possible[117] to isolate black crystals of a dimeric complex $[Mo_2(\mu\text{-}S)(S_2CNEt_2)_6]$.toluene, which has two pentagonal bipyramidal molybdenums axially linked through a single bridging sulphur. The Mo—S distances in the bridge average 2.19 Å, much shorter than those from the bidentate diethyldithiocarbamate ligands which fall in the range 2.496—2.570 Å. The seven-coordinate centres are slightly inclined to each other with Mo—S—Mo being 157.8°.

The synthesis has been reported[118] of three new molybdenum sulphide ions formed by the careful rupturing of the framework of $Mo_2S_{10}^{2-}$. The transformations are shown in Scheme 15, and the products were obtained in a microcrystalline form suitable for characterization by X-ray diffraction to be made.

The ion $W_3S_8^{2-}$ has been produced[119] by the interaction of $(NH_4)_2WS_4$ in sulphuric acid with a methanolic solution of $\{N(PPh_3)_2\}Cl$. A simple linear chain has been proposed for the structure (25), and reading along the chain from end to centre the average W—S bond lengths are 2.15, 2.24, and 2.34 Å. However, the interesting part of this is that while the outermost tungstens were found to be tetrahedral, the central metal takes a square-planar shape. Such a configuration is rather quite rare for tungsten. In the ion $[W_2S_{10}(SH)]^-$ each metal has pentagonal-bipyramidal symmetry with a double-bonded sulphur in an axial position (26). The ion was produced[120] from a work-up of an HCl–MeCN solution of $[PPh_4]WS_4$, and the crystals were

[116] M. Yu. Subbotin and L. A. Aslanov, *Russ. J. Inorg. Chem.*, 1986, **31**, 511.

[117] K. S. Jasim, C. Chieh, and T. C. W. Mak, *Angew. Chem., Int. Ed. Engl.*, 1986, **25**, 749.

[118] D. Coucouvanis and A. Hadjikyriacou, *Inorg. Chem.*, 1986, **25**, 4317.

[119] S. Bhaduri and J. A. Ibers, *Inorg. Chem.*, 1986, **25**, 3.

[120] F. Sécheresse, J. M. Manoli, and C. Potvin, *Inorg. Chem.*, 1986, **25**, 3967.

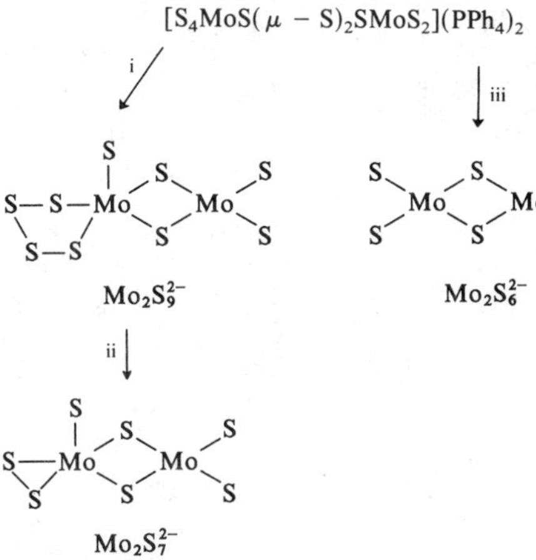

$$[S_4MoS(\mu - S)_2SMoS_2](PPh_4)_2$$

$Mo_2S_9^{2-}$

$Mo_2S_6^{2-}$

$Mo_2S_7^{2-}$

Reagents: i, DMF-NaBH$_4$-6h; ii, Ph$_3$P-DMF; iii, Ph$_3$P-DMF (6-fold excess)

Scheme 15

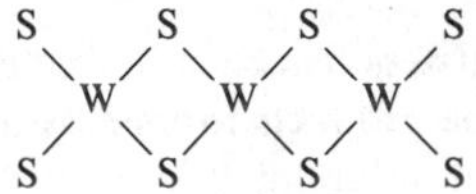

(25)

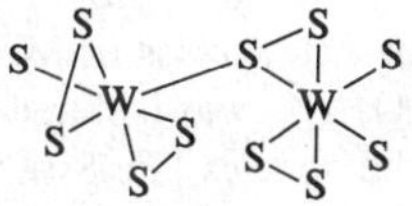

(26)

found to have a considerable degree of disorder so that it was not possible to locate the SH group.

To a large extent it is possible to view these sulphide ions as condensed versions of MS_4^{2-} ions and of course as well as binding to each other it is possible to use the sulphurs as ligands to other metals. Some of the most successful complexes of this type involve copper, and a few examples are shown in Scheme 16. These have been studied[121] for M = Mo and W and L = CN, SPh, S-tol, Cl, or Br. In particular the vibrational spectroscopic characteristics were determined through the use of infrared, Raman, and resonance Raman techniques. This approach was found to be a useful complement to the type of information available from n.m.r. spectroscopy. Particularly distinctive patterns and peak positions were noted for the type I and II complexes in the Scheme.

A copper–tungsten cluster related to type IV (Scheme 16) has been characterized[122] from the product of the reaction between $(NMe_4)_2WS_4$, CuCl, and KSCN in MeCN-CH$_2$Cl$_2$. The product has the empirical formula $(NMe_4)_2[(CuNCS)_4WS_4]$, but instead of having a trigonal arrangement each copper is tetrahedral. This is achieved by using the thiocyanates in bridging modes and constructing a polymer.

[121] R. J. H. Clark, S. Joss, M. Zvagulis, C. D. Garner, and J. R. Nicholson, *J. Chem. Soc., Dalton Trans.*, 1986, 1595.
[122] J. M. Manoli, C. Potvin, F. Sécheresse, and S. Marzak, *J. Chem. Soc., Chem. Commun.*, 1986, 1557.

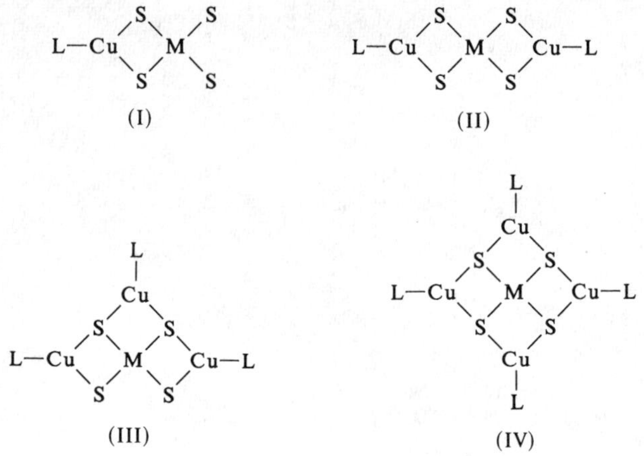

Scheme 16

Thus around any particular tungsten there are two coppers having one N-bonded and one S-bonded NCS, one copper with two S-bonded groups, and one with two N-bonded.

A member of a new class of Mo—Fe—S clusters has been described.[123] The ion $[MoOFe_5S_6(CO)_{12}]^{2-}$ was isolated as one of the products formed by adding $MoCl_5$ in MeCN to the species $[Fe_2S_2(CO)_6]^{2-}$ in thf solution. It is noteworthy that the molybdenum makes shorter (2.29 Å) Mo—S bonds with the two central sulphur atoms (27) than with the other two (2.46 Å). The construction of a classic cubane

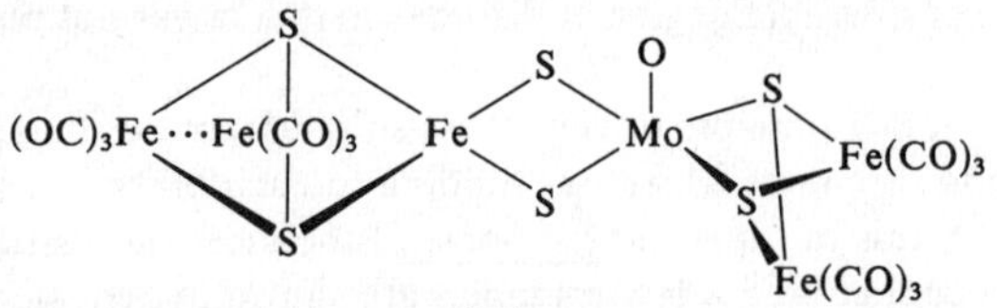

(27)

cluster using only simple starting materials has been announced.[124] The air-stable $[MoFe_3S_4(EtNCS_2)_5]^-$ ion was isolated from the mixture resulting from the addition of MoS_4^{2-} to Et_2NCS_2.Na, $FeCl_2$, and Et_4NBr in dimethylformamide. The structure is quite easy to describe, in that the central core is in the cubane style with alternating metals and sulphurs. Then, each metal has one chelating $Et_2NCS_2^-$ group and the fifth ligand acts as a bridge across the diagonal between the molybdenum and one of the irons. There are of course three Mo—Fe diagonals but the bridged one is found to be significantly shorter than the other two (2.624 Å compared to 2.85 Å).

These clusters have a potential interest as model compounds in discussions of bioinorganic chemistry, and hence the synthesis of a cubane having ammonia

[123] K. S. Bose, P. E. Lamberty, J. E. Kovacs, E. Sinn, and B. A. Averill, *Polyhedron*, 1986, **5**, 393.
[124] L. Qui-Tian, H. Liang-Ren, K. Bei-Sheng, L. Chun-Wan, W. Ling-Ling, and L. Jia-Xi, *Acta Chem. Sinica*, 1986, **44**, 343.

coordinated to the molybdenums is of considerable importance.[125] The ion $[Mo_4S_4(NH_3)_{12}]^{4+}$ was prepared as orange crystals by the addition of concentrated ammonia to an acid solution containing the aqua ion $[Mo_4S_4(H_2O)_{12}]^{5+}$. The aquo ion is regenerated by dissolving the solid in a mineral acid. The ion was shown to have a cubane core with each molybdenum being approximately octahedral from three sulphurs and three *mer*-ammonia ligands. The Mo—N bond length was found to be an average of 2.34 Å, which is longer than that formed by ammonia with other metals. A similar bonding style was found[126] also for $[Mo_3FeS_4(NH_3)_5(H_2O)]^{4+}$.

These latter complexes have considerable importance when it is realised that a glassy-carbon electrode modified by the addition of $[Mo_2Fe_6S_8(SPh)_9]^{3-}$ can act as a catalytic reductor for the conversion of NO_3^-, *via* NO_2^-, to ammonia.[127,128]

The rise to prominence over the recent years of dinuclear metal–metal compounds is documented in a number of publications and several additions have been made this year by leading practitioners in the field.

An account[129] of oxidative addition and reductive elimination at dinuclear centres forms also the centre of an article[130] on the $\sigma^2\pi^4$ triple bond found in Mo_2 and W_2. This discussion of the interrelationships between structure and reactivity lead to the triple bond being described as an inorganic functional group. As well as considering the dimolybdenum triple and quadruple bonded species, the chemistry of the equilateral triangular clusters has also been considered in a review[131] of the main advances from about 1981.

The strength of the metal–metal bond the Group 6 series $[M(CO)_3]_2(\mu$-fulvalene) has been probed by an electrochemical technique.[132] The use of cyclic voltammetry in aprotic media showed that all three could be oxidized at about +1V (*vs.* NHE) and reduced in a single wave to dianions. These dianions could then be readily reoxidized to the starting material. From a measurement of the potentials involved it was concluded that the metal–metal bond strength increased in the order Cr < Mo < W. This agrees with changes in the metal–metal bond length, which falls from 3.471 Å for the Cr compound to 3.347 Å for the tungsten. The Cr_2 triple-bonded entity $[(\eta^5$-$C_5Me_5)(CO)_2Cr]_2$ is cleaved by treatment with phosphorus in xylene at 140 °C. As well as the compound $(\eta^5$-$C_5Me_5)(CO)_2Cr(\eta^3$-$P_3)$, the triple-decker material shown in (28) is produced.[133] The chromium atoms are sandwiched between three planar, parallel, five-membered rings. The P—P distances are in the range 2.15—2.21 Å, with P—P—P angles *ca.* 108°. The linkages to the metals are quite uniform with P—Cr being 2.29—2.32 Å and there is a Cr to Cr separation of 2.727 Å.

The hydrostannolysis of $M_2(NMe_3)_6$, where M is Mo or W, gives mono- or di-substitution for molybdenum but only a mono-substituted product for the case of tungsten.[134] The reaction was controlled by varying the ratio of the complex to

[125] T. Shibahara, E. Kawano, M. Okano, M. Nishi, and H. Kuroya, *Chem. Lett.,* 1986, 827.
[126] T. Shibahara, H. Akashi, and H. Kuroya, *J. Am. Chem. Soc.,* 1986, **108**, 1342.
[127] S. Kuwabata, S. Uezumu, K. Tanaka, and T. Tanaka, *J. Chem. Soc., Chem. Commun.,* 1986, 135.
[128] S. Kuwabata, S. Uezumu, K. Tanaka, and T. Tanaka, *Inorg. Chem.,* 1986, **25**, 3018.
[129] M. H. Chisholm, *Polyhedron,* 1986, **5**, 25.
[130] M. H. Chisholm, *Angew. Chem., Int. Ed. Engl.,* 1986, **25**, 21.
[131] F. A. Cotton, *Polyhedron,* 1986, **5**, 3.
[132] R. Moulton, T. W. Weidman, K. P. C. Vollhardt, and A. J. Bard, *Inorg. Chem.,* 1986, **25**, 1846.
[133] O. J. Scherer, J. Schwalb, G. Wolmerhäuser, W. Kaim, and R. Gross, *Angew. Chem., Int. Ed. Engl.,* 1986, **25**, 363.
[134] P. R. Sharp and M. T. Rankin, *Inorg. Chem.,* 1986, **25**, 1508.

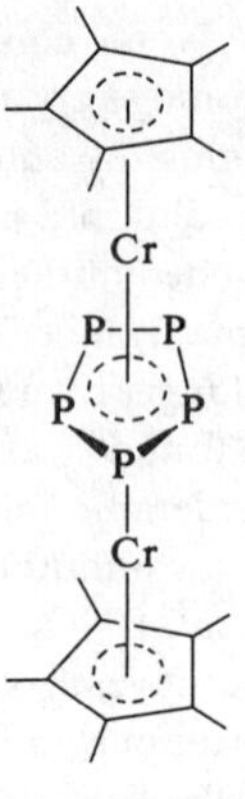

(28)

the reactant Ph_3SnH and the course of the substitution monitored by ^1H n.m.r. spectroscopy. The mono-substitution was shown to proceed by a mixed second-order process with a rate constant of $1 \times 10^{-4}\,mol^{-1}\,dm^3\,s^{-1}$.

Alkyne reactivity with similar materials has also been investigated.[135] It was found that in the presence of pyridine, alkynes will add across the triple bond in $W_2Cl_2(NMe_2)_4$ to give products of the type $W_2Cl_2(NMe_2)_4(\mu\text{-alkyne})(py)_2$. Except in the case of ethyne these decompose in hydrocarbon solvents with the release of $W_2(NMe_2)_6$ and the concomitant formation of $W_2Cl_x(NMe_2)_{6-x}(\mu\text{-alkyne})(py)_2$, where x is 3 or 4. These materials have been shown to possess a pseudo-tetrahedral core of $W_2(\mu\text{-}C_2)$.

The complex formulated as $W_2Et_2(NMe_2)_4$ exists in *anti* and *gauche* forms. If these are reacted[136] with 1,3-diaryltriazenes, ArNNNHAr, then the product $W_2Et_2(NMe_2)_2(ArNNNAr)_2$ can have three isomeric versions, shown as (A), (B), and (C) in Scheme 17. It was established by n.m.r. spectroscopy that isomer A, formed from the *gauche* reactant, slowly converts into a mixture of the isomers B

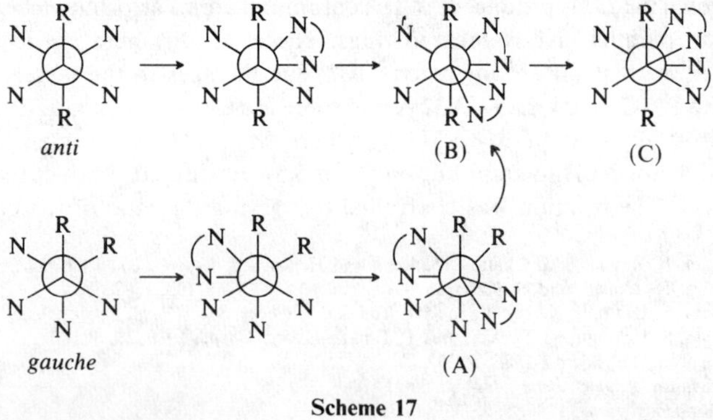

Scheme 17

[135] K. J. Ahmed, M. H. Chisholm, K. Folting, and J. C. Huffman, *Organometallics*, 1986, **5**, 2171.
[136] M. H. Chisholm, H. T. Chiu, J. C. Huffman, and R. J. Wang, *Inorg. Chem.*, 1986, **25**, 1092.

and C formed from the *anti* material. The exact position of equilibrium between these forms is dependent upon the nature of the aryl species.

Normal coordinate analysis of the vibrational spectra of molybdenum quadruple-bonded species has indicated a difference between the force constants found for isolated materials and those applicable to the solid state.[137] Thus for $Mo_2Cl_8^{4-}$ the value falls from $2.0 \, Nm^{-1}$ in the isolated state to $1.7 \, Nm^{-1}$ in the solid. For the tetraacetate the figures are 2.0 and $1.8 \, Nm^{-1}$ respectively.

As well as n.m.r. spectroscopy it is possible to use visible u.v. methods to assist in stereochemical transformations in these complexes. A simple model has been described[138] that enables the stereochemistry to be correlated to the signs in the circular dichroism spectrum of the $\delta \rightarrow \delta^*$ and the $\delta \rightarrow x^2 - y^2$ transitions. The examples discussed all relate to dimolybdenum quadruple-bonded compounds. The type of study that is possible is well illustrated[139] by a consideration of how the P—Mo—Mo—P torsion angle in the complex (29) affects the circular dichroic behaviour of the $\delta \rightarrow \delta^*$ band. Two forms of the chloro-complex were isolated with torsion angles of 22° and 24°, compared to 21.7° for the corresponding bromo-species.

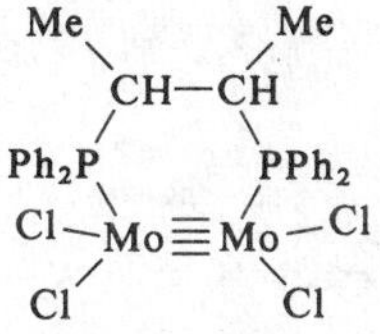

(29) 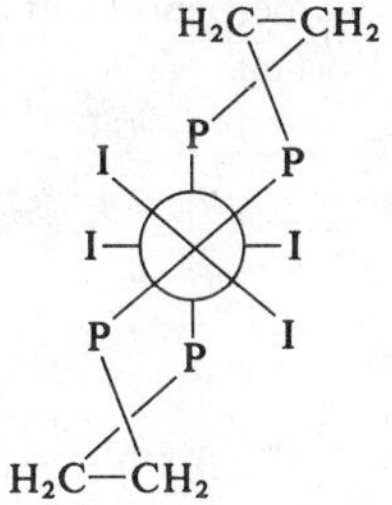   (30)

Using a slightly different phosphine,[140] both an eclipsed and a staggered form of the complex $Mo_2I_4(diphos)_2$ have been isolated. A projection of the staggered version is shown (30), and the torsion angle was determined as 27.9°. Two distinct peaks were evident in the visible spectra at 920 and 670 nm. The Mo—Mo bond length was found to be greater (2.180 Å) in the staggered form than that of the eclipsed (2.129 Å).

The trinuclear cluster ions $[M_3(\mu_3\text{-}O)_2(\mu\text{-}OOCMe)_6(H_2O)_3]^{2+}$ have been characterized for the cases of M = Mo and W, and the intermediate Mo_2W and MoW_2 cases have now been described.[141] The ions were separated by slow (*ca.* 1 month) elution from a cation exchange column. The series shows a graduation of colours from red (Mo_3) through to yellow(W_3). The main evidence for the postulation of mixed-metal clusters rather than mixtures of the trimetal complexes, comes from the ^1H n.m.r. spectra. The trimetal species show singlet methyl resonances whilst the mixed species show two resonances. There are also shifts in the resonance position that are in agreement with this formulation.

[137] E. M. Larson, T. M. Brown, and R. B. von Dreele, *Acta Crystallogr.*, 1986, **B42**, 533.
[138] I. F. Fraser, A. McVitie, and R. D. Peacock, *Polyhedron*, 1986, **5**, 39.
[139] P. A. Agaskar, F. A. Cotton, I. F. Fraser, L. Manojlovic-Muir, K. W. Muir, and R. D. Peacock, *Inorg. Chem.*, 1986, **25**, 2511.
[140] F. A. Cotton, K. R. Dunbar, and M. Matusz, *Inorg. Chem.*, 1986, **25**, 3641.
[141] B. Wang, Y. Sasaki, A. Nagasawa, and T. Ito, *J. Am. Chem. Soc.*, 1986, **108**, 6059.

The rest of this section on coordination chemistry deals with heterobimetallics. This is a somewhat arbitrary division, sliding as it does towards the next group of papers on organometallics.

An interesting set of compounds with formulae $(CO)_4\overline{M(\mu\text{-}PCy_2)_2M'}(PPh_3)$ for M = Mo or W, and M' = Ni, Pd, or Pt, have been prepared.[142] The reaction involved the effect of Bu^nLi and $M'Cl_2(PPh_3)_2$ on $cis\text{-}M(CO)_4(PCy_2H)_2$. A full structural analysis was undertaken for the Mo/Pd version and the central core was found to be non-planar with the dihedral angle between the PMoP and PPdP sections established as 15.4°. The bridging groups were identified also by the large downfield shift of the ^{31}P n.m.r. resonances. The production of cluster products has been aided by application of the isolobal principle. This has been used[143] to demonstrate the analogy between the ubiquitous $C_5H_5^-$ ligand and the d^6-metallate species $[\{\eta^5\text{-}Cp)M(CO)_3]^-$, for M = Cr, Mo, or W. These species have a piano-stool configuration and have indeed been shown[144] to take the place of Cp as for example in the reaction with the tetramethylbutadiene complex, $(MeC)_4.Co(CO)_3$. The interesting sandwich species shown (31) is produced and is notable for having the rings parallel to each other and perpendicular to the inter-metal axis. The bridging carbonyls show μ-CO at 1804 cm^{-1}.

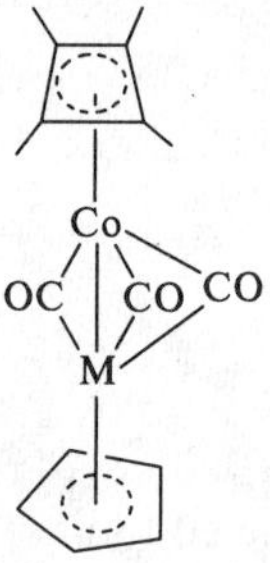

(31) 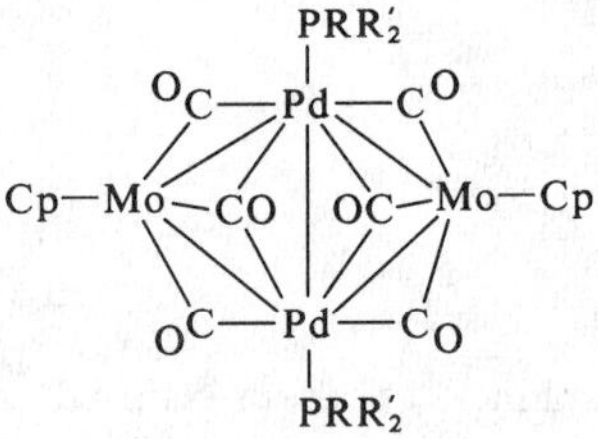   (32) 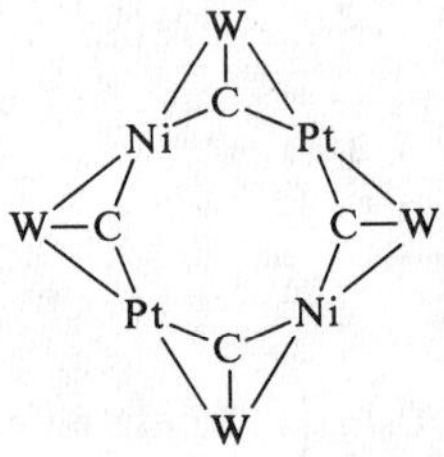   (33)

The four-metal cluster shown (32) was constructed[145] by the reaction of $Na[Mo(CO)_3Cp]$ with $[\mu\text{-}Cl\overline{-Pd(o\text{-}C_6H_4CH_2P}Ph_2)_2]$. The significant core distances were found to be Pd—Pd = 2.5 Å and Pd—Mo = 2.84 Å. In a similar vein the synthesis of the so-called 'star' clusters, one of which has the framework shown in (33), has been reported.[146] The full description of the star is $Ni_2Pt_2W_4(\mu_3\text{-}CPh)_4(CO)_8(\eta\text{-}Cp)_4$.

The replacement of a μ-Cr group in a tungsten–iron compound by a phosphine bridge has been reported.[147] Scheme 18 shows an outline of the procedure followed and the full paper is characterized by a discussion of the potential reaction pathways by which the change was accomplished.

[142] S. J. Loeb, H. A. Taylor, L. Gelmini, and D. W. Stephan, *Inorg. Chem.*, 1986, **25**, 1977.

[143] P. Hofmann and H. R. Schmidt, *Angew. Chem., Int. Ed. Engl.*, 1986, **25**, 837.

[144] P. Härter, H. Pfisterer, and M. L. Ziegler, *Angew. Chem., Int. Ed. Engl.*, 1986, **25**, 839.

[145] H.-P. Abicht, R. Barth, K. Peters, E. M. Peters, and H. G. von Schnering, *Z. Chem.*, 1986, 409.

[146] G. P. Elliott, J. A. K. Howard, T. Mise, C. M. Nunn, and F. G. A. Stone, *Angew, Chem., Int. Ed. Engl.*, 1986, **25**, 190.

[147] S. W. Hoskins, A. P. James, J. C. Jeffrey, and F. G. A. Stone, *J. Chem. Soc., Dalton Trans.*, 1986, 1709.

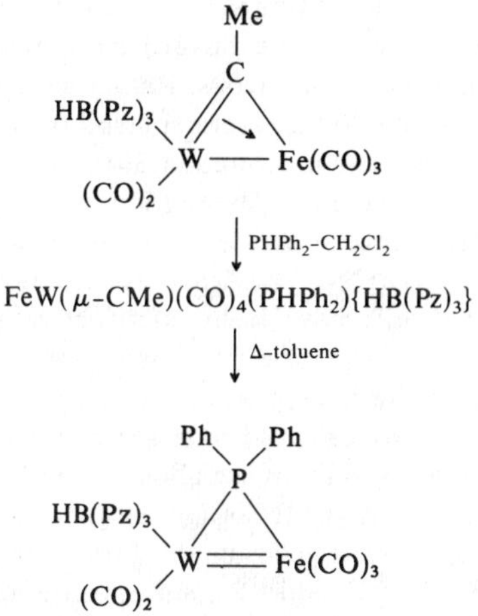

Scheme 18

A subtle differentiation of reactivity occurs[148] for the trimetal complex shown at the start of Scheme 19. Whereas the reaction of $W(CR)(CO)_2(Cp)$ leads to degradation of the trimetal complex for the case of $R = -C_6H_4Me-4$, a mixture of three trimetallates is given when $R = Me$. The structures were probed by a mixture of multinuclear n.m.r. spectroscopy and X-ray diffraction methods.

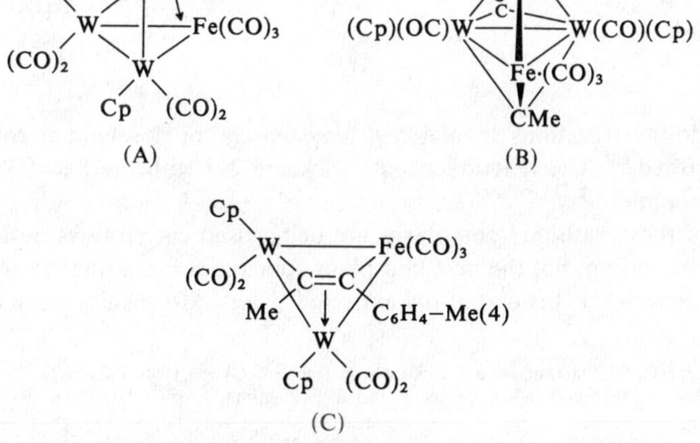

Scheme 19

[148] E. Delgado, J. C. Jeffrey, and F. G. A. Stone, *J. Chem. Soc., Dalton Trans.*, 1986, 2105.

Further results from u.v. photolytic reactions in the somewhat exotic solvent of liquid xenon have been reported.[149] The dihydrogen complex $M(CO)_5(H_2)$ has been characterized for all three Group 6 metals. Rather surprisingly the molybdenum version was found to have the lowest thermal stability, and indeed the ν (H—H) vibration for this compound was measured at 3080 cm^{-1} compared to that of the Cr (3030 cm^{-1}) and W at 2711 cm^{-1}. The chromium compound has been further studied in a series of exchange processes with deuterium, and in reaction with dinitrogen to form $Cr(CO)_5(N_2)$. Evidence is produced for the species *cis*-$Cr(CO)_4(H_2)_2$. Further details have been published[150] also on the dihydrogen complexes $M(CO)_3(PR_3)_2(\eta^2\text{-}H_2)$. The H—H vibration is at 2690 cm^{-1} and the H_2 atoms, but not those from the phosphines, readily exchange with deuterium. The use of ^{2}D n.m.r. spectroscopy on solid-state samples has indicated that a rapid rotation of the coordinated D_2 takes place about the metal—D_2 axis.

The complex $Mo(CO)_2(PPh_3)(MeCN)(\eta^2\text{-}SO_2)$ reacts[151] with PR_3 to give $Mo(PR_3)_3(CO)_2(SO_2)$. Three distinct isomeric forms of the product were distinguished by the use of i.r., ^{1}H, ^{13}C, and ^{31}P n.m.r. spectroscopy and X-ray diffraction methods. Isomers A and B are octahedral and have an S-bonded SO_2 ligand, differing in that while A has *mer*-phosphines and *trans*-carbonyls, B has *fac*-phosphines and *cis*-carbonyls. Isomer C however is seven-coordinate and has the SO_2 in a η^2 mode. The relative proportions of these isomers were seen to be dependent on the nature of the phosphine, for example with PMe_3 the sequence was C > B > A, for PMe_2Ph A ≃ B ≃ C, and for $PMePh_2$ A ≫ B ≃ C.

Photolysis of $M(CO)_6$, M = Cr, Mo, W, in the presence of excess of either of the compounds shown (34) forms[152] the complex $M(CO)_5L$. The molybdenum version has only low stability and decays to $Mo(CO)_4L$, but the others are more robust and the tungsten complexes were isolated. Tungsten complexes $W(CO)_5L$, have been produced[153] also for the ligand series L = R_3MSR' where R and R' are various alkyl groups and M is one of the Group 4 elements (C, Si, Ge, or Sn).

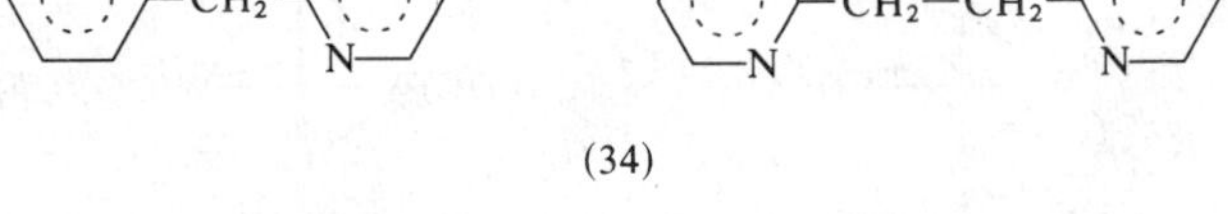

(34)

Cycloaddition reactions involving a tungsten carbonyl–selenium complex have been described.[154] The reaction occurs (Scheme 20) with the Se=C bond as the heterodienophile.

Many of these carbonyl complexes are being used currently as building blocks in cluster formation, but the best laid plans sometimes bring unexpected results as shown in Scheme 21. Instead of the expected[155] Ni—Mo species a di-molybdenum

[149] R. K. Upmacis, M. Poliakoff, and J. J. Turner, *J. Am. Soc. Chem.*, 1986, **108**, 3645.
[150] G. J. Kubas, C. J. Unkefer, B. I. Swanson, and E. Fukushima, *J. Am. Chem. Soc.*, 1986, **108**, 7000.
[151] F.-E. Baumann, C. Burschka, and W. A. Shenck, *Z. Naturforsch., Teil B*, 1986, **41**, 1211.
[152] D. E. Marx and A. J. Lees, *Organometallics*, 1986, **5**, 2072.
[153] C. R. Lucas, *Can. J. Chem.*, 1986, **64**, 1758.
[154] H. Fischer, U. Gerbing, J. Riede, and R. Benn, *Angew. Chem., Int. Ed. Engl.*, 1986, **25**, 78.
[155] D. Fenske and K. Merzweiler, *Angew, Chem., Int. Ed. Engl.*, 1986, **25**, 338.

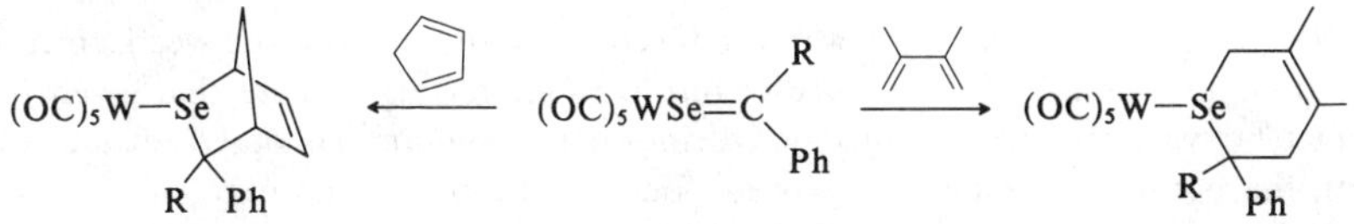

Scheme 20

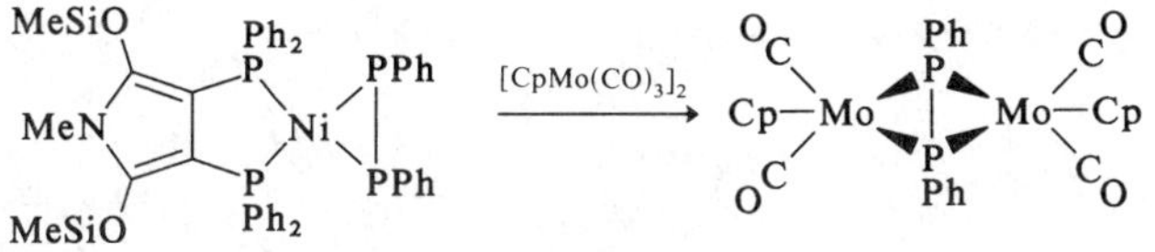

Scheme 21

compound has been formed. It shows a singlet in the ^{31}P n.m.r. spectrum and has two Mo—P bonds of 2.47 Å and two of 2.54 Å.

The Group 6 hydrogen isocyanide complexes $M(CO)_5CNH$ react[156] with aziridines $RNCHR'CH_2$ to give imidazolidin-2-ylidene products (35). A number of these were characterized for M = Cr, Mo, W including (R = H, R' = H, Me, Et; and R = Me, R' = H).

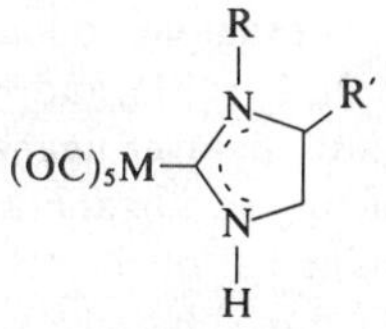

(35)

A homoleptic σ-bonded benzyl complex has been prepared[157] as dark red crystals by treating $MoCl_4(thf)_2$ with $Mg(CH_2Ph)_2$. The product was characterized as the dimetal complex $Mo_2(CH_2Ph)_6$ with an Mo—Mo bond length of 2.175 Å. The ligands take a staggered conformation with the Mo—C—Ph angles *ca.* 98.9°. The synthesis of an air-stable chromium(VI) σ-bonded aryl product was attained[158] by reacting together $(Bu^tN)_2Cr(OSiMe_3)_2$ and RMgBr in hexane. The product of this reaction is $(Bu^tN)_2CrR_2$ and it was shown to be monomeric with a distorted tetrahedral arrangement of ligands. The structure was determined for R = mesityl, and the angles C—Cr—C and N—Cr—N were found to be both greater than the tetrahedral angle at 121.3° and 114.7° respectively.

Evidence from infrared spectroscopy suggests that chromocene is converted into the ionic $[Cr(en)_3][Cp]_2$ by the action of excess ethylene diamine[159]. This view is supported by a change in the magnetic state from low-spin to high-spin.

[156] W. P. Fehlhammer, E. Bär, and B. Boyadjiev, *Z. Naturforsch., Teil B*, 1986, **41**, 1023.

[157] S. M. Beshouri, I. P. Rothwell, K. Folting, J. C. Huffman, and W. E. Streib, *Polyhedron*, 1986, **5**, 1191.

[158] M. B. Hursthouse, M. Motevalli, A. C. Sullivan, and G. Wilkinson, *J. Chem. Soc., Chem. Commun.*, 1986, 1398.

[159] J. Kalousova, L. Beneš, and J. Votinsky, *Coll. Czech. Chem. Commun.*, 1986, **51**, 314.

The compound shown (36) was produced[160] by the reaction between an iron complex and $Cr(CO)_3(NH_3)_3$ in dioxane. It is interesting to note that given the choice between the two five-membered rings the chromium has opted for the side with the boron. The planar rings were found to be parallel to each other and a carbonyl ligand almost eclipses the boron.

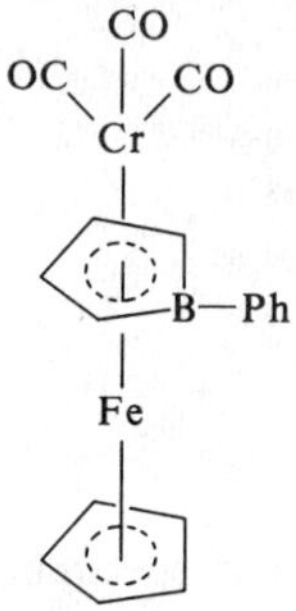

(36) 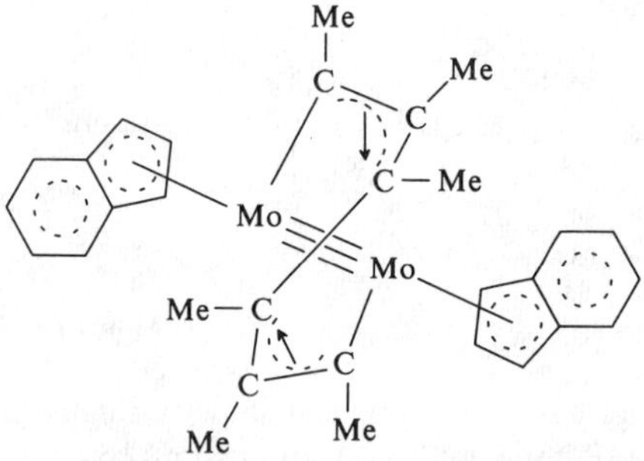   (37)

A review has been given[161] of the chemistry of bis-alkyne molybdenum species such as $[Mo(L)(\eta^2\text{-}RC_2R)_2(\eta\text{-}C_5H_5 \text{ or } \eta\text{-}C_9H_7)]^-$ ions. The reactions considered lead to alkyne–alkene complexes or to the linking of two molybdenum centres through 1e addition processes. For example, sodium or magnesium amalgam reduction of the ion having {L = MeCN, R = Me, and $\eta\text{-}C_9H_7$} forms[162] the three-alkyne 'fly-over' complex (37). The C_6Me_6 fly-over ligand can be considered as a pair of linked allylidene ligands each acting as a σ-donor to one metal and a η^3-donor to the other metal. From an examination of the ^1H and ^{13}C n.m.r. spectra it seems likely that the structure is retained in solution.

The alkyne complex $MoX[\eta\text{-}C_2(CF_3)_2]_2(Cp)$ reacts[163] with dienes to give $MoX_2(\eta\text{-diene})(Cp)$, for X = Cl, Br, or I. In the case of butadiene reacting with the chloro-variant, a small quantity of a $(\mu^3\text{-oxo})$ tri-molybdenum species was also obtained. Both products were characterized by X-ray diffraction methods. In the case of the mononuclear complex, the molybdenum was found to take roughly tetrahedral geometry with a mirror plane through the $MoCl_2$ moiety. The butadiene is in the endo-conformation with C_1 and C_4 slightly closer, (2.23 Å), to the metal than the two central atoms (2.32 Å). The Cp ring is not symmetric at all and displays slippage towards becoming a coordinated cyclic η^3-allyl fragment.

A wide range of sandwich compounds involving changes in substituents and in ring size have been investigated[164] systematically by ^{95}Mo and ^{13}C n.m.r. spectroscopy. The observed chemical shifts are interpreted on a ligand-field model, and it

[160] G. E. Herberich, B. Hessner, J. A. K. Howard, D. P. J. Köffer, and R. Saire, Angew. Chem., Int. Ed. Engl., 1986, 25, 165.
[161] M. Green, Polyhedron, 1986, 5, 427.
[162] L. Brammer, M. Green, A. G. Orpen, K. E. Paddick, and D. R. Saunders, J. Chem. Soc., Dalton Trans., 1986, 657.
[163] J. L. Davidson, K. Davidson, W. E. Lindsell, N. W. Murrall, and A. J. Welch, J. Chem. Soc., Dalton Trans., 1986, 1677.
[164] J. C. Green, R. A. Grieves, and J. Mason, J. Chem. Soc., Dalton Trans., 1986, 1313.

is noted that shielding appears to reduce in magnitude with either a decrease in stability or with an increase in reactivity of the complex. This point is then developed in the light of relevance to molybdenum as a catalytic centre.

Another use of n.m.r. is in the characterization of fluxionality in a molecule. The tungsten compound (38), with X = CO, has been shown[165] to have three distinct

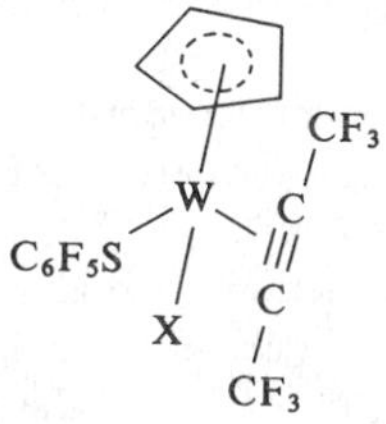

(38)

phases of fluxional behaviour in the temperature range from −90 to +80 °C, the rotations S—C_6F_5, W—SC_6F_5, and the propeller rotation (CF_3CCCF_3)—W. For the oxo-version (X = O) then only the W—SC_6F_5 rotation was recognized.

5 Manganese, Technetium, and Rhenium

The discussion of the chemistry of the elements of this group broadly follows the preceeding pattern in that it is led by the ligand structure rather than the oxidation number. It is noticeable that there is a marked disparity between the metals if this classification is followed. Manganese is poorly represented amongst the mononuclear complexes until those involving macrocyclic ligands are reached, whereas technetium seems to have little reported in the area of organometallic and metal cluster chemistry.

The synthesis and structure of $Mn(dmu)_3Br_2$ is a rare report[166] of a penta-coordinate manganese complex involving monodentate ligands. The ligands are in a slightly distorted trigonal-bipyramidal arrangement with both of the axial positions occupied by N,N'-dimethylurea molecules. The axial ligands are bound more weakly than that in the equatorial position. This is clear from both the longer Mn—O bond length (2.175 compared to 2.0353 Å) and from the weaker force constant for the Mn—O vibration (222 compared to 272 Nm^{-1}).

A manganese product that is of potential importance in discussions about manganese centres in biological areas such as photosystem II has been prepared[167] by reacting NaHsal with $[Mn_2O_2(bipy)_4]^{3-}$. The product is $Mn(sal)_2(bipy)$ and is a rare example of a six-coordinate Mn^{IV} that contains N and O donors but without either the porphyrin- or catecholate-type ligands which do not seem to be important in manganese centres of biological significance. The salicylate ligands are arranged so that the phenoxide linkages are *cis* to each other, while the carboxylates bind in a *trans* style.

The first example of two nitroxyl ligands bound to one metal has been produced[168] by the reaction of a cyclic nitroxyl radical species with the complex $Mn(hfaa)_2$,

[165] J. L. Davidson, *J. Chem. Soc., Dalton Trans.*, 1986, 2423.
[166] J. Delaunay and R. P. Hugel, *Inorg. Chem.*, 1986, **25**, 3957.
[167] P. S. Pavacik, J. C. Huffman, and G. Christou, *J. Chem. Soc., Chem. Commun.*, 1986, 43.
[168] M. H. Dickman, L. C. Porter, and R. J. Doedens, *Inorg. Chem.*, 1986, **25**, 2595.

where hfaa represents hexafluoroacetylacetonate. The products were formulated $Mn(hfaa)_2L_2$ and were shown to have monodentate O-bonded nitroxyls. The ligands chosen were formed from the cyclic amines piperidine and pyrrolidine, but where each amine carries two methyl groups at the positions adjacent to the nitrogen. Both complexes take similar *trans*-octahedral structures with the only major difference being that the piperidine derivative has an $Mn-O-N$ angle of 167.2° whereas the pyrrolidine has 145.3°.

The developing chemistry of technetium continues to produce examples of distinct behaviour from that of manganese. Whereas the reaction of either $NOPF_6$ or HNO_3–glacial acetic acid with $[M(CNBu^t)_6]^+$ gives $[Mn(CNBu^t)_6]^{2+}$ for manganese, a substitution takes place for technetium[169] to give $[Tc(NO)(CNBu^t)_5]^{2+}$.

Much of the interest in technetium chemistry is driven by the demands of nuclear medicine for a wide range of complex types of the isotope ^{99m}Tc. In brain imaging there is a need for neutral complexes for use in single-photon emission tomography so that cerebral blood flows can be measured. A range of technetium(V) oxime products has been produced[170] for this purpose and the members were shown to be five-coordinate taking a square-pyramidal shape and having an oxo-group in the apical position. The dimercaptosuccinic acid and its dimethylester shown (39) exist in *meso* and *racemic* forms. The technetium(V) complexes with these forms have been shown[171] to have quite different physiological effects in that while the complexes with the *meso* ligand are retained in the body and appear to have osteotropic properties, those formed from the racemate are rapidly eliminated.

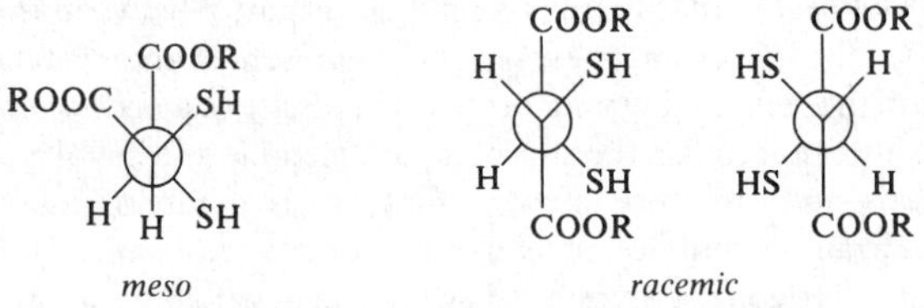

meso   *racemic*

(39)

The preparation and characterization of the complex ion $BaTcBr_6$ has been described.[172] Evidence for the existence of the mixed-ligand complexes $[TcNBr_{4-n}Cl_n]^-$, for $n = 1$, 2, or 3, was obtained from an examination[173] of the e.p.r. spectra of frozen solutions of mixtures between the chloro- and bromo-complex ions. There is a near linear dependence of various hyperfine coupling constants with the effective spin-orbit coupling constants of the equatorial donors which allows the different species to be distinguished. The chloro-complex can[174] be reduced to give a series of $Tc^V\equiv N$ complexes of which that formed with the ligand 8-quinolinethiol is particularly interesting. It adopts an approximately square-

[169] K. E. Linder, A. Davison, J. C. Dewan, C. E. Costello, and S. Maleknia, *Inorg. Chem.*, 1986, **25**, 2085.
[170] S. Jurisson, E. O. Schlemper, D. E. Troutner, L. R. Canning, D. P. Nowotnik, and R. D. Neirinckx, *Inorg. Chem.*, 1986, **25**, 543.
[171] H. Spies and D. Scheller, *Inorg. Chim. Acta*, 1986, **116**, 1.
[172] L. L. Zaitseva, A. A. Kruglov, and D. O. Bogdanov, *Russ. J. Inorg. Chem.*, 1986, **31**, 187.
[173] R. Kirmse, J. Stach, and U. Abram, *Inorg. Chim. Acta*, 1986, **117**, 117.
[174] J. Baldas, J. Bonnyman, and G. A. Williams, *Inorg. Chem.*, 1986, **25**, 150.

pyramidal arrangement with the apical (Tc≡N) distance being 1.60 Å. The planar bidentate ligands are arranged so that the sulphur atoms are mutually *cis*, but there is 19° twist between the two sides of the pyramid base. The structure of a six-coordinate Tc^V complex formed from quinolinol and the terdentate Schiff base (L) formed from salicylaldehyde and aminophenol has been investigated.[175] TcO(L)-(quin) has the Schiff base ligand in a *mer* position and the quinolinol binds so that the nitrogen atoms are *trans* to each other. There is a deal of distortion evident in this structure as the technetium is displaced *ca.* 0.08 Å away from the equatorial plane towards the oxo-group.

Some transformations of NSCl–rhenium complexes have been investigated.[176,177] Treatment of ReCl$_3$(NSCl)$_3$POCl$_3$ with a dichloromethane solution of pyridine lead to a complex with a thionitrile ligand, ReCl$_2$(NS)(NSCl)py$_2$. *X*-ray structural analysis revealed the presence of a slightly distorted octahedral environment about the metal with a near linear thionitrile group (Re—$\hat{\text{N}}$—S = 176.3°). The ion [ReCl$_4$(NSCl)$_2$]$^-$ can be converted into [ReCl$_4$(N$_2$S$_2$)]$^-$ by reaction with either S(NSiMe$_3$)$_2$ or PhC≡CPh. Addition of Me$_3$SiBr either to the initial reactant or to the former product, gives the bromo-version [ReBr$_4$(N$_2$S$_2$)]$^-$. The new ligand is a planar bidentate nitrogen donor to the octahedrally coordinated rhenium. There is little difference between the dimensions of the chloro and bromo cases with Re—N showing a change from 1.77 Å (Cl) to 1.84 Å (Br).

Tris-catecholates of rhenium have received some attention. Reaction of catechol with K$_2$ReCl$_6$ under basic, anaerobic conditions produces[178] [PPh$_4$]$_2$[Re(cat)$_3$].H$_2$O. A similar, but higher oxidation state, complex is formed by the photolysis[179] of Re$_2$(CO)$_{10}$ and 3,5-di-Bu^t-catechol in toluene. Dark purple crystals are obtained, in near quantitative yield, of Re(dbc)$_3$. The absence of carbonyls was indicated by consideration of the i.r. spectrum and a crystallographic structure determination showed the complex to be monomeric with an octahedral configuration. In contrast to the manganese analogue the complex is quite unaffected by traces of water.

The next set of papers concern manganese complexes with macrocyclic ligands, such as the manganese(IV) species that been recently characterized.[180] It was prepared by refluxing together salicylaldehyde, 1,3-dihydroxy-2-amino-2-methyl-propane, and Mn^{III} acetate with sodium hydroxide in dmf. The ligand species formed has a phenolate oxygen, two hydroxyls, and an imine nitrogen available for binding to the metal. It was found that a neutral Mn^{IV} complex was formed that had two terdentate ligands. These were arranged in a *mer* configuration with one of the hydroxyls unattached and with the attached hydroxyl in a deprotonated form thus giving an alkoxide. The structure adopted is broadly the slightly distorted octahedron that is expected for a d^3 ion. Electrochemical investigations show that this bis-Schiff base complex is stabilized by nearly 1V with respect to Mn(sal)$_2$(bipy). This may be a result of the alkoxide linkage, which could also account for the water stability shown by the complex. Oxidation of the porphyrin Mn^{III} (tpp)Cl by

[175] U. Mazzi, F. Refosco, F. Tisato, G. Bandoli, and M. Nicolini, *J. Chem. Soc., Dalton Trans.*, 1986, 1623.
[176] H.-G. Hauck, W. Willing, U. Müller, and K. Dehnicke, *Z. Anorg. Allg. Chem.*, 1986, **534**, 77.
[177] E. Conradi, H. G. Hauck, U. Müller, and K. Dehnicke, *Z. Anorg. Allg. Chem.*, 1986, **539**, 39.
[178] W. P. Griffiths, C. A. Pumphrey, and T. A. Rainey, *J. Chem. Soc., Dalton Trans.*, 1986, 1125.
[179] L. A. deLearie and C. G. Pierpont, *J. Am. Chem. Soc.*, 1986, **108**, 6393.
[180] D. P. Kessissoglou, W. M. Butler, and V. L. Pecoraro, *J. Chem. Soc., Chem. Commun.*, 1986, 1253.

$\overline{C_6H_4SC_6H_4O}(SbCl_6)$ leads[181] to the formation of the compound $Mn(tpp)Cl.SbCl_6$. This was formulated as a Mn^{III} porphyrin π-cation radical entity on the basis of the crystal structure, the visible spectrum, and the magnetic characteristics. In particular the u.v. spectrum was identical to that of the species obtained from a gold mini-grid after the one-electron oxidation of $Mn(tpp)Cl$. Treatment of the radical species with methanol gave a product tentatively identified as Mn^{IV} (tpp)-$(OMe)_2$, suggesting that in the presence of a strong π-donor such as $-OMe$ the site of oxidation switches from the ligand to the metal. Porphyrins can be assembled by refluxing together pyrrole and the appropriate aldehyde. Use of tolylaldehyde would produce tetratolylporphyrin (ttp), and if a mixture of tolylaldehyde and salicylaldehyde is used then a range of products with various combinations of tolyl and phenoxy substituents will be formed. From such a mixture the porphyrin having three tolyl groups and one phenoxy group was separated[182] by chromatography. Refluxing this ligand with Mn^{II} acetate, followed by aerial oxidation gave a dimeric Mn^{III} porphyrin. The species was characterized by various spectroscopic methods and the Curie Law plot was interpreted as showing minimal anti-ferromagnetic coupling between the two centres. A structure consistent with these results shows that the phenoxide substituent from one ring acts as an axial ligand to the other metal. Of necessity this means that the metals are displaced from each other and do not form any sort of sandwich.

The next group of papers are concerned with metal–metal bonded rhenium compounds. The complex anion $[Re_4S_{22}]^{4-}$ has been shown[183] to have a Re_4S_4 cubane core with six S_3^{2-} ligands bridging across each face of the cube. The rheniums across each face are bonded together ($Re-Re = ca.$ 2.76 Å) and the species is diamagnetic. The five-membered ring ($-Re-Re-S-S-S-$) formed at each face is non-linear and adopts an envelope configuration.

The classic metal–metal bonded species involving rhenium is probably the anion $Re_2Cl_8^{2-}$. It certainly forms a convenient starting point for the synthesis of a range of complexes such as the product[184] from the reaction with PMe_3, $Re_2X_4(PMe_3)_4$ where x is Cl or Br. The chloro complex proved to be sufficiently volatile for a gas-phase photoelectron spectrum to be measured and it was thus determined that the compound has a $\sigma^2\pi^4\delta^2\delta^{*2}$ configuration. This confirms suggestions that this species would indeed have this 'electron-rich' type of triple bond. The compounds can be oxidized by $NOPF_6$ to give a paramagnetic ion $Re_2X_4(PMe_3)_4^+$ the chloro version of which reacts with the diphosphines $Ph_2PCH_2PPh_2$ (dppm) or $Ph_2PNHPPh_2$ (dppa) to give Re_2Cl_4 $(PMe_3)_2$(diphos). Phosphorus n.m.r. was employed to show that these complexes have cis-PMe_3 ligands and bridging diphosphines. If the $Re_2Cl_8^{2-}$ species is reacted with the diphosphine $(PMe_2)_2CH_2$ in methanol solution then red–yellow crystals of $Re_2Cl_4(\mu$-diphos$)_3$ can be obtained.[185] All three ligands were found to be bridging with an orientation (40) consistent with

[181] L. O. Spreer, A. C. Maliyackel, S. Holbrook, J. W. Otvos, and M. Calvin, *J. Am. Chem. Soc.*, 1986, **108**, 1949.

[182] G. M. Godziela, D. Tilotta, and H. M. Goff, *Inorg. Chem.*, 1986, **25**, 2142.

[183] A. Müller, E. Krickemeyer, and H. Bögge, *Angew. Chem., Int. Ed. Engl.*, 1986, **25**, 272.

[184] D. R. Root, C. H. Blevins, D. L. Lichtenberger, A. P. Sattelberger, and R. A. Walton, *J. Am. Chem. Soc.*, 1986, **108**, 953.

[185] L. B. Anderson, F. A. Cotton, L. R. Falvello, W. S. Harwood, D. Lewis, and R. A. Walton, *Inorg. Chem.*, 1986, **25**, 3637.

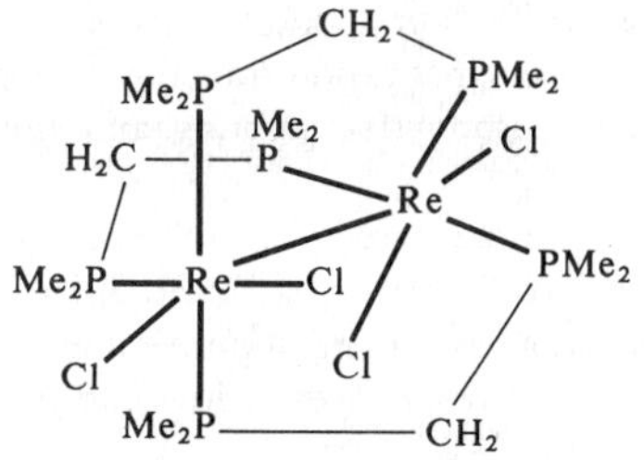

(40)

the observation of a 2:1 peak ratio in the ^{31}P n.m.r. spectrum. The Re—Re distance was established as 2.316 Å, and it would appear that this style of electron-rich bonding produces an entity that has a high degree of kinetic stability.[186]

A range of manganese carbonyls, $Me_{3-x}Cl_xSn[Mn(CO)_{5-y}L_y]_z$, has been prepared[187] and examined by different forms of spectroscopy. There is a good correlation between the ^{17}O n.m.r. chemical shift and the carbonyl stretching frequency, particularly after taking account of some deviations for some of the axial ligands. There was also a fair correlation between the ^{17}O and ^{55}Mn chemical shifts.

Electroreduction of a series of ionic carbonyls, $Mn(CO)_5L^+$, has been shown to give rise to broadly three types of product.[188] For L = CO the species $Mn(CO)_5^-$ is formed, probably *via* a 19-e transient radical species. With L = MeCN or py then $Mn_2(CO)_{10}$ is formed by a heterolytic coupling between the cation and the original anion, and for L = various phosphines, hydrides such as $HMn(CO)_3(phos)_2$ are produced.

An accurate redetermination of the metal–metal bond length in $MM'(CO)_{10}$ has been made.[189] The bond length of the mixed metal carbonyl (MnRe) was found to be 2.909 Å, much closer to that of the Mn_2 carbonyl (2.9038 Å) than to that of the Re_2 (3.0413 Å). Photolysis of $Mn_2(CO)_{10}$ is known to generate $Mn(CO)_5$ in matrix isolation, and has now been shown[190] to be formed during the equivalent gas-phase process. By the use of time-resolved i.r. spectroscopy it was demonstrated that the decay back to $Mn_2(CO)_{10}$ follows second-order kinetics with a value close to that expected for a gas-phase radical recombination process. Photolysis in the presence of ligands[191] gives a mixture of $Mn(CO)_5^-$ and $Mn(CO)_3L_3^+$, where L is one of a range of ligands. A 19-e intermediate is proposed for this disproportionation process, and depending on the nature of the ligand it is also possible to isolate a number of other substituted products. The nature of the product seems to be controlled by the ability of the incoming ligand to favour the formation of the intermediate through such factors as steric bulk and electron-donating power.

The complex ion $[Re_3(CO)_{10}H_3]^{2-}$ contains a range of carbonyl binding sites, bridging hydrides and metal–metal bonds. Even with this range of bonding types

[186] L. B. Anderson, T. J. Barder, F. A. Cotton, K. R. Dunbar, L. R. Falvello, and R. A. Walton, *Inorg. Chem.*, 1986, **25**, 3629.

[187] S. Onaka, T. Sugawara, Y. Kawada, Y. Yokoyama, and H. Iwamura, *Bull. Chem. Soc. Jpn.*, 1986, **59**, 3079.

[188] D. J. Kuchynka, C. Amatore, and J. K. Kochi, *Inorg. Chem.*, 1986, **25**, 4087.

[189] A. L. Rheingold, W. K. Meckstroth, and D. P. Ridge, *Inorg. Chem.*, 1986, **25**, 3706.

[190] T. A. Seder, S. P. Church, and E. Weitz, *J. Am. Chem. Soc.*, 1986, **108**, 1084.

[191] A. E. Stiegman, A. S. Goldman, C. E. Philbin, and D. R. Tyler, *Inorg. Chem.*, 1986, **25**, 2976.

it proved[192] possible to identify four fluxional processes by the use of a pulse sequence in conjunction with magnetization-transfer ^{13}C n.m.r. measurements. The processes identified include carbonyl scrambling and the rotation of a Re≡Re fragment.

N.m.r. ^1H spectra have been measured[193] in toluene for the paramagnetic complexes $(\eta^5\text{-}C_5H_4R)_2Mn$, where R = H, Me, or Et. All of the complexes showed evidence of rapid spin-exchange averaging at above ambient temperatures, but the R = H case showed two sets of resonances at low and high field positions at low temperatures.

The structure of the product of the reaction between $[Cp(CO)_2Re≡CPh]BBr_4$ and $(\mu\text{-}PhS)(\mu\text{-}LiS)Fe_2(CO)_6$ has been established (41). The Re=C bond length was found[194] to be 2.02 Å.

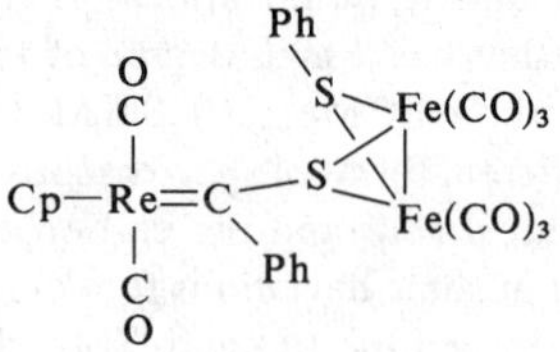

(41)

[192] T. Beringhelli, G. D'Alfonso, H. Molinari, B. E. Mann, B. T. Pickup, and C. M. Spencer, *J. Chem. Soc., Chem. Commun.*, 1986, 796.
[193] D. Cozak, F. Gauvin, and J. Demers, *Can. J. Chem.*, 1986, **64**, 71.
[194] C. Jia-Bi, L. Gui-Xin, Z. Ze-Ying, and T. You-Qi, *Acta Chim. Sinica*, 1986, **44**, 880.

9 Fe, Co, Ni

By B. W. FITZSIMMONS

Department of Chemistry, Birkbeck College, University of London, Malet Street, London, WC1E 7HX

1 General

The experimental excitation energies of the d-d bands for the ions V—Co inclusive have been analysed on the basis of the quantum mechanical virial theorem and the Hellman-Feynman theorem. The energy sequence is determined by the relative value of the electron-nuclear attraction energies in contrast to the Slater-Condon-Shortley picture in which d-d repulsion energy is the main feature.[1] Electronegativity values for the transitional elements have been deduced from bond energies and internuclear distances.[2]

2 Iron

Iron Atoms and Ions.—Condensation of iron atoms with C_6 aromatics or cyclodienes at 12—14 K leads to the formation of $[Fe(C_6H_6)]$, $[Fe(C_6H_6)_2]$, and $[Fe_2(C_6H_6)]$ complexes. 4-Cyclohexa-1,4-dienes yield $[Fe(C_6H_8)]$ and $[Fe_2(C_6H_8)]$ adducts whilst photolysis gives FeH_2 and benzene.[3] The ions Fe_2^+, $CoFe^+$, or Co_2^+ react with ethylene oxide by sequentially abstracting two oxygen atoms but $FeCo_2^+$ or Co_3^+ abstract up to three oxygens. In similar reactions $[Fe_2(CO)_4]^+$, $[CoFe(CO)_3]^+$, or $[Co_2(CO)_3]^+$ undergo carbon monoxide displacement forming $[MM^IO_2]^+$ exclusively. In reactions utilizing dioxygen as the source of oxygen atoms, $[FeCo_2(CO)_5]^+$ or $[Co_3(CO)_6]^+$ react with displacement of up to four CO groups and eventually yield $[MM^I_2O_4]^+$ and $[M^IM_2O_3]^+$. All relevant bond dissociation energies were determined in this FTMS study.[4] In their reactions with H_2, $Fe^+(^4F)$ reacts some 80 times faster than $Fe^+(^6F)$ despite the closeness in energy (0.25 eV) of these two states.[5]

The use of a crossed-beam study has proved effective in demonstrating the decarbonylation of acetaldehyde by either Fe^+ or Cr^+.[6] In gaseous-phase reactions of $FeOH^+$ or $COOH^+$ with alkanes, the cobalt species is the more reactive in processes which involve insertion followed by elimination of water leading to an activated Co^+-alkyl which decomposes further. The reactivity of $[MOH]^+$ is intermediate between that of MH^+ and MCH_3^+.[7] Gas-phase $ClFe^+$ is unreactive towards

[1] L. G. Vanquickenborne, P. Hoet, and K. Pierloot, *Inorg. Chem.*, 1986, **25**, 4228.
[2] R. T. Sanderson, *Inorg. Chem.*, 1986, **25**, 3518.
[3] D. W. Ball, Z. H. Kafafi, R. H. Hauge, and J. L. Margrave, *J. Am. Chem. Soc.*, 1986, **108**, 6621.
[4] D. B. Jacobson and B. S. Freiser, *J. Am. Chem. Soc.*, 1986, **108**, 27.
[5] J. L. Elkind and P. B. Armentrout, *J. Am. Chem. Soc.*, 1986, **108**, 2765.
[6] D. M. Sonnenfroh and J. M. Farrer, *J. Am. Chem. Soc.*, 1986, **108**, 3521.
[7] C. J. Cassady and B. S. Freiser, *J. Am. Chem. Soc.*, 1986, **108**, 5690.

alkanes.[8] Photodissociation of MCH_2^+ (M = Fe or Co) yields M^+, MC^+, and MCH^+ and from the photoappearance thresholds a number of bond energies were assigned.[9]

Iron Oxides and Compounds Containing Iron–Oxygen Bonds.—The oxygen content of each sample of the oxygen-deficient perovskite phase $[SrFe_{1-x}Sn_xO_{3-y}]$ ($0 \leqslant x \leqslant 0.7$) depends on its thermal history. Magnetic ordering in the ^{57}Fe Mössbauer spectrum sets in at low temperatures.[10] The intercalation compounds formed by contacting FeOCl with tetrathiafulvalene, tetramethyltetrahydrofuran, etc. have been studied using EXAFS and powder X-ray diffraction techniques. A structural model in which the intercalent is parallel to the interlayer axis[11] and neutron diffraction studies are consistent with a positive contribution by the tetrathiafulvalene to the diffraction pattern indicating long-range order.[12] Organotins interact with FeOCl by redox processes.[13] Iron(III) ion-exchanged A and B type crystalline zeolites give two mean signals in their e.p.r. spectra at $g = 2$ and $g \approx 4.3$ due, respectively, to ions *within* the zeolitic framework and ions *replacing* iron(III) in the zeolitic framework.[14] Open-framework iron(III) phosphates $[Fe_5P_4O_{20}H_{10}]$ and $[NaFe_3P_3O_{12}]$, analogues of minerals such as hureaulite or alluadite, are products of reaction of $Fe(CO)_5$ with phosphoric acid.[15]

The optical spectrum of the aqua ion $[Fe(H_2O)_6]^{3+}$ has been calculated with satisfactory accuracy using INDO/S methods.[16] Aerial oxidation of aqueous iron(III) hydroxide suspensions at pH 9 yields $Fe(OH)_2$ and Fe_3O_4 with γ-FeOOH as an intermediate. The $Fe^{II}Fe^{III}$ mol ratio in 'green rust' is $2:1$.[17] Oxygen transfer from percarboxylic acids to $[(edta)Fe^{III}]$ has been studied using 2,4,6-tri-butylphenol as a trapping agent for the iron-oxo intermediate.[18] The hydrolysis of iron(III) has been investigated at 25 °C in 1.0 mol dm^{-3} $NaClO_4$, KNO_3, or KCl and the product distribution analysed for each medium.[19] The kinetics of reduction of iron(III) in pyridine-2-carboxylate solutions are consistent with a single redox-active species $[Fe(pic)OH]$ but oxidation of iron(II) in picolinate solutions by $[Co(pic)_3]$ or $[Co(ox)_3]^{3-}$ (ox = oxalate) involve $[Fe(pic)_2]$.[20] Spin-labelled edta complexes of a range of transition metal ions including Fe^{III} and Ni^{II} have been studied in solution using e.p.r. Resolved spin–spin splitting was observed for the iron complex.[21] Rates and thermodynamic quantities for acetic acid exchange on iron(II), cobalt(II), nickel(II), or copper(II) ions have been determined in acetic acid or dichloromethane as solvent using ^{17}O n.m.r. methods.[22]

[8] M. L. Mandich, M. L. Steigerwald, and W. D. Reents, *J. Am. Chem. Soc.*, 1986, **108**, 6197.

[9] R. L. Hettich and B. S. Freiser, *J. Am. Chem. Soc.*, 1986, **108**, 2537.

[10] T. C. Gibb, *J. Chem. Soc., Dalton Trans.*, 1986, 1447.

[11] S. M. Kauzlarich, B. K. Teo, and B. A. Averill, *Inorg. Chem.*, 1986, **25**, 1209.

[12] S. M. Kauzlarich, J. L. Stanton, J. Faber, and B. A. Averill, *J. Am. Chem. Soc.*, 1986, **108**, 7946.

[13] J. E. Phillips and R. H. Herber, *Inorg. Chem.*, 1986, **25**, 3081.

[14] N. P. Evmiridis, *Inorg. Chem.*, 1986, **25**, 4362.

[15] D. R. Corbin, J. F. Whiteley, W. C. Fultz, G. D. Stucky, M. M. Eddy, and A. K. Cheetham, *Inorg. Chem.*, 1986, **25**, 2279.

[16] W. P. Anderson, W. D. Edwards, and M. C. Zerner, *Inorg. Chem.*, 1986, **25**, 2728.

[17] T. Kanzaki and T. Katsura, *J. Chem. Soc., Dalton Trans.*, 1986, 1243.

[18] P. N. Balasubramanian and T. C. Bruice, *J. Am. Chem. Soc.*, 1986, **108**, 5495.

[19] G. H. Khoe, P. L. Brown, R. N. Sylva, and R. G. Robins, *J. Chem. Soc., Dalton Trans.*, 1986, 1901.

[20] A. M. Lannon, A. G. Lappin, and M. G. Segal, *J. Chem. Soc., Dalton Trans.*, 1986, 619.

[21] K. M. More, G. R. Eaton, and S. S. Eaton, *Inorg. Chem.*, 1986, **25**, 2638.

[22] A. Hioki, S. Funahashi, M. Ishii, and M. Tanaka, *Inorg. Chem.*, 1986, **25**, 1360.

An unsymmetrical μ-oxo-di-iron(III) complex $[N_5Fe-O-FeCl_3]^+$ $\{N_5 =$ N,N,N'-tris-[(2-benzimidazoyl)methyl]-N'-(2-hydroxyethyl)-1,2-diaminoethane$\}$ has quite different magnetic properties from symmetric examples.[23] The crystal structure of the μ-oxo-compound $[Fe(Cp)_2][FeCl_3-O-FeCl_3]$, a product of reaction of ferrocene with iron(III) chloride, has been determined and the magnetic susceptibility and Mössbauer spectra have been analysed.[24] The complexes $[Fe_2L(OAc)_2]^-$ [L = N,N'-(2-hydroxy-5-methyl-1,3-xylene)bis(N-carboxymethylgylcine)] are the precursors of peroxo-complexes which are active catalysts in the decomposition of hydrogen peroxide.[25] The trinuclear complexes $[M_3O(Me_3N.BH_2CO_2)_6L_3]X$ (M = Cr^{III} or Fe^{III}; L = H_2O or MeOH; X = NO_3 or Cl) are true analogues of the better known carboxylato-complexes.[26] Solid-state ^2H n.m.r. has been applied to the μ-oxo-bridged $Fe^{II}Fe^{III}$ compound $[Fe_3O(O_2CMe)_6(4-MePy)_3]C_6H_6$ with a view to seeing whether the onset of rotational disorder of the benzene coincides with the onset of electron hopping. The authors put forward convincing arguments in favour of this.[27] It was heat-capacity measurements which originally enabled the detection of phase transitions taking place in these compounds at the temperature at which the Mössbauer spectra began to look peculiar. For $[Fe_3O(O_2CMe)_6Py_3]Py$, the phase transitions are at *ca.* 112 and 186 K, the lower one being critical in the Mössbauer context, and the authors think that the solvent pyridines begin to rotate at that temperature and that this induces electron hopping.[28]

EXAFS spectra of dinuclear μ-oxo- and μ-hydroxo-iron(III) complexes are highly sensitive to the presence or absence of a short μ-oxo Fe$-$O distance. This may be of value in assessing the stereochemistry of semimet forms of haemerythrin and related compounds.[29] There are three stages in the two-equivalent reduction of the dinuclear $Fe^{III}Fe^{III}$ μ-hydroxo-bridged active site of *Themiste zostericola* metmyo haemerythrin.[30] Proton n.m.r. spectroscopy has been used to investigate the three redox states of haemerythrin as well as the mechanism employed by this enzyme.[31] Resonance Raman studies of hydroxyhaemertythrin are consistent with hydrogen bonding of groups to the $[Fe-O-Fe]$ centre.[32]

Why is enterobactin the most efficient neutral binder of iron(III)? Analysis of conformational changes accompanying the association provides some answers.[33] Both iron(IV) and iron(V) have been detected in the pulse radiolysis of alkaline ferric and ferrate(VI) solutions. The hydroxyl radical HO$^{\cdot}$ or its conjugate base O$^{\cdot-}$ and the aquated electron e$^-_{aq}$ are oxidizer and reducer respectively.[34] EXAFS studies

[23] D. J. Jones, J. Roziere, and M. S. Lehmann, *J. Chem. Soc., Dalton Trans.*, 1986, 651.
[24] G. J. Bullen, B. J. Howlin, J. Silver, B. W. Fitzsimmons, I. Sayer, and L. F. Larkworthy, *J. Chem. Soc., Dalton Trans.*, 1986, 1937.
[25] B. P. Murch, F. C. Bradley, and L. Que, *J. Am. Chem. Soc.*, 1986, **108**, 5027.
[26] V. M. Norwood and K. W. Morse, *Inorg. Chem.*, 1986, **25**, 3690.
[27] S. E. Woeler, R. J. Wittebort, S. M. Oh, D. N. Hendrickson, D. Inniss, and C. E. Strouse, *J. Am. Chem. Soc.*, 1986, **108**, 2938.
[28] M. Sorai, K. Kaji, D. N. Hendrickson, and S. M. Oh, *J. Am. Chem. Soc.*, 1986, **108**, 702.
[29] B. Hedman, M. S. Co, W. H. Armstrong, K. O. Hodgson, and S. J. Lippard, *Inorg. Chem.*, 1986, **25**, 3708.
[30] G. D. Armstrong and A. G. Sykes, *Inorg. Chem.*, 1986, **25**, 3725.
[31] M. J. Maroney, D. M. Kurtz, J. M. Nocek, L. L. Pearce, and L. Que, *J. Am. Chem. Soc.*, 1986, **108**, 6871.
[32] A. K. Shiemke, T. M. Loehr, and J. Sanders-Loehr, *J. Am. Chem. Soc.*, 1986, **108**, 2437.
[33] A. Shanzer, J. Libman, S. Lifson, and C. E. Felder, *J. Am. Chem. Soc.*, 1986, **108**, 7609.
[34] J. D. Rush and B. H. J. Bielski, *J. Am. Chem. Soc.*, 1986, **108**, 523.

on horseradish peroxidase-(I) and -(II) are consistent with the ferryl formulation (Fe=O) with a bond distance of some 1.6 Å.[35]

Compounds Containing Iron–Nitrogen Bonds.—As in previous years' reports, the porphyrinato complexes have been allocated a separate section. The six-co-ordinate diazene complexes $[FeH(ArN=NH)L_4]BPh_4$ and $[Fe(ArN=NH)_2L_4][BPh_4]_2$ (L = $P(OEt)_3$; Ar = 4-MeC_6H_4, 4-MeOC_6H_4, or 4-FC_6H_4) are products of the reaction of hydrido-FeH_2L_4 with aryldiazonium cations at -80 °C.[36] The molecular structure of the high-spin compound (1) is as shown. It is decomposed by dioxygen but reacts smoothly with carbon monoxide.[37] Heterometallic trinuclear species of the type $M^{II}Fe^{II}_2(RL)_6$ [M = Co, Ni, or Zn; HRL = RC(=NOH)N=NPh] have been prepared and the X-ray structure of one example ($Fe^{II}Ni^{II}Fe^{II}R$ = Me) has been determined. It involves FeN_6 and NiO_6 chromophores. Alkaline earth or lanthanon complexes may also be synthesized.[38]

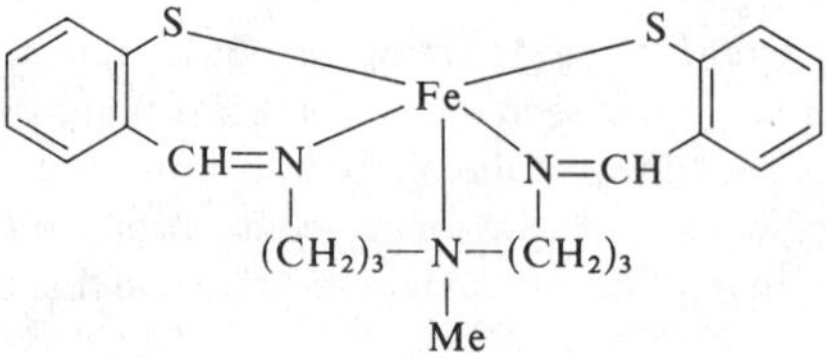

(1)

The structure of the *mer*-FeN_4O_2 iron(III) bis-complex of the Schiff base complex (2) has been determined so as to provide an example of a mononuclear complex containing both phenolate and imidazole ligands.[39] The octahedral complexes $[Fe(Pyz)_2X_2]$ and $[Fe(Pyz)Cl_2]$ (Pyz = pyrazine; X = Cl, Br, or I) are six-co-ordinate *trans*-FeX_2N_4 and *trans*-$FeN_2(\mu\text{-Cl}_2)$ polymers respectively. Both lack magnetic exchange interactions.[40] The dioxygen affinity of the lacunar iron(II) complex (3) has been analysed using a multiple equilibria model.[41] Kinetic results for axial ligand dissociation from low-spin *trans*-$[FeN_4XY]$ complexes $[N_4 = $ bis(benzoquinone dioxime) or bis(naphthoquinone dioxime); X, Y = methylimidazole, tributylphosphine, tributyl phosphite, benzyl isocyanide, or carbon monoxide] have been reported[42] and equilibria between H^+, Mn^{II}, Fe^{II}, Ni^{II}, Cu^{II}, or Zn^{II} and the polyamine (4) have been investigated by spectrophotometric and potentiometric methods and the co-ordination modes of this ligand elucidated.[43] The rates of axial ligand substitution have been determined for a wide range of iron(II) dimethylglyoximato-bis(ligand) (ligand = imidazole, R_3P, *etc.*) com-

[35] J. E. Penner-Hahn, K. Smith Eble, T. J. McMurry, M. Renner, A. L. Balch, J. T. Groves, J. H. Dawson, and K. O. Hodgson, *J. Am. Chem. Soc.*, 1986, **108**, 7819.
[36] G. Albertin, S. Antoniutti, G. Pelizzi, F. Vitali, and E. Bordigon, *J. Am. Chem. Soc.*, 1986, **108**, 6627.
[37] G. D. Fallon, P. J. Nichols, and B. O. West, *J. Chem. Soc., Dalton Trans.*, 1986, 2271.
[38] S. Pal, R. Mukherjee, M. Tomas, L. R. Falvello, and A. Chakravorty, *Inorg. Chem.*, 1986, **25**, 200.
[39] J. C. Davis, W. Kung, and B. A. Averill, *Inorg. Chem.*, 1986, **25**, 394.
[40] J. S. Haynes, J. R. Sams, and R. C. Thompson, *Inorg. Chem.*, 1986, **25**, 3740.
[41] K. A. Goldsby, B. D. Beato, and D. H. Busch, *Inorg. Chem.*, 1986, **25**, 2342.
[42] N. Siddiqui and D. V. Stynes, *Inorg. Chem.*, 1986, **25**, 1942.
[43] E. Garcia-Espana, M. Micheloni, P. Paoletti, and A. Bianchi, *Inorg. Chem.*, 1986, **25**, 1435.

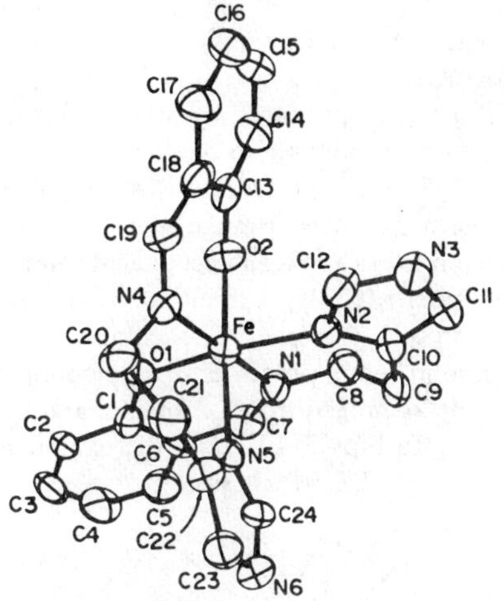

(2)

ORTEP diagram of the $[Fe(salhis)_2]^+$ *cation* (50% *probability ellipsoids*), *showing the atomic numbering scheme. Hydrogen atoms have been omitted for clarity*
(Reproduced by permission from *Inorg. Chem.*, 1986, **25**, 394)

plexes.[44] The mode of thiocyanate bonding in four low-spin iron(II) $[FeL(SCN)_2]^{2+}$ species (L = 2,3,9,10-tetramethyl-1,4,8,11-tetra-azacyclotetradeca-1,3,8,10-tetraene) has been studied. Whereas *S*-bonding obtains in these derivatives it is *N*-bonded in a monothiocyanato analogue.[45] A new hexadentate ligand (5) yields a series of

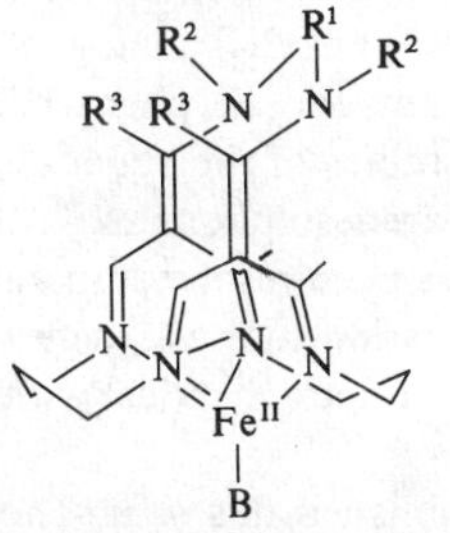

R^1, R^2, R^3 = alkyl or aryl

(3)

$$[H_2N(CH_2)_m]_2N-(CH_2)_n-N[(CH_2)_mNH_2]_2$$

$n = 2; m = 2$
$n = 2; m = 3$
$n = 3; m = 2$

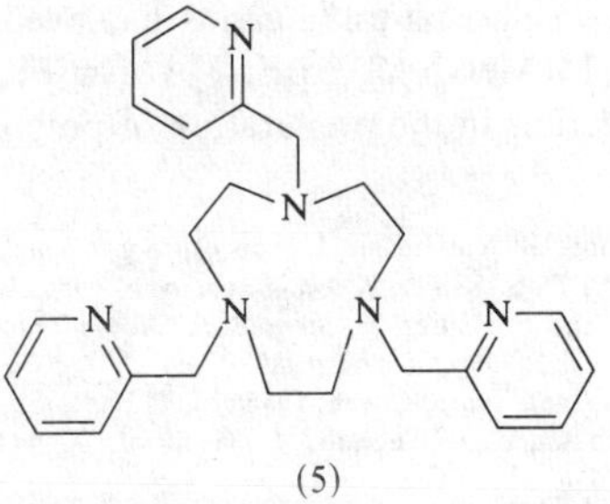

(5)

[44] X. Chen and D. V. Stynes, *Inorg. Chem.*, 1986, **25**, 1173.
[45] M. J. Marony, E. O. Fey, D. A. Baldwin, R. E. Stenkamp, L. H. Jensen, and N. J. Rose, *Inorg. Chem.*, 1986, **25**, 1409.

compounds $[MCl_2X(H_2O)]$ ($M = Mn^{II}$, Fe^{II}, Co^{II}, Ni^{II}, or Cu^{II}). Cobalt(III) and iron(III) complexes were obtained by oxidation. The metal ions adopt trigonally distorted $[MN_6]$ structures.[46]

Although (Δ)-$[Fe(bipy)_3]^{2+*}$ produced by means of laser photolysis has a short lifetime ($<10^{-9}$ s) it has been possible to record its c.d. spectrum and show that it is a $d-d$ excited state.[47] The rate of hydrolysis of $[Fe(phen)_3]^{2+}$ in dioxane–water mixtures has been determined over a range of compositions. Dissociation of the first ligand is a two-step process and is subject to acid catalysis.[48] Although solutions of the complex $[Fe(phen)_2(CN)_2]$ exhibit a bewildering variety of colours, only one species is present therein.[49] The compounds $[Fe(phen)_2(CN)_2]$ and $[\eta\text{-}CpFe(CO)_2CN]$ exhibit redox processes in acidic molten salts such as $AlCl_3$-n-butylpyridinium chloride as judged from i.r. spectra and cyclic voltammograms.[50] Complexes of the type $[Ru(bipy)FeL]$ [L = (6)] exhibit intermolecular $Ru \rightarrow Fe$ energy transfer after nitrogen laser excitation.[51]

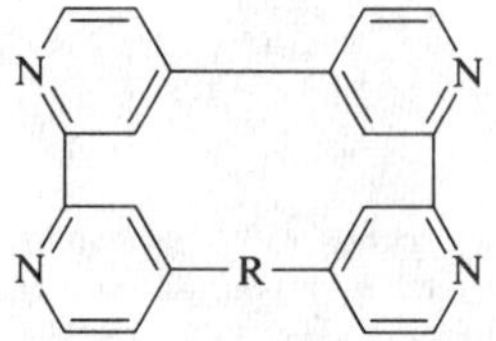

$R = (CH_2)_2$, $(CH_2)_5$, p-$(CH_2)_2(C_6H_4)(CH_2)_2$, or $(CH_2)_{12}$

(6)

There are only small micellar effects on the kinetics of aquation and base hydrolysis of $[Fe(phen)_3]^{2+}$ or $[Co(NH_3)_5Cl]^+$.[52] Orange $[Fe(bipy)Cl_2]$ is a chain of five-co-ordinate iron(II) centres with single chloro bridges. Two other $[FeLCl_2]$ analogues of this were included in this study.[53] Ion-pair formation constants for $[Fe(phen)_3]^{2+}$ and arenesulphonate ions have been obtained from kinetic studies of the aquation of the complex ion in aqueous solutions containing sodium arenesulphonates.[54]

Tris(bathophenanthrolinesulphanato)iron(II), $[FeL_3]^{2+}$, has been resolved on an optically active chromatographic column; this ligand dissociation rate was determined and compared with the rate of racemization. The two rates are identical with the same activation parameters.[55]

Relatively little work on phosphorus ligands has been published this year. The acetonitrile derivatives $[M(MeCN)_6]^{2+}(BF_4^-)_2$ ($M = $ Fe, Co, or Ni) have proved to be useful starting materials in the preparation of polyphosphine complexes such

[46] K. Wieghardt, E. Schoffmann, B. Nuber, and J. Weiss, *Inorg. Chem.*, 1986, **25**, 4877.
[47] S. J. Milder, J. S. Gold, and D. S. Kliger, *J. Am. Chem. Soc.*, 1986, **108**, 8295.
[48] M. Cazanga, J. G. Santos, and F. Ibanez, *J. Chem. Soc., Dalton Trans.*, 1986, 465.
[49] A. G. Maddock, *J. Chem. Soc., Dalton Trans.*, 1986, 2349.
[50] C. Woodcock and D. F. Shriver, *Inorg. Chem.*, 1986, **25**, 2137.
[51] R. H. Schmehl, R. A. Auerbach, W. F. Wacholtz, C. M. Elliott, R. A. Freitag, and J. W. Merkert, *Inorg. Chem.*, 1986, **25**, 2440.
[52] S. Tachiyashiki and H. Yamatera, *Inorg. Chem.*, 1986, **25**, 3043.
[53] F. F. Charron and W. M. Reiff, *Inorg. Chem.*, 1986, **25**, 2786.
[54] S. Tachiyashiki and H. Yamatera, *Inorg. Chem.*, 1986, **25**, 3209.
[55] A. Yamagishi, *Inorg. Chem.*, 1986, **25**, 55.

as $[FeL(MeCN)_2]^{2+}(BF_4^-)_2$ $[L = P(CH_2CH_2PPh_2)_2]$.[56] E.p.r. and i.r. spectra have been used in a characterization of five-co-ordinate bis(nitrosyl) cation radicals $[Fe(NO)_2L_3]^+$ $(L = e.g.$ $PPh_3)$.[57] The formally iron(IV) hydrido-complex $[FeH_4(PEtPh_2)_3]$ is better formulated as *cis*-$[H_2Fe(H_2)(PEtPh_2)_3]$, a six-co-ordinate iron(II) species with a non-classical H_2 occupying one co-ordination site.[58] A theoretical paper seems to be germane to this topic. Extended Hückel calculations have been applied to the interaction of d^6 ML_5 fragments with H_2 to form either a η^2-co-ordinated species or a dihydrido-complex. A *trans*-CO group favours the η^2-structure with respect to the case of a pure σ-donor.[59]

Normally, insertion of iron into sp^2 C—H bonds takes place only if the alkene is activated but the dihydro complex $[FeH_2L_2]$ $[L = 1,2$-bis(dimethylphos-phinoethane)] photolyses in benzene to yield a η^1-C_6H_5 product.[60]

Application of iron K-edge EXAFS to the relatively unstable $[Fe\{o$-$C_6H_4(PMe_2)_2\}Cl_4]$ $[BF_4]_n$ $(n = 0$—$2)$ not only enabled the structures to be established but also showed that the trends in bond parameters correlated well with the metal's formal oxidation state.[61] Iron(IV) complexes $[Fe(L-L)_2X_2]BF_4$ $[L-L = o$-$C_6H_4(PMe_2)_2$, o-$C_6H_4(PMe_2)(AsMe_2)$, or o-$C_6H_4(AsMe_2)_2$, X = Cl or Br; L-L = $Me_2PCH_2CH_2PMe_2$, X = Cl) have been isolated and the corresponding complexes $[Fe\{o$-$C_6H_4(PMe_2)_2\}_2X_2]^{2+}$ $(X = Cl$ or Br) and $[Fe(Me_2PCH_2CH_2PMe_2)_2Br_2]^{2+}$ have been obtained in solution. Iron K-edge EXAFS results for some of these provide the iron–ligand bond lengths.[62] The cationic nitrosyl allyls $[FeL(NO)_2(\eta$-allyl)]PF_6$ $[L = P(OPh)_3, P(OMe)_3$, or $PPh_3]$ are products of reaction of $[NO][PF_6]$ with $[Fe(CO)L(NO)$ η-allyl].[63]

Spin-crossover among Iron Compounds.—Although the spin isomers can be detected by appropriate experimental techniques, it has been shown that strong coupling between the two electronic states eliminates the individual potential energy wells leaving a single potential with some anharmonicity.[64] Germane to this theoretical work is the finding that octahedral $Fe^{III}N_4O_2$ $[FeL_2]PF_6$ $[HL = N$-(1-actyl-2-propyl-idine)-2-(pyridylmethyl)amine] exhibits fast electronic ' relaxation between $S = 1/2$ and $S = 5/2$. The spin interconversions are accompanied by continuous atomic displacements within the same lattice.[65]

Illumination of complexes prone to spin-crossover causes them to go over to the high-spin form, the change often being detected with the aid of Mössbauer spectroscopy and/or magnetic susceptibility measurements. Variable-temperature Fourier-transform i.r. spectroscopy has now been successfully applied to the classic example, $[Fe(phen)_2(SCN)_2]$.[66] A single-ion treatment of both spin equilibrium and relaxation

[56] D. G. Evans, *Inorg. Chem.*, 1986, **25**, 4602.
[57] D. Ballivet-Tkatchenko, B. Nickel, A. Rasset, and J. Vincent-Vaucquelin, *Inorg. Chem.*, 1986, **25**, 3497.
[58] R. H. Crabtree and D. G. Hamilton, *J. Am. Chem. Soc.*, 1986, **108**, 3124.
[59] Y. Jean, O. Eisenstein, F. Volatron, B. Maouche, and F. Sefta, *J. Am. Chem. Soc.*, 1986, **108**, 6547.
[60] M. V. Becker and L. D. Field, *J. Am. Chem. Soc.*, 1986, **108**, 7433.
[61] S. K. Harbron, S. J. Higgins, W. Levason, C. D. Garner, A. T. Steel, M. C. Feiters, and S. S. Hasnain, *J. Am. Chem. Soc.*, 1986, **108**, 526.
[62] S. K. Harbron, S. J. Higgins, W. Leavson, M. C. Feiters, and A. T. Steel, *Inorg. Chem.*, 1986, **25**, 1769.
[63] P. K. Baker, S. Clamp, N. G. Connelly, M. Murray, and J. B. Sheridan, *J. Chem. Soc., Dalton Trans.*, 1986, 459.
[64] M. Bacci, *Inorg. Chem.*, 1986, **25**, 2322.
[65] Y. Maeda, H. Oshio, Y. Takashima, M. Mikuriya, and M. Hidaka, *Inrog. Chem.*, 1986, **25**, 2948.
[66] R. Herber and L. M. Casson, *Inorg. Chem.*, 1986, **25**, 847.

has been applied to light-induced spin-state trapping in the case of $[Zn_{1-x}Fe_x(ptz)_6]^{2+}$, $(x \leqslant 0.1)$, ptz = 1-propyltetra-azole.[67]

The partial molar volumes of a range of spin-crossover iron(II), iron(III), and cobalt(II) complexes have been measured. The differences in the ionic radii were calculated and a minimum value was obtained for the change in the metal–ligand bond lengths between the spin isomers.[68] A variable-temperature single-crystal X-ray diffraction analysis has been carried out on the spin-crossover complex [Fe(2-pic)$_3$]Br$_2$.EtOH [2-pic = 2-(aminomethyl)pyridine]. Although there is a hysteresis in the fraction of the high-spin isomer, there is no structural phase transition.[69] In contrast, a first-order order–disorder transition associated with a change in space group sets in at 180 K in the spin-crossover benzene solvates [Fe(3-OEt-SalAPA)$_2$]ClO$_4$.C$_6$H$_6$, where 3-OEt-SalAPA is the Schiff base derived from 3-ethoxysalicylaldehyde and N-(3-aminopropyl)aziridine. The Mössbauer spectrum of this compound over the working temperature range is an unresolved doublet.[70,71] The iron(III) complexes [FeL$_2$(Salacen)]PF$_6$ [L = imidazole or N-methylimidazole; Salacen = ethylene(N-acetylacetonylideneiminate)] exhibit an $S = 1/2$, $S = 5/2$ spin-crossover. A stochastic model was developed and the conversion rates were calculated.[72]

Porphyrinato-iron Compounds.—The *trans*-complexation of nitrosylprotoporphyrin IX dimethylglyoxime iron(II) with various azolates has been studied using e.p.r. and electronic spectra and the results have been discussed in relation to the dissociation of the NH proton in proximal histidyl imidazole in haemoproteins.[73] Rate constants have been reported for the reduction by ascorbate of bromocyanogen-modified metmyoglobin reconstituted with 3,4-disubstituted deuterohaemin and with protohaemin dimethyl ester.[74] Equilibrium constants have been determined for the binding of phosphorus donor ligands [Bu$_3$P, (BuO)$_3$P] to methylimidazole complexes of iron protoporphyrin IX dimethyl ester and compared with similar results for ferrous phthalocyanin and ferrous dimethylglyoxime analogues.[75] An explanation has been put forward for the biological activity of sydnones in terms of their power to destroy cytochrome P-450, converting it into N-vinylprotoporphyrin IX.[76]

Although the lowest electronic state in iron(II) porphyrins is indeed $^3A_{2g}$, a $^3E_{2g}$ state is calculated to be some 240 cm^{-1} above it. The calculated electronic spectra arising from these states fit the experimental results satisfactorily.[77] Hydroporphyrin complexes [Fe(TPC)] and [Fe(TPiBC)] (TPC = 7,8-dihydro-5,10,15,20-tetra-phenylporphyrinate dianion; TPiBC = 2,3,7,8-tetrahydro-5,10,15,20-tetraphenyl-

[67] A. Hauser, P. Gutlich, and H. Spiering, *Inorg. Chem.*, 1986, **25**, 4245.
[68] R. A. Binstead and J. K. Beattie, *Inorg. Chem.*, 1986, **25**, 1481.
[69] L. Wiehl, G. Kiel, C. P. Kohler, H. Spiering, and P. Gutlich, *Inorg. Chem.*, 1986, **25**, 1565.
[70] M. D. Timken, C. E. Strouse, S. M. Soltis, S. A. Daverio, D. N. Hendrickson, A. M. Abdel-Mawgoud, and S. R. Wilson, *J. Am. Chem. Soc.*, 1986, **108**, 395.
[71] M. D. Timken, A. M. Abdel-Mawgoud, and D. N. Hendrickson, *Inorg. Chem.*, 1986, **25**, 160.
[72] Y. Maeda, Y. Takashima, N. Matsumoto, and A. Ohyoshi, *J. Chem. Soc., Dalton Trans.*, 1986, 1115.
[73] T. Yoshimura, *Inorg. Chem.*, 1986, **25**, 688.
[74] K. Tsukahara, T. Okazawa, H. Takahashi, and Y. Yamamoto, *Inorg. Chem.*, 1986, **25**, 4756.
[75] D. V. Stynes, D. Fletcher, and X. Chen, *Inorg. Chem.*, 1986, **25**, 3483.
[76] P. R. Ortiz de Montellano and L. A. Grab, *J. Am. Chem. Soc.*, 1986, **108**, 5584.
[77] W. D. Edwards, B. Weiner, and M. C. Zerner, *J. Am. Chem. Soc.*, 1986, **108**, 2196.

porphyrinate dianion) are magnetically different from [Fe(TPP)] despite the $S = 1$ ground state. They are rhombic rather than axial and show significant temperature-independent paramagnetism.[78] The iron(II) oxophlorin radical has been definitely characterized as being derivable from the iron(II) porphyrin bis(pyridine) complex plus hydrogen peroxide or from iron(III) *meso*-hydroxyoctaethylporphyrin.[79] A non-aggregating water-soluble tetraphenylporphyrin derivative proved useful in a study of the rate of decomposition of hydrogen peroxide and possible mechanisms for the general catalysis were advanced.[80] Both the free base and metal complexes of *mono*- and di-β-oxoporphyrins obtained by hydrogen peroxide oxidation of octaethylporphyrin have been prepared and the structure of one compound (7) has been determined.[81]   Iron–nitroxyl   electron–electron   spin–spin   splitting   in

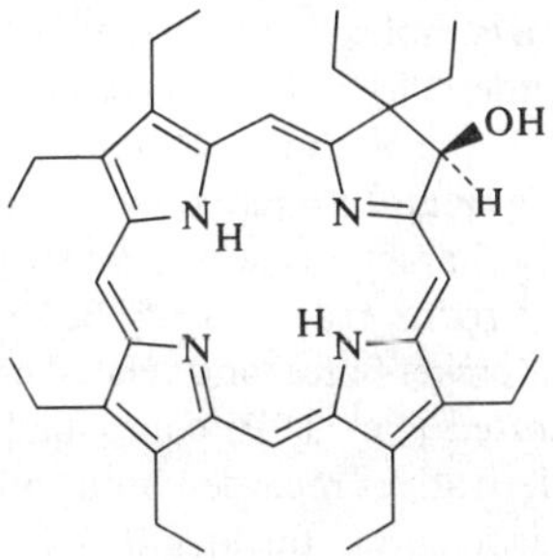

(7)

[(X)Fe(NO)] porphyrins (X = halogen) collapses as the temperature is raised, and the collapse temperature is in the order Br < Cl < F *i.e.* in the order of zero-field splitting.[82] The reaction of [$Fe^{III}(P)Cl$] or [$Fe^{III}P(NO_2)_n$] ($n = 1$ or 2; P = tetraphenylporphyrin, tetraphenylchlorin, or octaethylporphyrin) with nitrite leads to low-spin mono- and bis-nitrite complexes.[83] An iron bis-nitroporphyrin complex is implicated in the reduction of nitrite by [$Fe^{III}(P)Cl$] or [$Fe^{III}(P)(NO_3)$] (P = tetraphenylporphyrin, tetraphenylchlorin or octaethylporphyrin).[84] Rates of reaction of (tetraphenylporphinato)iron(III) azide with imidazole or *N*-methylimidazole have been determined in acetone or dichloromethane. A six-co-ordinate low-spin base adduct is an intermediate in these reactions in which azide ionization is rate limiting. Hydrogen bonding to the departing group is suggested.[85]

Iron(III) porphyrins suffer reduction to iron(II) bis-ligand species after addition of amines. This not only provides a useful synthetic route but also mimics cytochrome P-450 monoamine oxidase.[86] Iron(III) porphyrin autoreduction in the presence of

[78] S. H. Strauss and M. J. Pawlik, *Inorg. Chem.*, 1986, **25**, 1921.
[79] I. Morishima, H. Fujii, Y. Shiro, and S. Sano, *J. Am. Chem. Soc.*, 1986, **108**, 3858.
[80] M. F. Zipplies, W. A. Lee, and T. C. Bruice, *J. Am. Chem. Soc.*, 1986, **108**, 4433.
[81] A. M. Stolzenberg, P. A. Glazer, and B. M. Foxman, *Inorg. Chem.*, 1986, **25**, 983.
[82] L. Fielding, K. M. More, G. R. Eaton, and S. S. Eaton, *J. Am. Chem. Soc.*, 1986, **108**, 8194.
[83] L. A. Andersson, T. M. Loehr, C. Sotiriou, W. Wu, and C. K. Chang, *J. Am. Chem. Soc.*, 1986, **108**, 2908.
[84] J. B. Fernandes, D. Feng, A. Chang, A. Keyser, and M. D. Ryan, *Inorg. Chem.*, 1986, **25**, 2606.
[85] W. Byers, J. A. Cossham, J. O. Edwards, A. T. Gordon, J. G. Jones, E. T. P. Kenny, A. Mahmood, J. McKnight, D. A. Sweigart, G. A. Tondreau, and T. Wright, *Inorg. Chem.*, 1986, **25**, 4767.
[86] C. E. Castro, M. Jamin, W. Yokoyama, and R. Wade, *J. Am. Chem. Soc.*, 1986, **108**, 4179.

aliphatic amines may be inhibited by added cyanide which enables the spectral characteristics of Fe^{III} amine complexes to be established.[87]

The e.p.r. spectra of a series of seven side-chain nitroxyl spin-labelled low-spin iron(III) tetraphenylporphyrins require both electron–electron exchange and dipolar terms on the spin Hamiltonian for adequate simulation.[88] Electron nuclear double resonance has been effectively used to collect information on nitrogen and proton shifts in both natural and synthetic bis(imidazole)iron(III) porphyrins.[89] The strong g_{max} e.p.r. signals of low-spin iron(III) porphyrins can be correlated with perpendicularly aligned axial imidazole planes as based on a study of [(TPP)Fe(2-MeImH)$_2$]$^+$ (Im = imidazole) and related compounds.[90] An iron(III) porphyrin complex having a nitrene fragment inserted into an Fe—N bond is formed using alkene aziridination by [(tosylimido)iodo]benzene catalysed by iron(III) porphyrins.[91] Iron(II) porphyrin complexes such as $Fe^{II}P$ (P = TPP, TpClPP, etc.) react smoothly with iodonium ylide [PhI = X] to give Fe^{II}-bis-N-alkylporphyrin complexes.[92] An iron(III) porphyrin N-oxide [Fe^{III}(TMP)O] (TMP = 5,10,15,20-tetramesitylporphyrinato) prepared by m-chloroperoxobenzoic acid oxidation has been satisfactorily characterized.[93] There are large molecular environment effects due to side chains in redox reactions of [FeTPP] compounds as shown by an electrochemical study of 'picket fence' and related compounds.[94] Tetragonal and rhombic distortions of assorted [Fe^{III}TPP] compounds as detected through e.p.r. spectra of 'basket-handle' derivatives recorded both with and without added ligands are sufficiently different so as to permit the identification of the added axial ligand.[95] The association constants for nitrogenous bases such as pyridine or imidazole with synthetic iron(II) porphyrins having 'basket-handle' superstructures have been studied and the enhancement has been analysed.[96]

Hydroxide-ligated high-spin ($g_\perp = 5.9$, $g_\parallel = 2$) 5,10,15,20-tetrakis-(2,6-difluorophenyl)porphinatoiron(III) is a product of hydroxide displacement of chloride.[97] Oxygen affinity is an increasing function of iron(II)–base association according to an investigation based on electrochemistry of hanging-base 'basket-handle' iron porphyrins.[98] A similarity between the bonding properties of carbon monoxide and N-acyl isocyanides (RCONC) in ferrous porphyrin complexes has been demonstrated by use of i.r. and Mössbauer spectral data.[99]

Oxygen transfer from amine oxides, e.g. p-cyano-N,N-dimethylaniline N-oxide, is catalysed by meso-tetrakis(2,6-dimethylphenyl)porphinatoiron(III) chloride[100]

[87] Y. C. Hwang and D. W. Dixon, Inorg. Chem., 1986, 25, 3716.
[88] L. Fielding, K. M. More, G. R. Eaton, and S. S. Eaton, J. Am. Chem. Soc., 1986, 108, 618.
[89] C. P. Scholes, K. M. Falkowski, S. Chen, and J. Bank, J. Am. Chem. Soc., 1986, 108, 1660.
[90] F. A. Walker, B. H. Huynh, W. R. Scheidt, and S. R. Osvath, J. Am. Chem. Soc., 1986, 108, 5288.
[91] J. Mahy, P. Battioni, and D. Mansuy, J. Am. Chem. Soc., 1986, 108, 1079.
[92] J. Battioni, I. Artand, D. Dupre, P. Leduc, I. Akhrem, D. Mansuy, J. Fischer, R. Weiss, and I. Morgenstern-Bandarau, J. Am. Chem. Soc., 1986, 108, 5598.
[93] J. T. Groves and Y. Watanabe, J. Am. Chem. Soc., 1986, 108, 7836.
[94] C. Gueutin, D. Lexa, M. Momenteau, J. Saveant, and F. Xu, Inorg. Chem., 1986, 25, 4294.
[95] C. Schaeffer, M. Momenteau, J. Mispelter, B. Loock, C. Huel, and J. Lhoste, Inorg. Chem., 1986, 25, 4577.
[96] D. Lexa, M. Momenteau, J. M. Saveant, and F. Xu, J. Am. Chem. Soc., 1986, 108, 6937.
[97] T. C. Woon, A. Shirazi, and T. C. Bruice, Inorg. Chem., 1986, 25, 3845.
[98] D. Lexa, M. Momenteau, J. M. Saveant, and F. Xu, Inorg. Chem., 1986, 25, 4857.
[99] M. Le Plouzennec, A. Bondon, P. Sodano, and G. Simonneaux, Inorg. Chem., 1986, 25, 1254.
[100] T. C. Woon, C. M. Dicken, and T. C. Bruice, J. Am. Chem. Soc., 1986, 108, 7990.

and by the dichloro analogue.[101] Application of extended Hückel theory to this reaction indicates that the [Fe^{IV}—O] centres have radical character and are really RO⁻ radicals. Spin polarization is important in determining reactivity.[102]

The reaction of $Na_2Fe(CO)_4$ with [Porph(E)Cl₂] (E = Sn or Ge; Porph = tetra-arylporphyrin or octaethylporphyrin) yields [Porph(E)Fe(CO)₄].[103] ¹H n.m.r. spectra of low-spin iron(III) tetra-arylporphyrin complexes with axial η^1-aryl substituents have been recorded. Addition of base yields six-co-ordinate complexes. The spectra are similar to published spectra for the corresponding protein complexes derived from the arylhydrazine dioxygen reaction.[104]

meso-Substituents influence the exchange-coupled interaction in μ-oxo-iron(III) porphyrin dimers[105] and both the μ-oxo-dimer and the hydroxo-complex of tetrakis(pentafluorophenyl)porphinatoiron(III) have now been satisfactorily characterized.[106] The heterodinuclear complex RuCl₂Fe(Cl)L (8) involves six-co-ordinate [RuCl₂(N)₄] units. The redox properties of the system are different from those of the simple [Fe(TPP)] compounds.[107] The reactions of L_{aq}^- and CO_2^- with the monomers [Fe(TPPS)(H₂O)]³⁻ and [Fe(TMPyP)(OH)]⁴⁺ and with the dimers

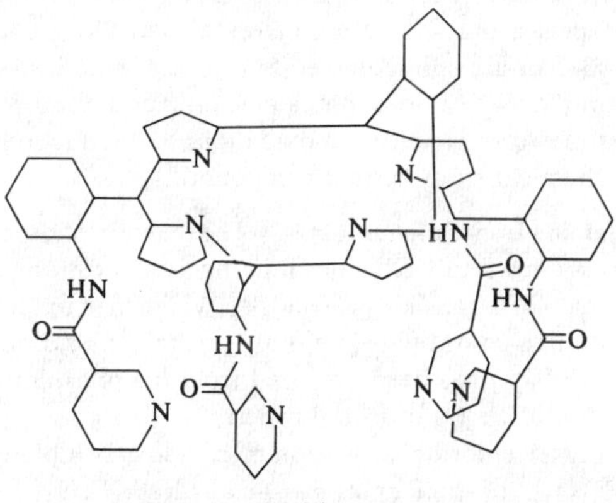

(8)

[(TPPS)Fe—O—Fe(TPPS)]⁸⁻ have been carried out and a product analysis has been attempted [H_2TPPS^{4-} = tetrakis-4-(sulphonatophenyl)porphine; H_2TMPyP^{4+} = tetrakis(N-methylpyrid-4-yl)porphine].[108] The absorption spectra of n-butanethiolate complexes of iron porphyrins and their carbon monoxide complexes have been recorded for comparison with those of haem-substituted cytochrome P-450.[109] A high-spin S = 5/2 antiferromagnetic ($J = -8.5$ cm⁻¹) porphyrin

101 C. M. Dicken, T. C. Woon, and T. C. Bruice, *J. Am. Chem. Soc.*, 1986, **108**, 1636.
102 A. Sevin and M. Fontecave, *J. Am. Chem. Soc.*, 1986, **108**, 3266.
103 K. M. Kadish, C. Swistak, B. Boisselier-Cocolios, J. M. Barbe, and R. Guilard, *Inorg. Chem.*, 1986, **25**, 4336.
104 A. L. Balch and M. W. Renner, *Inorg. Chem.*, 1986, **25**, 303.
105 J. H. Helms, L. W. ter Haar, W. E. Hatfield, D. C. Harris, K. Jayaraj, G. E. Toney, A. Gold, T. D. Mewborn, and J. R. Pemberton, *Inorg. Chem.*, 1986, **25**, 2334.
106 K. Jayaraj, A. Gold, G. E. Toney, J. H. Helms, and W. E. Hatfield, *Inorg. Chem.*, 1986, **25**, 3516.
107 C. M. Elliott and J. K. Arnette, *Inorg. Chem.*, 1986, **25**, 2867.
108 P. C. Wilkins and R. G. Wilkins, *Inorg. Chem.*, 1986, **25**, 1908.
109 E. M. Gaul and R. J. Kassner, *Inorg. Chem.*, 1986, **25**, 3734.

dimer with bridging *trans*-1,2-dicyanoethylenedithiolates has been prepared and fully characterized.[110] New sulphur-ligated iron porphyrins have been subjected to EXAFS and the Fe—N and Fe—S bond distances collected and correlated.[111] The direct measurement of an Fe^{III}/Fe^{IV} redox potential and the electrocatalytic epoxidation of alkanes in the absence of oxygen-donating reagents have been reported in a study of tetramesitylporphyrins.[112] The potential for the second-electron oxidation of tetrakis(2,4,6-trimethylphenyl)porphinatoiron(III) hydroxide has been confirmed.[113]

There can be no doubt that chloro(*meso*-tetraphenylporphyrinato)iron(III) hexachloroantimonate and bis(perchlorato)(*meso*-tetraphenylporphyrinato)iron(III) are iron(III) π-cation radical species having strong antiferromagnetic coupling leading to an overall $S = 2$ state. This non-semantic distinction is based on an impressive collection of physical evidence: it seems that authentic iron(IV) porphyrins are stable if terminal oxide or methoxide ligation is present.[114] The $S = 1$ iron(IV) TPP complexes are thermally unstable, undergoing clean reductive eliminations to form the iron(II) complex of the *N*-phenylporphyrin.[115]

Anionic six-co-ordinate $[Fe^{III}(Pc)L_2]$ (L = OH^-, OPh^-, NCO^-, N_3^-, or CN^-) are low-spin d^5 species but five-co-ordinate [XFePc] (X = Cl, Br, I, RCOO, or RSO_3) have $S = 3/2$ or are spin-admixed $S = 5/2$, $S = 3/2$. It seems as if there is only one high-spin $S = 5/2$ iron phthalocyanin, and that is the μ-oxo-dimer $[FePc]_2O$.[116] Lastly, two single-electron oxidations and reductions have been detected in an electrochemical study of that compound.[117]

Compounds Containing Iron–Sulphur Bonds.—The reaction of di-iodine with iron(III) dithiocarbamates is known to yield a bewildering array of products. Kinetic studies indicate the formation of a transient $[Fe(dtc)_3.I_2]$ which then oxidizes to $[Fe(dtc)_3]^+I_3$ in equilibrium with five-co-ordinate $[Fe(dtc)_2I]$, $(dtc)_2$, and di-iodine.[118] High-spin Fe 6(S) $[Fe\{S_2P(OPr^i)_2\}_3]$ has been prepared from the phosphorodithioic acid and a hexane suspension of Fe_2O_3 or iron powder.[119]

Reaction of iron(II) chloride with sodium pyridine-2-thiolate yields eventually $[Et_4N][Fe(SC_5H_4N)_3]$, or high-spin *mer*-octahedral $[FeN_3S_3]$ species, the first tris(pyridine-2-thiolato)metal(II) complex to be isolated and characterized.[120] Complexes of toluene-3,4-dithiolate with bivalent Mn, Fe, Co, or Ni have been prepared and satisfactorily characterized. An iron(III) compound $[Bu_4N]_2[Fe(TDT)_2]_2$ is an axially bridged dimer.[121] A cyclic voltammetric study in aprotic solvent of the

[110] C. M. Elliott, K. Akabori, O. P. Anderson, C. K. Schauer, W. E. Hatfield, P. B. Sczanieki, S. Mitra, and K. Spartalian, *Inorg. Chem.*, 1986, **25**, 1891.
[111] L. Kau, E. W. Svastits, J. H. Dawson, and K. O. Hodgson, *Inorg. Chem.*, 1986, **25**, 4307.
[112] J. T. Groves and J. A. Gilbert, *Inorg. Chem.*, 1986, **25**, 123.
[113] T. C. Calderwood and T. C. Bruice, *Inorg. Chem.*, 1986, **25**, 3722.
[114] P. Gans, G. Buisson, E. Duee, J. Marchon, B. S. Erler, W. F. Schloz, and C. A. Reed, *J. Am. Chem. Soc.*, 1986, **108**, 1223.
[115] A. L. Balch and M. W. Renner, *J. Am. Chem. Soc.*, 1986, **108**, 2603.
[116] B. J. Kennedy, K. S. Murray, P. R. Zwack, H. Homborg, and W. Kalz, *Inorg. Chem.*, 1986, **25**, 2539.
[117] L. A. Bottomley, C. Ercolani, J. Gorce, G. Pennesi, and G. Rossi, *Inorg. Chem.*, 1986, **25**, 2338.
[118] G. Crisponi, P. Deplano, and E. F. Trogu, *J. Chem. Soc., Dalton Trans.*, 1986, 365.
[119] M. G. Drew, W. A. Hopkins, P. C. H. Mitchell, and T. Colclough, *J. Chem. Soc., Dalton Trans.*, 1986, 351.
[120] S. G. Rosenfield, S. A. Swedberg, S. K. Arora, and P. K. Masharak, *Inorg. Chem.*, 1986, **25**, 2109.
[121] D. T. Sawyer, G. S. Srivatsa, M. E. Bodini, W. P. Schaefer, and R. M. Wing, *J. Am. Chem. Soc.*, 1986, **108**, 936.

compounds $[Fe_2(edt)_4]^{2-}$ and $[Co(edt)_2]$ (edt = ethane-1,2-dithiolato) has been reported.[122] Reaction of elemental sulphur with $[Fe(SR)_4]^{2-}$ yields $[Fe_2S_2(SR)_4]^{2-}$ and, in this way, the simplest alkanethiolate $[Fe_2S_2(SMe)_4]^{2-}$ was prepared for the first time.[123]

Tetrapeptide 2Fe-2S complexes $[Et_4N]_{2n}[Fe_2S_2(Z\text{-}Cys\text{-}X\text{-}Y\text{-}Cys\text{-}OMe)_2]_n$ (X-Y = Ala-Ala, Pro-Leu, Thr-Val, or Val-Val) have been synthesized from $(Et_4N)_2$-$[Fe_2S_2(SBu^t)_4]$ *via* ligand exchange reactions.[124]

A new inorganic ring system adopts a planar conformation for preference. It is $[Fe_3(SR)_3X_6]^{3-}$ (X = Cl or Br; R = Ph or $2,6\text{-}Me_2C_6H_3$) a new kind of Fe^{II} thiolate.[125]

The ^{13}C n.m.r. spectra of synthetic $[Fe_4S_4(SR)_4]^{2-}$ clusters have been recorded for R = Et, CH_2CH_2OH, Ph, $4\text{-}MeC_6H_4$, $3\text{-}NH_2C_6H_4$, or $4\text{-}NH_2C_6H_4$. Contact interactions dominate the shift mechanism as with the ^1H n.m.r. spectra[126] and the redox potentials have been determined for exactly analogous compounds but with R = $4\text{-}n\text{-}octyl\text{-}C_6H_4$.[127] The bond distances, Mössbauer spectra, and solid-state ^{13}C n.m.r. spectra have been examined for a series of $[Fe_4(\mu_3\text{-}S)_3(\mu_3\text{-}S_2)(\eta\text{-}C_5H_5)]^{n+}$ clusters in order to correlate core geometry with electronic distribution when oxidation level varies. It is concluded that unpaired spin density is localized on the sulphur atoms of the core.[128] A new cluster having a $[Fe_3CoS_4]$ core is formed in *Desulphovibriogigas ferrodoxin(II)* following purification of the protein and incubation with an excess of cobalt(II)[129] and a vanadium–iron–sulphur cluster $[VFe_3S_5Cl_3(dmf)_3]^-$ (dmf = dimethylformamide) has been assembled and characterized.[130] Tetrahedral $MoFe_3$ and square-pyramidal $MoFe_4$ clusters both give satisfactory simulations of the observed single-crystal EXAFS of nitrogenase.[131] A cubane having a $[Mo_3FeS_4]^{4+}$ aqua ion, *viz.* $[Mo_3FeS_4(NH_3)_9(H_2O)]Cl_4$, is a product of reaction of the $[Mo_3S_4]^{4+}$ aqua ion with iron wire.[132] Doping $[Fe_4S_4(SPh)_4]^{2-}$ with cobalt(II) enables the recording of an e.p.r. spectrum consistent with a polynuclear system $[CoFe_3S_4]^{2+}$ in which antiferromagnetic exchange leads to an $S' = 1/2$ ground state as in the parent.[133] E.p.r. and Mössbauer spectral results for $[NBu_4][Fe_4S_4(S\text{-}2,4,6\text{-}Pr^i_3C_6H_2)_4]$, a high-potential iron–protein model, indicate an $S = 1/2$ ground state with equivalent pairs of iron sites.[134] The core structure $[Fe_4S_4]^+$ can adopt different spin ground states, *e.g.* $S' = 1/2$ and $S' = 3/2$, in different but analogous situations. It further emerges that a mixture of spin states is possible and that nature provides one such in the reduced nitrogenase.[135] It has been possible to investigate the redox

[122] R. N. Mukerjee, C. P. Rao, and R. H. Holm, *Inorg. Chem.*, 1986, **25**, 2979.
[123] S. Han and R. C. Czernuszewicz, *Inorg. Chem.*, 1986, **25**, 2276.
[124] S. Veno, N. Veyama, A. Nakamura, and T. Tukihara, *Inorg. Chem.*, 1986, **25**, 1000.
[125] M. A. Whitener, J. K. Bashkin, K. S. Hagen, J. Girerd, E. Gamp, N. Edelstein, and R. H. Holm, *J. Am. Chem. Soc.*, 1986, **108**, 5607.
[126] T. J. Ollernshaw, S. Bristow, B. N. Anand, and C. D. Garner, *J. Chem. Soc., Dalton Trans.*, 1986, 2013.
[127] K. Tanaka, M. Moriya, and T. Tanaka, *Inorg. Chem.*, 1986, **25**, 835.
[128] N. Dupre, P. Aurie, H. M. J. Hendricks, and J. Jordanov, *Inorg. Chem.*, 1986, **25**, 1391.
[129] I. Moura, J. J. G. Moura, E. Munck, V. Papaefthymiou, and J. LeGall, *J. Am. Chem. Soc.*, 1986, **108**, 349.
[130] J. A. Kovacs and R. H. Holm, *J. Am. Chem. Soc.*, 1986, **108**, 340.
[131] A. M. Flank, M. Weininger, L. E. Mortenson, and S. P. Cramer, *J. Am. Chem. Soc.*, 1986, **108**, 1049.
[132] T. Shibahara, H. Akashi, and H. Kuroya, *J. Am. Chem. Soc.*, 1986, **108**, 1342.
[133] J. Gloux, P. Gloux, and G. Rius, *J. Am. Chem. Soc.*, 1986, **108**, 3541.
[134] V. Papaefthymiou, M. M. Millar, and E. Munck, *Inorg. Chem.*, 1986, **25**, 3010.
[135] M. J. Carney, R. H. Holm, G. C. Papaefthymiou, and R. B. Frankel, *J. Am. Chem. Soc.*, 1986, **108**, 3519.

behaviour of $[Fe_4S_4(SC_6H_4\text{-}n\text{-}C_8H_{17}\text{-}4)_4]^{2-}$ solubilized in a lecithin membrane.[136] Resonance Raman spectra have been reported for rubredoxin in frozen solutions with excitation by several lines. The three components of the $\nu_3(T_2)$ asymmetric Fe—S stretching mode of the FeS_4 tetrahedron were identified with the aid of ^{54}Fe substitution and vibrational coupling to other modes of the cysteine ligands, probably SCC bending. The SFeS bending modes were also assigned.[137]

The catalytic reduction of NO_3^- to NH_3 has been achieved with a glassy-carbon electrode modified with the trianion $[Mo_2Fe_6S_8(SPh)_9]^{3-}$ with an 80% current efficiency at pH 10. In contrast, NO_2^- yields N_2O but no NH_3.[138] The same electrode system has been applied to the reduction of alkyl azides. Ammonia and hydrazine were detected.[139] The odd-electron mono-capped prismane cluster (9) $[Fe_7S_6(PEt_3)_4Cl_3]$ is formed from the interaction of $FeCl_2$, PEt_3, and $(Me_3Si)_2S$.[140]

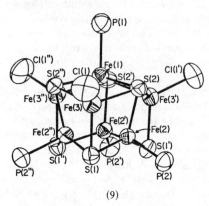

(9)

Structure of $[Fe_7S_6(PEt_3)_4Cl_3]$ *(C and H atoms omitted) showing 50% thermal ellipsoids and the atomic numbering scheme*
(Reproduced by permission from *Inorg. Chem.*, 1986, **25**, 3852)

Mononuclear η^5-Cyclopentadienyl and η^6-Benzene Derivatives.—The reaction of $[\eta\text{-}CpFe(CO_2Cl)]$ with $Me_2C{=}CHCO_2H$ in the presence of $AgSbF_6$ affords both $[\eta\text{-}CpFe(CO)_2Me_2C{=}CHCO_2H][SbF_6]$ and an orange HX_2 derivative $[\{\eta\text{-}CpFe(CO)_2(Me_2C{=}CHCOO)\}_2H][SbF_2]$. In contrast, the methyl ester yields just $[\eta\text{-}CpFe(CO)_2(Me_2C{=}CHCO_2Me)][SbF_6]$, which seems to involve an Fe—O=C$\subseteq$ fragment.[141] Metallacyclic compounds (10) are products of reaction of $[\eta\text{-}CpFe\text{-}(Me_2As)(CO)_2]$ with activated alkynes.[142] A comprehensive study of the photochemical reactions of (η-permethylcyclopentadienyl)dicarbonyliron alkyl and silyl compounds indicates that alkene hydroxilylation involves insertion into a M—Si

[136] K. Tanaka, M. Masanaga, and T. Tanaka, *J. Am. Chem. Soc.*, 1986, **108**, 5448.
[137] R. S. Czernuszewicz, J. LeGall, I. Moura, and T. G. Spiro, *Inorg. Chem.*, 1986, **25**, 696.
[138] S. Kuwabata, S. Vezumi, K. Tanaka, and T. Tanaka, *Inorg. Chem.*, 1986, **25**, 3018.
[139] S. Kuwabata, K. Tanaka, and T. Tanaka, *Inorg. Chem.*, 1986, **25**, 1691.
[140] I. Noda, B. S. Snyder, and R. H. Holm, *Inorg. Chem.*, 1986, **25**, 3851.
[141] D. C. Cupertino, M. M. Harding, D. J. Cole-Hamilton, H. M. Dawes, and M. B. Hursthouse, *J. Chem. Soc.*, *Dalton Trans.*, 1986, 1129.
[142] L. Carlton, J. L. Davidson, and M. Shiralian, *J. Chem. Soc.*, *Dalton Trans.*, 1986, 1577.

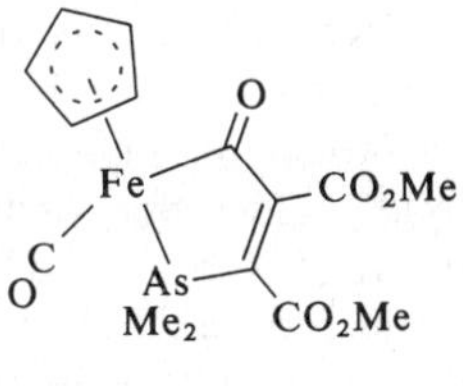

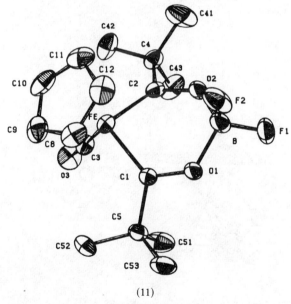

(10)

bond as a key step.[143] I.r. spectroscopy with microsecond time resolution reveals that the primary photoproducts in the u.v. flash photolysis of *trans*-[η-C$_5$Me$_5$Fe-(CO)$_2$]$_2$ is mainly [C$_5$Me$_5$Fe(CO)$_2$] with some [C$_5$Me$_5$Fe(μ-CO)$_3$Fe$_5$Me$_5$].[144] Photolysis of [η-CpFe(η^6-C$_8$H$_8$)]PF$_6$ at low temperatures in acetone or methylene chloride solutions of P(OMe)$_3$ or MeCN gives [η-CpFe(S)$_3$]PF$_6$ [S = MeCN or P(OMe)$_3$].[145]

The quantum yields for the photochemically induced arene release from [η-CpFe-(η^6-arene)]$^+$ (arene = C$_6$Me$_6$) is 0.007 but decreases by a factor of 2×10^4 if the Cp ring is permethylated.[146] The structure of a [(η-Cp)(CO)(ferra-β-diketonato)BF$_2$] complex has been determined and shown to involve a boat-shaped ferra chelate ring with the cyclopentadienyl on an axial position (11).[147] (Ferra-β-

(11)

(Reproduced by permission from *Inorg. Chem.*, 1986, **25**, 711)

[143] C. L. Randolph and M. S. Wrighton, *J. Am. Chem. Soc.*, 1986, **108**, 3366.
[144] B. M. Moore, M. Poliakoff, and J. J. Turner, *J. Am. Chem. Soc.*, 1986, **108**, 1819.
[145] J. L. Schrenk and K. R. Mann, *Inorg. Chem.*, 1986, **25**, 1906.
[146] J. L. Schrenk, A. M. McNair, F. B. McCormick, and K. R. Mann, *Inorg. Chem.*, 1986, **25**, 3501.
[147] D. Afzal, P. Galanlenhert, C. M. Lukehart, and R. Srinivasan, *Inorg. Chem.*, 1986, **25**, 710.

diketonato)BF_2 complexes (12) react as activated dienophiles in Diels–Alder cyclo-addition reactions, with some ten adducts being characterized.[148] Black [η-CpFeM(CO)$_6$PH(NPr$^i_2$)] is a product of photolysis of [η-CpFe(CO)$_2$PH(NPr$^i_2$)-M(CO)$_5$] (M = Cr or W).[149] The ferrocenium salt [Fe(C_5Me_5)$_2$]$^+$(BF$_4$) undergoes metathesis with K$^+$[C(CN)$_3$]$^-$ to give otherwise unremarkable [Fe(C_5Me_5)$_2$]$^+$[C(CN)$_3$]$^-$ in which the anion is definitely D_{3h}.[150]

This year has seen the publication of full details of the synthesis of the symmetrical perbridged [45]ferrocenophane (13).[151] Intramolecular electron hopping in the mixed-valence 1,6'-dihalogenobiferrocenium cations (14) is very strongly anion dependent.[152] The proton n.m.r. spectra of the ferrocenophanes (15) and (16) are consistent with conformational flips of the S_3 ring.[153] The anion in the salts formed

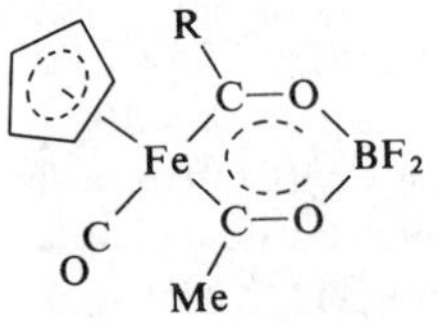

(12)   R = alkyl

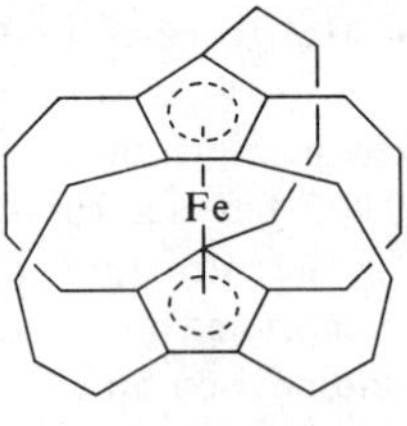

(13)

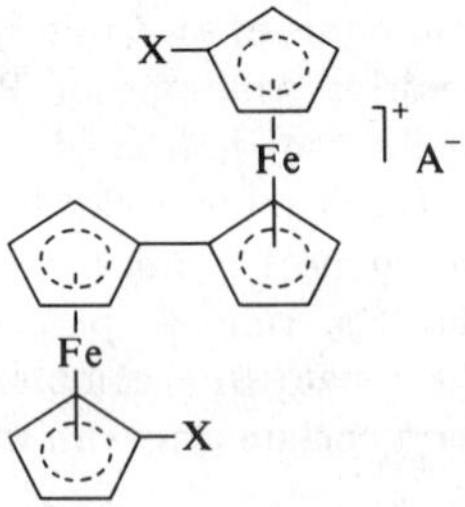

X = Cl, I, or Br
A = PF$_6$, Br$_2$I, *etc.*

(14)

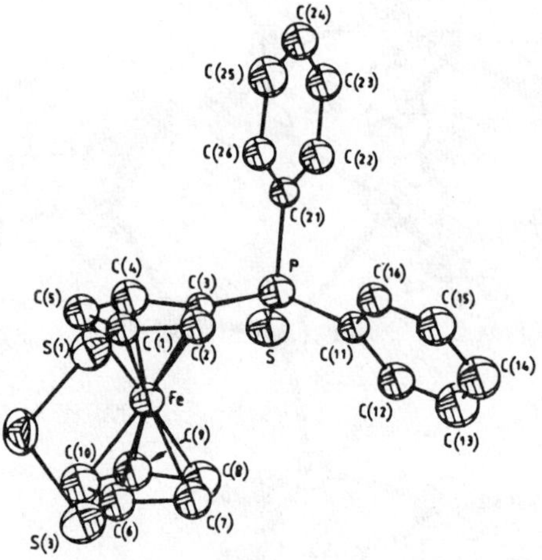

(15)

(Reproduced by permission from *Inorg. Chem.*, 1986, **25**, 4535)

[148] P. G. Lengert, C. M. Lukehart, and L. Sachsteder, *J. Am. Chem. Soc.*, 1986, **108**, 793.
[149] R. B. King, W. K. Fu, and E. M. Holt, *Inorg. Chem.*, 1986, **25**, 2394.
[150] D. A. Dixon, J. C. Calabrese, and J. S. Miller, *J. Am. Chem. Soc.*, 1986, **108**, 2582.
[151] M. Hisatome, J. Watanabe, K. Yamakawa, and Y. Iitaka, *J. Am. Chem. Soc.*, 1986, **108**, 1333.
[152] T. Dong, T. Kambara, and D. N. Hendrickson, *J. Am. Chem. Soc.*, 1986, **108**, 5857.
[153] I. R. Butler, W. R. Cullen, F. G. Herring, N. R. Jagannathan, F. W. B. Einstein, and R. Jones, *Inorg. Chem.*, 1986, **25**, 4534.

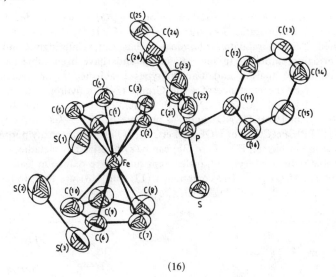

(16)

(Reproduced by permission from *Inorg. Chem.*, 1986, **25**, 4535)

from 2,3-dichloro-5,6-dicyanobenzoquinone and decamethylferrocene is a radical anion and not a singlet species as previously believed.[154] Bis(benzene)iron(II) cations are starting compounds for the preparation of cyclohexadienes as shown in Scheme 1. The substituents are *exo*.[155]

Scheme 1

[154] J. S. Miller, P. J. Krusic, D. A. Dixon, W. M. Reiff, J. H. Zhang, E. C. Anderson, and A. J. Epstein, *J. Am. Chem. Soc.*, 1986, **108**, 4449.
[155] D. Mandon, L. Toupet, and D. Astruc, *J. Am. Chem. Soc.*, 1986, **108**, 1320.

Mononuclear Iron Carbonyls.—The radical anion $[Fe(CO)_2]^-$ may be generated in a flowing afterglow apparatus. Reaction rates with H_2, H_2O, H_2S, or NH_3 have been recorded.[156] Structural data for terminal, linear, bent, semibridging, and symmetrically bridging carbonyls in some 47 compounds have been subjected to a detailed quantitative analysis and different types established.[157] I.r. evidence is offered as proof of the existence of some Group 18–19 dihydrogen complexes, $[Fe(CO)(NO)_2H_2]$ and $[Co(CO)_2(NO)H_2]$, generated by u.v. photolysis of $[Fe(CO)_2(NO)_2]$ or $[Co(CO)_3NO]$ under a pressure of H_2.[158] Tetracarbonyliron derivatives $[(TPP)EFe(CO)_4]$ or $[(OEP)EFe(CO)_4]$ (TPP = tetraphenylporphyrin, OEP = octaethylporphyrin; E = Sn or Ge) can be synthesized by metathesis. They show reversible reductive electron-transfer steps but electro-oxidation leads to the formation of $[Fe_3(CO)_{12}]$.[159] The compounds (17) and (18) are the products of reaction of $Na_2[Fe(CO)_4]$ with $Pr^i_2NPCl_2$.[160]

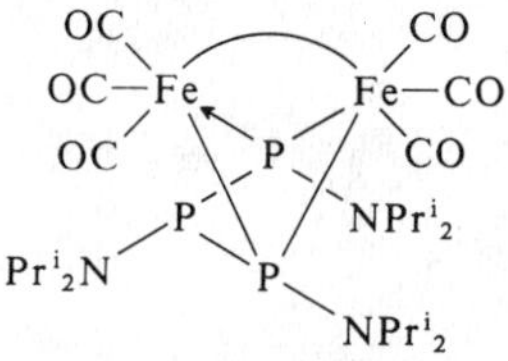

(17)

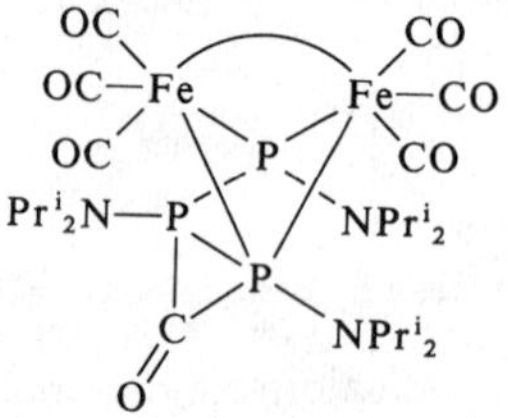

(18)

Steric inhibition of isomerization in tris(trifluoromethyl)phosphineiron(0) carbonyls, *e.g.* $[((CF_3)_3P)_nFe(CO)_{5-n}]$ (n = 1 or 2), is not observed. The structure of the n = 3 derivatives is all-equational.[161] Carbon monoxide substitution by pyridine in $[Fe(CO)_3L_2]^+$ (L = PR_3) is first-order in radical and first-order in entering nucleophile. These complexes then disproportionate, yielding iron(0) and iron(II) products.[162] Iron hydrido-complexes *trans*-$[FeH(CO)L_4][BPh_4]$ [L = P(OMe)$_3$, P(OEt)$_3$, or PhP(OEt)$_2$] have been synthesized by reaction of $[FeX(CO)L_4]^+$ (X = Cl or Br) with zinc dust in acetonitrile. The structure of *trans*-$[FeH(CO)\{P(OEt)_3\}_4][BPh_4]$ is as expected. It undergoes insertion reactions with aryldiazonium cations to give a diazone complex *trans*-$[Fe(ArNNH)(CO)L_4]$.[163] U.v. photolysis of $[Co(CO)_3NO]$ or $[Fe(NO)_2(CO)_2]$ dissolved in liquid xenon containing but-1-ene yields $[Co(CO)_2(NO)(\eta^2\text{-but-1-ene})]$ or $[Fe(CO)_{2-x}(NO)_2('\eta^2\text{-but-1-ene})_x]$ (x = 1 or 2).[164] Application of the SLCC MO method to the cyclodienyl complexes $[Fe(C_7H_9)(CO)_3]^+$ and $[Fe(C_8H_{11})(CO)_3]^+$ indicates non-planar rings

[156] R. N. McDonald, A. K. Chowdhury, and M. T. Jones, *J. Am. Chem. Soc.*, 1986, **108**, 3105.

[157] R. H. Crabtree and M. Lavin, *Inorg. Chem.*, 1986, **25**, 805.

[158] G. E. Gadd, R. K. Upmacis, M. Poliakoff, and J. J. Turner, *J. Am. Chem. Soc.*, 1986, **108**, 2547.

[159] K. M. Kadish, C. Swistak, B. Boisselier-Cocolios, J. M. Barbe, and R. Guiland, *Inorg. Chem.*, 1986, **25**, 4336.

[160] R. B. King, F. J. Wu, and E. M. Holt, *Inorg. Chem.*, 1986, **25**, 1733.

[161] A. B. Burg, *Inorg. Chem.*, 1986, **25**, 4751.

[162] M. J. Therien, C. Ni, F. C. Anson, J. C. Osteryoung, and W. C. Trogler, *J. Am. Chem. Soc.*, 1986, **108**, 4037.

[163] G. Albertin, S. Antoniutti, M. Lafranchi, G. Pelezzi, and E. Bordignon, *Inorg. Chem.*, 1986, **25**, 950.

[164] G. E. Gadd, M. Poliakoff, and J. J. Turner, *Inorg. Chem.*, 1986, **25**, 3604.

as in $[Fe(C_6H_7)(CO)_3]^+$ and a 'boat' conformation in the case of $[Fe(C_8H_{11})-(CO)_3]^+$.[165]

Proton spin–lattice relaxation time measurements on solid (cyclobutadiene)tricarbonyliron(0) have been interpreted in terms of two relaxation processes arising from inequivalent lattice sites. As yet, no crystal structure determination has been reported for this important molecule and the nature of the solid phase is speculative although the low-temperature FT i.r. and Raman results suggest four molecules per unit cell.[166] Iron(II) monocarbonyl compounds $[FeX(CO)L_4][BPh_4]$ [X = Cl or Br; L = P(OMe)$_3$, P(OEt)$_3$, or PhP(OEt)$_2$] and the dicarbonyl compound $[FeBr(CO)_2\{PhP(OEt)_2\}_3]BPh_4$ have been prepared and characterized.[167]

A straightforward high-yield route to $(Et_4N)_3[Fe(CN)_6]$ has been devised and the dependence of its redox potential on solvent well demonstrated. Variations of up to 1 V in aqueous binaries are observed, thus providing a tunable one-electron oxidizer.[168] The influence of closed-shell cations Li$^+$, Na$^+$, K$^+$, *etc.* on the ligand-to-metal charge-transfer transitions of pentacyanoferrate(III) complexes $[Fe^{III}(CN)_5L]$ [L = 4-(dimethylamino)pyridine or benzimidazole] has been correlated with hydration energies.[169] Low-spin tetracyanoferrates(II) and tetracyanoferrate(III) 1,3-diamine complexes have been prepared and characterized using ^{13}C n.m.r., electronic, and circular dichroic spectra,[170] and solute–solvent interactions in the sulphite–ferrocyanide reactions have been studied in a range of solvent mixtures.[171] New polynuclear tris(bipyrazine)ruthenium(II) pentacyanoferrate(II) complexes have been characterized in solution by means of electronic and ^1H n.m.r. spectroscopy.[172] Pentacyanonitrosylferrate(III) forms 1:1 complexes with aqua-cobalamin in which the linkage is of the [Fe—C—N—Co] type.[173] The intervalence charge-transfer bands in Prussian Blues have been reinvestigated making use of SnO$_2$ electrodes upon which partial oxidation or reduction was effected electrochemically.[174]

3 Cobalt

Low Valence State Compounds of Cobalt.—Transition metals, especially cobalt, are cocatalysts in NaBH$_4$ or LiAlH$_4$ reductions: *e.g.* black CoH plays an essential catalytic role in the reduction of alkyl halides; a radical mechanism seems likely.[175] Metallic cobalt as prepared from the reaction of anhydrous cobalt(II) salts with solutions of lithium or potassium naphthide is a useful catalyst (Fischer–Tropsch,

[165] D. A. Brown, N. J. Fitzpartick, and M. A. McGinn, *J. Chem. Soc., Dalton Trans.*, 1986, 701.
[166] P. D. Harvey, I. S. Butler, and D. F. R. Gilson, *Inorg. Chem.*, 1986, **25**, 1009.
[167] G. Albertin, D. Baldan, and E. Bordignon, *J. Chem. Soc., Dalton Trans.*, 1986, 329.
[168] P. K. Mascharak, *Inorg. Chem.*, 1986, **25**, 245.
[169] L. M. Warner, M. F. Hoq, T. K. Myser, W. W. Henderson, and R. E. Shepherd, *Inorg. Chem.*, 1986, **25**, 1911.
[170] M. Goto, H. Nakayabu, H. Ito, H. Tsubamoto, K. Nakabayashi, Y. Kuroda, and T. Sakai, *Inorg. Chem.*, 1986, **25**, 1694.
[171] A. Rodriguez, S. Lopez, M. C. Carmona-Guzman, F. Sanchez, and C. Piazza, *J. Chem. Soc., Dalton Trans.*, 1986, 1265.
[172] H. E. Toma and A. B. P. Lever, *Inorg. Chem.*, 1986, **25**, 176.
[173] A. R. Butler, C. Glidewell, A. S. McIntosh, D. Reed, and I. H. Sadler, *Inorg. Chem.*, 1986, **25**, 970.
[174] K. Itaya and I. Uchida, *Inorg. Chem.*, 1986, **25**, 389.
[175] J. O. Osby, S. W. Heinzman, and B. Ganem, *J. Am. Chem. Soc.*, 1986, **108**, 67.

nitrobenzene → azobenzene, *etc.*).[176] The dinitrogen complex [CoH(N$_2$)(PPh$_3$)$_3$] selectively cleaves C—O bonds in carboxylates R^{1}CO$_2$R but aldehydes dimerize to yield R^{1}CO$_2$CH$_2$R, and [MeCo(PPh$_3$)$_3$] converts esters into ketones RCOMe.[177]

Ligand exchange rates for cationic organocobalt complexes of the type (19) are similar to those of the cobaloximes. The structures of aniline complexes (R = Me or CH$_2$CO$_2$Me) are of the expected type.[178] Aqueous solutions of vitamin B$_{12}$ smoothly reduce maleic and fumaric acids to succinic acids in first-order processes accelerated by addition of H$^+$.[179] Reduction of the corresponding and analogous diesters is a two-stage process and, again, fastest in highest acidities.[180] A series of new phosphine-organocobalt vitamin B$_{12}$ model adducts has been synthesized and the structure of one example (20) determined.[181] Aqueous solutions of cob(I)alamin react rapidly with organic disulphides to yield the corresponding thiols and cob(II)alamin.[182]

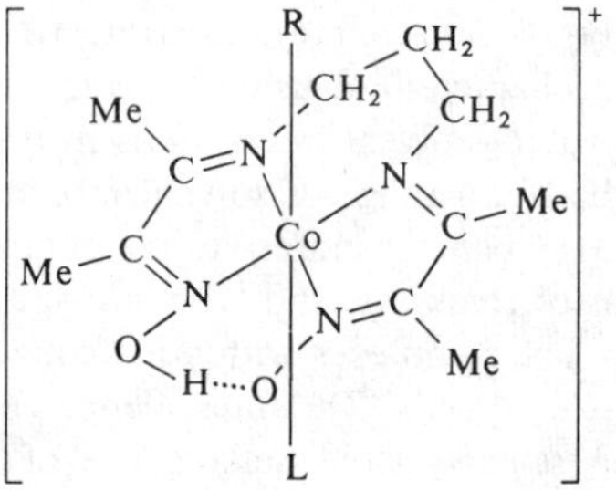

(19) 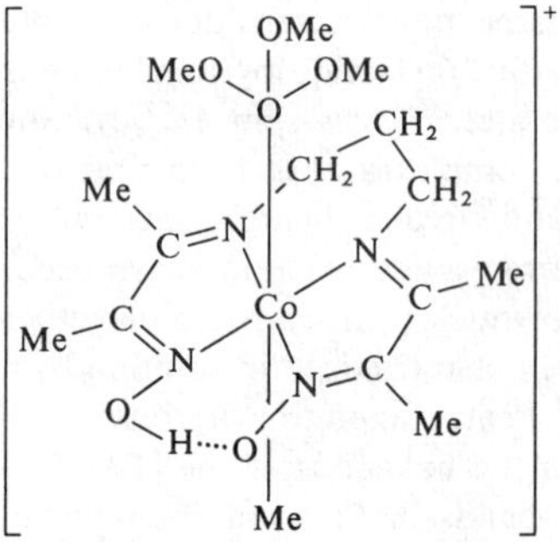 (20)

Low-spin five-co-ordinate Co^I and Ni^{II} complexes (21) have been synthesized and characterized and the structures of representative examples determined.[183] The variation with phosphite substituent has been studied in the cobalt(II) disproportion-ation to yield [CoL$_5$]$^+$ and [CoL$_6$]$^{3+}$.[184] 19-Electron d^7 mixed-sandwich compounds [Co(η-Cp)(arene)]$^+$ have the unpaired electron in an $e_g(d_{zx}d_{zy})$ orbital in a dynamically coupled Jahn–Teller ground state.[185] The organometallic anions (22) are ionophores with the ethyl derivative being specific for Li$^+$.[186] The reaction of octacarbonyldicobalt with C$_5$Ph$_5$Br and Zn dust in THF yields [Co(CO)$_2$(η-

[176] G. L. Rochfort and R. D. Rieke, *Inorg. Chem.*, 1986, **25**, 348.
[177] Y. Hayashi, T. Yamamoto, A. Yamamoto, S. Komiya, and Y. Kushi, *J. Am. Chem. Soc.*, 1986, **108**, 385.
[178] W. O. Parker, E. Zangrando, N. Bresciani-Pahor, L. Randaccio, and L. G. Marzilli, *Inorg. Chem.*, 1986, **25**, 3489.
[179] G. C. Pillai, J. W. Reed, and E. S. Gould, *Inorg. Chem.*, 1986, **25**, 4734.
[180] G. C. Pillai and E. S. Gould, *Inorg. Chem.*, 1986, **25**, 4740.
[181] W. O. Parker, N. Bresciani-Pahor, E. Zangrando, L. Randaccio, and L. G. Marzilli, *Inorg. Chem.*, 1986, **25**, 1303.
[182] G. C. Pillai and E. S. Gould, *Inorg. Chem.*, 1986, **25**, 3353.
[183] W. H. Hohman, D. J. Kountz, and D. W. Meek, *Inorg. Chem.*, 1986, **25**, 616.
[184] S. M. Socal and J. G. Verkade, *Inorg. Chem.*, 1986, **25**, 2658.
[185] B. L. Ramakrishna, A. K. Salzer, U. Rappli, J. H. Ammeter, and U. Koelle, *Inorg. Chem.*, 1986, **25**, 1364.
[186] H. Shinar, G. Navon, and W. Klaui, *J. Am. Chem. Soc.*, 1986, **108**, 5005.

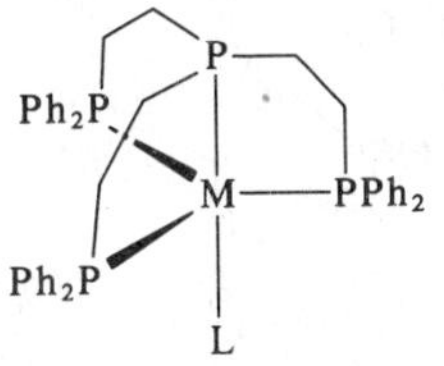

(21)

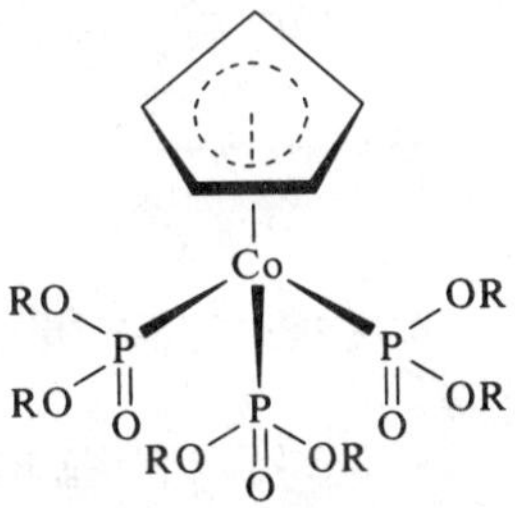

R = Et

(22)

$C_5Ph_5)]$.[187] There is a symmetric metal–cyclopentadienyl bonding in the electro-chemically generated 19-electron radical anions $[(\eta\text{-}C_5Ph_5)M(CO)_2]^-$ (M = Co or Rh) as opposed to the asymmetric situation in the 18-electron series.[188] Alkene bonding to $[\eta\text{-}CpCo(NO)_2]$ has been investigated theoretically in connection with the formation of intermediates in transition-metal catalysed synthesis of diols, amino-alcohols, and diamines.[189] The cobalt complexes $[(\eta\text{-}C_5H_5)Co\{P(OMe)_3\}_2]$ protonate with NH_4PF_6 in toluene–methanol. The electrochemical reduction of these acids has been achieved and the reduction products are catalysts for hydrogen production at -1.15 V on a mercury cathode in aqueous solution.[190]

Isocyanides add the co-ordinated alkynes as shown in Scheme 2.[191] Cobaltacyclo-pentenes may also be accessed by a similar reaction[192] (Scheme 3) and their structures have been determined.[193]

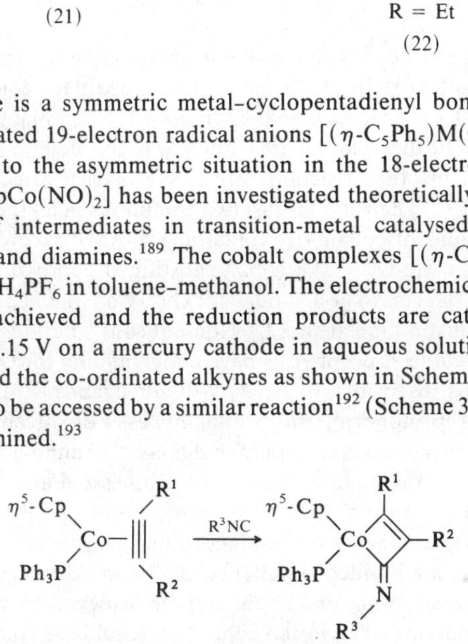

e.g. $R^1 = R^2 = R^3 = Ph$

Scheme 2

Compounds of Cobalt(II).—Electron transfer rates for the oxidation of $[Co(terpy)_2]^{2+}$ (terpy = terpyridyl) by $[Co(phen)_3]^{3+}$ or $[Co(bipy)_3]^{3+}$ in a range of solvents have been recorded and compared with those involving the electrically neutral oxidants such as the tris(phenanthroline-5-sulphonate)cobalt(III) complex. The results are interpreted using non-local electrostatic theory.[194] Oxidative dehydrogenation of 2-(aminomethyl)pyridine can originate in the formation and degradation of μ-

[187] N. G. Connelly and S. J. Raven, *J. Chem. Soc., Dalton Trans.*, 1986, 1613.
[188] N. G. Connelly, W. E. Geiger, G. A. Lane, S. J. Raven, and P. H. Rieger, *J. Am. Chem. Soc.*, 1986, **108**, 6219.
[189] K. A. Jorgensen and R. Hoffmann, *J. Am. Chem. Soc.*, 1986, **108**, 1867.
[190] U. Koelle and S. Ohst, *Inorg. Chem.*, 1986, **25**, 2689.
[191] Y. Wakatsuke, K. Aoke, and H. Yamazaki, *J. Chem. Soc., Dalton Trans.*, 1986, 1193.
[192] Y. Wakatuske, S. Miya, and H. Yamazaki, *J. Chem. Soc., Dalton Trans.*, 1986, 1201.
[193] Y. Wakatuske, S. Miya, and H. Yamazaki, *J. Chem. Soc., Dalton Trans.*, 1986, 1207.
[194] A. M. Kjaer and J. Ulstrup, *Inorg. Chem.*, 1986, **25**, 644.

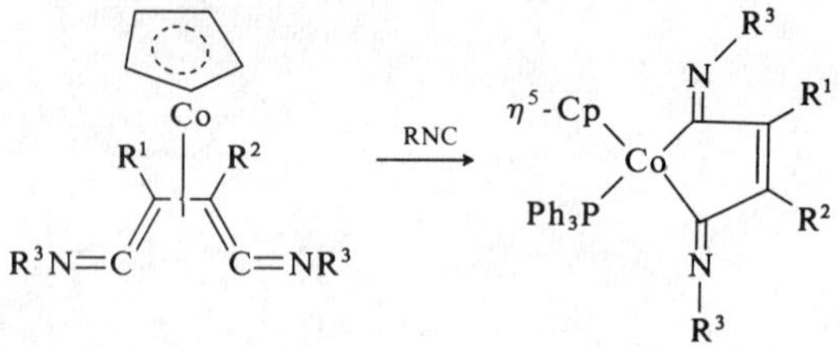

e.g. $R^1 = R^2 = R^3 = Ph$

Scheme 3

peroxo-cobalt(II) complexes involving diethylenetriamine as ligand.[195] The structures of the CoN_6 [Co(tacn))$_2$]I$_2$.2H$_2$O and the CoS_6 cobalt(III) complex [Co(ttcn)$_2$]-(ClO_4)$_3$ (tacn = 1,4,7-triazacyclononane, ttcn = 1,4,7-trithiacyclotonane) have been determined. A further part of this study was the determination of rates of electron exchange in the reaction of [(R-Metacn)Co(tacn)$_2$]$^{2+}$ and [(S-Metacn)Co(tacn)$_2$]$^{3+}$.[196] There are at least two complexes in temperature-dependent equilibrium in solutions of cobalt(III) containing 5'-AMP according to the results of a ^{31}P and ^1H n.m.r. study.[197] The cobalt-substituted cupro-zinc superoxide dismutase [Cu$_2$Co$_2$SOD] seems to be a 'model' for Cu^{II}–Co^{II} interactions with imidazolato bridges and tetrahedral metal sites.[198] Non-aqueous solutions of mixed-valence dicobalt [Co^{III}Co^{II}] cofacial porphyrins have a high affinity for O$_2$ which, although inhibited by water, is insensitive to axial N-donor ligands.[199] The mechanism of reaction of cobalt(III) protoporphyrin IX dimethyl ester dissolved in dipolar aprotic solvents has been shown to involve μ-peroxo dimers.[200] Amino-acid-derived salicylaldehydeimine cobalt(II) complexes have been synthesized and estimates made of their binding affinities with dioxygen.[201] Cobalt(II) and nickel(II) salts react with 3,6-bis(1'-pyrazolyl)pyradiazine to produce dinuclear derivatives in which two [MN$_4$(OH$_2$)$_2$] centres are bonded simultaneously between two adjacent ligands.[202]

Enrichment of oxygen in an air sample may be achieved by passing it through polystyrene which contains a supported cobalt(II) Schiff base complex.[203] Addition of the complexing agent 1,6-bis(2-hydroxyphenyl)-2,5-diazahexadiene to cobalt(II) in zeolite Y gives the cobalt(II) complex embedded in the pores of the zeolite with quite different properties from the bulk material: *e.g.* the dioxygen adduct is less prone to oxidation.[204] Cobalt(II), nickel(II), and copper(II) complexes of the ligands (23) and (24) have been characterized structurally. The co-ordination about the metal is essentially tetrahedral but weak M—O bonding induces pseudo five-co-

[195] A. K. Basak and A. E. Martell, *Inorg. Chem.*, 1986, **25**, 1182.
[196] H. Kuppers, A. Neves, C. Pomp, D. Ventur, K. Wieghardt, B. Nuber, and J. Weiss, *Inorg. Chem.*, 1986, **25**, 2400.
[197] J. Leroy and M. Gueron, *J. Am. Chem. Soc.*, 1986, **108**, 5753.
[198] I. Morgenstern-Badarau, D. Cocco, A. Desideri, G. Rotilio, J. Jordanov, and N. Dupre, *J. Am. Chem. Soc.*, 1986, **108**, 300.
[199] J. P. Collman, N. H. Hendricks, and L. McElwee-White, *J. Am. Chem. Soc.*, 1986, **108**, 533.
[200] D. Pavlovic, S. Asperger, and B. Domi, *J. Chem. Soc., Dalton Trans.*, 1986, 2535.
[201] L. Cassella and M. Guillotti, *Inorg. Chem.*, 1986, **25**, 1293.
[202] L. Rosenberg, L. K. Thompson, E. J. Gabe, and F. L. Lee, *J. Chem. Soc., Dalton Trans.*, 1986, 625.
[203] R. S. Drago and K. J. Balkus, *Inorg. Chem.*, 1986, **25**, 716.
[204] N. Herron, *Inorg. Chem.*, 1986, **25**, 4714.

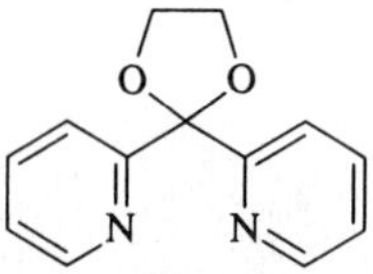

(23) 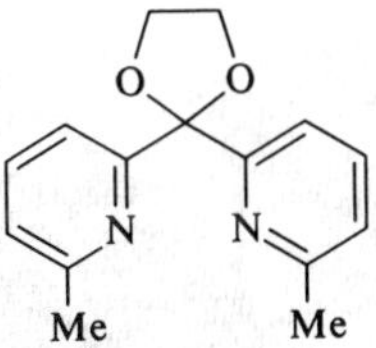   (24) 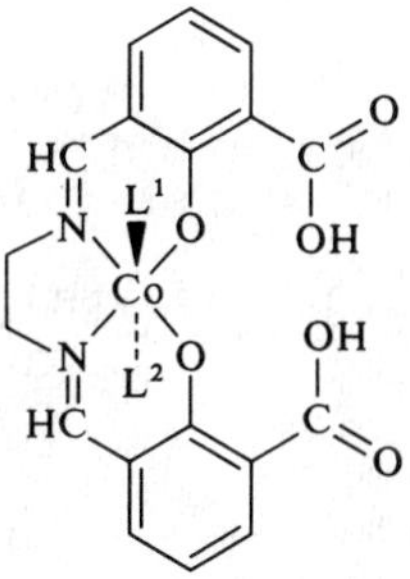   (25)

ordination.[205] The cobalt(II) complexes (25) are five-co-ordinate high-spin (L^1 = pyridine, L^2 absent) or low-spin (L^1 = 4-MePy, L^2 absent) or six-co-ordinate high-spin ($L^1 = L^2$ = 3-MePy, 3-NH_2Py, *etc.*) or spin-crossover ($L^1 = L^2$ = 4-MePy, 4-Bu^{t}Py).[206]

Complexes of the ligand (26) are either planar CoL or octahedral CoL(H_2O)$_2$ or tetrahedral [CoH$_2$LCl$_2$]. Solutions in DMF take up dioxygen reversibly or irreversibly depending on cases and conditions.[207] The six-co-ordinate [CoO$_2$N$_4$] complex (27) is distorted octahedral.[208] Studies of the use of Schiff base complexes in the catalytic oxidation by dioxygen of 3,5-di-t-butylcatechol to the *o*-benzoquinone have been published.[209] The structure of the low-spin cobalt(II) macrocyclic complex

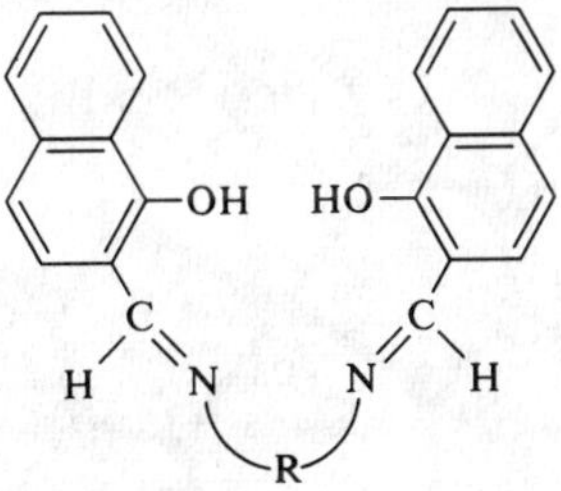

$R = -(CH_2)_n-$, n = 2—6

(26)

(28) has been solved by *X*-ray methods[210] and it has now been established that monocarboxylic acids are the major component of synthetic vitamin B_{12}.[211]

The formation constants of the complexes formed between H^+, Co^{2+}, or Ni^{2+} and 2-amino-*N*-hydroxyacetamide or 2-amino-*N*-hydroxypentamide have been measured at 25 °C using potentiometric techniques.[212] The new octadentate ligand

[205] G. R. Newkome, H. C. R. Taylor, F. R. Frorczek, and V. K. Gupta, *Inorg. Chem.*, 1986, **25**, 1149.
[206] P. Thuery and J. Zarembowitch, *Inorg. Chem.*, 1986, **25**, 2001.
[207] F. M. Ashmawy, R. M. Issa, S. A. Amer, C. A. McAuliffe, and R. V. Parish, *J. Chem. Soc., Dalton Trans.*, 1986, 421.
[208] A. E. Koziol, R. C. Palenik, and G. J. Palenik, *Inorg. Chem.*, 1986, **25**, 4461.
[209] S. Tsuruya, S. Yani, and M. Masai, *Inorg. Chem.*, 1986, **25**, 141.
[210] A. Bakac, M. E. Brynildson, and J. H. Espenson, *Inorg. Chem.*, 1986, **25**, 4108.
[211] L. G. Marzilli, W. O. Parker, R. K. Kohli, A. L. Carrell, and J. P. Glusker, *Inorg. Chem.*, 1986, **25**, 127.
[212] E. Leporati, *J. Chem. Soc., Dalton Trans.*, 1986, 2587.

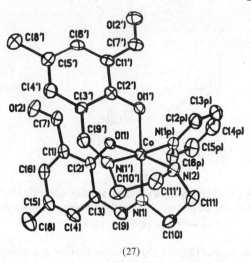

(27)

Structure of the [Co(3-formyl-5-Me-saldien)Py]$^+$ *cation, showing 30% probability thermal ellipsoids*
(Reproduced by permission from *Inorg. Chem.*, 1986, **25**, 4461)

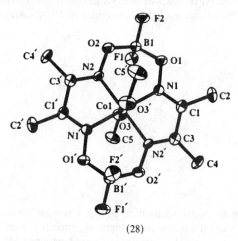

(28)

Full ORTEP drawing of the molecular structure of [(MeOH)$_2$Co(dmgBF$_2$)$_2$]
(Reproduced by permission from *Inorg. Chem.*, 1986, **25**, 4111)

N,N',N'',N'''-tetrakis(2-aminoethyl)-1,4,8,11-tetra-azacyclotetradecane has been synthesized as have dinuclear metal complexes M$_2$LX$_4$Y$_{5-n}$ where X and Y are anions Cl$^-$, Br$^-$, ClO$_4^-$, or OH$^-$ and M = Co, Ni, or Cu. The metal atoms are co-ordinated by two rings and two pendant nitrogen atoms.[213] A structural determination for the complex [Co$_2$\{H$_2$N(CH$_2$)OH\}$_3$\{H$_2$N(CH$_2$)$_2$O\}$_3$](ClO$_4$)$_3$.0.5H$_2$O using

213 I. Murase, M. Mikuriya, H. Sonoda, Y. Fukuda, and S. Kida, *J. Chem. Soc., Dalton Trans.*, 1986, 953.

neutron diffraction at 120 K points up the relative superiority of this technique over X-ray methods. Here the disorder in the water molecules and in the perchlorates was well resolved and some exceptionally short OHO hydrogen bonds were recorded.[214]

Reaction of alkynes with cobalt(II) octaethylporphyrins oxidized with iron(III) perchlorate yields N^{21},N^{22}-etheno-bridged porphyrins.[215] The presence of four electron-withdrawing groups in $[(CN)_4(TPP)Co^{II}]$ is responsible for extremely easy reductions, three such being detected using thin layer spectroelectrochemistry.[216] Cobalt(II) and copper(II) tetra-aza-annulenes based on (29) have been characterized and their redox properties studied.[217]

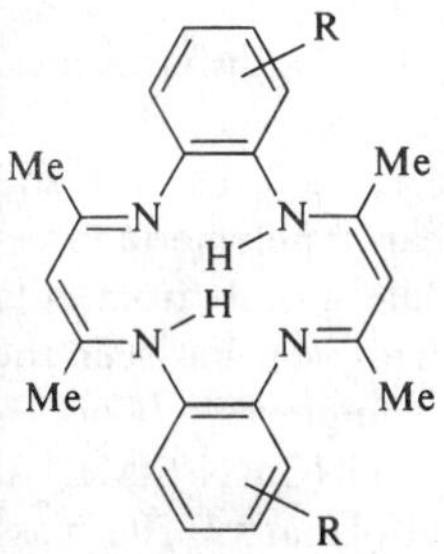

(29)

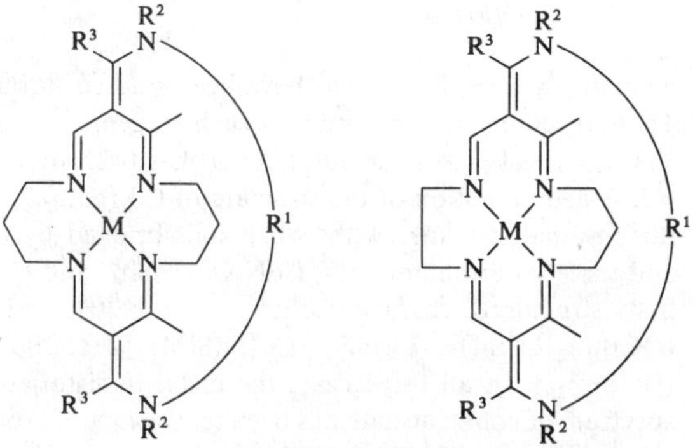

$R^1 = (CH_2)_{3-8 \text{ or } 12};$
$R^2 = Me; R^3 = Me$ or Ph

(30)

One-electron oxidation of the macrocyclic cyclidene complexes (30) of cobalt(II) or nickel(II) has been looked into with use of electronic and e.p.r. spectroscopy. The nature of the 3+ species depends on axial ligation and both ligand and metal oxidation were detected.[218] Reaction of cobalt(II) chloride with the ligand (31) yields the high-spin distorted tetragonal-pyramidal $[CoLCl]^+$ compounds, readily isolable as perchlorate salts.[219] An octahedral $[CoS_3N_3]$ cobalt(II) derivative $[CoL]$-$(ClO_4)_2$ (L = 1-methyl-3,13,16-triathia-6,8,10,19-tetra-azabicyclo[6.6.6]eicosane, azacapten) has been prepared and its structure determined. Strain energy minimization calculations reproduce the solid-state geometry quite well.[220]

The cobalt(II) macrocycle (32) catalyses the reaction between H^+ and Co^{II}, Eu^{II}, or V^{II}, liberating hydrogen.[221] Outer-sphere electron-transfer reactions of cobalt complexes $[CoL_3(BX)_2]^+$ (L = dimethylglyoximato, moximato, or diphenylgly-

[214] D. J. Jones, J. Roziere, and M. S. Lehmann, *J. Chem. Soc., Dalton Trans.*, 1986, 651.
[215] J. Stetsune, M. Ikeda, Y. Kishimoto, and T. Kitao, *J. Am. Chem. Soc.*, 1986, **108**, 1309.
[216] X. Q. Lin, B. Boisselier-Cocolios, and K. M. Kadish, *Inorg. Chem.*, 1986, **25**, 3242.
[217] C. L. Bailey, R. D. Bereman, and D. P. Rillema, *Inorg. Chem.*, 1986, **25**, 3149.
[218] M. Y. Chavan, T. J. Meade, D. H. Bush, and T. Kuwana, *Inorg. Chem.*, 1986, **25**, 414.
[219] C. Che, S. Mak, and T. C. W. Mak, *Inorg. Chem.*, 1986, **25**, 4705.
[220] T. W. Hambley and M. R. Snow, *Inorg. Chem.*, 1986, **25**, 1378.
[221] P. Connolly and J. H. Espenson, *Inorg. Chem.*, 1986, **25**, 2684.

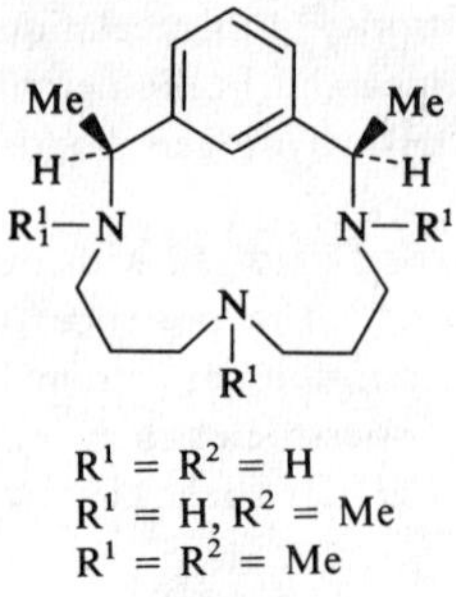

$$R^1 = R^2 = H$$
$$R^1 = H, R^2 = Me$$
$$R^1 = R^2 = Me$$

(31)

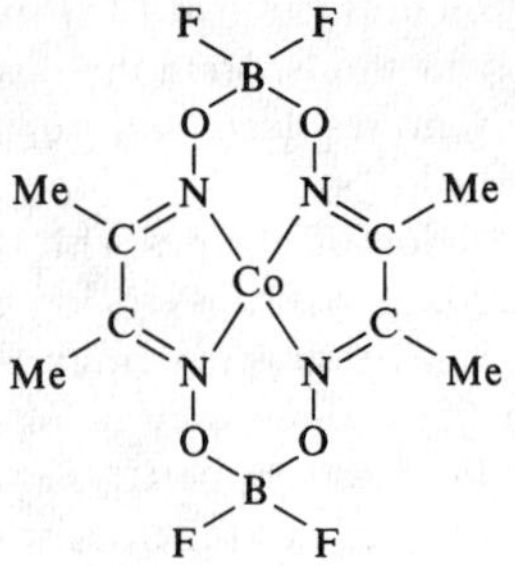

(32)

oximato; X = F, Bu, or Ph) have been studied through reactions with ferrocenes. These reactions are first-order in each reagent.[222]

The series [CoXNi$_2$-x(edta).2H$_2$O] has been prepared for x = 2.0, 1.5, 1.0, and 0.7. A determination of the structure of Co$_2$(edta).2H$_2$O reveals a polymeric three-dimensional structure with cobalt ions bridged by carboxylate groups from edta and three co-ordinate types, CoN$_2$O$_4$, CoO$_4$, and CoO$_6$.[223] This year has seen the first structural characterization of cobalt(I) alkoxide complexes, [Co(Cl)-(OCBu^t)$_2$]Li(thf)$_3$, Li(thf)$_{4.5}$[Co{N(SiMe$_3$)$_2$}(OCBu^t)$_2$], and Li[{Co(N(SiMe$_3$)$_2$)-(OCBu^t)$_2$}]. In all three cases the metal is distorted trigonal-planar.[224] The mass spectrum of cobalt acetate has been reinvestigated: some cluster ions with molecular masses in excess of 1000 were observed and several interesting ion molecule reactions established.[225] Oxygen-17 n.m.r. spectroscopy in [Co(H$_2$O)$_6$]$^{2+}$ recorded over a range of pressure indicates scalar coupling between the paramagnetic ion and water.[226] The kinetics and mechanism of the reaction of cobalt(II) or nickel(II) with 1-phenylhexane-1,3,5-trione to form the 1:1 and 2:2 complexes are consistent with a two-step mechanism, the first of which is second-order in complex and first-order in hydroxide.[227] The rates of exchange of *N,N*-diethylformamide on [Co(def)$_6$]$^{2+}$ or [Ni(def)$_6$]$^{2+}$ have been measured over a range of temperature using ^{1}H n.m.r. The same procedure using ^{13}C n.m.r. was followed in a study of dimethylacetamide exchange in analogous compounds.[228]

Redox properties and e.p.r. spectra for low-spin cobalt(II) bis(1,4,7-trithiacyclo-nonane) reveal three one-electron oxidation steps corresponding to cobalt(III)/(II), (II)/(I), and (I)/(0).[229] Cobalt(II) complexes of hexathia-18-crown-6 or 2,5,8-trithianonane have been fully characterized as tetrafluoroborates or picrates and their redox behaviour and e.p.r. spectra examined.[230]

[222] D. Borchardt and S. Wherland, *Inorg. Chem.*, 1986, **25**, 901.
[223] P. Gomez-Romero, G. B. Jameson, N. Cawen-Pastor, E. Coronado, and D. Beltran, *Inorg. Chem.*, 1986, **25**, 3171.
[224] M. M. Olmstead, P. P. Power, and G. Siegel, *Inorg. Chem.*, 1986, **25**, 1027.
[225] G. C. DiDonato and K. L. Busch, *Inorg. Chem.*, 1986, **25**, 1551.
[226] R. M. Nielson, H. W. Dodgen, J. P. Hunt, and S. E. Whertland, *Inorg. Chem.*, 1986, **25**, 582.
[227] J. Walsh and M. J. Hynes, *J. Chem. Soc., Dalton Trans.*, 1986, 2243.
[228] S. F. Lincoln, A. M. Hounslow, and A. N. Boffa, *Inorg. Chem.*, 1986, **25**, 1038.
[229] G. S. Wilson, D. D. Swanson, and R. S. Glass, *Inorg. Chem.*, 1986, **25**, 3827.
[230] Y. Wakatsuki, S. Miya, and H. Yamazaki, *J. Chem. Soc., Dalton Trans.*, 1986, 1207.

Compounds of Cobalt(III).—Flash photolysis of air equilibrated solutions of cobalt(III) hexammines yields cobalt(II) dioxygen complexes and, eventually, a μ-superoxo-dinuclear cobalt(III) complex and, lastly, a μ-superoxo-complex.[231] Discrepancies between theory and experiment for the $[Co(NH_3)_6]^{2+/3+}$ self-exchange have been resolved with the aid of an INDO calculation of the energy barrier.[232] Ligand-to-metal electron transfer in a range of eighteen cobalt(III) pentammines containing co-ordinated radicals has been studied using pulse radiolytic techniques.[233] Alkaline hydrolysis of p-$[Co(tren)(NH_3)SCN]^{2+}$ results in only p-$[Co(tren)(NH_3)(OH)]^{2+}$ but hydrolysis of t-$[Co(tren)(NH_3)(SCN)]^{2+}$ results in t-SCN-$[Co(tren)(OH)SCN]^+$ and t-$[Co(tren)(NH_3)(NCS)]^{2+}$ as well as p-$[Co(tren)(NH_3)(OH)]$ and t-$[Co(tren)(NH_3)(OH)]^{2+}$.[234] Pentakis(methyl-amine)cobalt(III) complexes $[Co(MeNH_2)_5L]^{3+}$ (L = urea, dimethyl sulphoxide, dimethylformamide, trimethyl phosphite, or acetonitrile), have been synthesized and the rates of their acid-catalysed aquation reactions measured over a range of temperature.[235] Flash photolysis of deoxygenated solutions of $[Co(sep)]^{3+}(X)_5^-$ (X = I$^-$ or Br$^-$) yields transient X_2^- species.[236] The kinetics for the reaction of percarboxylic acids or alkyl hydroperoxides with (tetraphenylporphinato)cobalt(III) chloride have been investigated and a mechanistic scheme advanced.[237] New *trans*-octahedral Co(Cl)(L) bis-dimethylglyoximato or -diphenylglyoximato complexes (L = *e.g.* 4-cyanopyridine) have been shown to undergo displacement reactions in the presence of cyanide ion to give cyano-derivatives.[238]

The photochemical formation of a sulphinato-O cobalt(III) complex has a volume of activation of 6 cm^3 mol^{-1} and proceeds by ring-opening of the sulphinate ligand. The thermal back reaction is associated with a decrease in volume of 13 cm^3 mol^{-1}.[239] Other kinetic work published this year is summarized in Table 1.[240-256]

[231] P. Ramamurthy and P. Natarajan, *Inorg. Chem.*, 1986, **25**, 3554.
[232] S. Larsson, K. Stahl, and M. C. Zerner, *Inorg. Chem.*, 1986, **25**, 3033.
[233] K. D. Whitburn, M. Z. Hoffman, N. V. Brezniak, and M. G. Simic, *Inorg. Chem.*, 1986, **25**, 3037.
[234] M. J. Gaudin, C. R. Clark, and D. A. Buckingham, *Inorg. Chem.*, 1986, **25**, 2569.
[235] N. J. Curtis and G. A. Lawrance, *Inorg. Chem.*, 1986, **25**, 1033.
[236] F. Pina, M. Maestri, R. Ballardine, Q. G. Mulazzani, M. D'Angelantonio, and V. Balzani, *Inorg. Chem.*, 1986, **25**, 4249.
[237] W. A. Lee and T. C. Bruice, *Inorg. Chem.*, 1986, **25**, 131.
[238] C. Lopez, S. Alvarez, X. Solans, and M. Font-Altaba, *Inorg. Chem.*, 1986, **25**, 2962.
[239] W. Weber, H. Maecke, and R. van Eldik, *Inorg. Chem.*, 1986, **25**, 3093.
[240] E. Baraniak, D. A. Buckingham, C. R. Clark, and A. M. Sargeson, *Inorg. Chem.*, 1986, **25**, 1956.
[241] R. van Eldik, Y. Kitamura, and C. P. Piriz Mac-Coll, *Inorg. Chem.*, 1986, **25**, 4252.
[242] I. Krack and R. van Eldik, *Inorg. Chem.*, 1986, **25**, 1743.
[243] E. Baraniak, D. A. Buckingham, C. R. Clark, B. H. Moynihan, and A. M. Sargeson, *Inorg. Chem.*, 1986, **25**, 3466.
[244] D. A. Buckingham and C. R. Clark, *Inorg. Chem.*, 1986, **25**, 3478.
[245] A. A. El-Awady and G. M. Harris, *Inorg. Chem.*, 1986, **25**, 1323.
[246] R. J. Balahura and A. Johnston, *Inorg. Chem.*, 1986, **25**, 652.
[247] H. Ogino, A. Masuko, S. Ito, N. Miura, and M. Shimura, *Inorg. Chem.*, 1986, **25**, 708.
[248] A. C. Dash and R. C. Nayak, *Inorg. Chem.*, 1986, **25**, 2237.
[249] V. K. Joshi, R. van Eldik, and G. M. Harris, *Inorg. Chem.*, 1986, **25**, 2229.
[250] N. J. Curtis, G. A. Lawrance, P. A. Lay, and A. M. Sargeson, *Inorg. Chem.*, 1986, **25**, 484.
[251] Y. Sasaki, K. Endo, A. Nagasawa, and K. Saito, *Inorg. Chem.*, 1986, **25**, 4845.
[252] S. K. Saha, M. C. Ghosh, and P. Banerjee, *J. Chem. Soc., Dalton Trans.*, 1986, 1301.
[253] M. H. M. Abou-El-Wafa, M. G. Burnett, and J. F. McCullagh, *J. Chem. Soc., Dalton Trans.*, 1986, 2083.
[254] A. Bakac, V. Butkovlic, J. H. Espenson, R. Marec, and M. Orhanovic, *Inorg. Chem.*, 1986, **25**, 341.
[255] E. Baraniak, D. A. Buckingham, C. R. Clark, and A. M. Sargeson, *Inorg. Chem.*, 1986, **25**, 1952.
[256] G. M. Miskelly, C. R. Clark, and D. A. Buckingham, *J. Am. Chem. Soc.*, 1986, **108**, 5202.

Table 1 *Kinetic studies of cobalt(III) compounds*

Complex	Reaction(s)	Ref.
cis-[Co(en)$_2$X(β-alaOPr^i)]$^{2+}$ X = halide	Hg^{2+}-catalysed removal of x, base hydrolysis	240
[Co(NH$_3$)$_4$(MeNH$_2$)Cl]$^{2+}$	Base-catalysed hydrolysis	2-+
[Co(NH$_3$)$_5$X]$^{3+}$ X = H$_2$O, Py, or Me$_2$SO	[Fe(CN)$_6$]$^{4-}$, ET reaction	242
[Co(en)$_2$(β-alaOPr^i)]$^{3+}$	Hydrolysis, aminolysis by glycine ethyl ester	243
		244
[Co(en)$_2$(OH$_2$)OH]$^{2+}$	SO$_2$ uptake	245
[NH$_3$)$_5$Co(NCacac)]$^{2+}$ NCacac = 3-cyanopentane-2,4-dionate	ET, Ti^{III}	246
[(NH$_3$)$_5$Co{(edta)Cr(H$_2$O)}]$^{2+}$	Ligand substitution	247
($\alpha\beta S$)-(tetren)(3NO$_2$-salicylato)Co^{III}	Base hydrolysis	248
cis-[Co(phen)$_2$(H$_2$O)$_2$]$^{3+}$ phen = 1,2-phenanthroline	SO$_2$ uptake	249
[Co(NH$_3$)$_5$OSO$_2$CF$_3$]$^{2+}$	Base hydrolysis, aquation	250
[Co(NH$_3$)$_5$H$_2$O]$^{3+}$	Ferrocyanide; ET, variable pressure study	251
Dodecatungstocobaltate(III)	Oxalate, ET	252
[Co(CN)$_5$N$_3$]$^{3-}$	Acid-catalysed substitution NCS$^-$	253
[Co(NH$_3$)$_5$(NC$_5$H$_4$X)]$^{3+}$ X = 3-CN, 3-Cl, 4-Me, or 4-CMe$_3$	ET; R·, V^{II}, Ru^{II}	254
[Co(en)$_2$(β-alao)]$^{2+}$	Hydrolysis	255
cis-[Co(en)$_2$(OC$_2$O$_3$)(H$_2$O)]PF$_6$	Hydrolysis	256

Three cobalt(III) complexes of L-histidyl glycylglicinate have been reported[257] and a seemingly stable complex containing an *N*-glycoside derived from D-mannose and 1,2-diaminoethane has been characterized by 400 MHz ^{1}H n.m.r. in course of which all the sugar signals were assigned and the conformation convincingly established.[258]

Cobalt(III) (oxalato)tetrammines or (α-amino-α-methylmalonato)tetrammines exhibit isotopic multiplets (^{13}C resonances shifted upfield owing to deuterium isotope effects) over three bonds. These shifts are dihedral angle-dependent which is what makes this technique useful.[259]

A ternary complex [Co(bipy)$_2$(PN-H)](ClO$_4$)$_2$ (PN-H represents an anionic pyridoxine) contains a [CoN$_4$O$_2$] centre[260] whilst Li[Co{(2-Py)$_3$CoH}$_2$]S$_2$O$_6$ is an octahedral [CoN$_6$] species.[261]

Methanolic [Co(NCS)$_3$(PMe$_3$)$_2$] undergoes disproportionation with NO at ambient temperature yielding *mer*-[Co(NCS)$_3$(PMe$_3$)$_3$] together with [Co(NO)$_2$(PMe$_3$)$_2$][Co(NCS)$_4$] from which latter a tetraphenylborate salt could be prepared.[262]

[257] C. J. Hawkins and J. Martin, *Inorg. Chem.*, 1986, **25**, 2146.
[258] K. Ishida, S. Yano, and S. Yoshikawa, *Inorg. Chem.*, 1986, **25**, 3552.
[259] M. Yashiro, S. Yano, and S. Yoshikawa, *J. Am. Chem. Soc.*, 1986, **108**, 1096.
[260] S. P. S. Rao, K. I. Varughese, and H. Manohar, *Inorg. Chem.*, 1986, **25**, 734.
[261] D. J. Szalda and F. R. Keene, *Inorg. Chem.*, 1986, **25**, 2795.
[262] O. Alnaji, Y. Peres, F. Dahan, M. Dartinguenave, and Y. Dartinguenave, *Inorg. Chem.*, 1986, **25**, 1383.

The structures of two (α-amino-oxime)(dinitro)cobalt(III) complexes, *trans*-[Co(ao-H-ao)(NO$_2$)$_2$] and *cis*-[Co(Hao)$_2$(NO$_2$)$_2$]NO$_3$ (Hao = 3-amino-3-methyl-butan-2-oneoxime), have been determined and the kinetic labilities of the nitro ligands measured by ^{15}N isotopic exchange. The rates did not correlate at all with the metal–nitrogen bond distances.[263]

A comparative study of [Co(CN)$_6$]$^{3-}$ and [Co(CN)$_5$OH]$^{3-}$ using Hartree–Fock calculations reveals a satisfactory correlation with the experimental ligand field spectrum.[264] The conformations of the chelate rings differ significantly as between *trans*-{Co(CN)$_2$L$_2$]$^+$Cl$^-$.H$_2$O and the trihydrate (L = propane-1,3-diamine), chair and skew boat conformations being detected by *X*-ray diffraction and further studied using electronic spectroscopy.[265]

A new type of atropisomerism (33) in metal complexes has been uncovered in the twin complex (3,4-diacetylhexane-2,5-dionato)bis(2,2',2''-triaminotriethyl-amine)cobalt(III).[266] The bis(μ-(+)-tartrato]diantimonate(III) anion is a useful eluent in ion-exchange chromatography of hexammine cobalt(III) cage complexes such as [Co(sep)]$^{3+}$ (sep = 1,3,6,8,10,13,16,19-octa-azabicyclo[6.6.6]eisocane) when the Λ-enantiomer elutes first.[267] Hydrogen- and π-bonding, important secondary interactions in complex chemistry, have been analysed in the isomorphous series [M(NH$_3$)Cl]Cl$_2$ (M = Co, Cr, Rh, Ir, Rn, or Os)[268] and ^{59}Co n.m.r. spectra have been recorded for some pentammines [Co(NH$_3$)$_5$L]$^{3+}$ (L = imidazole, *N*-methyl-imidazole, pyridine, or NH$_3$) and for [Co(por)L$_2$]$^+$ (por = dianion of tetraphenylpor-phyrin or octaethylphorphyrin). It is claimed that these spectral methods are more

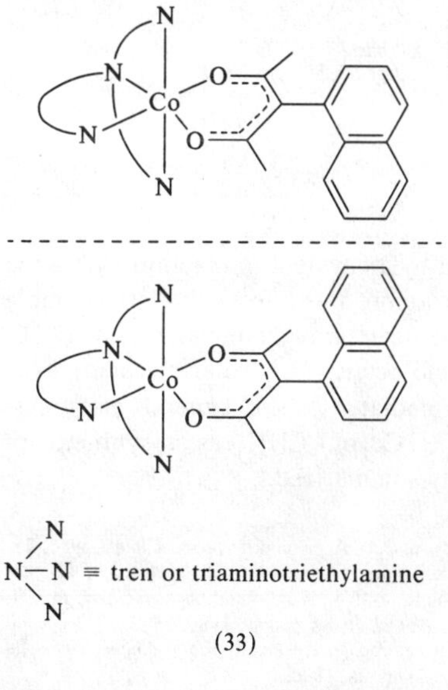

(34)

N—N ≡ tren or triaminotriethylamine

(33)

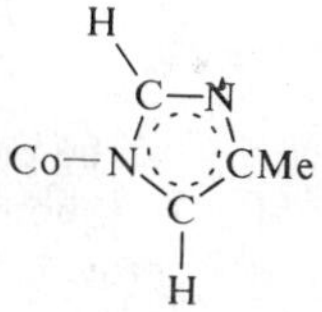

[263] D. E. Murray, E. O. Schlemper, S. Siripaisarnpipat, and R. K. Murmann, *J. Chem. Soc., Dalton Trans.*, 1986, 1759.
[264] L. G. Vanquickenborne, M. Hendrickx, I. Hyla-Kryspin, and L. Haspeslagh, *Inorg. Chem.*, 1986, **25**, 885.
[265] H. Kupka, J. Degan, A. Urushiyama, K. Angermund, and C. Kruger, *Inorg. Chem.*, 1986, **25**, 3294.
[266] Y. Nakano, Y. Yoshikawa, and H. Kondo, *J. Am. Chem. Soc.*, 1986, **108**, 7630.
[267] K. Miyoshi, S. Izumoto, K. Nakai, and H. Yoneda, *Inorg. Chem.*, 1986, **25**, 4654.
[268] T. W. Hambley and P. A. Lay, *Inorg. Chem.*, 1986, **25**, 4553.

sensitive than optical spectroscopy when it comes to placing ligands in the spectrochemical series.[269]

Controlled heating of the double salts $[CoH_2O(NH_3)_5][Co(CN)_5X]$ (X = CN, Cl, Br, *etc.*) yields dinuclear salts of the type $[(NH_3)_5Co(NC)Co(CN)_4X]$.[270] The absolute configurations of the isomers of triamminecobalt(III) adenosine triphosphate have been determined by an n.m.r. method.[271] A structure determination of [(4-methylimidazole)pentamminocobalt(III)] trichloride reveals 5-methylimidazole co-ordination (34).[272] Low-spin d^6 transition-metal complexes *e.g.* $[(Co(en)_3]^{3+}$, serve as photoactivated cleavers of the sugar-phosphate backbone of double stranded DNA.[273] The thermal reactions of solid, crystalline, *cis*-diammine bisethylenediamine salts, *e.g.* *cis*-$[Co(en)_2(NH_3)_2]^{3+}$, have been investigated. The PF_6 salt loses PF_5, forming *cis*-$[Co(en)_2(NH_3)_2]F(PF_6)_2$ and then *cis*-$[Co(en)_2(NH_3)F](PF_6)_2$.[274] Four *N*-substituted (H, Me, $HOCH_2CH_2$, benzyl) ethylenediaminetriacetato(thiocyanato)cobalt(III) complexes $[CoL(NCS)]^-$ have been prepared and satisfactorily characterized by ^1H and ^{13}C n.m.r. spectroscopy.[275] The preparation of some mixed-ligand complexes, *e.g.* *trans.*$[CoCl_2(Me_2tn)_2]^+X^-$ [X = ClO_4^- or $1/2ZnCl_4^{2-}$; tn = $NH_2(CH_2)_3H_2$] or *mer*-$[CoCl(Me_2tn)(tri)][tnCl_4]$ (tri = $NH_2CH_2CMe_2CH_2N=CHCMe_2CH_2NH_2$) has been achieved.[276] The tetra-anionic ligand (35) forms a cobalt(III) derivatives $Na[CoL].Me_2CO$ in which square-planar geometry has been forced upon the cobalt, leading to an $S = 1$ electronic ground state.[277]

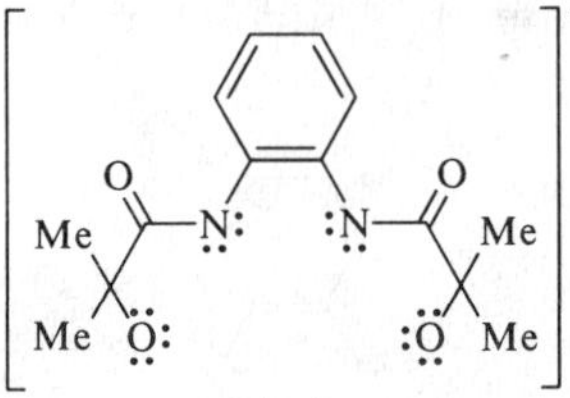

(35)

C—H Reactivity can be promoted by a covalent interaction with a metal to form a three-centre M—H—C bond. An example of this in a cobalt(III) complex–aqueous solution context is provided by the structural characterization of $[Co^{III}dacoda-C(2)(H_2O)]$ (dacoda = 1,5-diazacyclo-octane-*N,N*'-diacetic acid) as a deprotonation product of an agostic intermediate.[278] The general usefulness of *trans*-dichlorotetrakis(pyridine)cobalt(III), $[Co(py)_4Cl_2]^+$, as a synthetic precursor in cobalt(III) chemistry has been nicely summarized.[279]

[269] K. I. Hagen, C. M. Schwab, J. O. Edwards, and D. A. Sweigart, *Inorg. Chem.*, 1986, **25**, 978.

[270] J. Casabo, T. Flor, F. Texidor, and J. Ribas, *Inorg. Chem.*, 1986, **25**, 3166.

[271] D. C. Speckhard, V. L. Pecoraro, W. B. Knight, and W. W. Cleland, *J. Am. Chem. Soc.*, 1986, **108**, 4167.

[272] W. W. Henderson, R. E. Shepherd, and J. Abola, *Inorg. Chem.*, 1986, **25**, 3157.

[273] M. B. Fleischer, K. C. Waterman, N. J. Turro, and J. K. Barton, *Inorg. Chem.*, 1986, **25**, 3549.

[274] T. Fujiwara and J. C. Bailar, *Inorg. Chem.*, 1986, **25**, 1806.

[275] K. E. Rohly, B. E. Douglas, and C. Maricondi, *Inorg. Chem.*, 1986, **25**, 4119.

[276] D. A. House, *Inorg. Chem.*, 1986, **25**, 1671.

[277] T. J. Collins, T. G. Richmond, B. D. Santarsiero, and B. G. R. T. Treco, *J. Am. Chem. Soc.*, 1986, **108**, 2088.

[278] K. Kanamori, W. E. Broderick, R. F. Jordan, R. D. Willett, and J. I. Legg, *J. Am. Chem. Soc.*, 1986, **108**, 7122.

[279] W. L. Purcell, *Inorg. Chem.*, 1986, **25**, 4068.

4 Nickel

Low Valence State Compounds of Nickel.—The luminescence spectrum of tetracarbonylnickel following photolysis with an XeCl laser is strong enough to be visually detected. A three-step mechanism is advanced: luminescence is primarily from the $Ni(CO)_3$ fragment with its smaller HOMO–LUMO gap resulting in visible emission.[280] The dinickel complex $[\{\mu\text{-}Me_2PCH_2PMe_2\}Ni_2(CO)_4]$ is formed after reaction of nickel carbonyl with the phosphine.[281] A new series of organonickel complexes results from the interaction of alkyl halides with cationic nickel(I) $[NiN_4]$ macrocyclic complexes.[282]

In the isomerization of but-1-ene in bromobenzene as catalysed by $[(\eta^6\text{-}$arene)NiR_2]$ derivatives, (arene = benzene, toluene, or mesitylene; R = $SiCl_3$, SiF_3, or Ph) the intermediacy of a short-lived nickel hydride species from insertion of but-1-ene into the Ni—C bond has been established.[283] The nickelacyclopentene (36) undergoes an insertion reaction with CO_2.[284]

(36)

Compounds of Nickel(II).—The low-temperature phase of $Mg_2(NiD_4)$ is monoclinic with $Z = 8$ and contains tetrahedral NiD_4 units. This is another 18-electron hydrido complex to compare with $[FeD_6]^{4-}$ and $[CoD_5]^{4-}$.[285] It is suggested that the electronic and geometric structures of spin-paired nickel(II) complexes are derived from a Jahn–Teller distortion of the ^1E_g excited state of a regular octahedral complex.[286] The crystal structures of aromatic molecule inclusion compounds of $((\alpha,\omega\text{-}$diaminoalkane)cadmium(II) tetracyanonickelates(II)) have been determined. The cyano-groups bridge nickels to cadmiums giving sheets interconnected by ambident binding of the aminoalkanes. Cubic cells for the guest molecules are thereby generated.[287] The electronic, i.r., and resonance Raman spectra of the mixed-valence compounds $[Ni(en)_2][Pt(en)_2X_2](ClO_4)_2$ (X = Cl, Br, or I; en = 1,2-diaminoethane) have been recorded. These $[Ni^{II}Pt^{IV}]$ complexes have more localized valencies than their $[Pt^{II}Pt^{IV}]$ analogues.[288] The conformations of some nickel(II) tetra-aza-macrocyclic complexes (37) have been characterized using proton n.m.r.[289] Six stereoisomers of the nickel(II) complex of the ligand (38) have

[280] N. Rosch, M. Kotzian, H. Jorg, H. Schroder, B. Rager, and S. Metev, *J. Am. Chem. Soc.*, 1986, **108**, 4238.
[281] K. R. Porshke, Y. Tsay, and C. Kruger, *Inorg. Chem.*, 1986, **25**, 2097.
[282] M. S. Ram, A. Bakac, and J. H. Espenson, *Inorg. Chem.*, 1986, **25**, 3267.
[283] H. Kani, S. B. Choc, and K. J. Klabunde, *J. Am. Chem. Soc.*, 1986, **108**, 2019.
[284] E. Carmona, P. Plama, M. Paneque, M. L. Poveda, E. Gutierrez-Puebla, and A. Monge, *J. Am. Chem. Soc.*, 1986, **108**, 6424.
[285] P. Zolliker, K. Ivon, J. D. Jorgensen, and F. J. Rotella, *Inorg. Chem.*, 1986, **25**, 3590.
[286] R. J. Deeth and M. A. Hitchman, *Inorg. Chem.*, 1986, **25**, 1225.
[287] S. Nishikiori, and T. Iwamoto, *Inorg. Chem.*, 1986, **25**, 788.
[288] R. J. H. Clark and V. B. Croud, *Inorg. Chem.*, 1986, **25**, 1751.
[289] J. Chen, C. Chang, and C. Chung, *Inorg. Chem.*, 1986, **25**, 4794.

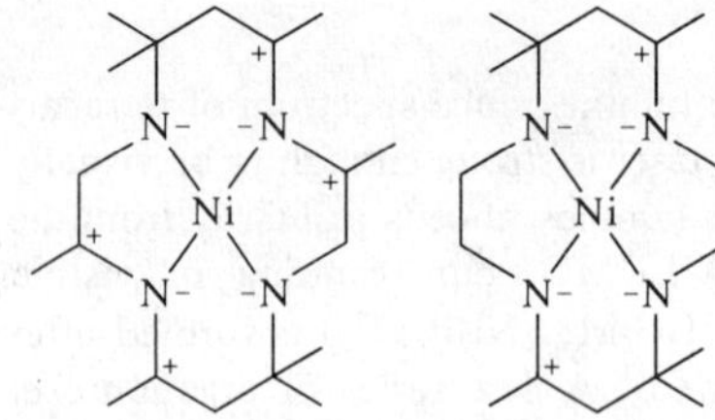

+, − indicate configurations at carbon atoms

(37)

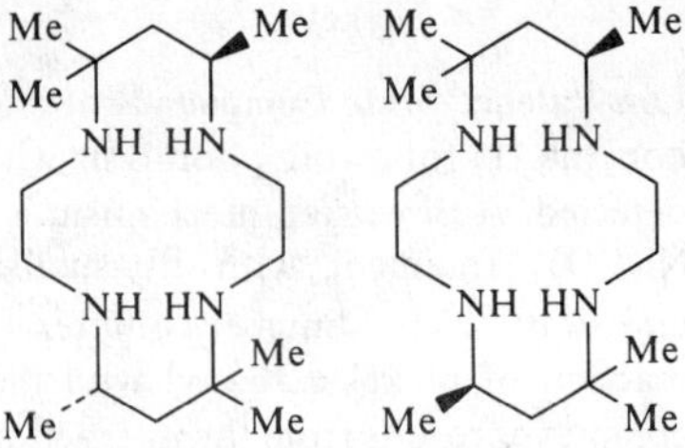

(38)

been characterized and their relative energies assessed.[290] The syntheses of a range of *cis*-octahedral nickel(II) complexes of the ligands (39), (40), and (41) have been achieved and their rates of base-catalysed isomerization determined.[291] Redox potentials for the oxidation and reduction of a range of nickel(II) complexes of the type

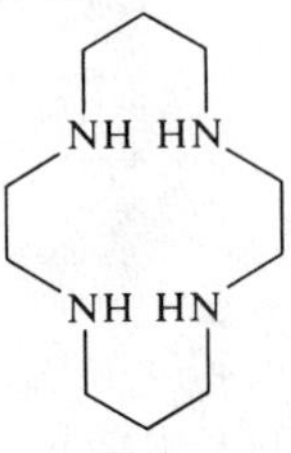

(39) 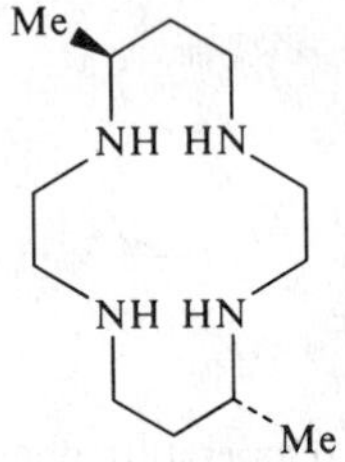   (40) 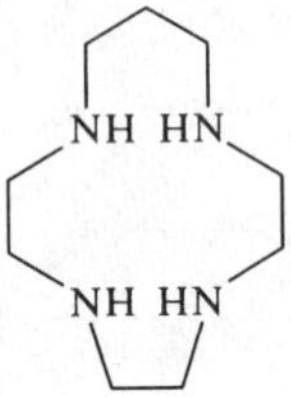   (41)

(42) have been determined and the structures of octahedral *trans* bis(aqua) and square-planar examples solved by X-ray diffraction.[292] The bis(macrocycle) ligand (43) forms a heterobimetallic $Ni^{II}Cu^{II}$ complex. Successive one-electron oxidations

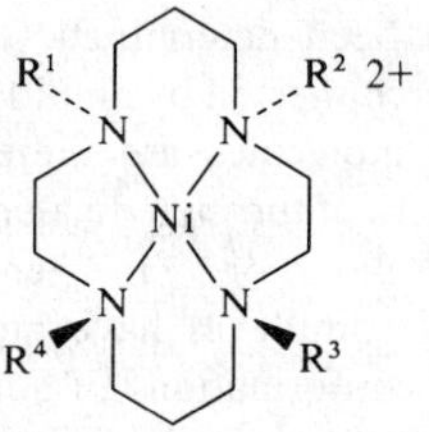

R^{1-4} = H, Me, CH_2Ph, *etc.*

(42)

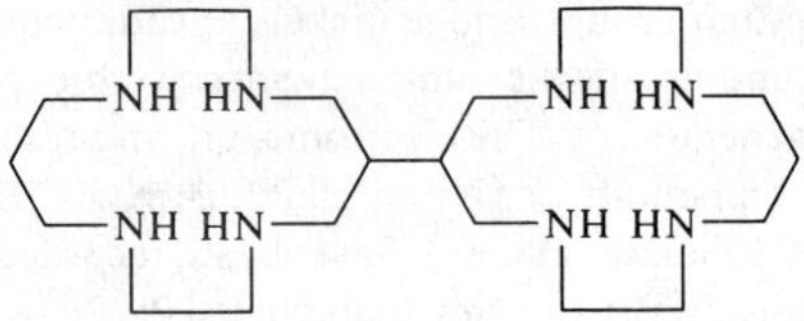

(43)

were detected, the nickel being oxidized first.[293] Potentiometric, spectrophotometric, kinetic, and electrochemical studies of nickel(II) or cobalt(II) complexes of the tetra-aza-macrocyclic species (44) and (45) have been reported.[294] A nickel(II)

[290] J. Chen and C. Chung, *Inorg. Chem.*, 1986, **25**, 2841.
[291] R. W. Hay and M. P. Pujari, *J. Chem. Soc., Dalton Trans.*, 1986, 1485.
[292] E. K. Barefield, G. M. Freeman, and D. G. Van Derveer, *Inorg. Chem.*, 1986, **25**, 552.
[293] L. Fabbrizzi, L. Montagna, A. Poggi, T. A. Kaden, and L. C. Siegfried, *Inorg. Chem.*, 1986, **25**, 2671.
[294] L. Fabbrizzi, T. A. Kaden, A. Perotti, B. Seghi, and L. Siegfried, *Inorg. Chem.*, 1986, **25**, 321.

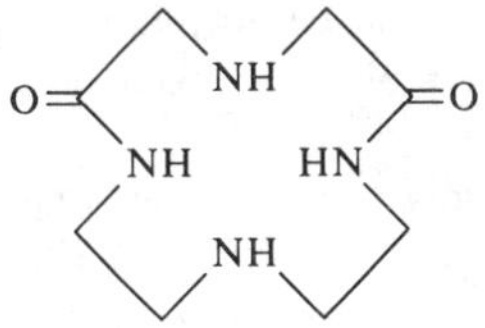

(44) 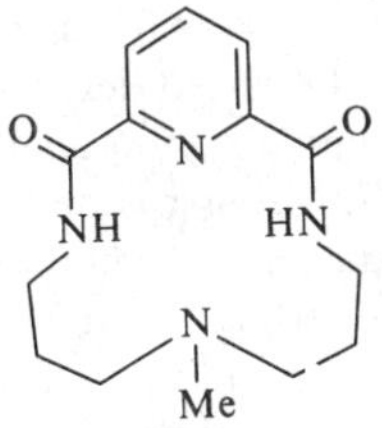   (45) 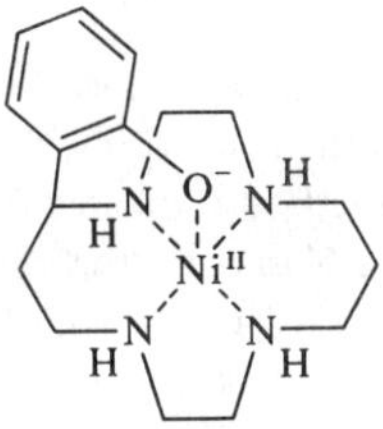   (46)

complex of the cyclam (46) has been prepared and its structure shown to be square pyramidal [NiN$_4$O].[295] Hydrolysis rates of the organonickel tetramethylcyclam complexes [RNi(tmc)]$^+$ (tmc = 1,4,8,11-tetramethyl-1,4,8,11-tetra-azacyclotetradecane) have been investigated for a range of alkyl groups and over a range of pH.[296] The nickel(I) macrocycle [Ni(1R,4R,8S,11S-tmc)]$^+$ reacts with primary alkyl halides in alkaline aqueous solution yielding alkyl nickel(II) complexes [RNi(tmc)]$^+$.[297] The reactivity of R^{1}X increases in the order methyl < primary < secondary and Cl < Br < I.[298] Single-crystal X-ray structures of two new nickel(III) cyclam complexes have been reported.[299] Complexing of nickel(III) with the macrocycles (47), (48), and (49) reveals a high-spin–low-spin conversion and some interesting redox reactions.[300] The reaction rates of chloride or bromide with [NiL(H$_2$O)$_2$]$^{3+}$ (L = dimethylcyclam) have been measured in aqueous perchloric acid and a mechanism has been advanced.[301]

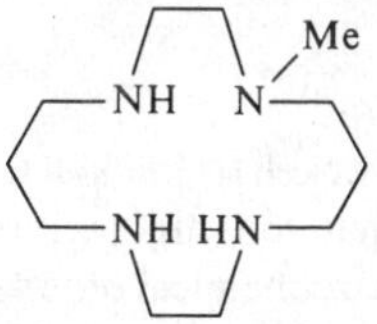

(47) 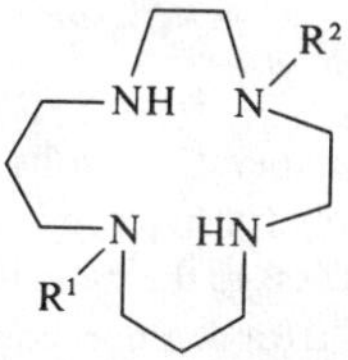   (48) R$_1$ = H; R$_2$ = Me
                        (49) R$_1$ = Me; R$_2$ = H

Carbon dioxide fixation continues to be a lively topic in applied co-ordination chemistry and [Ni(cyclam)]$^{2+}$ proves to be an efficient and selective electrocatalyst for the reduction of CO$_2$ to CO in aqueous solution.[302] The complexes [NiL][MCl$_4$] [L = (±)-5,7,7,12,14-hexamethyl-1,4,811-tetra-azacyclotetradecane; M = Mn, Co, Cu, or Zn] have been synthesized. The structure of the zinc derivative has been solved: it is a dinuclear [N$_4$NiCl$_2$ZnCl$_2$] *cis*-octahedral–tetrahedral species. The magnetic coupling in the other complexes is weak.[303] *cis*-Diaquanickel(II)cyclam goes over to the planar analogue at 25 °C *via* a transient species identified as a

[295] Y. Iitaka, T. Koike, and E. Kimura, *Inorg. Chem.*, 1986, **25**, 402.
[296] M. S. Ram, J. H. Espenson, and A. Bakac, *Inorg. Chem.*, 1986, **25**, 4115.
[297] A. Bakac and J. H. Espenson, *J. Am. Chem. Soc.*, 1986, **108**, 713.
[298] A. Bakac and J. H. Espenson, *J. Am. Chem. Soc.*, 1986, **108**, 719.
[299] E. K. Barefield, A. Bianchi, E. J. Billo, P. J. Connolly, P. Paoletti, J. S. Summers, and D. G. Van Derveer, *Inorg. Chem.*, 1986, **25**, 4197.
[300] M. Ciampolini, L. Fabbrizzi, M. Licchelli, A. Perotti, F. Pezzini, and A. Poggi, *Inorg. Chem.*, 1986, **25**, 4131.
[301] M. G. Fairbank and A. McAuley, *Inorg. Chem.*, 1986, **25**, 1233.
[302] M. Beley, J. Collin, R. Ruppert, and J. Sauvage, *J. Am. Chem. Soc.*, 1986, **108**, 7461.
[303] A. Bencini, A. Caneschi, A. Dei, D. Gatteschi, C. Zanchini, and O. Kahn, *Inorg. Chem.*, 1986, **25**, 1374.

configurational isomer of the planar species.[304] The complex [NiL]-$(CF_3SO_3)_2.Me_2CO.H_2O$ (L = 1,4,8,11-tetramethylcyclam) is four-co-ordinate but with distortions. Strain-energy minimization analysis by the molecular mechanics method indicates that the R,S,R,S-isomer is preferred.[305]

N.m.r. proton relaxation studies have been useful in a study of bimetallic complexes of Ni^{II}, Cu^{II}, and Sn^{II} of formula $[M^1M^2L_3]^{4+}$ [L = 2,5-di(2-pyridyl)-3,4-diazahexa-2,4-diene]. Magnetic coupling subsists in the bisnickel complex and to a lesser extent in the copper analogue.[306] The [4N] chelate ligand 7-methyl-3,7,11,17-tetra-azabicyclo[11.3.1]heptadeca-1(17),13,15-triene has been prepared and from it $[ML]X_2$ (M = Ni^{2+}, Cn^{2+}, or Zn^{2+}; X = ClO_4 or NO_3) have been isolated and characterized. Some products of reduction were isolated after reaction with H_2–Raney nickel. One of these, $[NiL^1](ClO_4)_2$, involves a ligand having a reduced pyridine ring.[307] A molecular mechanics study has been made of $[NiL]^{2+}$, $NiL(NO_2)(ONO)$, and $[NiL(en)]^{2+}$ [L = (50), en = 1,2-diaminoethane] and the steric energies have been calculated.[308]

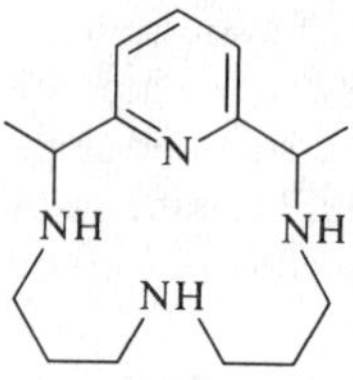

(50) 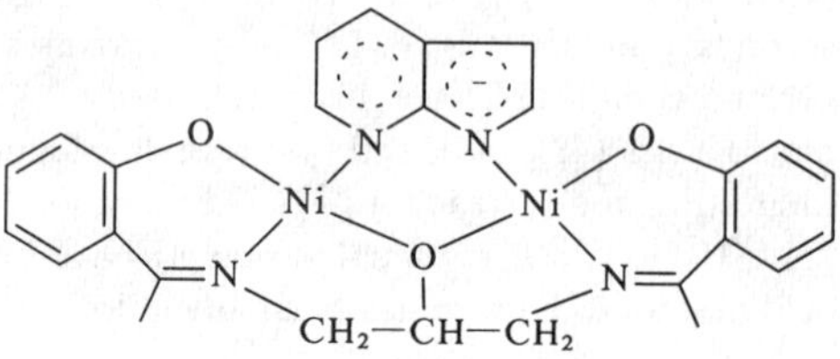   (51)

Variable-temperature voltammetry applied to the nickel(II)/nickel(III) redox couple in polyaza-macrocyclic complexes indicates that ion–dipole interactions control the access to the nickel(III) state.[309] Repeated electrochemical redox scanning of nickel(II) tetra-azalene complexes leads to deposition of polymeric films upon the electrodes. Characterization of the films and further studies led to the development of useful modified electrodes.[310] Electrochemical reducing paths for analogous dinuclear nickel(II) and copper(II) complexes, *e.g.* (51), have been shown to be quite different, two reversible processes for the nickel compounds being detected and the copper case being obscured by redox cross reactions.[311] The participation of macrocyclic nickel(II) complexes in new oscillating reactions has been demonstrated[312] and dinickel(II) bis(cyclidene) complexes are fluxional in solution and exhibit redox couples dependent upon the intermetallic distance.[313] Stopped-flow spectroscopy has been applied to the formation kinetics of ternary complexes of

[304] A. M. Martin, K. J. Grant, and E. J. Billo, *Inorg. Chem.*, 1986, **25**, 4904.
[305] T. W. Hambley, *J. Chem. Soc., Dalton Trans.*, 1986, 565.
[306] C. Owens, R. S. Drago, I. Bertini, C. Luchinat, and L. Banci, *J. Am. Chem. Soc.*, 1986, **108**, 3298.
[307] N. W. Alcock, P. Moore, and H. A. A. Omar, *J. Chem. Soc., Dalton Trans.*, 1986, 985.
[308] M. G. B. Drew, D. A. Rice, S. Silong, and P. C. Yates, *J. Chem. Soc., Dalton Trans.*, 1986, 1081.
[309] L. Fabbrizzi, A. Perotti, A. Profumo, and T. Soldi, *Inorg. Chem.*, 1986, **25**, 4256.
[310] C. L. Bailey, R. D. Bereman, D. P. Rillema, and R. Nowak, *Inorg. Chem.*, 1986, **25**, 933.
[311] W. Mazurek, A. M. Bond, M. J. O'Connor, and A. G. Wedd, *Inorg. Chem.*, 1986, **25**, 906.
[312] X. Ji-de and N. Shi-sheng, *Inorg. Chem.*, 1986, **25**, 1264.
[313] N. Hoshino, K. A. Goldsby, and D. H. Busch, *Inorg. Chem.*, 1986, **25**, 3000.

nickel(II), 2,2'-bipyridine, and adenosine 5'-triphosphate,[314] and an obscure assistive mechanism operates in ternary complex formation between terpyridine, nickel(II), and isomeric (pyridylazo)naphthol dyes.[315] Related to this is a study of the kinetics and equilibria of nickel(II) with a 1-(2-pyridylazo)-2-naphthol dye.[316]

High-resolution ^{13}C CPMAS n.m.r. has been applied to [Ni(pc)I], one of a number of phthalocyanin-based electrical conductors. Based on these results, a map of conduction electron hyperfine interaction was constructed.[317]

(Phthalocyaninato)nickel(II) bromide has been synthesized by the oxidation of Ni(pc) with Br_2[318] and nickel(II) and copper(II) *meso*-substituted chlorins have similar vibrational characteristics to the physiological chlorins.[319]

The reaction between $[(Ph_2PCH_2)_2NiX_2]$ (X = Cl or Br) and $K[C_2B_4H_7]$ has been examined between -35 and $-80\,°C$. When the bromo reagent was used, the only product was 1,1-[bis(diphenylphosphino)ethane]-2.3-dicarbon-1-nickel-*closo*-heptaborane even at $-80\,°C$. When the chloro reagent was used, positive evidence was obtained for the intermediate *nido*-species $4,5-\mu$-chloro-[bis(diphenylphosphino)ethane]nickel)-2,3-dicarba-*nido*-hexaborane. On warming, the *nido*-species loses HCl to give the *closo*-product.[320] The dinuclear complexes $[Ni(\mu\text{-}Bu^t_2As)\text{-}(PMe_3)]_2$ and $[Ni\mu\text{-}Bu^t_2As)(CN\text{-}p\text{-tol})_2]_2$ have been prepared from $[NiCl_2(PMe_3)_2]$ by reaction with $LiBu^t_2As$ in THF at $-78\,°C$ and characterized crystallographically.[321]

Over one hundred compounds crystallize with hexagonal NiAs structures. Distortion of triangular nets of atoms to a zig-zag chain leads to the MnP structure driven by a second-order Jahn–Teller distortion.[322] The complex $Ni_2(edta)(H_2O).2H_2O$ is a chain with two different nickel sites, $Ni4(H_2O)2(O)$ and $Ni4(H_2O)2(N)$. It is necessary to invoke alternating Landé factors in order that the chains' magnetic properties are satisfactorily fitted.[323] Little work on compounds having nickel-sulphur bonds has been published this year but the *cis*-[NiLCl₂] (52) has been prepared by a template synthesis.[324] Covalence is an elusive concept as applied to a molecule such as octahedral $Ni4(S)2(Cl)$ in $[Ni(tu)_4Cl_2]$ (tu = thiourea) and is often investigated with the aid of magnetic susceptibility and electronic spectroscopy. An *X*-ray diffraction study based on data collected at 140 K has enabled accurate charge density maps to be built up. The authors conclude that covalence is better probed by experiments sensitive to electron density than by those related to energies.[325]

High Oxidation State Compounds of Nickel.—The formation of nickel(III) bis-complexes with the ligands (53), (54), or (55) has been investigated electrochemically

[314] K. J. Butenhof, D. Cochenour, K. L. Banyasz, and J. E. Stuehr, *Inorg. Chem.*, 1986, **25**, 691.
[315] R. L. Reeves and J. A. Reczek, *Inorg. Chem.*, 1986, **25**, 4452.
[316] R. L. Reeves, *Inorg. Chem.*, 1986, **25**, 1473.
[317] P. J. Toscano and T. J. Marks, *J. Am. Chem. Soc.*, 1986, **108**, 437.
[318] S. M. Palmer, J. L. Stanton, B. M. Hoffman, and J. A. Ibers, *Inorg. Chem.*, 1986, **25**, 2296.
[319] L. A. Anderson, T. M. Loehr, C. Sotiriou, W. Wu, and C. K. Chang, *J. Am. Chem. Soc.*, 1986, **108**, 2908.
[320] L. Barton and P. K. Rush, *Inorg. Chem.*, 1986, **25**, 91.
[321] R. A. Jones and B. R. Whittlesey, *Inorg. Chem.*, 1986, **25**, 852.
[322] W. Tremel, R. Hoffmann, and J. Silvestre, *J. Am. Chem. Soc.*, 1986, **108**, 5174.
[323] E. Coronado, M. Drillon, A. Fuertes, D. Beltran, A. Mosset, and J. Galay, *J. Am. Chem. Soc.*, 1986, **108**, 900.
[324] E. C. Constable, J. Lewis, V. E. Marquez, and P. R. Raithby, *J. Chem. Soc., Dalton Trans.*, 1986, 1747.
[325] B. N. Figgis and P. A. Reynolds, *J. Chem. Soc., Dalton Trans.*, 1986, 125.

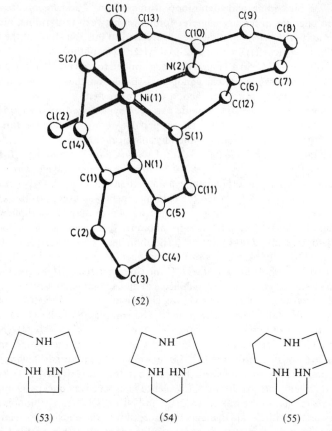

(52)

(53)   (54)   (55)

in acetonitrile, dimethyl sulphoxide, or water.[326] The structure of bis[bis(1,4,7-triazacyclonane)nickel(III)] dithionate has been shown to involve distorted octahedral [NiN$_6$] centres.[327] A high-spin–low-spin nickel(II)–nickel(III) pair based on deprotonated 1,4,7,10,13-penta-azacyclohexadecane-14,16-dione has been characterized.[328] The low-spin [Ni^{III}O$_6$] compound [Ni(bipyO$_2$)$_3$]$^{3+}$(bipyO$_2$ = 2,2'-bipyridyl 1,1'-dioxide) is the product of constant-potential electrolysis of the nickel(II) salt.[329] It is of interest to note that the nickel(III) state is more easily attained in the [N$_2$O$_2$S$_2$] donor-set compared with the [N$_2$O$_4$] family.[330] The structures of the two octahedral Ni^{IV} complexes [Ni{o-C$_6$H$_4$(EH$_2$)$_2$}$_2$Cl$_2$](ClO$_4$)$_n$ (E = P or As; n = 1 or 2) have been determined using EXAFS procedures.[331]

[326] A. Buttafava, L. Fabbrizzi, A. Perotti, A. Poggi, G. Poli, and B. Seghi, *Inorg. Chem.*, 1986, **25**, 1456.
[327] K. Weighardt, W. Walz, B. Nuber, J. Weiss, A. Ozarowdki, H. Stratemeier, and D. Reinen, *Inorg. Chem.*, 1986, **25**, 1650.
[328] R. Machida, E. Kimura, and Y. Kushi, *Inorg. Chem.*, 1986, **25**, 3461.
[329] S. Bhattacharya, R. Mukherjee, and A. Chakravorty, *Inorg. Chem.*, 1986, **25**, 3448.
[330] D. Ray, S. Pal, and A. Chakravorty, *Inorg. Chem.*, 1986, **25**, 2674.
[331] S. J. Higgins, W. Levason, M. C. Feiters, and A. T. Steel, *J. Chem. Soc., Dalton Trans.*, 1986, 317.

5 Cluster Compounds

Homonuclear Iron Clusters.—A molecular mechanics simulation of the ligand structures of transition-metal carbonyl clusters reveals that those with high steric energies have the most crowded ligand structures $\{[Fe_2(CO)_9], [Fe_3(CO)_{12}]\}$ as opposed to the $[M_4(CO)_{12}]$ clusters which have the lowest steric energies.[332] The Fenske–Hall molecular orbital method has been applied to the electronic structures of $[\eta\text{-}CpFe(NO)]_2$, $[\eta\text{-}CpCo(CO)]_2$, $[\eta\text{-}CpCo(NO)]_2$, and $[\eta\text{-}CpNi(CO)]_2$. There is a correlation between a simple valence bond description of the metal–metal interaction and a bond order derived from the electronic structures.[333]

The diverse chemistry of the iron dimer $[\eta\text{-}CpFe(CO)]_2$ in its interactions with CH_2, $C{=}CH_2$, or CH^+ at the bridging carbonyl is well reflected by use of this MO procedure.[334] The methylidyne compound (56), a relatively stable carbonium ion, is reactive towards nucleophiles such as trimethylamine, carbon monoxide, or ethane. In this last reaction the C—H bond of the carbocation adds across the double bond to give a μ-propylidyne complex: an example of a 'hydrocarbation' reaction.[335]

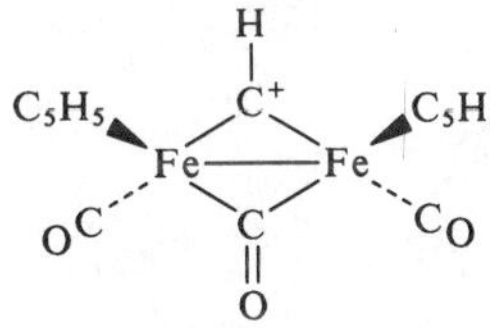

(56) 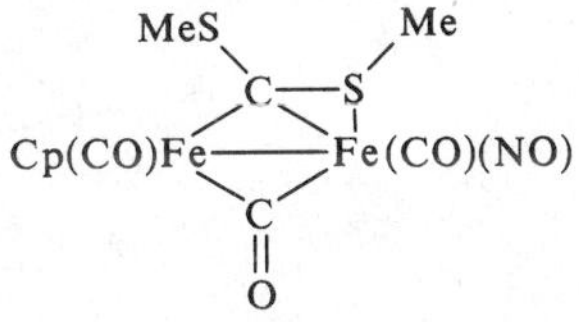   (57)

High nucleophilicity and basicity of the μ-CNR group in $[Fe_2(\eta\text{-}Cp)_2(CNR)_4]$ is well illustrated by this compound's reaction with alkyl halides to give $[Fe_2(\eta\text{-}Cp)_2L(CNR)(\mu\text{-}CNR)\{\mu\text{-}CN(R)(R^1)\}]I$ (L = CO or CNR).[336] The dinuclear complex (57) results from the reaction of $[Fe(CO)_3NO]^-$ with the carbene $[\eta\text{-}Cp(CO)(MeCN)C{=}C(SMe)_2]^+$. This is a general route: [Fe–Co] and [Ru–Co] examples were also described.[337] The electronic structures of the 'flyover' complexes $[Fe_2(CO)_6(R\text{-}C_2\text{-}R)_2Co]$ have been studied by u.v.-p.e. spectroscopy backed up by MO calculations. The Fe—C bonds to the organic part are essentially σ in nature.[338] EXAFS data have been used to determine the structures of the Fe_2E_2 cores of $[Fe_2E_2(CO)_6]^z$ ($z = 0$ or $2-$; E = S or Se) and $[Fe_2(\mu\text{-}ER)_2(CO)_4]$.[339] Reduction reactions of $[Fe_2(NO)_4(\mu\text{-}PPh_2)_2]$ with Na, Na/Hg, LiAlH$_4$, or redal have been shown to yield $[(NO)_2Fe(\mu\text{-}PPh_2)_2Fe(NO)_2]^{2-}$. Reduction using MBEt$_3$H* (M = Li, Na, or K; H* = H or D) takes place by transfer of H* to generate $[(NO)_2Fe(\mu\text{-}PPh_2)(\mu\text{-}NO)Fe(NO)(PPh_2H^*)]^-$. These anions react with CF$_3$CO$_2$H and alkyl iodides at temperatures of -78 to $25\,°C$, yielding mononuclear and dinuclear prod-

[332] J. W. Lauher, *J. Am. Chem. Soc.*, 1986, **108**, 1521.
[333] K. A. Schugard and R. F. Fenske, *J. Am. Chem. Soc.*, 1986, **108**, 5094.
[334] B. E. Bursten and R. H. Clayton, *J. Am. Chem. Soc.*, 1986, **108**, 8241.
[335] C. P. Casey, M. W. Meszaros, P. J. Fagan, R. K. Bly, S. R. Marder, and E. A. Austin, *J. Am. Chem. Soc.*, 1986, **108**, 4043.
[336] A. R. Manning and P. Murray, *J. Chem. Soc., Dalton Trans.*, 1986, 2399.
[337] J. R. Matachek and R. J. Angelici, *Inorg. Chem.*, 1986, **25**, 2877.
[338] M. Casarin, D. Ajo, A. Vittadini, G. Granozzi, R. Bertoncello, and D. Osella, *Inorg. Chem.*, 1986, **25**, 511.
[339] T. D. Weatherill, T. B. Rauchfuss, and R. A. Scott, *Inorg. Chem.*, 1986, **25**, 1466.

ucts.[340] The dianion $[Pb\{Fe(CO)_4\}_2\{Fe_2(CO)_8\}]$ is a product of reaction of $[Fe(CO)_5]$ with lead(II) acetate or lead chloride. The $[PbFe_4(CO)_{16}]^{2-}$ dianion is centred by a $[PbFe_4]$ unit.[341] Photolysis of matrix-isolated $[Fe_2(CO)_9]$ yields the CO-bridged isomer of $[Fe_2(CO)_8]$ which contains two bridging carbonyls as in $[Co_2(CO)_8]$.[342]

The photoelectron spectra of the compound $[Fe_2(CO)_6(Bu^tC_2Bu^t)]$ have been assigned on the basis of an SCF and CI calculation. A closed-shell ground state was assumed even though calculation suggested a triplet state. The HOMO is a back donation from both metals to a semi-bridging carbonyl group.[343] The FT i.r. spectrum of $[Fe_2(CO)_6(\mu\text{-}CO)_2(\mu\text{-}CH_2)]$ has been recorded and the methylene symmetric and asymmetric stretching frequencies have been shown to be consistent with a bond angle of $106°$.[344] The ferraborane (58), a by-product in the preparation of $[HFe_3(CO)_9BH_3]^-$, has a static structure on the 300 MHz time-scale at ambient

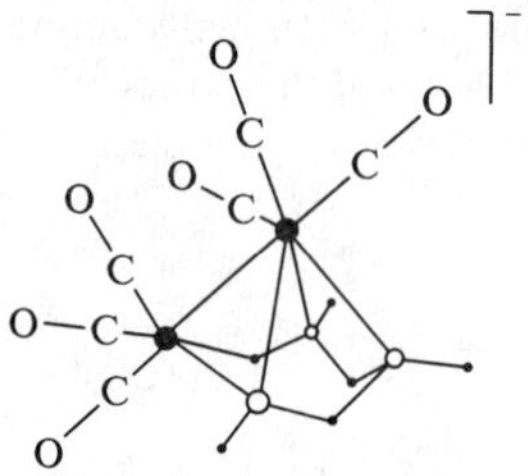

(58)

temperatures.[345] Ferraboranes, inorganic analogues of organometallic clusters, of formula $[(\mu\text{-}H)Fe_3(CO)_9BH_3R]$ and conjugate bases $[(\mu\text{-}H)Fe_3(CO)_9BH_2R]^-$ (R = Me or H) have now been prepared and characterized.[346] Reaction products of this last compound with Lewis bases such as CO, H_2O, NEt_3, or PMe_2P have been characterized by n.m.r. methods. There is a delicate balance between substitution and cluster degradation. Under mild conditions, dihydrogen elimination is observed.[347]

Relatively few iron carbonyl compounds of whatever nuclearity have been synthesized from iron(II) salts but a nice example has been reported this year. Iron(II) chloride in ethanolic NaSPh in a CO atmosphere yields $[Fe_3(SPh)_6(CO)_6]$, a linear sulphur-bridged complex with only terminal carbonyls. It is converted into $[Fe(SPh)_4]^{2-}$ in Me_2SO solution.[348] The bicapped tri-iron cluster $[Fe_3(\mu_3\text{-}PPh)_2(CO)_9]$ undergoes successive one-electron reductions to the anion radical A^- and then diamagnetic A^{2-}. The anion radical can be made to undergo carbon monoxide replacement by phosphorus-donor ligands.[349] The formation of

340 Y. Yu, C. Chau, and A. Wojcicki, *Inorg. Chem.*, 1986, **25**, 4098.
341 C. B. Lagrone, K. H. Whitmire, M. R. Churchill, and J. C. Fettinger, *Inorg. Chem.*, 1986, **25**, 2080.
342 S. C. Fletcher, M. Poliakoff, and J. J. Turner, *Inorg. Chem.*, 1986, **25**, 3597.
343 R. Bertoncello, G. Granozzi, P. Carsky, R. Wiest, and M. Benard, *J. Chem. Soc., Dalton Trans.*, 1986, 2581.
344 S. Chang, Z. H. Kafefi, R. H. Hauge, K. H. Whitmire, W. E. Billups, and J. L. Margrave, *Inorg. Chem.*, 1986, **25**, 4530.
345 C. E. Housecroft, *Inorg. Chem.*, 1986, **25**, 3108.
346 J. Vites, C. E. Housecroft, C. Eigenbrot, M. L. Buhl, G. J. Long, and T. P. Fehlner, *J. Am. Chem. Soc.*, 1986, **108**, 3304.
347 C. E. Housecroft and T. P. Fehlner, *Inorg. Chem.*, 1986, **25**, 404.
348 M. A. Walters, and J. C. Dewan, *Inorg. Chem.*, 1986, **25**, 4889.
349 H. H. Ohst and J. K. Kochi, *Inorg. Chem.*, 1986, **25**, 2066.

$[HeF_3(CO)_{11}]^-$ on hydroxylated alumina from $[Fe_3(CO)_{12}]$ is associated with CO evolution and disproportionation to $[Fe(CO)_5]$.[350] There are marked differences in reactivity towards dimethylamine between the iron and osmium analogues in $[M_3(CO)_9(\mu_2\text{-}S_2)]$ as revealed in an investigation which included characterization of the structures of two major products, $[Fe_3(CO)_8(Me_2NH)(\mu_3\text{-}S)_2]$ and $[Os_3(CO)_8(\mu_3\text{-}S)_2(\mu\text{-}Me_2NCO)(\mu\text{-}H)]$.[351] The reactions of $[(\mu\text{-}H)Fe_3(CO)_9BH_2R]^-$ (R = H or Me) with Lewis bases CO, $PhMe_2P$, NEt_3, or H_2O have been investigated. Elimination of H_2 and CO was established. For $PhMe_2P$, the rate is first-order both in substrate and in ligand.[352] Stepwise replacement of CO by $P(OMe)_3$ at three separate iron centres in the tri-iron cluster $[Fe_3(CO)_9(\mu_3\text{-}PPh)_2]$ is effected by electron-transfer catalysis.[353] Heterotrimetallic complexes $[Hg\{PtR(PPh_3)_2\}_2]$ (R = $2,3,4\text{-}C_6H_2Cl_3$, $2,4,6\text{-}C_6H_2Cl_3$, $2,3,5,6\text{-}C_6HCl_4$, or C_6Cl_5) are starting compounds for the preparation of the iron or cobalt compounds, e.g. $[\eta\text{-}Cp(CO)_3Fe\text{—}Hg\text{—}Pt(C_6Cl_5)(PPh_3)_2]$ or $[(CO)_4Co\text{—}Hg\text{—}Pt(C_6Cl_5)(PPh_3)_2]$.[354]

The structure of the iron–gold borido *arachno*-cluster $[Fe_4(CO)_{12}\{Au(PPh_3)\}_2BH]$ violates the $H/AuPR_3$ structural analogy.[355] The average Mössbauer organoiron clusters $[Fe_4(CO)_{13}]^{2-}$, $[Fe_4(CO)_{12}]^{2-}$, etc. are linearly related to the Slater charge experienced by the iron 4s electrons as calculated by Fenske–Hall methods.[356] The electron-deficient thallium Zintl-metal carbonylate $[Et_4N]_6[Tl_6Fe_{10}(CO)_{36}]$ consists of two thallium triangles held together by two $\eta_3\text{-}Fe(CO)_3$ units, the triangles being connected by two $Fe(CO)_4$ bridges.[357]

A hybrid Zintl iron–metal carbonyl having a 20-electron Bi_4 tetrahedron, $[Bi_4Fe_4(CO)_{13}]^{2-}$, contains a $Fe(CO)_4$ unit bonded to the apical bismuth and the three adjacent tetrahedral faces encrusted with $\eta_3\text{-}Fe(CO)_3$ groups, the remaining face of the tetrahedron being bare.[358]

Homonuclear Cobalt Clusters.—Ion-molecule reactions of $[Co(CO)_3NO]$ and $[Ni(CO)_4]$ yield ionic clusters of the type $[M_x(CO)_y(NO)_3]^+$. Relative reaction rates combined with the electron deficiency of the fragments were used to estimate the bond order of the fragments.[359]

The u.v. irradiation of $[(\eta\text{-}Cp)_2Co_2(CO)_3]$ in low-temperature hydrocarbon matrices results in rapid and clean loss of CO to produce the metal–metal double-bonded carbonyl-bridged product $[\eta\text{-}CpCo(\mu\text{-}CO)]_2$.[360] The e.p.r. spectrum of the radical anion $[\{MeN(PF_2)_2\}_3Co_2(CO)_4]^-$ is consistent with complete delocalization over the two cobalt sites even at 4.2 K.[361] The electron deformation density distribution in $[Co_2(CO)_6R_2C_2]$ (R = CMe_3) has been obtained from low-temperature X-ray

[350] B. E. Hanson, J. J. Bergmeister, J. T. Petty, and M. C. Connaway, *Inorg. Chem.*, 1986, **25**, 3089.
[351] R. D. Adams and J. E. Babin, *Inorg. Chem.*, 1986, **25**, 3418.
[352] C. E. Housecroft and T. P. Fehlner, *J. Am. Chem. Soc.*, 1986, **108**, 4867.
[353] H. H. Ohst and J. K. Kochi, *J. Am. Chem. Soc.*, 1986, **108**, 2897.
[354] O. Rossell, M. Secco, and I. Torra, *J. Chem. Soc., Dalton Trans.*, 1986, 1011.
[355] C. E. Housecroft and A. L. Rheingold, *J. Am. Chem. Soc.*, 1986, **108**, 6420.
[356] C. G. Benson, G. J. Long, J. S. Bradley, J. W. Kolis, and D. F. Shriver, *J. Am. Chem. Soc.*, 1986, **108**, 1898.
[357] K. H. Whitmire, R. R. Ryan, H. J. Wasserman, T. A. Albright, and S. Kang, *J. Am. Chem. Soc.*, 1986, **108**, 6831.
[358] K. H. Whitmire, T. A. Albright, S. Kang, M. R. Churchill, and J. C. Fettinger, *Inorg. Chem.*, 1986, **25**, 2799.
[359] D. Anderson Fredeen and D. H. Russell, *J. Am. Chem. Soc.*, 1986, **108**, 1860.
[360] F. R. Anderson and M. S. Wrighton, *Inorg. Chem.*, 1986, **25**, 112.
[361] F. Babonneau and J. Livage, *Inorg. Chem.*, 1986, **25**, 2741.

and neutron diffraction data backed up by theoretical wave-functions. The experimental results show no accumulation of electron density along the cobalt–cobalt line.[362]

I.r. spectroscopy and reactivity studies have been deployed in the elucidation of the species formed by the adsorption of $[Co_2(CO)_8]$ onto zeolites from pentane solution. The species assigned include $[Co(CO)_4]^-$, $(Co_4(CO)_2]$, $[Co(CO)_3L_2]^+$ (L = donor species in zeolite), and $[Co_x(CO)_y]$.[363]

Reaction of octacarbonyldicobalt with $(Pr^i_2N)_2PH$ at ambient temperature yields black, air-sensitive paramagnetic $[Pr^i_2NPCo_3(CO)_9]$ with a PCo_3 tetrahedral phosphinidene structure.[364] A new class of bicapped tricobalt clusters $[Co_3(\eta^5-C_5H_{5-x}Me_x)_3(\mu_3-X)(M_3Y)]^n$ (x = 0, 1, or 5; X = CO, NO; Y = NSiMe_3, NCONH_2, or NH) have been prepared and fully characterized.[365] The electron-deficient 46-electron tricobalt cluster $[Co_3(\eta^5-C_5Me_5)_3(\mu_3-CO)]_2$, isolobal with $[C_3H_3]^-$, being a spin-triplet is thus the first example of a multiply bonded trinuclear metal system containing only first-row transition metals.[366] Analysis of the structure and bonding in the tricobalt dicarbonyl monocation $[Co_3(\eta^6-C_6H_6)_3(\mu_3-CO)_2]^+$ indicates much anti-bonding character as expected from its 48-electron core.[367] Nitric oxide inserts into a cobalt–cobalt cluster bond as shown in Scheme 4.[368] The

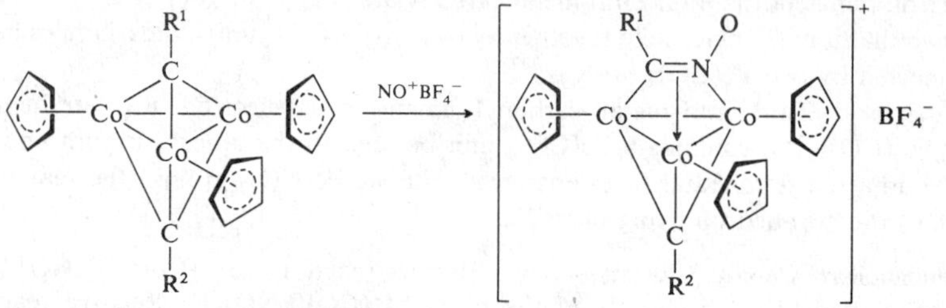

$$R^1 = R^3 = Bu^n$$

Scheme 4

geometries of the tetrahedral cobalt clusters $[Co_4(CO)_8(PMe_3)(tripod)]$, $[Co_4(CO)_7(dppm)]$, and $[Co_4(CO)_7(PMe_3)(tripod)]$ $[tripod = (PPh_2)_3CH, dppm = (Ph_2P)_2CH_2]$ have been determined.[369] The solution structure and dynamics of the molecules $[Co_4(CO)_8(L-L)_2]$ $[L-L = (Ph_2P)_2CH_2, (Me_2P)_2CH_2, or Me_2PCH_2PPh_2]$ have been investigated using ^{13}C and ^{31}P n.m.r. spectroscopy. The structure is related to that of $[Co_4(CO)_{12}]$ and all eight carbonyls exchange at the same rate indicating a concerted mechanism.[370] A variable-temperature solid-state ^{13}C CPMAS n.m.r.

[362] F. Baert, A. Guelzim, J. M. Poblet, R. Weist, J. Demuynck, and M. Benard, *Inorg. Chem.*, 1986, **25**, 1830.
[363] M. C. Connaway and B. E. Hanson, *Inorg. Chem.*, 1986, **25**, 1445.
[364] R. B. King, W. K. Fu, and E. M. Holt, *Inorg. Chem.*, 1986, **25**, 2390.
[365] R. L. Bedard and A. D. Rae, *J. Am. Chem. Soc.*, 1986, **108**, 5924.
[366] W. L. Olson, A. M. Stacy, and L. F. Dahl, *J. Am. Chem. Soc.*, 1986, **108**, 7646.
[367] W. L. Olson and L. F. Dahl, *J. Am. Chem. Soc.*, 1986, **108**, 7657.
[368] A. Goldhaber, K. P. C. Volhardt, E. C. Walborsky, and M. Wolfgruber, *J. Am. Chem. Soc.*, 1986, **108**, 516.
[369] D. J. Darensbourg, D. J. Zalewski, A. L. Rheingold, and R. L. Durney, *Inorg. Chem.*, 1986, **25**, 3281.
[370] E. C. Lisic and B. E. Hanson, *Inorg. Chem.*, 1986, **25**, 812.

study reveals unspecified fluxionality in solid $[Co_4(CO)_{12}]$.[371] The reduction potentials correlate satisfactorily with calculated ($SCFX_\alpha$) core changes for representative tetracobalt clusters such as $[Co_4(CO)_{12}]$. The new octahedral anion $[Co_6C(CO)_{13}]^{2-}$ has been prepared by refluxing a solution of $[Co_6C(CO)_{15}]^{2-}$ dissolved in THF.[372]

Nickel Clusters.—The cluster compounds $[H_{6-n}Ni_{34}(CO)_{38}C_4]^{n-}$ ($n = 5$ or 6) and $[Ni_{35}(CO)_{39}C_4]^{6-}$, both newly characterized, are possible molecular analogues of carbidized metal crystallites.[373] The cluster $[HNi_{38}(CO)_{42}C_6]^{5-}$ is an extended fragment of the $[Cr_{23}C_6]$ lattice neatly stabilized in a molecular nickel carbonyl cluster. Are new Ni–C binary phases to be prepared using this approach?[374] The non-centred icosahedral clusters $[Ni_9(AsPh)_3(CO)_{15}]^{2-}$ and $[Ni_{10}(AsMe)_2(CO)_{18}]^{2-}$ having AsR groups co-ordinated to five nickels are products of reaction of $AsPhCl_2$ or $AsMeBr_2$ on $[Ni_6(CO)_{12}]^{2-}$.[375]

Heterometallic Clusters.—Work published in this area is summarized in Table 2.[376]

Table 2 *Heterometallic clusters*

Metal framework	Other groups	Ref.
FeNi	Co, η-Cp, $C_7R^1R^2$	376
FeCo$_2$	CO, NPh	377
FeCo$_2$	CO, C=CHPh, Bu^{t}P	378
FeCu$_3$	CO	379
FeCo$_3$	CO, CS	380
Fe$_2$Pt$_2$	—	381
FeOs$_3$	H, CO	382
Ni$_2$Rn$_2$	η-Cp, CO, μ_5-PPh	383
NiOs$_3$	CO, H, HgBr	384
Fe$_2$CoBi$_2$	CO	385

[371] B. E. Hanson and E. C. Lisic, *Inorg. Chem.*, 1986, **25**, 715.
[372] V. G. Albano, D. Braga, and S. Martinengo, *J. Chem. Soc., Dalton Trans.*, 1986, 981.
[373] A. Ceriotti, A. Fait, G. Longoni, G. Piro, and L. Resconi, *J. Am. Chem. Soc.*, 1986, **108**, 5370.
[374] A. Ceriotti, A. Fait, G. Longoni, G. Piro, F. Demartin, M. Manassero, N. Masciocchi, and M. Sansoni, *J. Am. Chem. Soc.*, 1986, **108**, 8091.
[375] D. F. Rieck, R. A. Montag, T. S. McKechnie, and L. F. Dahl, *J. Am. Chem. Soc.*, 1986, **108**, 1330.
[376] F. W. B. Einstein, K. G. Tyers, A. S. Tracey, and D. Sutton, *Inorg. Chem.*, 1986, **25**, 1631.
[377] A. Basu, S. Bhaduri, H. Khwaja, P. G. Jones, K. Mayer-Base, and G. M. Sheldrick, *J. Chem. Soc., Dalton Trans.*, 1986, 2501.
[378] R. Mathieu, A. Caminade, J. Majoral, and J. Daran, *J. Am. Chem. Soc.*, 1986, **108**, 8007.
[379] G. Doyle, K. A. Eriksan, and D. Van Engen, *J. Am. Chem. Soc.*, 1986, **108**, 445.
[380] L. Busetto, V. Zanotti, V. G. Albano, D. Braga, and M. Monari, *J. Chem. Soc., Dalton Trans.*, 1986, 1791.
[381] D. M. L. Goodman, M. A. Hitchman, and B. Lippert, *Inorg. Chem.*, 1986, **25**, 2191.
[382] A. Choplin, L. Huang, A. Theolier, P. Gallezot, and J. M. Basset, *J. Am. Chem. Soc.*, 1986, **108**, 4224.
[383] M. Lanfranchi, A. Tiripicchio, E. Sappa, and A. J. Carty, *J. Chem. Soc., Dalton Trans.*, 1986, 2737.
[384] G. Predieri, A. Tiripicchio, C. Vignali, E. Sappa, and P. Braunstein, *J. Chem. Soc., Dalton Trans.*, 1986, 1135.
[385] M. R. Churchill, J. C. Fettinger, and R. F. See, *J. Am. Chem. Soc.*, 1986, **108**, 2778.

10   Ru, Os, Rh, Ir, Pd, Pt

By S. D. ROBINSON

Department of Chemistry, King's College (KQC), London WC2R 2LS

1 General

Topics of general interest to platinum metal chemists reviewed during 1986 include PF₃ complexes,[1a] coordination chemistry with alkanes,[1b] cyclometallation reactions of platinum group metals,[1c] the organometallic chemistry of transition metal–porphyrin complexes,[1d] bonding in metal clusters,[1e] alkyne-substituted metal clusters,[1f] cluster complexes containing opened transition metal polyhedra,[1g] and topology of osmium, rhodium, and platinum carbonyl clusters.[1h]

2 Ruthenium

The chemistry of substituted ruthenium carbonyl halides[2a] and of ruthenium carbene and carbyne complexes[2b] has been reviewed.

The boiling point of RuO_4 (129.6 ± 0.2 °C) has been measured directly and shown to be identical with that of OsO_4.[3a] The observed high E_f^0 value (1.25 V *vs.* NHE) for the *trans*-$[Ru(bipy)_2(O)_2]^{2+}$/*trans*-$[Ru(bipy)_2O(OH_2)]^{2+}$ couple indicates that *trans*-$[Ru(bipy)_2(O)_2]^{2+}$ is potentially useful for photocatalytic splitting of water.[3b] Paramagnetic *trans*-$[Ru(tmc)(O)_2]ClO_4$ (tmc = 1,4,8,11-tetramethyl-1,4,8,11-tetra-azacyclotetradecane) rapidly disproportionates into *trans*-$[Ru(tmc)(O)_2]^{2+}$ and *trans*-$[Ru(tmc)(O)(OH_2)]^{2+}$ upon dissolution in aqueous acid solution.[3c] Several other papers deal with the synthesis, and electrochemistry of ruthenium complexes with N^4-macrocyclic ligands.[3d–f] New work on ruthenium porphyrin

[1] (a) J. F. Nixon, *Adv. Inorg. Chem. Radiochem.*, 1985, **29**, 41; (b) M. L. Deem, *Coord. Chem. Rev.*, 1986, **74**, 101; (c) G. R. Newkome, W. E. Puckett, V. K. Gupta, and G. E. Kiefer, *Chem. Rev.*, 1986, **86**, 451; (d) P. J. Brothers and J. P. Collman, *Acc. Chem. Res.*, 1986, **19**, 209; (e) D. M. P. Mingos, *Chem. Soc. Rev.*, 1986, **15**, 31; (f) P. R. Raithby and M. J. Rosales, *Adv. Inorg. Chem. Radiochem.*, 1985, **29**, 169; (g) M. O. Albers, D. J. Robinson, and N. J. Coville, *Coord. Chem. Rev.*, 1986, **69**, 127; (h) R. B. King, *Inorg. Chim. Acta*, 1986, **116**, 99, 119, and 125.

[2] (a) N. C. Thomas, *Coord. Chem. Rev.*, 1986, **70**, 121; (b) M. A. Gallop and W. R. Roper, *Adv. Organomet. Chem.*, 1986, **25**, 121.

[3] (a) Y. Koda, *J. Chem. Soc., Chem. Commun.*, 1986, 1347; (b) C.-M. Che, K.-Y. Wong, W.-H. Leung, and C.-K. Poon, *Inorg. Chem.*, 1986, **25**, 345; (c) C.-M. Che and K.-Y. Wong, *J. Chem. Soc., Chem. Commun.*, 1986, 229; (d) C. Marzin, G. Tarrago, M. Gal, I. Zidane, T. Hours, D. Lerner, C. Andrieux, H. Gampp, and J. M. Savéant, *Inorg. Chem.*, 1986, **25**, 1775; (e) C. Marzin, G. Tarrago, I. Zidane, E. Bienvenue, P. Seta, C. Andrieux, H. Gampp, and J. M. Savéant, *Inorg. Chem.*, 1986, **25**, 1778; (f) C.-M. Che, K.-Y. Wong, and C.-K. Poon, *Inorg. Chem.*, 1986, **25**, 1809.

Abbreviations used: Cy = cyclohexyl; COD = cyclooctadiene (1,5 unless otherwise specified); dppm = Ph₂PCH₂PPh₂; dmpm = Me₂PCH₂PMe₂; dppe = Ph₂PCH₂CH₂PPh₂; PNP = Ph₃PNPPh₃; triphos = MeC(CH₂PPh₂)₃; *o*-phen = *ortho*-phenanthroline; bipy = 2,2'-bipyridyl; pzH = pyrazole; THF = tetrahydrofuran; OEP = octaethylporphyrin.

systems includes resonance Raman studies on back-bonding,[4a] photoinduced reduction of water[4b] addition of axial ligands (N_2, CO, Et_2O, THF) to the 14-electron complex Ru(5,10,15,20-tetramesitylporphyrin)[4c] and determination of the Ru—Et bond dissociation energy (21.7 ± 1.5 kcal mol^{-1}) in $RuEt_2(OEP)$.[4d] Halogen oxidation of [Ru(OEP)]$_2$ gives paramagnetic ruthenium(IV) complexes $RuX_2(OEP)$ which can be converted into diamagnetic organometallic derivatives $RuR_2(OEP)$ (R = Me or Ph).[4e] Isolation of (1) affords the first *in situ* template synthesis of a ruthenium(II) clathrochelate complex.[5] The spectroscopic and electrochemical properties of [Ru(bipy)$_3$]$^{2+}$ have been compared with those of the cyclometallated analogue Ru(bipy)$_2$(2-phenylpyridine-C,N)]$^+$.[6a]

(1)   (2)

Rotating-disc-electrode voltammetric studies on the ruthenium red analogue [(bipy)$_2$(H$_2$O)RuORu(bipy)$_2$ORu(H$_2$O)(bipy)$_2$]$^{6+}$ reveal seven distinct valence combinations from (II, II, II) to (IV, V, IV).[6b] Photochemical substitution of [Ru(CN)$_6$]$^{4-}$ affords the photosensitizer [Ru(bipy)(CN)$_4$]$^{2-}$.[6c] The spectroscopic and redox properties of the new heterodinuclear diimine chelates (2)[6d] and (3)[6e] have been described. The tris(bipyrazine) ruthenium(II) complex [Ru(bipz)$_3$][PF$_6$]$_2$ reacts with [Fe(CN)$_5$(H$_2$O)]$^{3-}$ anions to form a series of di- to hepta-nuclear bipyrazine-bridged complexes [Ru(bipz)$_3${Fe(CN)$_5$}$_n$]$^{(2-3n)+}$ ($n = 1$—6).[6f] Data obtained for ruthenium diimine complexes using bulky diimine ligands support the conclusion that dinuclear species are necessary to catalyse the oxidation of water.[6g] The electrochemical and photochemical properties of the complexes [Ru(dipy)$_2$(CO)L]$^{2+}$ (L = CO, MeCN, py, H$_2$O *etc.*) and [Ru(bipy)$_2$(CO)X]$^+$ (X = H, Cl, CNS) have been examined;[7a] Water-gas shift reactions catalysed by [Ru(bipy)$_2$(CO)$_2$]$^{2+}$ and [Ru(bipy)$_2$(CO)Cl]$^+$ yield the catalytic intermediates [Ru(bipy)$_2$(CO)(H$_2$O)]$^{2+}$, [Ru(bipy)$_2$(CO)(COOH)]$^+$, and [Ru(bipy)$_2$(CO)H]$^+$.[7b]

[4] (a) D. Kim, Y. O. Su, and T. G. Spiro, *Inorg. Chem.*, 1986, **25**, 3993; (b) W. Szulbinski and J. W. Strojek, *Inorg. Chim. Acta*, 1986, **118**, 91; (c) M. J. Camenzind, B. R. James, and D. Dolphin, *J. Chem. Soc., Chem. Commun.*, 1986, 1137; (d) J. P. Collman, L. McElwee-White, P. J. Brothers, and E. Rose, *J. Am. Chem. Soc.*, 1986, **108**, 1332; (e) C. Sishta, M. Ke, B. R. James, and D. Dolphin, *J. Chem. Soc., Chem. Commun.*, 1986, 787.

[5] J. G. Müller, J. J. Grzybowski, and K. J. Takeuchi, *Inorg. Chem.*, 1986, **25**, 2665.

[6] (a) E. C. Constable and J. M. Holmes, *J. Organomet. Chem.*, 1986, **301**, 203; (b) D. A. Geselowitz, W. Kutner, and T. J. Meyer, *Inorg. Chem.*, 1986, **25**, 2015; (c) C. A. Bignozzi, C. Chiorboli, M. T. Indelli, M. A. Rampi Scandola, G. Varani, and F. Scandola, *J. Am. Chem. Soc.*, 1986, **108**, 7872; (d) R. Sahai and D. P. Rillema, *Inorg. Chim. Acta*, 1986, **118**, L35; (e) R. Sahai, D. A. Baucom, and D. P. Rillema, *Inorg. Chem.*, 1985, **25**, 3843; (f) H. E. Toma and A. B. P. Lever, *Inorg. Chem.*, 1986, **25**, 176; (g) J. P. Collin and J. P. Sauvage, *Inorg. Chem.*, 1986, **25**, 135.

[7] (a) J. M. Kelly, C. M. O'Connell, and J. G. Vos, *J. Chem. Soc., Dalton Trans.*, 1986, 253; (b) H. Ishida, K. Tanaka, M. Morimoto, and T. Tanaka, *Organometallics*, 1986, **5**, 724.

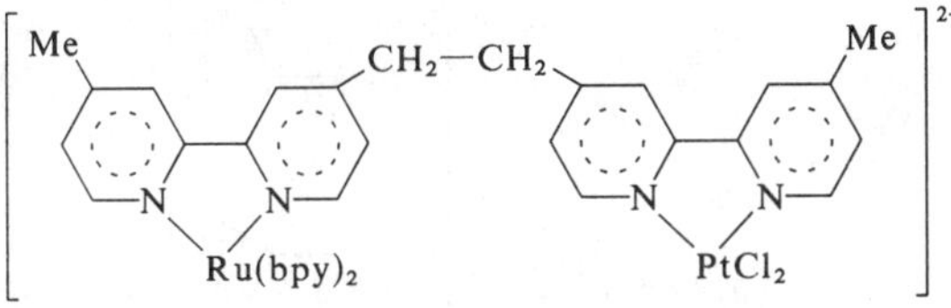

(3)

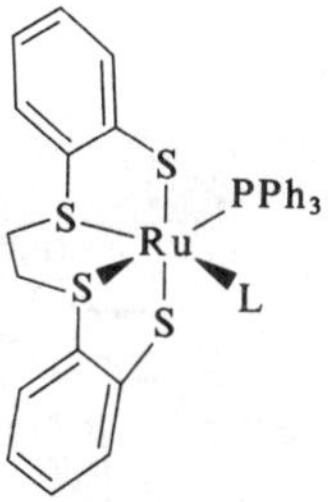

(4)

A very short Ru—N(heterocycle) bond [1.95(1) Å] has been found in salts of the pentaammine(N-methylpyrazinium)ruthenium(II) cation.[8] Oxidation of [RuCl(NO)(py)$_4$]$^{2+}$ with NaClO gives the oxoruthenium(IV) complex *trans*-[RuCl(O)(py)$_4$]$^+$ with an unusually long Ru=O bond distance [1.862(8) Å].[9] Trithiazyl chloride (NSCl)$_3$ reacts with RuCl$_3$ and PPh$_4$[RuCl$_4$(NO)] to form *cis*-RuCl$_4$(NS)$_2$ and PPh$_4$[RuCl$_4$(NO)(NSCl)] respectively, the latter product converts into [PPh$_4$]$_2$[Cl$_3$(NS)RuCl$_2$Ru(NS)Cl$_3$] on pyrolysis.[10] Treatment of RuCl$_4$(NS)$_2$ with [PPh$_4$]Br affords [PPh$_4$]$_2$[{RuCl$_4$(NS)}$_2$(μ-N$_2$S$_2$)] containing an N^1,N^3-N$_2$S$_2$ bridge.[10] The complex Ru(dttd)(PPh$_3$)$_2$ (4) undergoes S-alkylation[11a] and ligand substitution reactions[11b] to form [Ru(Me$_2$-dttd)I(PPh$_3$)]I and Ru(dttd)L(PPh$_3$) (L = NH$_3$, N$_2$H$_4$, N$_2$H$_3$Me, N$_2$H$_3$Ph) respectively. Oxidation of Ru(dttd)(N$_2$H$_4$)(PPh$_3$) affords {Ru(dttd)$_2$(PPh$_3$)}$_2$(μ-N$_2$H$_2$).[11b] The salt [AsPh$_4$][Ru(DL-MeSeCH$_2$CH$_2$SeMe)Cl$_4$] provides the first structurally characterized example of a diselenoether ligand adopting the DL conformation.[12a] The first example of a bis-transannular metallocenyl cyclophosphazene N$_4$P$_4$F$_4$[(η-C$_5$H$_4$)$_2$Ru]$_2$ (5) has been reported.[12b] Polythia[n](1,1′)ruthenocenophanes (6) bind as sulphur chelates to palladium and platinum dichlorides.[13a] 1,2,3-Triselena[3]ruthenocenophane (7) has been prepared from 1,1′-dilithioruthenocene and elemental selenium.[13b] A

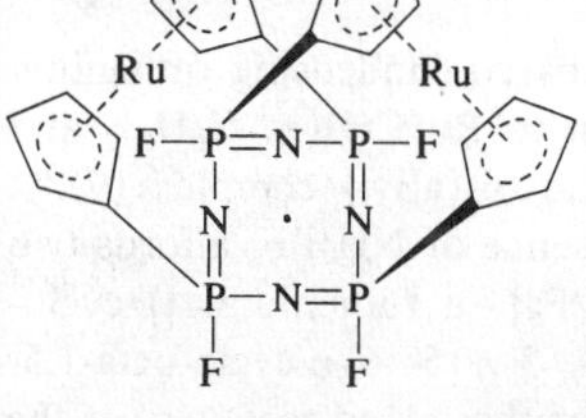

(5)

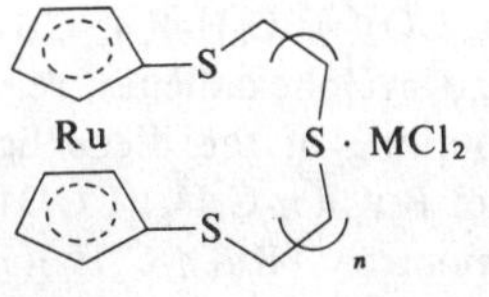

$n = 2$, M = Pt
$n = 2$, M = Pd
$n = 3$, M = Pt
$n = 3$, M = Pd

(6)

[8] J. F. Wishart, A. Bino, and H. Taube, *Inorg. Chem.*, 1986, **25**, 3318.

[9] K. Aoyagi, Y. Yukawa, K. Shimizu, M. Mukaida, T. Takeuchi, and H. Kakihana, *Bull. Chem. Soc. Jpn.*, 1986, **59**, 1493.

[10] U. Demant, W. Willing, U. Müller, and K. Dehnicke, *Z. Naturforsch.*, 1986, **41b**, 560; *Z. Anorg. Allg. Chem.*, 1986, **532**, 175.

[11] (a) D. Sellmann, M. Waeber, G. Huttner, and L. Zsolnai, *Inorg. Chim. Acta*, 1986, **118**, 49; (b) D. Sellmann and M. Waeber, *Z. Naturforsch.*, 1986, **41b**, 877.

[12] (a) E. G. Hope, H. C. Jewiss, W. Levason, and M. Webster, *J. Chem. Soc., Dalton Trans.*, 1986, 1479; (b) K. D. Lavin, G. H. Riding, M. Parvez, and H. R. Allcock, *J. Chem. Soc., Chem. Commun.*, 1986, 117.

[13] (a) S. Akabori, S. Sato, T. Tokuda, Y. Habata, K. Kawazoe, C. Tamura, and M. Sato, *Chem. Lett.*, 1986, 121; *Bull. Chem. Soc. Jpn.*, 1986, **59**, 3189; (b) A. J. Blake, R. O. Gould, and A. G. Osborne, *J. Organomet. Chem.*, 1986, **308**, 297.

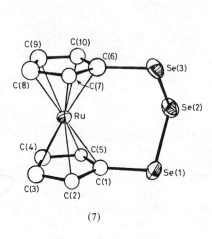

(7)

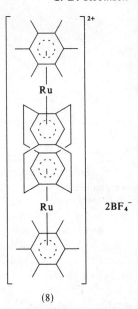

(8)

general method of synthesis for bis(η^6-[2_n]-cyclophane)ruthenium(II) derivatives has been described.[14a] Bis(η^6-hexamethylbenzene){n^4,n^6-[2_4](1,2,4,5)cyclophane}-diruthenium(0, II) bis(tetrafluoroborate) (8) undergoes reversible net two-electron intervalence transfer to form the corresponding $Ru_2^{0.0}$ complex.[14b] Treatment of [$Ru(H_2O)_6$][O_3SCF_3]$_2$ with a mixture of diene and arene in ethanol affords mixed sandwich compounds [$Ru^{II}(\eta^5$-dienyl)(η^6-arene)]$^+$ in high yield;[15a] use of cyclohexadiene (1,3 or 1,4 isomers) leads to formation of [$Ru(\eta$-$C_6H_6)(H_2O)_3$]$^{2+}$ a useful precursor for arene half-sandwich complexes.[15b] Amidine complexes [$Ru(\eta$-$C_6H_6)${NH=CMe(pz)}(Hpz)]$^{2+}$ are formed when pyrazoles (Hpz) react with [$RuCl_2(C_6H_6)$]$_2$ in acetonitrile.[16]

The complex $Ru(CO)(\eta^5$-$C_6H_7)(\eta^3$-$C_6H_9)$ formed by co-condensing ruthenium atoms with CO and cyclohexadienes, rearranges first to $Ru(CO)(\eta^4$-$C_6H_8)_2$ and then, by C—C coupling of the diene ligands, to the bis(allyl) complex (9).[17a] Dehydrogenation of $RuCl(\eta$-$C_5H_5)(COD)$ in the presence of NH_4PF_6 affords two isomers of stoicheiometry [$Ru(\eta$-$C_5H_5)(\eta$-$C_8H_{10})$][PF_6] – a ruthenium(II) cyclo-octa-1,3,5-triene complex and a novel ruthenium(IV) 1—3-η : 5—7-η cyclo-octa-1,5-dienediyl derivative.[17b] Several papers report on the synthesis and reactivity of the ruthenium(II) mixed sandwich cations [$Ru(\eta$-$C_5R_5)(arene)$]$^+$ (R = H or Me).[17c—e]

[14] (a) R. T. Swann, A. W. Hanson, and V. Boekelheide, *J. Am. Chem. Soc.*, 1986, **108**, 3324; (b) R. H. Voegeli, H. C. Kang, R. G. Finke, and V. Boekelheide, *J. Am. Chem. Soc.*, 1986, **108**, 7010.

[15] (a) M. Stebler-Röthlisberger, A. Salzer, H. B. Bürgi, and A. Ludi, *Organometallics*, 1986, **5**, 298; (b) M. Stebler-Röthlisberger, and A. Ludi, *Polyhedron*, 1986, **5**, 1217.

[16] C. J. Jones, J. A. McCleverty, and A. S. Rothin, *J. Chem. Soc., Dalton Trans.*, 1986, 109.

[17] (a) D. N. Cox and R. Roulet, *Organometallics*, 1986, **5**, 1886; (b) M. O. Albers, D. C. Liles, D. J. Robinson, and E. Singleton, *J. Chem. Soc., Chem. Commun.*, 1986, 1102; (c) N. Oshima, H. Suzuki, and Y. Moro-Oka, *Inorg. Chem.*, 1986, **25**, 3407; (d) J. L. Schrenk, A. M. McNair, F. B. McCormick, and K. R. Mann, *Inorg. Chem.*, 1986, **25**, 3501; (e) A. M. McNair and K. R. Mann, *Inorg. Chem.*, 1986, **25**, 2519.

(9)

(10)

(11)

Hydride abstraction by $[Ph_3C]PF_6$ converts $Ru(\eta\text{-}C_5H_5)(\eta^1\text{-}C_7H_7)(CO)_2$ into the carbene complex (10).[18a] The barrier to rotation about the $Ru{=}CH_2$ bond in $Ru({=}CH_2)(\eta\text{-}C_5H_5)(dppe)$ is 10.9 kcal mol^{-1}.[18b] The complexes $RuH(\eta\text{-}C_5H_5)$-(COD) and $RuX(\eta\text{-}C_5H_5)(COD)$ are useful synthetic precursors for 'open-face' ruthenium chemistry.[18c] Co-dimerization of two molecules of phenylacetylene at the ruthenium centre in $RuBr(\eta\text{-}C_5H_5)(COD)$ affords the first metallacyclopentatriene complex (11).[18d] The ruthenium(IV) allyl complex $RuBr_2(\eta\text{-}C_5H_5)(\eta^3\text{-}C_6H_9)$ formed by addition of 3-bromocyclohexene to $RuBr(\eta\text{-}C_5H_5)(COD)$ readily converts into the cation $[Ru(\eta\text{-}C_5H_5)(\eta\text{-}C_6H_6)]^+$.[18e] Convenient syntheses have been reported for $[RuX_3(\eta\text{-}C_5Me_5)]_n$ (X = Cl, Br, I),[19a] $Ru(\eta\text{-}C_5Me_5)_2$,[19b] and the trihydrides $RuH_3(\eta\text{-}C_5Me_5)(PR_3)$.[19c] The hydride $RuH_6(PCy_3)_2$ reacts with cyclopentene to afford $RuH_2(\eta^4\text{-}C_5H_6)(PCy_3)_2$ and $RuH(\eta\text{-}C_5H_5)(PCy_3)_2$.[19c] Protonation of $RuH(\eta\text{-}C_5H_5)(CNBu^t)(PPh_3)$ with HPF_6 yields the molecular hydrogen complex $[Ru(\eta\text{-}C_5H_5)(\eta^2\text{-}H_2)(CNBu^t)(PPh_3)][PF_6]$.[20a] N.m.r. relaxation times (T_1) favour reformulation of '$RuH_4(PPh_3)_3$' as $RuH_2(\eta^2\text{-}H_2)(PPh_3)_3$.[20b] Reductive amination of acetone by $RuH_2(\eta\text{-}C_6Me_6)(PMe_3)$ in the presence of $[NH_4]PF_6$ generates $[RuH(NH_2Pr^i)(\eta\text{-}C_6Me_6)(PMe_3)][PF_6]$.[20c] The ruthenium complex $RuMe_2(C_6Me_6)(PMe_3)$ reacts with $[Ph_3C][PF_6]$ in CH_2Cl_2 to form the ethylene (hydrido) product $RuH(C_2H_4)(\eta\text{-}C_6Me_6)(PMe_3)][PF_6]$; related reactions are reported for osmium and iridium.[20d] The bimetallic complex $[(dppm)_2Ru(\mu\text{-}H)_2Cu(MeCN)]BF_4$ reacts with CO and C_2H_4 or CO_2 to form (12) and (13) respectively.[21] The hydride $[RuH(1\text{---}6\text{-}\eta\text{-}C_8H_{10})(1,2;5,6\text{-}\eta\text{-}C_8H_{12})][BF_4]$ isomerizes first to $[RuH(\eta^5\text{-}C_8H_{11})_2][BF_4]$ and then to $Ru(1\text{---}5\text{-}\eta\text{-}C_8H_{11})(1\text{---}4\text{-}\eta\text{-}C_8H_{12})]$-$[BF_4]$.[22] The cyclopentadienyl complex $RuCl(\eta\text{-}C_5H_5)(PPh_3)_2$ is obtained quantitatively by refluxing $RuHCl(PPh_3)_3$ with penta-1,4-diene in methylethylketone.[23] The

18 (a) J. R. Lisko and W. M. Jones, *Organometallics*, 1986, **5**, 1890; (b) W. B. Studabaker and M. Brookhart, *J. Organomet. Chem.*, 1986, **310**, C39; (c) M. O. Albers, D. J. Robinson, A. Shaver, and E. Singleton, *Organometallics*, 1986, **5**, 2199; (d) M. O. Albers, D. J. A. de Waal, D. C. Liles, D. J. Robinson, E. Singleton, and M. B. Wiege, *J. Chem. Soc., Chem. Commun.*, 1986, 1680; (e) M. O. Albers, D. C. Liles, D. J. Robinson, A. Shaver, and E. Singleton, *J. Chem. Soc., Chem. Commun.*, 1986, 645.
19 (a) N. Oshima, H. Suzuki, Y. Moro-Oka, H. Nagashima, and K. Itoh, *J. Organomet. Chem.*, 1986, **314**, C46; (b) M. O. Albers, D. C. Liles, D. J. Robinson, A. Shaver, E. Singleton, M. B. Wiege, J. C. A. Boeyens, and D. C. Levendis, *Organometallics*, 1986, **5**, 2321; (c) T. Arliguie and B. Chaudret, *J. Chem. Soc., Chem. Commun.*, 1986, 985.
20 (a) F. M. Conroy-Lewis and S. J. Simpson, *J. Chem. Soc., Chem. Commun.*, 1986, 506; (b) R. H. Crabtree and D. G. Hamilton, *J. Am. Chem. Soc.*, 1986, **108**, 3124; (c) H. Werner, H. Kletzin, R. Zolk, H. Otto, *J. Organomet. Chem.*, 1986, **310**, C11; (d) H. Werner, H. Kletzin, A. Höhn, W. Paul, W. Knaup, M. L. Ziegler, and O. Serhadli, *J. Organomet. Chem.*, 1986, **306**, 227.
21 B. Delavaux, T. Arliguie, B. Chaudret, and R. Poilblanc, *Nouv. J. Chim.*, 1986, **10**, 619.
22 F. Bouachir, B. Chaudret, and I. Tkatchenko, *J. Chem. Soc., Chem. Commun.*, 1986, 94.
23 B. E. Mann, P. W. Manning, and C. M. Spencer, *J. Organomet. Chem.*, 1986, **312**, C64.

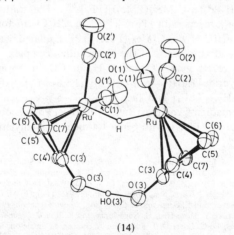

complexes $Ru(CH_2OH)(\eta\text{-}C_5Me_5)(CO)_2$ and $Ru(CHO)(\eta\text{-}C_5Me_5)(CO)_2$ have been synthesized and studied as models for possible intermediates in the transition metal-catalysed reduction of carbon monoxide.[24a] Studies on the isolable complexes $[M(CHO)(CO)(dppe)_2]^+$, $[M(CHOH)(CO)(dppe)]^{2+}$, and $M(CH_2OH)(CO)\text{-}(dppe)_2]^+$ (M = Ru or Os) suggest that homogeneous catalytic CO hydrogenation may involve successive intermolecular additions of H^- and H^+ to coordinated carbon monoxide.[24b] Hydrogen sulphide is cleaved by $Ru(CO)_2(PPh_3)_3$ to form $Ru(SH)_2(CO)_2(PPh_3)_2$ and liberate dihydrogen.[24c] Photoelectron spectra have been assigned for $RuH_2(PF_3)_4$ and $Ru(PF_3)_5$.[25] Formation of $RuCl\{C(C{\equiv}CPh){=}C(Ph)HgCl\}(CO)_2(PMe_2Ph)_2$ from $RuCl_2(CO)_2(PMe_2Ph)_2$ and $Hg(C{\equiv}CPh)_2$ is thought to involve *cis* addition of an Ru—HgCl bond across one of the triple bonds of a η^2-coordinated 1,4-diphenylbuta-1,3-diyne ligand.[26] A ruthenium cyclopentadienone complex, active in catalytic hydrogenation and dehydrogenation reactions has been shown to possess the novel dinuclear H-bonded structure (14).[27] The first ruthenium trimethylenemethane (tmm) complex $RuCl(NO)(PPh_3)(\eta^4\text{-tmm})$ has been reported.[28] The aluminohydride complex

(14)

[24] (a) G. O. Nelson and C. E. Sumner, *Organometallics*, 1986, **5**, 1983; (b) D. S. Barratt and D. J. Cole-Hamilton, *J. Organomet. Chem.*, 1986, **306**, C41; (c) C.-L. Lee, J. Chisholm, B. R. James, D. A. Nelson, and M. A. Lilga, *Inorg. Chim. Acta*, 1986, **121**, L7.
[25] J. F. Nixon, E. A. Seddon, R. J. Suffolk, M. J. Taylor, J. C. Green, and R. J. Clark, *J. Chem. Soc., Dalton Trans.*, 1986, 765.
[26] Z. Dauter, R. J. Mawby, C. D. Reynolds, and D. R. Saunders, *J. Chem. Soc., Dalton Trans.*, 1986, 433.
[27] Y. Shvo, D. Czarkie, Y. Rahamim, and D. F. Chodosh, *J. Am. Chem. Soc.*, 1986, **108**, 7400.
[28] M. D. Jones, R. D. W. Kemmitt, and A. W. G. Platt, *J. Chem. Soc., Dalton Trans.*, 1986, 1411.

$L_3HRu(\mu\text{-}H)_2AlH(\mu\text{-}H)_2AlH(\mu\text{-}H)_2RuHL_3$ reacts with tetramethylethylenediamine to eliminate AlH_3 and form $L_3HRu(\mu\text{-}H)_2AlH(\mu\text{-}H)_2RuHL_3$ (L = PMe_3).[29] New S-bonded thionitrosodimethylamine complexes of the platinum metals include the species $MClH(SNNMe_2)(CO)(PPh_3)_2$ (M = Ru, Os).[30] Unusually short Ru—B distances [2.08(3) and 2.12(3) Å] have been found in the complex $[(triphos)HRu(\mu\text{-}\eta^2\text{-}BH_4)RuH(triphos)][BPh_4]$.[31] The first alkyl complexes of ruthenium(VI), $[NBu_4][Ru(N)R_4]$ (R = Me or CH_2SiMe_3) and $[NBu_4]$-$[Ru(N)Me_2(CH_2SiMe_3)_2]$, have been reported.[32a] Other new ruthenium alkyls include the homoleptic species Ru_2R_6 and their oxygenation products $R_3Ru(\mu\text{-}O)_2RuR_3$ (R = CH_2CMe_3 or CH_2SiMe_3).[32b]

Detection of three new weak bands in the near infrared spectrum of $[Ru_2OCl_{10}]^{4-}$ has led to the reassignment of electronic transitions associated with the $\pi(Ru—O—Ru)$ skeleton; claims to the synthesis of $Cs_4[Ru_2OBr_{10}]$ have been shown to be in error.[33]

New dinuclear $Ru^{II/III}$ 'lantern' compounds include the polymeric carbonate $\{Na_3[Ru_2(O_2CO)_4]\cdot6H_2O\}_n$,[34a] and benzoate $\{Ru_2(O_2CPh)_4\cdot PhCO_2\cdot H\text{--}O_2CPh\}_n$,[34b] the chlorides $Ru_2(bridge)_4Cl$ (bridge = $Me_3CC(O)NH^{34c}$ and 2-oxo-6-fluoropyridine[34d]) and the axial acetylide $Ru_2(PhNpy)_4(C\equiv CPh)$.[34e] A novel $Ru^{II/II}$ mixed-bridge compound $[Ru_2Cl(dmpm)_2(PhNpy)_2][BPh_4]$ has also been characterized.[34f]

The triple thiolate-bridged complex $[(\eta\text{-}C_5Me_5)Ru(SPh)_3Ru(\eta\text{-}C_5Me_5)]Cl$ has a Ru—Ru distance [2.630(1) Å] compatible with a single bond.[35a] Nucleophilic substitution reactions on $[Ru(CO)_2(O_2CMe)]_\infty$ afford the dinuclear ruthenium(I) complexes $[Ru_2(CO)_n(CNR)_{10-n}][PF_6]_2$ (n = 0—2, R = Bu^t, Bz, $2,6\text{-}Me_2C_6H_3$).[35b] The ketene complex $Ru_2(CO)_2(\mu\text{-}CO)\{\mu\text{-}C(O)CH_2\}(\eta\text{-}C_5Me_5)_2$ formed by oxidation of the μ-vinylidene precursor $Ru_2(CO)_2(\mu\text{-}CO)(\mu\text{-}CCH_2)(\eta\text{-}C_5Me_5)_2$ decarbonylates to form $Ru_2(CO)_2(\mu\text{-}CO)(\mu\text{-}CH_2)(\eta\text{-}C_5Me_5)_2$ and is reduced by $BH_3\cdot THF$ to a mixture of the μ-ethylidene complex $Ru_2(CO)_2(\mu\text{-}CO)(\mu\text{-}CHMe)(\eta\text{-}C_5Me_5)_2$ and the ethylene complex $Ru_2(CO)(C_2H_4)(\mu\text{-}CO)_2(\eta\text{-}C_5Me_5)_2$.[36a] Treatment of $Ru_3(CO)_{10}(dppm)$ with CH_2N_2 affords the η^3-ketene complex $Ru_3(CO)_7\{\mu_3\text{-}\eta^3\text{-}C(O)CH_2\}(\mu\text{-}CH_2)(dppm)$ which readily carbonylates to form the dinuclear η^4-oxaallyl complex $Ru_2(CO)_7\{\mu\text{-}\eta^4\text{-}CH_2C(O)CH_2\}(dppm)$.[36b] Allenyl complexes

[29] A. R. Barron and G. Wilkinson, *J. Chem. Soc., Dalton Trans.*, 1986, 287.
[30] M. Herberhold and A. F. Hill, *J. Organomet. Chem.*, 1986, **315**, 105.
[31] L. F. Rhodes, L. M. Venanzi, C. Sorato, and A. Albinati, *Inorg. Chem.*, 1986, **25**, 3335.
[32] (a) P. A. Shapley and J. P. Wepsiec, *Organometallics*, 1986, **5**, 1515; (b) R. P. Tooze, G. Wilkinson, M. Motevalli, and M. B. Hursthouse, *J. Chem. Soc., Dalton Trans.*, 1986, 2711.
[33] D. Appleby, R. I. Crisp, P. B. Hitchcock, C. L. Hussey, T. A. Ryan, J. R. Sanders, K. R. Seddon, J. E. Turp, and J. A. Zora, *J. Chem. Soc., Chem. Commun.*, 1986, 483.
[34] (a) A. J. Lindsay, M. Motevalli, M. B. Hursthouse, and G. Wilkinson, *J. Chem. Soc., Chem. Commun.*, 1986, 433; (b) M. Spohn, J. Strähle, and W. Hiller, *Z. Naturforsch.*, 1986, **41B**, 541; (c) K. Ryde and D. A. Tocher, *Inorg. Chim. Acta*, 1986, **118**, L49; (d) A. R. Chakravarty, F. A. Cotton, and M. W. Extine, *Polyhedron*, 1986, **5**, 1821; (e) A. R. Chakravarty and F. A. Cotton, *Inorg. Chim. Acta*, 1986, **113**, 19; (f) A. R. Chakravarty, F. A. Cotton, and L. R. Falvello, *Inorg. Chem.*, 1986, **25**, 214.
[35] (a) M. Hidai, K. Imagawa, G. Cheng, Y. Mizobe, Y. Wakatsuki, and H. Yamazaki, *Chem. Lett.*, 1986, 1299; (b) M. O. Albers, D. C. Liles, E. Singleton, J. E. Stead, and M. M. de V. Steyn, *Organometallics*, 1986, **5**, 1262.
[36] (a) N. M. Doherty, M. J. Fildes, N. J. Forrow, S. A. R. Knox, K. A. MacPherson, and A. G. Orpen, *J. Chem. Soc., Chem. Commun.*, 1986, 1355; (b) J. S. Holmgren, J. R. Shapley, S. R. Wilson, and W. T. Pennington, *J. Am. Chem. Soc.*, 1986, **108**, 508; (c) D. Nucciarone, N. J. Taylor, and A. J. Carty, *Organometallics*, 1986, **5**, 1179.

$Ru_2(CO)_6(\mu\text{-}\eta^2\text{-}R_2C=C=CPh)(\mu\text{-}PPh_2)$ have been obtained by treatment of $Ru_2(CO)_6(\mu\text{-}\eta^2\text{-}C\equiv CPh)(\mu\text{-}PPh_2)$ with R_2CN_2 (R = H, Me or Ph).[36c] Photolysis of *concentrated* solutions of $Ru_3(CO)_{12}$ in THF or hydrocarbon solvents yields an insoluble purple red material which appears to be a new oligomer of stoicheiometry $\{Ru(CO)_4\}_n$.[37] Quinoline and other polynuclear heteroaromatic nitrogen compounds undergo ortho metallation reactions with $Ru_3(CO)_{12}$ to form a series of dimetal-laazacyclobutenes.[38] Thermolysis of $Ru_3(CO)_8(\mu\text{-}\eta^2\text{-}dppm)_2$ yields $Ru_3(\mu\text{-}H)\text{-}(CO)_7\{\mu_3\text{-}\eta^4\text{-}PhPCHPPh(C_6H_4)\}(dppm)$ which subsequently converts into $Ru_3(CO)_7(\mu_3\text{-}PPh)(\mu_3\text{-}\eta^2\text{-}CHPPh_2)(\mu\text{-}\eta^2\text{-}dppm)$ and, in the presence of dihydrogen, $Ru_3(\mu\text{-}H)_2(CO)_6(\mu_3\text{-}\eta^2\text{-}PhPCH_2PPh_2)_2$.[39a]

The oxo complex $Ru_3(\mu_3\text{-}O)(\mu_3\text{-}CO)(CO)_5(dppm)_2$ reversibly hydrogenates to $Ru_3(\mu_3\text{-}O)(\mu\text{-}H)_2(CO)_5(dppm)_2$.[39b] Cations of the complexes $[Ru(O_2CR)(dppm)_2]$-$[BPh_4]$ have mononuclear tris(chelate) rather than dinuclear carboxylate bridged structures.[39c] Azobenzenes react with $Ru_3(CO)_{12}$[40a] and $Ru_3(\mu_3\text{-}NAr)(CO)_{10}$[40b] to yield bis(imido) clusters $Ru_3(\mu_3\text{-}NAr)_2(CO)_9$ rather than the O-semidine derivatives previously proposed. Tetrahydrofuran solutions of $[PNP][Ru_3(\mu\text{-}NCO)(CO)_{10}]$ catalyse selective reduction of terminal, unactivated olefins, activated olefins are unaffected.[40c] Treatment of $[Ru_3H(CO)_{11}]^-$ with Bu^tCP gives the anionic $\mu_3\text{-}\eta^2$-phosphaalkyne complexes $[Ru_3H(CO)_9(Bu^tCP)]^-$.[41a] The metastable trinuclear complex $Ru_3(CO)_{11}(Ph_2P\cdot py)$ converts at ambient temperature into the η^2-aroyl complex $Ru_3(\mu\text{-}\eta^2\text{-}C(O)Ph)(\mu_3\text{-}\eta^2\text{-}PhP\cdot py)(CO)_9$.[41b] Photo conversion of the bridging methylidyne complex $Ru_3H(\mu\text{-}COMe)(CO)_{10}$ into the bridging acyl isomer $Ru_3H(\mu\text{-}C(O)Me)(CO)_{10}$ affords the first example of an oxygen-to-carbon alkyl migration.[41c] The reaction of $Ru_3(CO)_9(\mu_3\text{-}S)_2$ with excess Me_2NH at 25 °C yields $Ru_3(CO)_7(NHMe_2)(\mu\text{-}Me_2NC=O)(\mu_3\text{-}S)_2(\mu\text{-}H)$ and $Ru_3(CO)_6(NHMe_2)(\mu\text{-}Me_2NC=O)_2(\mu_3\text{-}S)_2$ on which the N,N-dimethylcarbamoyl ligands bridge non-bonding Ru—Ru pairs.[41d]

X-Ray structural analysis of the salts $[PNP][M_4H_3(CO)_{12}]$(M = Ru or Os) confirms the existence of discrete structural isomers with C_2 and C_{3v} symmetry in the solid state.[42a] The anion $[Ru_4H_3(CO)_{12}]^-$ catalyses the reductive C—N coupling of alkyl isocyanates to give N-formylureas.[42b] A synthetic programme for converting $Ru_5C(CO)_{15}$ *via* $[Ru_4H(C)(CO)_{12}]^-$ into $Ru_4H_2(C)(CO)_{12}$ (C—H and Ru—H isomers) and $Ru_4C(CO)_{13}$ has been described, the last product has a structure similar

[37] W. R. Hastings and M. C. Baird, *Inorg. Chem.*, 1986, **25**, 2913.

[38] R. H. Fish, T.-J. Kim, J. L. Stewart, J. H. Bushweller, R. K. Rosen, and J. W. Dupon, *Organometallics*, 1986, **5**, 2193.

[39] (a) C. Bergounhou, J.-J. Bonnet, P. Fompeyrine, G. Lavigne, N. Lugan, and F. Mansilla, *Organometallics*, 1986, **5**, 60; (b) A. Colombié, J.-J. Bonnet, P. Fompeyrine, G. Lavigne, and S. Sunshine; *Organometallics*, 1986, **5**, 1154; (c) E. B. Boyar, P. A. Harding, S. D. Robinson, and C. P. Brock, *J. Chem. Soc., Dalton Trans.*, 1986, 1771.

[40] (a) M. I. Bruce, M. G. Humphrey, O. B. Shawkataly, M. R. Snow, and E. R. T. Tiekink, *J. Organomet. Chem.*, 1986, **315**, C51; (b) J. A. Smieja, J. E. Gozum, and W. L. Gladfelter, *Organometallics*, 1986, **5**, 2154; (c) J. L. Zuffa, M. L. Blohm, and W. L. Gladfelter, *J. Am. Chem. Soc.*, 1986, **108**, 552.

[41] (a) M. F. Meidine, J. F. Nixon, and R. Mathieu, *J. Organomet. Chem.*, 1986, **314**, 307; (b) N. Lugan, G. Lavigne, and J. J. Bonnet, *Inorg. Chem.*, 1986, **25**, 7; (c) A. E. Friedman and P. C. Ford, *J. Am. Chem. Soc.*, 1986, **108**, 7851; (d) R. D. Adams and J. E. Babin, *Inorg. Chem.*, 1986, **25**, 4010.

[42] (a) M. McPartlin and W. J. H. Nelson, *J. Chem. Soc., Dalton Trans.*, 1986, 1557; (b) G. Süss-Fink and G. Herrmann, *Angew. Chem., Int. Ed. Engl.*, 1986, **25**, 570; (c) A. G. Cowie, B. F. G. Johnson, J. Lewis, and P. R. Raithby, *J. Organomet. Chem.*, 1986, **306**, C63; (d) M. I. Bruce, M. R. Snow, E. R. T. Tiekink, and M. L. Williams, *J. Chem. Soc., Chem. Commun.*, 1986, 701.

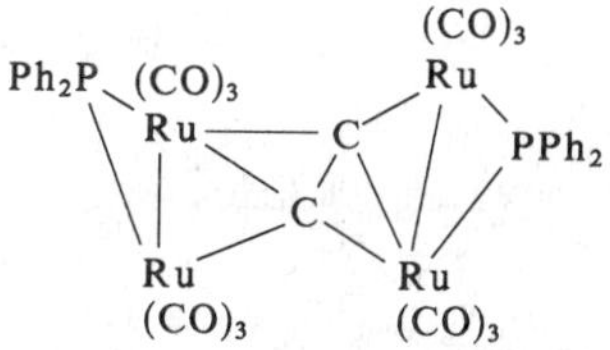

$$\begin{array}{c}
\text{(CO)}_3 \\
\text{Ph}_2\text{P} \diagdown \text{(CO)}_3 \qquad \text{Ru} \\
\text{Ru} \longrightarrow \text{C} \qquad \diagdown \text{PPh}_2 \\
\qquad\qquad \text{C} \\
\text{Ru} \qquad\quad \text{Ru} \\
\text{(CO)}_3 \qquad \text{(CO)}_3
\end{array}$$

(15)

to that of the iron analogue.[42c] Carbonylation of the diphenylphosphinoacetylide complex $Ru_5(\mu_5\text{-}C_2PPh_2)(\mu\text{-}PPh_2)(CO)_{15}$ affords (15) the first example of a μ_4-η^1,η^2-acetylide dianion.[42d] Protonation of $[Ru_4N(CO)_{12}]^-$ in the presence of diphenylacetylene yields $Ru_4(\mu_4\text{-}NH)(PhC\equiv CPh)(CO)_{11}$, the first example of a complex containing a μ_4-imido ligand.[43a] Reactions between tetracarbonylmetallate salts and $Ru_3(CO)_{10}(NCMe)_2$ provide a general route to the tetranuclear clusters $[PNP][Ru_3Co(CO)_{13}]$ and $[PNP][Ru_3MH(CO)_{13}]$ (M = Fe, Ru, or Os).[43b]

The heteropentametallic clusters $RuCo_3(CO)_{12}\{\mu_3\text{-}MPPh_3\}$ (M = Cu, Au) formed by treatment of the cluster anions $[RuCo_3(CO)_{12}]^-$ with $\{MCl(PPh_3)\}_n$ adopt trigonal bipyramidal structures with axial Ru and Cu or Au atoms.[44] The synthesis and dynamic behaviour of the cluster compounds $M_2Ru_4H_2\{Ph_2P(CH_2)_nEPh_2\}(CO)_{12}$ (n = 1 or 2, E = P or As) have been examined[45a,b] and X-ray crystal structures have been reported for $Ag_2Ru_4(\mu_3\text{-}H)_2(\mu\text{-}Ph_2PCH_2PPh_2)(CO)_{12}$ and $Au_2Ru_4(\mu\text{-}H)(\mu_3\text{-}H)(\mu\text{-}Ph_2PCH_2PPh_2)(CO)_{12}$ (16) and (17) respectively.[45b]

Oxidative addition of $PHPh_2$ to the 74 electron cluster $Ru_5(CO)_{13}(\mu_4\text{-}\eta^2\text{-}C_2Ph)(\mu\text{-}PPh_2)$ affords the 78 electron complex $Ru_5(CO)_{13}(\mu\text{-}H)(\mu_4\text{-}\eta^2\text{-}C_2Ph)(\mu\text{-}PPh_2)_2$

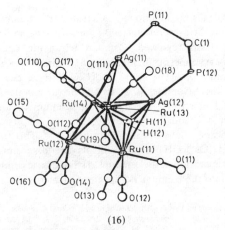

(16)

[43] (a) M. L. Blohm and W. L. Gladfelter, *Organometallics*, 1986, **5**, 1049; (b) G. A. Foulds, B. F. G. Johnson, J. Lewis, and R. M. Sorrell, *J. Chem. Soc., Dalton Trans.*, 1986, 2515.

[44] P. Braunstein, J. Rosé, A. Dedieu, Y. Dusausoy, J. P. Mangeot, A. Tiripicchio, and M. Tiripicchio-Camellini, *J. Chem. Soc., Dalton Trans.*, 1986, 225.

[45] (a) S. S. D. Brown, P. J. McCarthy, and I. D. Salter, *J. Organomet. Chem.*, 1986, **306**, C27; (b) P. A. Bates, S. S. D. Brown, A. J. Dent, M. B. Hursthouse, G. F. M. Kitchen, A. G. Orpen, I. D. Salter, and V. Šik, *J. Chem. Soc., Chem. Commun.*, 1986, 600.

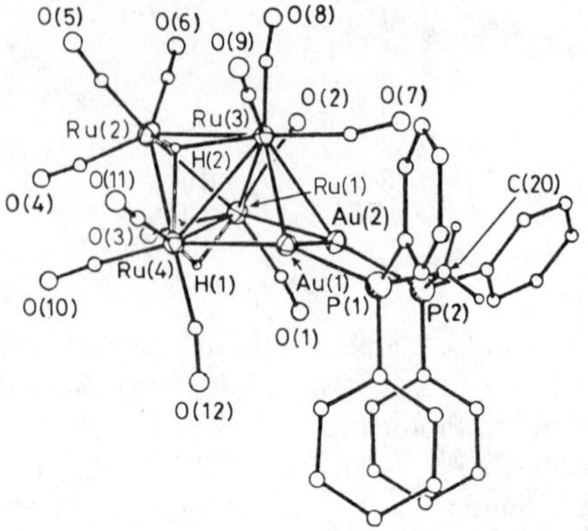

(17)

which adopts a 'bow-tie' structure.[46a] Products obtained from the reaction of $Ru_3(CO)_{12}$ with 3 moles of $PHPh_2$ in refluxing toluene include the distorted trig-onal prismatic clusters $Ru_6(\mu_4\text{-}PPh)_2(\mu_3\text{-}PPh)_2(CO)_{12}$ and $Ru_6(\mu_4\text{-}PPh)_3(\mu_3\text{-}PPh)_2(CO)_{12}$.[46b] When an equimolar amount of $PHPh_2$ is used the products include the encapsulated phosphido complex $Ru_8(\mu_8\text{-}P)(\mu_2\text{-}\eta^1,\eta^6\text{-}CH_2Ph)(\mu_2\text{-}CO)_2(CO)_{17}$.[46c] The novel cluster compound $Ru_{55}(PBu^t_3)_{12}Cl_{20}$ formed by treatment of $RuCl_3/PBu^t_3$ mixtures with B_2H_6 in THF or benzene[46d] can be electrochemically degraded to leave naked Ru_{13} clusters which aggregate into $(Ru_{13})_{13}$ super clusters.[46e]

3 Osmium

Some osmium complexes are mentioned alongside their ruthenium counterparts in the preceding section. Triosmium clusters[46f] and carbene and carbyne complexes of osmium[2b] have been reviewed. Osmium tetraoxide coordinated to chiral diamines catalyses asymmetric oxidation of olefins.[47a] Redox and spectral properties of *cis* and *trans*-[Os(bipy)_2(O)_2][ClO_4]_2 have been reported, the *cis* isomer affords the first example of a d^2 *cis* dioxo complex.[47b] The synthesis, electrochemistry and electronic spectra of the osmium(V) and (VI) complexes $[Os(tmc)(O)_2]^{n+}$ ($n = 1$ and 2 respectively; tmc = 1,4,8,11-tetramethyl-1,4,8,11-tetraazacyclotetradecane have been described.[47c,d] The oxo-bridged dimer $[(bipy)_2(H_2O)Os^{III}(O)Os^{IV}(OH)(bipy)_2]^{4+}$ displays redox chemistry extending from $Os^{V,V}$ to $Os^{II,II}$.[47e] Preparation of the osmium(VI) Schiff base complex $Os(salen)(O)_2$, and its conversion into the osmium(IV) species $Os(salen)(OR)_2$ (R = Me, Et) and

[46] (a) K. Kwek, N. J. Taylor, and A. J. Carty, *J. Chem. Soc., Chem. Commun.*, 1986, 230; (b) J. S. Field, R. J. Haines, and D. N. Smit, *J. Organomet. Chem.*, 1986, **304**, C17; (c) L. M. Bullock, J. S. Field, R. J. Haines, E. Minshall, D. N. Smit, and G. M. Sheldrick, *J. Organomet. Chem.*, 1986, **310**, C47; (d) G. Schmid and W. Huster, *Z. Naturforsch.*, 1986, **41B**, 1028; (e) G. Schmid and N. Klein, *Angew. Chem., Int. Ed. Engl.*, 1986, **25**, 922; (f) A. J. Deeming, *Adv. Organomet. Chem.*, 1986, **26**, 1.

[47] (a) T. Yamada and K. Narasaka, *Chem. Lett.*, 1986, 131; (b) J. C. Dobson, K. J. Takeuchi, D. W. Pipes, D. A. Geselowitz, and T. J. Meyer, *Inorg. Chem.*, 1986, **25**, 2357; (c) C.-M. Che and W.-K. Cheng, *J. Chem. Soc., Chem. Commun.*, 1986, 1519; (d) C.-M. Che and W.-K. Cheng, *J. Am. Chem. Soc.*, 1986, **108**, 4644; (e) J. A. Gilbert, D. Geselowitz, and T. J. Meyer, *J. Am. Chem. Soc.*, 1986, **108**, 1493; (f) C.-H. Che, W.-K. Cheng, and T. C. W. Mak, *Inorg. Chem.*, 1986, **25**, 703.

Os(salen)(SPh)$_2$ have been described.[47f] Square-pyramidal structures have been reported for the anion [OsN(CH$_2$SiMe$_3$)$_4$]$^-$ and its methylation product Os(NMe)-(CH$_2$SiMe$_3$)$_4$.[48] Other osmium alkyls include [Os(η^3-allyl)(CH$_2$CMe$_3$)$_2$]$_2$.[32b] With one exception staggered (D_{4d}) conformations with short Os—Os distances (*ca.* 2.182—2.196 Å) have been found for the anions in the salts [M]$_2$[Os$_2$X$_8$] [M = NBu$_4$ or PNP; X = Cl or Br] and triple (σ^2, π^4, δ^2, δ^{*2}) Os—Os bonds have been proposed.[49a,b] The dinuclear Os$_2^{5+}$ complexes Os$_2$Cl$_4$(Ph$_2$P·py)$_2$(O$_2$CMe)[50a] and Os$_2$(6-chloro-2-oxopyridine)$_4$Cl[50b] are the first structurally characterized compounds with an Os—Os bond order of 2.5. The trimethylsilyl reagents CH$_2$=C(OSiMe$_3$)CH$_2$Cl and Me$_3$SiCH$_2$C(=CH$_2$)CH$_2$· OSO$_2$Me react with Os(CO)$_2$(PPh$_3$)$_3$ to form Os$\overline{\{CH_2C(O)CH_2\}}(CO)_2$(PPh$_3$)$_2$[51] and Os$\{\eta^4$-C(CH$_2$)$_3\}$-(CO)$_2$(PPh$_3$)$_2$[28] respectively. Intra- and inter-molecular C—H bond activation reactions arising from the thermolysis of OsHR(PMe$_3$)$_4$ (R = CH$_2$CMe$_3$ or CH$_2$SiMe$_3$) appear to involve Os^{IV} rather than Os^0 intermediates.[52] The terminal phosphide ligands in the complexes OsCl(PRPh)(CO)$_2$(PPh$_3$)$_2$ (R = Me, I, Ph, OMe) display nucleophilic character in reactions with H$^+$, Me$^+$, and I$_2$ but add methoxide to form Os$\{$PR(OMe)Ph$\}$(CO)$_2$(PPh$_3$)$_2$.[53] Oxidation of OsCl(NO)(η^2-CH$_2$S)(PPh$_3$)$_2$ by 3-chloroperbenzoic acid affords OsCl(NO)(η^2-CH$_2$S=O)(PPh$_3$)$_2$ – the first example of a complex containing an unsubstituted sulphine ligand.[54a] New routes for conversion of coordinated CS$_2$ into CS ligands have permitted syntheses of OsX(NO)(CS)-(PPh$_3$)$_2$ (X = Cl, I) and [Os(NO)(CS)(PPh$_3$)$_2$L]$^+$ (L = CO, PPh$_3$).[54b] The carbyne complex Os($\equiv$CR)Cl(CO)(PPh$_3$)$_2$ reacts with HCl, Cl$_2$, chalcogens and group Ib metal halides to form OsCl$_2$(=CHR)(CO)(PPh$_3$)$_2$ OsCl$_2$($\overline{=CClR}$)(CO)(PPh$_3$)$_2$ OsCl$\{\eta^2$-C(X)R$\}$(CO)(PPh$_3$)$_2$ (X = S, Se, Te) and Os$\{\overline{=C(MY)R}\}$Cl(CO)(PPh$_3$)$_2$ (MY = CuI, AgCl, AuCl), the last named products appear to be dimetallocyclopropene derivatives.[55] Treatment of the complexes OsX(CH$_3$)(PPr$^i_3$)-(η-C$_6$H$_6$) (X = H or D) with [CPh$_3$]PF$_6$ affords the metallocycles [Os$\{\overline{CH_2CH(CH_3)PPr^i_2}\}$H(CH$_2$X)($\eta$-C$_6H_6$)][PF$_6$].[56] Iodide and/or carbonyl substitution or reduction reactions of OsI(CO)$_2$(η-C$_5$Me$_5$) afford a range of new hydride, alkyl, alkene, alkyne, acetylide, and vinylidene derivatives.[57] New high yield syntheses have been reported for Os(η-C$_5$R$_5$)$_2$ (R = H or Me).[19b] Os$\{$P(SiMe$_3$)$_2\}$-(CO)$_2$(η-C$_5$Me$_5$) reacts with aroyl chlorides ArCOCl to give mixtures of phosphaalkenyl complexes Os$\{$P=C(OSiMe$_3$)Ar$\}$(CO)$_2$(η-C$_5$Me$_5$) and diacylphosphido complexes Os$\{$P(C(O)Ar)$_2\}$(CO)$_2$(η-C$_5$Me$_5$).[58] The linear cluster Os$_3$(CO)$_{10}$(η-

[48] P. A. Shapley, Z.-Y. Own, and J. C. Huffman, *Organometallics*, 1986, **5**, 1269.
[49] (*a*) P. A. Agaskar, F. A. Cotton, K. R. Dunbar, L. R. Falvello, S. M. Tetrick, and R. A. Walton, *J. Am. Chem. Soc.*, 1986, **108**, 4850; (*b*) P. E. Fanwick, S. M. Tetrick, and R. A. Walton, *Inorg. Chem.*, 1986, **25**, 4546.
[50] (*a*) F. A. Cotton, K. R. Dunbar, and M. Matusz, *Polyhedron*, 1986, **5**, 903; (*b*) F. A. Cotton, K. R. Dunbar, and M. Matusz, *Inorg. Chem.*, 1986, **25**, 1585 and 1589.
[51] M. D. Jones, R. D. W. Kemmitt, J. Fawcett, and D. R. Russell, *J. Chem. Soc., Chem. Commun.*, 1986, 427.
[52] P. J. Desrosiers, R. S. Shinomoto, and T. C. Flood, *J. Am. Chem. Soc.*, 1986, **108**, 1346 and 7964.
[53] D. S. Bohle, T. C. Jones, C. E. F. Rickard, and W. R. Roper, *Organometallics*, 1986, **5**, 1612.
[54] (*a*) M. Herberhold and A. F. Hill, *J. Organomet. Chem.*, 1986, **309**, C29; (*b*) M. Herberhold, A. F. Hill, N. McAuley, and W. R. Roper, *J. Organomet. Chem.*, 1986, **310**, 95.
[55] G. R. Clark, C. M. Cochrane, K. Marsden, W. R. Roper, and L. J. Wright, *J. Organomet. Chem.*, 1986, **315**, 211.
[56] H. Werner and K. Roder, *J. Organomet. Chem.*, 1986, **310**, C51.
[57] D. B. Pourreau, G. L. Geoffroy, A. L. Rheingold, and S. J. Geib, *Organometallics*, 1986, **5**, 1337.
[58] L. Weber and D. Bungardt, *J. Organomet. Chem.*, 1986, **311**, 269.

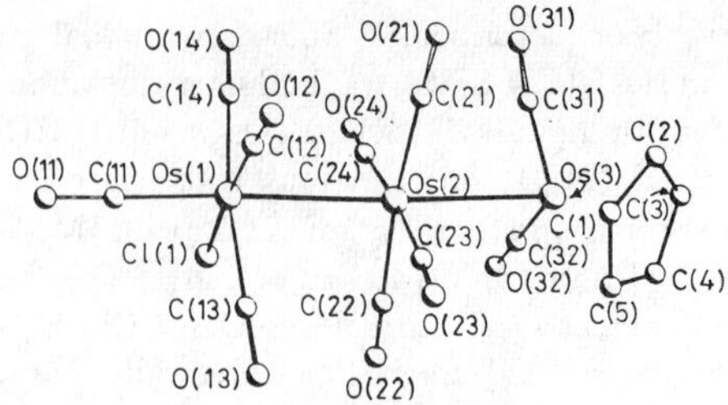

(18)

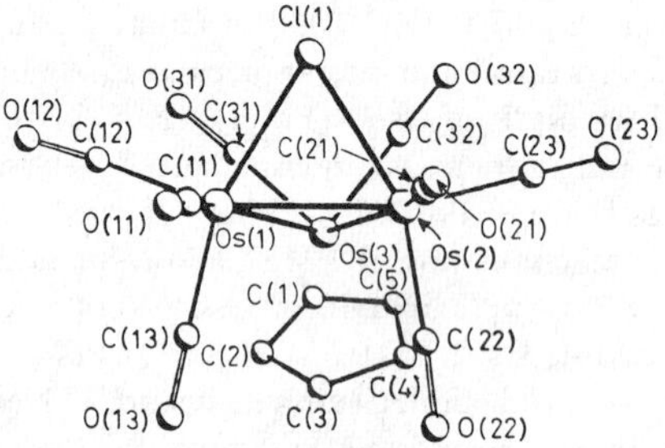

(19)

$C_5H_5)Cl$ (18) converts into the triangular cluster $Os_3(CO)_8(\eta\text{-}C_5H_5)(\mu\text{-}Cl)$ (19) on heating.[59] Nitrosoarenes, ArNO, react with $Os_3(CO)_{12}$ to afford mono and bis nitrene complexes $Os_3(\mu_3\text{-}NAr)(CO)_{10}$ and $Os_3(\mu_3\text{-}NAr)_2(CO)_9$.[60] Transfer of sulphur from ethylene sulphide to the methylene bridge of $Os_3(\mu\text{-}CH_2)(CO)_{11}$ yields the thioformaldehyde complexes $Os_3(\mu\text{-}HCHS)(CO)_{11}$ and $Os_3(\mu_3\text{-}HCHS)(CO)_{10}$.[61] Phenyl vinyl sulphide inserts into an Os—H bond of $Os_3H_2(CO)_{10}$ to give two diasteriomers of $Os_3H(\mu\text{-}PhSCHMe)(CO)_{10}$ but oxidatively adds to $Os_3(MeCN)_2(CO)_{10}$ to give $Os_3(\mu\text{-}SPh)(\mu\text{-}CH{=}CH_2)(CO)_{10}$.[62a] Other papers describe reaction of $Os_3(MeCN)_2(CO)_{10}$ with heterocyclic thioamides,[62b] aldehydes[62c] and 2-formyl derivatives of pyrrole, furan and thiophene.[62d] Nucleophilic attack by H_2O on the α- and β-carbon atoms of $Os_3H(\mu_3\text{-}C{\equiv}CH)(CO)_9$ leads to $Os_3H_3(\mu_3\text{-}CH)(CO)_9$ and $Os_3H_3(\mu_3\text{-}CCHO)(CO)_9$ respectively.[63] Addition of $SnCl_2$ to $Os_3(\mu\text{-}CH_2)(CO)_{11}$ gives the near-planar butterfly cluster (20).[64] Long wavelength photolysis of $Os_3(CO)_{12}$ and $RC{\equiv}CR$ (R = CO_2Me) affords $Os_2(CO)_8(\mu\text{-}\eta^1,\eta^1\text{-}RC{\equiv}CR)$ and $Os_2(CO)_6\{(RC{\equiv}CR)_4\}$ diffraction studies confirm a diosmacyclobutene structure for the first product and reveal a unique coupling of alkynes in the second.[65]

[59] M. A. Gallop, B. F. G. Johnson, J. Lewis, and P. R. Raithby, *J. Chem. Soc., Chem. Commun.*, 1986, 706.
[60] J. A. Smieja and W. L. Gladfelter, *Inorg. Chem.*, 1986, **25**, 2667.
[61] R. D. Adams, J. E. Babin, and M. Tasi, *Organometallics*, 1986, **5**, 1920.
[62] (a) E. Boyar, A. J. Deeming, K. Henrick, M. McPartlin, and A. Scott, *J. Chem. Soc., Dalton. Trans.*, 1986, 1431; (b) A. M. Brodie, H. D. Holden, J. Lewis, and M. J. Taylor, *J. Chem. Soc., Dalton Trans.*, 1986, 633; (c) B. F. G. Johnson, J. Lewis, and T. I. Odiaka, *J. Organomet. Chem.*, 1986, **307**, 61; (d) A. J. Arce, Y. De Sanctis, and A. J. Deeming, *J. Organomet. Chem.*, 1986, **311**, 371.
[63] E. Boyar, A. J. Deeming, and S. E. Kabir, *J. Chem. Soc., Chem. Commun.*, 1986, 577.
[64] N. Viswanathan, E. D. Morrison, G. L. Geoffroy, S. J. Geib, and A. L. Rheingold, *Inorg. Chem.*, 1986, **25**, 3100.
[65] M. R. Burke and J. Takats, *J. Organomet. Chem.*, 1986, **302**, C25.

$$
\begin{array}{c}
Cl_2 \\
Sn \\
(CO)_4Os \qquad\qquad Os(CO)_4 \\
\big| \\
H_2C \longrightarrow Os \\
(CO)_3
\end{array}
$$

(20)

Phosphido clusters $HOs_3(CO)_{10}(PRH)$ react with the dodecacarbonyls $M_3(CO)_{12}$ (M = Ru, Os) to yield phosphinidene clusters $M_3Os_3(CO)_{17}(PR)$, $M_2Os_3(CO)_{15}(PR)$ and $MOs_3H_2(CO)_{12}(PR)$.[66] The coordinated ketene ligand in $[PNP][Os_3(CO)_{10}(\mu\text{-}I)(\mu\text{-}CH_2CO)]$ is readily converted into η^1-enolate ligands by simple nucleophiles and into vinyl or acetyl ligands by electrophiles.[67] The unsaturated cluster $Os_4(CO)_{11}(\mu_4\text{-}S)(\mu_4\text{-}HC{\equiv}CCO_2Me)$ activates hydrogen and undergoes facile multiple addition reactions with $PhC{\equiv}CH$ and $CH_2{=}C{=}CH_2$.[68] The irregular planar cluster structure and fluxional character of $Os_4(CO)_{14}(PMe_3)$ have been rationalized in terms of $3c$, $2e$ metal–metal bonds.[69a] Addition of Me_2NH, H_2, and CO to $Os_4(CO)_{12}(\mu_3\text{-}S)$ affords $Os_4(CO)_{12}(NMe_2)(\mu_2\text{-}S)$, $Os_4(CO)_{12}H_2(\mu_3\text{-}S)$, and $Os_4(CO)_{13}(\mu_3\text{-}S)$ respectively.[69b] The 'raft' cluster $Os_6(CO)_{20}(MeCN)$[69c] reacts with terminal alkynes $RC{\equiv}CH$ (R = Me, Ph) to afford the clusters $Os_6(CO)_{20}(C{=}CHR)$ which on standing lose an $Os(CO)_5$ fragment to generate the edge-bridged tetrahedron clusters $Os_5(CO)_{15}(C{=}CHR)$.[69d] The complex $Os_3(CO)_8(NMe_3)(\mu\text{-}OH)$-$(\mu_3\text{-}S)(\mu\text{-}H)$ formed from $Os_3(CO)_{10}(\mu_3\text{-}S)$ and $Me_3NO{\cdot}2H_2O$ reacts with further $Os_3(CO)_{10}(\mu_3\text{-}S)$ to form the linear Os_6 open-chain complex $Os_6(CO)_{18}(\mu\text{-}OH)$-$(\mu_4\text{-}S)(\mu_3\text{-}S)(\mu\text{-}H)$.[70] Treatment of $Os_6(CO)_{18}$ with Me_3NO in the presence of $OsH_2(CO)_4$ affords new heptanuclear clusters $Os_7H_2(CO)_{22}$ and $Os_7H_2(CO)_{21}$.[71a] Other new osmium clusters include the species $Os_6H_2(CO)_{21}(\mu_3\text{-}PH)$ and $Os_6H_2(CO)_{20}(MeCN)(\mu_3\text{-}PH)$ in which Os_3 triangles are linked by μ_3-PH groups.[71b]

Heterometallic osmium carbonyl clusters have also received attention. The treatment of $OsH_4(PMe_2Ph)_3$ with $[Cu(OBu^t)]_4$ affords the novel trimeric product $Cu_3Os_3H_9(PMe_2Ph)_9$ with a triangular 'raft' structure (21).[72a] The clusters $Os_4(CO)_{12}(\mu_3\text{-}S)_2$ and $Os_5(CO)_{15}(\mu_4\text{-}S)$ react with $Pt(PMe_2Ph)_4$ and $Pt(C_2H_4)$-$(PPh_3)_2$ respectively to generate heterometallic clusters including $PtOs_4(CO)_{11}(PMe_2Ph)_2(\mu_3\text{-}S)_2$, $PtOs_3(CO)_9(PMe_2Ph)_2(\mu_3\text{-}S)_2$, $PtOs_4(CO)_{13}(PPh_3)$-$(\mu_4\text{-}S)$, $PtOs_5(CO)_{15}(PPh_3)(\mu_4\text{-}S)$, and $PtOs_5(CO)_{15}(PPh_3)_2(\mu_4\text{-}S)$.[72b,c] The syntheses

[66] S. B. Colbran, B. F. G. Johnson, J. Lewis, and R. M. Sorrell, *J. Chem. Soc., Chem. Commun.*, 1986, 525.

[67] S. L. Bassner, E. D. Morrison, G. L. Geoffroy, and A. L. Rheingold, *J. Am. Chem. Soc.*, 1986, **108**, 5358.

[68] R. D. Adams and S. Wang, *Organometallics*, 1986, **5**, 1272 and 1274.

[69] (a) L. R. Martin, F. W. B. Einstein, and R. K. Pomeroy, *J. Am. Chem. Soc.*, 1986, **108**, 338; (b) R. D. Adams and S. Wang, *Inorg. Chem.*, 1986, **25**, 2534; (c) R. J. Goudsmit, J. G. Jeffrey, B. F. G. Johnson, J. Lewis, R. C. S. McQueen, A. J. Sanders, and J.-C. Liu, *J. Chem. Soc., Chem. Commun.*, 1986, 24; (d) J. G. Jeffrey, B. F. G. Johnson, J. Lewis, P. R. Raithby, and D. A. Welch, *J. Chem. Soc., Chem. Commun.*, 1986, 318.

[70] R. D. Adams, J. E. Babin, and H. S. Kim, *Inorg. Chem.*, 1986, **25**, 1122.

[71] (a) B. F. G. Johnson, J. Lewis, M. McPartlin, J. Morris, G. L. Powell, P. R. Raithby, and M. D. Vargas, *J. Chem. Soc., Chem. Commun.*, 1986, 429; (b) C. J. Cardin, S. B. Colbran, B. F. G. Johnson, J. Lewis, and P. R. Raithby, *J. Chem. Soc., Chem. Commun.*, 1986, 1288.

[72] (a) T. H. Lemmen, J. C. Huffman, and K. G. Caulton, *Angew. Chem., Int. Ed. Engl.*, 1986, **25**, 262; (b) R. D. Adams, J. E. Babin, R. Mathab, and S. Wang, *Inorg. Chem.*, 1986, **25**, 4 and 1623; (c) R. D. Adams, I. T. Horváth, and S. Wang, *Inorg. Chem.*, 1986, **25**, 1617.

314   S. D. Robinson

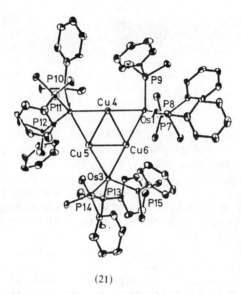

(21)

and crystal structures of the high nuclearity cluster anions,
$[Os_{10}C(CO)_{24}Cu(CNMe)]^-$, $[Os_{10}C(CO)_{24}Au(PPh_3)]^-$, $[Os_{10}C(CO)_{24}(AuBr)]^-$ (22)

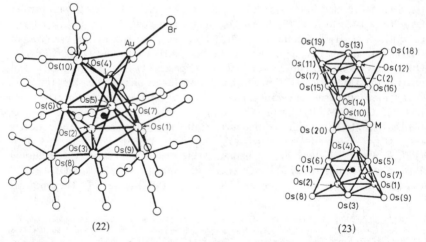

(22)   (23)

and $[Os_{20}M(C)_2(CO)_{48}]^{2-}$ (M = Ag or Hg) (23) have been described.[73a,b] Phos-
phide-centred trigonal-prismatic cores have been found in $[Os_6H(CO)_8P]^-$ and
$[Os_6(CO)_8P(AuPPh_3)]$.[73c] Reactions between tetracarbonylmetallate salts and
$Os_3(CO)_{10}(NCMe)_2$ provide a general route to the tetranuclear clusters [PNP]-
$[Os_3Co(CO)_{13}]$ and $[PNP][Os_3MH(CO)_{13}]$ (M = Fe, Ru or Os).[43b]

[73] (a) B. F. G. Johnson, J. Lewis, W. J. H. Nelson, M. D. Vargas, D. Braga, K. Henrick, and M. McPartlin,
J. Chem. Soc., Dalton Trans., 1986, 975; (b) S. R. Drake, K. Henrick, B. F. G. Johnson, J. Lewis, M.
McPartlin, and J. Morris, J. Chem. Soc., Chem. Commun., 1986, 928; (c) S. B. Colbran, C. M. Hay,
B. F. G. Johnson, F. J. Lahoz, J. Lewis, and P. R. Raithby, J. Chem. Soc., Chem. Commun., 1986, 1766.

4 Rhodium

The literature for 1983 has been reviewed.[74] Rhodium porphyrin complexes have continued to generate interest. Carbonylation of the octaethylporphyrin (OEP) complexes $RhH(OEP)$[75a] and $\{Rh(OEP)\}_2$[75b] afford the formyl complex $Rh(HCO)$-(OEP) and an equilibrium mixture of the adduct $\{Rh(OEP)\}_2CO$ and the dimetal-loketone $(OEP)Rh(\mu\text{-}CO)Rh(OEP)$ respectively. Arenes react with $RhCl(OEP)$ in the presence of $AgClO_4$ or $AgBF_4$ to give rhodium(III) aryl derivatives.[75c] $\{Rh(OEP)\}_2$ catalyses photo-assisted formation of $HCHO$ and $MeOH$ from H_2 and CO.[75d] The regioselective addition of alkyl iodides to $Rh(OEP)$ – $In(OEP)$ is consistent with a polar covalent $Rh(I)^- \rightarrow In(III)^+$ bond.[75e] Electrochemical reduction of the tetraphenylporphyrin (TPP) complexes $[Rh(TPP)(NHMe_2)_2]Cl$,[76a] $Rh(TPP)(O_2)$[76b] and $Rh(TPP)(COMe)$[76c] has been described. The mechanisms of oxidative addition of alkyl halides and dihalides to the nucleophilic rhodium(I) complex (24) have been investigated.[77] New di-, tri-, and tetranuclear 1,2,4-triazolate (tz) complexes including $Rh_2(\mu\text{-}tz)_2(L_2)_2$, $Rh_3(\mu\text{-}tz)_3(\eta^3\text{-}allyl)_6$, $Rh_2M(\mu_3\text{-}tz)(\mu\text{-}X)ClL_2(CO)_4$ and $Rh_2M_2(\mu_3\text{-}tz)_2Cl_2(L_2)_4$ (M = Rh, Ir, Pd, or Au, $L_2 = (CO)_2$, $(CO)(PPh_3)$ or diolefin, X = Cl or OH) have been synthesised.[78a] New 2,2'-bi-imidazolate(2-) bridged di- and tetranuclear Rh^{III}/Rh^I and Rh^{III}/Ir^I complexes have also been obtained.[78b] The cyclophosphorane complex (25) undergoes three-fragment oxidative addition with CH_2Cl_2 to yield the phosphoramide product (26).[78c]

(24)   (25)   (26)

Inclusion of $[RhCl(diolefin)]_2$ dimers by cyclodextrins[79a] and second sphere coordination of $[Rh(diolefin)(NH_3)_2]^+$ cations by dibenzo-3n-crown-n-ethers (n = 6—12)[79b] have been described. Solutions of $[RhCl(CO)_2]_2$ catalyse reduction of N_2O and NO by CO to N_2 at 100 °C.[80] Photochemical carbonylation of benzene to

[74] E. C. Constable, *Coord. Chem. Rev.*, 1986, **73**, 59.
[75] (a) M. D. Farnos, B. A. Woods, and B. B. Wayland, *J. Am. Chem. Soc.*, 1986, **108**, 3659; (b) B. B. Wayland, B. A. Woods, and V. L. Coffin, *Organometallics*, 1986, **5**, 1059; (c) Y. Aoyama, T. Yoshida, K.-I. Sakurai, and H. Ogoshi, *Organometallics*, 1986, **5**, 168; (d) H. W. Bosch and B. B. Wayland, *J. Chem. Soc., Chem. Commun.*, 1986, 900; (e) N. L. Jones, P. J. Carroll, and B. B. Wayland, *Organometallics*, 1986, **5**, 33.
[76] (a) J. E. Anderson, C.-L. Yao, and K. M. Kadish, *Inorg. Chem.*, 1986, **25**, 718; (b) 1986, **25**, 3224; (c) K. M. Kadish, J. E. Anderson, C.-L. Yao, and R. Guilard, *Inorg. Chem.*, 1986, **25**, 1277.
[77] J. P. Collman, J. I. Brauman, and A. M. Madonik, *Organometallics*, 1986, **5**, 215, 218, and 310.
[78] (a) L. A. Oro, M. T. Pinillos, C. Tejel, C. Foces-Foces, and F. H. Cano, *J. Chem. Soc., Dalton Trans.*, 1986, 1087, and 2193; (b) L. A. Oro, D. Carmona, M. P. Lamata, A. Tiripicchio, and F. J. Lahoz, *J. Chem. Soc., Dalton Trans.*, 1986, 15; (c) E. G. Burns, S. S. C. Chu, P. de Meester, and M. Lattman, *Organometallics*, 1986, **5**, 2383.
[79] (a) A. Harada and S. Takahashi, *J. Chem. Soc., Chem. Commun.*, 1986, 1229; (b) H. M. Colquhoun, S. M. Doughty, J. F. Stoddart, A. M. Z. Slawin, and D. J. Williams, *J. Chem. Soc., Dalton Trans.*, 1986, 1639.
[80] W. P. Fang and C. H. Cheng, *J. Chem. Soc., Chem. Commun.*, 1986, 503.

benzaldehyde is catalysed by square-planar rhodium(I) and iridium(I) complexes notably $MCl(CO)(PPh_3)_2$.[81a] Cleavage of the C—O bond in $RhBr(Me)(CH_2OMe)$-$(PMe_3)_3$ with Me_3SiBr provides C_2H_4, formed by methyl migration to a methylene entity and subsequent β-elimination.[81b] Facile conversion of (27) to (28) provides the first example of a spontaneous C—C bond breaking process at an (electrophilic) metal–metal double bond.[81c] Syntheses and oxidative addition reactions have been reported for a series of square-planar rhodium(I) and iridium(I) amide complexes

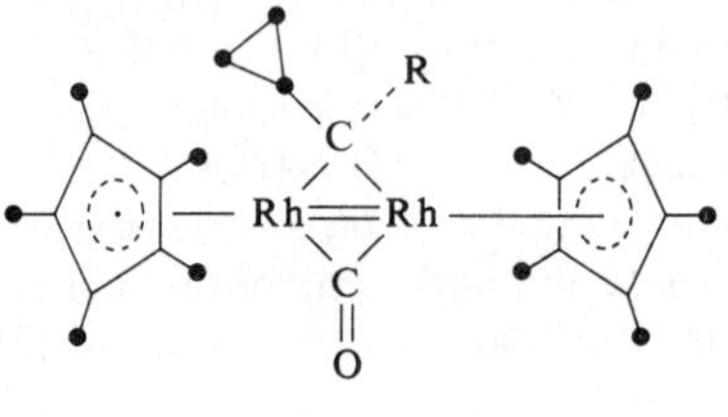

(27)

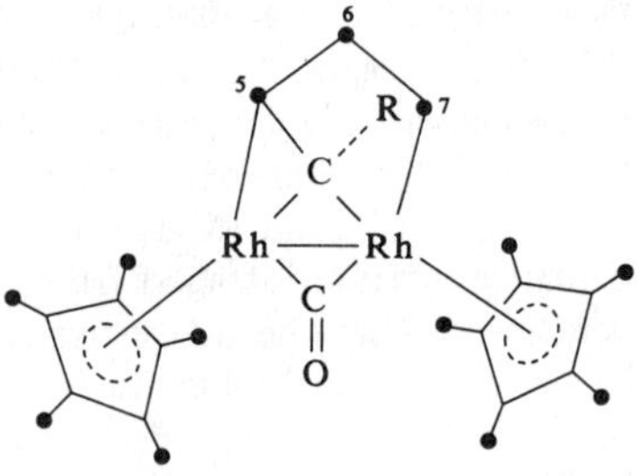

●— = Me; (a) R = Me; (b) R = —◁

(28)

$M(L)\{N(SiMe_2CH_2PR_2)_2\}$ (L = C_8H_{14}, C_2H_4, CO, PMe_3, PPh_3; R = Ph, Pr^i).[81d] New trimethylenemethane (tmm) complexes include $RhCl(\eta^4\text{-tmm})(PPh_3)_2$.[28] Rhodium–rhodium bonds of order 0, 1, and 2 have been found for $(COD)Rh(\mu$-$PBu^t_2)(\mu\text{-Cl})Rh(COD)$ (Rh-Rh 3.395 Å), $(OC)(Bu^t_2HP)Rh(\mu\text{-}Bu^t_2P)(\mu\text{-H})Rh$-$(PHBu^t_2)(CO)$ (Rh—Rh = 2.906 Å), and $[Rh(COD)]_2(\mu\text{-}Bu^tP(CH_2)_4PBu^t)$ respectively.[82a] In contrast to many previous examples of CO addition to metal clusters, reversible carbonylation of $Rh_3(\mu\text{-}Bu^t_2P)_3(CO)_3$ to $Rh_3(\mu\text{-}Bu^t_2P)_3(\mu\text{-CO})$-$(CO)_4$ does not involve metal–metal bond cleavage.[82b] Dinuclear oxidative addition of $RC{\equiv}CR$ (R = CO_2Me) to $[Rh(\mu\text{-}PPh_2)(COD)]_2$ and its iridium analogue yields the adducts $M_2(\mu\text{-}RC{\equiv}CR)(\mu\text{-}PPh_2)_2(COD)_2$.[82c] The synthesis, X-ray structure, and fluxional ring inversion of the dinuclear complex (29) have been described.[82d]

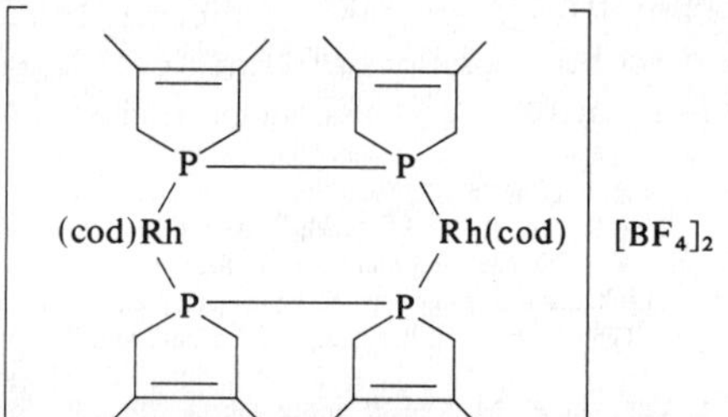

(29)

[81] (a) A. J. Kunin and R. Eisenberg, J. Am. Chem. Soc., 1986, **108**, 535; (b) D. L. Thorn, Organometallics, 1986, **5**, 1897; (c) W. A. Herrmann, E. Herdtweck, and C. Weber, Angew. Chem., Int. Ed. Eng., 1986, **25**, 563; (d) M. D. Fryzuk, P. A. MacNeil, and S. J. Rettig, Organometallics, 1986, **5**, 2469.

[82] (a) A. M. Arif, R. A. Jones, M. H. Seeberger, B. R. Whittlesey, and T. C. Wright, Inorg. Chem., 1986, **25**, 3943; (b) R. A. Jones and T. C. Wright, Inorg. Chem., 1986, **25**, 4058; (c) T. S. Targos, G. L. Geoffroy, and A. L. Rheingold, Organometallics, 1986, **5**, 12; (d) A. L. Crumbliss, R. J. Topping, J. Szewczyk, A. T. McPhail, and L. D. Quin, J. Chem. Soc., Dalton Trans., 1986, 1895.

The molecular and electronic structure of $RhCl(\eta^2\text{-}P_4)(PPh_3)_2$ has been examined.[83a] Displacement of ethylene from the complexes $M(\eta\text{-}C_5R_5)(C_2H_4)_2$ (M = Co, Rh, or Ir; R = H or Me) by $P\equiv CBu^t$ affords 1,3-diphosphacyclobutadiene derivatives $M(\eta\text{-}C_5R_5)\{\eta^4\text{-}\overline{P\text{-}C(Bu^t)\text{-}P\text{-}C}(Bu^t)\}$.[83b] $[Rh(\mu\text{-}S^tBu)(CO)(PPh_3)]_2$ is a much more active hydroformylation catalyst than the parent $RhH(CO)(PPh_3)_3$.[84] New chiral rhodium-based asymmetric hydrogenation[85a−d] and hydrosilylation[85e,f] catalysts have been reported. The complex $(Cy_3P)_2Rh(\mu\text{-}CO)_2Rh(CO)(PCy_3)$ contains square-planar $>Rh(CO)(PCy_3)$ and distorted tetrahedral $>Rh(PCy_3)_2$ groups.[86] Hydrogenation of $Rh_2(\mu\text{-}H)_2(RNC)\{P(OPr^i)_3\}_4$ generates $Rh_2(\mu\text{-}H)(\mu\text{-}RNCH_3)$- $\{P(OPr^i)_3\}_4$ by transfer of hydrogens to the isocyanide carbon atom.[87] $\{[RhH(Pr^i_2PCH_2CH_2CH_2PPr^i_2)]_2(\mu\text{-}H)_2(\mu\text{-}ClO_4)\}ClO_4$ affords the first example of rhodium atoms bridged by a perchlorate ligand.[88]

X-Ray diffraction studies have been reported for the paramagnetic Rh^{III} – Rh^{II} complex $[(triphos)Rh(\mu\text{-}H)_3Rh(triphos)][BPh_4]_2\cdot Me_2NCHO$[89a] and the related bimetallic complex $(triphos)Rh(\mu\text{-}H)_3Zn(OAr)_2$ (Ar = $2,4,6\text{-}Bu^t_3C_6H_2$).[89b] The former complex and some related bimetallic species $[(triphos)Rh(\mu\text{-}H)_3M(triphos)]$- $[ClO_4]_2$ (M = Ni, Co) have been shown to undergo reversible one-electron redox changes.[89c] Addition of LiMe to the acyl complex $Rh(COMe)(CO)(triphos)$ affords the acetone enolate complex $[Rh\{C(O)Me_2\}(CO)(triphos)]^-$ rather than leading to deprotonation of the acyl group and formation of a ketene complex.[89d] The complex $[(triphos)Rh(\mu\text{-}S)_2Rh(triphos)]^{2+}$ reacts with H_2 (reversibly) and with O_2 to form $[(triphos)HRh(\mu\text{-}SH)_2RhH(triphos)]^{2+}$ and $[(triphos)Rh(\mu\text{-}SO)_2Rh(triphos)]^{2+}$ respectively.[89e] Interaction of CO_2 with $RhCl\{(Me_2PCH_2CH_2CH_2)_2PBu^t\}$ affords $Rh(O_2CO)Cl\{(Me_2PCH_2CH_2CH_2)_2PBu^t\}$ and CO.[89f]

The complex $Rh_2(\mu\text{-}CO)(\mu\text{-}CF_3C\equiv CCF_3)(\eta\text{-}C_5H_5)_2$ adds alkenes to form bis(alkenyl) complexes $Rh_2(alkene\text{-}H)(CF_3C=CHCF_3)(\eta\text{-}C_5H_5)_2$[90a] and reacts with alkynes $RC\equiv CR'$ to form μ-pentadienone complexes $Rh_2\{C_4(CF_3)_2RR'CO\}(\eta\text{-}C_5H_5)_2$, or the metallodiene complexes $Rh_2\{C_4(CF_3)_2RR'\}(\eta\text{-}C_5H_5)_2$ and the dicarbonyl $Rh_2(CF_3C\equiv CCF_3)(CO)_2(\eta\text{-}C_5H_5)_2$.[90b] Photolysis of $Rh(C_2H_4)_2(\eta\text{-}C_5H_5)$ in

[83] (a) A. P. Ginsberg, W. E. Lindsell, K. J. McCullough, C. R. Sprinkle, and A. J. Welch, *J. Am. Chem. Soc.*, 1986, **108**, 403; (b) P. B. Hitchcock, M. J. Maah, and J. F. Nixon, *J. Chem. Soc., Chem. Commun.*, 1986, 737.

[84] P. Escaffre, A. Thorez, P. Kalck, B. Besson, R. Perron, and Y. Colleuille, *J. Organomet. Chem.*, 1986, **302**, C17.

[85] (a) F. Alario, Y. Amrani, Y. Colleuille, T. P. Dang, J. Jenck, D. Morel, and D. Sinou, *J. Chem. Soc., Chem. Commun.*, 1986, 202; (b) M. Yamashita, M. Kobayashi, M. Sugiura, K. Tsunekawa, T. Oshikawa, S. Inokawa, and H. Yamamoto, *Bull. Chem. Soc. Jpn.*, 1986, **59**, 175; (c) A. Kinting and H.-W. Krause, *J. Organomet. Chem.*, 1986, **302**, 259; (d) U. Nagel and E. Kinzel, *Chem. Ber.*, 1986, **119**, 1731; (e) I. D. McKay and N. C. Payne, *Acta Crystallogr.*, 1986, **C42**, 307; (f) H. Brunner, G. Riepl, I. Bernal, and W. H. Ries, *Inorg. Chim. Acta*, 1986, **112**, 65.

[86] M. A. Freeman and D. A. Young, *Inorg. Chem.*, 1986, **25**, 1556.

[87] S. T. McKenna, R. A. Andersen, and E. L. Muetterties, *Organometallics*, 1986, **5**, 2233.

[88] K. Tani, T. Yamagata, Y. Tatsuno, T. Saito, Y. Yamagata, and N. Yasuoka, *J. Chem. Soc., Chem. Commun.*, 1986, 494.

[89] (a) C. Bianchini, C. Mealli, A. Meli, and M. Sabat, *J. Chem. Soc., Chem. Commun.*, 1986, 777; (b) R. L. Geerts, J. C. Huffman, and K. G. Caulton, *Inorg. Chem.*, 1986, **25**, 590; (c) C. Bianchini, A. Meli, and P. Zanello, *J. Chem. Soc., Chem. Commun.*, 1986, 628; (d) G. G. Johnston and M. C. Baird, *J. Organomet. Chem.*, 1986, **314**, C51; (e) C. Bianchini, C. Mealli, A. Meli, and M. Sabat, *Inorg. Chem.*, 1986, **25**, 4617; (f) L. Dahlenburg, C. Prengel, and N. Hock, *Z. Naturforsch.*, 1986, **41b**, 718.

[90] (a) R. S. Dickson, G. D. Fallon, S. M. Jenkins, B. W. Skelton, and A. H. White, *J. Organomet. Chem.*, 1986, **314**, 333; (b) C. W. Baimbridge, R. S. Dickson, G. D. Falcon, I. Grayson, R. J. Nesbit, and J. Weigold, *Aust. J. Chem.*, 1986, **39**, 1187.

N_2 or CO[91a] doped xenon at 173 °C affords $Rh(C_2H_4)(N_2)(\eta\text{-}C_5H_5)$ and $Rh(C_2H_4)$-$(CO)(\eta\text{-}C_5H_5)$ respectively, the latter product loses CO to form $Rh(C_2H_4)(\eta\text{-}C_5H_5)$ and C_2H_4 to form $Rh(CO)(\eta\text{-}C_5H_5)$ both of which activate C—H and Si—H bonds.[91b] Isomerization of hydridoalkylrhodium(III) complexes formed by oxidative addition of alkane C—H bonds across $Rh(\eta\text{-}C_5Me_5)(PMe_3)$ provides evidence of intermediate $\eta^2\text{-}$C—H alkane complexes.[91c] The cyclopropane derivative $RhH(\text{cyclopropyl})(\eta\text{-}C_5Me_5)(PMe_3)$ rearranges readily to the rhodacyclobutane $Rh(CH_2CH_2CH_2)(\eta\text{-}C_5Me_5)(PMe_3)$.[91c] Rhodacyclobutanes $Rh\{CH_2E(Me)_2CH_2\}$-$(\eta\text{-}C_5Me_5)(PPh_3)$ (E = C or Si) have also been obtained from $RhCl_2(\eta\text{-}C_5Me_5)$-$(PPh_3)$ and $Mg(CH_2EMe_3)Cl$.[91d]

Electrochemical oxidation of $[Rh_2(1,8\text{-diisocyanomenthane})_2(dppm)_2]^{2+}$ generates a $d^7\text{-}d^8$ radical cation.[92] Dinuclear Rh^{II}—Rh^{II} carboxylates and their derivatives are again a focus of attention. New examples synthesized and characterized include derivatives of R- and S-mandelic acid,[93a,b] the substituted species $[Rh_2(O_2CMe)_2(MeCN)_4L_2][BF_4]_2$ (L = MeCN or py)[93c] and the cyclometallated product $Rh_2\{PhP(C_6H_4)(C_6F_4Br)\}(O_2CMe)_3\{PPh_2(C_6F_4Br)\}$ which contains the structural skeleton shown in (30).[93d] Cavity-shaped tri- and tetra-aza ligands span axial and bridging sites in the complexes $[Rh_2(O_2CMe)_3L]^+$ [L = (31) and related

(30)

X = Y = N
X = N, Y = CH

(31)

species];[93e] electrochemical reduction to $[Rh_2(O_2CMe)_3L]$ and $[Rh_2(O_2CMe)_3L]^-$ is described.[93f] An e.s.r. study of $[Rh_2(O_2CMe)_4(H_2O)_2]^+$ places the odd electron in a degenerate π^*_{RhRh} orbital.[93g] Resonance Raman data for isotopically labelled $Rh_2(O_2CMe)_4(PPh_3)_2$ indicate that $\nu(Rh\text{—}Rh)$ occurs at ca. 289 cm^{-1} rather than 150—170 cm^{-1}.[93h] Dinuclear lantern structures are also found in the new Rh^{II}—Rh^{II}

[91] (a) D. M. Haddleton, R. N. Perutz, S. A. Jackson, R. K. Upmacis, and M. Poliakoff, J. Organomet. Chem., 1986, 311, C15; (b) D. M. Haddleton, J. Organomet. Chem., 1986, 311, C21; (c) R. A. Periana and R. G. Bergman, J. Am. Chem. Soc., 1986, 108, 7332 and 7346; (d) L. Andreucci, P. Diversi, G. Ingrosso, A. Lucherini, F. Marchetti, V. Adovasio, and M. Nardelli, J. Chem. Soc., Dalton Trans., 1986, 477.

[92] D. C. Boyd, P. A. Matsch, M. M. Mixa, and K. R. Mann, Inorg. Chem., 1986, 25, 3331.

[93] (a) P. A. Agaskar, F. A. Cotton, L. R. Falvello, and S. Han, J. Am. Chem. Soc., 1986, 108, 1214; (b) F. Pruchnik, B. R. James, and P. Kvintovics, Can. J. Chem., 1986, 64, 936; (c) G. Pimblett, C. D. Garner, and W. Clegg, J. Chem. Soc., Dalton Trans., 1986, 1257; (d) F. Barcelo, F. A. Cotton, P. Lahuerta, R. Llusar, M. Sanau, W. Schwotzer, and M. A. Ubeda, Organometallics, 1986, 5, 808; (e) R. P. Thummel, F. Lefoulon, D. Williamson, and M. Chavan, Inorg. Chem., 1986, 25, 1675; (f) J. L. Bear, L. K. Chau, M. Y. Chavan, F. Lefoulon, R. P. Thummel, and K. M. Kadish, Inorg. Chem., 1986, 25, 1514; (g) T. Kawamura, H. Katayama, and T. Yamabe, Chem. Phys. Lett., 1986, 130, 20; (h) R. J. H. Clark, A. J. Hempleman, and C. D. Flint, J. Am. Chem. Soc., 1986, 108, 518.

species $Rh_2(2\text{-PhNpy})_4$[94a] and $[Rh_2\{MeC(O)NH\}_4\cdot 2H_2O]3H_2O$,[94b] the Rh^{II}—Rh^{III} salts $[Rh_2\{MeC(O)NH\}_4(\text{theophylline})_2]NO_3H_2O$,[94c] $[Rh_2(2\text{-HNpy})_4]X$ (X = Cl, Br)[94d] and $K_3[Rh(SO_4)_4]2H_2O$,[94e] and the Rh^{III}—Rh^{III} dication $[Rh_2(2\text{-PhNpy})_4]^{2+}$.[94a]

New dinuclear rhodium complexes with bridging 2-Ph_2Ppy ligands include the dimetallated olefin derivatives $Rh_2Cl_2(\mu\text{-RC}=CR)(\mu\text{-2-}Ph_2Ppy)_2$[95a] (R = CF_3, CO_2CH_3) and the nitrate-bridged species $Rh_2(\mu\text{-}NO_3)(\mu\text{-2-}Ph_2Ppy)_2(CO)Cl_3$.[95b] New rhodium complexes containing monodentate (*S*-coordinated), chelate and bridging (*S*- or *N*,*S*-coordinated) 2-pyridine thiolate ligands have been characterized.[96a–c] Tri-, di-, and mono-nuclear rhodium complexes with 1,8-naphthyridine-2-onate ligands include salts of the linear cations (32) [L_2 = $(CO)_2$, (CO)-(PPh$_3$), or COD].[97] Rhodium compounds involving the tridentate ligands

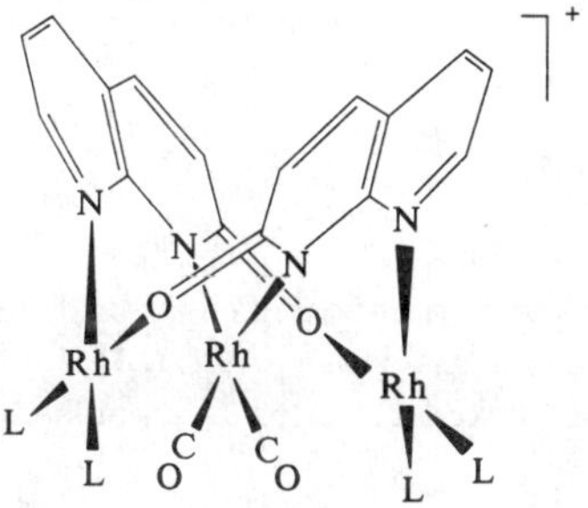

(32)

$Ph_2PCH_2P(Ph)CH_2PPh_2$(dpmp), $Me_2PCH_2P(Me)CH_2PMe_2$(dmmm), and $Ph_2PCH_2As(Ph)CH_2PPh_2$(dpma) have received considerable attention. Trinuclear products include the dimetallated olefin complex $[Rh_3(\mu\text{-dpmp})_2(\mu\text{-CO})(CO)(\mu\text{-Cl})(Cl)(\mu\text{-MeO}_2CC=CCO_2Me)][BPh_4]$[98a] and the dihydride $[Rh_3(\mu\text{-dpmp})_2(H)_2(\mu\text{-Cl})_2(CO)_2][BPh_4]$.[98b] Dinuclear species include the remarkably crowded cation $[Rh_2(dmmm)_2(CO)_2]^{2+}$ (33)[98c] and the 12-membered metallocyclic $Rh_2(dpma)_2Cl_2(CO)_2$(34).[98d] The latter product and its iridium analogue are readily converted into trimetallic chain complexes $[M_2M'(\mu\text{-dpma})_2(CO)_3(\mu\text{-Cl})Cl][BPh_4]$ (M and M' = Rh or Ir) by addition of a Rh^I or Ir^I ions to the central cavity.[98d] Heterobimetallic complexes of rhodium with platinum, silver, and gold have been obtained using the branched tridentate $2(Ph_2P)_2CHpy$ ligand.[98e] The phosphido

94 (a) D. A. Tocher and J. H. Tocher, *Polyhedron*, 1986, **5**, 1615; (b) M. Q. Ahsan, I. Bernal, and J. L. Bear, *Inorg. Chem.*, 1986, **25**, 260; (c) K. Aoki, M. Hoshino, T. Okada, H. Yamazaki, and H. Sekizawa, *J. Chem. Soc., Chem. Commun.*, 1986, 314; (d) M. Zuber, *Transition Met. Chem.*, 1986, **11**, 5; (e) I. B. Baranovskii, A. N. Zhilyaev, and A. V. Rotov, *Russ. J. Inorg. Chem.*, 1985, **30**, 1822.

95 (a) J. T. Mague, *Organometallics*, 1986, **5**, 918; (b) L. J. Tortorelli, C. A. Tucker, C. Woods, and J. Bordner, *Inorg. Chem.*, 1986, **25**, 3534.

96 (a) A. J. Deeming and M. N. Meah, *Inorg. Chim. Acta*, 1986, **117**, L13; (b) L. Oro, M. A. Ciriano, F. Viguri, A. Tiripicchio, M. Tiripicchio-Camellini, and F. J. Lahoz, *Nouv. J. Chim.*, 1986, **10**, 75; (c) A. J. Deeming, M. N. Meah, H. M. Dawes, and M. B. Hursthouse, *J. Organomet. Chem.*, 1986, **299**, C25.

97 M. A. Ciriano, B. E. Villarroya, and L. A. Oro, *Inorg. Chim. Acta*, 1986, **120**, 43.

98 (a) A. L. Balch, L. A. Fossett, J. C. Linehan, and M. M. Olmstead, *Organometallics*, 1986, **5**, 691; (b) A. L. Balch, J. C. Linehan, and M. M. Olmstead, *Inorg. Chem.*, 1986, **25**, 3937; (c) A. L. Balch, M. M. Olmstead, and D. E. Oram, *Inorg. Chem.*, 1986, **25**, 298; (d) A. L. Balch, L. A. Fossett, M. M. Ormstead, and P. E. Reedy, *Organometallics*, 1986, **5**, 1929; (e) R. J. McNair and L. H. Pignolet, *Inorg. Chem.*, 1986, **25**, 4717.

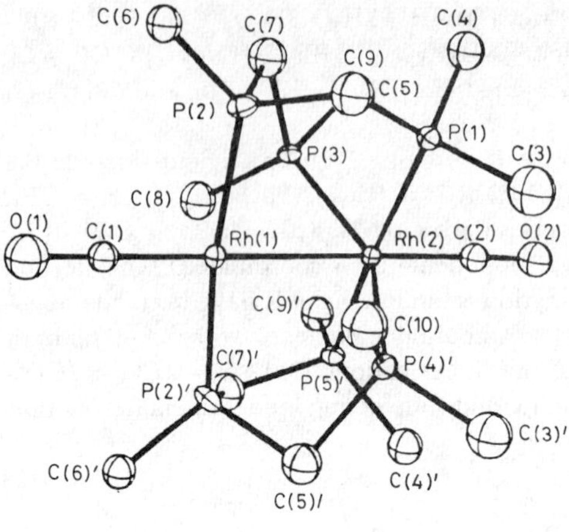

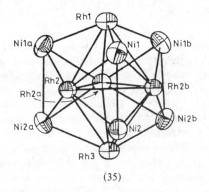

(33)

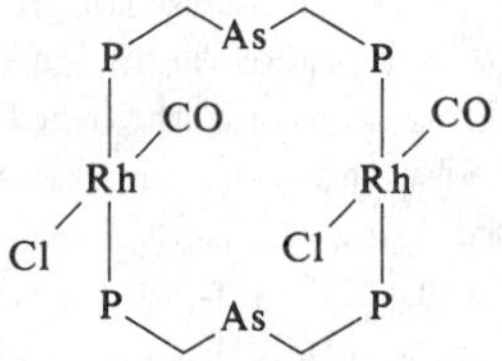

(34)

complex $Rh_6(\mu\text{-}Bu^t_2P)_4(CO)_6(\mu\text{-}CO)_2(\mu\text{-}H)_2$ has an unusual Rh_6 core consisting of an Rh_4 tetrahedron with two opposing edges bridged by rhodium atoms.[99a] The 11-atom heterometallic species $[Rh_5Ni_6(CO)_{21}H_x]^{3-}$ ($x = 1$ *or* 3) possesses a hitherto unknown close-packed 11-vertex D_{3h} polyhedral core (35).[99b]

(35)

The anion $[Rh_{10}(CO)_{21}]^{2-}$ contains a novel decanuclear skeleton which is a fragment of hexagonal close-packed lattice formed by three superimposed layers (3/4/3 metals).[99c] The dodecanuclear skeleton of $[Rh_{12}H(N)_2(CO)_{23}]^{3-}$ contains N atoms encapsulated in trigonal-prismatic and distorted-square anti-prismatic holes.[99d] Variable temperature, multinuclear n.m.r. studies on the clusters $[Rh_{13}(CO)_{24}H_x]^{(5-x)-}$ ($x = 1$—4) and $[Rh_{14}(CO)_{25}H_y]^{(4-y)-}$ ($y = 0$ or 1) show interstitial H-migration at room temperature with partial randomization of carbonyls

⁹⁹ (*a*) A. M. Arif, D. E. Heaton, and R. A. Jones, *J. Chem. Soc., Chem. Commun.*, 1986, 1506; (*b*) D. A. Nagaki, J. V. Badding, A. M. Stacy, and L. F. Dahl, *J. Am. Chem. Soc.*, 1986, **108**, 3825; (*c*) S. Martinengo, G. Ciani, and A. Sironi, *J. Chem. Soc., Chem. Commun.*, 1986, 1282; (*d*) 1986, 1742; (*e*) C. Allevi, B. T. Heaton, C. Seregni, L. Strona, R. J. Goodfellow, P. Chini, and S. Martinengo, *J. Chem. Soc., Dalton Trans.*, 1986, 1375.

in the former and complete randomization in the latter.[99e] Reduction of $RhCl(PPh_3)_3$ with B_2H_6 in THF affords the cluster $Rh_{55}(PPh_3)_{12}Cl_6$.[46d]

5 Iridium

Some iridium complexes are mentioned alongside their rhodium analogues in the preceding section. The literature for 1983 has been reviewed.[100] Carbene and carbyne complexes of iridium have been surveyed.[2b] New aqua ions of iridium (IV) and (V) have been prepared by electrochemical oxidation of $[Ir(H_2O)_6]^{3+}/HClO_4$ solutions.[101a] The X-ray crystal structure of Δ,Λ-$[(H_2O)(en)_2Ir(OH)Ir(en)_2(OH)]$-$(S_2O_6)_{3/2}(ClO_4)\cdot2.75\ H_2O$ has been reported.[101b] Oxidative addition of H_2E (E = O or S) to $[Ir(PMe_3)_4][PF_6]$ affords the thermally stable products $[IrH(EH)(PMe_3)_4]$-$[PF_6]$.[101c] The iridium(II) octaethylporphyrin dimer $[Ir(OEP)]_2$, prepared by photolysis of $IrMe(OEP)$, reacts with toluene and $CH_2=CHOEt$ to form $IrH(OEP)/Ir(CH_2Ph)(OEP)$ and $(OEP)IrCH_2-CH(OEt)Ir(OEP)$ respectively.[102a] The *cis*-bicyclo(3,3,0)oct-1-yl derivative $Ir(C_8H_{13})(OEP)$ provides the first example of an iridium porphyrin crystal structure and undergoes reversible oxidation without cleavage of the axial Ir—C bond.[102b]

The complexes $[Ir(S-N-S-N)(Ph_2PCH=CHPPh_2)_2][BPh_4]^{103a}$ and *fac*-$Ir(S-N-S-N-S-N-S-N)Cl(CO)(PPh_3)^{103b}$ prepared by the reactions of S_4N_4 with $[IrPh_2PCH=CHPPh_2)_2][BPh_4]$ and $IrCl(CO)(PPh_3)_2$ respectively have been characterized by diffraction methods. An Ir—Br distance of 2.479(2) Å has been reported for the P,Br chelate $Ir\{PPh_2(o-C_6H_4Br)\}Cl_3(PPh_3)$.[103c] New trimethylene methane (tmm) complexes include $IrX(CO)(EPh_3)(\eta^4$-tmm) (X = Cl or Br, E = P or As).[28] The 'push–pull' precursor $Me_3SiCH_2C(O)CH_2Cl$ reacts with $IrH(CO)(PPh_3)_3$, $IrH(CO)_2(PPh_3)$ or $IrCl(CO)(PPh_3)_2$ to generate the metallocyclobutan-3-one $Ir\{CH_2C(O)CH_2\}Cl(CO)(PPh_3)_2$.[51] The alkoxides *trans*-$Ir(OR)$-$(CO)(PPh_3)_2$ (R = Me, Pr^i, Pr^n) decompose by β-elimination in the presence of PPh_3 to give $IrH(CO)(PPh_3)_3$ and an aldehyde or ketone.[104a] $Ir\{C(O)CH_2OH\}HCl(PMe_3)_3$, a model compound for CO hydrogenation catalysis, has been characterized by diffraction methods.[104b] The cation $[Ir(np_3)]^+$ [$np_3 = N(CH_2CH_2PPh_2)_3$] activates an sp^2 C—H bond of cyclo-octa 1,5-diene to give the *cis* σ-cyclooctadienyl hydride $[IrH(\sigma$-$C_8H_{11})(np_3)][BPh_4]$.[104c]

Oxidative addition of simple and functionalized s-alkyl iodide to $IrX(CO)L_2$ (X = Cl, I; L = PMe_3, $PPhMe_2$, PPh_2Me) and $IrR(CO)(PMe_3)_2$ is accompanied by rearrangement of the *simple* s-alkyl iridium(III) products to the corresponding

[100] E. C. Constable, *Coord. Chem. Rev.*, 1986, **73**, 113.

[101] (a) S. E. Castillo-Blum, D. T. Richens, and A. G. Sykes, *J. Chem. Soc., Chem. Commun.*, 1986, 1120; (b) F. Galsbøl, S. Larsen, B. Rasmussen, and J. Springborg, *Inorg. Chem.*, 1986, **25**, 290; (c) D. Milstein, J. C. Calabrese, and I. D. Williams, *J. Am. Chem. Soc.*, 1986, **108**, 6387.

[102] (a) K. J. Del Rossi and B. B. Wayland, *J. Chem. Soc., Chem. Commun.*, 1986, 1653; (b) J.-L. Cornillon, J. E. Anderson, C. Swistak, and K. M. Kadish, *J. Am. Chem. Soc.*, 1986, **108**, 7633.

[103] (a) C. A. Ghilardi, S. Midollini, S. Moneti, and A. Orlandini, *J. Organomet. Chem.*, 1986, **312**, 383; (b) F. Edelmann, H. W. Roesky, C. Spang, M. Noltemeyer, and G. M. Sheldrick, *Angew. Chem., Int. Ed. Engl.*, 1986, **25**, 931; (c) F. A. Cotton, F. Lahuerta, M. Sanau, W. Schwotzer, and I. Solana, *Inorg. Chem.*, 1986, **25**, 3526.

[104] (a) K. A. Bernard, W. M. Rees, and J. D. Atwood, *Organometallics*, 1986, **5**, 390; (b) D. Milstein, W. C. Fultz, and J. C. Calabrese, *J. Am. Chem. Soc.*, 1986, **108**, 1336; (c) C. Bianchini, D. Masi, A. Meli, M. Peruzzini, M. Sabat, and F. Zanobini, *Organometallics*, 1986, **5**, 2557.

n-alkyl derivatives.[105a] Photolysis of $Ir(\eta\text{-}C_5H_5)(C_2H_4)_2$ in low temperature matrices generates first $Ir(\eta\text{-}C_5H_5)H(CH=CH_2)(C_2H_4)$ and then $Ir(\eta\text{-}C_5H_5)(H)_2(C=CH_2)$.[105b] 1,1-Dimethylcyclopentane reacts with $[IrH_2(Me_2CO)_2L_2]$-$[SbF_6]$ [L = $P(C_6H_4F\text{-}p)_3$] to give first $[Ir(5,5\text{-}Me_2C_5H_4)IrL_2][SbF_6]$ and then $[Ir(\eta^5\text{-}MeC_5H_4)(Me)L_2][SbF_6]$.[105c] Carbon dioxide extrusion converts $Ir\{C(Ar)=N-O-C(O)\}(\eta\text{-}C_5H_5)(PPh_3)$ (Ar = $p\text{-}C_6H_4X$, X = Cl, F) to the η^2-nitrile complexes $Ir(\eta^2\text{-}ArCN)(\eta\text{-}C_5H_5)(PPh_3)$.[105d] Conversion of $(\eta^5$-indenyl)$Ir(C_8H_{14})_2$ into $(\eta^3$-indenyl)IrL_3 (L = PMe_3, PMe_2Ph) affords the first example of $\eta^5 \rightarrow \eta^3$-ring slippage by an indenyl ligand.[105e] Thermolysis of $IrH(Cy)$-$(\eta\text{-}C_5Me_5)(PMe_3)$ in benzene generates cyclohexane and $IrH(Ph)(\eta\text{-}C_5Me_5)$-$(PMe_3)$.[105f] The X-ray crystal structures of $[Ir(\mu\text{-}Cl)(COD)]_2$ and its aerobic oxidation product $Ir_2(\mu\text{-}OH)_2(\mu\text{-}O)Cl_2(COD)_2$ have been reported.[106a] A carbon–carbon coupling reaction between $CH_2(CN)_2$ and CO_2 in the presence of $IrH(PMe_3)_4$ affords $[IrH_2(PMe_3)_4][O_2CCH(CN)_2]$.[106b] The pentahydride $IrH_5(PPr^i_3)_2$ catalyses H–D exchange between benzene and methane under mild conditions.[106c] The stable paramagnetic iridium(IV) hydride $IrH_2Cl_2(PPr^i_3)_2$ has an all *trans* structure.[106d] Inversion-recovery n.m.r. spectra of $[IrH(H_2)(benzoquinolinate)(PPh_3)_2][SbF_6]$ show that the dihydrogen protons relax much faster than the terminal hydride.[106e] The reaction of $[Ir(\mu\text{-}pz)(CO)_2]_2$ with CH_2I_2 affords $Ir_2(\mu\text{-}pz)_2I(CH_2I)(CO)_4$, $Ir_2(\mu\text{-}CH_2)(\mu\text{-}pz)_2I_2(CO)_4$, and $[Ir(\mu\text{-}pz)I(CO)_2]_2$.[107] Addition of AgO_3SCF_3 to $IrH_3(PPh_3)_3$ in acetone produces $[(Ph_3P)_3Ir(\mu_3\text{-}H)(\mu\text{-}H)_2Ag_2(OSO_2CF_3)$-$(H_2O)][O_3SCF_3]$ a complex with a bent Ag—Ir—Ag chain.[108a] The anions $[Ir_3(\mu_3\text{-}E)_2(CO)_6]^-$ prepared from $Ir_4(CO)_{12}$ or $Ir_6(CO)_{16}$ and ECN^- (E = S, Se) contain Ir_3 triangles bicapped by triply bridging chalcogenide atoms.[108b] Planar unsaturated $[Ir(\mu\text{-}AsBu^t)]_4(\mu\text{-}CO)_2(CO)_2(Ir=Ir)$ and tetrahedral saturated $[Ir(\mu\text{-}PCy_2)$-$(CO)]_4(\mu\text{-}CO)_2$ provide the first examples of arsenido- or phosphido-stabilized tetranuclear iridium clusters.[108c]

6 Palladium

The very extensive palladium literature for 1982[109a] and 1983[109b] has been surveyed. Other reviews cover the electronic structure and reactivity of palladium com-

[105] (a) M. A. Bennett and G. T. Crisp, *Organometallics*, 1986, **5**, 1792 and 1800; *Aust. J. Chem.*, 1986, **39**, 1363; (b) D. M. Haddleton and R. N. Perutz, *J. Chem. Soc., Chem. Commun.*, 1986, 1734; (c) R. H. Crabtree, R. P. Dion, D. J. Gibboni, D. V. McGrath, and E. M. Holt, *J. Am. Chem. Soc.*, 1986, **108**, 7222; (d) P. A. Chetcuti, C. B. Knobler, and M. F. Hawthorne, *Organometallics*, 1986, **5**, 1913; (e) J. S. Merola, R. T. Kacmarcik, and D. Van Engen, *J. Am. Chem. Soc.*, 1986, **108**, 329; (f) J. M. Buchanan, J. M. Stryker, and R. G. Bergman, *J. Am. Chem. Soc.*, 1986, **108**, 1537.
[106] (a) F. A. Cotton, P. Lahuerta, M. Sanau, and W. Schwotzer, *Inorg. Chim. Acta*, 1986, **120**, 153; (b) A. Behr, E. Herdtweck, W. A. Herrmann, W. Keim, and W. Kipshagen, *J. Chem. Soc., Chem. Commun.*, 1986, 1262; (c) C. J. Cameron, H. Felkin, T. Fillebeen-Khan, N. J. Forrow, and E. Guittet, *J. Chem. Soc., Chem. Commun.*, 1986, 801; (d) P. Mura, *J. Am. Chem. Soc.*, 1986, **108**, 351; (e) R. H. Crabtree, M. Lavin, and L. Bonneviot, *J. Am. Chem. Soc.*, 1986, **108**, 4032.
[107] D. G. Harrison and S. R. Stobart, *J. Chem. Soc., Chem. Commun.*, 1986, 285.
[108] (a) P. Braunstein, T. M. G. Carneiro, D. Matt, A. Tiripicchio, and M. Tiripicchio Camellini, *Angew. Chem., Int. Ed. Eng.*, 1986, **25**, 748; (b) R. Della Pergola, L. Garlaschelli, S. Martinengo, F. Demartin, M. Manassero, and M. Sansoni, *J. Chem. Soc., Dalton Trans.*, 1986, 2463; (c) A. M. Arif, R. A. Jones, S. T. Schwab, and B. R. Whittlesey, *J. Am. Chem. Soc.*, 1986, **108**, 1703.
[109] (a) P. A. Chaloner, *Coord. Chem. Rev.*, 1986, **71**, 235; (b) 1986, **72**, 1; (c) O. V. Gritsenko, A. A. Bagatur'Yants, I. I. Moiseev, and V. B. Kazanskii, *Russ. Chem. Rev.*, 1985, **54**, 1151; (d) N. K. Eremenko, E. G. Mednikov, and S. S. Kurasov, *Russ. Chem. Rev.*, 1985, **54**, 394; (e) J. Tsuji, *J. Organomet. Chem.*, 1986, **300**, 281.

pounds,[109c] palladium carbonyl/phosphine clusters[109d] and a personal account by J. Tsuji of 25 years in the organic chemistry of palladium.[109e] A basket shaped structure with four monodentate nitrate ligands coordinated to square planar Pd^{II} has been reported for $K_2[Pd(NO_3)_4]$.[110a] Palladous acetate cleaves P—C bonds in PPh_3 to form $Pd_2(\mu\text{-}O_2CMe)_2Ph_2(PPh_3)_2$[110b] and reacts with cupric acetate to generate polynuclear mixed acetates $Cu_2Pd(O_2CMe)_6$ and $Cu_2Pd_4(O_2CMe)_{12}$ for which triangular and octahedral arrangements of metal atoms are proposed.[110c] The stereodynamics of the cyclic selenoether complexes $PdX_2\{Se(CH_2)_n\}_2$ and $PdX_2\{SeCH_2CMe_2CH_2\}_2$ (X = Cl, Br or I, n = 4—6) have been investigated by variable temperature ^1H n.m.r. spectroscopy.[111a] Orange red crystals of $(PPh_4)_2(NH_2Me_2)(NH_4)[Pd_2(S_7)_4]$, obtained from the reaction of $Pd(acac)_2$ in $MeCN/Me_2NCHO$ with an ethanolic solution of ammonium polysulphide in the presence of the cations PPh_4^+ and $NH_2Me_4^+$, have been shown to contain a

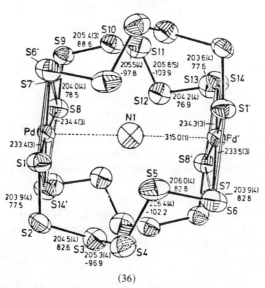

(36)

30-membered cage anion containing a trapped NH_4^+ cation (36).[111b] Reactions of halide complexes $PdCl_2(PR_3)_2$ with $Na(S_3N_3)$ afford N,S-chelates $Pd(SNSN)$-$(PR_3)_2$.[111c] Palladous chloride reacts with S_7NH in the presence of [PNP]OH to form the complex salt $[PNP][Pd(SSNS)(S_5)]$.[111d] Bis(amidino)palladium complexes $Pd\{R'NC(R)NR'\}_2$ (R' = Me, Ph, p-tolyl) are monomeric (R = Me) or dimeric (R = Ph) in the solid state but monomeric in benzene solution.[111e] The cation

[110] (a) L. I. Elding, B. Norén, and A. Oskarsson, *Inorg. Chim. Acta*, 1986, **114**, 71; (b) M. B. Hursthouse, O. D. Sloan, P. Thornton, and N. P. C. Walker, *Polyhedron*, 1986, **5**, 1475; (c) O. D. Sloan and P. Thornton, *Inorg. Chim. Acta*, 1986, **120**, 173.

[111] (a) E. W. Abel, T. E. MacKenzie, K. G. Orrell, and V. Šik, *J. Chem. Soc., Dalton Trans.*, 1986, 205; (b) A. Müller, K. Schmitz, E. Krickemeyer, M. Penk, and H. Bögge, *Angew. Chem., Int. Ed. Eng.*, 1986, **25**, 453; (c) P. A. Bates, M. B. Hursthouse, P. F. Kelly, and J. D. Woollins, *J. Chem. Soc., Dalton Trans.*, 1986, 2367; (d) J. Weiss, *Z. Anorg. Allg. Chem.*, 1986, **542**, 137; (e) J. Barker, N. Cameron, M. Kilner, M. M. Mahoud, and S. C. Wallwork, *J. Chem. Soc., Dalton Trans.*, 1986, 1359; (f) Z. Taira and S. Yamazaki, *Bull. Chem. Soc. Jpn.*, 1986, **59**, 649.

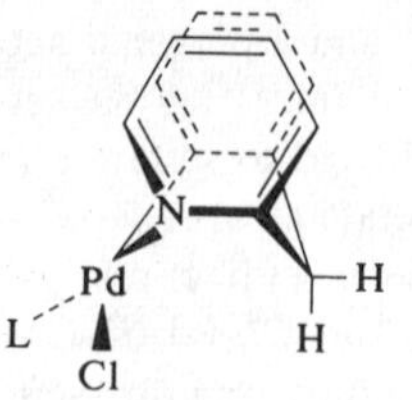

(37)

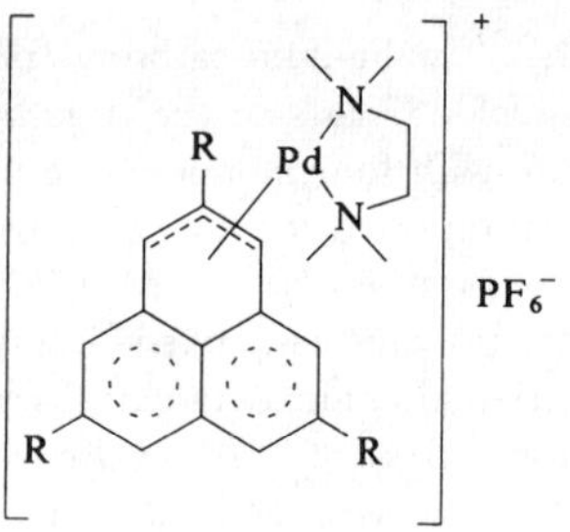

(a) R = H; (b) R = Me

(38)

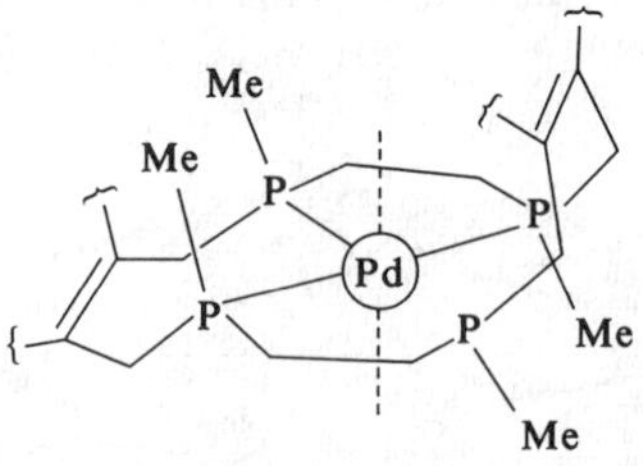

(39)

$[PdCl(o\text{-phen})(PPh_3)_2]^+$ has a distorted square-pyramidal structure with a long apical Pd—N distance [2.68 (4) Å].[111f] New palladium complexes containing polydentate pyrazolyl ligands include $Pd\{HB(pz)_3\}_2$, $[Pd\{HC(pz)_3\}_2][BF_4]_2$,[112a] $PdCl_2\{H_2C(pz)_2\}$,[112b] and $[(\eta^3\text{-allyl})Pd(\mu\text{-pz})_2B(\mu\text{-pz})_2Pd(\eta^3\text{-allyl})][PF_6]$.[112c] Variable temperature n.m.r. spectroscopy has been used to demonstrate inversion of the six-membered palladocycle in (37)[113a] and migration of the $Pd(Me_2NCH_2CH_2NMe_2)^+$ unit around the phenalenyl skeleton in (38).[113b] Cyclization of $HMePCH_2CH_2PMeH$ with o-xylylene dichloride within a Pd^{II} template affords three diastereomeric cations, one of which is shown in (39).[114a] The low steric requirements of $Pr^iHPCH_2PHPr^i$ lead to formation of tris(diphosphine)-bridged complexes $[Pd_2Cl_2(\mu\text{-}Pr^iHPCH_2PHPr^i)_3]X_2$ (X = Cl, PF_6) and their platinum analogues.[114b] Insertions of CO, MeNC, SO_2, and CS_2 into the Pd—Pd bond of $Pd_2Cl_2(\mu\text{-dmpm})_2$ have been reported.[114c] The dimetallated olefin complex $Pd_2Cl_2(\mu\text{-dppm})_2(\mu\text{-HC}=CH)$, obtained by acid-catalysed addition of acetylene to $Pd_2Cl_2(\mu\text{-dppm})_2$, has been compared with the isomeric 'vinylidene derivative $Pd_2Cl_2(\mu\text{-dppm})_2(\mu\text{-C}=CH_2)$, prepared from $Pd(PPh_3)_4$, dppm and

112 (a) A. J. Canty, N. J. Minchin, L. M. Engelhardt, B. W. Skelton, and A. H. White, J. Chem. Soc., Dalton Trans., 1986, 645; (b) G. Minghetti, M. A. Cinellu, A. L. Bandini, G. Banditelli, F. Demartin, and M. Manassero, J. Organomet. Chem., 1986, 315, 387; (c) J. Bielawski, T. G. Hodgkins, W. J. Layton, K. Niedenzu, P. M. Niedenzu, and S. Trofimenko, Inorg. Chem., 1986, 25, 87.

113 (a) V. A. Polyakov and A. D. Ryabov, J. Chem. Soc., Dalton Trans., 1986, 589; (b) K. Nakasuji, M. Yamaguchi, and I. Murata, J. Am. Chem. Soc., 1986, 108, 325.

114 (a) D. J. Brauer, F. Gol, S. Hietkamp, H. Peters, H. Sommer, O. Stelzer, and W. S. Sheldrick, Chem. Ber., 1986, 119, 349; (b) W. Radecka-Paryzek, A. J. McLennan, and R. J. Puddephatt, Inorg. Chem., 1986, 25, 3097; (c) M. L. Kullberg and C. P. Kubiak, Inorg. Chem., 1986, 25, 26; (d) S. J. Higgins and B. L. Shaw, J. Chem. Soc., Chem. Commun., 1986, 1629; (e) A. M. Herring, S. J. Higgins, G. B. Jacobsen, and B. L. Shaw, J. Chem. Soc., Chem. Commun., 1986, 882; (f) G. Besenyei, C.-L. Lee and B. R. James, J. Chem. Soc., Chem. Commun., 1986, 1750.

$Cl_2C=CH_2$.[114d] Palladium chloride and particularly $Pd(O_2CMe)_2$ highly activate coordinated $(Ph_2P)_2C=CH_2$ towards Michael addition.[114e] The 'A' frame complexes $Pd_2Cl_2(\mu\text{-Se})(\mu\text{-Ph}_2PCHRPPh_2)_2$ (R = H or Me) are converted into the corresponding μ-SeO derivatives by Bu^tOOH.[114f] Insertion of one alkyne molecule into the Pd—C bond of orthopalladated 2-benzylpyridine affords an eight-membered organometallic ring, a further insertion generates the spiro-compound (40).[115a] The bis(acetylene) complex cis-$Pd(C_6H_5)_2(PhC\equiv CPh)_2$ and its platinum analogue are stable even though they appear to lack π-back bonding.[115b] $Pd(C_6F_5)_2(dppa)_2$ (dppa = $Ph_2PNHPPh_2$) reacts with $Pd_2(dibenzylideneacetone)_3$ and $AgClO_4$ to give the palladium(I) complex $Pd_2(C_6F_5)_2(\mu\text{-dppa})_2$ and the heterobimetallic salt $[(C_6F_5)_2Pd(\mu\text{-dppa})_2Ag]ClO_4$ respectively.[115c] The synthesis, molecular structure and some reactions of the dinuclear palladium(I) complex $Pd_2(C_6Cl_5)_2(\mu\text{-dppm})_2$ have been described.[115d] The dinuclear complex $[Pd(\mu\text{-Se}_2CPMe_3)(PMe_3)]_2$ (41)

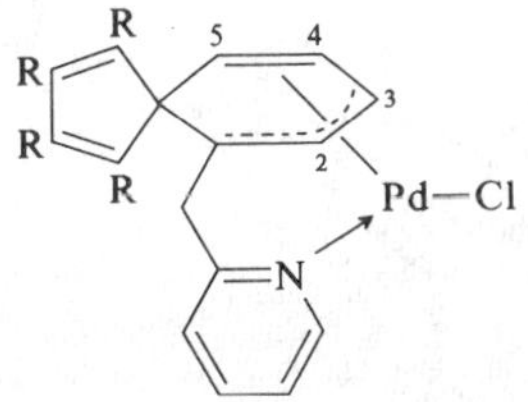

(40)

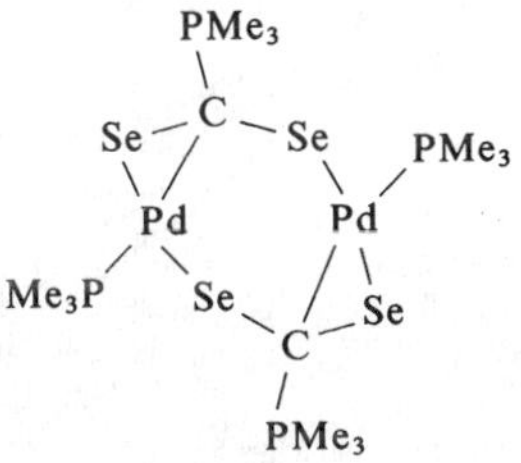

(41)

contains bridging zwitterionic $\overset{+}{P}Me_3C\overset{-}{Se}_2$ units coordinated in an unprecedented η^1,η^2-mode.[116a] α-Diazomethyl palladium complexes $Pd\{C(N_2)R'\}X(PR_3)_2$ have been obtained from the halides $PdX_2(PR_3)_2$ and $Hg[C(N_2)R']_2$ or $LiC(N_2)R'$ (X = Cl, Br, I; R = Et, Ph; R' = CO_2Et, COMe, Ph, Pr^i, Bu^t).[116b] The novel infinite-chain mixed-metal carbene complex $>Pd=C(PPh_2)_2Pt(\mu\text{-Cl})_2Pd=C(PPh_2)_2Pt<$ has been synthesized from $PdCl_2(PhCN)_2$ and $Pt(Ph_2PCHPPh_2)_2$.[117] Oxidative addition of MeI to $PdMe_2(bipy)$ affords *fac*-$PdMe_3I(bipy)$ the first hydrocarbyl palladium(IV) complex.[118a] The pallado or platinocyclosulphones $M\{PhCHS(O)_2CHPh\}L_2$ have been obtained by treatment of the dichlorides *cis*- or *trans*-MCl_2L_2 (L = PMe_3, PEt_3, PPh_3, or $AsPh_3$) with salts of the dianion $[PhCHS(O)_2CHPh]^{2-}$.[118b] The palladium complexes $Pd(R_2PCH_2CH_2PR_2)(butadiene)$ and their nickel analogues contain fluxional η^2-bonded butadiene ligands, the corresponding platinum complexes have a similar arrangement when R = Bu^t or Cy but adopt a platinocyclopen-

115 (a) F. Maassarani, M. Pfeffer, and G. Le Borgne, *J. Chem. Soc., Chem. Commun.*, 1986, 488; (b) R. Usón, J. Forniés, M. Tomás, B. Menjón, and A. J. Welch, *J. Organomet. Chem.*, 1986, **304**, C24; (c) R. Usón, J. Fornies, R. Navarro, and J. I. Cebollada, *J. Organomet. Chem.*, 1986, **304**, 381; (d) P. E. J. Fornies, C. Fortuño, G. Hidalgo, F. M. M. Tomas, and A. J. Welch, *J. Organomet. Chem.*, 1986, **317**, 105.

116 (a) H. Otto, M. Ebner, and H. Werner, *J. Organomet. Chem.*, 1986, **311**, 63; (b) S.-I. Murahashi, Y. Kitani, T. Uno, T. Hosokawa, K. Miki, T. Yonezawa, and N. Kasai, *Organometallics*, 1986, **5**, 356.

117 S. I. Al-Resayes, P. B. Hitchcock, and J. F. Nixon, *J. Chem. Soc., Chem. Commun.*, 1986, 1710.

118 (a) P. K. Byers, A. J. Canty, B. W. Skelton, and A. H. White, *J. Chem. Soc., Chem. Commun.*, 1986, 1722; (b) K. W. Chiu, J. Fawcett, W. Henderson, R. D. W. Kemmitt, and D. R. Russell, *J. Chem. Soc., Chem. Commun.*, 1986, 41; (c) R. Benn, P. W. Jolly, T. Joswig, R. Mynott, and K.-P. Schick, *Z. Naturforsch.*, 1986, **41b**, 680.

tene structure when R = Pr^i.[118c] Theoretical calculations[119a] and experimental evidence for the existence of Pd(η^1-H$_2$) and Pd(η^2-H$_2$) in frozen Kr and Xe matrices[119b] have been reported. Treatment of PdX$_2$(PPh$_3$)$_2$ (X = Cl, Br, or I) with LiR (R = Me, Ph, or CH=CHBu^t) in THF gives LiXPd(PPh$_3$)$_2$, Li$_2$X$_2$Pd(PPh$_3$)$_2$, and aggregates thereof rather than the expected 'Pd(PPh$_3$)$_2$'.[120a] A tetrahedral structure with Pd—P = 2.283(3) Å has been reported for Pd{P(C≡CPh)$_3$}$_4$.[120b] The Pd^0 complex Pd{N(CH$_2$CH$_2$PPh$_2$)$_3$} displays out-of-plane trigonal coordination (N not coordinated) and reacts with RI (R = Me or Et) to form salts [PdR{N(CH$_2$CH$_2$PPh$_2$)$_3$}]I containing trigonal-bipyramidal cations.[120c] The cationic cluster [Pd$_3$(μ_3-CO)(μ-dppm)$_3$]$^{2+}$ reacts with thiocyanate to form [Pd$_3$(SCN)(μ_3-CO)(μ-dppm)$_3$]$^+$ which rearranges to [Pd$_3$(μ_3-S)(CN)-(μ-dppm)$_3$]$^+$.[121a] The tetranuclear structure (42) has been established for the imidoyl complex [Pd$_2${μ-C(C$_6$F$_5$) = NMe}$_2$(μ-Cl)(μ-O$_2$CMe)]$_2$.[121b] Treatment of Pd$_{10}$(CO)$_{12}$(PEt$_2$)$_6$ with

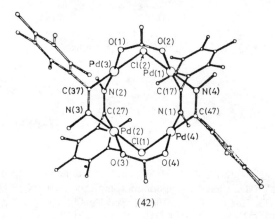

(42)

Pd(O$_2$CMe)$_2$ under argon at room temperature affords the Pd$_{23}$(CO)$_{22}$(PEt$_3$)$_{10}$ cluster (43).[121c]

7 Platinum

A number of platinum compounds are discussed along with their palladium analogues in the preceding section. The platinum literature for 1982[109a] and 1983[109b] has been surveyed. Considerable attention has been directed towards complexes containing platinum in the 3+ oxidation state. Deep blue [NBu$^t_4$][Pt(C$_6$Cl$_5$)$_4$] affords the first example of a fully characterized mononuclear Pt^{III} complex, Pt—C distances

[119] (a) C. Jarque, O. Novaro, M. E. Ruiz, and J. Garcia-Prieto, J. Am. Chem. Soc., 1986, 108, 3507; (b) G. A. Ozin and J. Garcia-Prieto, J. Am. Chem. Soc., 1986, 108, 3099.

[120] (a) E. Negishi, T. Takahashi, and K. Akiyoshi, J. Chem. Soc., Chem. Commun., 1986, 1338; (b) L. G. Kuz'mina, Yu. T. Struchkov, L. Yu. Ukhin, and N. A. Dolgopolova, Sov. J. Coord. Chem., 1986, 11, 973; (c) C. A. Ghilardi, S. Midollini, S. Moneti, and A. Orlandini, J. Chem. Soc., Chem. Commun., 1986, 1771.

[121] (a) G. Ferguson, B. R. Lloyd, L. Manojlović-Muir, K. W. Muir, and R. J. Puddephatt, Inorg. Chem., 1986, 25, 4190; (b) R. Uson, J. Fornies, P. Espinet, E. Lalinde, A. Garcia, P. G. Jones, K. Meyer-Bäse, and G. M. Sheldrick, J. Chem. Soc., Dalton Trans., 1986, 259; (c) E. G. Mednikov, N. K. Eremenko, Yu. L. Slovokhotov, and Yu. T. Struchkov, J. Organomet. Chem., 1986, 301, C35.

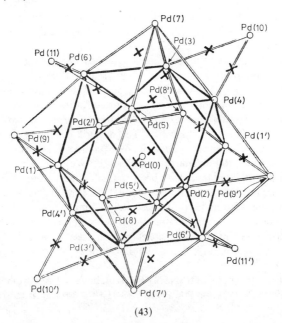

(43)

in the square-planar anion average 2.094(8) Å.[122a] New dinuclear examples with bridging α-pyridonate ligands include $Pt_2Me_4(\mu\text{-}C_5H_4NO)_2(py)_2$[122b] and $[Pt_2(en)_2(C_5H_4NO)_2(NO_2)(NO_3)][(NO_3)_2 \cdot 0.5H_2O$.[122c] Dinuclear platinum(III) 1-methyluracil (MeU·H) derivatives $[Cl(NH_3)_2Pt(\mu\text{-}MeU)_2Pt(NH_3)_2Cl]$-$Cl_2 \cdot 3.5H_2O$[123a] and $(NO_3)(NH_3)_2Pt(\mu\text{-}MeU)_2Pt(NH_3)_2(OH_2)][NO_3]_3 \cdot 2H_2O$[123b] contain head-to head and head-to-tail arrangements of the MeU ligands respectively. A third example, $[(NH_3)_2Pt(\mu\text{-}MeU)_2Pt(NH_3)_2(MeU)]^{3+}$, contains an axial C(5)-coordinated 1-methyluracil anion,[123c] and this along with the cation $[(NO_2)$-$(NH_3)_2Pt(\mu\text{-}MeU)_2Pr(NH_3)_2]^{3+}$ affords the first examples of dinuclear Pt^{III} complexes with one six coordinate and one five coordinate platinum centre.[123d] Platinum(III) complexes containing N,S or S,S donors include $IPt(\mu\text{-}2\text{-}TU)_4PtI$ (2-TUH = 2-thiouracil)[123e] and $IPt(S_2CR)_4PtI$ (R = Bz or Pr^i).[123f]

The tetranuclear platinum(III) α-pyrrolidonate complex $[Pt_4(C_4H_6NO)_4$-$(NH_3)_8]^{8+}$ is reduced by water to the corresponding 6+ cation with liberation of dioxygen.[123g]

The labile water molecules in the diplatinum(III) complex $Na_2[Pt_2(\mu\text{-}PO_4H)_4(H_2O)_2]$ have been sequentially replaced by halide, amine, thioether, and

[122] (a) R. Usón, J. Forniés, M. Tomás, B. Menjón, R. Bau, K. Sünkel, and E. Kuwabara, *Organometallics*, 1986, **5**, 1576; (b) D. P. Bancroft, F. A. Cotton, L. R. Falvello, and W. Schwotzer, *Inorg. Chem.*, 1986, **25**, 763; (c) T. V. O'Halloran, M. M. Roberts, and S. J. Lippard, *Inorg. Chem.*, 1986, **25**, 957.
[123] (a) B. Lippert, H. Schöllhorn, and U. Thewalt, *Inorg. Chem.*, 1986, **25**, 407; (b) H. Schöllhorn, P. Eisenmann, U. Thewalt, and B. Lippert, *Inorg. Chem.*, 1986, **25**, 3384; (c) H. Schöllhorn, U. Thewalt, and B. Lippert, *J. Chem. Soc., Chem. Commun.*, 1986, 258; (d) *J. Am. Chem. Soc.*, 1986, **108**, 525; (e) D. M. L. Goodgame, R. W. Rollins, A. M. Z. Slawin, D. J. Williams, and P. W. Zard, *Inorg. Chim. Acta*, 1986, **120**, 91; (f) C. Bellitto, M. Bonamico, G. Dessy, V. Fares, and A. Flamini, *J. Chem. Soc., Dalton Trans.*, 1986, 595; (g) K. Matsumoto and T. Watanabe, *J. Am. Chem. Soc.*, 1986, **108**, 1308.

thiolate ligands,[124a] the guanine (GuH$_2$) derivative reacts with aqueous NaOH to afford Na$_{10}$[Pt$_2$(μ-PO$_4$)$_4$(Gu)$_2$].22H$_2$O in which the guanine dianion is coordinated through N(9).[124b]

The electronic spectra[125a,b] and electrochemical behaviour[125c] of dinuclear platinum(III) pyrophosphito complexes [NBu$_4$]$_2$[Pt(P$_2$O$_5$H$_2$)$_4$L$_2$] (L = MeCN or py) and K$_4$[Pt$_2$(P$_2$O$_5$H$_2$)$_4$X$_2$] (X = Cl, Br, I, or SCN) have been examined. X-Ray crystal structure determinations have been reported for the latter complex (X = Br) and for the platinum(II/III) complex K$_4$[Pt$_2$(P$_2$O$_5$H$_2$)$_4$Cl]3H$_2$O which contains linear —Pt—Pt—Cl—Pt—Pt—Cl— chains.[125d] The dimeric platinum(II) pyrazole complex [Pt(pz)$_2$(Hpz)$_2$]$_2$ has square-planar platinum centres linked by four symmetrical pz—H—pz bridges.[126] Methylamine analogues of Wolffram's red and Reihlen's green platinum(II,IV) complexes have been characterized.[127] Dioxocyclam (44) acts as a selective sequestering agent for platinum(II) in aqueous metal ion mixtures.[128] Interest in platinum antitumour drugs has catalysed much new work on the interaction of platinum(II) and (IV) ammine or amine complexes with DNA fragments notably 5'-guanosine monophosphate,[129a,b] various nucleoside 5'-monophosphates,[129c] nucleic acid bases,[129d] and oligonucleotides[129e,f] including d(GpG)[129g,h] and d(GpTpG).[129i]

Platinum(II) chelates Pt(SNSN)(PR$_3$)$_2$ have been obtained from PtCl$_2$(PR$_3$)$_2$/Na(S$_3$N$_3$)[111c] and Pt(PPh$_3$)$_4$/S$_4$N$_4$H$_4$.[130a] Crystal structures of the com-

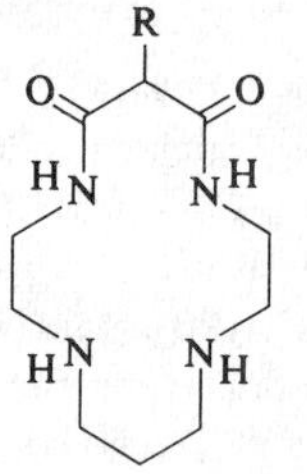

(44) (R = H or C$_{16}$H$_{33}$)

[124] (a) R. El.-Mehdawi, F. R. Fronczek, and D. M. Roundhill, Inorg. Chem., 1986, 25, 1155; (b) Inorg. Chem., 1986, 25, 3714.
[125] (a) C.-M. Che, T. C. W. Mak, V. M. Miskowski, and H. B. Gray, J. Am. Chem. Soc., 1986, 108, 7840; (b) A. E. Stiegman, V. M. Miskowski, and H. B. Gray, J. Am. Chem. Soc., 1986, 108, 2781; (c) S. A. Bryan, R. H. Schmehl, D. M. Roundhill, J. Am. Chem. Soc., 1986, 108, 5408; (d) R. J. H. Clark, M. Kurmoo, H. M. Dawes, and M. B. Hursthouse, Inorg. Chem., 1986, 25, 409.
[126] W. Burger and J. Strähle, Z. Anorg. Allg. Chem., 1986, 539, 27.
[127] R. J. H. Clark, V. B. Croud, H. M. Dawes, and M. B. Hursthouse, J. Chem. Soc., Dalton Trans., 1986, 403.
[128] E. Kimura, Y. Lin, R. Machida, H. Zenda, J. Chem. Soc., Chem. Commun., 1986, 1020.
[129] (a) J. L. van der Veer, A. R. Peters, and J. Reedijk, J. Inorg. Biochem., 1986, 26, 137; (b) Y.-T. Fanchiang, J. Chem. Soc., Dalton Trans., 1986, 135; (c) O. Yamauchi, A. Odani, R. Shimata, and Y. Kosaka, Inorg. Chem., 1986, 25, 3337; (d) H. Basch, M. Krauss, W. J. Stevens, and D. Cohen, Inorg. Chem., 1986, 25, 684; (e) A. L. Pinto, L. J. Naser, J. M. Essigmann, and S. J. Lippard, J. Am. Chem. Soc., 1986, 108, 7405; (f) J. Kozelka, G. A. Petsko, G. J. Quigley, and S. J. Lippard, Inorg. Chem., 1986, 25, 1075; (g) K. Inagaki, C. Ninomiya, and Y. Kidani, Chem. Lett., 1986, 233; (h) K. Inagaki, Y. Kidani, Inorg. Chem., 1986, 25, 1; (i) J. L. van der Veer, G. J. Ligtvoet, H. van den Elst, and J. Reedijk, J. Am. Chem. Soc., 1986, 108, 3860.
[130] (a) T. Chivers, F. Edelmann, U. Behrens, and R. Drews, Inorg. Chim. Acta, 1986, 116, 145; (b) R. Jones, P. F. Kelly, C. P. Warrens, D. J. Williams, and J. D. Woollins, J. Chem. Soc., Chem. Commun., 1986, 711.

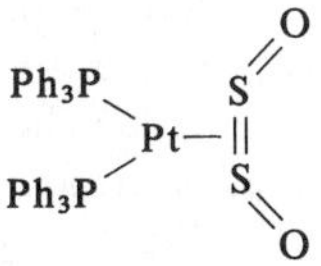

(45)

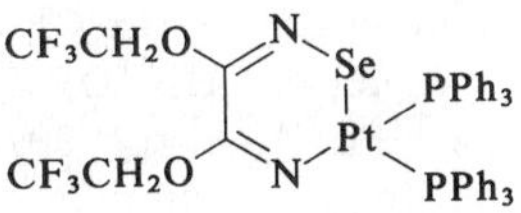

(46)

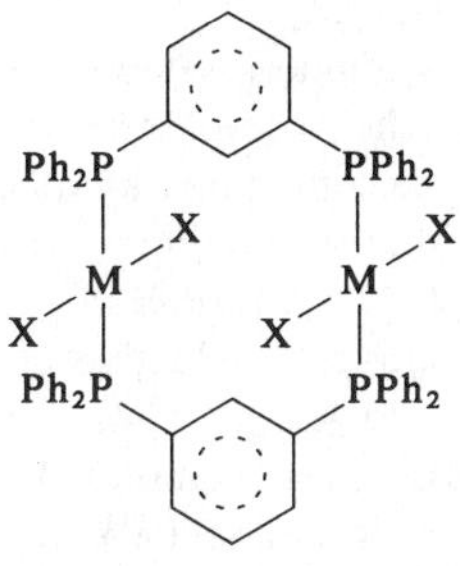

(47)

plex salts [Pt(SNSNH)(PMe$_2$Ph)$_2$][BF$_4$] and [Pt(SNSNH)(PMe$_3$)$_2$][PF$_6$] reveal stacking of the metallocycles.[130b] The thiirane *S*-oxide complex Pt{S(O)CH$_2$CH$_2$}-(PPh$_3$)$_2$ undergoes thermal fragmentation at 110 °C to give the η^2-S$_2$O$_2$ derivative (45).[131a] Insertion of platinum(0) into 1,2,5-selenadiazole rings affords metallocyclic products including (46).[131b] Stepwise selenium extrusion converts [Pt-{Ph$_2$P(Se)CH$_2$P(Se)Ph$_2$}$_2$](O$_3$SCF$_3$)$_2$ into [Pt{Ph$_2$P(Se)CH$_2$PPh$_2$}$_2$][O$_3$SCF$_3$]$_2$.[131c] The structure of the *m*-bis(diphenylphosphino)benzene complex (47) has been established by *X*-ray diffraction methods.[132a] A short Pt—Th distance [2.984(1) Å] in the phosphido-bridged complex (C$_5$Me$_5$)$_2$Th(μ-PPh$_2$)$_2$Pt(PMe$_3$) is indicative of a Pt—Th bond.[132b] Alkyl iodides RI (R = Pr^i, Bu^t) react with PtMe$_2$(*o*-phen) in the presence of alkenes CH$_2$=CHX (X = CN, CHO, or COMe) to give platinum(IV) species PtIMe$_2$(CHX·CH$_2$R)(*o*-phen).[132c] The facile rearrangement of Pt(CH$_2$Cl)-(CH$_3$)(CO-1,4-D) to Pt(CH$_2$CH$_3$)Cl(CO-1,4-D) may model the key step in transition metal-catalysed diazomethane polymerization.[132d] For the complexes PtEt$_n$(OMe)$_{2-n}$(dppe) (*n* = 0, 1, or 2) thermolysis *via* β-elimination from the ethyl ligands is energetically easier than the comparable process from the methoxoligand by 0.3 kcal/mole.[132e] The abnormally long Pt—C bonds in the complexes PtBr{C(R)-CH$_2$}(PEt$_3$)$_2$ (R = adamant-1-yl and naphth-1-yl) insert CO under mild conditions.[132f] New platinacyclobutylcarbinyl esters Cl$_2$L$_2$Pt{CH$_2$CR^1(CHR^{2}OR^3)CH$_2$} (R^1 = H, Me, or Ph; R^2 = H or Me, R^3 = methanesulphonyl or 4-nitrobenzoyl,

[131] (*a*) I.-P. Lorenz and J. Kull, *Angew. Chem., Int. Ed. Engl.*, 1986, **25**, 261; (*b*) H. W. Roesky, T. Gries, H. Hofmann, J. Schimkowiak, P. G. Jones, K. Meyer-Bäse, and G. M. Sheldrick, *Chem. Ber.*, 1986, **119**, 366; (*c*) P. Peringer and J. Schwald, *J. Chem. Soc., Chem. Commun.*, 1986, 1625.

[132] (*a*) N. W. Alcock, J. Ludd, and P. G. Pringle, *Inorg. Chim. Acta*, 1986, **113**, L13; (*b*) P. J. Hay, R. R. Ryan, K. V. Salazar, D. A. Wrobleski, and A. P. Sattelberger, *J. Am. Chem. Soc.*, 1986, **108**, 313; (*c*) P. K. Monaghan and R. J. Puddephatt, *Organometallics*, 1986, **5**, 439; (*d*) R. McCrindle, G. J. Arsenault, R. Farwaha, M. J. Hampden-Smith, and A. J. McAlees, *J. Chem. Soc., Chem. Commun*, 1986, 943; (*e*) H. E. Bryndza, J. C. Calabrese, M. Marsi, D. C. Roe, W. Tam, and J. E. Bercaw, *J. Am. Chem. Soc.*, 1986, **108**, 4805; (*f*) C. J. Cardin, D. J. Cardin, H. E. Parge, and A. C. Sullivan, *J. Chem. Soc., Dalton Trans.*, 1986, 2315; (*g*) J. T. Burton and R. J. Puddephatt, *Organometallics*, 1986, **5**, 1312.

L = pyridine or L_2 = bipy) undergo solvolysis in aqueous acetone to give ring-expanded products $Cl_2L_2PtCHR^2CR^1(OH)CH_2CH_2$.[132g] Synthesis of the platinacyclobut-3-one complex $Pt\{CH_2C(O)CH_2\}(PPh_3)_2$ has been reported.[51] The complexes $PtH_2\{Cy_2P(CH_2)_nPCy_2\}$ (n = 2, 3, or 4) show evidence of a dynamic cis-dihydride $\rightleftharpoons \eta^2$-dihydrogen equilibrium.[133a] E.s.r. studies of $PtH_2(PBu^t_2Bu^n)_2$/acetylene reaction mixtures reveal the presence of persistent bis(phosphine)hydridoplatinum(II)(π-acetylene) radicals in which the unpaired electron is delocalized over an unsaturated metallocyclic ring.[133b]

The chemistry of dinuclear platinum(I) systems continues to develop apace. Rapid, reversible carbonylation of the complexes $[HPt(\mu-PP)_2Pt(CO)]^+$ ($P-P$ = $R_2PCH_2PR_2$, R = Et or Ph) is accompanied by intramolecular disproportionation to yield the first reported platinum(0)-platinum(II) species $[HPt(\mu-PP)_2Pt(CO)_2]^+$ and $[H(CO)Pt(\mu-PP)_2Pt(CO)_2]^+$.[134a] A further Pt^0-Pt^{II} example, $[Pt(\mu-PP)_3PtH]^+$, has been obtained by ligand-induced reductive elimination of dihydrogen from $[HPt(\mu-H)(\mu-PP)_2PtH]^+$ at −70 °C.[134b] Short- and long-range n.m.r. trans-influences in dinuclear platinum(I) complexes have been investigated.[134c,d] Definitive ^1H, ^{31}P, and ^{195}Pt n.m.r. data have been reported for the dinuclear complexes $XPt(\mu-dppm)_2PtX$ (X = Cl, Br, or I) and a range of 'A'-frame derivatives.[134e] Addition of $HgCl_2$ to $ClPt(\mu-dppm)_2PtCl$ affords $Pt_2(\mu-HgCl_2)Cl_2(\mu-dppm)_2$ an 'A' frame complex with an apical $HgCl_2$ group.[134f]

The platinum(0) complex $Pt(MesP=CPh_2)(PPh_3)_2$ (Mes = mesityl) provides the first example of a directly observable equilibrium between η^1 and η^2 coordination modes for a phosphaalkene ligand.[135] The electrochemical generation and chemical reactivity of $Pt(PEt_3)_2$[136a] and $Pt(PPh_3)_2$[136b] have been reported. The 1,2,3-triphenyl-3-vinylcycloprop-1-ene ring opens with $Pt(C_2H_4)(PPh_3)_2$ to give (48), the first example of a 1-platinacyclohexa-2,4-diene complex.[136c] The reactions of CH_2ClI with $Pt(PPh_3)_4$ and $Pt(C_2H_4)(PPh_3)_2$ afford the ylide cis-$[Pt(CH_2PPh_3)Cl(PPh_3)_2]I$ and the oxidative addition product $PtI(CH_2Cl)(PPh_3)_2$ respectively.[136d] The inferred intermediate $Pt(Cy_2PCH_2CH_2PCy_2)$, formed by elimination of neopentane from $PtH(CH_2CMe_3)(Cy_2PCH_2CH_2PCy_2)$, undergoes oxidative addition reactions with activated and non-activated C−H bonds.[136e] The energetics and structures of the most stable intermediates formed during the successive hydrogenation steps of Pt_n clusters (n = 2−12) have been investigated by the extended Hückel method.[137]

[133] (a) H. C. Clark and M. J. Hampden Smith, J. Am. Chem. Soc., 1986, **108**, 3829; (b) H. C. Clark, G. Ferguson, A. B. Goel, E. G. Janzen, H. Ruegger, P. Y. Siew, and C. S. Wong, J. Am. Chem. Soc., 1986, **108**, 6961.

[134] (a) A. J. McLennan and R. J. Puddephatt, Organometallics, 1986, **5**, 811; (b) A. J. McLennan and R. J. Puddephatt, J. Chem. Soc., Chem. Commun., 1986, 422; (c) S. S. M. Ling and R. J. Puddephatt, Polyhedron, 1986, **5**, 1423; (d) R. J. Blau and J. H. Espenson, Inorg. Chem., 1986, **25**, 878; (e) M. C. Grossel, J. R. Batson, R. P. Moulding, and K. R. Seddon, J. Organomet. Chem., 1986, **304**, 391; (f) P. R. Sharp, Inorg. Chem., 1986, **25**, 4185.

[135] J. G. Kraaijkamp, G. van Koten, T. A. van der Knaap, F. Bickelhaupt, and C. H. Stam, Organometallics, 1986, **5**, 2014.

[136] (a) J. A. Davies and C. T. Eagle, Organometallics, 1986, **5**, 2149; (b) J. A. Davies, C. T. Eagle, D. E. Otis, and U. Venkataraman, Organometallics, 1986, **5**, 1264; (c) N. A. Grabowski, R. P. Hughes, B. S. Jaynes, and A. L. Rheingold, J. Chem. Soc., Chem. Commun., 1986, 1694; (d) C. Engelter, J. R. Moss, L. R. Nassimbeni, M. L. Niven, G. Reid, and J. C. Spiers, J. Organomet. Chem., 1986, **315**, 255; (e) M. Hackett, J. A. Ibers, P. Jernakoff, and G. M. Whitesides, J. Am. Chem. Soc., 1986, **108**, 8094.

[137] C. Minot, B. Minot, and A. Hariti, J. Am. Chem. Soc., 1986, **108**, 196; Nouv. J. Chim., 1986, **10**, 461.

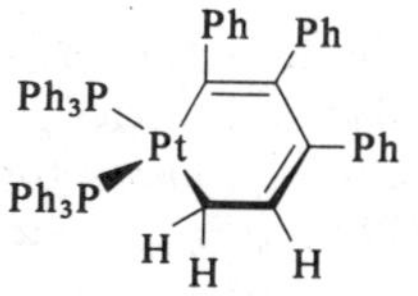

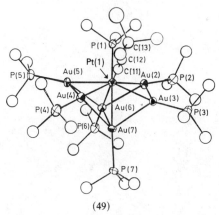

(48)

Sulphur and carbon disulphide react with $Pt_3(CNR)_6$ (R = 2,4,6-Bu$^t_3$C$_6$H$_2$) to afford the dinuclear Pt—Pt bonded products $(RNC)_2Pt(\mu\text{-S})Pt(CNR)_2$ and $(RNC)_2Pt(\mu\text{-CS}_2)Pt(CNR)_2$ respectively.[138a] The reaction of $Pt(C_2H_4)(PPh_3)_2$ with $[Pt(CNR)_4]^{2+}$ (R = 2,6-Me$_2$C$_6$H$_3$) affords dinuclear $[Ph_3P\{Pt(CNR)_2\}_2PPh_3]^{2+}$ and linear trinuclear $[Ph_3P\{Pt(CNR)_2\}_3PPh_3]^{2+}$.[138b] Use of Me_3NO facilitates substitution of $Pt_3(\mu\text{-SO}_2)_3(PCy_3)_3$ to yield $Pt_3(CO)_3(PCy_3)_3$ and $[Pt_3(\mu\text{-X})(\mu\text{-}SO_2)_2(PCy_3)_3]^-$ (X = Cl, Br).[138c] The cluster complex $[Pt_3(\mu_3\text{-CO})(\mu\text{-dppm})_3]$-$(O_2CCF_3)_2$ has been obtained in 97% yield by carbonylation of $Pt(O_2CCF_3)_2(dppm)$ in aqueous methanol.[138d] The complexes $[Pt_3H(\mu_3\text{-SMe})(\mu\text{-dppm})_3]^{2+}$, $[Pt_3(SPh)\text{-}(\mu_3\text{-SPh})(\mu\text{-dppm})_3]^{2+}$ and $[Pt_3H(\mu_3\text{-S})(\mu\text{-dppm})_3]^+$ formed in reactions of $[Pt_3(\mu_3\text{-}CO)(\mu\text{-dppm})_3]^{2+}$ with MeSH, PhSH, and H_2S respectively model important features of Pt (111) surface poisoning by these reagents.[138e] Acetylene reacts with $[Pt_3(\mu_3\text{-H})(\mu\text{-dppm})_3][PF_6]$ *via* a weakly bound fluxional adduct $[Pt_3(\mu_2\text{-H})\text{-}(HCCH)(\mu\text{-dppm})_3][PF_6]$ to give the vinylidene complex $Pt_3H[\mu_3\text{-}\eta^2\text{-C}=CH_2)(\mu\text{-}dppm)_3][PF_6]$.[138f]

The platinum blue analogue, $Pt_4(NH_3)_8(C_4O_4)_4H_2O$, prepared from hydrolysed $PtCl_2(NH_3)_2$ and squaric acid, has a linear tetranuclear chain structure with bridging squarate ligands.[139a] Partial oxidation of $[Pt(NH_3)_4][PtCl_4]$ with H_2O_2 affords new $Pt^{II/III}$ mixed valence complexes $[Pt_6(NH_3)_{14}Cl_{10}]X_4$ (X = ClO$_4$, BF$_4$, or PF$_6$) containing one dimensional chains with direct Pt—Pt contacts.[139b] High resolution

(49)

[138] (a) Y. Yamamoto and H. Yamazaki, *J. Chem. Soc., Dalton Trans.*, 1986, 677; (b) C. E. Briant, D. I. Gilmour, D. M. P. Mingos, *J. Organomet. Chem.*, 1986, **308**, 381; (c) M. F. Hallam and D. M. P. Mingos, *J. Organomet. Chem.*, 1986, **315**, C35; (d) G. Ferguson, B. R. Lloyd, and R. J. Puddephatt, *Organometallics*, 1986, **5**, 344; (e) M. C. Jennings, N. C. Payne, and R. J. Puddephatt, *J. Chem. Soc., Chem. Commun.*, 1986, 1809; (f) M. Rashidi and R. J. Puddephatt, *J. Am. Chem. Soc.*, 1986, **108**, 7111.
[139] (a) G. Bernardinelli, P. Castan, and R. Soules, *Inorg. Chim. Acta*, 1986, **120**, 205; (b) M. Tanaka and I. Tsujikawa, *Bull. Chem. Soc. Jpn.*, 1986, **59**, 2773.

transmission electron microscopic investigations of $Pt_{55}(AsBu^t_3)_{12}Cl_{20}$ show 55 closest-packed Pt atoms arranged in the expected full-shell cuboctahedron.[46d] New gold–platinum cluster compounds including $[Pt_2Au_2(PPh_3)_4(CNC_6H_3Me_2-2,6)_4]$-$[PF_6]_2$,[140a] $[Pt_3Au(\mu\text{-}CO)_3(PCy_3)_4][PF_6]$, and $[Pt_3Au(\mu\text{-}CO)_2(\mu\text{-}SO_2)$-$(PCy_3)_4][PF_6]$[140b] have been characterized by diffraction methods. The bonding in these and some related gold–platinum clusters has been analysed using semi-empirical MO calculations.[140c] The cation (49) of the novel gold–platinum cluster $[Au_6Pt(C{\equiv}CBu^t)(PPh_3)_7][Au(C{\equiv}CBu^t)_2]$ displays fluxional behaviour thought to involve rotation of the $Pt(PPh_3)(C{\equiv}CBu^t)$ unit with respect to the Au_6 group, and interconversion of bridgehead and peripheral gold atoms.[140d]

[140] (a) C. E. Briant, D. I. Gilmour, and D. M. P. Mingos, J. Chem. Soc., Dalton Trans., 1986, 835; (b) D. M. P. Mingos and R. W. M. Wardle, J. Chem. Soc., Dalton Trans., 1986, 73; (c) D. I. Gilmour and D. M. P. Mingos, J. Organomet. Chem., 1986, **302**, 127; (d) D. E. Smith, A. J. Welch, I. Treurnicht, and R. J. Puddephatt, Inorg. Chem., 1986, **25**, 4616.

11   Cu, Ag, Au; Zn, Cd, Hg

By J. SILVER

Department of Chemistry, University of Essex, Colchester CO4 3SQ

1 Copper, Silver, and Gold

Copper.—*Copper(I)*. Fused salts based on CuCl/MeEtImCl mixtures and their reactions with dioxygen have been examined.[1] The thermodynamics of formation of Cu^I and Ag^I halide and thiocyanate complexes in pyridine have been determined by potentiometric and calorimetric measurements.[2] Reactions of Cu^I and Zn^{II} halides with phenylacetylene have been reported.[3] Di-μ-chloro-bis[bis(2-methyl-pyridine)]Cu^I contains two pseudo-tetrahedrally coordinated Cu atoms.[4] Cu^I is approximately tetrahedral in $[CuCl(C_3H_4O)]$.[5] The characterization of $(PPh_4)_2Cu_2Cl_2WS_4$ has been reported.[6] Formation constants of CuCl complexes with *cis,cis*-1,5-cyclooctadiene have been presented.[7] The $[Cu_4Cl_4(C_4H_6O)_4]$ tetramer is centrosymmetric.[8] Tetranuclear $(DENC)_3Cu_3M(NS)X_4 \cdot DENC$ or dimeric $(DENC)_4M_2Cl_4$ complexes (DENC = N,N-diethylnicotinamide; X = Cl, Br; M = Co, Ni, Cu, Zn; NS is a monoanionic Schiff base) have been studied.[9a] Other transmetallation reactions of both Cu^I and Cu^{II} complexes have been reported.[9b-e] $[NMe_4][Cu_2Cl_3]$ contains the $[Cu_2Cl_3]$-ion in an infinite double chain[10] and in $[NPr_4][CuCl_2]$ the dichlorocuprate(I) ion is a nearly linear monomer.[11] In $[Cu_6(oxl)_4Br_2]$ (oxl = 4,4-dimethyl-2-oxazoline-4-methylphenyl) the six Cu atoms form a distorted octahedron with four two-electron three centre-bonded bridging aryl groups and two bridging Br atoms.[12] In $[Cu_2Br_2(C_5H_5N)_2\{PPh_3\}]$ the Cu atoms

[1] S. A. Bolkan and J. T. Yoke, *Inorg. Chem.*, 1986, **25**, 3587.
[2] S. Ahrland, S.-I. Ishiguro, and I. Persson, *Acta Chem. Scand., Ser. A*, 1986, **40**, 418.
[3] W. Hiller, *Acta Crystallogr.*, 1986, **C42**, 149.
[4] T. A. Emokpae and A. E. Ukwueze, *Inorg. Chim. Acta*, 1986, **112**, L17.
[5] S. Andersson, M. Hakansson, S. Jagner, M. Nilsson, C. Ullenius, and F. Urso, *Acta Chem. Scand., Ser. A*, 1986, **40**, 58.
[6] F. Secheresse, M. Salis, C. Potvin, and J. M. Manoli, *Inorg. Chim. Acta*, 1986, **114**, L19.
[7] I. Rencken and S. W. Orchard, *Inorg. Chem.*, 1986, **25**, 1972.
[8] S. Andersson, M. Hakansson, S. Jagner, M. Nilsson, and F. Urso, *Acta Chem. Scand., Ser. A*, 1986, **40**, 194.
[9] (a) G. Davis, M. A. El-Sayed, A. El-Toukhy, T. R. Gilbert, and K. Nabih, *Inorg. Chem.*, 1986, **25**, 1929; (b) G. Davis, M. A. El-Sayed, and A. El-Toukhy. *Inorg. Chem.*, 1986, **25**, 1925; (c) G.-Z. Cai, G. Davis, M. A. El-Sayed, A. El-Toukhy, T. A. Gilbert, and K. Nabih, *Inorg. Chem.*, 1986, **25**, 1935; (d) G. Davis, M. A. El-Sayed, and A. El-Toukhy, *Inorg. Chem.*, 1986, **25**, 2269; (e) G. Davis, M. A. El-Sayed, A. El-Toukhy, M. Henary, and T. R. Gilbert, *Inorg. Chem.*, 1986, **25**, 2373.
[10] S. Andersson and S. Jagner, *Acta Chem. Scand., Ser. A*, 1986, **40**, 177.
[11] S. Andersson and S. Jagner, *Acta Chem. Scand., Ser. A*, 1986, **40**, 52.
[12] E. Wehman, G. Van Koten, and J. T. B. H. Jastrzebski, *J. Organomet. Chem.*, 1986, **302**, C35.

are tetrahedrally coordinated to pyridine, two Br atoms, and triphenylphosphine.[13] [P(Bu)Ph$_3$][CuBr$_2$] contains monomeric anions.[14] The linear anionic dicoordinate species CuI$_2^-$ is found in CuI$_2$K(dicyclohexano-18 crown-6) (1) and CuI$_2$K(18-crown-6).[15]

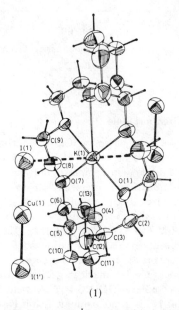

(1)

The solution usually described as Cu^I borohydride contains Cl$^-$ which continues to be associated with the Cu^I, even through thermal decomposition or complexation.[16] In Eu$_2$CuS$_3$ the Cu atoms have tetrahedral environments.[17] [Cu$_8$(SC$_5$H$_{11}$)$_4$(S$_2$CSC$_5$H$_{11}$)$_4$] has a structure based on an unusual C$_8$S$_{12}$ cage.[18] Cu(N,N'-dimethylthiourea)$_2$NO$_3$ contains infinite chains of edge-sharing CuS$_4$ tetrahedra.[19] Spectroscopic studies of Cu^{I}tetrathio-molybdate(VI) and tungstate(VI) anions have been reported[20] and (NMe$_4$)$_2$[(CuNCS)$_4$WS$_4$] has been characterized.[21] Spontaneous reduction to Cu^I derivatives has been observed when Cu^{II} salts are reacted with some potentially dinucleating tetradentate phthalazine-hydrazone ligands.[22] The reaction of Cu^I with 4,5-dicyanoimidazole has been described.[23] The

[13] J. Zukerman-Schpector, E. E. Castellano, A. E. Maruo, and M. R. Roveri, *Acta Crystallogr.*, 1986, **C42**, 302.
[14] S. Andersson and S. Jagner, *Acta Chem. Scand., Ser. A*, 1986, **40**, 210.
[15] N. P. Rath and E. M. Holt, *J. Chem. Soc., Chem. Commun.*, 1986, 311.
[16] R. J. Spokas and B. D. James, *Inorg. Chim. Acta*, 1986, **118**, 99.
[17] P. Lemoine, D. Carré, and M. Guittard, *Acta Crystallogr.*, 1986, **C42**, 390.
[18] R. Chadha, R. Kumar, and D. G. Tuck, *J. Chem. Soc., Chem. Commun.*, 1986, 188.
[19] E. Dubler and W. Bensch, *Inorg. Chim. Acta*, 1986, **125**, 37.
[20] R. J. H. Clark, S. Joss, M. Zvagulis, G. D. Garner, and J. R. Nicholson, *J. Chem. Soc., Dalton Trans.*, 1986, 1595.
[21] J. M. Monoli, C. Potvin, F. Secheresse, and S. Marzak, *J. Chem. Soc., Chem. Commun.*, 1986, 1557.
[22] L. K. Thompson and T. C. Woon, *Inorg. Chim. Acta*, 1986, **111**, 45.
[23] P. G. Rasmussen, L. Rongguang, W. M. Butler, and J. C. Bayon, *Inorg. Chim. Acta*, 1986, **118**, 7.

stability constant of the triammine Cu^I complex has been reported.[24] The activation of O_2 by Cu^I complexes containing 2,2'-bipyridine (biby) or 1,10-phenanthroline (phen) and 2-methylimidazole has been presented.[25] Methylviologen has been photo-catalytically reduced by Cu^I complexes.[26] Cu^I complexes of a series of acyclic Schiff base ligands have been prepared, and the structure of a dinuclear complex containing coordinated and uncoordinated alkene groups has been reported,[27] as has the high yield synthesis of multiring catenates by acetylenic oxidative coupling.[28]

$[Cu(Ph_2P(CH_2)_2PEt_2)_2]Cl$ has been characterized and its rate of inversion compared with those of the isostructural Ag^I and Au^I complexes.[29] ^{63}Cu n.m.r. studies of tris(triethylphosphite)Cu^I chloride in non-aqueous solution have been presented.[30] In $PbCuAsO_4$ the Cu^I atom has a linear coordination.[31] A stable monomeric Cu^I phenolate complex has been described,[32] as have some Cu^I phenoxide complexes.[33]

Evidence for the existence of Cu^+ and $Cu(CN)_2^-$ in supported CuCN reagents has been found.[34] Cu^I abstracts phosphine from $(\eta\text{-}C_5H_5)(CO)(PPh_3)FeCOMe$.[35] The reactions of methylcopper and its complexes with trialkylaluminiums have been studied,[36] as have those of ethoxyacetylene and organo-Cu^I species.[37] α-(Methylene)alkanoic esters have been prepared from Grignard reagents in the presence of Cu^I salts.[38] Regioselectivity of α-additions of organo-copper reagents has been described.[39] Formation constants of Cu^I olefin complexes with biby and phen have been presented.[40] Resonance Raman studies of vinyl modes in Cu^I-protohaem π-complexes have been described.[41]

The raft structures of $Cu_3Fe_3(CO)_{12}^{3-}$, $Cu_5Fe_4(CO)_{16}^{3-}$, and $[Cu_3Os_3H_9(PMe_2Ph)_9]$ have been discussed.[42a,b]

Copper(II). The e.p.r. spectrum of a tetragonally compressed CuF_6^{4-} guest species in K_2ZnF_6 has been studied.[43] Distinguishable sites in tetranuclear oxo-Cu^{II} complexes $(py)_3Cu_4Cl_4O_2$ and $(Denc)_3Cu_3M(H_2O)Cl_4O_2$ (M = Cu, Zn) have been observed.[44] Weak Cl^- complex formation by Cu^{II} in aqueous Cl^- solutions have

[24] J. Bjerrum, *Acta Chem. Scand., Ser. A*, 1986, **40**, 233.

[25] S. Bhaduri, N. Y. Sapre, and A. Basu, *J. Chem. Soc., Chem. Commun.*, 1986, 197.

[26] S. Sakaki, G. Koga, and K. Ohkubo, *Inorg. Chem.*, 1986, **25**, 2330.

[27] S. M. Nelson, A. Lavery, and M. G. B. Drew, *J. Chem. Soc., Dalton Trans.*, 1986, 911.

[28] C. O. Dietrich-Buchecker, A. Khemiss, and J. P. Sauvage, *J. Chem. Soc., Chem. Commun.*, 1986, 1376.

[29] S. J. Berners Price, C. Brevard, A. Pagelot, and P. J. Sadler, *Inorg. Chem.*, 1986, **25**,596.

[30] S. Kitagawa, M. Munakata, and M. Sasaki, *Inorg. Chim. Acta*, 1986, **120**, 77.

[31] F. Pertlik, *Acta Crystallogr.*, 1986, **C42**, 774.

[32] T. N. Sorrell, A. S. Borovik, and C.-C. Shen, *Inorg. Chem.*, 1986, **25**, 589.

[33] P. Fiaschi, C. Floriani, M. Pasquali, A. Chiesi-Villa, and C. Guastini, *Inorg. Chem.*, 1986, **25**, 462.

[34] J. H. Clark, C. V. A. Duke, J. M. Miller, and S. J. Brown, *J. Chem. Soc., Chem. Commun.*, 1986, 877.

[35] S. A. Levitre, C. C. Tso, and A. R. Cutler, *J. Organomet. Chem.*, 1986, **308**, 253.

[36] S. Pasynkiewicz and J. Poplawska, *J. Organomet. Chem.*, 1986, **302**, 269.

[37] P. Wijkens and P. Vermeer, *J. Organomet. Chem.*, 1986, **301**, 247.

[38] H. Amri, M. Rambaud, and J. Villieras, *J. Organomet. Chem.*, 1986, **308**, C27.

[39] V. Calò, L. Lopez, and G. Pesce, *J. Chem. Soc., Chem. Commun.*, 1986, 1252.

[40] M. Munakata, S. Kitagawa, S. Kosome, and A. Asahara, *Inorg. Chem.*, 1986, **25**, 2622.

[41] S. S. Sibbert, T. M. Loehr, and J. K. Hurst, *Inorg. Chem.*, 1986, **25**, 307.

[42] (*a*) G. Doyle, K. A. Eriksen, and D. Van Engen, *J. Am. Chem. Soc.*, 1986, **108**, 445; (*b*) T. H. Lemmen, J. C. Huffman, and K. G. Caulton, *Angew. Chem., Ent. Ed. Engl.*, 1986, **25** 262.

[43] M. A. Hitchman, R. G. McDonald, and D. Reinen, *Inorg. Chem.*, 1986, **25**, 519.

[44] G. Davis, M. A. El-Sayed, A. El-Toukhy, M. Henary, and C. A. Martin, *Inorg. Chem.*, 1986, **25**, 4479.

been discussed[45a,b] as have highly electro-conductive tetrathiafulvalenium salts of Cu^{II} halides.[46] Halogeno-Cu^{II} complexes of phenyl-2-picolyl ketone hydrazone have been reported.[47] The structures of bis-2,2′,*N*,*N*′-bipyridylaminechloro-Cu^{II} chloride,[48] dichloro(2,5-dithiahexane)Cu^{II},[49] dichloro(4,7-dithiadecane)Cu^{II} [49] and bis(μ-pyridine *N*-oxide)bis[dichloro(dimethylsulphoxide)Cu^{II}][50] have been determined. In dichloro(ethylenediamine)Cu^{II} the Cu atom is square planar.[51] The structures of $[Cu_2(C_2H_7NO_2P)_2Cl_2]$,[52] $CuCl_2C_{13}H_{12}N_2O_2$,[53] and $CuCl_2C_{15}H_{16}N_2O_2$,[53] have also been described, as has that of $[Me_4P][Cu_2Cl_5]$, in which the $[Cu_2Cl]_5{}^-$ anions are present as interconnected chains.[54] Et_4NCuCl_3 contains discrete $Cu_4Cl_{12}{}^{4-}$ anions.[55] The structures of a series of $ACuCl_3$ salts containing uniform bi-bridged chains have been presented,[56] as has a cellular ligand-field study of the $CuCl_4{}^{2-}$ in $Cs_2[CuCl_4]$[57]. The results of spectrophotometric studies of Cu^{II}-chloride–trimethyl phosphate solutions have been reported[58] and studies of paraquat hexabromodicuprate(II) presented.[59] $[Cu_2L(Br)]Br_2$ [HL = 2,6-bis{*N*-(2-pyridylmethyl)formidoyl}-4-methylphenol] is shown to be a dinuclear complex.[60] The structures of $[CuBr_2(C_{15}H_{16}N_4)]$ and $[CuBr(C_{16}H_{18}N_4)]Br$ have been reported[61] while those of two five-coordinate mononuclear Cu^{II} complexes are similar (2),[62] (3)[63] have been shown to be similar.

(2)

[45] (*a*) J. Bjerrum and L. H. Skibsted, *Inorg. Chem.*, 1986, **25**, 2479; (*b*) R. W. Ramtte, *Inorg. Chem.*, 1986, **25**, 2481.
[46] M. B. Inoue, M. Inoue, Q. Fernando, and K. W. Nebesny, *Inorg. Chem.*, 1986, **25**, 3976.
[47] A. El-Dissouky, A. M. Hindawey, and A. Abdel-Salem, *Inorg. Chim. Acta*, 1986, **119**, 207.
[48] R. A. Jacobson and W. P. Jensen, *Inorg. Chim. Acta*, 1986, **114**, L9.
[49] W. E. Hatfield, L. W. ter Haar, M. M. Olmstead, and W. K. Musker, *Inorg. Chem.*, 1986, **25**, 558.
[50] A. Bencini, D. Gatteschi, and C. Zanchini, *Inorg. Chem.*, 1986, **25**, 2211.
[51] D. A. Harvey and C. J. L. Lock, *Acta Crystallogr.*, 1986, **C42**, 799.
[52] T. Glowiak, *Acta Crystallogr.*, 1986, **C42**, 62.
[53] G. R. Newkome, H. C. R. Taylor, F. R. Fronczek, and V. K. Gupta, *Inorg. Chem.*, 1986, **25**, 1149.
[54] W. G. Haije, J. A. L. Dobbelaar, and W. J. A. Maaskant, *Acta Crystallogr.*, 1986, **C42**, 1485.
[55] R. D. Willett and U. Geiser, *Inorg. Chem.*, 1986, **25**, 4558.
[56] U. Geiser, R. M. Gaura, R. D. Willett, and D. X. West, *Inorg. Chem.*, 1986, **25**, 4203.
[57] R. J. Deeth and M. Gerloch, *J. Chem. Soc., Dalton Trans.*, 1986, 1531.
[58] C. Amuli, M. Elleb, J. Meullemeestre, M.-J. Schwing, and F. Vierling, *Inorg. Chem.*, 1986, **25**, 856.
[59] R. D. Willett, *Inorg. Chem.*, 1986, **25**, 1918.
[60] C. J. O'Connor, D. Firmin, A. K. Pane, B. R. Baby, and E. D. Stevens, *Inorg. Chem.*, 1986, **25**, 2300.
[61] A. Pajunen and S. Pajunen, *Acta Crystallogr.*, 1986, **C42**, 53.
[62] A. W. Hanson, *Acta Crystallogr.*, 1986, **C42**, 501.
[63] J. B. Veldhuis, W. L. Driessen, and J. Reedijk, *J. Chem. Soc., Dalton Trans.*, 1986, 537.

(3)

Six-coordinated Cu^{II} has been found in $[CuBr_2(C_{16}H_{20}N_4O)]\cdot H_2O$,[64] and $[CuBr_2(C_{12}H_{25}N_4O_2)]\cdot 1.5H_2O$.[65] $[Me_3NH]_2Cu_4Br_{10}$ contains $Cu_4Br_{10}^{2-}$ oligomers.[66]

Anisotropic exchange and magneto-structural correlations have been studied in a number of bridged Cu^{II} dimers[67a—f] and the interactions between Cu^{II} ions through azido bridges have been studied.[68]

A number of reports of studies of compounds containing Cu^{II} ions bound to imidazole ligands have appeared[69a—d] and the structures of two 1:1 mixed-ligand complexes of Cu^{II} azide with substituted pyridines discussed.[70] The pillaring of layers in a $Zr(HPO_4)_2(EtOH)_2$ by $[Cu(phen)_2]^{2+}$ has been observed.[71] The heats of reaction between TAPEN and Cu^{II} and Zn^{II} ions have been reported.[72] $[Cu(nn)Cl]ClO_4$ (nn = N,N'-bis(8-quinolilethylenediamine) contains a CuN_4Cl chromophore.[73] The structures of two Cu^{II} complexes prepared from 2-aminomethyl-pyridine have been described[74] as has the complex formation of Cu^{II} with mono- and di-ethanolamines in aqueous solution.[75] The binding of acetamide oxime with Cu^{II} has been studied,[76] as has that of 2-aminoacetamidoxime and its N-methylated

[64] A. Pajunen and S. Pajunen, *Acta Chem. Scand., Ser. A*, 1986, **40**, 190.
[65] A. Pajunen and S. Pajunen, *Acta Chem. Scand., Ser. A*, 1986, **40**, 413.
[66] U. Geiser, R. D. Willett, M. Lindbeck, and K. Emerson, *J. Am. Chem. Soc.*, 1986, **108**, 1173.
[67] (a) M.-F. Charlot, Y. Journaux, O. Kahn, A. Bencini, D. Gatteschi and C. Zanchini, *Inorg. Chem.*, 1986, **25**, 1060; (b) L. K. Thompson and B. S. Ramaswamy, *Inorg. Chem.*, 1986, **25**, 2664; (c) R. Chiari, J. H. Helms, O. Piovesana, T. Tarantelli, and P. F. Zanazzi, *Inorg. Chem.*, 1986, **25**, 2408; (d) P. Chaudhuri, K. Oder, K. Wieghardt, B. Nuber, and J. Weiss, *Inorg. Chem.*, 1986, **25**, 2818; (e) E. G. Bakalbassis, J. Mrozinski, and C. A. Tsipis, *Inorg. Chem.*, 1986, **25**, 3684; (f) A. Ozarawski and D. Reinen, *Inorg. Chem.*, 1986, **25**, 1704.
[68] M. F. Charlot, O. Kahn, M. Chaillet, and C. Larrieu, *J. Am. Chem. Soc.*, 1986, **108**, 2574.
[69] (a) K. Miki, S. Kaida, M. Saeda, K. Yamatoya, N. Kasai, M. Sato, and J.-I. Nakaya, *Acta Crystallogr.*, 1986, **C42**, 1004; (b) A. Bencini, C. Benelli, D. Gatteschi, and C. Zanchini, *Inorg. Chem.*, 1986, **25**, 398; (c) M. Sato, S. Nagae, K. Ohmae, J.-I. Nakaya, K. Miki, and N. Kasai, *J. Chem. Soc., Dalton Trans.*, 1986, 1949; (d) S. Siddiqui and R. E. Shepherd, *Inorg. Chem.*, 1986, **25**, 3869.
[70] T. C. W. Mak and M. A. S. Goher, *Inorg. Chim. Acta*, 1986, **115**, 17.
[71] C. Ferragina, M. A. Massucci, P. Patrona, A. La Ginestra, and A. A. G. Tomlinson, *J. Chem. Soc., Dalton Trans.*, 1986, 265.
[72] M. Micheloni, P. Paoletti, A. Bianchi, and E. Garcia-España, *Inorg. Chim. Acta*, 1986, **117**, 165.
[73] J. Ribas, C. Diaz, M. Monfort, M. Corbella, X. Solans, and M. Font-Altaba, *Inorg. Chim. Acta*, 1986, **117**, 49.
[74] C. J. O'Connor, E. E. Eduok, J. W. Owens, E. D. Stevens, and C. L. Klein, *Inorg. Chim. Acta*, 1986, **117**, 175.
[75] R. Tauler, E. Casassas, and B. M. Rode, *Inorg. Chim. Acta*, 1986, **114**, 203.
[76] T. Hirotsu, S. Katoh, K. Sugasaka, M. Seno, and T. Itagaki, *J. Chem. Soc., Dalton Trans.*, 1986, 1609.

derivatives.[77] Square-planar Cu ions are found in diformatobis(2-methylbenz-imidazole)Cu^{II}.[78] Cu^{II}-promoted hydrolysis of 4-nitrophenyl glycinate has been reported[79] as has the electrochemical reduction of pentadentate dinuclear Cu^{II} compounds in DMF.[80] The Cu^{II} complex of the ligand 1-(2'-aminophenyl)-6-methyl-2,5-diazanona-1,6-diene-8-one has been prepared[81] and the nature of the Cu^{II} com-plexes of picoline [82a] and pyridine N-oxides[82b,c] have been discussed. In the structures of $Cu[C_{11}H_8N_2(OH)_2]_2Cl_2\cdot4H_2O$ and $Cu[C_{11}H_8N_2(OH)_2]_2(NO_3)_2\cdot2H_2O$ the pyridyl nitrogens are strongly coordinating to the metal in the equitorial plane.[83] Interactions in $Cu^{II}Cu^{II}Cu^{II}$, $Cu^{II}Zn^{II}Cu^{II}$, and $Cu^{II}Ni^{II}Cu^{II}$ trinuclear species have been studied.[84] A number of studies of Cu^{II} with a variety of Schiff bases have been described.[85a—g] Cu^{II} and Ni^{II} trinuclear species with dithiooxamide ligands have been prepared[86] as have some substituted acetato-Cu^{II} complexes.[87a—c] In $Na_2[Cu(C_{14}H_{12}N_2O_8)]\cdot4H_2O$ the coordination geometry of Cu is distorted from O_h and twisted towards that of a trigonal prism owing to the planarity of the phenyl-enediamine chelate ring (4).[88]

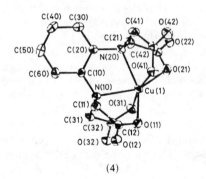

(4)

[77] H. Saarinen, M. Orama, T. Raikas, and J. Korvenranta, *Acta Chem. Scand., Ser. A*, 1986, **40**, 396.

[78] T. S. Schilperoort, F. J. Reitmeyer, R. A. G. de Graaf, and J. Reedijk, *Acta Crystallogr.*, 1986, **C42**, 1491.

[79] R. W. Hay and A. K. Basak, *J. Chem. Soc., Dalton Trans.*, 1986, 39.

[80] W. Mazurek, A. M. Bond, M. J. O'Connor, and A. G. Wedd, *Inorg. Chem.*, 1986, **25**, 906.

[81] E. Kwiatkowski and M. Kwiatkowski, *Inorg. Chim. Acta*, 1986, **117**, 145.

[82] (a) D. X. West and C. A. Nipp, *Inorg. Chim. Acta*, 1986, **118**, 157; (b) G. Ertem, J. C. Severns, and D. X. West, *Inorg. Chim. Acta*, 1986, **115**, 141; (c) P. Knuuttila and H. Knuuttila, *Acta Crystallogr.*, 1986, **C42**, 989.

[83] S.-L. Wang, J. W. Richardson, jun., S. J. Briggs, R. A. Jacobsen, and W. P. Jensen, *Inorg. Chim. Acta*, 1986, **111**, 67.

[84] Y. Journaux, J. Sletten, and O. Kahn, *Inorg. Chem.*, 1986, **25**, 439.

[85] (a) S. P. S. Rao, K. I. Varughese, and H. Manohar, *Inorg. Chem.*, 1986, **25**, 734; (b) I. Sasaki, A. Gaudemer, A. Chiaroni, and C. Riche, *Inorg. Chim. Acta*, 1986, **112**, 119; (c) M. G. M. Reyes, P. Gili, P. M. Zarza, A. M. Ortega, and M. C. Diaz Gonzalez, *Inorg. Chim. Acta*, 1986, **116**, 153; (d) N. A. Bailey, D. E. Fenton, and J. R. Tate, *Inorg. Chim. Acta*, 1986, **112**, 17; (e) M. N. Patel, M. M. Patel, F. E. Cassidy, and J. W. Fitch, *Inorg. Chim. Acta*, 1986, **118**, 33; (f) U. Casellato, P. Guerriero, S. Tamburini, F. A. Vigato, and R. Graziani, *Inorg. Chim. Acta*, 1986, **119**, 215; (g) F. Nepveu, F.-J. Bormuth, and L. Walz, *J. Chem. Soc., Dalton Trans.*, 1986, 1213.

[86] R. Veit, J.-J. Girerd, O. Kahn, F. Robert, and Y. Jeannin, *Inorg. Chem.*, 1986, **25**, 4175.

[87] (a) A. El-Dissouky, A. M. Hindawey, and A. Abdel-Salem, *Inorg. Chim. Acta*, 1986, **118**, 109; (b) R. Nakon, C. R. Krishnamoorthy, S. Townsend, and J. Grayson, *Inorg. Chim. Acta*, 1986, **124**, L5; (c) S.-B. Teo, S.-G. Teoh, T. W. Hambley, and M. R. Snow, *J. Chem. Soc. Dalton Trans.*, 1986, 553.

[88] N. Nakasuka, S. Azuma, and M. Tanaka, *Acta Crystallogr.*, 1986, **C42**, 1482.

Five-coordinate Cu^{II} ions are found in substituted benzoato[89] and carboxylato[90] complexes, whereas four-coordinate planar Cu^{II} ions are found in benzoato,[91] butyrato,[92] and phenolato[93] complexes. A number of structures containing substituted salicylaldiminato ligands bound to Cu^{II} have been determined.[94a—e]

$[Cu(C_7H_5O_2)(C_{10}H_8N_2)]ClO_4$ consists of a dimeric unit involving two (2',2'-bipyridyl)salicylaldehydato Cu^{II} cations[95] and the structures of three complexes of $[Cu_2(Fdmem)(X)]ClO_4 \cdot nH_2O$ (X = OH^-, 1,1-N_3^-, and 1,1-OCN^-) have been compared.[96] The structure of aqua(L-aspartato)(1,10-phenanthroline)$Cu^{II} \cdot 4H_2O$ has been determined and is as shown in (5).[97]

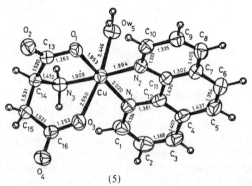

(5)

The structure of a tetradentate amine oxime of Cu^{II} has been described.[98] The preparation of complexes of Cu^{II} with pyrazine-2,3-dicarboxylic and pyridine-2,3-dicarboxylic acids have been reported[99] as has the structure of a N-(2-pyridinyl)ketoacetamide of Cu^{II} (6).[100]

The weak exchange interactions between nitroxides and Cu^{II} ions have been monitored by e.p.r. spectroscopy.[101] ZSM-5 zeolites containing Cu^{2+} ions showed high activity for the catalytic decomposition of NO.[102]

$[(AE)_3(Cu_3OH)](ClO_4)_2$ (AEH = 7-amino-4-methyl-5-aza-3-hepten-2-one) contains a Cu_3OH core.[103] ^{17}O-n.m.r. studies on hexakis(methanol)Cu^{II} ions have been

[89] M. Gawron, A. C. Palenik, and G. J. Palenik, *Inorg. Chim. Acta*, 1986, **112**, 71.
[90] H. M. Haendler, *Acta Crystallogr.*, 1986, **C42**, 147.
[91] S. M. Boudreau and H. M. Haendler, *Acta Crystallogr.*, 1986, **C42**, 980.
[92] G. Oliva, E. E. Castellano, J. Zukerman-Schpector, and R. Calvo, *Acta Crystallogr.*, 1986, **C42**, 19.
[93] J. Marengo-Rullàn and R. D. Willett, *Acta Crystallogr.*, 1986, **C42**, 1487.
[94] (a) G. M. Gray, A. L. Zell, and H. Einspahr, *Inorg. Chem.*, 1986, **25**, 2923; (b) L. Daizheng, Z. J. Zhong, H. Okawa, and S. Kida, *Inorg. Chim. Acta*, 1986, **118**, 21; (c) R. Cini, *Acta Crystallogr.*, 1986, **C42**, 550; (d) M. T. Garland, J. Y. Le-Marouille, and E. Spodine, *Acta Crystallogr.*, 1986, **C42**, 299; (e) A. Bencini, C. Benelli, A. Caneschi, A. Dei, and D. Gatteschi, *Inorg. Chem.*, 1986, **25**, 572.
[95] M. T. Garland, J. Y. Le Marouille, and E. Spodine, *Acta Crystallogr.*, 1986, **C42**, 1518.
[96] T. Mallah, M.-L. Boillat, O. Kahn, J. Gouteron, S. Jeannin, and Y. Jeannin, *Inorg. Chem.*, 1986, **25**, 3058.
[97] L. Antolini, L. P. Battaglia, A. B. Corradi, G. Marcotrigiano, L. Menabne, G. C. Pellacani, M. Saladini, and M. Sola, *Inorg. Chem.*, 1986, **25**, 2901.
[98] L. Pal, R. K. Murmann, E. O. Schlemper, C. K. Fair, and M. S. Hussain, *Inorg. Chim. Acta*, 1986, **115**, 153.
[99] R. Bucci, V. Carunchio, and A. M. Girelli, *Inorg. Chim. Acta*, 1986, **111**, 1.
[100] P. Iliopoulos, G. D. Fallon, and K. S. Murray, *J. Chem. Soc., Dalton Trans.*, 1986, 437.
[101] C. Benelli, D. Gatteschi, C. Zanchini, J. M. Latour, and P. Rey, *Inorg. Chem.*, 1986, **25**, 4242.
[102] M. Iwamoto, H. Furukawa, Y. Mine, F. Uemura, S. Mikuriya, and S. Kagawa, *J. Chem. Soc., Chem. Commun.*, 1986, 1272.
[103] J.-P. Costes, F. Dahan, and J.-P. Laurent, *Inorg. Chem.*, 1986, **25**, 413.

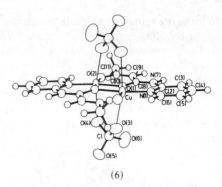

(6)

reported.[104] Tetrakis[μ-3,3′-dimethylacrylato]bis(ethanol)dicopper(II) contains a Cu···Cu separation of 2.594(2) Å.[105] Square-pyramidal Cu^{II} ions are found in $Cu(bpy)DBCat)\cdot1.5CH_2Cl_2 0.5H_2O$.[106] $C_{28}H_{56}Cu_4N_8O_8\cdot CHCl_3$ contains a Cu_4O_4 core,[107] whereas $[Cu_6(C_4H_{10}NO)_8](ClO_4)_4$ contains a 'bicapped-cubane' Cu_6O_8 cluster.[108] The structure of tris{di-μ-methoxobis[4,4,4-trifluoro-1-(2-thienyl)butane-1,3-dionato]dicopper(II)} has been reported[109] as has that of a linear trinuclear Cu^{II} cluster complex with a single oxygen bridged between neighbouring Cu^{II} ions.[110]

A number of Cu^{II} organic acid structures based on the Cu^{II} acetate monohydrate type have been described.[111a—e]. The structure of a linear-chain Cu^{II} compound with bridging acetate and oxamidate ligands has been presented.[112] $Cu_2L_2(MeCOO)_2\cdot 2H_2O$ contains Cu in (4 + 1) square-pyramidal coordination.[113] In (hexafluoroacetylacetonato)(acetylacetonato)Cu^{II} square-planar geometry is found at each Cu^{II} for the four oxygens but the planes of the chelate rings are at an angle of 163.5° with each other.[114] The structure of $[Cu_2(cdta)]\cdot4H_2O$ contains octahedral sites bridged by carboxylate group.[115] The magneto chemistry of $[Cu(NH_3)_2(MeCOO)Br]$ has been studied.[116] Three structures containing phenoxy-alkanoic acids bonded to Cu^{II} have been described[117a,b] and spectroscopic studies have been carried out on mixed Cu^{II}–Pd^{II} acetates[118] and bis(acetylacetonato)Cu^{II}.[119]

[104] L. Helm, S. F. Lincoln, A. E. Merbach and D. Zbinden, *Inorg. Chem.*, 1986, **25**, 2550.

[105] W. Clegg, I. R. Little, and B. P. Straugham, *Acta Crystallogr.*, 1986, **C42**, 1489.

[106] R. M. Buchanan, C. Wilson-Blumenberg, C. Trapp, S. K. Larsen, D. L. Greene, and C. G. Pierpont, *Inorg. Chem.*, 1986, **25**, 3070.

[107] L. Schwabe and W. Haase, *Acta Crystallogr.*, 1986, **C42**, 667.

[108] H. Muhonen, W. E. Hatfield, and J. H. Helms, *Inorg. Chem.*, 1986, **25**, 800.

[109] Z. Olejnik, B. Jezowska-Trzebiatowska, and T. Lis, *J. Chem. Soc., Dalton Trans.*, 1986, 97.

[110] H. Muhonen and W. E. Hatfield, *Acta Chem. Scand., Ser. A*, 1986, **40**, 41.

[111] (a) G. R. Newkowe, K. J. Theriot, V. K. Gupta, R. N. Baltz, and F. R. Fronczek, *Inorg. Chim. Acta*, 1986, **114**, 21; (b) O. W. Steward, R. C. McAfee, S.-C. Chang, S. R. Piskor, W. J. Schreiber, C. F. Jury, C. E. Taylor, J. F. Pletcher, and C.-S. Chen, *Inorg. Chem.*, 1986, **25**, 771; (c) L. C. Porter, M. H. Dickman, and R. J. Doedens, *Inorg. Chem.*, 1986, **25**, 678; (d) L. P. Battaglia, A. Bonamartini Corradi, and L. Menabue, *J. Chem. Soc., Dalton Trans.*, 1986, 1653; (e) K. Smolander, *Inorg. Chim. Acta*, 1986, **114**, 1.

[112] A. Bencini, C. Benelli, A. C. Fabretti, G. Franchini, and D. Gatteschi, *Inorg. Chem.*, 1986, **25**, 1063.

[113] B. Chiari, J. H. Helms, O. Piovesana, T. Tarantelli, and P. F. Zanazzi, *Inorg. Chem.*, 1986, **25**, 870.

[114] P. C. Le Brun, W. D. Lyon, and H. A. Kuska, *Inorg. Chem.*, 1986, **25**, 3106.

[115] A. Fuertes, C. Miravitlles, E. Escriva, E. Coronado, and D. Beltran, *J. Chem. Soc., Dalton Trans.*, 1986, 1795.

[116] R. L. Carlin, K. Kapinga, O. Kahn, and M. Verdaguer, *Inorg. Chem.*, 1986, **25**, 1786.

[117] (a) C. H. L. Kennard, G. Smith, E. J. O'Reilly, *Inorg. Chim. Acta*, 1986, **112**, 47; (b) T. C. W. Mak, W.-H. Yip, G. Smith, E. J. O'Reilly, and C. H. L. Kennard, *Inorg. Chim. Acta*, 1986, **112**, 53.

[118] O. D. Sloan and P. Thornton, *Inorg. Chim. Acta*, 1986, **120**, 173.

[119] Y. L. Chow and G. E. Buono-Core, *J. Am. Chem. Soc.*, 1986, **108**, 1234.

Cu, Ag, Au; Zn, Cd, Hg 341

Dinuclear Cu^{II} 1,3,5-triketonates undergo sequential two-electron transfer.[120] A Cu^{II} complex with two thiolato and two imino donor atoms has been characterized[121] and the formation of Cu^{II}-cyclic polythia ether complexes have been described.[122] The coordination of thiocyanate in tetraamine and diimine-diamine complexes of Cu^{II} have been discussed.[123]

Macrocyclic Systems containing Copper(II). Cu^{II} and Ag^{II} porphyrins and their one-electron oxidation products have been characterized.[124] The oxidative microelectrode voltammetry of CuTPP has been studied in toluene.[125] The structure of the dicopper complex of a biphenylene pillared cofacial diporphyrin has been described (7).[126]

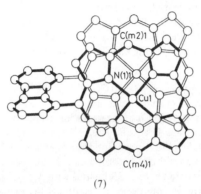

(7)

The preparation of a soluble Cu phthalocyanine with crown ether moieties has been described.[127] Two Cu^{II} structures containing the [1*SR*, 4*RS*, 7*RS*, 8*SR*, 11*RS*, 14*RS*)-5,5,7,12,12,14-hexamethyl-1,4,8,11-tetraazacyclotetradecane have been determined[128a,b] as have two other related macrocycle Cu^{II} structures from the same workers.[129a,b] Similar macrocyclic Cu^{II} complexes have continued to attract much attention from both the structural[130a—d] and chemical[131a—e] viewpoints. The kinetics

[120] R. L. Lintvedt, B. A. Schoenfelner, and K. A. Rupp, *Inorg. Chem.*, 1986, **25**, 2704.
[121] O. P. Anderson, J. Becher, H. Frydendahl, L. F. Taylor, and H. Toftlund, *J. Chem. Soc., Chem. Commun.*, 1986, 699.
[122] I. R. Young, L. A. Ochrymowcyz, and D. B. Rorabacher, *Inorg. Chem.*, 1986, **25**, 2576.
[123] D.-T. Wu and C.-S. Chung, *Inorg. Chem.*, 1986, **25**, 3584.
[124] G. M. Godziela and H. M. Goff, *J. Am. Chem. Soc.*, 1986, **108**, 2237.
[125] L. Geng and R. W. Murray, *Inorg. Chem.*, 1986, **25**, 3115.
[126] J. P. Fillers, K. G. Ravichandran, I. Abdalmuhdi, A. Tulinsky, and C. K. Chang, *J. Am. Chem. Soc.*, 1986, **108**, 417.
[127] A. R. Koray, V. Ahsen, and O. Bekaroglu, *J. Chem. Soc., Chem. Commun.* 1986, 932.
[128] (a) T.-H. Lu, W.-C. Liang, D.-T. Wu, and C.-S. Chung, *Acta Crystallogr.*, 1986, **C42**, 801; (b) T.-H. Lu, D.-T. Wu, and C.-S. Chung, *J. Chem. Soc., Dalton Trans.*, 1986, 1999.
[129] (a) T.-J. Lee, T.-Y. Lee, C.-Y. Hong, D.-T. Wu, and C.-S. Chung, *Acta Crystallogr.*, 1986, **C42**, 999; (b) J. W. Chen, D.-S. Wu, and C.-S. Chung, *Inorg. Chem.*, 1986, **25**, 1940.
[130] (a) N. W. Alcock, K. P. Balakrishnan, and P. Moore, *J. Chem. Soc., Dalton Trans.*, 1986, 1743; (b) N. W. Alcock, P. Moore, and H. A. A. Omar, *J. Chem. Soc., Dalton Trans.*, 1986, 985; (c) R. Schneider, A. Riesen, and T. A. Kaden, *Helv. Chim. Acta*, 1986, **68**, 53; (d) D. P. Comba, N. F. Curtis, G. A. Lawrence, A. M. Sargeson, B. W. Skelton, and A. H. White, *Inorg. Chem.*, 1986, **25**, 4260.
[131] (a) D. K. Geiger and G. Ferraudi, *Inorg. Chem.*, 1986, **117**, 139; (b) D. E. Whitmoyer, D. P. Rillema, and G. Ferraudi, *J. Chem. Soc., Dalton Trans.*, 1986, 677; (c) L. Fabbrizzi, T. A. Kaden, A. Perotti, B. Seghi, and L. Siegfried, *Inorg. Chem.*, 1986, **25**, 321; (d) I. Murase, S. Ueno, and S. Kida, *Inorg. Chim. Acta*, 1986, **111**, 57; (e) N. Aoi, G.-E. Matsubayashi, and T. Tanako, *Inorg. Chim. Acta*, 1986, **114**, 25.

of reduction of a Cu^{II} complex with a tetrabenzo(b,f,j,n)(1,5,9,13)tetraacyc-lohexadecine by ascorbic acid has been investigated[132] and a Cu^{II} complex of a pentaazamacrobicycloalkane has been described.[133] The ordering of metal chelates on the basis of bilayer assembly has been discussed.[134] Cascade halide binding by multiprotonated Cu^{II} BISTREN cryptrates has been reported.[135] Reports on three crystal structures of macrocycles containing two Cu^{II} ions have been presented,[136a—c] and in a fourth two Cu^{II} ions are shown to be bound by a novel octadentate macrocycle.[137] The electrochemical behaviour of acyclic and macrocyclic complexes of Cu^{II} ions has been studied.[138] The n.m.r. spectra of a Cu^{II} macrocyclic peptide containing histidyl residues has been observed.[139]

Copper(III). The dinuclear, hydroxo-bridged, five-coordinate Cu^{II} complex of a hexadentate (N_6) phthalazine ligand undergoes two-electron oxidation to form a dinuclear Cu^{III} derivative.[140] Three forms of a Cu^{III} tripeptideamide have been studied.[141] Cu^{III} is discussed in the kinetics of formation of the methyl–Cu^{II} complex in aqueous solution.[142]

Mixed Valence Complexes. $[Cu_5(CN)_6(dmf)_4]$,[143]$[Cu(C_6H_{15}N_3)_2][Cu(CN)_3] \cdot 2H_2O$[144] and a variety of cyanide containing complexes of copper[145] have all been shown to contain both Cu^I and Cu^{II}. The steric constraints of mixed valence Cu^{II}–Cu^I macrocyclic complexes has been discussed.[146] Cu^{II} salts react with the bidentate hybrid ligands 1-thiophenyl-2-diphenylphosphinoethane (PSPh) and 1-thioethyl-2-diphenylphosphinoethane (PSEt) to form a mixed-valence species $Cu^{II}(OPSEt)_2Y_2 \cdot Cu^I(PSR)_2Y$. The interaction of the PSR ligands with Ag^I and Au^{III} is also reported.[147] Cu^{II} and Cu^I geometries in complexes of hexathia-18-crown-6,[148] a di-Schiff base[149] and macrocyclic and related acyclic ligands[150] have been discussed.

[132] J. Labuda and J. Sima, *Inorg. Chim. Acta*, 1986, **112**, 59.

[133] M. Ciampolini, M. Micheloni, F. Vizza, F. Zanobini, S. Chimichi, and P. Dapporto, *J. Chem. Soc., Dalton Trans.*, 1986, 505.

[134] T. Kunitake, T. Ishikawa, M. Shimoumura, and H. Okawa, *J. Am. Chem. Soc.*, 1986, **108**, 327.

[135] R. J. Motekaitis, A. E. Martell, and I. Murase, *Inorg. Chem.*, 1986, **25**, 938.

[136] (a) H. Adams, N. A. Bailey, D. E. Fenten, S. Moss, C. O. Rodriguez de Barbarin, and G. Jones, *J. Chem. Soc., Dalton Trans.*, 1986, 693; (b) N. A. Bailey, D. E. Fenton, R. Moody, C. O. Rodriguez de Barbarin, I. N. Sciambarella, J.-M. LaTour, and D. Limosin, *Inorg. Chim. Acta*, 1986, **125**, L41; (c) L. R. Gahan, K. E. Hart, C. L. Kennard, M. A. Kingston, G. Smith, and T. C. W. Mak, *Inorg. Chim. Acta*, 1986, **116**, 5.

[137] I. Murase, M. Mikuriya, H. Sonoda, Y. Fukuda, and S. Kida, *J. Chem. Soc., Dalton Trans.*, 1986, 953.

[138] P. Zanello, A. Cinquantini, P. Guerriero, S. Tamburini, and P. A. Vigata, *Inorg. Chim. Acta*, 1986, **117**, 91.

[139] J.-P. Laussac, A. Robert, R. Haran, and B. Sarkar, *Inorg. Chem.*, 1986, **25**, 2760.

[140] L. K. Thompson, S. K. Mandal, E. J. Gabe, and J.-P. Charland, *J. Chem. Soc., Chem. Commun.* 1986, 1537.

[141] J. P. Hinton and D. W. Margerum, *Inorg. Chem.*, 1986, **25**, 3248.

[142] H. Cohen and D. Meyerstein, *Inorg. Chem.*, 1986, **25**, 1505.

[143] S.-M. Peng and D.-S. Liaw, *Inorg. Chim. Acta*, 1986, **113**, L11.

[144] P. Chauduri, K. Oder, K. Wieghardt, J. Weiss, J. Reedijk, W. Hinrichs, J. Wood, A. Ozarowski, H. Stratemaier, and D. Reinan, *Inorg. Chem.*, 1986, **25**, 2951.

[145] J. A. Connor, D. Gibson, and R. Price, *J. Chem. Soc., Dalton Trans.*, 1986, 347.

[146] S. K. Mandal, B. Adhikary, and K. Nag, *J. Chem. Soc., Dalton Trans.*, 1986, 1175.

[147] M. Bressan and A. Morvillo, *Inorg. Chim. Acta*, 1986, **120**, 33.

[148] J. A. R. Hartman and S. R. Cooper, *J. Am. Chem. Soc.*, 1986, **108**, 1202.

[149] M. Linss and U. Weser, *Inorg. Chim. Acta*, 1986, **125**, 117.

[150] M. G. B. Drew, P. C. Yates, B. P. Murphy, J. Nelson, and S. M. Nelson, *Inorg. Chim. Acta*, 1986, **118**, 37.

Copper Compounds of Biological Interest. Cu^{II} complexes of a number of amino acids have once again been studied.[151a—f] The catalytic activity of a Cu^{II}-oxidized glutathione complex on aqueous superoxide ion dismutation has been investigated.[152] The structure of bis(S-methyl-L-cysteinato)Cu^{II} has appeared.[153] Some binary Cu^{II} complexes with N-dansylglycine and their amine adducts with pyridine have been synthesized.[154] The first stable Cu^{II} complex that incorporates 'biological' S(cys) ligation has been reported[155] as have some Cu^{II} complexes of cinnamic acids.[156] Several studies of Cu^{II} dipeptides have been published[157a—e] as has one of a Cu^{II} tetrapeptide.[158] A Cu^{II} complex of an N-protected aspartic acid salt contains two Cu^{II} ions.[159]

A D-glucosamine complex of Cu^{II} has been studied[160] as has the reaction of CuO with L-lysine.[161] Cu^{II} complexes of L-proline amides have been prepared[162] and the Cu^{II}-promoted hydrolysis of 4-nitrophenyl L-leucinate studied.[163] Cu^{I} and Cu^{II} complexes of the histamine H$_2$-antagonist cimetidine have been isolated[164] as have the Cu^{II} complexes of adenine N-oxides[165] and phytic acid.[166]

The complexation of Cu^{II} with the biochemical buffer HEPES[167] with ε-ATP[168] has been observed. The solution properties of Cu^{II}-L-α-alaninehydroxamic acid were studied[169] as were those of the Cu^{II}-prometone-D-glucosamine ternary system.[170] Cu^{II} complexes of a series of dihydroxamine acids have been structurally characterized.[171]

[151] (a) B. Henry, J.-C. Boubel, and J.-J. Delpuech, *Inorg. Chem.*, 1986, **25**, 623; (b) R. Basoli, G. Valensin, E. Gaggelli, W. Froncisz, M. Pasenkiewicz-Gierula, W. E. Antholine, and J. S. Hyde, *Inorg. Chem.*, 1986, **25**, 3006; (c) E. Leporati, *J. Chem. Soc., Dalton Trans.*, 1986, 199; (d) B. Radomska, E. Matczak-Jon, and W. Wojciechowski, *Inorg. Chim. Acta*, 1986, **124**, 83; (e) M. T. Leal, S. Duarte, M. A. A. F. de C. T. Carrondo, M. L. S. S. Goncalves, M. B. Hursthouse, N. P. C. Walker, and H. M. Dawes, *Inorg. Chim. Acta*, 1986, **124**, 41; (f) A. Hanaki and H. Yokoi, *Inorg. Chim. Acta*, 1986, **123**, L7.

[152] M. Jouini, G. Lapluye, J. Huet, R. Julien, and C. Ferradini, *J. Inorg. Biochem.*, 1986, **26**, 269.

[153] E. Dubler, N. Cathomas, and G. B. Jameson, *Inorg. Chim. Acta*, 1986, **123**, 99.

[154] L. Antolini, L. Menabue, M. Sola, L. P. Battaglia, and A. Bonmartini Corradi, *J. Chem. Soc., Dalton Trans.*, 1986, 1367.

[155] P. K. Bhardwaj, J. A. Potenza, and H. J. Schugar, *J. Am. Chem. Soc.*, 1986, **108**, 1351.

[156] M. A. Cabras and M. A. Zoroddu, *Inorg. Chim. Acta*, 1986, **117**, 5.

[157] (a) R. P. Bonomo, R. Cali, V. Cucinotta, G. Impellizzeri, and E. Rizzarelli, *Inorg. Chem.*, 1986, **25**, 1641; (b) T. Glowiak and I. Podgórska, *Inorg. Chim. Acta*, 1986, **125**, 83; (c) M. Sato, S. Matsuki, and J. Nakaya, *Inorg. Chim. Acta*, 1986, **120**, L19; (d) M. Sato, S. Matsuki, M. Ideka, and J. Nakaya, *Inorg. Chim. Acta*, 1986, **125**, 49; (e) M. Aihara, Y. Nakanura, Y. Nishida, and K. Noda, *Inorg. Chim. Acta*, 1986, **124**, 169.

[158] B. Decock-Le Reverend, A. Lebkiri, C. Livera, and L. D. Pettit, *Inorg. Chim. Acta*, 1986, **124**, L19.

[159] L. Antolini, L. P. Battaglia, A. Bonamartini Corradi, L. Menabue, G. Micera, and M. Saladini, *Inorg. Chem.*, 1986, **25**, 3301.

[160] J. Lerivrey, B. Dubois, P. Decock, G. Micera, J. Urbanska, and H. Kozlowski, *Inorg. Chim. Acta*, 1986, **125**, 187.

[161] Y. Yorkovsky, M. Kapon, and Z. Dori, *Inorg. Chim. Acta*, 1986, **124**, 149.

[162] T. Murakami and M. Hatano, *Inorg. Chim. Acta*, 1986, **125**, 111.

[163] R. W. Hay and A. K. Basak, *Inorg. Chim. Acta*, 1986, **123**, 237.

[164] E. Kimura, T. Koike, Y. Shimizu, and M. Kodama, *Inorg. Chem.*, 1986, **25**, 2242.

[165] W. Nerdal and E. Sletten, *Inorg. Chim. Acta*, 1986, **114**, 41.

[166] C. J. Martin and W. J. Evans, *J. Inorg. Biochem.*, 1986, **28**, 39.

[167] K. Hegetschweiler and P. Saltman, *Inorg. Chem.*, 1986, **25**, 107.

[168] T. A. Kaden, K. H. Scheller, and H. Sigel, *Inorg. Chem.*, 1986, **25**, 1313.

[169] B. Kurzak, K. Kurzak, and J. Jezierska, *Inorg. Chim. Acta*, 1986, **125**, 77.

[170] J. Lerivrey, P. Decock, B. Dubois, J. Urbánska, and H. Kozlowski, *Inorg. Chim. Acta*, 1986, **124**, L11.

[171] S. J. Barclay and K. N. Raymond, *Inorg. Chem.*, 1986, **25**, 3561.

Cu^{II} complexes have been investigated for antitumour activity[172a,b] and bleomycin-like properties.[173a,b] A Cu^{II} complex of the anti-inflammatory drug nictindole has been reported.[174] The reaction of D-penicillamine with Cu-metal has been discussed.[175]

The Cu–Mo antagonism in ruminants[176a,b] and the biological significance of Cu^{II} complexes of tetracycline[177] have been studied.

Reports on dinuclear copper complexes as models for multicopper enzymes have appeared[178a,b] and model complexes for haemocyanins,[179a,b] 'non-blue' copper proteins,[180] blue copper proteins,[181a—e] Cu/Zn superoxide dismutase,[182] and galactose oxidase[183] have been described.

A number of enzymes containing copper have been studied, these include bovine ceruloplasmin,[184a,b] haemocyanin,[185a,b] Cu/Zn superoxide dismutase,[186a,b] azurin,[187] and plantacyanin.[188] Cu^{II} binding sites have been investigated in yeast inorganic pyrophosphatase,[189] myoglobin,[190] and dopamine β-monooxygenase.[191] The secondary structures of copper proteins have been discussed.[192] Papers on Cu^{II} interactions with blood plasma[193] and bovine serum albumin[194] have also been published.

Silver.—*Silver–heterometallic Clusters.* The cluster $[Au_2Ag_2(C_2Ph)_4(PPh_3)_2]$ has been characterized.[195] $[(Ph_3P)_3Ir(\mu_3\text{-}H)(\mu\text{-}H)_2Ag_2(OSO_2CF_3)(H_2O)](CF_3SO_3)$ contains

[172] (a) M. Mohan, P. Sharma, M. Kumar, and N. K. Jha, *Inorg. Chim. Acta*, 1986, **125**, 9; (b) H. Beraldo and L. Tosi, *Inorg. Chim. Acta*, 1986, **125**, 173.

[173] (a) S. J. Brown, X. Tao, D. W. Stephan, and P. K. Mascharak, *Inorg. Chem.*, 1986, **25**, 3377; (b) I. Sasaki, M. N. DuFour, A. Graudemer, A. Chiaroni, and C. Riche, *Inorg. Chim. Acta*, 1986, **112**, 129.

[174] A. E. Underhill, R. P. Blundell, P. S. Gomm, and M. E. Jacks, *Inorg. Chim. Acta*, 1986, **124**, 133.

[175] S. H. Laurie, *Inorg. Chim. Acta*, 1986, **123**, L15.

[176] (a) J. D. Allen and M. Gawthorne, *J. Inorg. Biochem.*, 1986, **27**, 95; (b) S. H. Laurie, D. E. Pratt, and J. B. Raynor, *Inorg. Chim. Acta*, 1986, **123**, 193.

[177] M. Brion, L. Lambs, and G. Berthon, *Inorg. Chim. Acta*, 1986, **123**, 61.

[178] (a) T. Glowiak, I. Podgórska, and J. Baranowski, *Inorg. Chim. Acta*, 1986, **115**, 1; (b) A. Evers, R. D. Hancock, and I. Murase, *Inorg. Chem.*, 1986, **25**, 2160.

[179] (a) N. A. Bailey, D. E. Fenton, R. Moody, P. J. Scrimshire, and J.-M. Latour, *Inorg. Chim. Acta*, 1986, **124**, L1; (b) R. J. Butcher, G. Diven, G. Erickson, A. M. Mockler, and E. Sinn, *Inorg. Chim. Acta*, 1986, **123**, L17.

[180] G. L. Bailey, R. D. Bereman, and D. P. Rillema, *Inorg. Chem.*, 1986, **25**, 3149.

[181] (a) J. Whelen and B. Bosnich, *Inorg. Chem.*, 1986, **25**, 3671; (b) H. K. Baek, K. D. Karlin, and R. A. Holwerda, *Inorg. Chem.*, 1986, **25**, 2347; (c) J. M. Berg and K. O. Hodgson, *Inorg. Chem.*, 1986, **25**, 1800; (d) L. Casella, M. Gullotti, and R. Vigano, *Inorg. Chim. Acta*, 1986, **124**, 121; (e) E. John, P. K. Bharadwaj, J. A. Potenza, and H. J. Schugar, *Inorg. Chem.*, 1986, **25**, 3065.

[182] M. Rosi, A. Sgamellotti, F. Tarantelli, I. Bertini, and C. Luchinat, *Inorg. Chem.*, 1986, **25**, 1005.

[183] N. Kitajima, K. Whang, Y. Moro-oka, A. Uchida, and Y. Sasada, *J. Chem. Soc., Chem. Commun.*, 1986, 1504.

[184] (a) T. Sakurai and A. Nakahara, *Inorg. Chim. Acta*, 1986, **123**, 217; (b) T. Sakurai and A. Nakahara, *J. Inorg. Biochem.*, 1986, **27**, 85.

[185] (a) K. Lerch, M. Huber, H.-J. Schneider, R. Drexel, and B. Linzen, *J. Inorg. Biochem.*, 1986, **26**, 213; (b) J. Lorosch, W. Haase, and P. V. Huong, *J. Inorg. Biochem.*, 1986, **27**, 53.

[186] (a) D. M. Dooley and M. A. McGuirl, *Inorg. Chem.*, 1986, **25**, 1261; (b) I. Morgenstern-Badarau, D. Cocco, A. Desideri, G. Rotilio, J. Jordanov, and N. Dupre, *J. Am. Chem. Soc.*, 1986, **108**, 300.

[187] G. E. Norris, B. F. Anderson, and E. N. Baker, *J. Am. Chem. Soc.*, 1986, **108**, 2784.

[188] T. Sakurai, S. Sawada, and A. Nakahara, *Inorg. Chim. Acta*, 1986, **123**, L21.

[189] A. Banerjee, R. LoBrutto, and B. S. Cooperman, *Inorg. Chem.*, 1986, **25**, 2417.

[190] O. Baffa, J. C. Say, M. Tabak, and O. R. Nascimento, *J. Inorg. Biochem.*, 1986, **26**, 117.

[191] C. Syvertsen, R. Gaustad, K. Schröder, and T. Ljones, *J. Inorg. Biochem.*, 1986, **26**, 63.

[192] M. Lundeen, *J. Inorg. Biochem.*, 1986, **27**, 151.

[193] G. Berthen, B. Hacht, M.-J. Blais, and P. M. May, *Inorg. Chim. Acta*, 1986, **125**, 219.

[194] A. Zatòn, M. Trueba, and C. Abad, *Inorg. Chim. Acta*, 1986, **124**, 219.

[195] O. M. Abu-Salah and C. B. Knobler, *J. Organomet. Chem.*, 1986, **302**, C10.

a bent Ag—Ir—Ag chain.[196] A triangular Rh_2Ag framework was found in [$Rh_2(\eta$-$C_5H_5)_2(\mu$-CO)(μ-$Ph_2PCH_2PPh_2$)(μ-AgO_2CMe)].[197] Ag— and Cu—Ru cluster compounds containing $Ph_2As(CH_2)_nPPh_2$ (n = 1 or 2) have been reported.[198] The structures of three Pt—Ag compounds have been described[199a] as has the chemistry of the bis(diphenylphosphino)amine ligand with Pt—Ag and Pd—Ag complexes.[199b]

Silver(I). The structure of [NMe_4][Ag_2Br_3] has been discussed.[200] The coagulation of AgI solutions in water is accelerated by $RNMe_3^+$ ions if R comprises five or more carbon atoms.[201] The structure and electrical properties of ($BEDT$-TTF)$_3Ag_xI_8$ ($x \sim 6.4$) [BEDT-TTF is bis(ethylenedithio)tetrathiafulvene] has been studied.[202] Discussion has appeared on the composition and ionic conductivity of $Ag_{8-x}Nb_{16-x}W_{12+x}O_{80}$[203] and the structures of $Ag_2Li(NO_2)_3$ and $Hg_2(NO_2)_2$ have been compared.[204] The stability constant of the triamine Ag^I complex has been determined.[205] A report on the structure of the dinuclear $Ag_2[S_2C_2(CN)_2)_2][PPh_3]_4$ complex has appeared.[206] A series of Ag^I and Cu^I nonvolatile isocyanide complexes have been studied using SIMS.[207]

The reactions of $Fe_2[(\eta$-$C_5H_5)_2(CO)_2(CNMe)\{CN(Me)H\}]^{+208}$ and diphosphaferrocenes[209] with Ag^I salts have been studied. The new heterobimetallic phosphido-bridged compounds [$AgL(\mu$-$PR_2)M(CO)_5$] (L = phen or tricyclohexylphosphine; M = Cr, Mo, W) have been prepared[210] as has the bis(phenylethynyl)-argentate(I) and chlorophenylethynylargentate(I) complex anions.[211] The structures of [$(Ph_3P)(O_3ClO)AgN(Ph_2PAuPPh_2)_2NAg(OClO_3)(PPh_3)$] (8) and

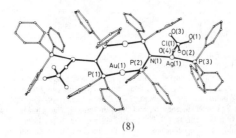

(8)

[196] P. Braunstein, T. M. G. Carneiro, D. Matt, A. Tiripicchio, and M. Tiripicchio Camellini, *Angew. Chem., Int. Ed. Engl.*, 1986, **25**, 748.

[197] S. Lo Schiavo, G. Bruno, P. Piraino, and F. Faraone, *Organometallics*, 1986, **5**, 1400.

[198] S. S. D. Brown, P. J. McCarthy, and I. D. Salter, *J. Organomet. Chem.*, 1986, **306**, C27.

[199] (a) R. Uson, J. Forniés, M. Tomás, J. M. Cusas, F. A. Cotton, and L. R. Falvello, *Inorg. Chem.*, 1986, **25**, 4519; (b) R. Uson, J. Forniés, R. Navarro, and J. I. Cebollada, *J. Organomet. Chem.*, 1986, **304**, 381.

[200] S. Jagner, S. Olson, and R. Stomberg, *Acta Chem. Scand., Ser. A*, 1986, **40**, 230.

[201] F. M. Menger, S. Richardson, and U. V. Venkataram, *J. Chem. Soc., Chem. Commun.*, 1986, 1015.

[202] U. Geiser, H. H. Wang, K. M. Donega, B. A. Anderson, J. M. Williams, and J. F. Kwak, *Inorg. Chem.*, 1986, **25**, 401.

[203] S. Frostäng, B.-O. Marinder, and M. Nygren, *Acta Chem. Scand., Ser. A*, 1986, **40**, 361.

[204] S. Ohba, F. Matsumoto, M. Ishihara, and Y. Saito, *Acta Crystallogr.*, 1986, **C42**, 1.

[205] J. Bjerrum, *Acta Chem. Scand., Ser. A*, 1986, **40**, 392.

[206] D. D. Heinrich, J. F. Fackler, jun., and P. Lahuerta, *Inorg. Chim. Acta*, 1986, **116**, 15.

[207] L. D. Detter, R. G. Cooks, and R. A. Walton, *Inorg. Chim. Acta*, 1986, **115**, 55.

[208] B. Callan and A. R. Manning, *J. Organomet. Chem.*, 1986, **306**, C61.

[209] R. M. G. Roberts, J. Silver, and A. S. Wells, *Inorg. Chim. Acta*, 1986, **119**, 165.

[210] D. Obendorf and P. Peringer, *J. Organomet. Chem.*, 1986, **299**, 127.

[211] O. M. Abu-Salah, A. R. Al-Ohaly, and H. A. Al-Qahtani, *Inorg. Chim. Acta*, 1986, **117**, L29.

$[(C_6F_5)AuCH(Ph_2PAuPPh_2)_2CHAu(C_6F_5)]$ have been determined.[212] The coordination of triphenylphosphine with $AgNO_3$[213] and the synthesis of trifluoromethylalkyls of Ag^I and Au^{III} have been described.[214]

Macrocyclic Systems containing Silver Ions. The chemistry of $[(p-x)TPP]Ag$ complexes were investigated in an electrochemical study. Ag^{III}, Ag^{II}, and an Ag^I complexes were generated.[215] The peroxodisulphate oxidation of an Ag^I complex of 5,10,15,20-tetra-(3'-N-methylpyridinio)porphyrin to the Ag^{II} complex has been studied[216] and the kinetics and equilibria of Ag^I cryptates in dimethyl sulphoxide reported.[217]

Silver Compounds of Biological Interest. The structure of silver pefloxacin, an antibiotic related to nalidixic acid, has been described.[218] Sugar–Ag^I interactions have been studied[219a,b] as have some Ag^I dipeptides.[220a,b]

Gold.—Two reviews of gold chemistry have appeared. The first covers recent development in arylgold chemistry[221] and the second classifies gold compounds on the basis of X-ray structural and Mössbauer spectroscopic data.[222]

Gold Heterometallic Complexes. The number of papers concerning Au atoms in heterometallic cluster continues to grow[223-233] (see Table). The structure of the anion $[Os_{20}M(C)_2(CO)_{48}]^{2-}$ (M = Au or Hg) is shown (9).[229]

Molecular orbital calculations have been carried out on Pt—Au heterometallic clusters.[234]

[212] R. Usón, A. Laguan, M. Laguna, M. C. Gimeno, P. G. Jones, C. Fittschen, and G. M. Sheldrick, *J. Chem. Soc., Chem. Commun.*, 1986, 509.

[213] P. F. Barron, J. C. Dyason, P. C. Healy, L. M. Engelhardt, B. W. Skelton, and A. H. White, *J. Chem. Soc., Dalton Trans.*, 1986, 1965.

[214] M. A. Guerra, T. R. Bierschenk, and R. J. Lagow, *J. Organomet. Chem.*, 1986, **307**, C58.

[215] K. M. Kadish, X. Q. Lin, J. Q. Ding, Y. T. Wu, and C. Araullo, *Inorg. Chem.*, 1986, **253**, 3236.

[216] J. M. Okoh and M. Krishnamurthy, *J. Chem. Soc., Dalton Trans.*, 1986, 449.

[217] B. G. Cox, J. Garcia-Rosas, H. Schneider, and Ng van Truong, *Inorg. Chem.*, 1986, **25**, 1165.

[218] N. C. Baenziger, C. L. Fox, jun., and S. L. Modak, *Acta Crystallogr.*, 1986, **C42**, 1505.

[219] (a) H. A. Tajmir-Riahi, *J. Inorg. Biochem.*, 1986, **27**, 205; (b) H. A. Tajmir-Riahi, *Inorg. Chim. Acta*, 1986, **125**, 43.

[220] (a) L. D. Pettit and A. Q. Lyons, *J. Chem. Soc., Dalton Trans.*, 1986, 499; (b) E. Benedetti, A. Bavoso, B. Di Blasio, V. Pavone, C. Pedone, and F. Rossi, *Inorg. Chim. Acta*, 1986, **116**, 31.

[221] R. Uson and A. Laguna, *Coord. Chem. Rev.*, 1986, **70**, 1.

[222] M. Melnik and R. V. Parish, *Coord. Chem. Rev.*, 1986, **70**, 157.

[223] A. A. Aitchison and L. J. Farrugia, *Organometallics*, 1986, **5**, 1103.

[224] P. A. Bates, S. S. D. Brown, A. J. Dent, M. B. Hursthouse, G. F. M. Kitchen, A. G. Orpen, I. D. Salter, and V. Sik, *J. Chem. Soc., Chem. Commun.*, 1986, 600.

[225] P. D. Boyle, B. J. Johnson, A. Buchler, and L. H. Pignolet, *Inorg. Chem.*, 1986, **25**, 5.

[226] T. J. Henly, J. R. Shapley, and A. L. Rheingold, *J. Organomet. Chem.*, 1986, **310**, 55.

[227] A. J. Deeming, S. Donovan-Mtunzi, and K. Hardcastle, *J. Chem. Soc., Dalton Trans.*, 1986, 543.

[228] B. F. G. Johnson, J. Lewis, W. J. H. Nelson, M. D. Vargas, D. Braga, K. Henrick, and M. McPartlin, *J. Chem. Soc., Dalton Trans.*, 1986, 975.

[229] S. R. Drake, K. Henrick, B. F. Johnson, J. Lewis, M. McPartlin, and J. Morris, *J. Chem. Soc., Chem. Commun.*, 1986, 928.

[230] J. N. Nicholls, P. R. Raithby, and M. D. Vargas, *J. Chem. Soc., Chem. Commun.*, 1986, 1617.

[231] W. Bos, J. J. Bour, P. P. J. Schlebos, P. Hageman, W. P. Bosman, J. M. M. Smits, J. A. C. van Wietmarschen, and P. T. Beurskens, *Inorg. Chim. Acta*, 1986, **119**, 141.

[232] C. E. Briant, D. I. Gilmour, and D. M. P. Mingos, *J. Chem. Soc., Dalton Trans.*, 1986, 835.

[233] D. M. P. Mingos and R. W. M. Wardle, *J. Chem. Soc., Dalton Trans.*, 1986, 73.

[234] D. I. Gilmour and D. M. P. Mingos, *J. Organomet. Chem.*, 1986, **302**, 127.

Table *Gold-containing cluster complexes*

Complex	Structure of Cluster	Ref.
$CoFe_2Au(\mu_3\text{-COMe})(\mu_3\text{-CO})(CO)_6(PPh_3)(\eta\text{-}C_5H_5)$	$CoFe_2$ triangle, μ-AuPPh$_3$ bridge	223
$Ru_2Au_4(\mu\text{-H})(\mu_3\text{-H})(\mu\text{-Ph}_2PCH_2PPh_2)(CO)_{12}]$	Capped square-based pyramidal metal core structure, with 2 Au and 3 Ru atoms defining the pyramid and an Ru atom capping the 3 Ru atom face	224
$[RhAu_3(H)(CO)(PPh_3)_5](PF_6)$	Au$_3$Rh tetrahedron	225
$[ReAu_5(H_4)_4(PPh_3)_7](PF_6)_2$	Distorted edge-shared bitetrahedron	225
$[Ph_4As]_2[Re_7C(CO)_{21}Au(PPh_3)]$	Metal cluster has bicapped octahedral geometry	226
$[Os_3(C{\equiv}CPh)\{Au(PMe_2Ph)\}(CO)_{10}]$	Butterfly Os$_3$Au group with a wing tip Au atom	227
$[PPh_3Me]^+[Os_{10}C(CO)_{24}Au(PPh_3)]^-$	Au ligand added to capping tetrahedra of $[Os_{10}C(CO)_{24}]^{2-}$ group	228
$[Os_{10}C(CO)_{24}AuBr]^-$	As for ref. 228	229
$[Os_{20}Au(C)_2(CO)_{48}]^{2-}$	See (9) for structure of core	229
$[Ir_4(CO)_{11}PhPPPhIr_4(CO)_9(AuPEt_3)_2]$	Ir$_4$ and Ir$_4$Au$_2$ units linked by a diphosphane ligand	230
$[Pt_2(PPh_3)_4(\mu\text{-SAuCl})_2]\cdot 2CH_2Cl_2$	Both bridging S atoms of the nearly planar Pt$_2$S$_2$ core are bonded to AuCl	231
$[Pt_2(PPh_3)_4(\mu\text{-S})(\mu\text{-SAuPPh}_3)][NO_3]\cdot 0.5H_2O$	AuPPh$_3$ is bonded to one of the bridging S atoms of the hinged square planar Pt$_2$S$_2$ core	231
$[Pt_2Au_2(PPh_3)_4(CNC_6H_3Me_2\text{-}2,6)_4][PF_6]_2$	Distorted flattened butterfly core	232
$[Pt_3Au(\mu\text{-CO})_2(\mu\text{-SO}_2)\{P(C_6H_{11})_3\}_4PF_6$	Tetrahedral Pt$_3$Au cluster	233
$[Pt_3Au(\mu\text{-SO}_2)_2(\mu\text{-Cl})\{p(C_6H_{11})_3\}_3\{P(C_6H_4F\text{-}p)_3\}]$	Tetrahedral Pt$_3$Au cluster	233
$[(C_6F_5)Au\{Ph_2P\overline{CH(PPh_2Me)}\}Au(C_6F_5)]$	Contains two separate Au atoms	235a
$[Au(CH_2)_2PPh_2]_2S_8$	Ring containing two units of Au—Au [distance 2.662(1) Å] with two S$_4$ units bridging the two Au$_2$ units	235b
$[Au(CH_2)_2PPh_2]_2S_9$	Ring containing two units of Au—Au [distances 2.649(9) Å] with S$_5$ and S$_4$ units bridging the two Au$_2$ units	235b
$[Au(PhNNNPh)]_4$	Contains the four Au atoms at the corner of a rhombus	236
$[Au_6(PPh_3)_6][NO_3]_2$	Pair of tetrahedra sharing a common edge	237

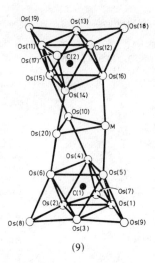

(9)

Gold Clusters. The structures of several gold clusters have been reported[235] (Table) including those shown in (10)[235b] and (11).[236] Novel modification of Au, Rh, and Ru—M_{13} clusters as building blocks of 'superclusters' have been reported.[238] The topology of gold clusters has been discussed.[239]

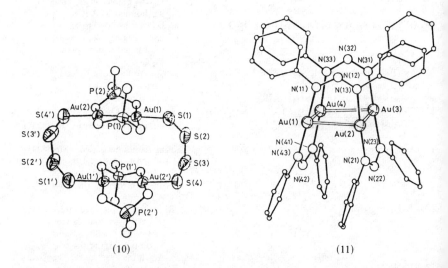

(10)   (11)

[235] (a) R. Usón, A. Laguna, M. Laguna, T. Lázaro, A. Morata, P. G. Jones, and H. M. Sheldrick, *J. Chem. Soc., Dalton Trans.*, 1986, 669; (b) J. P. Fackler, jun. and L. C. Porter, *J. Am. Chem. Soc.*, 1986, **108**, 2750.
[236] J. Beck and J. Strähle, *Angew. Chem., Int. Ed. Engl.*, 1986, **25**, 95.
[237] C. E. Briant, K. P. Hall, D. M. P. Mingos, and A. C. Wheeler, *J. Chem. Soc., Dalton Trans.*, 1986, 687.
[238] G. Schmid and N. Klein, *Angew. Chem., Int. Ed. Engl.*, 1986, **25**, 922.
[239] R. B. King, *Inorg. Chim. Acta*, 1986, **116**, 109.

Gold(I). The structures of the tetra-n-butylammonium salts of the $AuCl_2^-$, $AuBr_2^-$, and AuI_2^- ions have been discussed.[240] A report of a single-crystal luminescence study of the layered compound $KAu(CN)_2$[241] has appeared. The competition of thiols and cyanide for Au^I has been investigated.[242] The structure of (t-butyl-isocyanide)chlorogold(I) has been reported[243] as has the structure of a mononuclear Au^I complex containing a covalently bound phosphonium ylide ligand.[244] The complexes $R_3P-Au-Cl$ (R_3P = $TolPh_2P$ or Me_2PhP) have been studied.[245] The structures of (isocyanato)triphenylphosphine)gold(I),[246] (fulminato)(triphenylphosphine)gold(I),[246] and cyano(triethylphosphine)gold(I)[247] have been investigated. Halogeno(ylide)gold(I) complexes have been synthesized.[248] Exchange kinetics of Au^I phosphine complexes have been studied using $^{31}P-\{^1H\}$ n.m.r. spectra.[249] In (η^5-(C_5H_4CHO)Cr(CO)$_3$(Au(PPh$_3$)) a linear $Cr-Au-P$ geometry is found.[250] $\xi(Et)_2AuBr_2$ has been investigated.[251] The heterotrimetallic species $[FeL(PPh_2)_2C(AuPPh_3)_2(CNPh)_3]^{n+}$ (L = I, n = 1; L = CNPh, n = 2) have been obtained.[252]

Gold(II). A number of papers on dinuclear gold ylide complexes have appeared, including a vibrational study,[253] structural studies,[254a,b] and reactivity.[255] Mono- and di-nuclear Au^I, Au^{II}, and Au^{III} perhalogenoaryl complexes and the structure of an Au^{II} complex have been presented.[256] $[Au_2(C_7H_5O_2)_2\{P(CH_2)_2Ph_2\}_2]C_4H_8O$ is reported as the first dinuclear Au^{II} complex possessing bonds to oxygens (12).[257]

Gold(III). The reduction of $Au(NH_3)_4^{3+}$ by iodide has been investigated[258] as has the recovery of Au^{III} in chloride media.[259] Leaving group selectivity in reductive elimination from organogold(III) complexes takes place when R is alkenyl, aryl, and furyl.[260] The structures of trichloro(diethylenetriamine)gold(III)[261]

[240] P. Braunstein, A. Müller, and H. Bogge, *Inorg. Chem.*, 1986, **25**, 2104.
[241] N. Nagasundaram, G. Roper, J. Biscoe, J. W. Chai, H. H. Patterson, N. Blom, and A. Ludi, *Inorg. Chem.*, 1986, **25**, 2947.
[242] G. Lewis and C. F. Shaw III, *Inorg. Chem.*, 1986, **25**, 58.
[243] D. S. Eggleston, D. F. Chodosh, L. L. Webb, and L. L. Davis, *Acta Crystallogr.*, 1986, **C42**, 36.
[244] L. C. Porter, H. Knachel, and J. P. Fackler, jun., *Acta Crystallogr.*, 1986, **C42**, 1125.
[245] U. Schubert and J. Meyer, *J. Organomet. Chem.*, 1986, **303**, C5.
[246] W. P. Bosman, W. Bos, J. M. M. Smits, P. T. Beurskens, J. J. Bour, and J. J. Steggerda, *Inorg. Chem.*, 1986, **25**, 2093.
[247] A. L. Hormann, C. F. Shaw III, D. W. Bennett, and W. M. Reiff, *Inorg. Chem.*, 1986, **25**, 3953.
[248] R. Usón, A. Laguna, M. Laguna, A. Usón, and M. C. Gimeno, *Inorg. Chim. Acta*, 1986, **114**, 91.
[249] S. Al-Baker, W. E. Hill, and C. A. McAuliffe, *J. Chem. Soc., Dalton Trans.*, 1986, 1297.
[250] F. Edelmann, S. Töfke, and U. Behrens, *J. Organomet. Chem.*, 1986, **309**, 87.
[251] E. Amberger, K. Polborn, and H. Fuchs, *Angew. Chem., Int. Ed. Engl.*, 1986, **25**, 727.
[252] V. Riera and J. Ruiz, *J. Organomet. Chem.*, 1986, **310**, C36.
[253] R. J. H. Clark, J. H. Tocher, J. P. Fackler, jun., R. Neira, H. H. Murray, and H. Knackel, *J. Organomet. Chem.*, 1986, **303**, 437.
[254] (a) H. H. Murray, J. P. Fackler, jun., A. M. Mazany, L. C. Porter, J. Shain, and L. R. Falvello, *Inorg. Chim. Acta*, 1986, **114**, 171; (b) H. H. Murray III, J. P. Fackler, jun., L. C. Porter, and A. M. Mazany, *J. Chem. Soc., Dalton Trans.*, 1986, 321.
[255] H. H. Murray and J. P. Fackler, jun., *Inorg. Chim. Acta*, 1986, **115**, 207.
[256] R. Usón, A. Laguna, M. Laguna, M. N. Fraile, P. G. Jones, and G. M. Sheldrick, *J. Chem. Soc., Dalton Trans.*, 1986, 291.
[257] L. C. Porter and J. P. Fackler, jun., *Acta Crystallogr.*, 1986, **C42**, 1128.
[258] L. I. Elding and L. H. Skibsted, *Inorg. Chem.*, 1986, **25**, 4048.
[259] E. Borgarello, N. Serpone, G. Emo, R. Harris, E. Pelizzetti, and €. Minero, *Inorg. Chem.*, 1986, **25**, 4449.
[260] S. Komiya, S. Ozaki, and A. Shibue, *J. Chem. Soc., Chem. Commun.*, 1986, 1555.
[361] R. C. Elder and J. W. Watkins II, *Inorg. Chem.*, 1986, **25**, 223.

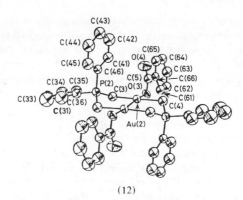

(12)

$[(2\text{-}Ph\overline{N=NC_6H_4})Au(2\text{-}C_6H_4CH_2\overline{NMe_2})][Au_4Cl_4]$ (13)[262] and guanidinium tetrachloroaurate(III)[263] have been described. Au^I and Au^{III} *ortho*-nitrophenyl complexes[264] and methylene-bridged dinuclear Au^{III} complexes with terminal and bridging ylide ligands[265] have been reported.

Mössbauer Spectroscopy. A number of studies on ^{197}Au Mössbauer spectroscopy have appeared.[266—271] These include studies on mixed-metal cluster compounds,[267] Au^I derivatives of five- or six-membered nitrogen-containing heterocycles,[268] three-coordinate Au^I complexes,[269] isocyanide, carbene, or methanide derivatives of gold.[270]

Gold Compounds of Biological Interest. Studies related to anti-tumour activity involving $Au^{I\,272a}$ and $Au^{III\,272b}$ complexes have been discussed. Studies on the anti-arthritic gold drugs 'Myocrisin'[273] and 'Auranafin'[274—276] have appeared. Reactions of triethylphosphine Au^I complexes with haem proteins have been reported.[277] The accumulation of elemental gold on the alga *Chlorella vulgaris* has been studied[278] as has the characterization of protein-bound gold in rat urine following aurothio-

[262] J. Vicente, M.-T. Chicote, M. D. Bermudez, M. J. Sanchez-Santano, P. G. Jones, C. Fittschen, and G. M. Sheldrick, *J. Organomet. Chem.*, 1986, **310**, 401.

[263] H. Kiriyama, N. Matsushita, and Y. Yamagata, *Acta Organomet. Chem.*, 1986, **C42**, 277.

[264] J. Vicente, A. Arcas, M. Mora, X. Solans, and M. Font-Altaba, *J. Organomet. Chem.*, 1986, **309**, 369.

[265] H. Schmidbaur and C. Hartmann, *Angew. Chem., Int. Ed. Engl.*, 1986, **25**, 575.

[266] S. S. D. Brown, L. S. Moore, R. V. Parish, and I. D. Salter, *J. Chem. Soc., Chem. Commun.*, 1986, 1453.

[267] A. J. Batchelor, T. Birchall, and R. C. Burns, *Inorg. Chem.*, 1986, **25**, 2009.

[268] F. Bonati, A. Burini, B. R. Pietroni, S. Calogero, and F. E. Wagner, *J. Organomet. Chem.*, 1986, **309**, 363.

[269] R. Usón, A. Laguna, A. Navarro, R. V. Parish, and L. S. Moore, *Inorg. Chim. Acta*, 1986, **112**, 205.

[270] G. Banditelli, F. Bonati, S. Calogero, G. Valle, F. E. Wagner, and R. Wordel, *Organometallics*, 1986, **5**, 1346.

[271] H. Schmidbaur, Th. Pollok, R. Herr, F. Wagner, R. Bau, J. Riede, and G. Müller, *Organometallics*, 1986, **5**, 566.

[272] (*a*) C. F. Shaw III, A. Beery, and H. C. Stocco, *Inorg. Chim. Acta*, 1986, **123**, 213; (*b*) Y. Mizuno and S. Komiya, *Inorg. Chim. Acta*, 1986, **125**, L13.

[273] A. A. Isab, *J. Chem. Soc., Dalton Trans.*, 1986, 1049.

[274] D. J. Ecker, J. C. Hempel, B. M. Sutton, R. Kirsch, and S. T. Crooke, *Inorg. Chem.* 1986, **25**, 3139.

[275] C. F. Shaw III and J. E. Laib, *Inorg. Chim. Acta*, 1986, **123**, 197.

[276] M. T. Coffer, C. F. Shaw III, M. K. Eidsness, J. W. Watkins II, and R. C. Elder, *Inorg. Chem.*, 1986, **25**, 333.

[277] M. C. Grootveld, G. Otiko, P. J. Sadler, and R. Cammack, *J. Inorg. Biochem.*, 1986, **27**, 1.

[278] M. Hosea, B. Greene, R. McPherson, M. Henzl, M. D. Alexander, and D. W. Darnall, *Inorg. Chim. Acta*, 1986, **123**, 161.

malate administration.[279] The model application of a thiol-reactive gold cluster for the specific labelling of cysteine residues in proteins has been suggested.[280]

2 Zinc, Cadmium, and Mercury

As large a covering of the 1986 papers dealing with Zn, Cd, and Hg as possible is presented here, but papers covering more than one element in the group are cited in the first occasion only, *viz.* a paper covering Zn, Cd, and Hg is cited only in the zinc section.

Zinc.—Neutral and ionic metal methyl bond energies of zinc have been investigated.[281] Studies of propan-2-ol[282] and CO[283] on single-crystal surfaces of ZnO have been presented. The halide ion reactivity in $[Zn_2Cr(OH)_6]X \cdot nH_2O$ has been studied.[284] Fourteen new ZnS polytypes were identified[285] as was $ZnH_{10}(AsO_4)_4$.[286] The molecular structures of $ZnCl_2$, $ZnBr_2$, and ZnI_2 were reinvestigated using electron diffraction.[287] In $[C_6H_9N_2S_3][ZnCl_3(H_2O)]$ the $ZnCl_3(H_2O)^-$ ions are distorted.[288] Rechargeable $Zn/ZnSO_4/MnO_2$-type cells are reported.[289] The structures of $Na_6[Zn(SO_4)_4(H_2O)_2]$[290] and $ZnSeO_3$[291] have been reported, the latter contains trigonal-bipyramidal Zn.

The mass spectrum of zinc acetate has been reinvestigated.[292] The structures of some dinuclear (13) and trinuclear zinc carboxylate complexes have been reported[293] as have those of $Zn(C_4H_8N_2O_6)$[294] and $[Zn(C_4H_8NO_2)_2(H_2O)_2]$-$2[Zn(C_4H_8NO_2)_2(H_2O)]$.[295] Quasi-tetrahedral Zn^{II} has been found in $Zn(OAr)_2(THF)_2$ (Ar = 2,4,6-tri-t-butylphenyl).[296] Nearly regular octahedral coordination is found for the Zn atom in $[Zn(H_2O)_5][Cu(C_{14}H_{18}N_2O_8)]H_2O$.[297] A $Zn\cdots Zn$ separation of 3.247(3) Å has been established in $[Zn_2(\mu-O_2CCH:CHMe)_3(O_2CCH:CHMe)]$.[298]

The structure of $[Zn(\text{nic-}N)_2(H_2O)_4]$ (14) has been reported.[299] ^{13}C N.m.r. studies involving Zn^{II} with triethylenetetramine[300a] and 2,2'2''-triaminotriethylamine[300b] have been carried out and the effect of changes in the ligands backbone moiety on

[279] C. F. Shaw III, N. Schaeffer-Memmel, and D. Krawczak, *J. Inorg. Biochem.*, 1986, **26**, 185.
[280] D. Safar, L. Bolinger, and J. S. Leigh, jun., *J. Inorg. Biochem.*, 1986, **26**, 77.
[281] R. Georgiadis and P. B. Armentrout, *J. Am. Chem. Soc.*, 1986, **108**, 2120.
[282] P. Berlowitz and H. H. Kung, *J. Am. Chem. Soc.*, 1986, **108**, 3532.
[283] A. B. Anderson and J. A. Nichols, *J. Am. Chem. Soc.*, 1986, **108**, 1385.
[284] K. J. Martin and T. J. Pinnavaia, *J. Am. Chem. Soc.*, 1986, **108**, 541.
[285] V. Medizadeh and S. Mardix, *Acta Crystallogr.*, 1986, **C42**, 518.
[286] D. Tran Qui and M. Chiadmi, *Acta Crystallogr.*, 1986, **C42**, 391.
[287] M. Hargittai, J. Tremmel, and I. Hargittai, *Inorg. Chem.*, 1986, **25**, 3163.
[288] M. Kubiak and T. Glowiak, *Acta Crystallogr.*, 1986, **C42**, 419.
[289] T. Yamamoto and T. Shoji, *Inorg. Chim. Acta*, 1986, **117**, L27.
[290] M. J. Heeg, W. Redman, and R. Frech, *Acta Crystallogr.*, 1986, **C42**, 949.
[291] F. C. Hawthorne, T. S. Ercit, and L. A. Groat, *Acta Crystallogr.*, 1986, **C42**, 1285.
[292] G. C. Di Donato and K. L. Busch, *Inorg. Chem.*, 1986, **25**, 1551.
[293] W. Clegg, I. R. Little, and B. P. Straughan, *J. Chem. Soc., Dalton Trans.*, 1986, 1283.
[294] R. E. Marsh, *Acta Crystallogr.*, 1986, **C42**, 1327.
[295] E. E. Castellano, G. Oliva, J. Zukerman-Schpector, and R. Calvo, *Acta Crystallogr.*, 1986, **C42**, 21.
[296] R. L. Geerts, J. C. Huffman, and K. G. Caulton, *Inorg. Chem.*, 1986, **25**, 1803.
[297] A. Fuertes, C. Miravitlles, E. Molins, E. Escriva, and D. Beltran, *Acta Crystallogr.*, 1986, **C42**, 421.
[298] W. Clegg, I. R. Little, and B. P. Straughan, *Acta Crystallogr.*, 1986, **C42**, 919.
[299] W. E. Broderick, M. R. Pressprich, U. Geiser, R. D. Willett, and J. I. Legg, *Inorg. Chem.*, 1986, **25**, 3372.
[300] (*a*) S. P. Dagnall, D. N. Hague, and A. D. Moreton, *J. Chem. Soc., Dalton Trans.*, 1986, 1499; (*b*) ibid., 1986, 1505.

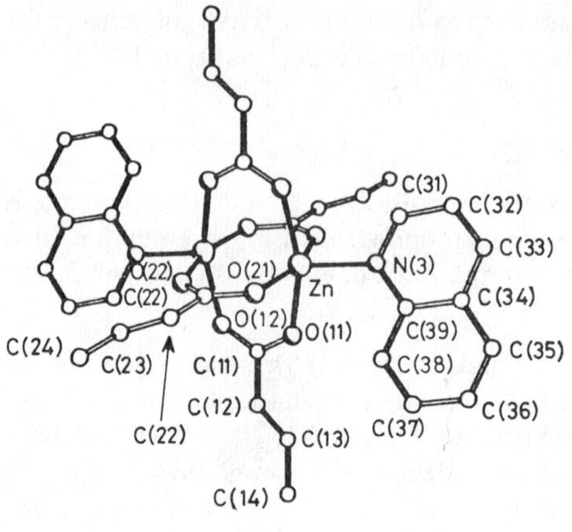

(13)

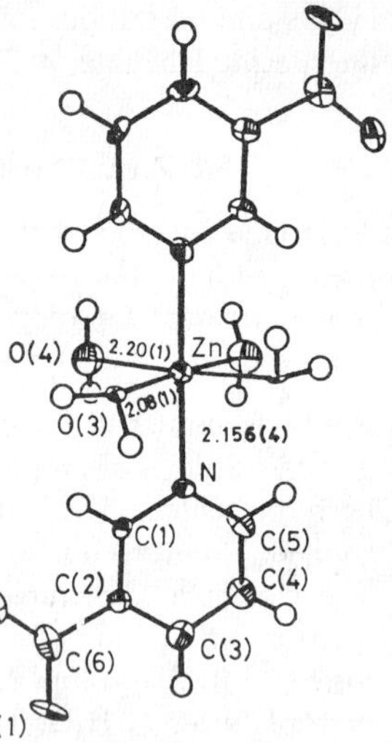

(14)

the formation constants of 1:1 Zn, Cu, and Cd chelates have been studied.[301] Zn^{II} salts of *N*-isopropyl- and *N*-cyclohexyl-2-pyrrolidinone have been studied.[302] The structures of bis(*S*-methylbenzylidenehydrazinecarbodithioata)zinc(II) and bis(*S*-methylisopropylidenehydrazinecarbodithioato)zinc(II) have been reported.[303] High spin → low spin relaxation kinetics have been found in [Zn$_{1-x}$Fe$_x$(ptz)$_6$]-(BF$_4$)$_2$(ptz = 1-propyltetrazole).[304]

The structures of [Ph$_4$As]$_2$[Zn(mnt)$_2$],[305a] [Ph$_4$As]$_2$[(ZnC$_2$O$_2$S$_2$)$_2$],[305b] [Zn$_4${S$_2$P(OEt)$_2$}$_6$S],[305c] Zn[S$_2$P(OPr^i)$_2$]$_2$·biby (15),[305d] [Zn{S$_2$P(OPr^i)$_2$}$_2$]· H$_2$NCH$_2$CH$_2$NH$_2$ (16),[305e] and [Zn{S$_2$P(OPr^i)$_2$}$_2$]NC$_5$H$_5$[305e] were described in the review year.

In [Ph$_4$As][Zn(mnt)(Et$_2$dtc)] the ZnS$_4$ unit is nearly tetrahedral.[306]

The preparations and molecular structures of bis(pentamethylcyclopentadienyl)zinc and bis(trimethylsilylcyclopentadienyl)zinc have been presented.[307] A number of papers describing aspects of Zn^{II}-catalysed reactions have appeared.[308a—e].

[301] J. E. Powell, D. R. Ling, and P. K. Tse, *Inorg. Chem.*, 1986, **25**, 587.

[302] B. A. Wilson and S. K. Madan, *Inorg. Chim. Acta*, 1986, **114**, 9.

[303] K. D. Onan, G. Davis, M. A. El-Sayed, and A. El-Toukhy, *Inorg. Chim. Acta*, 1986, **113**, 109.

[304] A. Hauser, P. Gutlich, and N. Spierling, *Inorg. Chem.*, 1986, **25**, 4245.

[305] (*a*) J. Stach, R. Kirmse, J. Sieler, U. Abram, W. Dietzsch, R. Böttcher, L. K. Hansen, H. Vergoosen, M. C. M. Gribnau, and C. P. Keijzers, *Inorg. Chem.*, 1986, **25**, 1369; (*b*) L. Golic, N. Bulc, W. Dietzsch, *Acta Crystallogr.*, 1986, **C42**, 811; (*c*) P. G. Harrison, M. J. Begley, T. Kikabhai, and F. Killer, *J. Chem. Soc., Dalton Trans.*, 1986, 925; (*d*) ibid., 1986, 929; (*e*) M. G. B. Drew, M. Hasan, R. J. Hobson, and D. A. Rice, *J. Chem. Soc., Dalton Trans.*, 1986, 1161.

[306] I. Leban, L. Golic, R. Kirmse, J. Stach, U. Abram, H.-J. Sieler, W. Dietzsch, H. Vergoosen, and K. P. Keijzers. *Inorg. Chim. Acta*, 1986, **112**, 107.

[307] R. Blom, J. Boersma, P. H. M. Budzelaar, B. Fischer, A. Haaland, H. V. Volden, and J. Weidlein, *Acta Chem. Scand., Ser. A*, 1986, **40**, 113.

[308] (*a*) B. Szpoganicz and A. E. Martell, *Inorg. Chem.*, 1986, **25**, 327; (*b*) R. Csuk, A. Furstner, B. I. Glanzer, and H. Weidmann, *J. Chem. Soc., Dalton Trans.*, 1986, 1149; (*c*) J. Suh, O. Han, and B. Chang, *J. Am. Chem. Soc.*, 1986, **108**, 1839; (*d*) P. Knochel and J. F. Normant, *J. Organomet. Chem.*, 1986, **309**, 1; (*e*) Y. Wakita, T. Yasunaga, M. Akita, and M. Kojima, *J. Organomet. Chem.*, 1986, **301**, C17.

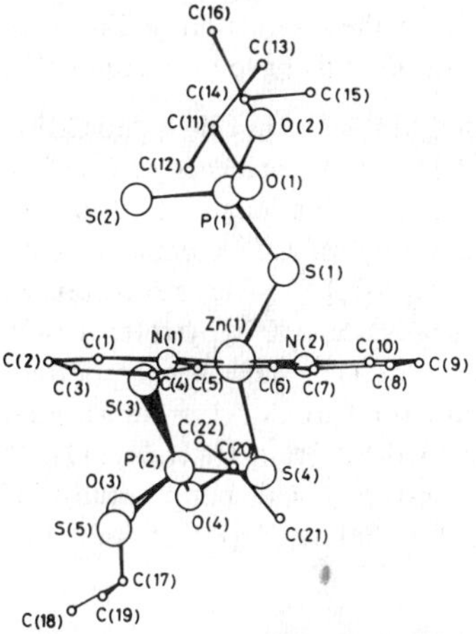

(15)   (16)

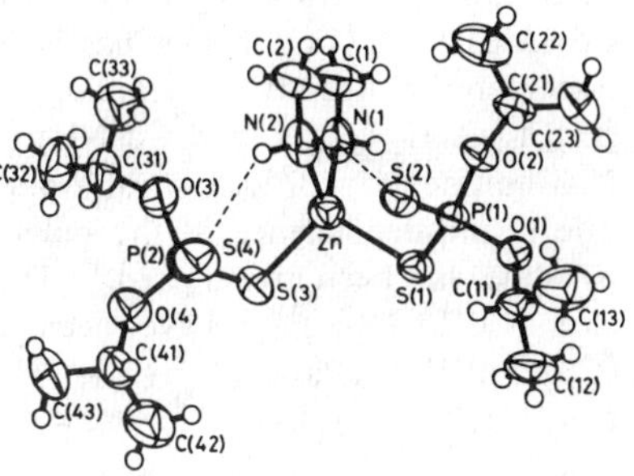

Macrocyclic Zinc Compounds. The structures of ZnTPP[309] and AgTPP[309] and spectroscopic studies on ZnTPP[310] and oxygen donor ligands bonding to ZnTPP[311] have been reported. Complex formation between viologen and Zn—TPPS$_3$$^{3-}$ by the addition of surfactant micelles has been studied.[312] The interaction of Zn^{II} and Cd^{II} with dibenzo-substituted macrocyclic and open-chain tetramines has been described[313] as has the catalytic hydrolysis of a phosphate triester by a tetrazamacrocyclic Zn^{II} complex.[314] The redox properties of Zn^{II} tetra-N-methyl-2,3-pyridinoporphyrazine in aqueous solution have been investigated.[315]

Zinc Compounds of Biological Interest. The interactions of Zn^{II} and Cu^{II} with ε-ATP and ATP have been discussed.[316] The coordination of Zn^{II} to vincristine,[317] cimetidine sulphoxide,[318] and phytic acid[319] and also to the sugars D-glucuronate[320a] and L-arabinose[320b] has been reported. The binding of Zn^{II} to mercaptosuccinate has been studied.[321] ZnTPP models for cytochrome c oxidase have been suggested.[322] Rates of π-electron oxidation and reduction of Zn^{II} porphyrins, chlorins, and

[309] W. R. Scheit, J. U. Mondal, C. W. Eigenbrot, A. Adler, L. R. Radonovich, and J. L. Hoard, *Inorg. Chem.*, 1986, **25**, 795.

[310] D. Kim, J. Terner, and T. G. Spiro, *J. Am. Chem. Soc.*, 1986, **108**, 2097.

[311] J. V. Nardo and J. H. Dawson, *Inorg. Chim. Acta,* 1986, **123**, 9.

[312] I. Okura, T. Kita, S. Aono, and N. Kaji, *Inorg. Chim. Acta*, 1986, **116**, L53.

[313] K. R. Adam, C. W. G. Ansell, K. P. Dancey, L. A. Drummond, A. J. Leong, L. F. Lindoy, and P. A. Tasker, *J. Chem. Soc., Dalton Trans.*, 1986, 1011.

[314] S. A. Gellman, R. Petter, and R. Breslow, *J. Am. Chem. Soc.*, 1986, **108**, 2388.

[315] M.-C. Richoux and Z. M. Abou-Gamra, *Inorg. Chim. Acta*, 1986, **118**, 115.

[316] V. Scheller-Krattiger and H. Sigel, *Inorg. Chem.*, 1986, **25**, 2628.

[317] K. Burger, M. Veber, P. Sipos, Z. Galbacs, I. Horvath, S. Szepesi, G. T. Nagy and J. Siemroth, *Inorg. Chim. Acta*, 1986, **124**, 175.

[318] E. Freijanes and G. Berthon, *Inorg. Chim. Acta*, 1986, **124**, 141.

[319] G. J. Martin and W. J. Evans, *J. Inorg. Biochem.*, 1986, **26**, 169.

[320] (a) H.-A. Tajmir-Riahi, *J. Inorg. Biochim.*, 1986, **26**, 23; (b) H.-A. Tajmir-Riahi, *J. Inorg. Biochim.*, 1986, **27**, 65.

[321] M. Filella, A. Izquierdo, and E. Casassas, *J. Inorg. Biochem.*, 1986, **28**, 1.

[322] C. T. Brewer and G. Brewer, *J. Inorg. Biochem.*, 1986, **26**, 247.

isobacteriochlorins[323] have been reported. Models for the interaction of Zn^{II} with DNA have been proposed.[324] [ZnFe^{III}] haemoglobin hybrids have been studied.[325]

Cadmium.—The optical spectra of exchange-coupled Mn^{II} pairs in $CdCl_2$ and $CdBr_2$ has been reported.[326] The photo-decomposition of H_2S in the presence of CdS has been studied.[327] The Cd_2^{2+} ion has been found to have a bond length of 2.576 Å in $Cd_2(AlCl_4)_2$.[328] ^{113}Cd N.m.r. has been used on a variety of Cd^{II} complexes with N-containing ligands[329] including [Cd(NO$_3$)$_2$(C$_6$H$_8$N$_2$)$_2$].[330] The structures of [CdCl$_2$(C$_{16}$H$_{28}$N$_6$)][331] and Cd[NH$_2$(CH$_2$)$_4$NH$_2$]Ni(CN)$_4$X$_n$ (X = pyrrole, n = 1; X = aniline, n = 3/2; X = N,N-dimethylaniline, n = 1)[332] have been reported. The ligand-dependence of the polymeric structure for CdLCl$_2$ (L = substituted pyridine) has been investigated.[333] The structures of Cd(S$_3$C$_5$H$_9$)$_2$·C$_{10}$H$_8$N$_2$ (17)[334a] and [Cd{Et$_2$PS$_2$}$_2$][334b] have been determined and a ^{113}Cd n.m.r. study of [Cd$_4$(SePH)$_x$(TePh)$_{10-x}$]$^{2-}$ (x = 7—10), [Cd$_4$(SPh)$_x$(SePh)$_{10-x}$]$^{2-}$ (x = 0—10), and [Cd$_x$Zn$_{4-x}$(SPh)$_{10}$]$^{2-}$ (x = 2—4)[334c] reported.

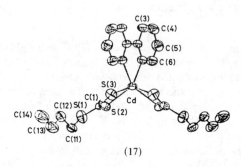

(17)

Cadmium Compounds of Biological Interest. ^{113}Cd N.m.r. study of Cd(C$_5$H$_{12}$N$_2$S)$_2$(NO$_3$)$_2$ has been reported.[335] The isolation and characterization of metallothionein from guinea-pig liver has been presented.[336]

[323] S. H. Strauss and R. G. Thompson, *J. Inorg. Biochem.*, 1986, **27**, 173.
[324] S. K. Miller, L. G. Marzilli, S. Dorre, P. Kollat, R.-D. Stigler, and J. J. Stezowski, *Inorg. Chem.*, 1986, **25**, 4272.
[325] S. E. Peterson-Kennedy, J. L. McGourty, J. A. Kalweit, and B. M. Hoffman, *J. Am. Chem. Soc.*, 1986, **108**, 1739.
[326] P. J. McCarthy and H. U. Gudel, *Inorg. Chem.*, 1986, **25**, 838.
[327] E. Borgarello, N. Serpone, M. Gratzel, and E. Pelizzetti, *Inorg. Chim. Acta*, 1986, **112**, 197.
[328] R. Faggiani, R. J. Gillespie, and J. E. Vekris, *J. Chem. Soc., Chem. Commun.*, 1986, 517.
[329] M. Munakata, S. Kitagawa, and F. Yagi, *Inorg. Chem.*, 1986, **25**, 964.
[330] P. F. Rodesiler, N. G. Charles, E. A. H. Griffith, K. Lewinski, and E. L. Amma, *Acta Crystallogr.*, 1986, **C42**, 538.
[331] F. Paap, A. Erdonmez, W. L. Driessen, and J. Reedijk, *Acta Crystallogr.*, 1986, **C42**, 783.
[332] S.-I. Nishikiori and T. Iwamoto, *Inorg. Chem.*, 1986, **25**, 788.
[333] M. Goodgame and J. N. Okey, *Inorg. Chim. Acta*, 1986, **114**, 179.
[334] (a) J. J. Black, F. W. B. Einstein, P. C. Hayes, R. Kumar, and D. G. Tuck, *Inorg. Chem.*, 1986, **25**, 4181; (b) H. Wunderlich, *Acta Crystallogr.*, 1986, **C42**, 631; (c) P. A. W. Dean and J. J. Vittal, *Inorg. Chem.*, 1986, **25**, 514.
[335] R. S. Honkonen, P. S. Marchetti, and P. D. Ellis, *J. Am. Chem. Soc.*, 1986, **108**, 912.
[336] M. J. Stillman, A. J. C. Law, E. M. K. Lui, and M. G. Cherian, *Inorg. Chim. Acta*, 1986, **124**, 29.

Mercury.—*Mercury(I).* The structures of three basic Hg^I nitrates have been described.[337] Hg$_2$(caffeine)$_2$(NO$_3$)$_2$ contains Hg—Hg ions linearly bonded to N-9.[338]

Mercury(II). The solvation thermodynamics of the neutral Hg$^{II}_n$ halides in different solvents has been studied.[339] The structures of [Hg$_2$PbI$_2$S$_2$],[340a] [HgI$_2$(C$_7$H$_7$NS)]$_2$,[340b] [C$_{12}$H$_{24}$O$_6$·HgI$_2$],[340c] and [C$_{12}$H$_{24}$O$_6$]·2[C$_{13}$H$_{13}$S]$^+$· [Hg$_2$I$_6$]$^{2-}$ [340d] have been reported. In NNMo(CO)$_3$(HgCl)Cl (NN = biby, phen, or dmp) the Mo—Hg bond is retained.[341] Mercury halides have been shown to react with tricarbonyl(fulvene)chromium,[342a] Mo^{II} dicarbonyl complexes[342b] and ruthenium clusters.[342c] Octahedral and monocapped trigonal prismatic coordinations are found for Hg^{II} in HgTeO$_3$.[343] The structure of Hg(NH$_4$)$_2$Na$_2$(P$_3$O$_9$)$_2$ has been reported.[344] The reaction of Hg^{II} trifluoroacetate with disilenes has been reported[345a] as have the structures of [M{NC$_5$H$_4$C(SiMe$_3$)$_2$-2}$_2$] (M = Hg, Cd, or Zn).[345b] The rate and extent of Si—C bond cleavage by Hg^{II} species has been studied.[346] Several studies on methyl- and phenyl-Hg^{II} derivatives have appeared[347a—g] including the structure of [Ru(CO)$_2${C(C≡CPh)=C(Ph)HgCl}-Cl(PMe$_2$Ph)$_2$] (18).[347g] Spiroacetals are obtained from dienones and hydroxyenones

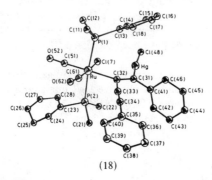

(18)

[337] B. Kamenar, D. Matkovic-Calogovic, and A. Nagl, *Acta Crystallogr.*, 1986, **C42**, 385.

[338] M. A. Romero-Molina, E. Calacio-Rodriguez, J. Ruiz-Sanchez, J. M. Salas-Peregrin, and F. Nieto, *Inorg. Chim. Acta*, 1986, **123**, 133.

[339] I. Persson, M. Landgren, and A. Marton, *Inorg. Chim. Acta*, 1986, **116**, 135.

[340] (a) R. Blachnik, W. Buchmeier, and H. A. Dreisbach, *Acta Crystallogr.*, 1986, **C42**, 515; (b) W. Hiller, A. Castineiras, A. Arquero, and J. R. Masaguer, *Acta Crystallogr.*, 1986, **C42**, 151; (c) D. A. Pears, J. F. Stoddart, J. Crosby, B. L. Allwood, and D. J. Williams, *Acta Crystallogr.*, 1986, **C42**, 51; (d) *ibid.*, 1986, **C42**, 804.

[341] A. Lopez, M. Panizo, and M. Cano, *J. Organomet. Chem.*, 1986, **311**, 145.

[342] (a) F. Edelmann, P. Behrens, S. Behrens, and U. Behrens, *J. Organomet. Chem.*, 1986, **309**, 109; (b) M. A. Lobo, M. F. Perpinan, M. P. Pardo, and M. Cano, *J. Organomet. Chem.*, 1986, **299**, 197; (c) E. Rosenberg, D. Ryckman, I-N. Hsu, and R. W. Gellert, *Inorg. Chem.*, 1986, **25**, 194.

[343] V. Kramer and G. Brandt, *Acta Crystallogr.*, 1986, **C42**, 917.

[344] M. T. Averbuch-Pouchot and A. Durif, *Acta Crystallogr.*, 1986, **C42**, 932.

[345] (a) C. Zybill and R. West, *J. Chem. Soc., Dalton Trans.*, 1986, 857; (b) M. J. Henderson, R. I. Papasergio, C. L. Raston, A. H. White, and M. F. Lappert, *J. Chem. Soc., Chem. Commun.*, 1986, 672.

[346] J. D. Nies, J. M. Bellama, and N. Ben-Zvi, *Inorg. Chim. Acta*, 1986, **118**, 1.

[347] (a) A Castineiras, W. Hiller, J. Strahle, J. Bravo, J. S. Casas, M. Gayoso, and J. Sordo, *J. Chem. Soc., Dalton Trans.*, 1986, 1945; (b) J. Mink, D. K. Breitinger, and W. Kress, *J. Organomet. Chem.*, 1986, **301**, 1; (c) E. R. T. Tiekink, *Inorg. Chim. Acta*, 1986, **112**, L1; (d) J. Bravo, J. S. Casas, Y. P. Mascarenhas, A. Sanchez, C. de O. P. Santos, and J. Sordo, *J. Chem. Soc., Chem. Commun.*, 1986, 1100; (e) S. Fukuzumi, S., Kurodo, and T. Tanaka, *J. Chem. Soc., Chem. Commun.*, 1986, 1553; (f) N. A. A. Al-Jabar, J. Bowen Jones, D. S. Brown, and A. G. Massey, *Acta Crystallogr.*, 1986, **C42**, 425; (g) Z. Dauter, R. J. Mawby, C. D. Reynolds, and D. R. Saunders, *J. Chem. Soc., Dalton Trans.*, 1986, 433.

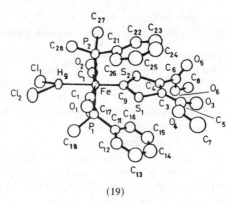

(19)

by Hg^{II} cyclization.[348] A dinuclear complex forms in the reaction of [$(H_2O)_5CrCH_2CN$]$^{2+}$ with Hg^{II}.[349] The structures of [$Cl_2HgFe(CO)_2(PMe_2Ph)_2$-$\{CS_2C_2(CO_2Me)_2\}$] (19)[350a] and [$Hg_9Co_6(CO)_{18}$[350b] have been reported as have those of [$HgCu(\mu\text{-}SCN)_3(dmtp)_2$]^{350c} and Hg($C_4H_8Se$)]$Cl_2$.[350d] A ^{31}P CP-MAS n.m.r. study of phosphine complexes of Hg^{II} has appeared.[351] Mixed Hg^{II}-tetraphos-phine complexes of the type [Hg(tripod)L]$^{+ and 2+}$ (tripod = MeC(CH_2PPh_2)$_3$ (L = anionic or neutral P ligand) have been discussed.[352]

Mercury Compounds of Biological Interest. A mixed-ligand thymidine–Hg–quanosine complex has been suggested as a putative Hg^{II} interstrand cross-linking structure of DNA.[353] The structures of [$(MeHg)_5Ad_2$](NO_3)$_2$·$3H_2O$ and [MeHg)$_3Ad$)]-(NO_3)$_2$·H_2O both contain [$(MeHg)_3Ad$]$^{2+}$ ions.[354]

[348] W. Kitching, J. A. Lewis, M. T. Fletcher, J. J. De Voss, R. A. I. Drew, and C. J. Moore, *J. Chem. Soc., Chem. Commun.* 1986, 855.

[349] M. J. Sisley and R. B. Jordon, *Inorg. Chem.*, 1986, **25**, 3547.

[350] (a) D. V. Khasnis, H. Le Bozec, P. H. Dixneuf, and R. D. Adams, *Organometallics*, 1986, **5**, 1772; (b) J. M. Ragosta and J. M. Burlitch, *Organometallics*, 1986, **5**, 1517; (c) M. B. Cingi, A. M. Manotti Lanfredi, A. Tiripicchio, J. Reedijk, and J. Haasnoot, *Acta Crystallogr.*, 1986, **C42**, 427; (d) C. Stalhandske and F. Zintl, *Acta Crystallogr.*, 1986, **C42**, 1449.

[351] T. Allman and R. E. Lenknski, *Inorg. Chem.*, 1986, **25**, 3202.

[352] P. Peringer and M. Lusser, *Inorg. Chim. Acta*, 1986, **117**, L25.

[353] E. Buncel, C. Boone, and H. Joly, *Inorg. Chim. Acta*, 1986, **125**, 167.

[354] J.-P. Charland, J. F. Britten, and A. L. Beauchamp, *Inorg. Chim. Acta*, 1986, **124**, 161.

12 Sc, Y, the Lanthanoids and the Actinoids

By J. D. MILLER

Department of Applied Chemistry, University of Aston, Aston Triangle, Birmingham B4 7ET

1 Introduction

The pattern and style of this chapter will be similar to that adopted last year. Thus, scandium will be discussed separately, yttrium and the lanthanoids will be considered together, and the last section will be devoted to the actinoids. Since the number of published papers relating to the chemistry of these elements is so large some areas must, of necessity, be excluded or seriously under-represented. For example, in spite of the extremely promising findings of superconduction at unusually high temperatures for some lanthanoid-containing systems, the large bodies of information relating to electrical, magnetic, or luminescent properties of solid compounds are not considered here. Similarly, the large literature dealing with ternary oxides, phosphides, etc., as well as that for the determination of stability constants for complex formation, is almost completely neglected. By contrast, a few aspects that appear to be of increasing interest or of genuine novelty receive much fuller coverage.

Several new publications and further volumes in existing texts relating to the lanthanoids and actinoids have appeared during 1986. Newly published volumes and supplements to Gmelin's Handbook dealing with compounds formed between Se and Sc, Y, and the Rare Earth Elements; Uranium, UO_2, physical properties and electrochemical behaviour; Compounds of thorium with S, Se, Te, and B; and Co-ordination compounds of thorium have appeared.[1] The most important book appearing during 1986 was clearly the second edition of 'The Chemistry of the Actinide Elements'. This two volume work is an essential library text for anyone interested in the field of f-block chemistry.[2] A more specialized double volume entitled 'Actinides—Chemistry and Physical Properties' has been published. It deals primarily with free atoms, the metals, and a few binary compounds, and so will be of much less interest to most chemists.[3]

The collected papers from three separate conferences held in 1984 or 1985 have now been published in book form. Thus 'Americium and Curium Chemistry and Technology' has come from the International Chemical Congress in Honolulu in 1984.[4] The papers from the International Conference on Rare Earth Developments

---

[1] Gmelin's Handbook of Inorganic Chemistry', 8th Edn., Springer-Verlag, Berlin, 1986.
[2] 'The Chemistry of the Actinide Elements', 2nd Edn., ed. J. J. Katz, G. T. Seaborg, and L. R. Morss, Chapman & Hall, London, 1987.
[3] 'Structure and Bonding', Vol. 59/60, 'Actinides—Chemistry and Physical Properties', ed. L. Manes, Springer-Verlag, Heidelberg, 1986.
[4] 'Americium and Curium Chemistry and Technology', ed. N. M. Edelstein, J. Navratil, and W. W. Schulz, D. Reidel Publ. Co., Dordrecht, 1985.

and Applications are now available,[5] and the volume resulting from the International Symposium on Hydrogen Systems[6] contains many relevant articles concerning those lanthanoid alloys capable of taking up hydrogen.

However, the most timely articles concerning this chapter appear in Volume 4 of 'Advances in Inorganic and Bioinorganic Mechanisms', in the form of lengthy review chapters on the kinetics and mechanisms of actinide redox and complexation reactions,[7a] and on the solvation, solvent exchange, and ligand substitution reactions of the trivalent lanthanide ions.[7b] The investigation of the nature of lanthanoid and actinoid species in solution and of the mechanisms of their reactions is currently the most exciting aspect of research on these elements. These reviews should help underpin the next generation of studies.

Several other reviews of aspects of actinoid chemistry appeared during 1986 and are cited below under the appropriate headings. In the following sections of this chapter, studies of inorganic materials in the solid state are reported first. Accounts of investigations of the behaviour of solutions and of reagents in solution are then presented. Discussion then turns to the properties of co-ordination compounds, and ends with an account of organometallic chemistry.

2 Scandium

There has been little research activity connected with scandium, and some of that has treated the element as an extension of the lanthanoids. The reader should therefore refer also to Section 3 for information on this element.

Ab initio calculations of potential energy surfaces have been used to investigate the activation of H_2 by Sc^+. Various states of Sc^+ were considered as also were two different interaction geometries, from which it was concluded that the formation of both ScH^+ and ScH_2^+ is possible.[8] By contrast, an experimental investigation of the interaction between carbon monoxide and metal oxide surfaces, where the metal is Sc, Y, or La, shows that paramagnetic polyatomic surface compounds are formed. The presence of adsorbed $(OCCO)^{\cdot-}$ radical anions has been advanced as an explanation of this finding.[9]

Scandium cluster compounds can be made in solid-state, high-temperature reactions between the metal, its halide, and the appropriate additive. Products of formula $Sc_7X_{12}Z$ (X = Br or I; Z = B or C) have been isolated[10] as also have $Sc_7I_{12}T$ (T = Co or Ni).[11] The crystallographic investigation of one of these products shows it to be more correctly written as $Sc(Sc_6I_{12}C)$. One scandium atom occurs as Sc^{III} in an approximately trigonal antiprismatic environment of six I atoms from two different clusters, and the clusters are distorted from D_{3d} to C_3 symmetry. Y, Pr, or Gd can be used in place of Sc to give the transition metal encapsulated clusters.

[5] 'New Front. Rare Earth Sci. Appl.', Proc. Int. Conf. Rare Earth Dev. Appl., ed. G. Xu and J. Xiao, Sci. Press, Beijing, 1985.
[6] 'Hydrogen Syst., Pap. Int. Symp.', ed. T. N. Vesiroglu, Y. Zhu, and D. Bao, China Acad. Publ., Beijing, 1986.
[7] 'Advances in Inorganic and Bioinorganic Mechanisms', ed. A. G. Sykes, Academic Press, London, 1986, Vol. 4 (*a*) K. L. Nash and J. C. Sullivan, p. 185; (*b*) S. F. Lincoln, p. 217.
[8] T. H. Upton, *J. Chem. Phys.*, 1986, **85**, 4400.
[9] K. V. Topchieva, S. E. Spiridonov, and A. Yu. Loginov, *J. Chem. Soc., Chem. Commun.*, 1986, 636.
[10] D. S. Dudis, J. D. Corbett, and S.-J. Hwu, *Inorg. Chem.*, 1986, **25**, 3434.
[11] T. Hughbanks, G. Rosenthal, and J. D. Corbett, *J. Am. Chem. Soc.*, 1986, **108**, 8289.

The same research group has also reported the synthesis and structural analysis of double-metal-layered chloride carbides and chloride nitrides, M_2Cl_2Z (M = Sc or Y; and Z = C or N). XPS results show that the C is carbidic while there is a large degree of Sc—C covalency.[12]

X-Ray crystallographic investigations show that the two compounds $M(Cp)_3$ (M = Sc or Lu; Cp = cyclopentadienyl) are isomorphous because of the similar sizes of the metal ions.[13] The structures are best written as $[(\eta^5\text{-}Cp)_2M(\mu\text{-}\eta^1:\eta^1\text{-}Cp)]_\infty$. Notice the difference from the structure of $La(Cp)_3$ discussed below.

Finally under scandium we note the two newly reported methods for the spectrophotometric determination of Sc at below 1 p.p.m. concentrations.[14,15]

3 Yttrium and the Lanthanoids

The ordering of material within this section is the same as that adopted in previous years. So also is the integration of information on yttrium with that for the lanthanoids. When this year's topics are compared with those of earlier Reports, only one stands out as attracting increasing attention. A considerable number of reports have appeared, mainly in Japanese patents, of the use of $LaNi_5$-type alloys as hydrogen-absorbing anodes. We can expect to see more on this research topic in the next few years. There have been fewer reviews and books dealing with lanthanoid chemistry this year than in 1985. The three main items have already been mentioned in the Introduction,[5-7] with the review of solvation, solvent exchange, and ligand-substitution reactions of the Ln^{3+} ions being the most welcome.[7b]

Over the past few years there has been an increasing awareness of the problems associated with the disposal of radioactive waste. Some papers with obvious chemical relevance are mentioned in Section 4. Research on this problem is likely to spread over to the lanthanoids. A recent comparison of the energy characteristics of valence orbitals and of $E°$ data demonstrates the similarity between the stabilities of M^{3+} species of the first half of the lanthanoid series to those of the second half of the actinoids.[16] It has been argued that Am and Cm will be present in repositories as An^{III}, so that environmental studies of lanthanoids can be used to determine the consequences of a breach in a repository. Data for La and Nd from an area in Brazil have been used to show that there would be little escape of activity after such a breach.[17] It seems clear that more environmental studies of the early lanthanoids will be undertaken in the future.

In last year's Report the announcement of the $C_{60}La^+$ cluster vaporized from graphite was discussed. Its formation and detection have been confirmed, but the 'soccerball' shape and the claimed high stability were queried.[18] However, quantum calculations indicate that that shape is more stable than would be a planar graphitene

[12] S.-J. Hwu, R. P. Ziebarth, S. von Winbush, J. E. Ford, and J. D. Corbett, *Inorg. Chem.*, 1986, **25**, 283.
[13] S. H. Eggers, H. Schultze, J. Kopf, and R. D. Fischer, *Angew. Chem.*, 1986, **25**, 656.
[14] I. Mori, Y. Fujita, K. Fujita, A. Usami, H. Kawabe, Y. Koshiyama, and T. Tanaka, *Bull. Chem. Soc. Jpn.*, 1986, **59**, 1623.
[15] C. D. Sharma, S. G. Nagarkar, and M. C. Eshwar, *Bull. Chem. Soc. Jpn.*, 1986, **59**, 1662.
[16] N. B. Mikheev, V. I. Spitsyn, G. V. Ionova, L. N. Auerman, and B. G. Korshunov, *Radiokhimiya*, 1986, **28**, 82.
[17] K. B. Krauskokf, *Chem. Geol.*, 1986, **55**, 323.
[18] D. M. Cox, D. J. Trevor, K. C. Reichmann, and A. Kaldor, *J. Am. Chem. Soc.*, 1986, **108**, 2457.

structure. A spheroidal C_{70} should be even more stable.[19] This area will interest physical and theoretical chemists for some time to come.

Hydrogen absorption by alloys continues to attract attention, and to produce new uses. The distribution of H over the available sites in LnH_2 (Ln = Y, La, or Ce) has been investigated by neutron inelastic scattering and diffraction. Over the temperature range 15—200 K the fraction in octahedral sites remains constant, but it then drops as the temperature increases further.[20] There have been several reports of studies of the rate of hydrogen absorption by alloys. When $LaNi_5$ is coated with oxide there are two separate rate-determining steps for the initial activation. Both the dissociation of H_2 on the surface and the permeation of H atoms through the oxide layers are involved.[21] Only one slow step occurs for suspensions of $LaNi_5$ or $LaNi_{4.7}Al_{0.3}$ in hydrocarbons. In both cases it is the dissolution of H_2 in the liquid phase.[22] For the solids it has been found that the addition of another metal can increase the rates of desorption[23] and adsorption[24] of H by $LaNi_5$, or misch-metal systems such as $MmNi_{5-x}Al_x$. The effect of gas pressure is more important than the hydride content in determining the rates. Kinetic studies have been carried out on the isotope exchange of H with H_2O using a $LaNi_5$/PTFE-based catalyst. A first-order rate law was observed, with an activation energy of 36 kJ mol^{-1}, implying diffusion control.[25] The substitution of Si for some of the Ni reduces the hydriding capacity of $LaNi_5$ but raises the stability of the hydride as the plateau pressure rises.[26]

A promising development for organic synthesis is the finding that Ln_2Co_7 (Ln = Nd or Sm) can be used to dehydrogenate organics under milder conditions than are usually used, *i.e.* in the 300—423 K range. In the case of the dehydrogenation of propan-2-ol by Nd_2Co_7 at low temperatures there is good evidence that hydride formation is the driving force of the reaction; *e.g.* the activation energy of 62 kJ mol^{-1} is very similar to the 69 kJ mol^{-1} for hydrogen absorption.[27] Other reported uses of the hydriding intermetallics include an attempt to develop a cooling system using hydrogen storage in two different alloys,[28] their use in the purification and analysis of H_2,[29,30] and their use in the recovery of H_2 during NH_3 production.[31]

Thin layers of copper can be electrodeposited onto powdered $LaNi_5$ and the material then pressed to form compacted pellets which have better mechanical properties than has the alloy, while still retaining good hydriding behaviour.[32] Such pelletized discs using Cu, Ag, or Au as coating have been examined for use as hydrogen storage electrodes in molar NaOH media. The results look moderately encouraging.[33] The use of non-encapsulated $LaNi_5$ powder as the H electrode in

[19] M. D. Newton and R. E. Stanton, *J. Am. Chem. Soc.*, 1986, **108**, 2469.
[20] J. A. Goldstone, J. Eckert, P. M. Richards, and E. L. Venturini, *Physica B + C*, 1986, **136**, 183.
[21] H. Uchida and M. Ozawa, *Z. Phys. Chem.*, 1986, **147**, 77.
[22] J. R. Johnson and J. J. Reilly, *Z. Phys. Chem.*, 1986, **147**, 263.
[23] H. Chui, Q. Li, and J. Chui, *Hydrogen Syst., Pap. Int. Symp.*, 1985, **1**, 451.
[24] Y. Komazaki and S. Suda, *Kagaku Kogaku Ronbunshu*, 1986, **12**, 447.
[25] B. M. Andreev, E. P. Magomedbekov, A. A. Firer, and I. Yu. Nagaev, *Kinet. Katal.*, 1986, **27**, 224.
[26] J. Liang and J. Zhao, *Hydrogen Syst., Pap. Int. Symp.*, 1985, **1**, 459.
[27] H. Imamura, K. Yamada, K. Nukui, and S. Tsuchiya, *J. Chem. Soc., Chem. Commun.*, 1986, 367.
[28] S. Li, S. Ge, X. Jin, W. Wang, and X. Zhang, *Hydrogen Syst., Pap. Int. Symp.*, 1985, **2**, 313.
[29] D. Bao, L. Sou and F. Lui, *Hydrogen Syst., Pap. Int. Symp.*, 1985, **2**, 335.
[30] E. Yin, X. Yang, W. Lan, and F. Yu, *Hydrogen Syst., Pap. Int. Symp.*, 1985, **2**, 339.
[31] Q. Way, J. Wu, C. Chen, and W. Luo, *Hydrogen Syst., Pap. Int. Symp.*, 1985, **2**, 319.
[32] H. Ishikawa, K. Oguro, A. Kato, H. Suzuki, and E. Ishii, *J. Less-Common Met.*, 1986, **120**, 123.
[33] K. Machida, M. Enyo, G. Adachi, H. Sakaguchi, and J. Shiokawa, *Bull. Chem. Soc. Jpn.*, 1986, **59**, 925.

an alkaline fuel cell has also been reported. Limiting current densities up to 250 mA cm^{-2} were measured.[34] A considerable number of accounts of hydrogen-absorbing anodes have appeared this year, usually in patent disclosures. A typical example of anode construction involves the mixing and melting together of the metals to make alloy powders of composition such as $Ca_{1-x}La_xNi_{5-y}Al_y$. The alloy is then pressed with small amounts of acetylene black and a fluoropolymer at 280 °C to form discs.[35] There is also one report of the use of $LaNi_5$ as a cathode. During the electrolysis of organic sewage an alloy cathode was continuously fed into the system and exposed to ultrasonics. Oxygen produced at the anode is consumed by the sewage, while metal hydride is formed from the cathode.[36]

The structure of Y_5Si_3 has been described. At elevated temperatures it reversibly takes up hydrogen; reaction starts at 250 °C and 1 atm pressure. Activation parameters were determined.[37] The ternary compounds LnCoC (Ln = Y, Gd—Tm, or Lu) have been prepared. They have simple structures, *cf.* CsCl, with no variable positional parameters. On hydrolysis with HCl they yield mainly CH_4, C_3H_8, and C_2H_6.[38] By contrast there are distinct C_2 dimers in $Gd_2C_2Cl_2$; the C—C bond length of 1.36 Å is consistent with C_2^{4-}. The structure is accounted for by the absence of Gd—Gd bonds but the presence of strong Gd—C π-bonding, while the dispersion of the C_2 bands through this Gd—C interaction results in metallic behaviour.[39]

Various LnP samples (Ln = Y, La, Nd, or Sm) have been prepared and examined by optical, electrical, and photoelectrical methods. The valence band is composed of the $3p$ band of P, while the conduction band comes from the empty $5d$ and $6s$ bands of Ln. There is also a Fermi energy band close to the conduction band, which results in very low resistivities.[40] High catalytic activity for the hydrogenation of CO to C_1 and C_2 hydrocarbons has been observed for the amorphous alloy $Ni_{78}P_{19}La_3$. The alloy is pre-treated with dilute HNO_3, O_2, and H_2; the temperature of the gaseous pre-treatment affects the turnover frequency achieved.[41]

Once again in 1986 there was considerable research activity on the lanthanoid oxides. The calculation of the values of low-lying energy levels, and hence the LnO dissociation energies, has been described. They were discussed in terms of a $4f^{n+1}5d^0 \rightarrow 4f^n5d^1$ promotion model.[42] A thermodynamic method which accurately represents the experimental data has been used to recalculate the data base for the dependence of the non-stoicheiometric CeO_{2-x} on temperature and the chemical potential of oxygen. The method appears to be of general applicability.[43] The electron diffraction patterns of Ln_nO_{2n-2} ($n > 3$) have been observed for Ln = Ce, Pr, and Tb. They only occur under an electron beam, which presumably induces reduced metal-containing species in the vapour phase with the corresponding disproportionation leaving higher oxides as solid residues.[44] The oxidation–reduction processes

[34] H. Tanaka, N. Kaneki, H. Hora, K. Shimada, and T. Takeuchi, *Can. J. Chem. Eng.*, 1986, **64**, 267.
[35] S. Furukawa, S. Murakami, and T. Matsumoto, Japan. Patent 61 19 061.
[36] K. Inoue, Japan. Patent 61 101 401.
[37] I. J. McColm, V. Kotroezo, T. W. Button, N. J. Clark, and B. Bruer, *J. Less-Common Met.*, 1985, **115**, 113.
[38] H. M. Gerss and W. Jeitschko, *Z. Naturforsch., B: Anorg. Chem. Org. Chem.*, 1986, **41**, 946.
[39] G. J. Miller, J. K. Burdett, C. Schwarz, and A. Simon, *Inorg. Chem.*, 1986, **25**, 4437.
[40] Y. Ren and J. Meng, *Zhongguo Xitu Xuebao*, 1986, **4**, 1 (*Chem. Abstr.*, 1986, **105**, 163779m).
[41] H. Yamashita, M. Yoshikawa, T. Funabiki, and S. Yoshida, *J. Catal.*, 1986, **99**, 375.
[42] M. Dulick, E. Murad, and R. F. Barrow, *J. Chem. Phys.*, 1986, **85**, 385.
[43] T. B. Lindemer, *Chem. Abstr.*, 1986, **105**, 50013p.
[44] M. Gasgier, G. Schiffmacher, P. Caro, and L. Eyring, *J. Less-Common Met.*, 1986, **116**, 31.

involving Pr oxides have been investigated at temperatures below 523 K. Oxidation proceeds in steps involving Pr_7O_{12}, Pr_9O_{16}, $Pr_{10}O_{18}$, and $Pr_{12}O_{22}$. At 513 K and pressures below 4 kPa, oxidation beyond $Pr_{13}O_{30}$ does not occur.[45]

When a sample of La_2O_3 is quenched from 650 °C in liquid oxygen, an e.p.r. signal corresponding to surface O_2^- can be detected. It presumably arises from the reaction between O_2 and electron–hole pairs or with surface O_2^{2-}.[46] Radical anions, $(CO)_2^{2-}$, have also been detected on the surface of the oxides of Sc, Y, and La after exposure to CO.[9] Reduced Ln^{2+} ions (Ln = La, Nd, or Dy) are formed in the reduction of Ln_2O_3 by H_2 at temperatures above 500 °C. The maximum concentration of the lower oxidation state for neodymium occurs at 700 °C, and the state is stable at oxygen pressures below 10^{-6} volume %. Also for neodymium there is a direct correlation between $[Nd^{2+}]$ and the catalytic activity in the hydrogenation of C_2H_4.[47] The catalytic activity of $La_{1-x}M_xMnO_3$ catalysts (M = Sr, Ce, or Hf) towards the oxidation of propane also varies with the reducibility of the surface, and also with the amount of reversibly adsorbed oxygen.[48] Lastly on the subject of oxides, it has been reported that the process of heating with $LnCl_3$ for one or two weeks at 730–850 °C enables oxygen to be selectively removed from lanthanoid metals as $LnOCl$.[49]

X-Ray photoelectron spectroscopy has been applied to the LnX_3 halides. For X = Cl, Br, and I, the binding energies of the $3d$ and $4d$ peaks of the lanthanoid are related to the atomic charge as calculated from their Pauling electronegativity values; with the exception of Y and Lu those for the fluorides are lower.[50] As is the case for Sc, so also Y and some of the lanthanoids have been shown to yield cluster compounds in which $3d$ metals can be encapsulated, for example $Pr_7I_{12}M$ (M = Mn, Fe, Co, or Ni).[11]

In last year's Report, attention was drawn to the advances being made in the investigation of the nature of lanthanoid ions in solution. This encouraging development is again apparent in the 1986 literature, which contains accounts of some promising new approaches.

A method using measurements of luminescence lifetimes has been described for the investigation of the co-ordination environment of europium ions in solution. There is a difference in the decay constants for solutions of $EuCl_3$ dissolved in ^1H_2O and ^2H_2O. The hydration number for $EuCl_3$ in water is already known from X-ray diffraction studies, and so this value can be used to calibrate the luminescence decay data. The luminsecence experiment can then be repeated using previously unstudied solutes such as $Eu_2(SO_4)_3$. In this latter case investigation led to the deduction that the ratio of inner-sphere to outer-sphere sulphato-complexes increases with temperature over the range 5—45 °C.[51] Fourier-transform i.r. spectroscopy has been used to study $ClO_4^- $-$Ln^{III}$ interactions in anhydrous CH_3CN solution. The findings are in agreement with those obtained from conductivity data, namely that two of

[45] K. Otsuka, M. Kunitomi, and T. Saito, *Inorg. Chim. Acta*, 1986, **115**, L31.

[46] J. X. Wang and J. H. Lunsford, *J. Chem. Phys.*, 1986, **90**, 3890.

[47] K. I. Slovetskaya, Yu. S. Khodakov, A. M. Rubinstein, and Kh. M. Minachev, *Izv. Akad. Nauk SSSR., Ser. Khim.*, 1986, 1958.

[48] T. Nitadori, S. Kurihara, and M. Misono, *J. Catal.*, 1986, **98**, 221.

[49] J. D. Corbett, J. D. Smith, and E. Garcia, *J. Less-Common Met.*, 1986, **115**, 343.

[50] Y. Uwamino, A. Tsuga, T. Ishizuka, and H. Yamatera, *Bull. Chem. Soc. Jpn.*, 1986, **59**, 2263.

[51] F. Tanaka, Y. Kawasaki, and S. Yamashita, *Bull. Chem. Soc. Jpn.*, 1986, **59**, 3389.

the perchlorate groups are ionic, yielding a 2:1 electrolyte, whereas the third perchlorate is co-ordinated predominantly in a unidentate fashion.[52]

Further information has been published on the stabilization of Pr^{IV} and Tb^{IV} in aqueous solution, the species being formed by ozone oxidation. The standard electrode potentials in alkaline pyrophosphate are quoted as 1.01 and 0.93 V respectively, and absorption maxima and extinction coefficients are also given. The rates of reduction of both species are pseudo-first order.[53] Ce^{IV} is claimed to be unique in catalysing the oxidation of thioethers to sulphoxides by molecular oxygen.[54] Kinetic and stoicheiometric data are in agreement with the scheme shown in equations (1)—(4), in which a steady-state treatment is applied to all the sulphur-containing intermediates.

$$SR_2 + Ce^{IV} \rightleftharpoons R_2S^{+\cdot} + Ce^{III} \tag{1}$$

$$R_2S^{+\cdot} + O_2 \rightleftharpoons R_2S^+OO^\cdot \tag{2}$$

$$R_2S^+OO^\cdot + Ce^{III} \rightleftharpoons R_2S^+OO^- + Ce^{IV} \tag{3}$$

$$R_2S^+OO^- + R_2S \rightarrow 2R_2SO \tag{4}$$

Most ligand substitution reactions on lanthanoid ions in solution are too fast to be studied easily; therefore it is pleasing to find some interesting reports of such studies in the year's literature. An intriguing mechanistic interpretation of some ligand exchange results has been made. The exchange between free 1,1,3,3-tetramethylurea (tmu) and the same ligand co-ordinated in $[Tb(tmu)_6]^{3+}$ in CD_3CN was studied by ^1H n.m.r. Although a dissociative mechanism was deduced, as one might expect for such a ligand, a more active role was postulated for the second co-ordination sphere than is found in the better known reactions of the analogous first-row transition-metal complexes. The investigators even suggested that the molecules in the second co-ordination sphere adopt a tricapped trigonal prismatic arrangement about the metal ion in the transition state of the reaction.[55] A stopped-flow study of the ligand exchange reaction between Tb^{3+} and $Ca(edta)^{2-}$ over the range $4.4. < pH < 6.0$ has been carried out by following the increase in Tb^{3+} luminescence on binding to edta.[56] The processes involved are shown in equations (5)—(9).

$$Ca(edta)^{2-} + H^+ \rightleftharpoons Ca(edtaH)^- \qquad K_H \tag{5}$$

$$Ca(edta)^{2-} + Tb^{3+} \rightleftharpoons Ca(edta)Tb^+ \qquad K_T \tag{6}$$

$$Ca(edta)Tb^+ \rightarrow Tb(edta)^- + Ca^{2+} \qquad k_1 \tag{7}$$

$$Ca(edtaH)^- \rightarrow Ca^{2+} + Hedta^{3-} \qquad k_2 \tag{8}$$

$$Hedta^{3-} + Tb^{3+} \rightleftharpoons Tb(edta)^- + H^+ \qquad fast \tag{9}$$

The dissociation of the more complicated ligand K21DA from its complexes with Ln^{3+} has been investigated.[57] The ligand is an N_2O_3 macrocycle with two pendant

[52] J.-C. Bunzli and C. Mabillard, *Inorg. Chem.*, 1986, **25**, 2750.
[53] W. Dong and R. Yang, *Huaxue Xuebao*, 1986, **44**, 563.
[54] D. P. Riley and P. E. Correa, *J. Chem. Soc., Chem. Commun.*, 1986, 1097.
[55] S. F. Lincoln and A. White, *Polyhedron*, 1986, **5**, 1351.
[56] P. J. Breen, W. De W. Horrocks, and K. A. Johnson, *Inorg. Chem.*, 1986, **25**, 1968.
[57] V. C. Sekhar and C. A. Chang, *Inorg. Chem.*, 1986, **25**, 2061.

CH$_2$CO$_2$H groups, giving seven-co-ordinate complexes. Again in this investigation a second metal is present, this time Cu^{2+}, to act as a scavenger for the ligand. There is a clear distinction between the behaviour of the early lanthanoids and that of the later members of the series. For the early members catalysis by both H$^+$ and HCO$_2^-$ is observed. The rate of reaction shows a first-order dependence on each of these ions. The later members show a dependence only on [H$^+$], which is between zero and first order. The difference is probably due to the co-ordination of a water molecule only to the lighter, larger ions. This water ligand can be substituted quickly by a carboxylate, which can then take part in the main dissociation reaction. In that reaction the partial dissociation of the ligand is the slow step, which is then followed by its rapid scavenging by Cu^{2+}.

In circularly polarized and total luminescence studies of complexes formed between Ln^{3+} and 2,6-pyridinedicarboxylate ligands it was found that optical activity can be generated from the racemic solutions, and observed, because of the kinetic stability of the complexes formed.[58] When the ligand 4(−)-methyloxy]-1-(p-tolyl)-1,3-butanedionate forms complexes with Ln^{3+}, inter-ligand interactions cause high stereoselectivity, as is shown in the c.d. bands of f–f transitions.[59] Luminescence techniques have also been applied to the investigation of Eu^{III} complexes with an 18-crown-6 ether in various solvents. Again the data provide indications both of the species present in solution and of the exchange rates operating.[60] Thermodynamic studies of other dicarboxylate ligands show that the stabilities of their complexes drop markedly as the chelate ring size increases from 5 to 7, but change little from 7 to 9. The entropy change becomes more negative with increasing size due to the loss in configurational entropy in the alkyl chain.[61]

The conformation of antipyrine complexes with Ln^{3+} in aqueous media has been studied by ^{1}H and ^{13}C n.m.r. The complexes are not the usual inner-sphere type, but rather the organic ligand is found in the second co-ordination sphere of the metal ion. Eight different metals were studied and only in the case of Tm^{3+} did the complex not possess effective axial symmetry.[62] When ^{13}C spectra of hydroxycarboxylate complexes in aqueous solution are obtained, it is found that Gd^{III} induces enhanced relaxation rates. This ion was considered as a model for Ca^{2+} and used to determine the differing modes of co-ordination assumed by various structural units.[63]

Lanthanoid complexes of the ligand nota, where nota = 1,4,7-triaza-cyclononane-N,N',N''-triacetate, have been examined by ^{1}H and ^{13}C n.m.r. spectroscopy at 25 and 70 °C. Usually shifts are dominated by contact interactions. More than one chelate structure is present for Pr(nota), while the addition of Cl$^-$ to the solutions shows that the earlier metals can form mixed complexes, with both five- and six-co-ordinate nota present. The smaller, later Ln^{3+} ions fit the ligand cavity better and so give more rigid, and purely six-co-ordinated, complexes.[64] Similar investigations have been carried out on complexes of the 14-membered macrocyclic

[58] G. L. Hilmes and J. P. Riehl, Inorg. Chem., 1986, 25, 2617.

[59] H. Okawa, M. Nakamura, Y. Shuin, and S. Kida, Bull. Chem. Soc. Jpn., 1986, 59, 3658.

[60] D. H. Metcalf, R. C. Carter, R. G. Ghirardwlli, and R. A. Palmer, Inorg. Chem., 1986, 25, 2175.

[61] G. R. Choppin, A. Dadgar, and E. N. Rizkalla, Inorg. Chem., 1986, 25, 3581.

[62] A. L. du Preez and R. J. P. Williams, J. Chem. Soc., Dalton Trans., 1986, 1425.

[63] C. A. M. Vijverberg, J. A. Peters, A. P. G. Kieboom, and H. van Bekkum, Tetrahedron., 1986, 42, 167.

[64] A. D. Sherry, M. Singh, and C. F. G. C. Geraldes, J. Magn. Reson., 1986, 66, 511.

ligand teta; where teta = 1,4,8,11-tetra-azacyclotetradecane-N,N',N'',N'''-tetra-acetic acid. ^{1}H studies of Yb(teta) and ^{13}C for Lu(teta) show that exchange between two equivalent dodecahedral geometries occurs. The dynamic behaviour depends on the radius of the Ln^{3+} ion. In the case of lutetium the free energy of activation for the exchange is 64 kJ mol^{-1}.[65] ^{1}H N.m.r. spectra have also been described for several complexes of the ligand whose ring is one member smaller *i.e.* the 1,4,8,10-cyclotridecane derivative. Here the ring cavity is too small to accommodate the metal ion, and so the ligand folds to give co-ordination with four N-donors on one side of Ln and four O-donors on the other.[66]

Once again this year the properties of the lanthanoid ions have been used to probe the behaviour of biochemically important species. ^{31}P N.m.r. has been used to examine the properties of cation–proton exchange systems in biological membranes. The transport of Pr^{3+} ions across phosphatidylcholine vesicles is accelerated in the joint presence of an ionophore (X-537A) and a fatty acid, because of the coupling of the flow of Pr^{3+} with the opposing flows of H^+ and Na^+.[67] N.m.r. data show that Dy^{3+} attaches to a carboxylate group rather than to a sulphonate, as does Ca^{2+}, on the binding sites of the bile salts glycocholate and taurocholate.[68] The binding constants for Nd^{3+} and Sm^{3+} with human serum transferrin have also been determined.[69] A γ-carboxyglutamic acid-containing heptapeptide, LOOP, which corresponds to a section of Bovine Prothrombin residues, has been synthesized. It forms a 2 : 1 Tb^{3+} : LOOP complex. When the luminescence decay is analysed in the presence of an excess of La^{3+} it is seen that metal exchange is slow compared with the lifetime of the excited state. Co-operative effects between the two co-ordination sites may account for this slowness.[70]

Several reports this year have described the use of the lanthanoid metals themselves as reagents in the formation of complexes. Sometimes the products have uncommon properties. Complexes which are at least formally Ln^0 species have been synthesized by the co-condensation of Ln and RN=CH—CH=NR vapours (R = Pr^i or Bu^t; Ln = Y, Nd, Sm, or Yb). These species are 'molecular' rather than salts. ^{1}H N.m.r. spectroscopy confirms this, as there is no exchange between the free ligand and the co-ordinated diimine.[71] We can reasonably hope that this preliminary report signals the start of a new research effort. Various Ln metals have also been used as reagents in transmetallation reactions with $[(Me_3Si)_2N]_2Hg$ in dimethoxyethane. Ligand exchange with alcohols and thiols can subsequently be effected for $[(Me_3Si)_2N]_2Yb$ in pentane.[72]

In the presence of a trace of iodine, organocerium compounds, RCeI, can be formed from alkyl, allyl, or aryl iodides. As well as showing Grignard-like behaviour when used *in situ*, these materials show reductive effects attributable to Ce^{II} or a Ce^{III}–H compound. For example, both reductive and reductively coupled products

[65] J. F. Desraux and M. F. Loncin, *Inorg. Chem.*, 1986, **25**, 69.
[66] J. R. Ascenso, R. Delgado, and J. J. R. Frauto de Silva, *J. Chem. Soc., Dalton Trans.*, 1986, 2395.
[67] J. Grandjean and P. Laszlo, *J. Am. Chem. Soc.*, 1986, **108**, 3483.
[68] E. Mukidjan, S. Barnes, and G. A. Elgavish, *J. Am. Chem. Soc.*, 1986, **108**, 7082.
[69] W. R. Harris, *Inorg. Chem.*, 1986, **25**, 2041.
[70] H. C. Marsh, R. A. Hoke, D. W. Deerfield, L. G. Pedersen, R. G. Hiskey, and K. A. Koehler, *Inorg. Chem.*, 1986, **25**, 4503.
[71] F. G. N. Clarke, H. C. de Lemos, and A. A. Sameh, *J. Chem. Soc., Chem. Commun.*, 1986, 1344.
[72] Yu F. Rad'kov, E. A. Fedorova, S. Ya. Khorshev, G. S. Kalinina, M. N. Bochkarev, and G. A. Razuvaev, *Zh. Obsch. Khim.*, 1985, **55**, 2153.

as well as the tertiary alcohol are obtained from cyclohexanone.[73] Similarly, in the presence of iodine, the metals suspended in thf have been used to mediate an easy organic preparation from methyl β-bromopropionate, presumably *via* an organometallic intermediate.[74] NOBF$_4$ in CH$_3$CN can be used to oxidize metallic Eu. The [Eu(CH$_3$CN)$_3$(BF$_4$)$_3$]$_x$ product is a 1:2 electrolyte in acetonitrile, and x is determined as 2. The co-ordinated fluoroborate ligands are easily displaced. The compound is capable of initiating the polymerization of alkenes both in acetonitrile and nitromethane, the rate being much greater in the latter.[75]

Each year there are a great number of papers published describing complexes of the lanthanoids with various multidentate ligands, and 1986 was no exception. Some of the more original of these concern the nature of species in solution and have already been mentioned above. A brief summary of some others follows. Highly purified lanthanoid ions can be obtained by using newly synthesized chelating resins, which carry pendant groups similar to edta and dtta.[76] Conditions have been described which enable the formation of two series of complexes of a macrocyclic hexa-aza-ligand with all the Ln^{3+} ions to be achieved. The complexes appear to be inert with respect to macrocycle substitution, and to be soluble in both water and organic solvents. The authors suggest that these complexes may be of use as n.m.r. shift reagents.[77] The structure of the product of the template condensation between ethylenediamine and 2,6-diacetylpyridine on Lu^{3+} has been determined crystal-lographically. The C$_{22}$H$_{26}$N$_6$ ligand is non-planar with a dihedral angle of 114°. This is in marked contrast to the same ligand in complexes with the earlier lanthanoids, even though the spectra, stabilities, and substitution inertness of all the complexes are comparable.[78] Stability constants of lanthanoid complexes with some polyethyleneglycols, EO$_n$, show unusual trends. Whereas the maximum stability for $n = 3$ occurs at Eu, the stability constants drop with increasing atomic number for $n = 4$ and 5. This is attributed to the fact that the ligands form ring-like structures with a cavity in the region of 2.7 Å.[79]

The reaction between YbCl$_2$ and crown ethers leads to the formation of new complexes whose luminescence has been studied.[80] I.r.-raman and luminescence spectra have also been reported and interpreted for Eu^{III} complexes,[81] and the luminescence of a Tb^{III} complex with a 2,2,1-cryptate has been investigated both in aqueous solution and in the solid state.[82]

The redox potentials of the couples Ln^{3+}/Ln^{2+} and LnL^{3+}/LnL^{2+} (Ln = Eu, Yb, or Sm; L is either the 2,2,1- or 2,2,2-cryptand) have been compared as the solvent is changed. The variations with the solvent have been used to study the degree of shielding of the Ln ion from the solvent by the cryptand. The approach seems to

[73] S.-I. Fukuzawa, T. Fujinami, and S. Sakai, *J. Organomet. Chem.*, 1986, **299**, 97.
[74] S. Fukuzawa, T. Fujirami, and S. Sakai, *J. Chem. Soc., Chem. Commun.*, 1986, 475.
[75] R. R. Thomas, V. Chebolu, and A. Sen, *J. Am. Chem. Soc.*, 1986, **108**, 4096.
[76] K. Taneda, M. Akiyama, F. Kawakami, and M. Sasaki, *Bull. Chem. Soc. Jpn.*, 1986, **59**, 2225.
[77] L. De Cola, D. L. Smailes, and L. M. Vallarino, *Inorg. Chem.*, 1986, **25**, 1729.
[78] G. Bombieri, F. Benetollo, A. Polo, L. De Cola, D. L. Smailes, and L. M. Vallarino, *Inorg. Chem.*, 1986, **25**, 1127.
[79] Y. Hirashima, K. Kanetsuki, I. Yonezu, N. Isobe, and J. Shiokawa, *Bull. Chem. Soc. Jpn.*, 1986, **59**, 25.
[80] W. Li, H. Fujikawa, G. Adachi, and J. Shiokawa, *Inorg. Chim. Acta*, 1986, **117**, 87.
[81] J. C. G. Bunzli, G. A. Leonard, D. Plancherel, and G. Chapius, *Helv. Chim. Acta*, 1986, **69**, 288.
[82] N. Sabbatini, S. Dellonte, and G. Blasser, *Chem. Phys. Lett.*, 1986, **129**, 541.

offer some useful guidance.[83] The crown ether derivative $[Y(OH_2)_8]Cl_3$(15-crown-5) has been prepared and its structure determined. The hydrogens of the co-ordinated water molecules are hydrogen-bonded to other species in the solid. Three are attached to the chloride ions and the other five to the small crown ether.[84]

There has been considerable research activity on phthalocyanin complexes this year, much of it concerned with species containing more than one Pc ligand per Ln. A range of complexes of forms LnPcCl and LnPc(OAc) have been prepared and examined. They undergo both reversible one-electron oxidation to $[LnPcX]^{+\cdot}$ and irreversible oxidation to $[LnPcX]^{2+}$. The absorption spectra of the singly charged products do not vary with Ln.[85] The blue form of $Lu(Pc')_2$, where Pc′ is the derivative obtained from 4-t-butylphthalodinitrile, has been separated chromatographically from the green form and examined. It has a sandwich structure and its electrochemistry has been reported.[86] An investigation into the preparation of complexes of general form Ln_xPc_y has shown that the main product varies across the rare earth series. Ln_2Pc_3 is the dominant form for La and Nd, becoming less so for Eu and Gd, whereas $LnPc_2H$ predominates for Dy, Yb, and Lu.[87] Spectral comparisons of the Ln_2Pc_3 species show that the Q-band undergoes a blue shift as the radius of Ln^{3+} decreases, whereas the X-band undergoes a red shift.[88] Three separate species can be found after the column chromatography of a crude mixture of $Nd^{III}Pc$; they are [PcNdPcH], [PcNdPc˙], and [PcNdOAc]. For Ln = Y, La, or Nd, [PcLnPcH] can be oxidized by *p*-benzophenone to [PcLnPc˙].[89]

Some photochemical results have been reported for the irradiation of [PcLnPcH] species in CH_2Cl_2–CH_3CN mixture above 320 nm.[90] The steps described are shown in equations (10) and (11). The quantum yields can approach unity (at above 80% v/v methylene dichloride in the case of Nd). For a given solvent ratio the quantum yields increase in the order Nd (0.64), Y (0.81), Lu (1.00). The radical intermediate can be detected by the absorption band which appears in the otherwise clear region of 470 nm, and by its intense e.s.r. signal at $g = 2.00$.[90]

$$[PcLnPcH] \rightarrow [PcLnP\dot{c}H]^+ + \dot{C}H_2Cl + Cl^- \qquad (10)$$

$$[PcLnP\dot{c}H]^+ \rightarrow [PcLnP\dot{c}] + H^+ \qquad (11)$$

Complexes containing porphyrin have also been discussed during the year. One, a triple-decker sandwich, includes both a porphyrin and phthalocyanin as [(Por)Ln(Pc)Ln(Pc)]. Here the *meso*-tetra(4-methoxyphenyl)porphyrin was used.[91] Triple-decker complexes of Ce^{III} and double-deckers of Ce^{IV} have been reported for octaethyl-(OEP) and other porphyrin derivatives. $Ce_2(OEP)_3$ is a brown paramagnetic material which is scarcely soluble.[92] By contrast $Ce(OEP)_2$ is a reddish-brown,

[83] J. Tabib, J. T. Hupp, and M. J. Weaver, *Inorg. Chem.*, 1986, **25**, 1916.
[84] R. D. Rogers and H. G. Brittain, *Inorg. Chim. Acta*, 1986, **116**, 163.
[85] N. B. Subbotin, L. G. Tomilova, N. A. Kostromina, and E. A. Luk'yanets, *Zh. Obshch. Khim.*, 1986, **56**, 397.
[86] V. I. Gavrilov, V. A. Vazhnina, and K. Konstantinov, *Electrokhimiya*, 1986, **22**, 1112.
[87] M. M'Sadak, J. Roncali, and F. Garnier, *J. Chim. Phys., Phys.-Chim. Biol.*, 1986, **83**, 211.
[88] K. Kasuga, M. Ando, H. Morimoto, and M. Isa, *Chem. Lett.*, 1986, 1095.
[89] K. Kasuga, M. Ando, and H. Morimoto, *Inorg. Chim. Acta*, 1986, **112**, 99.
[90] K. Kasuga, H. Morimoto, and M. Ando, *Inorg. Chem.*, 1986, **25**, 2478.
[91] M. Moussavi, A. De Cian, J. Fischer, and R. Weiss, *Inorg. Chem.*, 1986, **25**, 2107.
[92] J. W. Buchler, A. De Cian, J. Fischer, M. Kihn-Botulinski, H. Paulus, and R. Weiss, *J. Am. Chem. Soc.*, 1986, **108**, 3652.

is diamagnetic, and can be dissolved in $1,2,4$-$C_6H_3Cl_3$. It can be oxidized to $[Ce(OEP)_2]^+$ which can then be isolated as a solid salt.[92,93]

Almost all the references to the organometallic chemistry of the lanthanoids deal with cyclopentadienyl derivatives. They are covered in the final paragraphs of this section, but before turning to them we consider a small number of other topics.

Amalgamated lanthanoids (La, Sm, Yb) can be used to insert a Ln atom into a metal–halide bond in an organometallic compound. The reactions are carried out in THF at or below room temperature, and are typified by the reactions (12)—(14).[94,95]

$$Ln/Hg + Co(CO)_4I \rightarrow (CO)_4CoLnI \qquad (12)$$

$$Ln/Hg + CpMo(CO)_3Br \rightarrow Cp(CO)_3MoLnBr \qquad (13)$$

$$CpMo(CO)_3LnBr + Cp(CO)_3MoTl \rightarrow [CpMo(CO)_3]_2Ln(THF)_n \qquad (14)$$

$$LaCl_3 + 3K[(Ph_2P)_2CH] \rightarrow [La\{(Ph_2P)_2CH\}_3].C_7H_8 \qquad (15)$$

Diphosphinomethanide ligands have been attached to lanthanum [equation (15)] giving a π-allyl-like ligand geometry (1). The central carbon atom of each ligand is slightly closer to the lanthanum than are the phosphorus atoms. The orientations of the phenyl rings show that the lone pairs on P are not arranged to maximize La–P interactions. The ^{31}P n.m.r. spectrum in benzene is very similar to that of the potassium salt of the ligand, and so the bonding to La is judged to be strongly ionic in character.[96]

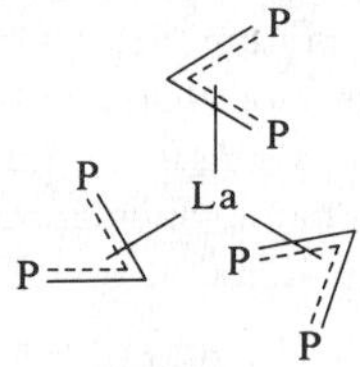

(1)

The crystal structures of the two compounds $[Cp'_2YN(SiMe_3)_2]$ and $[Cp'_2YCH(SiMe_3)_2]$ have been described; Cp' is used to represent C_5Me_5 throughout this chapter. There are agostic interactions between Y and methyl-H in both compounds. These interactions are judged to be stronger in the second species, as the solid-state ^{13}C n.m.r. spectra show two and three different Si—Me groups respectively. Both the n.m.r. and X-ray data show that there is also an α-agostic Y$\cdots$CH interaction in the substituted methanido-complex.[97] Thermolysis of the related yttrium compounds $Cp''_2Y(CH_2SiMe_3).THF$ and $[Cp''_2YMe]_2$ (Cp'' always

[93] J. W. Buchler, K. Elsasser, M. Kihn-Botulinski, and B. Scharbert, *Angew. Chem.*, 1986, **25**, 286.
[94] I. P. Beletskaya, G. Z. Suleimanov, R. R. Shifrin, R. Yu. Mekhdiev, T. A. Agdamskii, V. N. Khandozhko, and N. E. Kolobova, *J. Organomet. Chem.*, 1986, **299**, 239.
[95] I. P. Beletskaya, G. Z. Suleimanov, R. Yu. Mekhdiev, T. A. Agdamskii, V. N. Khandozhko, P. V. Petrovskii, and N. E. Kolobova, *Dokl. Akad. Nauk SSSR*, 1986, **289**, 96.
[96] H. H. Karsch, A. Appelt, and G. Muller, *Angew. Chem.*, 1986, **25**, 823.
[97] K. H. den Haan, J. L. de Boer, J. H. Teuben, A. L. Spek, B. Kojic-Prodic, G. R. Hays, and R. Huis, *Organometallics*, 1986, **5**, 1727.

represents C_5H_4Me), leads to the formation of an enolate-bridged dimer of formula $[Cp''_2Y(\mu\text{-}OCH=CH_2)]_2$. The molecular structure of this compound is reported. Similar compounds can be obtained from reaction (16). The preparative route is not restricted to yttrium as the analogous compounds containing Yb or Lu were prepared, as were the derivatives with Cp in place of Cp''.[98] It is also worth noting that the first arene complex of a lanthanoid, $Sm(\eta^6\text{-}C_6H_6)(\eta^2\text{-}AlCl_4)_3$, has been prepared and structurally characterized.[99]

$$2[Cp''_2LnCl]_2 + 2LiOCH=CH_2 \rightarrow [Cp''_2Ln(\mu\text{-}OCH=CH_2)]_2 \qquad (16)$$

Several structures of binary lanthanoid cyclopentadienyls were described during 1986. They cover a wide range of shapes. The electron diffraction results for the compound Cp'_2Yb show that it has a bent structure in the gas phase. An angle of 158° is subtended by the two Yb—Cp centroid axes, and all the parameters are very similar to those of the equivalent calcium compound.[100] Full details have now been given of the structure of Cp'_2Sm, reported in 1984, together with those of the isomorphous Cp'_2Eu. Again we have bent molecules, now with 140° angles. It is not clear whether the change from 140° and 158° reflects anything other than the difference between the states in which the measurements were made.[101] In the solid state, sublimed Cp_3Ln (Ln = Er or Tm), occur as discrete molecules aligned in chains by van der Waals interactions.[102] That is in contrast to the more complicated structures found for Cp_3Ln (Ln = Sc,[13] La,[103] or Lu[13]). The compounds of the smaller Sc and Lu are isomorphous, and can be represented as $[(\eta^5\text{-}Cp)_2Ln(\mu\text{-}\eta^1\text{:}\,\eta^1\text{-}Cp)]_\infty$. With lanthanum zig-zag chains of form $[(\eta^5\text{-}Cp)_2Ln(\mu\text{-}\eta^5\text{:}\,\eta^2\text{-}Cp)]_\infty$ are found. Obviously the concept of co-ordination number is inappropriate in such circumstances, but it is amusing to note that there are 17 carbon atoms within 3.035 Å of each La atom. Structure (2) shows the arrangement around the bridging ligand in Cp_3La.

(2)

The oxidative addition of alkyl and aryl halides, RX, to the Yb^{II} compound $Cp'_2Yb.(OEt_2)$ has been investigated and found to be between 10^3 and 10^6 times faster than equivalent reactions for d-block compounds. The reaction leads to a complicated mixture of products including Cp'_2YbX, $Cp'YbX_2$, $Cp'R$, R—R and RH alkanes, and $R(-H)$ alkenes. A suggested mechanistic scheme has been devised.[104] X-Ray diffraction studies have also been carried out on the Yb^{III}

---

[98] W. J. Evans, R. Dominguez, and T. P. Honusa, *Organometallics*, 1986, **5**, 1291.

[99] F. A. Cotton and W. Schwotzer, *J. Am. Chem. Soc.*, 1986, **108**, 4657.

[100] R. A. Andersen, J. M. Boncella, C. J. Burns, R. Blom, A. Haaland, and H. V. Volden, *J. Organomet. Chem.*, 1986, **312**, C49.

[101] W. J. Evans, L. A. Hughes, and T. P. Hanusa, *Organometallics*, 1986, **5**, 1285.

[102] S. H. Eggers, W. Hinrichs, J. Kopf, W. Jahn, and R. D. Fischer, *J. Organomet. Chem.*, 1986, **311**, 313.

[103] S. H. Eggers, J. Kopf, and R. D. Fischer, *Organometallics*, 1986, **5**, 383.

[104] R. G. Finke, S. R. Keenan, D. A. Schiraldi, and P. L. Watson, *Organometallics*, 1986, **5**, 598.

compound $Cp_2YbMe(THF)$. It, and other compounds of the same family, *e.g.* $Cp_2LnR(THF)$, $[Cp_2LnR]_2$, and the equivalent Cp' species, have been compared in their reactivity in hydrogenolysis reactions. In this instance Ln = Y, Er, Yb, or Lu, and R = H or Me. Size factors appear to be of great importance as small changes in $r(Ln)$, or in the steric crowding caused by R or the solvent, cause a large change in reactivity.[105] During the thermolysis of $[Cp'_2YH]_2$ at 100 °C only half as much hydrogen was evolved as was expected, giving a stoicheiometry of $1:1$ $H_2:Y_2$. The product (3) is believed to be formed.[106] Diffraction studies have been described for

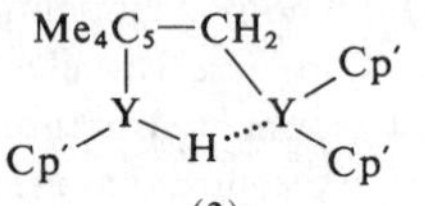

$$Me_4C_5-CH_2$$
$$| \qquad \backslash \qquad Cp'$$
$$\diagdown \quad Y \quad \diagup \quad Y \diagup$$
$$Cp' \diagup \quad \diagdown H \cdots \diagdown Cp'$$

$$(3)$$

other hydride species, this time of lutetium.[107] These are the products of the reactions (17)—(19) in THF.

$$Cp_2LuR.THF + H_2 \rightarrow THF.Cp_2Lu(\mu\text{-}H)_2LuCp_2.THF \qquad (17)$$

$$Cp_2LuCl + Na \rightarrow [Na(THF)_6][(Cp_2LuH)_3H] \qquad (18)$$

$$Cp_3Lu + NaH \rightarrow [Na(THF)_6][(Cp_3Lu)_2H].2THF \qquad (19)$$

4 The Actinoids

Yet again this year the chemistry of the actinoids has been the subject of several useful and informative reviews and books,[1—4] with the publication of the second edition of 'The Chemistry of the Actinide Elements' being the most important event.[2] A number of interesting pieces of research have been described in the primary literature during the year, but no aspect can be recognized as one that is about to blossom.

The preparations and properties of the trans-plutonium elements in their metallic state has been reviewed.[108] An analysis of the products, NH_3, N_2, N_2O, and HN_3, has been used as the basis for an interpretation of the mechanism of dissolution of Pu metal in HNO_3-HF-N_2H_4.[109]

Studies, centred solely on the initial oxidation reactions of clean polycrystalline uranium surfaces, have also been announced. The behaviour of the surfaces under atmospheres of dry O_2, dry CO, and pure H_2O vapour over the temperature range 85—295 K was examined using a variety of physical techniques. In all cases a nearly stoicheiometric UO_2 layer was rapidly formed, and no UO was detected. In the initial stages the active oxide surface promotes UO_{2+x} production.[110] When uranium metal is exposed to HCl or HBr an adherent hydride layer is formed. If the specimen

[105] W. J. Evans, R. Dominguez, and T. P. Hanusa, *Organometallics*, 1986, **5**, 263.
[106] K. H. den Haan and J. H. Tenben, *J. Chem. Soc., Chem. Commun.*, 1986, 682.
[107] H. Schumann, W. Genthe, and E. Hahn, *J. Organomet. Chem.*, 1986, **299**, 67.
[108] A. G. Seleznev, V. A. Stupin, V. D. Shushakov, and N. S. Kosulin, Radiokhimiya, 1986, **28**, 235.
[109] D. G. Karraker, *J. Less-Common Met.*, 1986, **122**, 337.
[110] K. A. Winer, *Diss. Abstr. (B)*, 1986, **46**, 3987.

is then heated under vacuum an activated metal surface is produced that reacts 'immediately' with molecular hydrogen at room temperature.[111]

A direct preparative route to the hydride CfH_{2+x} is available. The product has the same f.c.c. structure as have the MH_2 hydrides formed by the lanthanoids and the earlier actinoids. The lattice parameters, together with the observation that CfH_3 can not be formed from Cf under the usual conditions, indicate the increasing trend to the oxidation state II in the actinoid series.[112] The gas-phase X-ray photoelectron spectra of several compounds, including $An(BH_4)_4$ and $An(BH_3CH_3)_4$ (An = Th or U), have been recorded.[113,114] Comparison of these data with the results of molecular orbital calculations for the BH_4 species suggest that the $5f$ orbitals are important to covalent An—L bonding.[115]

The results of thermogravimetric studies of the reduction of UF_4 by Si under N_2 to form nitrides have been reported. UN at greater than 99.9% purity can be obtained when reaction (23) is carried out under vacuum.[116]

$$4UF_4 + Si \rightarrow 4UF_3 + SiF_4 \qquad 940\ K \qquad (20)$$

$$2UF_3 + Si + N_2 \rightarrow 2UNF + SiF_4 \qquad 1360\ K \qquad (21)$$

$$4UNF + Si + N_2 \rightarrow 2U_2N_3 + SiF_4 \qquad 1640\ K \qquad (22)$$

$$U_2N_3 \rightarrow 2UN + \tfrac{1}{2}N_2 \qquad 1370\ K \qquad (23)$$

The air-stable phosphide ThP_7 can be prepared directly from the elements in the presence of I_2. It decomposes under vacuum at 650 K to yield Th_2P_{11}.[117] Further reports have appeared of the introduction of foreign species into the layered uranyl phosphates. Hydrogen uranyl phosphate can undergo complete exchange with Eu^{3+} to produce a material of approximate composition $Eu(UO_2)_3(PO_4)_3$. This is a single phase with an interlamellar spacing of *ca.* 10.2 Å. When only the uranyl group is initially irradiated, excited-state energy transfer to Eu^{3+} of almost unit efficiency occurs; the rate is approximately $2 \times 10^5\ s^{-1}$.[118] Butylamine intercalates of the layered phosphates are capable of retaining divalent metal ions. This occurs as a result of exchange between M^{2+} and $BuNH_3^+$.[119]

Other than a review of the recent developments in the field of oxidic compounds of the actinoids,[120] there have been few reports of note in this area. A perovskite containing U^{4+} and U^{5+} can be formed in the $CaO-UO_2$ system in wet air.[121] Single crystals of $Na_2U_2O_7$ have been grown from the $Nb_2O_5-UO_2-Na_2CO_3$ mixture, and their structure has been determined. The material is isotypic with $CaUO_4$ with O vacancies located upon the O(1) sites of the uranyl-type bond. Therefore the formula

[111] G. L. Silver, *J. Radioanal. Nucl. Chem.*, 1986, **103**, 199.
[112] J. K. Gibson and R. G. Haire, *Radiochim. Acta*, 1985, **38**, 193.
[113] D. B. Beach, K. D. Bomben, N. M. Edelstein, D. C. Eisenberg, W. L. Jolly, R. Shinomoto, and A. Streitweiser, *Inorg. Chem.*, 1986, **25**, 1735.
[114] J. C. Green, R. Shinomoto, and N. Edelstein, *Inorg. Chem.*, 1986, **25**, 2719.
[115] D. Hohl and N. Rosch, *Inorg. Chem.*, 1986, **25**, 2711.
[116] R. Vengkataramani, Y. J. Bhatt, N. Krishnamurthy, and S. P. Garg, *J. Less-Common Met.*, 1986, **118**, 281.
[117] H. G. Von Schnering and D. Vu, *J. Less-Common Met.*, 1986, **116**, 259.
[118] M. M. Olken, C. M. Verschoor, and A. B. Ellis, *Inorg. Chem.*, 1986, **25**, 81.
[119] R. Pozas-Tormo, L. Moreno-Real, M. Martinez-Lara, and S. Brusque-Gamez, *Can. J. Chem.*, 1986, **64**, 30.
[120] C. Keller, *J. Less-Common Met.*, 1986, **121**, 15.
[121] J. Holc and D. Kolar, *J. Solid State Chem.*, 1986, **61**, 260.

is $Na(UO_{1.5})O_2$ and not $Na(UO_2)O_{1.5}$.[122] Further information about peroxy-uranate(VI) species has been published. Whereas the peroxo-group is bridging in $(M^+)_2[UO_2(O_2)SO_4(H_2O)]$, which also contains unidentate sulphate, both peroxo and oxalato ligands are bidentate in $(M^+)_2[UO_2(O_2)(ox)].H_2O$.[123] An X-ray crystallographic study of the mixed-valence compound $Cs_7[Np^VO_2][Np^{VI}O_2]_2Cl_{12}$ has been made. Only one type of Np position occurs.[124] The preparation of $Cs_4[(NpO_2)_3Cl_6(NO_3)H_2O]$ has also been described.[125]

Synthetic routes to high purity, polycrystalline binary and tertiary compounds, free of magnetic impurities, have been described which start from the oxides of U, Pu, and Np. The products include borides and oxychalcogenides.[126] Uranium oxides can also be used as oxidation catalysts. Such oxides, supported on SiO_2, TiO_2, Al_2O_3, or MgO have been prepared and characterized. Thermoanalytical techniques have been used in order to investigate their mechanisms of formation and stabilization. The dispersion of the active phase was also studied. When the SiO_2- and TiO_2-based catalysts are calcined at 900 °C encapsulation occurs. By contrast segregation probably occurs on Al_2O_3 and MgO.[127]

The halides have not attracted much attention this year. A review of the preparation, structures, and properties of the complex halides of uranium has been published.[128] The hexafluorides of Pu, U, and Np, but not of Am, can be prepared at ambient temperature from their dioxides using KrF_2 as the fluorinating agent.[129] This is only the second low-temperature preparation known, and so is of obvious value. The X-ray photoelectron spectrum of gaseous UF_6 has been measured. The core binding energies show that the bonding is surprisingly ionic.[113] The mixed halide UF_5Cl has been isolated and studied in matrices of Ar, N_2, and CO. It is considerably more reactive than is UF_6 and undergoes photodissociation of the chloride around 500 nm.[130]

The most interesting research area this year for the actinoids, as with the lanthanoids, has been the study of solutes and of their reactions in solution. Most of the relevant papers are reviewed in the following paragraphs, but a small number have been delayed for inclusion in the section that covers cyclopentadienyl complexes of Th and U to which they are more appropriate. The whole area has been reviewed this year.[7a]

An ionic model has been used to predict the thermodynamic parameters of An^{n+} aquo ions, where $n = 2$—4.[131] Th^{4+} only forms mononuclear hydroxo-complexes at pH <4, and relatively low thorium concentrations. The first hydrolysis constant at 20 °C and an ionic strength of 0.1 in nitrate or perchlorate media is $(4.4 \pm 1.2) \times 10^{-5}$. At pH >4 pseudocolloids are formed.[132] The oxidative breakdown of Pu^{IV}

[122] M. Gasperin, *J. Less-Common Met.*, 1986, **119**, 83.
[123] M. Bhattacharjee, M. K. Chaudhuri, and R. N. Purkayastha, *Inorg. Chem.*, 1986, **25**, 2354.
[124] N. W. Alcock, D. J. Flanders, and D. Brown, *J. Chem. Soc., Dalton Trans.*, 1986, 1403.
[125] Yu. F. Volkov, R. F. Melkaya, V. I. Spiryakov, S. V. Tomilin, and I. I. Kapshukov, *Radiokhimiya*, 1986, **28**, 311.
[126] J. Larroque, R. Chipaux, and M. Beauvy, *J. Less-Common Met.*, 1986, **121**, 487.
[127] H. Collette, S. Maroie, J. Riga, J. J. Verbist, Z. Gabelicia, J. B. Nagy, and E. G. Derouane, *J. Catal.*, 1986, **98**, 326.
[128] J. Drozdzynski, *Wied. Chem.*, 1986, **38**, 855.
[129] L. B. Asprey, P. G. Eller, and S. A. Kinkead, *Inorg. Chem.*, 1986, **25**, 671.
[130] A. J. Downs and C. J. Gardner, *J. Chem. Soc., Dalton Trans.*, 1986, 1289.
[131] S. G. Bratsch and J. J. Lagowski, *J. Phys. Chem.*, 1986, **90**, 307.
[132] Yu. P. Davidov and I. G. Toropov, *Zh. Neorg. Khim.*, 1986, **31**, 351.

colloid by Ce^{IV} has been investigated.[133] The rate equation is

$$rate = k[Pu][Ce^{IV}]^{0.5}[H^+]^{-1} \tag{24}$$

In mildly alkaline solutions both Pu^{VI} and Np^{VI} can be reduced by HS^- to the corresponding An^V at stopped-flow rates. These reactions have been investigated kinetically, and are deduced to proceed *via* the complexes $O_2An^{VI}(OH)(SH)$. Apparent stability constants were determined, as were the activation parameters for Pu. The reaction for Np is roughly five times faster than that for Pu.[134] Np^V can be reduced further by H_2O_2 in acidic media, in a reaction which is first order in peroxide but varies between first and second order in Np^V. The rate-determining step is deduced to be the one-electron reaction between NpO_2^+ and H_2O_2 which produces HO_2 radicals.[135]

The photochemistry of some $[UO_2L_5]^{2+}$ complexes in solution has been described. When the ligand L is dimethylformamide, and the complex is dissolved in DMF or acetone, irradiation at 430 nm yields a U^V product which can be spectrally character-ized under oxygen- and acid-free conditions. The process occurs *via* a U^{VI} excited state whose luminescence at 522 nm can be measured. This undergoes an intramolecular transfer to produce $U^VO_2L_4$ and $L^{\cdot}$.[136] N.m.r. studies show that the paramagnetic shifting of dimethyl sulphoxide by U^V is almost entirely due to pseudo-contact rather than contact terms. That is, there is very little U–DMSO orbital overlap. Not surprisingly then, the photochemical reaction of $[UO_2(DMSO)_5]^{2+}$ has a much smaller quantum yield than is found for the DMF complex. However, visible irradiation does still produce a U^V complex which is stable in the absence of O_2 and H^+, but one which rapidly disproportionates when acid is added.[137]

Acetylacetonate-ligand exchange on $[UO_2(acac)_2DMF]$ in an inert solvent has been investigated by n.m.r. line broadening. At low enol concentrations the rate depends on that concentration, but a limiting rate can be reached in the high region. The dissociation of one end of an acac ligand is involved in the rate-determining step.[138] The same research group has extended this work to include other neutral unidentate ligands both in place of DMF and as solvent. A pronounced kinetic isotope effect ($k_2^H/k_2^D = 2.6$) was measured. The rate-determining step for the whole series of reactions is deduced to be proton transfer from a singly co-ordinated entering group in its enol form to the leaving group.[139] A very similar mechanism is proposed for the ligand exchange on the lower IV oxidation state. Thus, when the rates of ligand exchange between $[U(acac)_4]$ and free Hacac are studied, it is found that they change with solvent. A mechanistic scheme which involves alternative nine-co-ordinate transition states is proposed. It can be summarized as shown in reactions (25)—(31). In this scheme the asterisk is used as an identifier for the

[133] D. J. Savage and T. W. Kyffin, *Polyhedron*, 1986, **5**, 743.
[134] K. L. Nash, J. M. Cleveland, J. C. Sullivan, and M. Woods, *Inorg. Chem.*, 1986, **25**, 1169.
[135] N. N. Malkova, V. A. Matyukha, T. V. Afanas'eva, and N. N. Krot, *Radiokhimiya*, 1986, **28**, 328.
[136] H. Fukutomi, T. Harazono, and T. Kojima, *Bull. Chem. Soc. Jpn.*, 1986, **59**, 2387.
[137] H. Fukutomi and T. Harazono, *Bull. Chem. Soc. Jpn.*, 1986, **59**, 3678.
[138] H. Fukutomi and Y. Ikeda, *Inorg. Chim. Acta*, 1986, **115**, 223.
[139] W.-S. Jung, H. Tomiyasu, and H. Fukutomi, *Inorg. Chem.*, 1986, **25**, 2582.

entering ligand, unidentate ligands are indicated by the prime, and B represents a unidentate base.[140]

$$(acac)_4U + enol^* \leftrightharpoons (acac)_4U(enol^*)' \qquad (25)$$

$$(acac)_4U(enol^*)' \rightarrow (acac)_3(acac^*)U(enol)' \qquad (26)$$

$$(acac)_3(acac^*)U(enol)' \rightarrow (acac)_3(acac^*)U + Hacac \qquad (27)$$

$$(acac)_4U + B \leftrightharpoons (acac)_4UB \qquad (28)$$

$$(acac)_4UB \leftrightharpoons (acac)_3UB(enolate)' \qquad (29)$$

$$(acac)_3UB(enolate)' + enol^* \rightarrow (acac)_3UB(enol)'(enolate^*)' \qquad (30)$$

$$(acac)_3UB(enol)'(enolate^*)' \rightarrow (acac)_3(acac^*)UB + Hacac \qquad (31)$$

Equilibria and rates of exchange have also been measured for the interactions between $[UO_2(CO_3)_3]^{4-}$ and several dithiocarbamate ligands. For the acyclic entering ligands complex formation follows a dissociative pathway, whereas complex formation with a macrocyclic dithiocarbamate was found to be associative. This difference is believed to be due to the topological requirements of the entering group, which can be met in one transition state but not in the other. When once the complex with the macrocycle has been formed, a 'macrocyclic protection effect' is noted. Its presence prevents further substitution by OH^- or HCO_3^-, but not by CO_3^{2-}.[141] ^1H and ^{13}C n.m.r. spectroscopy have been used to study the formation of complexes between UO_2^{2+} and tartaric acid. More different complexes can be detected for (D)-tartaric acid than for the *meso*-form, and (D,L)-tartaric acid gives still one more, a cyclic trimer.[142]

The behaviour of the actinoids in the environment is of obvious interest and importance. The migration of the normal forms of these elements involves their solution in water; since Th^{IV} is a good analogue for Pu^{IV}, the usual form in sub-surface geological environments, this more amenable ion is often investigated. So also are the earlier Ln^{III} ions as models for Am and Cm.[17] The migrations of Th^{IV}, Pa^V, U^{VI}, Np^V, Pu^{IV}, and Am^{III} through a compacted bentonite have been investigated under laboratory conditions. Added metallic Fe and $Fe_3(PO_4)_2$ slowed down Np movement, but only Fe did so for U. Neither HCO_3^- nor humic acid dissolved in the aqueous phase affected the uranyl migration.[143] That probably reflects strong humic–clay interactions, as large conditional stability constants have been determined for UO_2^{2+}–humate in soft, acid waters.[144] The rate of Th dissociation from humic acid has been measured in the range $4.2 < pH < 5.94$. A detailed analysis of the data shows evidence of thorium binding to seven different territorial, surface, or internal sites within the extremely complicated organic material.[145]

The final report of solution studies to be noted here is a sensitive colorimetric method to determine either Th^{IV} or Sm^{III}. The presence of a trace of either metal

[140] H. Fukutomi, H. Ohno, and H. Tomiyasu, *Bull. Chem. Soc. Jpn.*, 1986, **59**, 2303.
[141] I. Tabushi and A. Yoshizawa, *Inorg. Chem.*, 1986, **25**, 1541.
[142] M. T. Nunes and V. M. S. Gil, *Inorg. Chim. Acta*, 1986, **115**, 107.
[143] B. Torstenfelt, *Radiochim. Acta*, 1986, **39**, 105.
[144] J. P. Giesy, R. A. Geiger and N. R. Kevern, *J. Environ. Radioactivity*, 1986, **4**, 39.
[145] G. R. Choppin and W. P. Cacheris, *J. Less-Common Met.*, 1986, **122**, 551.

ion can be used to catalyse a colour reaction involving the other. This could prove to be a useful analytical tool.[146]

Although there has been continued activity in the area of the co-ordination chemistry of the actinoids, little excitement has been generated, and only a small number of reports are mentioned here. Raman and i.r. studies in the CN stretching region have been described for solutions of $[U(NCS)_8]^{4-}$ in CH_3NO_2 and CH_3CN. No free SCN^- was detected and so the configuration was deduced to be D_{4d} rather than O_h or D_{2d} as have been suggested previously.[147] An X-ray crystallographic study of thiophosphinate complexes of thorium, $[Th(R_2P(S)O)_4(solv)_4]$ (R = Me or Ph; solv = H_2O or EtOH), has revealed unexpectedly short Th—O bonds at 2.29 Å.[148] A good synthetic route to complexes of form UI_4L_2, *via* precursors with L = Ph_2CO or MeCN, has been described. It involves the reaction of uranium metal with I_2 and L in dry, oxygen-free CH_2Cl_2 and gives 60% yields of the precursors.[149]

Most new results for co-ordination compounds are concerned with multidentate ligands, often with newly synthesized ligands. One such carries a *vic*-dioxime ligand site on the side of a 15-crown-5. This ligand can attach to UO_2^{2+}, but only as an (NO)-bidentate with the crown well away from U.[150] Another newly derived uranophile comes from calix[6]arene. It has six carboxylate groups in a pseudo-planar array.[151] In a different system the uranyl group forms a seven-co-ordinate complex with the common distorted pentagonal bipyramidal geometry when treated with a dibasic, cyclic Schiff base compartmental ligand. The ligand, which offers two phenolate-O, four imine-N, and two thioether-S donor atoms, co-ordinates in an (O_2N_2S)-manner.[152]

The extracted complex from (18-crown-6)-UO_2Cl_2-HCl-H_2O has been prepared and characterized and its structure determined. The uranium is present as $UO_2Cl_4^{2-}$, whose charge is balanced by H_3O^+ ions. Two of these hydroxonium ions are anchored in the crown's cavity, each by three hydrogen bonds.[153] There is a pronounced difference in the interactions between U^{VI} and the A and B isomers of dicyclohexano-18-crown-6. Not only do the extracted species have different stoicheiometries, 1:2 and 2:3 respectively, but the A isomer bonds more strongly.[154] The final macrocycle to be mentioned is octaethylporphyrin. Both Th and U undergo the reaction sequence (32) with this ligand. The crystal structure of the thorium-containing product has been determined. As expected, the Th is situated above the N_4 plane with Th—N separations of *ca.* 2.5 Å.[155]

$$AnCl_4 + Por \rightarrow (Por)AnCl_2 \xrightarrow{\text{Na(acac)}} (Por)An(acac)_2 \qquad (32)$$

[146] I. Mori, Y. Fujita, K. Fujita, S. Kitano, I. Ogawa, H. Kawabe, Y. Koshiyama, and T. Tanaka, *Bull. Chem. Soc. Jpn.*, 1986, **59**, 955.

[147] W. P. Griffith and M. J. Muchford, *J. Chem. Soc., Dalton Trans.*, 1986, 1057.

[148] R. Mattes and L. Nieland, *Inorg. Chim. Acta*, 1986, **112**, 215.

[149] J. G. H. du Preez and B. Zeelie, *J. Chem. Soc., Chem. Commun.*, 1986, 743.

[150] A. Gul, A. I. Okur, A. Cihan, N. Tan, and O. Bekaroglu, *J. Chem. Res.*, 1986, 90.

[151] S. Shinkai, H. Koreishi, K. Ueda, and O. Moinabe, *J. Chem. Soc., Chem. Commun.*, 1986, 233.

[152] U. Casellato, S. Sitran, S. Tamburini, P. A. Vigato, and R. Graziani, *Inorg. Chim. Acta*, 1986, **114**, 111.

[153] W. J. Wang, B. Chen, P. Zheng, B. Wang, and M. Wang, *Inorg. Chim. Acta*, 1986, **117**, 81.

[154] W. J. Wang, Q. Sun, and B. Chen, *J. Radioanal. Nucl. Chem.*, 1986, **98**, 11.

[155] A. Dormond, B. Belkalen, P. Charpin, M. Lance, D. Vigner, G. Folcher, and R. Guilard, *Inorg. Chem.*, 1986, **25**, 4785.

There have been very few papers dealing with organo-actinoid chemistry that have not been based on cyclopentadienyl ligands. One such included details of the gas-phase X-ray photoelectron spectrum of $U(C_8H_8)_2$, which shows that although the bonding can be described as covalent the $5f$-electrons are essentially non-bonding.[113] That finding is pertinent to some of the information obtained for other organo-uranium compounds. Two other papers follow up last year's report of the synthesis and structural characterization of an η^6-C_6H_6 complex of uranium. The ligand has now been changed to C_6Me_6, which yields the otherwise identical $[U_2Cl_4(\mu\text{-}Cl)_3(\eta^6\text{-}C_6Me_6)_2]AlCl_4$. The solid-state ^{13}C n.m.r. spectrum of this dimer gives reasonable linewidths in relatively short times despite containing U^{IV}. At room temperature only a single resonance is seen, but this splits into two unequal peaks as the temperature is reduced, with the peak due to aromatic C shifting.[156] By contrast the U^{III} trimer, $[U_3(\mu^3\text{-}Cl)_2(\mu^2\text{-}Cl)_3(\mu^1, \eta^2\text{-}AlCl_4)_3(\eta^6\text{-}C_6Me_6)_3]AlCl_4$, prepared by the same research group with the same ligand, shows only very broad lines in its ^{13}C n.m.r. spectrum.[157]

All the papers still to be discussed in this chapter are concerned with mixed-ligand complexes of Th or U that contain cyclopentadienyl ligands. The order in which these are discussed is decided mainly by the position of the other donor atoms in the Periodic Table. That is a reasonable approach as the properties of Th and U complexes are usually comparable. For example, when the bond disruption enthalpies of $Cp'_2An(R)X$ pairs (An = Th or U; R = H, alkyl, or aryl; X = alkyl, Cl, or OR) are determined the approximate relationship (33) holds.[158] There are some differences of course, for example the relative stabilities of differing oxidation states, but such factors can be recognized. Thus, when comparisons of bond lengths are made between $[Cp_3U^{OX}(Bu^n)]$ (OX = III or IV), the U—C σ-bond lengths are seen to change from 2.557 to 2.426 Å respectively; the U—Cp bonds are also approximately 0.1 Å shorter for IV than III.[159]

$$D(\text{Th}-\text{R}) - D(\text{U}-\text{R}) = 40 \text{ kJ mol}^{-1} \tag{33}$$

$$K[L_2ReH_6] + Cp_3UCl \rightarrow L_2ReH_6UCp_3 \tag{34}$$

A heterometallic compound can be made by reaction (34) where L is a triarylphosphine. At 30 °C all the hydride ligands are apparently equivalent in the ^1H n.m.r. spectrum, and a slow limiting spectrum could not be obtained. Nevertheless, at least three bridging protons are considered to be present.[160] The first preparation at a normal temperature of a molecular carbonyl complex of uranium has been reported, and the evidence so far available appears to be satisfactory for the formulation $(Me_2SiC_5H_4)_3U(CO)$.[161] Very good yields of cationic uranium complexes containing isocyanide ligands have been described in the cases of $[Cp_3U(CNR')_2]^+$ and $[Cp_3U(NCR)(CNR')]^+$.[162]

[156] G. C. Campbell, F. A. Cotton, J. F. Haw, and W. Schwotzer, *Organometallics*, 1986, **5**, 274.

[157] F. A. Cotton, W. Schwotzer, and C. Q. Simpson, *Angew. Chem.*, 1986, **25**, 637.

[158] J. W. Bruno, H. A. Stecher, L. R. Morss, D. C. Sonnenberger, and T. J. Marks, *J. Am. Chem. Soc.*, 1986, **108**, 7275.

[159] L. Arnaudet, P. Charpin, G. Folcher, M. Lance, M. Nierlich, and D. Vigner, *Organometallics*, 1986, **5**, 270.

[160] D. Baudry and M. Ephritikhine, *J. Organomet. Chem.*, 1986, **311**, 189.

[161] J. G. Brennan, R. A. Andersen, and J. L. Robbins, *J. Am. Chem. Soc.*, 1986, **108**, 335.

[162] H. Aslan and R. D. Fischer, *J. Organomet. Chem.*, 1986, **315**, C64.

A range of Cp'_2ThR_2 complexes has been synthesized and characterized. Some of these complexes undergo a unimolecular cyclometallation reaction in solution in saturated hydrocarbons. For example, reaction (35) has enthalpy and entropy of activation values of $90 \, kJ \, mol^{-1}$ and $-67 \, J \, K^{-1} \, mol^{-1}$ respectively. The rate of reaction increases, both parameters becoming more positive ($105 \, kJ \, mol^{-1}$ and $-44 \, J \, K^{-1} \, mol^{-1}$), when the alkyl group is changed to CH_2SiMe_3. When both these alkyl groups are on the same Th atom cyclometallation proceeds exclusively on the silicon-containing ligand. The main determinant here is likely to be the increase in size on going from C to Si. Some speculations on mechanistic routes are offered.[163] Other studies of Cp'_2AnR_2 for both Th and U have also been reported. Double carbonylation and C=C bond formation, leading to the production of complexes containing the *cis*-enediolato ligand, $RC(O)=C(O)R$, occur. Experiment shows that diolate formation always occurs on one metal centre. A theoretical treatment supports the idea that the obvious intermediate, $Cp'_2An(\eta^2\text{-}COR)_2$, is involved. The coupling of carbamoyl ligands in $Cp'_2An(\eta^2\text{-}CONR)_2$ is, as expected, more difficult.[164] A carbonyl group can be inserted into a U—C multiple bond, as shown in reaction (36). It is retained on heating to 90 °C, at which temperature the isomerization (4a)—(4b) occurs.[165]

$$Cp'_2Th(CH_2CMe_3)_2 \rightarrow Cp'_2Th\overline{CH_2CMe_2CH_2} + CMe_4 \qquad (35)$$

$$Cp_3U=CHP(PhRMe) + W(CO)_6 \rightarrow (OC)_5WC(OUCp_3)-CHP(PhRMe) \qquad (36)$$

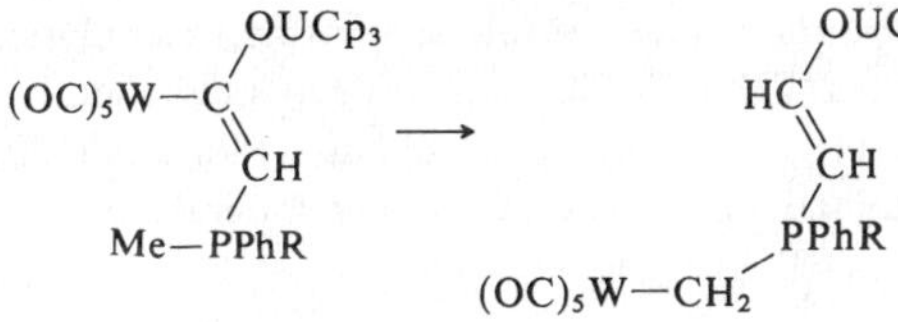

(4a)   (4b)

The product of reaction (37), giving an η^4-butadiene complex, is very air- and moisture-sensitive. Its n.m.r. spectrum shows it to be a fluxional molecule, undergoing a ring inversion (5a) $\rightleftharpoons$ (5b) with a free energy of activation of $65 \, kJ \, mol^{-1}$ at 23 °C.[166]

$$Cp'_2ThCl_2 + Mg(C_4H_6) \rightarrow Cp'_2Th(\eta^4 - C_4H_6) \qquad (37)$$

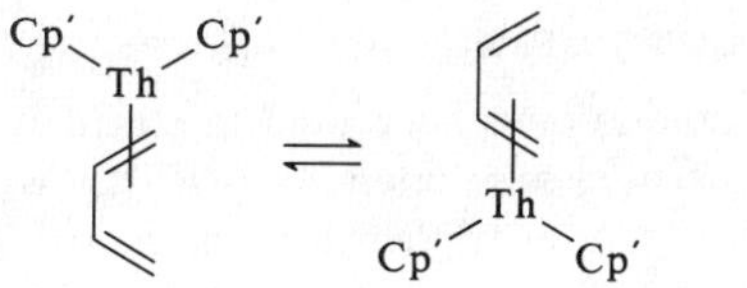

(5a)   (5b)

[163] J. W. Bruno, G. M. Smith, T. J. Marks, C. K. Fair, A. J. Schultz, and J. M. Williams, *J. Am. Chem. Soc.*, 1986, **108**, 40.
[164] K. Tatsumi, A. Nakamura, P. Hofmann, R. Hoffmann, K. G. Moloy, and T. J. Marks, *J. Am. Chem. Soc.*, 1986, **108**, 4467.
[165] R. E. Cramer, J. H. Jeong, and J. W. Gilje, *Organometallics*, 1986, **5**, 2555.
[166] G. Erker, T. Muhlenbernd, R. Benn, and A. Rufinska, *Organometallics*, 1986, **5**, 402.

High-yielding preparative routes to the various diethylamine derivatives indicated in reaction (38) have been described; n does not take the value 1 for Th. The reactions are under stoicheiometric and kinetic control.[167]

$$An(NEt_2)_4 + Cp \rightarrow Cp_nAn(NEt_2)_{4-n} \qquad n = 1, 2, \text{ or } 3 \qquad (38)$$

The uranium(IV) compounds $Cp_2U(NEt_2)_2$ and $Cp_3U(NEt_2)$ undergo U—N bond cleavage reactions (39) with alcohols and phenols. The product yields are controlled by the bulk of the new ligand, and the bis-alkoxide can disproportionate or decompose to give the mono-derivative.[168]

$$Cp_nU(NEt_2)_{4-n} + ROH \rightarrow Cp_nU(OR)_{4-n} \qquad n = 2 \text{ or } 3 \qquad (39)$$

Mass spectral patterns have been studied for the Cp_3UX species (X = η^1-NEt$_2$; η^2-CMe=N-cyclohexyl, or η^2-CMe=NPr$_n$). Some hydride species were detected, and a mechanism for the β-hydride rearrangement was discussed.[169]

There have been some reports concerned with cyclopentadienyl compounds containing terminal or bridging diphenylphosphido ligands. More detailed information has been given about the $Cp'_2Th(PPh_2)_2$-type compounds. They are highly air-sensitive, with intense purple–red colours attributed to phosphide–metal CTTM transitions. The average Th—P bond length is 2.87 Å for the cited compound, *i.e.* showing no sign of significant multiple bonding. The P(cyclohexyl)$_2$-containing compound is very soluble in non-polar organic solvents. Its $^{31}P(^1H)$ n.m.r. spectrum is an unbroadened singlet even down to $-120\,°C$, indicating that the solid and solution structures are the same.[170] Molecular orbital calculations on the Th—Ni interaction in $Cp_2Th(\mu\text{-}PH_2)_2Ni(CO)_2$ have been published.[171] This species is a simplified version of last year's reported $Cp_2Th(\mu\text{-}PPh_2)_2Ni(CO)_2$. A new bridged product can be obtained in toluene at room temperature by reaction (40). The

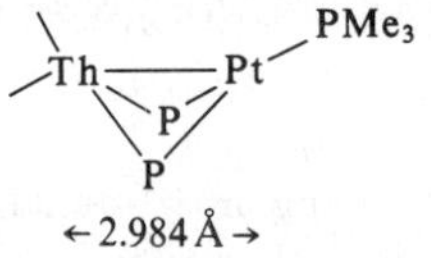

$$\leftarrow 2.984\,\text{Å} \rightarrow$$

(6)

bridging geometry is as shown in (6), with a surprisingly short Th—Pt separation. N.m.r. evidence suggests that the structure is unchanged in toluene solution.[172]

$$Cp'_2Th(PPh_2)_2 + Pt(COD)_2 + PMe_3 \rightarrow Cp'_2Th(\mu\text{-}PPh_2)_2Pt(PMe_3) + 2COD \qquad (40)$$

^1H and ^{31}P n.m.r. studies have been used to elucidate the structure and follow the ligand exchange reactions in solution of $CpUCl_3L_2$, where L is hexamethylphosphoramide, tetrahydrofuran, or triphenylphosphine oxide. In CDCl$_3$ two isomers

[167] F. Ossola, G. Rossetto, P. Zanella, and R. D. Fischer, *J. Organomet. Chem.*, 1986, **309**, 55.

[168] A. Berton, M. Porschia, G. Rossetto, and P. Zanella, *J. Organomet. Chem.*, 1986, **302**, 351.

[169] G. Paducci, S. Daolio, and P. Traldi, *J. Organomet. Chem.*, 1986, **309**, 283.

[170] D. A. Wrobleski, R. R. Ryan, H. J. Wasserman, K. V. Salazar, R. T. Paine, and D. C. Moody, *Organometallics*, 1986, **51**, 90.

[171] J. V. Ortiz, *J. Am. Chem. Soc.*, 1986, **108**, 551.

[172] P. J. Hay, R. R. Ryan, K. V. Salazar, D. A. Wrobleski, and A. P. Sattelberger, *J. Am. Chem. Soc.*, 1986, **108**, 313.

occur as shown in (7). The equatorial ligand L in the *cis-mer*-isomer exchanges associatively with free L, which the *trans-mer*-ligands do not. This gives considerable guidance about the transition state, which is obviously pentagonal bipyramidal with the cyclopentadienyl ligand axial, and with steric constraints on the equatorial groups.[173]

$$
\begin{array}{c}
\text{Cp} \\
\text{Cl} \diagdown \,\big|\, \diagup \text{Cl} \\
\diagup \,\big|\, \diagdown \\
\text{Cl} \;\;\big|\;\; \text{L} \\
\text{L}
\end{array}
\;\rightleftharpoons\;
\begin{array}{c}
\text{Cp} \\
\text{L} \diagdown \,\big|\, \diagup \text{Cl} \\
\diagup \,\big|\, \diagdown \\
\text{Cl} \;\;\big|\;\; \text{L} \\
\text{Cl}
\end{array}
$$

$$(7)$$

During a carbonylation reaction in the presence of phosphines, enedionediolate- and ylide-containing complexes can be formed according to equations (41) and (42). Kinetic studies of reaction (42) showed that the rate-determining step has a second-order rate law and is the coupling of CO to the η^2-acyl ligand to yield an intermediate which is either a ketene or is ketene-like. The initial carbonylation is regiospecific to the α-position of the ylide, and isocyanides attack in a similar fashion.[174]

$$
\text{Cp}'_2\text{Th}(\eta^2\text{-COCH}_2\text{Bu}^t)\text{Cl} \rightarrow \text{Cp}'_2\text{Th}[\overline{\text{OC(CH}_2\text{Bu}^t)\text{C(PR}_3)\text{O}}]\text{Cl} \qquad (41)
$$

$$
\text{Cp}'_2\text{Th}(\eta^2\text{-COCH}_2\text{Bu}^t)\text{Cl} \rightarrow \{\text{Cp}'_2\text{Th}[\overline{\text{OC(CH}_2\text{Bu}^t)\text{CO}}]\text{Cl}\}_2 \qquad (42)
$$

More detail of the compound $\text{Cp}'_3\text{ThS}_5$ has now been given,[175] and new compounds containing S co-ordinated to U have been described. When UCp''_3 (where Cp'' represents a monosubstituted cyclopentadienyl ligand such as $\text{C}_5\text{H}_4\text{Me}$) is treated with carbon disulphide, electron-transfer reactions occur leading to the formation of uranium(IV) dimers, $[\text{Cp}''_3\text{U}]_2[\mu\text{-}\eta^1, \eta^2\text{-CS}_2]$. No U—U magnetic interactions occur in these dimers, even down to 5 K.[176] Related species of formula $[\text{Cp}''_3\text{U}]_2\text{E}$ can be made where E = COS, SPPh$_3$, SePPh$_3$, or TePBu$^n_3$.[177]

Finally we note that for $\text{Cp}'' = \text{C}_5\text{H}_3(\text{SiMe}_3)_2$ and with a halide as X, a complete series of dimers $[\text{Cp}''_2\text{UX}]_2$ has been prepared and characterized.[178]

[173] J. F. Le Marechal, M. Ephritikhine, and G. Folcher, *J. Organomet. Chem.*, 1986, **299**, 85.

[174] K. G. Moloy, P. J. Fagan, J. M. Manriquez, and T. J. Marks, *J. Am. Chem. Soc.*, 1986, **108**, 56.

[175] D. A. Wrobleski, D. T. Cromer, J. V. Ortiz, T. B. Rauschfuss, R. R. Ryan, and A. P. Sattelberger, *J. Am. Chem. Soc.*, 1986, **108**, 174.

[176] J. G. Brennan, R. A. Andersen, and A. Zalkin, *Inorg. Chem.*, 1986, **25**, 1756.

[177] J. G. Brennan, R. A. Andersen, and A. Zalkin, *Inorg. Chem.*, 1986, **25**, 1761.

[178] P. C. Blake, M. F. Lappert, R. G. Taylor, J. L. Atwood, W. E. Hunter, and H. Zhang, *J. Chem. Soc., Chem. Commun.*, 1986, 1394.

13 Radiochemistry

By D. S. URCH

Chemistry Department, Queen Mary College, London E1 4NS

and

C. E. URCH

Department of Medicine, University of Southampton, Southampton SO9 5NH

1 Introduction

Following the style and precedent of previous years this section of *Annual Reports* is devoted to a review of recent (roughly mid-85 to mid-86) work in those aspects of radiochemistry where the radioactivity itself is of importance or interest; the chemistry of those elements which happen to be radioactive is excluded. Three new Nuclear Chemistry text-books[1] have appeared.

2 Isotope Production

General aspects of the production of radioisotopes in Belgium[2] and South Africa[3] have been reviewed as have the more specific topics of the production of short-lived positron emitters of biomedical interest using cyclotrons[4a—4d] or linear accelerators[4e] and of the use of radionuclide generators.[5]

Tritium, the radioactive isotope of hydrogen, can be produced by the laser,[6] (CO_2)[6b] irradiation of suitable compounds (*e.g.* CF_3C^3HClF),[6c] or separated from other isotopes using a thermal diffusion column[7] or photoassisted catalysis $(Pt—TiO_2)$.[8] ^{18}F can be made by a variety of nuclear reactions such as that of

[1] (a) A. Willi, 'Advanced Course in Nuclear Chemistry', Diesterweg, Frankfurt am Main, F. R. Germany, 1986; (b) O. Andrysek, 'Radiological Physics', Avicenum, Prague, Czechoslovakia, 1984; (c) H. J. Arniker, 'Essentials of Nuclear Chemistry', Halsted Press, New York, U.S.A., 1982.

[2] J. M. Alardin, *Rev. IRE*, 1986, **10**, 35.

[3] 'Annual Report–National Accelerator Centre', Council for Scientific and Industrial Research, Faure, South Africa, 1985, p. 152.

[4] (a) A. P. Wolf and W. B. Jones, *Radiochim. Acta*, 1983, **34**, 1; (b) S. M. Qaim and G. Stoecklin, *ibid.*, p. 25; (c) H. Guratzsch, *Radiol. Diagn.*, 1986, **27**, 296; (d) T. Bjoernstad, Thesis (Ph.D.), Bergen Univ., Bergen, Norway, 1986; (e) J. McKeown, *IEEE Trans. Nucl. Sci.*, 1985, **NS-32**, 3292.

[5] R. M. Lambrecht, *Radiochim. Acta*, 1983, **34**, 9.

[6] (a) K. Takeuchi, Y. Makide, and I. Inoue, *J. Chem. Eng. Jpn.*, 1985, **18**, 1; (b) K. Takeuchi, S. Satooka, and Y. Makido, *J. Nucl. Sci. Technol.*, 1984, **21**, 959; (c) O. Kuihara, K. Takeuchi, S. Sakae, and Y. Makido, *J. Nucl. Sci. Technol.*, 1983, **20**, 617.

[7] A. Neubert, H. Heimbach, and H. R. Ihle, Proc. 13th Symposium on Fusion Technology, Pergamon Press, Oxford, U.K., 1984, Vol. 1, p. 417.

[8] K. Watanabe, K. Ichimura, N. Inoue, and I. Matsuura, *J. Phys. Chem.*, 1986, **90**, 866.

accelerated protons with ^{18}O-enriched water,[9] [^{18}O(p, n)^{18}F], or the bombardment of ordinary water with ^{3}He ions,[10] [^{16}O(^{3}He, p)^{18}F]. Improvements have also been reported[11] in methods for ^{18}F production in nuclear reactors utilizing the reaction of recoil tritons, produced by the neutron irradiation of lithium, with oxygen.

A method has been described[12] for the concentration and separation of the transition-metal isotopes ^{46}V, ^{51}Cr, ^{54}Mn, ^{59}Fe, ^{58}Co, and ^{60}Co by precipitation with ferric hydroxide. More specifically techniques for making the short-lived isotope ^{55}Co, [^{54}Fe(d-12 MeV, n)^{55}Co],[13] and the isotopes of copper ^{64}Cu, [^{63}Cu(n, γ)^{64}Cu],[14] and ^{67}Cu, [^{68}Zn(p-193 MeV, 2p) or ^{70}Zn(p-193 MeV, α) or ^{67}Zn(n-fast, p)^{67}Cu],[15] have been reported. Other isotopes, e.g. ^{7}Be, ^{46}Sc, ^{52}Mn, ^{59}Fe, and ^{67}Ga, are routinely isolated from the spent zinc oxide targets[16] after their bombardment with protons of energies up to 800 MeV.

The bombardment of nickel with 72.5 MeV ^{12}C ions leads to the formation of ^{66}Ge and ^{67}Ge. These isotopes can be separated[17] from other reaction products in the helium gas flow by the addition of silver chloride at 600 °C. ^{77}As can be made[18] by the more conventional neutron irradiation of germanium oxide. As many of the isotopes of bromine have radiopharmaceutical potential,[19] interest continues in their production, [^{76}Se(p-28 MeV, 2n)^{75}Br], from copper selenide[20] and in their final chemical state.[21] The isotopes of rubidium have also received attention, ^{81}Rb as a source[3] of ^{81m}Kr, and ^{85}Rb—^{87}Rb as a problem in isotope separation. This problem has been solved[22] by repeated ion electromigration on a cation-exchange membrane. ^{85}Sr can be made by the deuteron bombardment of rubidium chloride[23] followed by separation on an ion-exchange column. Similar methods of purification have been used to separate ^{89}Sr from irradiated uranium (column with polyantimonic acid),[24] ^{90}Sr from ^{90}Y (column of ceric phosphate),[25] and ^{89}Sr from ^{91}Y and ^{147}Nd.[26] A very different approach to isotope separation has been used to concentrate ^{91}Zr

[9] (a) M. Vogt, I. Huszar, M. Argentini, H. Oehninger, and R. Weinreich, *Appl. Radiat. Isot.*, 1986, **37**, 448; (b) J. Keinonen, A. Fontell, and A. L. Kairento, *ibid.*, p. 631; (c) O. T. Dejesus, J. A. Martin, N. J. Yasillo, S. J. Gatley, and M. D. Cooper, *ibid.* p. 397.

[10] E. J. Kunst, H. J. Machulla, and W. Roden, in 'Radioactive Isotopes in Clinic and Research, Gastein International Symposium', ed. R. Hoefer and H. Bergmann, Egermann, Vienna, Austria, 1986, Vol. 17, p. 45.

[11] B. W. Wessels, W. R. Yuosoff, and D. Ercegovic, *Radional. Nucl. Chem.*, 1985, **92**, 27.

[12] A. I. Novikov, V. F. Samojlova, and A. A. Shaffert, *Radiokhimiya*, 1985, **27**, 508.

[13] H. Sharma, J. Zweit, A. M. Smith, and S. Downey, *Appl. Radiat. Isot.*, 1986, **37**, 105.

[14] C. Nakamshi, C. P. G. da Silva, Proc. 1st Gen. Congress Nuc. Energy, Ass. Brasileira de Energia Nuclear, Rio de Janeiro, Brasil, 1986, Vol. 2, p. 213.

[15] S. Mirzadeh, L. F. Mausner, and S. C. Srivastava, *Appl. Radiat. Isot.*, 1986, **37**, 29.

[16] K. E. Thomas, *Radiochim. Acta*, 1986, **39**, 69.

[17] J. Ma, X. Sun, W. Mou, X. Xiujun, C. Zhou, and Z. Zhao, *J. Nucl. Radiochem.*, 1985, **7**, 96.

[18] A. R. Byrne, Abstracts, Int. Conf. on Analytical Chemistry in Nuclear Technology, Pub. Kernforschungs- zentrum, Karlsruhe, B.R.D., 1985, p. 205.

[19] M. J. Welch and K. D. McElvany, *Radiochim. Acta*, 1983, **34**, 41.

[20] W. Vaalburg, A. M. J. Paans, J. W. Terpstra, T. Wiegman, K. Dekens, A. Rijskamp, and M. G. Woldring, *Int. J. Appl. Radiat. Isot.*, 1985, **36**, 961.

[21] P. M. Wanek and P. M. Grant, *J. Radioanal. Nucl. Chem.* 1986, **99**, 187.

[22] M. Hosoe and T. Nagumo, *Nippon Kagaku Kaishi*, 1985, **5**, 877.

[23] P. M. Smith-Jones, F. W. E. Strelow, and R. G. Boehmer, *Appl. Radiat. Isot.*, 1986, **37**, 240.

[24] R. N. Varma, K. L. N. Rao, G. N. Chavan, K. R. Balasubramanian, and T. S. Murthy, Proc. Radio- chemistry and Radiation Chemistry Symposium, Pune, India, 1982, paper SC-9.

[25] D. K. Bhattacharyya and A. De, ref. 24, paper SC-3.

[26] R. N. Varma, C. Mathai, V. I. Dhiwar, and K. R. Balasubramanian, ref. 24, paper SC-8.

selectively. It has been suggested[27] that the magnetic properties of a nucleus with an odd number of nucleons will facilitate electronic relaxation processes and thereby alter the rates at which ^{91}Zr reacts chemically.

Whilst most attention focuses on ^{99}Mo as a source of ^{99m}Tc the production of the isotope ^{93}Mo has been reported[28] from the proton or deuteron irradiation of niobium. If α-particle bombardment (16 MeV) is used[29] then isotopes of technetium result, ^{95m}Tc and ^{96}Tc. Whether ^{99}Mo results from nuclear fission or from the irradiation of a molybdenum compound, if impurities are present they will compromise the quality of the daughter technetium. Methods have therefore been developed to ensure the removal of ^{132}Te from fission-produced ^{99}Mo,[30] and of rhenium from molybdenum trioxide[31] (traces of tungsten which give rise to ^{188}Re are still a problem though). Various types of new technetium generators have been described,[32,33] insoluble zirconium molybdate gel being quite popular.[33]

Methods for the preparation of ^{102}Rh,[34] ^{103}Ru,[35] and ^{111}In[36] have been described. The more extensively used isotope of indium ^{113m}In is usually made by decay of ^{113}Sn; a suitable generator is tin oxide mounted on silica gel.[37] A new method for the production of ^{125}I by the neutron irradiation of solid xenon difluoride has been reported[38] and the lighter isotope ^{123}I, which is also of interest in nuclear medicine, can be made either by the proton irradiation of tellurium oxide[39] [^{124}Te(p-27 MeV, 2n)^{123}I] or by the decay of ^{123}Xe following the γ-irradiation of ^{124}Xe[40a] [^{124}Xe(γ, n)^{123}Xe]. A ^{122}I generator has been reported[40b] based on the decay of ^{122}Xe. Lighter still isotopes of iodine (^{117}I, ^{118}I, and ^{119}I) are produced[41] when a silver target is bombarded with 67.4 MeV ^{12}C ions.

Details have been given for the extraction of ^{161}Tb from neutron irradiated gadolinium[42] and of ^{167}Tm from ytterbium sulphate targets after γ-irradiation.[43] A wide range of rare earth, together with some hafnium, isotopes was found to result[44]

[27] D. C. Phillips and S. H. Peterson, Calif. Patent 1 173 054/A, U.S.A., 1984.
[28] A. I. Silantev, A. G. Maklachkov, and B. Z. Iofa, *Radiokhimiya*, 1985, **27**, 679.
[29] N. Ramamoorthy, M. K. Das, B. R. Sarkar, and R. S. Mani, *J. Radioanal. Nucl. Chem.*, 1986, **98**, 121.
[30] S. A. C. Mestnik and C. P. G. daSilva, ref. 14, Vol. 2, p. 263.
[31] E. J. Lee, P. J. Sorby, R. K. Barnes, and R. E. Boyd, Report AAEC/E-615, Australian Atomic Energy Commission Research Establishment, Lucas Heights, Australia, 1985.
[32] (*a*) N. A. Morcos, G. A. Bruno, and T. A. Haney, Norwegian Patent 149 926/C, 1984; (*b*) K. Svoboda, F. Melichar, M. Tympl. J. Lezama, and J. Tendilla, *J. Radioanal. Nucl. Cham.*, 1985, **96**, 405; (*c*) O. P. D. Noronha, *Isotopenpraxis*, 1986, **22**, 53.
[33] (*a*) R. E. Boyd, *Nucl. Spectrom.*, 1986, **2**, 18; (*b*) J. G. Wilson and R. E. Boyd, Report AAEC/E-616, Australian Atomic Energy Commission Research Establishment, Lucas Heights, Australia, 1985.
[34] B. Gorski and W. Heinig, *Isotopenpraxis*, 1986, **22**, 179.
[35] V. L. Kochetkov, V. I. Plotnikov, and T. I. Taurbaeva, *Radiokhimiya*, 1985, **27**, 257.
[36] A. G. daSilva, F. C. M. Teixeira, and A. S. F. deSouza, Ann. Tech. Report of Nuclear Engineering Institute, Rio de Janeiro, Brazil, 1984, p. 38.
[37] A. Galik, *Radiochim. Acta*, 1985, **38**, 149.
[38] P. Hradilek, L. Kronrad, and K. Kopicka, *Radioisotopy*, 1985, **26**, 28.
[39] (*a*) P. Kopecky, P. Hradilek, and L. Kronrad, *Radioisotopy*, 1985, **26**, 38; (*b*) J. L. Q. deBritto, ref. 36, p. 39.
[40] (*a*) I. K. Kikoin, S. S. Yakimov, and N. A. Chernoplekov, *J. Radioanal. Nucl. Chem.*, 1986, **103**, 27; (*b*) C. A. Mathis, T. Sargent, Y. Yano, A. Vuletich, M. C. Lagunas-Solar, L. J. Harris, and M. H. Grant, *Appl. Radiat. Isot.*, 1986, **37**, 258.
[41] T. Zhang, M. Fu, and D. Wu, *J. Nucl. Radiochem.*, 1985, **7**, 101.
[42] A. Subhodaya, C. R. Biswas, B. S. Kullolli, and V. C. Nair, ref. 24, paper SC-10.
[43] N. Toda, Y. Shirota, and K. Sakamoto, *J. Radioanal. Nucl. Chem.*, 1985, **92**, 333.
[44] K. E. Thomas, *Radiochim. Acta*, 1983, **34**, 135.

from the proton bombardment of tantalum. Conversely hafnium, when subjected to α-particle irradiation, gives rise[45] to isotopes of tantalum (^{179}Ta and ^{182}Ta) and also some of tungsten (^{178}W and ^{181}W). Details of an electrochemical procedure for the preparation of tungsten-181 oxide have been reported.[46] Another isotope of tungsten (^{188}W) is used as a source of ^{188}Re; adsorption of compounds of this isotope of tungsten onto alumina makes[47] a convenient generator of ^{188}Re. As platinum complexes are used in anti-tumour therapy there is interest in the production of radioactive isotopes of platinum for use as tracers. Neutron irradiation initiates many (n, γ) reactions,[48] giving rise to isotopes with masses 191, 193, 195, and 197. Radioactive gold also has radiopharmaceutical interest and details have recently been given of the preparation of ^{194}Au from ^{194}Hg[49] and of the construction of a gold 'generator' in which the mercury 'parent' is adsorbed onto a thiolated kieselghur support.[50] New methods have been reported for the preparation of ^{201}Tl from proton irradiated lead[51] or thallium[52] targets.

Optimization of the conditions for the preparation[53a] and extraction[53b] of ^{211}At produced by the bombardment of bismuth with 26 MeV α-particles [^{209}Bi(α, 2n)^{211}At] has been reported. A recent patent[54] describes how ^{226}Ra can be separated from solution by precipitation. Interest is still focused on methods for the extraction of ^{235}U from natural uranium by laser irradiation[55] (of UF$_6$[56]), by distribution between two liquid phases[57] or by non-equilibrium gas diffusion using a rotating cylindrical slit nozzle.[58] Extended high neutron flux irradiation of ^{252}Cf leads to the production[59] of microgram quantities of ^{253}Es. This isotope, when bombarded with 26 MeV α-particles,[60] gives rise to ^{256}Md. Methods for the production of these and other trans-plutonium elements, up to $Z = 109$, have recently been reviewed.[61]

[45] S. M. Gasita, A. I. Silant'ev, A. G. Maklachkov, and B. Z. Iofa, *Radiokhimiya*, 1985, **27**, 236.

[46] I. V. Matveev, V. A. Pisanko, A. A. Sverdlina, and Y. N. Simirskij, *Radiokhmiya*, 1985, **27**, 270.

[47] N. Botros, M. El-Garhy, S. Abdulla, and H. F. Aly, *Isotopenpraxis*, 1986, **22**, 368.

[48] T. R. Sykes, L. G. Stephens-Newsham, and A. A. Noujaim, *Appl. Radiat. Isot.*, 1986, **37**, 231.

[49] B. Ajkhler and V. P. Domanov, Soviet Union Patent 1 119 981/A, 1984.

[50] A. Rosevear, U.K. Patent 2 151 599/A, 1985.

[51] (*a*) J. L. Q. deBritto, A. M. S. Braghirolli, and A. G. deSilva, *J. Radioanal. Nucl. Chem.*, 1985, **96**, 181; (*b*) V. A. Ageev, A. A. Kljucnikov, A. F. Linev, V. A. Chalkin, and N. G. Zajceva, East German Patent 3 402 348/A, 1984.

[52] P. Kopecky, K. Zdrazil, and P. Svihal, *Radioisotopy*, 1984, **25**, 767.

[53] (*a*) F. Roesch, R. Dreyer, J. Henniger, and G. J. Beyer, *J. Radioanal. Nucl. Chem.*, 1985, **96**, 319; (*b*) C. Luo, Q. Duan, J. Zhang, Y. Zhang, L. Xiao, B. Liu, Y. Jín, Z. Liu, and Z. Tang, *J. Nucl. Radiochem.*, 1984, **6**, 243.

[54] P. M. Huck, R. C. Andrews, and W. B. Anderson, South African Patent 84/8197/A, 1985.

[55] I. Bernstein, *Isotopenpraxis*, 1986, **22**, 164.

[56] H. Johansen and K. Johst, *Isotopenpraxis*, 1986, **22**, 157.

[57] J. Stamberg, J. Cabicar, and K. Stamberg, Czech Patent 224 844/B, 1984.

[58] M. Fiebig, W. Schwan, and N. K. Mitra, Proc. 14th Int. Symp. on Rarefied Gas Dynamics, ed. H. Oguchi, Univ. of Tokyo Press, Tokyo, Japan, 1984, p. 665.

[59] S. A. Kalyukhin, L. N. Auehrman, V. L. Novichenko, N. B. Mikheev, I. A. Rumer, A. N. Kamenskaya, L. A. Goncharov, and A. I. Smirnov, *Radiokhimiya*, 1985, **27**, 833.

[60] N. B. Mikheev, V. L. Novichenko, L. N. Auehrman and A. N. Kamenskaya, *Radiokhimiya*, 1985, **27**, 835.

[61] C. Keller, *Chem. Ztg.*, 1986, **110**, 233.

3 Labelled Compounds

Introduction.—General reports from Russia,[62] East Germany,[63] and Czechoslovakia[64] give details of the production of labelled compounds in those countries. More specific reviews have considered the preparation of radiopharmaceuticals[65] in South Africa[66] and Czechoslovakia.[67] The particular importance of compounds labelled with short-lived positron emitting isotopes was also the subject of many recent publications.[68] The specific requirements for labelled compounds for use as tracers in an industrial environment have also been reviewed.[69]

Tritium, ^{3}H.—New methods have been reported for the preparation of tritiated alcohols, by using[70a] a beam of tritium atoms, and tritiated proteins and peptides by bombarding the compounds to be labelled with $[^3H_3]^+$, $[^3H_2]^+$ ions and fast $^3H_2$ molecules.[70b] More conventional procedures to achieve the same ends have been reviewed.[71] Many compounds are labelled with tritium simply by exposure to tritium gas; the efficiency of this process can be enhanced[72] by heating the gas with a tungsten filament. Catalytic reduction of unsaturated centres using tritium gas is widely used for the introduction of a radioactive label at a specific site within a molecule. This method has been reported in the preparation of ^{3}H-labelled amino-acids,[73] 2-fluoro-L-histidine,[74] 1,3-dipropylxanthine,[75] benzamide dopamine-D_2 ligands,[76] dihydro-1H-pyrrolizine-3,5(2H,5H)-dione,[77] steroids,[78] retinoidal benzoic acids,[79] and hexadecyl phosphocholine derivatives (PAF).[80] Reduction effected at triple bonds has been used in the preparation of tritium-labelled eicosatrienoic

[62] Committee for uses of Atomic Energy, Report, in Proc. 2nd, Socialist Countries Symp. Uses of Atomic Energy for Peacful Purposes–Part 1, Radioisotope-labelled Organic Compounds, 1982, p. 3.

[63] M. Bubner, K. H. Haize, E. Mittag, and J. Remer, ref. 62, p. 9.

[64] Abstracts, 3rd National Seminar on Labelled Compounds, applied in Biochemistry, Medicine and Biochemical Analysis, Srni, Czechoslovakia, 1984.

[65] H. Vera-Ruiz, *Int. At. Energy Agency Bull.*, 1985, **27**, 48.

[66] B. Mutch, *S. Afr. Pharm. J.*, 1984, **51**, 493.

[67] F. Melichar, K. Svoboda, and M. Hradil, *Jad. Energ.*, 1985, **30**, 425; *ibid.*, 1985, **31**, 57, 229, 306.

[68] (a) M. J. Welch, M. R. Kilbourn, and M. A. Green, *Radioisotopes (Tokyo)*, 1985, **34**, 170; (b) A. P. Wolf and J. S. Fowler, 2nd, Int. Symp. on the Synthesis and Application of Isotopically Labelled Compounds, Kansas City, U.S.A., 1985; (c) G. W. Kabalka, Report, DOE/EV/10363-8, 1985; (d) J. Hanus, ref. 64, p. 13; (e) H. H. Coenen, S. M. Moerlein, and G. Stoecklin, *Radiochim. Acta*, 1983, **34**, 47.

[69] R. Otto, Report No. 109, Central Institute for Isotope and Radiation Research, Leipzig, D.D.R., 1985, 125.

[70] G. A. Bush, Thesis, Georgia Inst. of Tech., Atlanta, U.S.A., 1981 (Univ. Microfilm 82-08786).

[71] (a) C. T. Peng and R. L. Hua, *Fusion Technol.*, 1985, **8**, 2265; (b) D. E. Brundish and R. Wade, *J. Labelled Compd. Radiopharm.*, 1986, **23**, 9.

[72] L. A. Nejman, V. S. Smolyakov, and L. P. Antropova, ref. 62, p. 33.

[73] (a) D. Ego, J. P. Beaucourt, and L. Pichat, *J. Labelled Compd. Radiopharm.*, 1986, **23**, 229; (b) D. Ego and J. P. Beaucourt, *ibid.* p. 553.

[74] K. Takahashi, K. L. Kirk, and L. A. Cohen, *J. Labelled Compd. Radiopharm.*, 1986, **23**, 1.

[75] K. A. Jacobson, K. Ukena, J. W. Daly, and K. L. Kirk, *J. Labelled Compd. Radiopharm.*, 1986, **23**, 519.

[76] L. Gawell, H. Hall, and C. Koehler, *J. Labelled Compd. Radiopharm.*, 1985, **22**, 1033.

[77] J. D. Hartman, J. H. Dodd, J. L. Hicks, F. M. Hershenson, C. C. Huang, and D. E. Butler, *J. Labelled Compd. Radiopharm.*, 1985, **22**, 583.

[78] (a) P. N. Rao and K. M. Damodaran, *Steriods*, 1984, **43**, 343; (b) J. Roemer and H. Wagner, ref. 62, p. 193; (c) F. V. Tatarkina, N. A. Ignat'eva, L. N. Kaklyushkina, I. F. Tupitsyn, V. I. Mishin, V. N. Kramerov, M. M. Moskvina, N. S. Pigoreva, L. E. Shabunevich, and L. I. Dovedova, ref. 62, p. 209.

[79] S. W. Rhee, J. I. DeGraw, and H. H. Kaegi, *J. Labelled Compd. Radiopharm.*, 1985, **22**, 843.

[80] S. D. Wyrick, J. S. McClanahan, R. L. Wykle, and J. T. O'Flaherty, *J. Labelled Compd. Radiopharm.*, 1985, **22**, 1169.

acid[81] and cholecystokinin derivatives.[82] If reduction is carried out in the presence of a chiral catalyst then labelled optically active peptides can be made.[83] Other studies of this tritium labelling reaction have included an investigation of the expected isotope effect[84] and the use of nickel saturated with pre-adsorbed tritium in place of platinum catalyst.[85] Another popular labelling reaction involving tritium gas is catalytic dehalogenation,[86]

$$RX + {}^3H_2 \rightarrow R^3H + {}^3HX \quad (X = \text{halogen})$$

which again has the advantage of placing the label at a specific site in the molecule. Uracil has been labelled at both the 5- and 6-positions,[87a] and pseudouracil has also been labelled[87b] as have many nucleic acids and nucleosides.[88] Palladium-based catalysts have been reported[89] to be particularly effective. Other compounds produced by this method were labelled phosphorane derivatives,[90] 5-succinimide-4-oxopent-2-enoic acid,[91] and thioridazine.[92] Gaseous tritium can also be used in catalysed exchange reactions involving hydrogen atoms; again palladium catalysts seem very efficacious in promoting the labelling of amino-acids,[93] acetylene derivatives,[94] and gossypol.[95]

Tritiated water can be used to produce labelled compounds, usually by catalysed exchange, *e.g.* pseudouracil,[87b] steroids,[96] aromatic compounds,[97] and tomatidin.[98] Tritium can be introduced by the use of specific reagents such as sodium [³H]hydride (acetylenes[99]), sodium [³H]borohydride (sugar derivatives with CrO_3 catalyst,[100] juvenile hormone,[101] and noradrenaline neurotoxin[102]) or lithium [³H]aluminohy-

[81] L. P. Dukat, V. M. Fedoseev, T. Y. Lazurkina, G. I. Myagkova, N. F. Myasoedov, and V. P. Shevchenko, *Sov. Radiochem.*, 1986, **27**, 281.
[82] N. A. Sasaki, S. Funakoshi, P. Potier, J. L. Morgat, G. Gacel, B. Charpentier, and B. P. Roques, *J. Labelled Compd. Radiopharm.*, 1985, **22**, 1123.
[83] H. Parnes and E. J. Shelton, *Int. J. Pept. Protein Res.*, 1986, **27**, 239.
[84] Z. Li, *J. Nucl. Radiochem.*, 1985, **7**, 160.
[85] G. Y. Cao and C. T. Peng, *Trans. Am. Nucl. Soc.*, 1983, **45**, 18.
[86] C. T. Peng and Z. L. Xue, *Trans. Am. Nucl. Soc.*, 1983, **45**, 19.
[87] (a) J. Filip and L. Bohacek, *Radioisotopy*, 1984, **25**, 483; (b) K. Kusama, in Report of Meeting on Design of the Facility for High Flux and Low Temperature Irradiations, ed. T. Shibata, H. Yoshida and M. Nakagawa, Kumatori, Osaka, Japan, 1984, p. 116.
[88] (a) L. S. Gordeeva, N. A. Patokina, N. A. Korsakova, L. N. Rumyantseva, V. K. Dedva, L. F. Chernysheva, A. G. Neopikhanova, and Y. L. Kaminskij, ref. 62, p. 127; (b) L. A. Yakovleva, Y. L. Kaminskij, and L. P. Sosnova, *Radiokhimiya*, 1985, **27**, 455.
[89] N. M. Anikeev, V. V. Ivanov, V. N. Kramerov, F. V. Tatarkina, L. B. Kudryavtseva, V. I. Mishin, N. A. Ignateva, and L. N. Kaklyushkina, ref. 62, p. 137.
[90] Ya. Kozel and Ya. Ganush, ref. 62; p. 88.
[91] A. L. Gutman and J. B. Campbell, *J. Labelled Compd. Radiopharm.*, 1985, **22**, 649.
[92] S. D. Wyrick, D. M. Niedzwiecki, and R. B. Mailman, *J. Labelled Compd. Radiopharm.*, 1986, **23**, 95.
[93] B. V. Petrenik, Yu. A. Zolotarev, and N. F. Myasoedov, ref. 62, p. 156.
[94] L. P. Dukat, V. M. Fedoseev, N. F. Myasoedov, V. P. Shevchenko, T. Y. Larurkina, and G. I. Myagkova, *Radiokhimiya*, 1985, **27**, 301.
[95] Z. Yang, Z. Guo, and L. Lixin, *J. Nucl. Radiochem.*, 1985, **7**, 119.
[96] K. Veres, J. Fajkos, and P. Kocovsky, Czech. Patent 217 028/B, 1984.
[97] J. L. Garnett, M. A. Long, and C. A. Lukey, *J. Labelled Compd. Radiopharm.*, 1985, **22**, 641.
[98] D. Doller and E. G. Gros, *J. Labelled Compd. Radiopharm.*, 1986, **23**, 109.
[99] S. A. Lermontov, S. E. Tkachenko, and R. G. Gafurov, *Izv. Akad. Nauk SSSR, Ser. Khim.*, 1985, 1921.
[100] S. Usuki and Y. Nagai, *Anal. Biochem.*, 1986, p. 172.
[101] F. C. Baker and C. A. Schooley, *J. Labelled Compd. Radiopharm.*, 1986, **23**, 533.
[102] C. Sahlberg and L. Gawell, *J. Labelled Compd.*, *Radiopharm.*, 1985, **22**, 1143.

dride (psilocin[103]), and even by methyl exchange using [³H]methyl iodide (proteins).[104]

Carbon, ¹¹C.—The short half-life (20 min) of this positron-emitting isotope makes it ideal for a host of nuclear medicine applications,[105] provided that it can be rapidly incorporated into specific molecules. The challenge to produce ever more complicated radiopharmaceuticals ever more rapidly continues to be met. The first stage is to select a reactive ¹¹C-labelled starting material, the second to devise very fast preparative techniques. If [¹¹C]methyl iodide is used it is possible to produce (usually in less than 40 minutes) labelled compounds such as fluoromethane,[106] choline,[107] N-methylspiperone,[108] quinidine and tamoxifen,[109] suricione,[110] enkephalins,[111] indolealkylamines,[112] and even coenzyme Q₁₀.[113] [¹¹C]Carbon dioxide is another very useful starting material from which the following labelled compounds have recently been made: a range of amino-acids,[114] pyruvic[115] and palmitic[116] acids, ethanol, butanol,[117] and diphenyl methanol,[118] and N-(4-methyl)imipramine.[119] [¹¹C]Carbon dioxide can also be used to prepare cyclopropane[¹¹C]carbonyl chloride[120] and [1-¹¹C]iodoethane,[121] from which many radiopharmaceuticals can easily be made. Another very convenient starting material is ¹¹CN⁻, leading easily to ¹¹C-labelled methylamine,[122] putrescine,[123] [9-¹¹C]heptadecan-9-one,[124] and 2-

[103] G. Poon, C. Yun-Cheung, and F. C. P. Law, *J. Labelled Compd. Radiopharm.* 1986, **23**, 167.

[104] W. S. Hancock, D. R. K. Harding, P. M. Barling, and J. T. Sparrow, *Int. J. Pept. Protein Res.*, 1985, **26**, 105.

[105] C. Crouzel and D. Comar, Proc. 2nd. ACOMEN Int. Conf. Biarritz, France, (CEA-Conf 8387), 1986.

[106] S. Stone-Elander, P. Roland, J. E. Litton, L. Widen, L. Eriksson, and P. Johnstroem, *Eur. J. Nucl. Med.*, 1986, **12**, 236.

[107] M. A. Rosen, R. M. Jones, Y. Yano, and T. F. Budinger, *J. Nucl. Med.*, 1985, **12**, 1424.

[108] O. Hiroyoshi, A. Tanaka, M. Iio, Y. Nishihara, O. Inoue, and T. Yamazaki, *Radioisotopes (Tokyo)*, 1985, **34**, 480.

[109] D. VanHaver, T. Vandewalle, G. Slegers, and C. Vandecasteele, *J. Labelled Compd. Radiopharm.*, 1985, **22**, 535.

[110] C. Boullais, F. Oberdorfer, J. Sastre, C. Prenant, and C. Crouzel, *J. Labelled Compd. Radiopharm.*, 1985, **22**, 1081.

[111] K. Naagren, B. Laagstroem, and U. Ragnarsson, *Appl. Radiat. Isot.*, 1986, **37**, 537.

[112] T. Toshihiro, K. Takahashi, T. Ido, K. Yanai, I. Ren, R. Iwata, K. Ishiwata, and S. Nozoe, *Int. J. Appl. Radiat. Isot.*, 1985, **36**, 965.

[113] T. Takahashi, T. Ido, R. Iwata, K. Ishiwata, K. Hamamura, and K. Kogure, *J. Labelled Compd. Radiopharm.*, 1985, **22**, 565.

[114] (a) C. Haldin, Doctoral Dissertation, Uppsala Univ., Uppsala, Sweden, 1984; (b) C. Halldin and B. Langstroem, *J. Labelled Compd. Radiopharm.*, 1985, **22**, 631.

[115] T. Hara, M. Iio, R. Izuche, T. Tsukiyama, and F. Yokoi, *Eur. J. Nucl. Med.*, 1985, **11**, 275.

[116] D. M. Jewett, R. L. Ehrenkaufer, and S. Ram., *Int. J. Appl. Radiat. Isot.*, 1985, **36**, 672.

[117] G. Del Fiore, J. M. Peters, L. Quaglia, M. C. Pardon, J. L. Piette, R. Cantineau, Ch. de Landsheere, P. Rigo, and R. Boudjelida, *J. Radioanal. Nucl. Chem.*, 1985, **104**, 301.

[118] D. D. Dischino, thesis (Ph.D.), Washington Univ. St. Louis, MO, U.S.A., 1983 (microfilm 83-20554).

[119] S. Ram, R. E. Ehrenkaufer, and D. M. Jewett, *Appl. Radiat. Isot.*, 1986, **37**, 391.

[120] (a) S. K. Luthre, V. W. Pike, and F. Brady, *J. Chem. Soc., Chem. Commun.*, 1985, p. 1423; (b) D. W. McPherson, D. R. Hwang, J. S. Fowler, A. P. Wolf, R. M. MacGregor, and C. D. Arnett, *J. Labelled Compd. Radiopharm.*, 1986, **23**, 505.

[121] G. Slegers, J. Sambre, P. Goethals, C. Vandecasteele, and D. vanHaver, *Appl. Radiat. Isot.*, 1986, **37**, 279.

[122] R. Amano and C. Crouzel, *Appl. Radiat. Isot.*, 1986, **37**, 541.

[123] D. W. McPherson, J. S. Fowler, A. P. Wolf, C. D. Arnett, J. D. Brodie, and N. Volkow, *J. Nucl. Med.*, 1985, **10**, 1186.

[124] P. J. Kothari, M. M. Vora, T. E. Boothe, A. M. Emran, R. D. Finn, and G. W. Kabalka, *Appl. Radiat. Isot.*, 1986, **37**, 471.

388 *D. S. Urch and C. E. Urch*

deoxy-D-[1-^{11}C]glucose.[125] A new synthesis has been reported for [^{11}C]phosgene[126] from which [2-^{11}C]-5,5-dimethyl-oxazolidine-2,4-dione[127] and 4-(3-t-butylamino-2-hydroxypropoxy)benzimidazol-2[^{11}C]one[128] have been prepared. A new method for making [^{11}C]urea has been described.[129]

Carbon, ^{14}C.—For the most part ^{14}C-labelled compounds can be produced by standard preparative methods, so only a selection of new techniques will be considered here. [^{14}C]-films can be made[130] by the pyrolysis of labelled methane and the optimum conditions for the preparation of [^{14}C]oxalates from Ba^{14}CO$_3$ have been determined.[131] A new method for making high specific activity ^{14}C-labelled methyl iodide and L-methylmethionine has been described.[132] The discovery[133] that malonic acid salts apparently catalyse the exchange of ^{14}CO$_2$ with carboxylic acids facilitates their labelling. New methods for the preparation of ^{14}C-labelled 1-pyrenecarboxaldehyde,[134] acetylcholine iodide,[135] fluoromethylstyrenes,[136] steroids,[137] and sugars[138] have all been described recently. Improved ways of preparing labelled hydrocarbons[139] and derivatives of 5-nitrofuran[140] on a micro-scale have been reported as well as the use of [^{14}C]ethanolamine as a precursor in the manufacture[141] of labelled drugs. Some other compounds for which modified preparations have been given include [^{14}C]methylphosphonic difluoride,[142] [^{14}CH$_3$]bis(dimethylphosphino)ethane,[143] 2,5-dihydroxy[carboxy-^{14}C]benzoic acid,[144] and histamine H$_1$[145] and H$_2$[146] antigens.

Nitrogen, ^{13}N.—Labelled alkylamines can be prepared by allowing [^{13}N]ammonia to react with the corresponding trialkylborane in tetrahydrofuran.[147] Other

[125] D. vanHaver, N. A. Rabi, M. Vandewalle, P. Goethals, and C. Vandecasteele, *J. Labelled Compd. Radiopharm.*, 1985, **22**, 657.
[126] P. Landais, Thèse (3ème cycle), Univ. Paris–11 Paris, France, 1985.
[127] C. Boullais, C. Crouzel, and A. Syrota, *J. Labelled Compd. Radiopharm.*, 1986, **23**, 565.
[128] J. Z. Ginos, *Int. J. Appl. Radiat. Isot.*, 1985, **36**, 793.
[129] A. M. Emran, T. E. Boothe, R. D. Finn, M. M. Vora, P. J. Kothari, and J. T. Wooten, *Int. J. Appl. Radiat. Isot.*, 1985, **36**, 739.
[130] A. S. Dem'yanova, B. G. Novatskij and A. G. Obaztsov, *Prib. Tekh. Ekhsp.*, 1985, **5**, 196.
[131] M. F. Barakat, A. N. Farag, and M. T. Ragab, *Isotopenpraxis*, 1986, **22**, 215.
[132] (a) T. Elbert, *Radioisotopy*, 1985, **26**, 153; (b) I. Kleinmann and V. Svoboda, *ibid.*, p. 12.
[133] Ya. Sammer, J. Vejs, and L. Ehtvjosh, ref. 62, p. 56.
[134] G. H. Posner, I. Barness, W. H. Biggley, H. H. Seliger, and C. D. Kaplan, *J. Labelled Compd. Radiopharm.*, 1985, **22**, 1023.
[135] B. Anjaneyulu, *J. Labelled Compd. Radiopharm.*, 1985, **22**, 745.
[136] H. M. Ali, C. A. Barson, and P. L. Coe, *J. Labelled Compd. Radiopharm.*, 1985, **22**, 559.
[137] F. V. Tatarkina, L. N. Kaklyushkina, N. A. Ignat'eva, E. V. Kochetkova, N. P. Klimenko, I. F. Tupitsyn, V. I. Mishin, V. N. Kramerov, E. E. Komova, M. M. Moskivina, L. E. Shabunevich, L. I. Dovedova, and V. P. Segeeva, ref. 62, p. 180.
[138] (a) L. Skala, *Radioisotopy*, 1984, **25**, 331, 455; (b) L. Skala, *ibid.*, 1985, **26**, 18.
[139] U. Kosakowsky and R. Russow, *Isotopenpraxis*, 1986, **22**, 280.
[140] R. I. Duclos, jun. and B. A. Hoener, *J. Labelled Compd. Radiopharm.*, 1986, **23**, 103.
[141] J. C. Madelmont, D. Parry, D. Godeneche, and J. Duprat, *J. Labelled Compd. Radiopharm.*, 1985, **22**, 851.
[142] W. E. Bechtold and A. R. Dahl, *J. Labelled Compd. Radiopharm.*, 1985, **22**, 1181.
[143] P. Maeding, D. Kloetzer, and R. Muenze, *Isotopenpraxis*, 1986, **22**, 353.
[144] J. Varghese, L. J. Vennerstrom, and T. J. Holmes, jun., *J. Labelled Compd. Radiopharm.*, 1986, **23**, 313.
[145] (a) M. A. Armitage and D. Saunders, *J. Labelled Compd. Radiopharm.*, 1985, **22**, 1075; (b) M. M. Cashyap, M. B. Mitchell, D. C. Osborne, and D. Saunders, *ibid.*, p. 1239.
[146] S. M. Stefanick, C. F. Kasulanis, S. D. Levine, and A. C. Fabian, *J. Labelled Compd. Radiopharm.*, 1985, **22**, 677.
[147] P. J. Kothari, M. M. Vora, T. E. Boothe, A. M. Emran, R. D. Finn, and G. W. Kabalka, *Appl. Radiat. Isot.*, 1986, **37**, 469.

methods for the production of labelled amines[105] and enzymes[68c] have also been reported.

Oxygen, ^{15}O.—Despite its short half-life ^{15}O is routinely incorporated into dioxygen,[105] carbon monoxide, and water,[148] and these compounds are then used in the preparation of radiopharmaceuticals. ^{15}O-Labelled alcohols can be made by a variety of methods,[68c.105] including the reaction of $H_2^{15}O$ with alkyl boranes[149] (see above).

Fluorine, ^{18}F.—A new method for the introduction of ^{18}F *via* the cleavage of a carbon–silicon bond with [^{18}F]fluorine gas has been developed and applied to the preparation of labelled 3,4-dihydroxy-6-fluoro-L-phenylalanine (6-fluoro-L-dopa)[150] and 4-fluoroantipyrine.[151] This latter compound has also been made[152] by an alternative route using labelled acetyl hypofluorite. The more conventional approach to the problem of the rapid introduction of ^{18}F into a molecule is to utilize the nucleophilicity of the fluoride anion. Thus tetra-alkylammonium salts [which can be made from ^{18}F(CH$_3$)$_3$Si][153] have been used to initiate exchange reactions leading to labelled fluoroperidol,[154] fluorocarboxylic acids,[153] 2-deoxy-2-fluoro-D-mannose,[155] and 16α-fluoro-17β-o-estradiol.[156] Enhanced nucleophilic activity has been reported[157] when ^{18}F$^-$ is supported on an aminopolyether in a dipolar aprotic solvent. This reagent has been successfully used in the production of labelled fluoroalkanes and fluorocarboxylic acids.[158] It has also been demonstrated[159] that aromatic fluorination with [^{18}F]fluoride proceeds more efficiently with iodides than with azides. Hydrogen [^{18}F]fluoride in the presence of antimony(III) oxide has been proposed[160] as a new selective radiofluorinating reagent and applied to the labelling of trihalogenomethyl groups.

Phosphorus, ^{32}P.—Labelled nucleotides and polyphosphoinositides can be prepared from acyl[161] and free [^{32}P]orthophosphate[162] respectively. The optimum conditions for the manufacture of desoxycytidine-5'-[α-^{32}P]triphosphate have also been determined.[163] Although most routes to labelled proteins from ^{32}P-nucleoside triphosphates require the intermediacy of a suitable enzyme (*e.g.* to produce

[148] M. J. Welch and M. R. Kilbourn, *J. Labelled Compd. Radiopharm.*, 1985, **22**, 1193.
[149] G. W. Kabalka, S. A. Kunda, G. W. McCollum, R. M. Lambrecht, M. Sajjad, J. S. Fowler, and R. MacGregor, *Int. J. Appl. Radiat.*, 1985, **36**, 853.
[150] M. Diksic and S. J. Farrokhzad, *J. Nucl. Med.*, 1985, **11**, 1314.
[151] P. Di Raddo and M. Diksic, *Int. J. Appl. Radiat. Isot.*, 1985, **36**, 953.
[152] M. Diksic and P. Di Raddo, *Int. J. Appl. Radiat. Isot.*, 1985, **36**, 643.
[153] A. L. Bosh and T. R. Degrado, *Appl. Radiat. Isot.*, 1986, **37**, 305.
[154] S. Farrokhzad and M. Diksic, *J. Labelled Compd. Radiopharm.*, 1985, **22**, 721.
[155] A. Luxen, N. Satyamurthy, G. T. Bida, and J. R. Barrio, *Appl. Radiat. Isot.*, 1986, **37**, 409.
[156] J. W. Brodack, M. R. Kilbourn, and M. J. Welch, *Appl. Radiat. Isot.*, 1986, **37**, 217.
[157] D. Block, B. Klatte, A. Knoechel, R. Beckmann, and U. Holm. *J. Labelled Compd. Radiopharm.*, 1986, **23**, 467.
[158] H. H. Coenen, M. Scheller, G. Stoecklin, B. Klatte, and A. Knoechel, *J. Labelled Compd. Radiopharm.*, 1986, **23**, 455.
[159] M. S. Berridge, C. Crouzel, and D. Comar, *J. Labelled Compd. Radiopharm.*, 1985, **22**, 687.
[160] G. Angelini, M. Speranza, C.-Y. Shiue, and A. P. Wolf, *J. Chem. Soc., Chem. Commun.*, 1986, p. 924.
[161] J. J. DiMeo, U.K. Patent 2 161 484/A, 1986.
[162] O. B. Tysnes, G. M. Aarbakke, A. J. Verhoeven, and H. Holmsen, *Thromb. Res.*, 1985, **3**, 329.
[163] M. Havranek, *Radioisotopy*, 1985, **26**, 5.

[^{32}P]vitellogenin[164]) it has been shown[165] that non-specific labelling can also occur, especially in the presence of alcohols.

Sulphur, ^{35}S.—Methods have been described for the O-sulphonation of hydroxyamino-acids and peptides[166] from labelled sulphuric acid. [^{35}S]Sulphuric acid can also be used to produce[167] labelled 3'-phosphoadenosin-5'-phosphosulphate. Methods for the preparation of dimethyl[^{35}S] sulphide by isotopic exchange[168] as well as labelled letosteine[169] and omeprazole[170] have been reported.

Chlorine, ^{36}Cl.—The optimum conditions for the preparation of labelled chloramine-B by exchange with ^{36}Cl$^-$ anions have been determined.[171]

First-row Transition Metals, ^{59}Fe, ^{60}Co.—The glycoprotein transferrin has been labelled with just ^{59}Fe[172] and also double-labelled with ^{3}H as well.[173] ^{60}Co has been incorporated into phthalocyanines[174] and also into a mixed-valence complex containing both bi- and tri-valent cobalt.[175]

Gallium, ^{67}Ga, ^{68}Ga.—The positron emitter ^{68}Ga can be incorporated[176] into complexes with 8-hydroxyquinoline, tropolone, *etc.* as well as being used to label proteins.[177] The production of other radiopharmaceuticals, but labelled with ^{67}Ga, has also been reported[178] (*e.g.* albumen, lactoferrin).

Arsenic, ^{76}As.—A new preparation of labelled arsenic trichloride has been reported[179] in which neutron irradiated arsenic(III) oxide is reacted with sulphur monochloride. This reagent was then used in the preparation of [^{76}As]dihydrophenarsazine.

Selenium, ^{73}Se, ^{75}Se.—Exchange reactions have been used to produce both ^{73}Se-labelled 2-selenouracil[180] and ^{75}Se-labelled dimethylselenium.[181] [^{75}Se]Di-

164 L. Opresko and H. S. Wiley, *Anal. Biochem.*, 1984, **140**, 372.
165 M. C. Schmidt and M. M. Hanna, *FEBS Lett.*, 1986, **194**, 305.
166 (*a*) S. Pongor, M. Brownlee, and A. Cerami, *Arch. Biochem. Biophys.*, 1985, **2**, 458; (*b*) T. Nakahara, M. Waki, and H. Uchimura, *Anal. Biochem.*, 1986, **154**, 194.
167 D. M. Delfert and H. E. Conrad, *Anal. Biochem.*, 1985, **2**, 303.
168 E. V. Sharvadze, E. E. Grinberg, L. V. Vasil'eva, and V. I. Morozov, *Radiokhimiya*, 1985, **27**, 681.
169 C. Nicolas, F. Gachon, and M. Faurie, *J. Labelled Compd. Radiopharm.*, 1985, **22**, 711.
170 A. M. Crowe, R. J. Ife, M. B. Mitchell, and D. Saunders, *J. Labelled Compd. Radiopharm.*, 1986, **23**, 21.
171 M. Subhashini, M. S. Subramanian, and V. R. S. Rao, *J. Radioanal. Nucl. Chem.*, 1986, **103**, 261.
172 J. M. Legendre, M. P. Moineau, A. Turzo, and J. F. Menez, *Pathol. Biol.*, 1985, **33**, 741.
173 T. Kishimoto and M. Tavassoli, *Anal. Biochem.*, 1986, **153**, 324.
174 I. C. S. Jardim, J. C. de Andrade, C. H. Collins, and E. Collins, *J. Radioanal. Nucl. Chem.*, 1985, **96**, 621.
175 G. Albarran and C. Archundia, *Rev. Soc. Quim. Mex.*, 1984, **28**, 354.
176 Y. Yano, T. F. Budinger, S. N. Ebbe, C. A. Mathis, M. Singh, K. M. Brennan, and B. R. Moyer, *J. Nucl. Med.*, 1985, **12**, 1429.
177 (*a*) W. Wolf, J. F. Harwig, A. G. Janoki, and W. Chanachai, Int. Symp. on the Developing Role of Short-Lived Radionuclides in Nuclear Medical Practice, ed. P. Paras and J. W. Thiessen, Dept. of Energy, Washington, DC, U.S.A., 1982, p. 484; (*b*) B. Maziere, C. Loc'h, M. Steinling, and D. Comar, *Appl. Radiat. Isot.*, 1986, **37**, 360.
178 (*a*) A. G. Janoki, J. F. Hawig, J. B. Slater, and W. Wolf, *J. Radioanal. Nucl. Chem.*, 1985, **91**, 115; (*b*) C. Motta-Hennessy, G. Coghlan, S. A. Eccles, and C. Dean, *Eur. J. Nucl. Med.*, 1985, **11**, 240; (*c*) R. E. Weiner, G. J. Schreiber, P. B. Hoffer, and J. T. Bushberg, *J. Nucl. Med.*, 1985, **26**, 908.
179 A. M. Emran, N. M. Shanbaky and R. P. Spencer, *Appl. Radiat. Isot.*, 1986, **37**, 545.
180 K. Ogawa and K. Taki, *Appl. Radiat. Isot.*, 1986, **37**, 315.
181 E. V. Sharvadze, E. E. Grinberg, and L. V. Morozov, *Radiokhimiya*, 1985, **27**, 683.

selenosalicylic acid has also been prepared[182] as an intermediate in the synthesis of labelled ebselen.

Bromine, [77]Br, [82]Br.—The cleavage of aryl—M bonds (M = Si, Ge, or Sn) by bromide in the presence of an oxidizing agent such as chloramine-T enables radioactive bromine to be located at specific sites on aromatic rings.[183] Other compounds that have been labelled with radiobromine recently have included, *p*-bromo-3-*N*-alkylspiperone,[184] 18-bromo-5-telluraoctadec-17-enoic acid[185] and 2-deoxy-2-bromo-D-mannose.[186]

Yttrium, [90]Y.—Methods for the preparation of ^{90}Y-labelled antibodies have been reported.[187]

Technetium, [99m]Tc.—The preparation and production of technetium radiopharmaceuticals in Australia has been reviewed[33b] and the use of specific chelating agents critically considered.[188] A new development has been the concentration on 'bifunctional' ligands,[189] which can form a stable complex with technetium at one site (thiosemicarbazide) and bind to a protein or confer water solubility at another.[190] Further studies with sulphur-based ligands have shown[191] that toluene-3,4-dithiol complexes can be characterized with technetium in valence states III, IV, and V. The preparation of the technetium tetrasulphophthalocyanine complex has been reported.[192] A series of technetium complexes with ligands based on substituted phosphonate groups has been prepared and used for myocardial imaging[193] or in bone scanning.[194] The chromatographic isolation and purification of these complexes has also been studied.[195] The complex $[Tc^{VI}NCl_4]^-$ has proved a useful starting point for the preparation of new technetium compounds and radiophar-

[182] R. Cantineau, G. Tihange, A. Plenevaux, L. Christiaens, M. Guillaume, A. Welter, and N. Dereu, *J. Labelled Compd. Radiopharm.*, 1986, **23**, 59.
[183] (a) S. M. Moerlein, *J. Chem. Soc., Perkin Trans.* 1, 1985, 1687; (b) S. M. Moerlein and H. H. Coenen, *ibid.*, p. 1941; (c) M. J. Adam, T. J. Ruth, Y. Homma, and B. D. Pate, *Int. J. Appl. Radiat. Isot.*, 1985, **36**, 935.
[184] S. M. Moerlein, P. Laufer, and G. Stoecklin, *J. Labelled Compd. Radiopharm.*, 1985, **22**, 1007.
[185] P. C. Srivastava, F. F. Knapp, jun., A. P. Callahan, B. A. Owen, G. W. Kabalka, and K. A. Sastry, *J. Med. Chem.*, 1985, **28**, 408.
[186] Y. Zhou, C. Shiue, A. P. Wolf, and C. D. Arnett, *Eur. J. Nucl. Med.*, 1985, **11**, 252.
[187] A. T. M. Vaughan, A. Keeling, and S. C. S. Yankuba, *Int. J. Appl. Radiat. Isot.*, 1985, **36**, 803.
[188] V. Vrana and I. Kleisner, *Radioisotopy*, 1984, **25**, 621.
[189] (a) F. Grases and C. Genestar, *Radiochim. Acta*, 1985, **39**, 43; (b) T. Hosotani, A. Yokoyama, Y. Arano, K. Horiuchi, H. Wasaki, H. Saji, and K. Torizuka, *Int. J. Nucl. Med. Biol.*, 1986, **12**, 431; (c) Y. Arano, A. Yokohama, Y. Magata, K. Horiuchi, H. Saji, and K. Torizuka, *ibid.*, p. 425.
[190] A. Davison, B. V. DePamphilis, A. G. Jones, R. Faggiani, C. J. L. Lock, and C. Orvig, *Can. J. Chem.*, 1985, **63**, 319.
[191] F. Grases, P. March, and G. Far, *Appl. Radiat. Isot.*, 1986, **37**, 201.
[192] J. Rousseau, H. Ali, G. Lamoureux, E. Lebel, and J. E. van Lier, *Int. J. Appl. Radiat. Isot.*, 1985, **36**, 709.
[193] (a) R. Berger, R. Syhre, S. Seifert, D. Kloetzer, P. Maeding, and R. Muenze, *Appl. Radiat. Isot.*, 1986, **37**, 197; (b) J. Simon, D. A. Wilson, and W. A. Volkert, South African Patent 84/7275/A, 84/7379/A, 1986.
[194] (a) L. J. Degrossi, P. Oliveri, H. Garcia del Rio, R. Labriola, D. Artagaveytia, and E. B. Degrossi, *J. Nucl. Med.*, 1985, **10**, 1135; (b) R. Ollivier, G. Sturtz, J. M. Legendre, G. Jacolot, and A. Turzo, *Eur. J. Med. Chem.-Chim. Ther.*, 1986, **21**, 103; (c) G. J. deGroot, H. A. Das, J. Kroesbergen, W. J. Gelsema, and C. L. De Ligny, *Int. J. Nucl. Med. Biol.*, 1986, **12**, 419.
[195] G. J. deGroot, H. A. Das, and C. L. De Ligny, *Appl. Radiat. Isot.*, 1986, **37**, 23.

maceuticals[196] based on Tc^V: the chlorines can be replaced by a variety of groups, (*e.g.* isothiocyanate,[196b,c] 8-quinolinethiol,[196c] gluconate[196a]). Technetium is also quinquevalent in the complex formed when pertechnate ions react with L-cysteine.[197] A similar reduction of Tc^{VII} is effected by tetrahydroxy-1,4-quinone[198] but the valency of the technetium in the final complex was not determined. Other complexes that have been reported recently are those of 1,7-bis(2-pyridyl)-2,6-diazaheptane,[199] 4-*p*-*N*-butylphenylcarbamoylethyliminodiacetic acid,[200] and a series of β-diketonates.[201] Recipes for technetium–sulphur[202] and technetium–tin phosphate colloids[203] have been published as well as details of technetium-labelled polymers.[204]

Indium, ^{111}In, ^{113m}In.—The preparation of dialkyldithiocarbamate complexes of ^{113m}In has been reported[205] together with methods for the labelling of proteins,[177a] platelets, leucocytes,[206] and antibodies.[207]

Iodine, ^{121}I, ^{122}I, ^{123}I, ^{125}I, ^{131}I.—Current interest in the preparation of radio-iodine-labelled compounds centres on their use in nuclear medicine. The use of ^{125}I compounds for tumour imaging has been discussed[208] and general methods for making iodine-labelled compounds[209] and iodine-labelled plasma proteins[210] have been reviewed. Reactions involving organometallic intermediates have been advocated[211] as useful in locating a radioactive atom at a specific site in a molecule, especially at vinyl or phenyl sites where the *in vivo* half-life of the label is greatest.

The short half-lives of the light isotopes of iodine have led to the development of high-speed variations of conventional labelling methods based on exchange reactions[212] and the use of chloramine-T. Only three minutes were required to

[196] (*a*) J. Baldas and J. Bonnyman, *Int. J. Appl. Radiat. Isot.*, 1985, **36**, 919; (*b*) U. Abram, H. Spies, S. Abram, R. Kirmse, and J. Stach, *Z. Chem.*, 1986, **26**, 140; (*c*) J. Baldas, J. Bonnyman, and G. A. Williams, *Inorg. Chem.*, 1986, **25**, 150.

[197] F. Grases, J. Palou, and E. Amat, *Transition Met. Chem.*, 1986, **11**, 253.

[198] F. Grases and G. Far, *Radiochim. Acta*, 1986, **39**, 81.

[199] Z. Proso and P. Lerch, *J. Radioanal. Nucl. Chem.*, 1985, **92**, 237.

[200] E. S. Hamada, M. F. de Barbosa, M. A. T. M. de Almeida, M. T. Colturato, C. P. G. da Silva, and E. Muramoto, ref. 14, Vol. 2, p. 134.

[201] K. Hashimoto, T. Sekine, T. Omori, and K. Yoshihara, *J. Radioanal. Nucl. Chem.*, 1986, **103**, 19.

[202] (*a*) J. Steigman, N. A. Solomon, and L. L. Y. Hwang, *Appl. Radiat. Isot.*, 1986, **37**, 223; (*b*) A. M. Al-Hilli, N. H. Agha, I. F. Al-Jumaili, H. M. Al-Azzawi, M. H. S. Al-Hissoni, and M. N. Jassim, *J. Radioanal. Nucl. Chem.*, 1986, **104**, 273.

[203] Faiz-Ur-Rehman, Shamas-Uz-Zaman, M. A. Shahid, S. L. Imran, M. Ashraf, and M. W. Akhtar, *Appl. Radiat. Isot.*, 1986, **37**, 249.

[204] S. J. Douglas and S. S. Davis, *J. Labelled Compd. Radiopharm.*, 1986, **23**, 495.

[205] S. Abram, U. Abram, H. Spies, and R. Muenze, *Int. J. Appl. Radiat. Isot.*, 1985, **36**, 653.

[206] M. R. Hardeman, E. G. J. Eitjes-Van-Ovebeek, A. J. M. vanVelzen, and M. H. Roevekamp, Proc. Symp. 'Radio-Labelled Cells', ed. M. R. Hardman and Y. Najean, Martinus Nijhoff, The Hague, Netherlands, 1984, p. 17.

[207] C. H. Paik, J. J. Hong, M. A. Ebbert, S. C. Heald, R. C. Reba, and W. C. Eckelman, *J. Nucl. Med.*, 1985, **26**, 482.

[208] M. E. Van Dort, Thesis (Ph.D.) Lehigh Univ., Bethlehem, PA, U.S.A., 1983, (microfilm, 83-17143).

[209] L. Kronrad, P. Hradilek, and K. Kopicka, *Radioisotopy*, 1984, **25**, 781.

[210] E. Regoeczi, 'Iodine-Labelled Plasma Proteins', Vol. 1, CRC Press, Boca Raton, FL, U.S.A., 1984.

[211] P. C. Srivastava, M. M. Goodman, and F. F. Knapp, jun., ref. 68(*b*).

[212] (*a*) H. J. Sinn, H. H. Schenk, and W. Maier-Borst, *Appl. Radiat. Isot.*, 1986, **37**, 17; (*b*) D. M. Wieland, ref. 177, p. 384.

prepare a *m*-dimethoxy-*N,N*-dimethyliodophenylisopropylamine labelled with either ^{122}I ^{213a} or ^{125}I.^{213b} It having been found^{214} that long-chain fatty acids tagged with radio-iodine are particularly effective at imaging the heart many new preparations of compounds such as 17-[^{123}I]iodoheptadecanoic^{215,216a} and 15-(*p*-[^{123}I]iodophenyl)pentadecanoic acid^{216} (including a telluro-derivative^{217}) have been reported. Methods for the preparation of *o*-iodohippuric acid and its salts labelled with ^{123}I,^{218a} ^{125}I, or ^{131}I have been described^{218} and the catalytic roles of copper and cobalt salts in this preparation investigated.^{219} Other radiopharmaceuticals that have been labelled with ^{123}I recently include 6-iodonicotinic acid diethylamide (also copper-catalysed exchange),^{220} *m*-iodobenzylguanidine,^{221} and *N*-isopropyl-p-iodoamphetamine.^{222}

The simplest way of making radio-iodine compounds is by exchange with labelled iodide anions. This method has been used to prepare labelled iodothyrosines and iodothyronines,^{223} iodosulphophthalein,^{224} and iodo-derivatives of nucleosides.^{225} When solutions of iodine and labelled sodium iodide are used nucleophilic attack can produce radioactive iodine compounds of thyroxine^{226} and aminodarone.^{227} The chloramine-T method of introducing radio-iodine has been used to make 4-[^{131}I]iodoantipyrine^{228} and [^{125}I]tri-iodothyronine,^{229} as well as labelled humic acid,^{230} porcine insulin,^{231} and monoclonal antibodies.^{232} Enzymatic assistance has often been found advantageous as in the lactoperoxidase–glucose oxidase method used to label thyroglobulin,^{233} vasopressin,^{234} a DNA ligand,^{235} and a polypeptide

[213] (*a*) C. A. Mathis, T. Sargent III, and A. T. Shulgin, *J. Nucl. Med.*, 1985, **11**, 1295; (*b*) C. A. Mathis, A. T. Shulgin, and T. Sargent III, *J. Labelled Compd. Radiopharm.*, 1986, **23**, 115.

[214] F. F. Knapp, *Int. At. Energy Agency Bull.*, 1986, **28**, 26.

[215] (*a*) J. Steinbach and G. J. Beyer, *Radiol. Diagn.*, 1986, **27**, 297; (*b*) J. Mertens, W. Vanryckeghem, and A. Bossuyt, *Eur. J. Nucl. Med.*, 1986, **11**, 361.

[216] (*a*) F. F. Knapp, jun., M. M. Goodman, D. R. Elmaleh, R. Okada, and H. W. Strauss, ref. 177(*a*), p. 289; (*b*) C. A. Otton, H. Lee, T. J. Mangner, and D. M. Wieland, *Appl. Radiat. Isot.*, 1986, **37**, 205.

[217] M. M. Goodman, A. P. Callahan, F. F. Knapp, jun. H. Strauss, D. Elmaleh, P. Richards, and L. F. Mausner, ref. 177(*a*), p. 488.

[218] (*a*) J. L. Q. de Britto, ref. 36, p. 46; (*b*) P. Hradilek, L. Kronard, and K. Kopicka, Czech. Patent 202 202/B, 1983.

[219] (*a*) T. Li, Y. Wang, and Q. Ouyang, *J. Nucl. Radiochem.*, 1985, **7**, 85; (*b*) H. Y. Aboul-Enein and D. F. Wolczak, *J. Radioanal. Nucl. Chem.*, 1986, **104**, 359.

[220] E. J. Knust, H. J. Machulla, P. L. Zabel, C. Astfalk, and U. Schmidt, *J. Radioanal. Nucl. Chem.*, 1985, **91**, 285.

[221] P. A. van Doremalen and A. G. M. Janssen, *J. Radioanal. Nucl. Chem.*, 1985, **96**, 97.

[222] G. Wunderlich and W. G. Franke, *Radiobiol. Radiother.*, 1986, **27**, 88.

[223] P. Hradilek, L. Kronrad, M. Becicova, and K. Kopicka, *Radioisotopy*, 1984, **25**, 675.

[224] K. Kopicka, P. Hradilek, and L. Kronrad, *Radioisotopy*, 1984, **25**, 726.

[225] A. Verbruggen, C. Julien, E. De Clercq, and M. De Roo, *Appl. Radiat. Isot.*, 1986, **37**, 355.

[226] P. Hradilek, L. Kronrad, and K. Kopicka, Czech. Patent 221 884/B, 1985.

[227] R. Sion, *J. Labelled Compd. Radiopharm.* 1985, **22**, 799.

[228] T. E. Boothe, R. D. Finn, M. M. Vora, P. J. Kothari, and A. M. Emran, *J. Labelled Compd. Radiopharm.*, 1986, **23**, 479.

[229] M. M. Hamada, C. H. de Mesquita, and C. P. G. da Silva, ref. 14, vol. 2, p. 98.

[230] J. C. Lobartini, N. R. Curvetto, and G. A. Orioli, *Appl. Radiat. Isot.*, 1986, **37**, 237.

[231] I. T. de Toledo e Souza, D. Giannella Neto, and B. L. Wajchenbeg, ref. 14, vol. 2, p. 102.

[232] S. Matzku, H. Kirchgessner, W. G. Dippold, and J. Brueggen, *Eur. J. Nucl. Med.*, 1985, **11**, 260.

[233] C. A. Spencer, B. W. Platler, R. B. Guttler, and J. T. Nicoloff, *Clin. Chim. Acta*, 1985, **151**, 121.

[234] R. P. Sequeira and K. S. Rahavan, *Indian J. Exp. Biol.*, 1982, **20**, 824.

[235] R. F. Martin and M. Pardee, *Int. J. Appl. Radiat. Isot.*, 1985, **36**, 745.

hormone.[236] Various antibodies have been labelled using the Iodo-gen reagent.[237]

A new group of methods for introducing radioactive iodine atoms at specific sites is based on iodide attack at an R—M bond where R is an organic group and M can be boron,[238] silicon,[183b] germanium,[183a] or tin.[183a] [^{123}I or ^{125}I]-iodobenzyl bromide can be prepared by the iodide displacement of a trimethylsilyl group[239] generating a reagent which can be used to make labelled quaternary ammonium and phosphonium salts. [125]ICl has been used to label large molecules such as equine fibrinogen.[240] Other techniques that have been reported recently have included apparent heterogeneous catalysis on silica gel[241] (to produce 4-[^{131}I]antipyrine), exchange in melts,[242] and the use of u.v. irradiation to effect the labelling of di-iodofluorescein.[243]

Mercury, [203]Hg.—The preparation of methyl [[203]Hg]mercuric chloride of high specific activity has been reported.[244]

Radiochemical Integrity.—Two fundamental problems beset the production of labelled compounds. The first concerns the initial radiochemical purity of the product. To what extent is it the compound it should be, and to what extent is it contaminated by minute traces of other radioactive species? The problem is exacerbated by the fact that such species would be carrier-free and present with the highest possible specific activities. The second problem is that a radioactive substance will irradiate itself during storage, giving rise to labelled and unlabelled impurities.

The most widely used methods of purification and for the determination of impurities are based on chromatography (paper, thin layer) or electrophoresis.[245] The specific problem of the nature of the eluant from ^{99}Mo/^{99m}Tc generators (*i.e.* how much ^{99}Mo and ^{99}Tc are also present) has been investigated[246] and the different commercial generators have been compared.[247] With ^{77m}Se/^{77}Br generators it was found[21] that after a few weeks about 0.5% of the bromine activity is present as bromate. Similarly [^{35}S]sulphide solutions produced, on standing, a wide range of labelled oxy-sulphur anions.[248] Labelled iodine compounds have been studied to determine their long-term stability by measuring the inorganic iodide build-up from

[236] J. Silberring and A. Dubin, *Endokrynol. Pol.*, 1981, **32**, 353.
[237] (*a*) P. M. Kleveland, S. E. Haugen, and H. L. Waldum, *Scand. J. Gastroenterol.*, 1985, **20**, 569; (*b*) P. G. Burhol, R. Jorde, J. Florholmen, T. G. Jenssen, and B. Vonen, *ibid.*, p. 466; (*c*) P. Motte, P. Vauzelle, F. Troalen, D. Bellet, G. F. Alberici, and C. Bohuon, *J. Immunol. Methods*, 1986, **87**, 223.
[238] G. W. Kabalka, ref. 177(*a*), p. 377.
[239] A. A. Wilson, R. F. Dannals, H. T. Ravert, H. D. Burns, and H. N. Wagner, jun., *J. Labelled Compd. Radiopharm.*, 1986, **23**, 83.
[240] C. P. Coyne, W. J. Hornof, A. B. Kelly, T. R. O'Brien, and S. J. de Nardo, *Am. J. Vet. Res.*, 1985, **12**, 2572, 2578.
[241] T. E. Boothe, R. D. Finn, M. M. Vora, A. Emran, P. J. Kothari, and G. W. Kabalka, *J. Labelled Compd. Radiopharm.*, 1985, **22**, 1109.
[242] B. Liu, Y. Jin, and T. Li, *J. Nucl. Radiochem.*, 1985, **7**, 29.
[243] M. Raieh and E. M. Mikhail, *Isotopenpraxis*, 1986, **22**, 276.
[244] T. Y. Toribara, *Int. J. Appl. Radiat. Isot.*, 1985, **37**, 903.
[245] (*a*) A. Hammermaier, E. Reich, and W. Boegl, *Nuc. Compact.*, 1985, **16**, 200; (*b*) J. Bonnyman, R. Lauder, and B. Van Every, Report, TR-073, Australian Radiation Lab., Melbourne, Australia, 1986.
[246] A. Hammermaier, E. Reich, and E. Boegl, *Eur. J. Nucl. Med.*, 1986, **12**, 41.
[247] M. E. Holland, Thesis (Ph.D.), Cincinnati Univ., Ohio, U.S.A., 1984 (microfilm 85-09485).
[248] A. Kandasamy, S. N. Muddukrishna, and S. Mukherji, ref. 24, paper AR-1.

[^{131}I]Rose Bengal and [^{131}I]Bromosulphthalein^{249} and the use of t.l.c. to determine^{250} the labelled products formed from *o*-iodohippuric acids labelled with ^{131}I, ^{125}I, or ^{123}I. Greatest decomposition was found with ^{131}I. Photo-induced decomposition of labelled methyl iodide has been investigated^{251} to elucidate basic reaction mechanisms.

Standardization, Detection.—A method has been described for determining separately the H^3H and H^3HO content of air containing both tritiated species.^{252} Standardization techniques for both ^{111}In ^{253} and ^{207}Bi ^{254} have been reported.

Automation.—Microcomputers are being increasingly applied in radiochemical laboratories for a variety of tasks. The automation of specific manipulations coupled with information fed to the computer from detectors enables^{255} simple labelling procedures to be carried out without human intervention.^{256} A robot system for the production of ^{125}I-labelled proteins has been described.^{257} Other applications of computers have included their use in unravelling g.c.–m.s. data of ^{14}C-compounds,^{258} in calculating the specific activity of fission-produced ^{99}Mo,^{259} in organizing radiochemical laboratory records and inventories of radiochemicals,^{260} and in keeping track of specific isotopes present in waste from nuclear reactors.^{261}

4 Chemistry of Nuclear Transformations

The chemistry of recoil atoms in solids has been critically reviewed^{262} as has the use of hot-atom reactions in making suitable starting materials for the rapid preparation of radiopharmaceuticals labelled with short-lived positron emitting isotopes.^{263} Since it seems reasonable to assume that the chemical reactions that take place in space will be for the most part hot-atom reactions it is of interest to report^{264} that the laboratory experiments to simulate such reactions involving the impact of carbon and nitrogen atoms with frozen water, ammonia, and methane have been carried out. Another aspect of the chemistry associated with nuclear transformations that

^{249} I. T. de Toledo e Souza, N. P. S. Pereira, and C. P. G. Silva, Report, No. 79, Insituto de Pesquisas Energeticas e Nucleares, Sao Paulo, Brazil, 1985.
^{250} A. Hammermaier, E. Reich, and W. Boegl, Report, No. 48, Inst. für Strahlenhygiene, Bundesgesundheitsamt, Neuherberg, B.R.D., 1984.
^{251} H. Noguchi, H. Matsui, and M. Murata, *Nippon Genshiryoku Gakkai-Shi*, 1983, **25**, 658.
^{252} M. Sakaguchi, M. Ohta, K. Uematsu, and A. Wakamoto, *Radioisotopes (Tokyo)*, 1985, **34**, 207.
^{253} A. J. Kolb and D. L. Horrocks, in Proc. Internat. Conf. 'Advances in Scintillation Counting', ed. S. A. McQuarrie, C. Ediss, and L. I. Wiebe, Univ. of. Alberta, Edmonton, Alberta, Canada, 1983, p. 208.
^{254} E. A. Aristov, V. A. Bakhshi-Zade, S. V. Matve, N. M. Samykin, G. A. Sichkar', and M. D. Stepanov, *Meas. Tech. (Engl. Transl)*, 1985, **27**, 1059.
^{255} (*a*) G. Toeroek, L. Gy. Nagy, J. Ruip, M. Vodicska, N. Vajda, and J. Solymosi, *Period. Polytech. Chem. Eng.*, 1984, **28**, 233; (*b*) S. C. Sweet and H. C. Griffin, *Int. J. Appl. Radiat. Isot.*, 1985, **36**, 908.
^{256} T. Hiroishi, Y. Adachi, and Y. Nishihara, *Sumitomo Jukikai Giho*, 1985, **33**, 28.
^{257} W. M. Hurni, W. J. Miller, E. H. Wasmuth, and W. J. McAleer, *Appl. Radiat. Isot.*, 1986, **37**, 623.
^{258} H. Kanamaru, I. Nakatsuka, and A. Yoshitake, *Radioisotopes (Tokyo)*, 1985, **34**, 401.
^{259} Anon., Report, No. E-625, Australian Atomic Energy Commission Research Establishment, Lucas Heights, Australia, 1986.
^{260} D. J. Wilkinson and I. Ward, *Appl. Radiat. Isot.*, 1986, **37**, 2.
^{261} J. N. Vance, J. S. Herrin, and R. T. Hope, in Proc. Conf. 'Waste Management 84', American Nuclear Society, La Grange Park, IL, U.S.A., 1984, p. 651.
^{262} H. Mueller, *Radiat. Phys. Chem.*, 1986, **27**, 25.
^{263} R. A. Ferrieri and A. P. Wolf, *Radiochim. Acta*, 1983, **34**, 69.
^{264} K. Roessler, in Proc. Conf. 'Properties and Interactions of Interplanetary Dust', ed. R. H. Giese and P. Lamy, D. Reidel, Dordrecht, Netherlands, 1985, p. 357.

continues to excite some attention is the effect that chemical environment might have on the nuclear transformation process itself, in particular the half-life. The effect is small but discernible and has been the subject of a recent review.[265]

Tritium.—Studies of the reactions of recoil tritium atoms with aliphatic (n-hexane),[266] aromatic (naphthalenes),[267] and heterocyclic (pyrroles, uracil,[268] pseudouridine[87b]) compounds have all been reported recently. Gas-phase investigations, which give more direct information about primary processes, have been extended[269] to ethyl alcohol. The formation of $C_2{}^3HH_4OH^*$ or $C_2{}^3HH_5{}^*$ in highly excited states as a result of the reaction with hot tritium atoms was postulated to explain the observed pressure dependence of [^3H]ethene. Other gas-phase studies have looked[270] at the reaction of recoil tritium with para-hydrogen in hydrogen–deuterium mixtures and claim that low-energy recoil atoms react preferentially with the ortho-isomer. Further reactions of low-energy recoil atoms, with ethene and hydrogen sulphide, have been studied[271] by carrying out experiments in the presence of large excesses of bath gases such as carbon tetrafluoride or krypton. Low-energy tritium atoms can also made by heating[272a] (and allowed to react with polythene), and by microwave discharge.[272b]

The investigation of the chemistry of ions activated by tritium decay by using doubly labelled molecules has been continued. [^3H]Phenylium ions produced by the decay of one of the tritium atoms in [$^3H_2$]benzene were allowed to react with a range of aromatic compounds in the gas phase in the presence of an excess of carbon monoxide.[273] Both benzoylation and phenylation of the substrate were observed.

Group III.—The reactions of recoil ^{116}In produced by neutron irradiation in indium iodate and indium sulphate have been studied.[274]

Group IV.—The reaction of recoil-^{11}C with water vapour leads to the formation of ^{11}C-labelled carbon monoxide (98% at the lowest dose), carbon dioxide, and methane.[275] As the yield of the former compound decreases with radiation dose it is postulated that carbon monoxide is the primary product. The importance of singlet silylene in understanding the chemistry of recoil silicon atoms has been critically reviewed.[276]

Group V.—Ice, when bombarded with recoil ^{13}N at 77 K, reacts[277] to produce ^{13}N-labelled ammonium (51%), nitrite (37%), and nitrate (9%) ions. With increasing

[265] M. R. Harston, N. C. Pyper, and A. G. Maddock, ref. 18, p. 108.
[266] A. A. Karim Al-Dharir, *J. Indian Chem. Soc.*, 1982, **59**, 632.
[267] Y. Muramatsu, G. Izawa, and K. Yoshihara, *Radiochim. Acta*, 1985, **38**, 5.
[268] Y. Murano, J. Akimoto, and K. Yoshihara, *Radiochim. Acta*, 1985, **38**, 11.
[269] L. Jiunnguang and N. Jongchen, *J. Radioanal. Nucl. Chem.*, 1986, **97**, 237.
[270] J. G. Hawke, A. A. Suweda, and C. A. Lukey, *J. Chem. Soc., Chem. Commun.*, 1986, p. 499.
[271] N. Y. Wang, S. Iyer, and F. S. Rowland, *J. Phys. Chem.*, 1986, **90**, 931.
[272] (a) G. A. Badun, A. I. Kostin, and Eh. S. Filatov, *Radiokhimiya*, 1985, **27**, 222; (b) R. W. Ehrekaufer, W. C. Hembree, and A. P. Wolf, *J. Labelled Compd. Radiopharm.*, 1985, **22**, 819.
[273] G. Occhiucci, F. Cacace, and M. Speranza, *J. Am. Chem. Soc.*, 1986, **108**, 872.
[274] B. Nebeling, K. Roessler, and G. Stoecklin, *Radiochim. Acta*, 1985, **38**, 15.
[275] S. P. Mishra, A. Patnaik, R. A. Singh, and D. P. Wagley, ref. 24, paper NC-14.
[276] P. P. Gaspar, S. Konieczny, and S. H. Mo, *J. Am. Chem. Soc.*, 1984, **106**, 424.
[277] K. Roessler and K. Schurwanz, Report, No. 1990, Kernforschungsanlage, Jülich, B.R.D., 1985.

radiation dose the yield of ammonium rises to 95%, indicating the importance of radiation-induced secondary reactions in its formation. Some of the reactions of recoil antimony have been investigated[278] in the solid state using ^{128}Sb produced by the decay of ^{128}Sn. β-Decay energy was suggested as being a factor determining the final oxidation state of the antimony.

Group VI.—When alkaline chloride solutions are irradiated with thermal neutrons the recoil sulphur atoms that are produced react initially to produce [^{35}S]sulphite,[279] but subsequent reactions lead to the formation of labelled species in both higher and lower oxidation states. The reactions of recoil ^{35}S in chloroform have also been reported.[280] When diphenyl selenide is irradiated with thermal neutrons the (n, γ) reaction generates ^{75}Se, ^{81m}Se, and ^{83}Se. As only low yields of labelled diphenyl selenide were observed it was concluded[281] that the selenium–carbon bonds were usually ruptured by the nuclear reaction. Reactions in the solid state of recoil tellurium (^{129}Te), made either by the β-decay of antimony[278] or by isomeric transition from ^{129m}Te,[282] have been reported.

Group VII.—*Chlorine.* Recoil chlorine, ^{38}Cl, has been shown to replace chlorine in molecules such as chloropropionyl chloride by a Walden inversion mechanism. This reaction has now been studied[283] in the liquid phase and the effect of solvent changes investigated. When methyl chloride, either as a frozen solid or in frozen aqueous solution, reacts with recoil chlorine[284] the main products are labelled chloride anions and labelled methyl chloride. The reactions of recoil chlorine with chloro-substituted toluenes,[285] anilines,[286] phenols and acetanilides,[287] as well as more complex heterocyclic compounds[288] have been reported. Work has also continued in aromatic systems to determine[289] the degree of selectivity that recoil chlorine atoms can exhibit. Many of the above studies have been extended to investigate the effect of added solvents such as methanol, cyclohexane,[285] benzene,[286] or tetrahydrofuran.[288] Related experiments have studied binary mixtures of alcohols with carbon tetra-chloride[290] or tetrachloroethene.[291] The analysis of results from such experiments is usually by the Urch–Kontis method which enables relative reactivities to be determined.

In inorganic systems very low yields of ^{38}ClO$_3^-$ were found[292] following the neutron irradiation of alkaline earth chlorates (but increased yields were observed in similar

[278] H. Moriyama and I. Fujiwara, *Radiochim. Acta*, 1983, **34**, 117.
[279] S. N. Muddukrishna and S. Mukherji, ref. 24, paper NC-17.
[280] M. F. de Ramirez, M. Jimenez-Reyes, and S. Bulbulian, *Rev. Soc. Quim. Mex.*, 1985, **27**, 246.
[281] C. Beltran, S. Bulbulian, and C. Archundia, *J. Radioanal. Nucl. Chem.*, 1985, **96**, 59.
[282] S. I. Bondarevskij and S. A. Timofeev, *Radiokhimiya*, 1985, **27**, 75; Engl. Trans., *Sov. Radiochem.*, 1986, **27**, 68.
[283] N. Borkar, A. A. Latifi, and E. P. Rack, *J. Chem. Phys.*, 1986, **85**, 3125.
[284] L. R. Opelanio-Buencamino and E. P. Rack, *Radiochim. Acta*, 1985, **38**, 87.
[285] N. Chandrasekhar, R. N. Bhave, and B. S. M. Rao, *Radiochim. Acta*, 1985, **39**, 5.
[286] V. G. Dedgaonkar, S. Mitra, and S. Waghmare, ref. 24, paper NC-8.
[287] V. G. Dedgaonkar, S. Mitra, and S. Waghmare, *Radiat. Phys. Chem.*, 1986, **27**, 375.
[288] V. G. Dedgaonkar, S. Mitra, and S. Waghmare, *J. Radioanal. Nucl. Chem.*, 1985, **96**, 79.
[289] K. Berei, Z. Kardos, and L. Vasaros, *Radiochim. Acta*, 1985, **38**, 83.
[290] M. Pertessis-Keis, *Radiochim. Acta*, 1985, **38**, 79.
[291] A. S. Agrawal and B. S. M. Rao, ref. 24, paper NC-11.
[292] V. G. Dedgaonkar, R. Harnesswala, R. S. Lokhande, M. B. Chaudari, and K. A. Bhagwat, ref. 24, paper NC-5.

reactions involving bromates and iodates, see below). The reactions of recoil chlorine in mixed crystals, $K_2[OsCl_6]-K_2[OsBr_6]$, have shown most interesting results[293] by the formation of mixed complex ions containing both bromine and inactive chlorine as well as ^{38}Cl. A mechanistic model to rationalize these observations has been proposed.

Bromine. In the gas phase the reactions of ^{76}Br and ^{77}Br, produced by the β^+ decay of the corresponding krypton isotopes, with propane and cyclopropane have been studied.[294] The pattern of labelled products suggested that bromine positive ions as well as neutral atoms were important. This idea is strongly supported[295] by related work which studied the reactions of the same bromine species with simple benzenoid compounds. The question of the charged state of recoil bromine has also been investigated[296] for ^{80}Br, ^{80m}Br, and ^{82}Br produced by the (n, γ) reaction using the charged plate technique. In bromobenzene enhanced yields were found at the anode. The reactions of recoil bromine in solid methyl bromide have also been studied.[284]

The reactions of recoil bromine in bromates have been extensively studied to determine the efficiency of production of labelled bromate anions[292,297] and the effect of the counter cation[298] as well as to investigate the importance of radiation damage.[299]

Iodine. The reaction of ^{128}I with methyl iodide has been studied in the solid phase, in liquid and frozen aqueous solutions,[284] and in benzene.[300] Neutron irradiation of iodine pentoxide leads to 44% of the ^{128}I being found[301] as labelled pentoxide. Pre-irradiation heating increased this yield. Conversely a similar pre-treatment for lithium iodate diminished[302] the amount of radio-iodine found as labelled iodate. Other investigations of iodates (potassium and indium,[275] alkaline earths[292]) have also been reported. An interesting example of an isotope effect would appear to have been discovered[303] as a result of the neutron irradiation of some periodate compounds which had been labelled with ^{129}I. This isotope undergoes the $^{129}I(n, \gamma)^{130}I$ reaction so that the recoil chemistry of ^{128}I and ^{130}I could be compared directly. Differences in Auger relaxation patterns seem to provide the best rationalization for the observed differences in reactivity.

Transition Metals and Actinoids.—Investigations of the reaction of recoil ^{56}Mn, produced by neutron capture, with the permanganate anion both in solution[304] and in mixed crystals with perchlorate[305] have shown that small yields of $^{56}MnO_4^-$ are produced. The extent to which a recoil atom might be able to replace an atom

[293] H. Mueller, P. Obergfell, and I. Hagenlocher, *J. Phys. Chem.*, 1986, **90**, 3418.
[294] D. de Jong, G. A. Brinkman, and B. W. van Halterren, *Radiochim. Acta*, 1983, **34**, 93.
[295] S. M. Moerlein, Thesis (Ph.D.), Washington Univ., St. Louis, MO, U.S.A., 1982 (microfilm 82-23805).
[296] B. M. Shukla, S. P. Mishra, and C. Pandey, *Met. Miner. Rev.*, 1983, **22**, 170.
[297] T. Tamai, S. Nishikawa, Y. Tanaka and H. Takemi, *Ann. Rep. Res. React. Inst. Kyoto Univ.*, 1984, **17**, 150.
[298] V. C. Dedgaonkar and D. A. Bhagwat, ref. 24, paper NC-4.
[299] B. M. Shukla, S. P. Mishra, and D. P. Wagley, ref. 24, paper NC-13.
[300] B. M. Shukla and C. Panday, ref. 24, paper NC-15.
[301] S. P. Mishra and A. Patnaik, *J. Radioanal. Nucl. Chem.*, 1986, **103**, 63.
[302] S. P. Mishra and A. Patnaik, ref. 24, paper NC-12.
[303] E. M. Batalha and A. V. Bellido, *J. Radioanal. Nucl. Chem.*, 1985, **91**, 251.
[304] V. G. Dedgaonkar and S. Mitra, ref. 24, paper NC-6.
[305] S. P. Mishra and J. Singh, *J. Radioanal. Nucl. Chem.*, 1986, **103**, 241.

closely surrounded by ligands, as in permanganate, has also been studied for the reactions of recoil chromium [^{51}Cr] in mixed iron and iron-chromium 4β-diketonate complexes[306] and for recoil chromium, molybdenum, and tungsten in the hexacarbonyl complexes of those metals.[307] In the former system direct replacement of the central atom of the complex was found to be a more important reaction than that of ligand collection (labelled complexes with mixed ligands have low yields) but for the carbonyl complexes the experimental evidence indicates that radical intermediates are important. Not unrelated are the annealing studies[308] using ^{51}Cr-doped K$_3$[Cr(NCS)$_6$].4H$_2$O in which it was found that most of the radio-chromium ended up as [^{51}Cr(NCS)$_5$(H$_2$O)]$^{2-}$ (goodness knows how). Similar studies[309] were carried out with potassium chromate to which trivalent ^{51}Cr had been added. The yields of labelled chromate were followed as a function of the annealing conditions. Annealing following the neutron irradiation of chromates has revealed[310] curious oscillatory behaviour.

Neutron irradiation has been used to initiate the reactions of recoil ^{64}Cu, ^{97}Ru, and ^{103}Ru in copper[14] and ruthenium[311] complexes. For ruthenium an optically active complex was chosen and it was shown that more of the labelled complex was in the original than in the enantiomeric form.

α-Decay of ^{241}Am leads to recoil ^{237}Np which in a cerium aluminate matrix ends up mostly as Np^{4+} (70%).[312] The valence state was determined using Mössbauer spectroscopy which also showed the remainder to be Np^{3+}, replacing trivalent cerium in the host lattice.

5 The Environment

Most recent work that might reasonably fall under this general heading is concerned, in some way or other, with nuclear reactors, with the sources of their fuel (uranium), with the control of radioisotopes made by them, with the disposal of radioactive waste from them, and with their ultimate decommissioning.

Uranium.—The intense effort to develop economic methods for the extraction of uranium from the sea continues unabated, and has been reviewed[313] by an IAEA technical committee. Reviews of Japanese[314] and Korean[315] progress on specific projects have also been published. It has been suggested[316] that uranium exists in the sea in its hexavalent state as hydroxyuranyl anions, mostly [UO$_2$(OH)$_4$]$^{2-}$.

[306] K. Yoshihara and T. Sekine, *Nippon Kagaku Kaishi*, 1984, 1873.

[307] Y. Muto and H. Ebihara, *Radiochim. Acta*, 1985, **39**, 11.

[308] F. M. Lancas, K. E. Collins, and C. H. Collins, *Radiochim. Acta*, 1985, **38**, 189.

[309] M. I. Stamouli, *J. Radioanal. Nucl. Chem.*, 1985, **91**, 35.

[310] P. N. Dimotakis and B. D. Symeopoulos, *Radiochim. Acta*, 1986, **39**, 65.

[311] M. Tanaka, *Radiochim. Acta*, 1983, **34**, 109.

[312] V. M. Filin, V. F. Gorbunov, and S. A. Ulanov, *Radiokhimiya*, 1985, **27**, 86; Engl. Trans., *Sov. Radiochem.*, 1985, **27**, 79.

[313] J. Bitte, in Proc. Tech. Committee 'Advances in Uranium Ore Processing and Recovery from Non-conventional Resources', IAEA, Vienna, Austria, 1985, p. 299.

[314] (a) M. J. Driscoll, in Proc. Internat. Meeting on Recovery of Uranium from Seawater, Atomic Energy Society of Japan, Tokyo, Japan, 1983, p. 1; (b) M. Kanno, *ibid.*, p. 12; (c) N. Ogata, *Nippon Kaisui Gakkai-Shi*, 1983, **37**, 33.

[315] C. K. Yun and C. K. Choi, ref. 314(a), p. 23.

[316] O. T. Krylov, P. D. Novikov, and M. P. Nesterova, *Okeanologiya*, 1985, **25**, 242.

Whereas one proposed extraction method relies on co-precipitation (with mag-nesium hydroxide)[317] many seek to adsorb the uranium onto some suitable substrate. To this end hydrated titanium oxides[318] and also ferric oxyhydroxides[319] have been the subject of continued study. A composite material of titanium and activated carbon has also been shown[320] to be effective in adsorbing uranium from sea-water. Most effort though continues in the development of chelating polymers and resins, usually with an amino-residue somewhere, *e.g.* polyethenimine,[321] polyacryl-amidoxime,[322] and hydroxybenzamide polymers.[323] Polymers based on dihy-droxamic acid[324] and sulphinamide[325] have also been shown to be effective at chelating with and thus sequestering the elusive uranyl. The electrical properties of amine-adsorbed uranium species have been investigated[326] so that optimum condi-tions for desorption can be found. Extraction processes based on liquid membranes have also been proposed.[327]

Coal often contains a significant amount of uranium (0.02%) which with care can be extracted from the ash after combustion.[328] The advantage here is two-fold, an economically useful source of uranium and a reduction in the amount of radioactive material discharged into the environment.

Trans-uranic Elements.—The general behaviour of long-lived trans-uranic elements in marine environments has been reviewed[329] and the particular problem of the oxidation states of neptunium, plutonium, and americium in ground-water has been investigated[330] in some detail, (oxidation states V and VI being detected). Pu^{4+} is expected to be strongly adsorbed on solid surfaces, and recent studies have shown[331] that the same is true of the plutonyl(V) cation. Furthermore some mineral surfaces seem capable of catalysing the interconversion of the plutonium valence states Pu^{IV}, Pu^{V} and Pu^{VI}. It is therefore not clear whether the observation of oxidized plutonium

[317] T. Fujinaga, T. Kuwamoto, E. Nakayama, and K. Isshiki, *Nippon Kaisui Gakkai-Shi*, 1984, **38**, 50.
[318] (a) D. Alexandre and P. P. Vistoli, ref. 313, p. 289; (b) S. Senoh, *Nippon Kaisui Gakkai-Shi*, 1984, **38**, 218.
[319] (a) C. H. Ho and D. C. Doern, *Can. J. Chem.*, 1985, **63**, 1100; (b) H. D. Chingkuo, *Geochim. Cosmochim. Acta*, 1985, **49**, 1931.
[320] T. Hirotsu, A. Fujii, K. Sakane, S. Katoh, and K. Sugasaka, *Shikoku Kogyo Gijutsu Shikenjo Hokoku*, 1984, **16**, 45, 83.
[321] S. Usami, K. Hasegawa, K. Takata, R. Naito, and H. Uchida, ref. 314(a), p. 147.
[322] (a) T. Hirotsu, N. Takagi, S. Katoh, K. Sugasaka, N. Takai, M. Seno, ref. 314(a), p. 110; (b) S. Katoh, K. Sugasaka, T. Hirotsu, N. Takai, T. Itagaki, and H. Ouchi, *ibid.*, p. 138.
[323] M. Sakuragi, K. Ichimura, Y. Suda, T. Marumo, T. Iwaki, Y. Abe, and T. Misono, *Kogyo Gijutsuin Seni Kobunshi Zairyo Kenkyusho Kenkyu Hokoku*, 1985, **148**, 33.
[324] (a) T. Hirotsu, S. Katoh, and K. Sugasaka, *Kogyo Gijutsuin Seni Kobunshi Zairyo Kenkyusho Kenkyu Hokoku*, 1985, **148**, 21; (b) M. Sakuragi, K. Ichimura, Y. Suda, Y. Hirotsu, S. Katoh, K. Sugasaka, M. Fujishima, Y. Abe, and T. Misono, *ibid.*, p. 29.
[325] M. Sakuragi, K. Ichimura, and Y. Suda, *Kogyo Gijutsuin Seni Kobunshi Zairyo Kenkyusho Kinkyu Hokoku*, 1985, **148**, 17.
[326] (a) S. Senho, Y. Oda, and H. Konishi, *Nippon Kaisui Gakkai-Shi*, 1984, **38**, 212; (b) Y. K. Litsis, *Radiokhimiya*, 1985, **27**, 614.
[327] W. S. Kim, Y. J. Kyun, K. Youm, H. Kyung, and H. G. Tack, *Yonsei Eng. Rep.* 1985, **17**, 17.
[328] (a) J. Slivnik, A. Stergarsek, Z. Beslin, Z. Krempl, P. Marijanovic, and V. Valkovic, ref. 313, p. 263; (b) G. Morales, P. J. Nadal, J. L. Merino, and P. Gasos, ref. 313, p. 275.
[329] D. N. Edgington and D. M. Nelson, in 'Behaviour of Long-lived Radionuclides Associated with Deep Sea Disposal of Radioactive Wastes' (IAEA Tech. Doc. 368), IAEA, Vienna, Austria, 1986, p. 41.
[330] J. M. Cleveland, K. L. Nash, and T. F. Rees, *Nucl. Technol.*, 1985, **69**, 380.
[331] W. L. Keeney-Kennicutt and J. W. Morse, *Geochim. Cosmochim. Acta*, 1985, **49**, 2577.

species in solution near test sites[332] is a 'memory' of the oxidation state of the ^{239}U or ^{240}U which β-decayed to ^{239}Pu or ^{240}Pu or is due to subsequent contact with minerals such as goethite or calcite.

Nuclear Waste.—A method for calculating the chemical hazard of radioactive waste has been described[333] which takes account of such factors as intrinsic toxicity, persistence with time, and the build-up of degradation products. A theoretical approach has also been used[334] to assess the solubility of uranium dioxide from nuclear fuel as a function of pH, temperature, anions, *etc*. A method has been proposed[335] for the recovery of uranium from waste-water dams by co-precipitation with calcium phosphate. Other topics considered in the recent literature are the disposal of aluminium nitrate solutions of fission waste products[336] and quantitative methods for the detection of ^{131}I.[337]

Decontamination.—The procedures to be followed following radiochemical or nuclear accidents or in the decommissioning of a nuclear reactor or similar facility are considered in a recent book edited by Elder.[338] A computer simulation of such problems and strategies to be adopted in their solution has also been reported.[339]

6 Miscellaneous

Muon Chemistry.—The quasiresonant formation of μ^2H_2 and μ^2H^3H molecules when muons pass through deuterium or deuterium–tritium mixtures has been reported[340] and their possible importance in fusion reactions discussed.

Radon.—Evidence has been reported to suggest the formation of radon cations[341] and the existence (briefly) of radon trioxide.[342]

Astatine.—A detailed account of the chemistry of astatine is contained in the recent volume of 'Gmelin' devoted to this element.[343] On a slightly less monumental scale the chemistry of cationic (+1) astatine has been reviewed[344] and evidence produced for complex formation with alkenes.[345]

[332] E. A. Bondietti, *J. Radioanal. Nucl. Chem.*, 1985, **91**, 221.

[333] L. E. Wickham, in 'Waste Isolation in the US Technical Programmes and Public Education', American Nuclear Society, La Grange Park, IL, U.S.A., 1984, p. 655.

[334] F. Garisto and N. C. Garisto, *Nucl. Sci. Eng.*, 1985, **90**, 103.

[335] S. Komoto, *Donen Giho*, 1985, **54**, 90.

[336] G. Bernhard, W. Boessert, and O. Hladik, *Kernenergie*, 1986, **29**, 108.

[337] J. Leib, R. Crecelius, H. Pfeiff, and G. L. Faengewisch, *mt Medizintechnik*, 1985, **105**, 33.

[338] H. K. Elder, 'Technology, Safety and Costs of Decommissioning Reference Nuclear Fuel Cycle and Non-fuel Cycle Facilities following Postulated Accidents', Pacific Northwest Labs, Richland, WA, U.S.A., 1985 (NUREG/CR-3293).

[339] J. J. Tawil and D. L. Strenge, in 'Computer Applications in Health Physics', ed. R. L. Kathren, D. P. Higby, and M. A. McKinney, Health Physics Society, Richlands, WA, U.S.A., 1984, p. 6001.

[340] L. I. Menshikov and L. I. Ponomrev, *Phys. Lett., B*, 1986, **167**, 141.

[341] L. Stein, *J. Chem. Soc., Chem. Commun.*, 1985, p. 1631.

[342] V. V. Avrorin, R. N. Krasikova, V. D. Nefedov, and M. A. Toropova, *Radiokhimiya*, 1985, **27**, 511.

[343] H. K. Kugler, C. Keller, K. Berei, L. Vasaros, S. H. Eberle, H. W. Kirby, H. Murnzel, H. Roessler, and A. Seidel, 'Astatine—Gmelin Handbook of Iorganic Chemistry', Gmelin Inst., Frankfurt-am-Main, B.R.D., 1985.

[344] (*a*) R. Dreyer, I. Dreyer, S. Fischer, H. Hartmann, and F. Roesch, *J. Radioanal. Nucl. Chem.*, 1985, **96**, 333; (*b*) R. Dreyer, I. Dreyer, W. Doberenz, S. Fischer, W. A. Chalkin, and F. Roesch, *Isotopenpraxis*, 1986, **22**, 81.

[345] Y. V. Norseev, D. D. Nyan, N. K. Khuan, and L. Vasaros, Report, R6-85-625, Joint Inst. of Nuclear Research, Dubna, U.S.S.R., 1986.

Author Index

Gadd, G. E., 276
Gafurov, R. G., 386
Gaggelli, E., 343
Gahan, L. R., 240, 342
Gaines,, D. F., 43
Gais, H.-J., 24
Gal, M., 301
Galanlenhert, P., 273
Galay, J., 293
Galazka, R. R., 193
Galbacs, Z., 353
Galik, A., 383
Gallezot, P., 299
Gallop, M. A., 301, 312
Galsbøl, F., 321
Gambarott, S., 220
Gamp, E., 271
Gampp, H., 301
Ganapathi, L., 161
Gandour, R. D., 31
Ganem, B., 40, 277
Ganis, P., 142
Gano, D. R., 135
Gans, P., 270
Gantar, D., 212
Ganush, Ya., 386
Ganzer, G. A., 105
Garbauskas, M. F., 39, 40
Garcia, A., 326
Garcia, E., 362
Garcia, S. G., 230
Garcia del Rio, H., 391
Garcia-España, E., 262, 337
Garcia-Fernandez, M. E., 102
Garcia-Prieto, J., 326
Garcia-Rosas, J., 346
Gard, G. L., 200
Gard, P., 9
Gardner, C. J., 372
Garg, S. P., 371
Garisto, F., 401
Garisto, N. C., 401
Garland, M. T., 339
Garlashelli, L., 322
Garner, C. D., 176, 240, 243, 265, 271, 318, 334
Garnett, J. L., 386
Garnier, F., 367
Garreau, F. B., 169
Garrigues, B., 73
Garst, J. F., 17
Gasdaska, J. R., 144
Gasgier, M., 361
Gasita, S. M., 384
Gasos, P., 400
Gaspar, P. P., 119, 120, 121, 396
Gasparin, M., 85, 372
Gasymov, V. A., 174
Gates, P. N., 198
Gatley, S. J., 382
Gatteschi, D., 291, 336, 337, 339, 340
Gatto, V. J., 22, 31
Gattow, G., 116, 117, 175

Gaudemer, A., 338
Gaudin, M. J., 285
Gaul, E. M., 269
Gaune-Escard, M., 174
Gaura, R. M., 336
Gaustad, R., 344
Gautheron, B., 192
Gauvin, F., 258
Gavalyan, V. B., 203
Gavrilov, V. I., 367
Gawell, L., 385, 386
Gawron, M., 339
Gawthorne, M., 344
Gay, J. G., 192
Gayoso, M., 183, 355
Ge, S., 360
Geantet, C., 170
Gebicki, J., 11
Gedanken, A., 69
Geerts, R. L., 317, 351
Geib, S. J., 143, 311, 312
Geiger, D. K., 341
Geiger, R. A., 374
Geiger, W. E., 279
Geilich, K., 64
Geiser, U., 214, 237, 336, 337, 345, 351
Geissler, M., 20
Gelb, R. I., 175
Gellert, R. W., 355
Gellman, S. A., 353
Gelmini, L., 248
Gelmont, B. L., 193
Gelsema, W. J., 391
Genestar, C., 391
Genet, R., 386
Geng, L., 341
Genthe, W., 370
Genzel, L., 188
Geoffroy, G. L., 143, 311, 312, 313, 316
George, T. A., 233
Georgiadis, R., 351
Geraldes, C. F. G. C., 364
Gerbing, U., 190, 250
Gereke, R., 145
Gerets, R., 94
Gerloch, M., 336
Germeshausen, J., 5
Gerner, B., 188
Gerry, M. C. L., 75, 198, 202
Gerss, H. M., 361
Gerstein, B. C., 218
Geselowitz, D. A., 160, 302, 310
Gessner, W., 83
Geurink, P. J. A., 19
Ghilardi, C. A., 27, 149, 321, 326
Ghirardelli, R. G., 30, 33, 364
Ghosal, B., 87
Ghosh, M. C., 285
Ghosh, P. K., 89
Giannella Neto, D., 393
Gibb, T. C., 260
Gibb, W. R., 118

Gibboni, D. J., 322
Gibson, D., 342
Gibson, J. F., 179
Gibson, J. K., 371
Giesy, J. P., 374
Gil, V. M. S., 177, 374
Gilbert, J. A., 270, 310
Gilbert, T. R., 182, 333
Gil-Bortnowska, R., 233
Gilday, J. P., 17
Gileadi, E., 77
Gili, P., 338
Gilje, J. W., 377
Gill, G. B., 4
Gill, W. R., 24
Gillespie, R. J., 79, 168, 189, 190, 204, 354
Gillet, J. P., 15
Gillies, C. W., 109
Gilmour, D. I., 331, 332, 346
Gilson, D. F. R., 277
Gimarc, B. M., 45, 54
Gimeno, M. C., 346, 349
Ginos, J. Z., 388
Ginsberg, A. P., 317
Girelli, A. M., 339
Girerd, J., 271
Girerd, J.-J., 338
Girolami, G. S., 46
Girreser, U., 15
Giusti, J., 30
Gladfelter, W. L., 308, 309, 312
Glanzer, B. I., 352
Glaser, B., 66
Glaser, J., 103
Glass, R. S., 284
Glaunsinger, W. S., 171
Glazer, P. A., 267
Gleiter, R., 112
Glidewell, C., 135, 210, 277
Gloede, J., 207
Gloux, J., 271
Gloux, P., 271
Glowacki, A., 143
Glowiak, T., 336, 343, 344, 351
Glushkova, M. A., 238
Glusker, J. P., 281
Gobbi, G. C., 137
Goddard, J. P., 65, 143
Godeneche, D., 388
Godfrey, P. D., 112, 116
Godfrey, S., 116
Godziela, G. M., 256, 341
Goedken, V., 17
Goedken, V. L., 91, 219, 241
Goel, A. B., 330
Goerlich, M., 118, 216
Goethals, P., 387, 388
Goeva, L. V., 44
Goff, D. A., 37
Goff, H. M., 256, 341
Goh, L. Y., 182
Goher, M. A. S., 337
Goiffon, A., 84

Author Index 415

Gokel, G. W., 22, 31
Gol, F., 324
Gold, A., 269
Gold, J. S., 264
Goldberg, D. E., 130
Goldhaber, A., 298
Goldman, A. S., 257
Goldsby, K. A., 160, 262, 292
Goldstone, J. A., 360
Golic, L., 352
Gollinger, W., 76
Goltsos, W. C., 194
Gomez-Romero, P., 284
Gomm, P. S., 344
Goncalves, M. L. S. S., 343
Goncharov, L. A., 384
Gonsalves, K., 17
Gonzalez-Elipe, A. R., 174
Goode, M. J., 48, 95
Goodfellow, R. J., 320
Goodgame, D. M. L., 327
Goodgame, M., 354
Goodman, D. M. L., 299
Goodman, E. M., 10
Goodman, M. M., 392, 393
Gopal, E. S. R., 192
Gopalakrishnan, J., 161
Gopinathan, S., 66
Gorbunov, V. F., 399
Gorce, J., 270
Gordeeva, L. S., 386
Gordon, A. T., 267
Gordon, M. S., 135, 146
Goren, Z., 171
Gorochov, O., 194
Gorski, B., 383
Goswamin, S., 184
Goto, M., 277
Goudsmit, R. J., 313
Gougeon, P., 170
Gould, E. S., 182, 232, 278
Gould, I. A., 144
Gould, R. O., 175, 303
Gourdon, A., 147
Goursof, A., 41
Gouteron, J., 339
Gozum, J. E., 308
Grab, L. A., 266
Grabowski, N. A., 330
Graddon, D. P., 142
Grady, G., 135
Grady, J. K., 224
Grätzel, M., 170, 354
Grandjean, J., 365
Grange, P., 88
Granger, P., 141, 195
Grankin, V. M., 216
Granozzi, G., 295, 296
Grant, K. J., 292
Grant, M. H., 383
Grant, P. M., 382
Grases, F., 391, 284
Grasselli, R. K., 194
Graudemer, A., 344

Graule, T., 15
Gray, G. M., 399
Gray, H. B., 328
Grayson, I., 317
Grayson, J., 338
Graziani, R., 143, 215, 338, 375
Greatrex, R., 43
Grebille, D., 82
Green, J. C., 48, 252, 306, 371
Green, M., 69, 252
Green, M. A., 385
Green, M. L. H., 239
Greenblatt, M., 9, 13
Greene, B., 350
Greene, D. L., 340
Greenwood, N. N., 43, 45, 51, 53
Grein, F., 72, 189, 198, 204
Grev, R. S., 120
Grey, I. E., 213
Gribnau, M. C. M., 352
Griend, L. V., 147
Gries, T., 186, 329
Grieves, R. A., 232, 252
Griffin, H. C., 395
Griffith, E. A. H., 354
Griffith, W. P., 238, 255, 375
Griffiths, L., 157
Griller, D., 139
Grim, S. O., 166
Grimblot, J., 88, 169
Grimmer, A.-R., 10, 157
Grimsrud, E. P., 113
Grinberg, E. E., 390
Gritsenko, O. V., 322
Groat, L. A., 351
Grobe, J., 155, 205
Gronchi, G., 150
Grootveld, M. C., 350
Gros, E. G., 386
Gross, R., 149, 245
Grossel, M. C., 330
Grotemeyer, J., 37
Groth, Th., 213
Groves, J. T., 262, 268, 270
Gruberger, J., 77
Gruehn, R., 99
Grunwald, M., 145
Grutzner, J. B., 24
Grzybek, R., 233
Grzybowski, J. J., 302
Guastini, C., 27, 220, 225, 335
Güdel, H. U., 241, 354
Guelzim, A., 298
Günther, H., 18
Guerney, P. J., 168
Gueron, M., 280
Guerra, M. A., 138, 346
Guerriero, P., 338, 342
Gueutin, C., 268
Guilard, R., 92, 99, 134, 269, 276, 315, 375
Guillaume, M., 391
Guinand, G., 32
Guittard, M., 93, 174, 334

Guittet, E., 322
Gui-Xin, L., 258
Gul, A., 375
Gulloti, M., 160, 280, 344
Gummin, D. D., 185
Guo, B.-S., 15
Guo, Z., 386
Gupta, K. D., 115, 209
Gupta, S. H., 19, 138
Gupta, V. K., 151, 281, 301, 336, 340
Guratzsch, H., 381
Gustowski, D. A., 31
Gutierrez-Puebla, E., 189, 235, 289
Gutlich, P., 266, 352
Gutman, A. L., 386
Gutowsky, H. S., 42
Gutsev, G. L., 208
Guttler, R. B., 393
Guy, J. J., 213
Guyot, D., 132
Guzzo, F., 32

Ha, T.-K., 66
Haaland, A., 130, 352, 369
Haas, A., 210
Haase, M., 68, 340, 344
Haasnoot, J., 356
Habata, Y., 303
Habben, C., 69, 75
Hachgenei, J., 149
Hacht, B., 344
Hackert, M. L., 33
Hackett, M., 330
Hackett, S., 14
Haddleton, D. M., 318, 322
Hadel, L. M., 106
Hadjikyriacou, A., 176, 242
Häberle, K., 136
Haendler, H. M., 339
Händler, V., 157
Härter, P., 248
Haeuseler, H., 173
Hageman, P., 346
Hagen, K., 190, 215, 229
Hagen, K. I., 288
Hagen, K. S., 271
Hagenlocher, I., 398
Hagenmuller, P., 100
Haggstrom, L., 103, 190
Hague, D. N., 351
Hahn, E., 129, 143, 370
Hahn, J., 152
Haiden, I., 167
Haiduc, I., 139, 140, 158
Haije, W. G., 214, 336
Haines, R. E., 142
Haines, R. J., 310
Haire, R. G., 371
Haize, K. H., 385
Hajdasz, D. J., 135
Hakansson, M., 333

424   *Author Index*